An Introduction to Distributed Optical Fibre Sensors

Series in Fiber Optic Sensors

Series Editor: Alexis Mendez

An Introduction to Distributed Optical Fibre Sensors
Arthur H. Hartog

Fothcoming Titles:

Fiber-Optic Fabry-Perot Sensors: An Introduction
Yun-Jiang Rao, Zeng-Ling Ran, and Yuan Gong

Fiber Optic Structural Health Monitoring (SHM) and Smart Structures
Alfredo Güemes and Julian Sierra Perez

Introduction to Fiber Bragg Grating Sensors
John Canning and Cicero Martelli

An Introduction to Distributed Optical Fibre Sensors

Arthur H. Hartog

CRC Press
Taylor & Francis Group
Boca Raton London New York

CRC Press is an imprint of the
Taylor & Francis Group, an **informa** business

Cover Image: Courtesy of Statoil
Back Cover Image: Oil painting courtesy of Christine Maltby

CRC Press
Taylor & Francis Group
6000 Broken Sound Parkway NW, Suite 300
Boca Raton, FL 33487-2742

First issued in paperback 2018

ISBN 13: 978-1-138-08269-4 (pbk)
ISBN 13: 978-1-4822-5957-5 (hbk)

Visit the Taylor & Francis Web site at
http://www.taylorandfrancis.com

and the CRC Press Web site at
http://www.crcpress.com

To the memory of my parents, Dolf Hartog and Joyce Whitney.

Arthur H. Hartog

Contents

Series Preface xvii
Foreword xix
Preface xxi
Author xxiii
List of Symbols xxv
List of Abbreviations xxix

PART ONE Fundamentals **1**

1 Introduction **3**
1.1 The Concept of Distributed Optical Fibre Sensors 3
1.1.1 Optical Fibre Sensors 3
1.1.2 The Advent of Multiplexed Sensors 6
1.1.3 Distributed Optical Fibre Sensors 7
1.2 Historical Development of DOFS 9
1.2.1 Optical TDR: An Enabling Technology 9
1.2.2 Distributed Temperature Sensors 11
1.2.3 Elastic and Inelastic Scattering 11
1.2.4 Raman OTDR 13
1.2.5 Brillouin-Based Distributed Sensing 14
1.2.6 Distributed Sensing with Rayleigh Backscatter 15
1.3 Performance Criteria in Distributed Fibre-Optic Sensors 16
1.3.1 Four Key Criteria 16
1.3.1.1 Measurand resolution 17
1.3.1.2 Range 17
1.3.1.3 Spatial resolution 17
1.3.1.4 Sampling resolution 19
1.3.1.5 Measurement time 19
1.3.2 Interplay between Performance Criteria 20
1.3.3 Estimation of Key Performance Parameters 21
1.3.4 Sensor Accuracy and Cross-Talk 22
1.4 History of Commercialisation: Successes and Failures 22
1.4.1 Raman DTS 22
1.4.2 Commercial Brillouin-Based Distributed Strain and Temperature Sensors 24
1.4.3 Distributed Vibration Sensing 24
1.5 Other Technologies 25
1.6 Distributed Sensing and Dense Point Sensor Arrays 25
1.7 Organisation of the Book 26
References 26

2 Optical Fibre Technology **31**
2.1 Propagation in Optical Fibres 31
2.1.1 Types of Optical Fibre 33
2.1.1.1 Step-index multimode fibres 33
2.1.1.2 Graded-index multimode fibres 34
2.1.1.3 Single-mode optical fibres 36
2.1.2 Loss Mechanisms 37
2.1.2.1 Absorption 37
2.1.2.2 Scattering 37
2.1.2.3 Other loss mechanisms 39
2.1.3 Bandwidth of Optical Fibres 40
2.2 Construction of Optical Fibres 41
2.3 Principal Components Used in Distributed Sensing Systems 42
2.3.1 Lasers 42
2.3.1.1 Broad contact semiconductor lasers 42
2.3.1.2 Fabry-Pérot single-mode laser diodes 43
2.3.1.3 Semiconductor distributed feedback lasers 44
2.3.1.4 External cavity feedback semiconductor lasers 44
2.3.1.5 Semiconductor DFB lasers with active phase/frequency feedback 44
2.3.1.6 Fibre DFB lasers 44
2.3.1.7 Tunable lasers 45
2.3.1.8 Symbols for optical sources 45
2.3.2 Optical Amplifiers 45
2.3.2.1 Doped fibre amplifiers 45
2.3.2.2 Semiconductor optical amplifiers 46
2.3.2.3 Symbol for optical amplifiers 46
2.3.3 Fused-Taper Couplers 46
2.3.4 Micro-Bulk Optics 47
2.3.5 Isolators, Circulators and Faraday Rotation Mirrors 47
2.3.6 Fibre Bragg Gratings 47
2.3.7 Modulators 48
2.3.7.1 Acousto-optic modulators 48
2.3.7.2 Electro-optic modulators 49
2.3.8 Switches 49
2.3.9 Polarisation Controllers 50
2.3.10 Connectors and Splices 50
2.3.11 Detectors 50
2.3.12 Other Symbols 52
References 52

3 Principles of Optical Time-Domain Reflectometry (OTDR) for Distributed Sensing **55**
3.1 Optical Time-Domain Reflectometry 55
3.1.1 Single-Ended Measurement 56
3.1.2 Double-Ended Measurement 60
3.1.3 Capture Fraction 64
3.1.4 Coherence Effects 66
3.1.5 Signal Levels 72
3.2 OTDR System Design 73
3.2.1 Sources and Detectors 73
3.2.2 Common System Designs 74
3.2.3 Noise in Optical Receivers 75

3.2.3.1 Worked example 1: Multimode, short-wavelength OTDR 78
3.2.3.2 Worked example 2: Single-mode, long-wavelength OTDR 85
3.2.4 Overload Considerations 85
3.2.5 Acquisition and Further Processing 86
3.2.6 Limitations to Power Launched 87
3.3 Enhancing the OTDR Range 88
3.3.1 Improved Detection 88
3.3.1.1 Coherent detection systems 88
3.3.2 Photon Counting 91
3.3.3 Spread-Spectrum Techniques 92
3.3.3.1 OFDR 92
3.3.3.2 Coherent OFDR 92
3.3.3.3 Incoherent OFDR 94
3.3.3.4 Pulse compression coding 94
3.3.3.5 Frequency-diverse C-OTDR 99
3.3.4 Amplified OTDR Systems 101
3.4 Trends in OTDR 102
3.5 Applying Fibre Reflectometry to Distributed Sensing 102
References 103

PART TWO Distributed Sensing Technology 107

4 Raman-Based Distributed Temperature Sensors (DTSs) 109
4.1 Principles of Raman OTDR 109
4.1.1 Spontaneous Raman Scattering 109
4.1.2 Temperature Sensitivity of the Raman Scattering 111
4.1.3 Raman Backscatter Signals 114
4.2 Raman OTDR Design 115
4.2.1 Wavelength, Laser and Fibre Selection 117
4.2.2 Reference Coil Design 119
4.2.3 Receiver Design 120
4.2.4 Principal Types of Raman OTDR Systems 120
4.2.4.1 Short-range systems 120
4.2.4.2 High-performance systems operating at around 1064 nm 121
4.2.4.3 Long-range systems 121
4.3 Performance of Raman OTDR DTS Systems 121
4.3.1 Spatial Resolution 121
4.3.2 Temperature Resolution 122
4.3.3 Fundamental Limitations 122
4.3.4 Signal-to-Noise Ratio 125
4.3.5 Performance Comparisons: A Figure of Merit and Typical Specifications 125
4.4 Alternative Approaches to Raman OTDR 128
4.4.1 Optical Frequency-Domain Reflectometry (OFDR) 128
4.4.1.1 Data acquisition in Raman OFDR 129
4.4.1.2 Signal recovery 130
4.4.1.3 Comparison with OTDR methods 130
4.4.2 Pulse Compression Coding (PCC) 131
4.4.3 Systems for Very High Spatial Resolution 132

4.4.3.1 Time-correlated single-photon counting 134
4.4.3.2 Multichannel photon counting 136
4.4.3.3 Correction for dead time and other detector imperfections 136
4.4.3.4 Recent improvements in photon counting 136
4.4.3.5 Comparison with analog direct detection 137
4.5 Practical Issues and Solutions 137
4.5.1 Loss Compensation 137
4.5.1.1 Fibre degradation 138
4.5.1.2 Doubled-ended measurements 143
4.5.1.3 'J-fibre' configuration 144
4.5.1.4 Multi-wavelength measurements 145
4.5.1.5 Raman ratio with separate sources 145
4.5.1.6 Anti-Stokes Raman with Rayleigh loss correction 147
4.5.1.7 Dual sources separated by two Stokes shifts 148
4.5.1.8 Summary of the loss compensation methods 148
4.5.2 Calibration 149
4.5.3 Probe Power Limitations 150
4.5.3.1 Addressing the linear power limitation 151
4.5.3.2 The role of wavelength selection in extending linearity 152
4.5.4 Insufficient Filter Rejection 152
4.5.5 Forward Raman Scattering 153
4.5.6 High-Temperature Operation of Raman Sensors: Addressing the Reduction in Sensitivity 154
4.5.7 Brillouin Backscatter 154
4.6 Summary 155
References 155

5 Brillouin-Based Distributed Temperature and Strain Sensors 161
5.1 Spontaneous Brillouin Scattering 162
5.1.1 The Effect 162
5.1.2 Frequency Shift 162
5.1.3 Spectral Width 163
5.1.4 Sensitivity to Temperature and Strain 164
5.1.4.1 Sensitivity coefficients 164
5.1.4.2 Sensitivity of the intensity to temperature 165
5.1.4.3 Sensitivity of intensity to strain 166
5.1.4.4 Loss compensation for intensity measurements 166
5.1.4.5 Sensitivity of the Brillouin frequency shift 167
5.1.4.6 Coating and coupling effects 168
5.2 Stimulated Brillouin Scattering 169
5.3 Systems Based on SpBS 171
5.3.1 Optical Separation 171
5.3.2 Electrical Separation 177
5.3.3 Estimation of Frequency, Gain and Linewidth 183
5.3.4 Limitations and Benefits 184
5.3.4.1 Launched power 185
5.3.4.2 Spatial resolution 186
5.4 Recent Advances in Distributed Strain Sensing Based on SpBS 187
5.4.1 Remote Amplification for Range Extension 187

5.4.1.1 Distributed Raman amplification 188
5.4.1.2 Remotely pumped rare-earth-doped fibre amplifiers 190
5.4.1.3 Hybrid Raman/EDFA amplification 192
5.4.2 Fast Measurement Techniques 192
5.4.2.1 Frequency discriminators 192
5.4.2.2 Parallel multi-frequency acquisition 192
5.4.2.3 Signal tracking 193
5.4.3 Pulse Compression Coding 194
5.4.4 Simultaneous Multi-Frequency Measurement 194
5.4.5 Dual-Pulse Method in BOTDR 195
5.4.6 Spectral Efficiency 197
5.5 System Based on SBS 197
5.5.1 BOTDA 197
5.5.2 Optical Arrangements for BOTDA 199
5.5.3 Sensitivity and Spatial Resolution 203
5.5.4 Trade-Offs in BOTDA 203
5.5.5 Composite Pulse Methods in BOTDA 203
5.5.5.1 Single shaped pump interrogation 203
5.5.5.2 Differential pump interrogation 205
5.5.5.3 Simultaneous gain/loss interrogation 206
5.5.6 Pulse Compression Coding 206
5.5.6.1 Pulse compression coding for range extension 206
5.5.6.2 Pulse compression coding as an extension of composite pulse techniques 207
5.5.7 Slow Light Correction 207
5.5.8 BOFDA 208
5.5.9 Performance and Limitations 208
5.5.9.1 Parallel (multi-tone) acquisition approaches 208
5.5.9.2 Slope-assisted fast acquisition 210
5.5.9.3 Distance enhancements in BOTDA through remote amplification 210
5.5.9.4 Power limitations due to SPM 211
5.5.10 Phase Measurement in BOTDA 211
5.6 Brillouin Sensors Based on Coherence-Domain Reflectometry: BOCDA and BOCDR 213
5.6.1 BOCDA 213
5.6.1.1 Simplified BOCDA approach 215
5.6.1.2 Range extension 215
5.6.1.3 Background noise suppression 216
5.6.1.4 BOCDA using phase coding 216
5.6.1.5 Low coherence BOCDA 217
5.6.2 BOCDR 217
5.7 Dynamic Bragg Gratings 218
5.8 Polarisation Effects 220
5.9 Summary 221
References 222

6 Rayleigh Backscatter: Distributed Vibration Sensors and Static Measurements 231
6.1 Distributed Vibration Sensors 232
6.1.1 Recall of Coherent Rayleigh Backscatter 232
6.1.2 Timescales in Distributed Vibration Sensing 232
6.1.3 Modelling of Coherent OTDR 234
6.2 Amplitude-Based Coherent Rayleigh Backscatter DVS 235

6.3 Differential Phase-Measuring DVS (dΦ-DVS) 239
6.3.1 Techniques for dΦ-DVS 240
6.3.2 Phase Unwrapping 240
6.3.3 Dual-Pulse DVS 241
6.3.4 Interferometric Phase Recovery dΦ-DVS 246
6.3.5 Heterodyne DVS (hDVS) 249
6.3.6 Homodyne DVS 254
6.4 Comparison and Performance of the DVS Techniques 256
6.5 Issues and Solutions 257
6.5.1 Fading 257
6.5.1.1 Requirements for statistically independent signals 257
6.5.1.2 The effect of diversity on signal quality 259
6.5.1.3 Specific arrangements for frequency diversity 260
6.5.1.4 Alternative approaches to diversity in DVS 261
6.5.2 Reflections 262
6.5.3 Dynamic Range of the Signal 262
6.5.4 Dynamic Range of the Measurand 263
6.5.5 Linearity in dΦ-DVS Systems 264
6.5.6 Optical Non-Linearity 266
6.6 Dynamic Measurements Converted to Strain 267
6.7 Static Measurements with Coherent Rayleigh Backscatter 267
6.7.1 Static vs. Dynamic Measurements 267
6.7.2 Coherent Rayleigh Backscatter in Static Measurements 267
6.7.2.1 Illustration of the coherent Rayleigh response to static measurements 267
6.7.2.2 Relationship between the backscatter power spectrum and the refractive index variations 271
6.7.3 Obtaining Location-Dependent Data from Frequency-Domain Rayleigh Backscatter 271
6.7.3.1 Frequency-domain-only interrogation 271
6.7.3.2 Combined time-/frequency-domain interrogation 272
6.7.4 Limitations in Frequency-Scanned Coherent Rayleigh Backscatter 272
6.7.5 Quasi Static Strain Measurements as a Vibration Sensor 273
6.8 Linearisation of the Dynamic Strain Measurement in Intensity Coherent OTDR 274
6.9 Summary 274
References 275

PART THREE Applications of Distributed Sensors 279

7 Applications of Distributed Temperature Sensors (DTSs) 281
7.1 Energy Cables: Dynamic Rating of Carrying Capacity 281
7.2 Transformers 282
7.3 Generators and Circuit Breakers 283
7.4 Fire Detection 283
7.5 Process Plant 284
7.6 Cryogenic Measurements 285
7.7 Pipelines: Leak Detection and Flow Assurance 286
7.8 Borehole Measurement (Geothermal, Oil and Gas) 288
7.8.1 Completion of Oil and Gas Wells 288
7.8.2 Deployment of Optical Fibres in Wells 290

7.8.3 Steam-Assisted Heavy Oil Production 294
7.8.4 Conventional Oil 295
7.8.5 Thermal Tracer Techniques 297
7.8.6 Injector Wells: Warm Back and Hot Slugs 297
7.8.7 Intervention Services 298
7.8.8 Geothermal Energy Extraction 299
7.8.9 CO_2 Sequestration 299
7.8.10 Gas Wells 299
7.8.11 Gas Lift Valves 300
7.8.12 Electric Submersible Pumps (ESP) 300
7.8.13 Well Integrity Monitoring 301
7.8.14 Hydrate Production 301
7.8.15 Subsea DTS Deployment 302
7.9 Cement and Concrete Curing 302
7.10 Hydrology 303
7.10.1 Dams and Dykes 303
7.10.2 Near-Surface Hydrology, Streams and Estuaries 304
7.11 Environment and Climate Studies 305
7.12 Building Efficiency Research 305
7.13 Nuclear 306
7.14 Mining 307
7.15 Conveyors 307
7.16 Measurements Inferred from Temperature 308
7.16.1 Thermal Measurement of Air Flow 308
7.16.2 Liquid-Level Measurement 309
7.16.3 Performance of Active Fibre Components 310
7.17 Summary 310
References 310

8 Distributed Strain Sensors: Practical Issues, Solutions and Applications 315
8.1 Separation of Temperature and Strain Effects 315
8.1.1 Separation through Cable Design 315
8.1.2 Hybrid Measurement 316
8.1.2.1 Spontaneous Brillouin intensity and frequency measurement 316
8.1.2.2 Raman/Brillouin hybrid approach 316
8.1.2.3 Coherent Rayleigh/Brillouin hybrid 317
8.1.2.4 Brillouin/fluorescence 317
8.1.3 Discrimination Based Solely on Features of the Brillouin Spectrum 317
8.1.3.1 Independent optical cores 317
8.1.3.2 Multiple Brillouin peaks with distinct properties 318
8.1.3.3 Polarisation-maintaining fibres and photonic crystal fibres 318
8.1.3.4 Few-mode fibres 318
8.1.3.5 Multi-wavelength interrogation 319
8.1.3.6 Direct measurement of the Brillouin beat spectrum 319
8.1.3.7 Fibres designed for Brillouin measurand discrimination 320
8.2 Inferring Other Measurands from Temperature or Strain 322
8.2.1 Pressure 322
8.2.2 Liquid Flow 322
8.2.3 Chemical Detection 323
8.3 Polymer Fibre Based Strain Measurement 324

8.4 Applications of Distributed Strain Measurements 325
8.4.1 Mechanical Coupling and Limitations on Strain Measurements 325
8.4.2 Energy Cables 326
8.4.3 Telecommunications 326
8.4.4 Civil Engineering 327
8.4.4.1 Tunnels 327
8.4.4.2 Crack detection 327
8.4.4.3 Damage in offshore structures 328
8.4.5 Nuclear 328
8.4.6 Shape Sensing 328
8.4.7 Composite Structural Health Monitoring 329
8.4.8 Casing Deformation 329
8.4.9 Pipeline Integrity Monitoring 329
8.4.10 Compaction and Subsidence in Subterranean Formations 330
8.4.11 Ground Movement and Geomechanics 330
8.4.12 River Banks and Levees 332
8.4.13 Security/Perimeter Fence 332
8.5 Summary 333
References 333

9 Applications of Distributed Vibration Sensors (DVSs) 339
9.1 Active Seismic Applications 340
9.1.1 Surface Seismic Acquisition 340
9.1.2 Borehole Seismic Surveying 341
9.1.2.1 Introduction to borehole seismic acquisition 341
9.1.2.2 Optical borehole seismic acquisition on wireline cable 343
9.1.2.3 Borehole seismic acquisition using permanently installed optical cables 345
9.1.3 Response of Distributed Vibration Sensors to Acoustic or Seismic Waves 346
9.1.3.1 Distance resolution in differential phase DVS systems 346
9.1.3.2 Nature of the measurement 347
9.1.3.3 Gauge length effects 348
9.1.3.4 Angular and directional response 353
9.1.3.5 Interaction of wavenumber filtering and angular response 354
9.1.3.6 Polarity of the seismic signal captured with DVS 356
9.1.3.7 Pulse duration 356
9.1.3.8 Sampling rates 358
9.1.3.9 Diversity in borehole seismic acquisition with DVS 358
9.1.4 Examples of Borehole Seismic DVS Data 359
9.1.5 Coupling in Borehole Seismic 361
9.1.6 DVS in Surface Seismic 363
9.2 Security – Perimeter, Pipeline Surveillance 364
9.3 Transportation 370
9.4 Energy Cables 371
9.4.1 Subsea Cables 371
9.4.2 Detection of Faults and Partial Discharge 372
9.4.3 Wind and Ice Loading in Overhead Cables 372
9.5 Production and Integrity Applications in Boreholes 373
9.5.1 Flow Monitoring 373
9.5.1.1 Total vibration energy 374
9.5.1.2 Spectral discrimination in DVS analysis 374

9.5.1.3 Tube wave velocity 376
9.5.1.4 Turbulent structure velocity 376
9.5.1.5 Ultra-low frequency 376
9.5.1.6 Sensor placement 377
9.5.1.7 Conveyance methods for DVS 377
9.5.1.8 Applications of DVS flow monitoring in boreholes 377
9.5.2 Hydrofracturing Monitoring Using Microseismic Events 378
9.5.3 Artificial Lift Monitoring 378
9.5.4 Perforation 380
9.5.5 Well Abandonment 381
9.6 Multimode Fibre in Distributed Vibration Sensing 381
9.7 Interpretation of DVS Data 382
References 382

10 Other Distributed Sensors 387
10.1 Loss-Based Sensors 387
10.1.1 Micro-Bending-Based Sensors 387
10.1.1.1 Safety barriers 389
10.1.1.2 Linear heat detection 389
10.1.1.3 Moisture/hydrocarbon sensing 390
10.1.2 Radiation Detection 390
10.1.3 Loss of Numerical Aperture 391
10.1.4 Rare-Earth-Doped Fibre Temperature Sensing 392
10.2 Fluorescence 394
10.3 Changes in Capture Fraction 394
10.4 Polarisation Optical Time-Domain Reflectometry (POTDR) 395
10.4.1 Basic Principles of POTDR 395
10.4.2 POTDR in High-Birefringence Fibres 396
10.4.3 Polarisation Optical Frequency-Domain Reflectometry 398
10.4.4 Dual-Wavelength Beat Length Measurements 398
10.4.5 Full Polarimetric Analysis of POTDR Signals 399
10.4.6 Alternative Approaches to Estimating the Distribution of the State of Polarisation 400
10.4.7 Applications of POTDR 400
10.4.7.1 Distributed pressure sensing 400
10.4.7.2 Current/magnetic fields 401
10.4.7.3 Polarisation mode dispersion 402
10.4.7.4 Vibration sensing 402
10.5 Blackbody Radiation 403
10.6 Forward-Scattering Methods 403
10.6.1 Mode Coupling in Dual-Mode Fibres 404
10.6.2 Modalmetric Sensors 405
10.6.3 Distributed Interferometers 406
10.7 Non-Linear Optical Methods 408
10.7.1 Stimulated Raman 409
10.7.2 The Kerr Effect in Distributed Temperature Sensing 409
10.7.3 The Kerr Effect for POTDR in High-Birefringence Fibres 410
10.7.4 Two-Photon Fluorescence 411
10.8 Closing Comments 413
References 413

11 Conclusions **419**
11.1 Selection Process 421
11.2 Choice of Distributed Sensors vs. Multiplexed Point Sensors 423
11.3 Trends in DOFS 424
References 426

Index 427

Series Preface

Optical fibers are considered among the top innovations of the twentieth century, and Sir Charles Kao, a visionary proponent who championed their use as a medium for communication, received the 2009 Nobel Prize in Physics. Optical fiber communications have become an essential backbone of today's digital world and internet infrastructure, making it possible to transmit vast amounts of data over long distances with high integrity and low loss. In effect, most of the world's data flows nowadays as light photons in a global mesh of optical fiber conduits. As the optical fiber industry turns fifty in 2016, the field might be middle aged, but many more advances and societal benefits are expected of it.

What has made optical fibers and fiber-based telecommunications so effective and pervasive in the modern world? It is its intrinsic features and capabilities make it so versatile and very powerful as an enabling and transformative technology. Among their characteristics we have their electromagnetic (EM) immunity, intrinsic safety, small size and weight, capability to perform multi-point and multi-parameter sensing remotely, and so on. Optical fiber sensors stem from these same characteristics. Initially, fiber sensors were lab curiosities and simple proof-of-concept demonstrations. Nowadays, however, optical fiber sensors are making an impact and serious commercial inroads in industrial sensing, biomedical applications, as well as in military and defense systems, and have spanned applications as diverse as oil well downhole pressure sensors to intra-aortic catheters.

This transition has taken the better part of thirty years and has now reached the point where fiber sensor operation and instrumentation are well understood and developed, and a variety of diverse variety of commercial sensors and instruments are readily available. However, fiber sensor technology is not as widely known or deeply understood today as other more conventional sensors and sensing technologies such as electronic, piezoelectric, and MEMS devices. In part this is due to the broad set of different types of fiber sensors and techniques available. On the other hand, although there are several excellent textbooks reviewing optical fiber sensors, their coverage tends to be limited and do not provide sufficiently in-depth review of each sensor technology type. Our book series aims to remedy this by providing a collection of individual tomes, each focused exclusively on a specific type of optical fiber sensor.

The goal of this series has been from the onset to develop a set of titles that feature an important type of sensor, offering up-to-date advances as well as practical and concise information. The series encompasses the most relevant and popular fiber sensor types in common use in the field, including fiber Bragg grating sensors, Fabry-Perot sensors, interferometric sensors, distributed fiber sensors, polarimetric sensors, polymer fiber sensors, structural health monitoring (SHM) using fiber sensors, biomedical fiber sensors, and several others.

This series is directed at a broad readership of scientists, engineers, technicians, and students involved in relevant areas of research and study of fiber sensors, specialty optical fibers, instrumentation, optics and photonics. Together, these titles will fill the need for concise, widely accessible introductory overviews of the core technologies, fundamental design principles, and challenges to implementation of optical fiber-based sensors and sensing techniques.

This series has been made possible due to the tenacity and enthusiasm of the series manager, Ms. Luna Han, to whom I owe a debt of gratitude for her passion, encouragement and strong support – from the initial formulation of the book project and throughout the full series development. Luna has been a tremendous resource and facilitator, and a delight to work with in the various stages of development for each of the series volumes.

Information, as the saying goes, is knowledge. And thanks to the dedication and hard work of the individual volume authors as well as chapter co-authors, the readers have enriched their knowledge on the subject of fiber optic sensors. I thank all the authors and extend my deep appreciation for their interest and support for this series and for all the time and effort they poured into its writing.

To the reader, I hope that this series is informative, fresh and of aid in his/her ongoing research, and wish much enjoyment and success!

Alexis Méndez, PhD
Series Editor
President
MCH Engineering, LLC
Alameda, California

Foreword

Fibre optic sensor technology was initially suggested about half a century ago. By the early 1980s, the thought that this basic principle could be used in a so-called distributed architecture began to emerge. It is an exciting idea – lay out a perfectly normal fibre and look out for the way the environment changes the detail of the fibre transmission characteristics. Detect these changes by using a variant on optical radar – the optical time-domain reflectometer (OTDR) – and relate the detected changes to changes in the environment located around the fibre at the point to which the OTDR is tuned. Thence appears a map of what is happening as function of position along the fibre itself! It all seems too good to be true!

Indeed, whilst the system designers were convinced, the prospective users of this magical device were initially far from convinced. It took until the late 1990s for the quite amazing capabilities of this idea to become appreciated, and since then, distributed optical fibre sensing has emerged as the preferred approach to sensing events along long, thin structures like tunnels, pipelines and oil wells. The system has also become progressively more versatile, both in performance capability and in parameters that can be addressed.

Arthur Hartog is most certainly amongst the very best placed to record this history, to enlighten readers on the principles and limitations of the technique and to shed insight on the ever-expanding portfolio of applications. Arthur has been closely involved from the very beginning as both an active researcher and as a very successful industrial protagonist, improving on the basics, appreciating the many faceted needs in the practical environment and ensuring that these diverse needs are effectively satisfied.

This book reflects this wealth of experience and expertise, giving the most authoritative account available of an important and ever expanding technology. The text encompasses basic principles and established and emerging applications and insights into ongoing development. It will appeal to the student as well as the practising engineer and also presents insight into current and prospective applications.

Distributed systems are the predominant success story in fibre-optic sensing. This book presents an authoritative account of this important topic benefitting from the author's unique experiences and insights. It is an invaluable addition to the technical literature and promises to become the essential reference text for many, many years to come!

Brian Culshaw
Glasgow

Preface

The field of distributed optical fibre sensors (DOFSs) has grown from an intriguing concept in the early 1980s to a technology that has widespread usage today, and it continues to be the subject of very active research. Distributed sensors have become a *de facto* standard way of operating, for example, in the monitoring of high-voltage transmission cables, certain fire detection applications and, increasingly, in the monitoring of boreholes in geothermal and hydrocarbon production. These practical applications have, in turn, led to the development of a number of industrial teams dedicated to the supply of distributed fibre sensors and the interpretation of the results.

The research into DOFS is documented in a large, and growing, set of scientific papers, conference presentations and patent applications. DOFSs are also described in single chapters in books on the more general subject of optical fibre sensors. However, there is no single volume dedicated to the subject of DOFS that provides the detail that students or engineers wishing to enter the field need to become effective contributors. I am frequently asked by new engineers or interns for background reading on DOFS and I have usually provided them with references to review articles. However, I usually feel that the available documents are unsatisfactory because they cover only selected parts of the field whilst assuming too much prior knowledge on the part of the reader.

The aim of this book is therefore to provide, in a single volume, sufficient background on the technology and applications to allow people new to the field to gain rapidly a view of the entire subject; I hope it will also be a useful reference for those of us who have worked on this topic for some time.

On a personal note, the field of distributed sensing has held my interest, in one way or another, for most of my professional life, with ever more fascinating science to explore whilst refining the design of interrogation systems and discussing applications with clients and partners. The many applications of the technology have also opened doors into vast swathes of engineering that can benefit from DOFS technology. This has led me to electrical substations and transformer manufacturing sites, down coal mines, onto dam construction sites, oil production rigs and many more. As someone with a keen interest in science and technology, DOFS has provided me with a wonderful insight into many fields of science and engineering.

The writing of this book was prompted by an invitation of the Series Editor Alexis Méndez (MCH Engineering LLC) and Luna Han, Senior Publishing Editor at Taylor & Francis, and I am grateful to them for their initial comments on the proposal, and to Luna for her suggestions and support throughout this project. I am also grateful to Adel Rosario at the Manila Typesetting Company (MTC) who oversaw the production phase of the publication process most effectively and to Anna Grace de Castro, Lisa Monette and Mamta Jha at MTC for their careful copyediting, proofreading and indexing.

The outline of this work was reviewed by several anonymous reviewers. I am grateful to all of them for their supportive and constructive comments that have helped shape the way in which the material is presented.

I thank Brian Culshaw (University of Strathclyde) for kindly writing a foreword to this volume.

The first demonstration of a distributed temperature sensor with my co-inventor, David Payne, would not have occurred without the doctoral training that he, Mike Adams and Alec Gambling provided within the Optical Fibre Research Group at Southampton University and the support of members of that group, including Alan Conduit, Frédérique de Fornel, Martin Gold, Bob Mansfield, Steve Norman, Cathy Ragdale, Francis Sladen and Eleanor Tarbox.

This book is heavily influenced by my interaction with friends and colleagues at York Sensors over the time that this distributed sensing technology has developed, whether on the technology or the commercial aspects. I would like to thank Glen Fowler, Roger Hampson, Richard Kimish, Adrian Leach, Darryl Marcus-Hanks, Richard Marsh, Iain Robertson, Pat Ross, Jane Rowsell, Robert Theobald,

Peter Travis, Peter Wait and Val Williams amongst those who contributed in one way or another to the success of this technology. As the ownership of York Sensors moved to Sensor Highway, George Brown, Nigel Leggett and Glyn Williams were instrumental in moving the technology to the oilfield, ultimately within Schlumberger. In more recent years, I have also had the pleasure of working at the Schlumberger Fibre-Optic Technology Centre with a number of colleagues, including Dom Brady, Yuehua Chen, Alexis Constantinou, Matt Craig, Theo Cuny, Tim Dean, Florian Englich, Alireza Farahani, Max Hadley, Will Hawthorne, Graeme Hilton, Ian Hilton, Kamal Kader, Gareth Lees, Adam Stanbridge, Paul Stopford and many others (some of whom were previously at York Sensors). The initial demonstrations of oil-field applications of distributed vibration sensors were greatly facilitated by collaboration with colleagues within Schlumberger, including William Allard, Mike Clarke, Richard Coates, Bernard Frignet, Duncan Mackie, Doug Miller and Merrick Walford and many colleagues at the Schlumberger Gould Research Centre (Cambridge, United Kingdom), Schlumberger Doll Research Centre (Cambridge, Massachusetts, USA) and the Schlumberger Moscow Research Centre.

The development of distributed temperature sensor (DTS) technology at York Sensors was considerably assisted by the close collaboration with our then Japanese partners, initially part of Nippon Mining and later becoming Y.O. Systems and now YK Gikken. The support and friendship of their founder Osamu Yasuda (now deceased) and of Shunsuke Kubota is an enduring contribution to the field.

I have the great pleasure of working with a team at the Peter the Great St. Petersburg Polytechnical University led initially by Oleg I. Kotov and now by Leonid Liokumovich together with their colleagues Andrey Medvedvev, Nikolai Ushakov, Artem Khlybov and (at St Petersburg State University) Mikhail Bisyarin. Their contributions, spanning more than 10 years, to Raman distributed sensing and more recently to distributed vibration sensing are greatly appreciated as is their continuing collaboration on new topics. Other aspects of my understanding of the subject were enhanced by collaborations with Trevor Newson and his students at Southampton University, particularly Yuh Tat Cho, Mohammad Belal and Mohamed Alahbabi. More recently, I have appreciated the support of Dimitris Syvridis and his colleagues at the National and Kapodistrian University of Athens, Greece, and I have also enjoyed a fruitful collaboration with Daniela Donna (Ecole des Mines – Paritech) and James Martin (then at the Institut de Physique du Globe de Paris).

Although this work is a personal project, I am happy to acknowledge the support and encouragement of Simon Bittleston (Schlumberger Vice President, Research) and Frédérique Kalb (then Centre Manager at the Schlumberger Fibre-Optic Technology Centre). I am also keen to recognise the assistance of the Schlumberger librarians, particularly Clare Aitken and Jacqui Wright, in finding a number of obscure references.

A special word of thanks is due to my friends and colleagues Will Hawthorne, Dom Brady, George Brown, Alexis Constantinou and Paul Dickenson, as well as to Jennifer Hartog, for their comments on various parts of the manuscript.

Finally, and most importantly, I want to thank my wife, Christine Maltby, in addition to her careful proofreading of the manuscript, for her unwavering and unquestioning support, love and patience for the several years that I have hidden away working on this project and for our many happy years together.

Arthur H. Hartog
Martyr Worthy, Winchester, United Kingdom

Author

Arthur H. Hartog earned a BSc in Electronics at the University of Southampton, where he was first exposed to the world of optical fibres. He followed up with a PhD in the Optical Fibre Research Group in a team led by Prof. W.A. Gambling at the same university. In his doctoral and post-doctoral research, he focused on measurements of light propagation in optical fibres, including research in optical time-domain reflectometry. This work led to the first demonstration of a distributed optical fibre sensor in 1982.

In 1984, Arthur moved to York Ltd., a start-up providing instrumentation for manufacturers of optical fibre preforms, fibres and cables as well as speciality optical fibres (these businesses still continue as part of Photon Kinetics, Inc for the instrumentation and as Fibercore Ltd. for the speciality fibre). After initial work on optical time-domain reflectometers (OTDRs) and other measurements at York Ltd., he set up a team to design and commercialise the first Raman distributed temperature sensor, a business that was eventually separated as York Sensors Ltd. During his time at York and York Sensors, he dealt with most aspects of the product life, from research to manufacturing, including applications development and working with strategic overseas partners.

After several changes of ownership, York Sensors became part of Schlumberger, the leading oilfield services company. As a Schlumberger Fellow, Arthur continues to research optical fibre sensors and their applications to the oilfield whilst providing advice on wider matters within the company.

Arthur Hartog is a senior member of the Institute of Electrical and Electronics Engineers and of the Optical Society of America and a member of the European Association of Geoscientists & Engineers.

List of Symbols

a	core radius
a_e	amplitude of element electric field (in model of coherent scattering)
A_e	amplitude of electric field summation (in model of coherent scattering)
A_{eff}	effective area of a mode
A_V	forward gain of a voltage amplifier
B	backscatter capture fraction
B'	normalised backscatter capture fraction
$B_{as,s}$	backscatter capture fraction for the anti-Stokes and Stokes Raman signals
B_T	isothermal compressibility
B_w	noise bandwidth of a receiver
c	speed of light in vacuum
$C_{\nu_B\varepsilon}$, C_{ν_BT}	coeffecient of the Brillouin frequency shift due to strain and temperature, respectively
$C_{I_B\varepsilon}$, C_{I_BT}	coeffecient of the Brillouin intensity due to strain and temperature, respectively
C_T	total capacitance at the input node of an optical receiver
D_a	peak displacement due to a seismic wave
e	electron charge
E_b	electric field of backscatter signal
E_{LO}	electric field of local oscillator
E_m	electric field distribution of mode m
E_p	probe energy
E_{ph}	photon energy
E_s	electric field summation in modelling of coherent backscatter
E_{tot}	sum of local oscillator and backscatter field in coherent-detection OTDR
E_Y	Young's modulus
f_b	beat frequency relating to a location in OFDR
$f_{b_{max}}$	maximum value of the beat frequency in OFDR
f_d	sampling frequency
f_m	modulation frequency (in BOCDA)
$f_p(r)$	refractive index profile function
f_R	pulse repetition frequency
$\mathcal{F}$	finesse (of a Fabry-Pérot interferometer)
$F(M_G)$	excess noise multiplication factor for an APD
g_B	Brillouin gain
g_m	transconductance of a field-effect transistor
h	Planck's constant
H_{SBS}	transfer function of the Brillouin gain process
i_{fe}	channel thermal noise of an FET
i_{het}	heterodyne signal term in the output of a coherent OTDR
i_n	receiver current noise spectral density
i_{RF}	Johnson noise current density
I_{as}	intensity of the anti-Stokes Raman backscatter signal
I_{du}, I_{dm}	unmultiplied and multiplied (respectively) detector dark current
I_G	gate leakage current of an FET

I_s	intensity of the Stokes Raman backscatter signal
k_A	acoustic wavenumber
K_{as}, K_s	constants used in the expressions for the anti-Stokes Raman and Stokes Raman intensities (respectively) to account for losses in the optical system
k_B	Boltzmann's constant
L	fibre length
L_{eff}	effective fibre length for non-linear interactions
L_G	gauge length
m_d	modulation depth (in BOCDA)
M	code length in pulse-compression OTDR
M_{DTS}	figure of merit for DTS systems
M_f	number of degrees of freedom for coherent Rayleigh backscatter statistics
M_G	gain of an avalanche photodiode
n_1, n_2	refractive index in core and cladding, respectively
n_{eff}	effective index for a mode
n_x, n_y	effective index for the principal modes of a birefringent optical fibre
N_g	group refractive index
N_s	number of independently sampled points
NA	numerical aperture
p_{11}, p_{12}	photoelastic constants
$P_{BS}(t)$	backscatter power vs time; suffices $BS1$ and $BS2$ are used to distinguish power measured from End 1 and End 2 of a fibre loop
P_d	probability that one photon will be detected in photon counting DTS
P_{LO}	local oscillator power (coherent detection)
P_p	probe pulse power
$P_s(z)$	signal power reaching an optical receiver
$\Pr(P_{BS}(z))$	probability distribution for coherent Rayleigh backscatter power
q	electron charge
Q	detector quantum efficiency
R_d	detector responsivity
R_F	feedback resistor value (Transimpedance amplifier)
$R(T(z))$	anti-Stokes/Stokes Raman ratio; suffices $E1$ and $E2$ are used to distinguish power ratios measured from End 1 and End 2 of a fibre loop
$\bar{R}(T(z))$	normalised anti-Stokes/Stokes Raman ratio
S_{LE}	temperature sensitivity coefficient of the natural logarithm of the anti-Stokes/Stokes Raman ratio
T	temperature
T_A	ambient temperature
T_C	temperature of the channel of an FET
T_f	fictive temperature (of a glass)
T_r	sweep duration in OFDR
T_{ref}	temperature of the reference coil
v_g	group velocity
V	normalised frequency
V_A	acoustic velocity
V_{sw}	voltage swing available at the output of a transimpedance preamplifier
z	distance variable along an optical fibre
z_{sc}^m	distance to the mth scatterer in coherent backscatter model
α	fibre attenuation per unit length
α_p, α_{as}, α_s	attenuation at the probe, anti-Stokes and Stokes Raman wavelengths, respectively

α_{ras}	anti-Stokes Raman scattering coefficient at the probe wavelength
α_{ray}	Rayleigh scattering coefficient at the probe wavelength
α_{rs}	Stokes Raman scattering coefficient at the probe wavelength
β_m	propagation constant of mode m
γ	sweep rate in OFDR
Γ	excess channel noise factor of an FET
Δ	relative index difference
Δf	frequency sweep range in C-OFDR
ΔL	path imbalance in an interferometer
ΔT	temperature differential
$\Delta\alpha$	differential attenuation between the anti-Stokes and Stokes wavelengths in Raman DTS systems
$\Delta\omega$	angular Frequency offset (coherent detection)
δT	temperature resolution
δz	spatial resolution
δz_{PC}	bin size in photon counting DTS
$\delta\zeta^m$	random variable used in the modelling of coherent Rayleigh scattering and representing the location of the mth scatterer within a distance interval
$\delta\tau_{GI}$	range of transit times between modes of a graded-index optical fibre
$\delta\tau_{SI}$	range of transit times between modes of a step-index optical fibre
ε_{as}	relative sensitivity of the Raman anti-Stokes intensity to temperature
ε_f	fibre strain
ε_R	relative sensitivity of the Raman intensity ratio to temperature
ε_s	relative sensitivity of the Raman Stokes intensity to temperature
η	backscatter factor; subscripts *as* and *s* refer to anti-Stokes and Stokes, respectively
$\theta_1, \theta_2, \theta_c, \theta_e$	angles used in the construction of Figure 2.1b
λ	optical wavelength
λ_A	acoustic (seismic) wavelength
λ_{as}	anti-Stokes Raman wavelength
λ_c	cut-off wavelength of the second mode of an optical fibre
λ_e	fluorescence wavelength
λ_p	probe wavelength
λ_s	Stokes Raman wavelength
$\Lambda(z)$	loss distribution measured with the double-ended OTDR method
ν	optical frequency
ν_B	Brillouin frequency shift
ν_p	Poisson's ratio
ν_R	Raman frequency shift
ξ	photoelastic correction factor for the phase change with strain (Chapter 6)
ρ	density (of silica)
τ_f	impulse response of the fibre
τ_p	probe pulse duration
τ_{rx}	impulse response of the receiver
τ_R	time constant at front end of a transimpedance preamplifier
Φ_e	phase of electric field summation (in model of coherent scattering)
ϕ_e	phase of element electric field (in model of coherent scattering)
ω	carrier angular frequency
ω_0	carrier angular frequency (coherent detection)

Note: Certain symbols, used only locally in the text, have been omitted from this table.

List of Abbreviations

A/D, A/D C	analog-to-digital converter
AOM	acousto-optic modulator
APD	avalanche photodiode detector
ASE	amplified spontaneous emission
BOCDA	Brillouin optical coherence domain analysis
BOCDR	Brillouin optical coherence domain reflectometry
BOFDA	Brillouin optical frequency-domain analysis
BOTDA	Brillouin optical time-domain analysis
BOTDR	Brillouin optical time-domain reflectometry
C-OFDR	coherent optical frequency-domain reflectometry
C-OTDR	coherent optical time-domain reflectometry
CD	chromatic dispersion
CPOTDR	computational POTDR
CT	coiled tubing
dΦ-DVS	differential phase distributed vibration sensor(s)
DAS	distributed acoustic sensor(s)
DFB	distributed feedback
DOFS	distributed optical fibre sensor(s)
DSTS	distributed strain and temperature sensor(s)
DTS	distributed temperature sensor(s)
DVS	distributed vibration sensor(s)
EDFA	erbium-doped fibre amplifier
EOM	electro-optic modulator
EOR	enhanced oil recovery
ESP	electrical submersible pump
FBE	frequency band energy
FBG	fibre Bragg grating
FET	field-effect transistor
FMCW	frequency-modulated, continuous wave
FPI	Fabry-Pérot interferometer
FRM	Faraday rotation mirror
FSR	free spectral range
GLV	gas lift valve
GOR	gas-to-oil ratio
HEMT	high electron-mobility transistor
HFM	hydraulic fracture monitoring
IF	intermediate frequency
IR	infrared
LED	light-emitting diode
LNG	liquefied natural gas
LPR	Landau-Placzek ratio
MC-PC	multi-channel photon counting
MCVD	modified chemical vapour-phase deposition

MEMS	micro-electro-mechanical systems
MI	modulation instability
MOPA	master oscillator, power amplifier optical source
MZI	Mach-Zehnder interferometer
NA	numerical aperture
OFDR	optical frequency-domain reflectometry
OLO	optical local oscillator
OPGW	optical ground wire
OTDR	optical time-domain reflectometry
OVPO	outside vapour phase oxidation
PCC	pulse-compression coding
PCF	photonic crystal fibre
PD	photo-diode
PMT	photomultiplier tube
POF	polymer optical fibres
POFDR	polarisation optical frequency-domain reflectometry
POTDR	polarisation optical time-domain reflectometry
PRF	pulse repetition frequency
PRM	permanent reservoir monitoring
r.m.s.	root-mean square
RF	radio frequency
RIA	radiation-induced attenuation
SA-BOTDA	slope-assisted BOTDA
SAGD	steam-assisted, gravity drainage method for heavy oil recovery
SBS	stimulated Brillouin scattering
SF-BOTDA	sweep-free BOTDA
SNR	signal-to-noise ratio
SOA	semiconductor optical amplifier
SOP	state of polarisation
SPAD	single photon avalanche detector
SpBS	spontaneous Brillouin scattering
SPM	self-phase modulation
SRS	stimulated Raman scattering
SNSPD	superconducting nanowire single photon detector
SWE	short-wavelength edge
TAC	time-to-amplitude converter
TC-SPC	time-correlated single photon counting
TDC	time-to-digital converter
TDR	time-domain reflectometry
TE	thermo-electric (device)
UV	ultraviolet
VAD	vapour-phase axial deposition
VDA	viscosity-diverting acid
VIT	vacuum-insulated tubing
VIVSP	vertical-incidence VSP
VSP	vertical seismic profiling
ZOVSP	zero-offset VSP

PART ONE

Fundamentals

Introduction 1

Sensors are the means by which inputs about the real world are provided to electronic systems that archive and process the data collected; often, these systems also make decisions based upon that information. Sensors convert measurands,* such as temperature, flow, pressure or strain, from the physical world into signals that can be read in electronic form.

A stylised illustration of the industrial use of sensors is shown in Figure 1.1, where a petro-chemical process taking place in the bulbous vessel (which is typically a refractory-lined pressure housing, shown to the right) is monitored by a number of sensors that penetrate the housing through apertures that preserve the pressure seal and report their readings to the control room. Here, the data are not only normalised and archived but also passed to a control computer that monitors the process and issues signals to control valves and actuators that keep the process within intended parameters.

1.1 THE CONCEPT OF DISTRIBUTED OPTICAL FIBRE SENSORS

1.1.1 Optical Fibre Sensors

Optical fibre sensors are devices that use light to convey the information which they sense. A typical optical fibre sensor system (Figure 1.2) consists of the sensor itself that is *probed* by the input light which it modulates in accordance with the value of the measurand. The system also includes an optical fibre within a *transit cable* that conveys the probe light to the sensor and returns the modulated light from the sensor to the interrogator through a *patch panel* that houses the connectors or splices. The interrogator is the opto-electronic system that emits the probe light and converts the returned light into an electrical signal which is processed to create the output of the system. The sensor is designed to respond to the intended measurand and ideally not at all to any other external influence. Likewise, the transit cable, that carries the probe and the returning optical signal, should protect the fibres it contains from external influence and from damage caused by the environment through which it passes.

This book is about a particular family of sensors, *distributed optical fibre sensors* (DOFSs), that determine the spatial distribution of a measurand along a section of fibre, often many kilometres long, rather than measuring the physical parameter of interest in one location only. DOFSs use optical fibres both as the sensing element and as the means of carrying the optical signals used for this purpose.

In the early 1970s, rapid advances were made in the technology of optical fibres for telecommunications, including low-loss fibres [1], reliable laser diodes operating at room temperature [2] and an understanding of the design of transmitting and receiving electronics. This progress led to pioneering thoughts about applying this same technology to physical [3–6] and even chemical [7] sensing. A number of concepts appeared in the scientific literature at the time, demonstrating that optical fibres could be used

* Throughout this book, the term *measurand* is used to denote the input to a sensor that is converted into an optical or electronic signal.

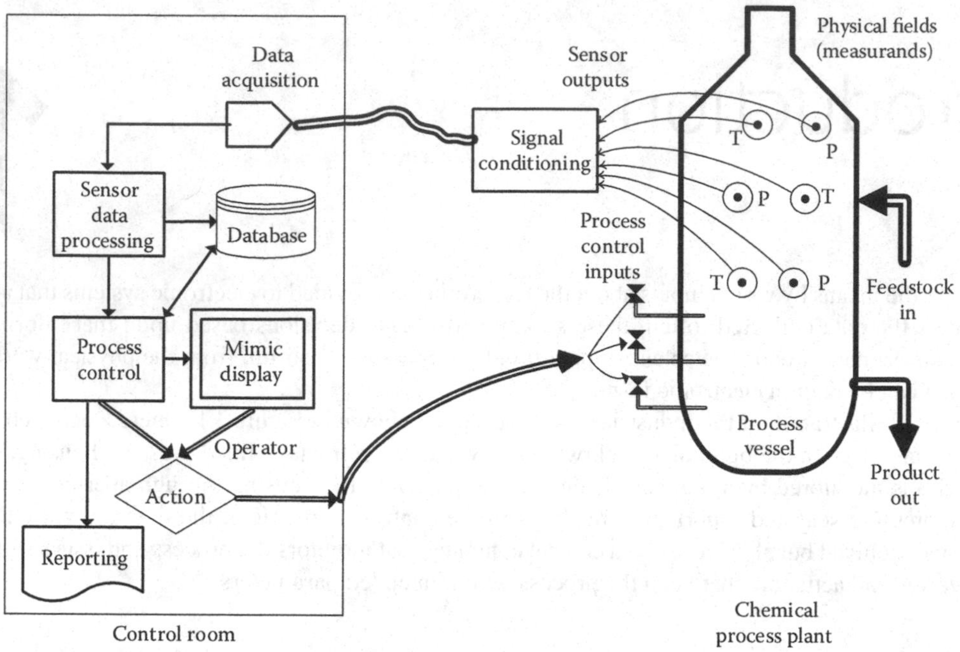

FIGURE 1.1 Simplified representation of a chemical process plant, its sensor systems and process control.

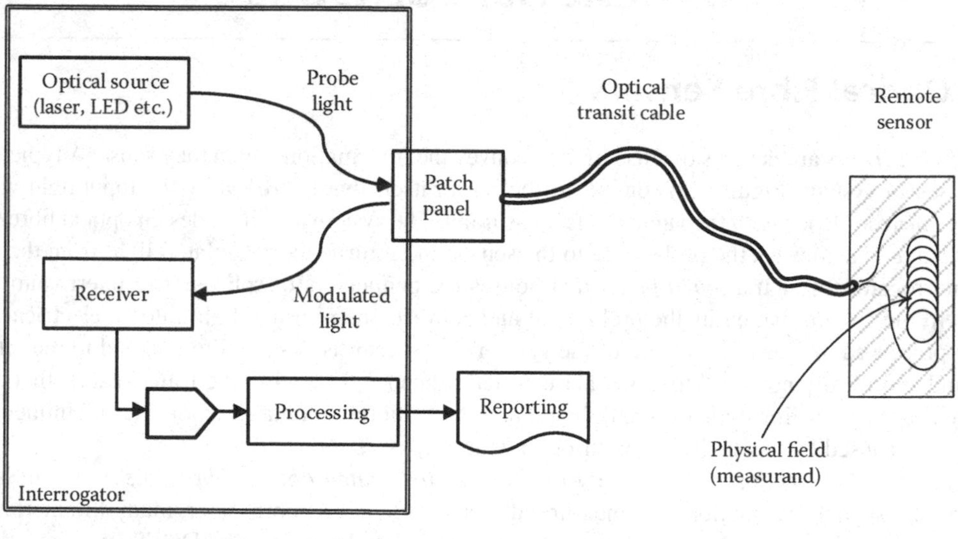

FIGURE 1.2 General arrangement of an optical fibre sensor system.

as sensors for a wide variety of measurands. The literature also showed that many attributes of the light travelling in the fibre could be applied to the sensing task, including its intensity [8], polarisation [9], phase [10], propagation time [6], optical spectrum [11] and coherence [12].

Thus, a highly versatile sensing technology emerged, able to provide compact, lightweight sensors that could be shaped to match a wide variety of space, power and environmental constraints. In particular, the ability of optical sensors to convey the information in optical form from a hostile environment to the interrogating electronics located in a more benign site enabled the development of sensors operating in very high-temperature conditions, for example, in borehole applications.

INTRINSIC VS. EXTRINSIC OPTICAL FIBRE SENSORS

At this point, we should make the distinction (illustrated in Figure 1.3) between *intrinsic* (Figure 1.3a) and *extrinsic* (Figure 1.3b) optical fibre sensors. In intrinsic sensors, the light stays within an optical fibre throughout the system and is modulated within the fibre, for example, by induced loss, changes to its spectrum or polarisation and so on. In contrast, in extrinsic sensors, the sensor is a bulk-optic device, such as an electro-optic crystal, or a strain-birefringent component placed between crossed polarisers; in another example, it might be a fluorescent element that absorbs probe light from the fibre and re-emits fluorescent light at a different wavelength that is collected by the transit fibre and taken back to the interrogator. The illustrations of Figure 1.3 capture only a small fraction of the many intrinsic and extrinsic sensor concepts that have been explored. In each of the examples, the measurand that is captured is indicated above or below as is (to the right) the physical process invoked in the transduction process and how the information is encoded onto the input optical signal.

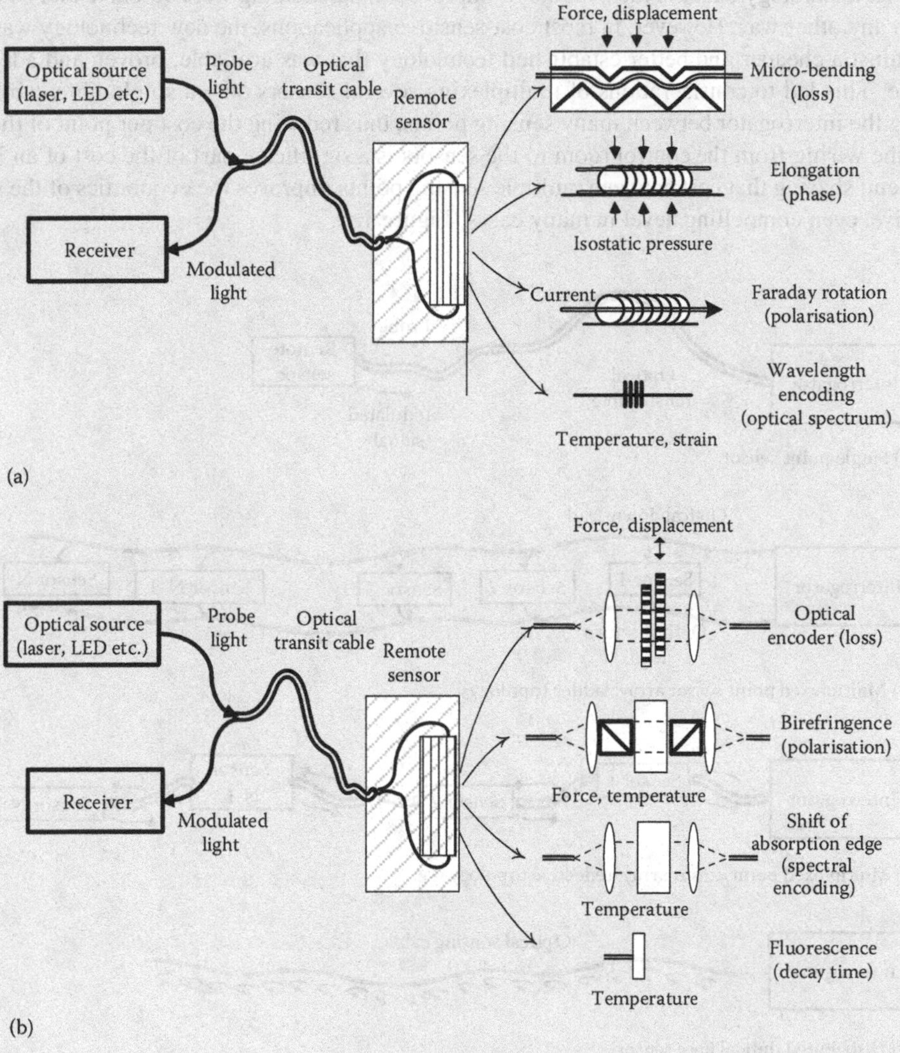

FIGURE 1.3 (a) Some examples of intrinsic optical fibre sensors. (b) Some examples of extrinsic optical fibre sensors.

Likewise, the electrically inert fibres allowed sensors to be installed within high-voltage transformers and energy cables and thus enabled low-cost monitoring of these expensive assets. In this context, the cost of a sensor system must be viewed as the complete cost of ownership, including installation of the cable, the need to provide for intrinsic safety in hazardous environments and any maintenance or re-calibration that may be required. On this scale, optical fibre sensors are frequently relatively expensive as individual items but are cost-effective when all aspects of the system are considered.

Optical fibre sensors can be designed to be impervious to electro-magnetic interference, and yet, in different forms, they are used as current or voltage sensors.

1.1.2 The Advent of Multiplexed Sensors

As more concepts and applications for optical fibre sensors emerged, the issue of their cost became more pressing: some applications, such as sensing in extremely hostile environments, could bear the expense of this specialist technology because the benefits of improved understanding were valuable and could not be obtained in any other way. However, in most cost-sensitive applications, the new technology was uncompetitive against a cheaper and better established technology that was available, proven and adequate for the purpose. This led to considerations of multiplexing several sensors onto a single fibre cable [13–16] and sharing the interrogator between many sensing points, thus reducing the cost per point of the system. Installing the wiring from the control room to the sensors is a significant part of the cost of an industrial sensor system; sharing that cost between multiple sensing points improves the economics of the system to a competitive, even compelling, level in many cases (Figure 1.4).

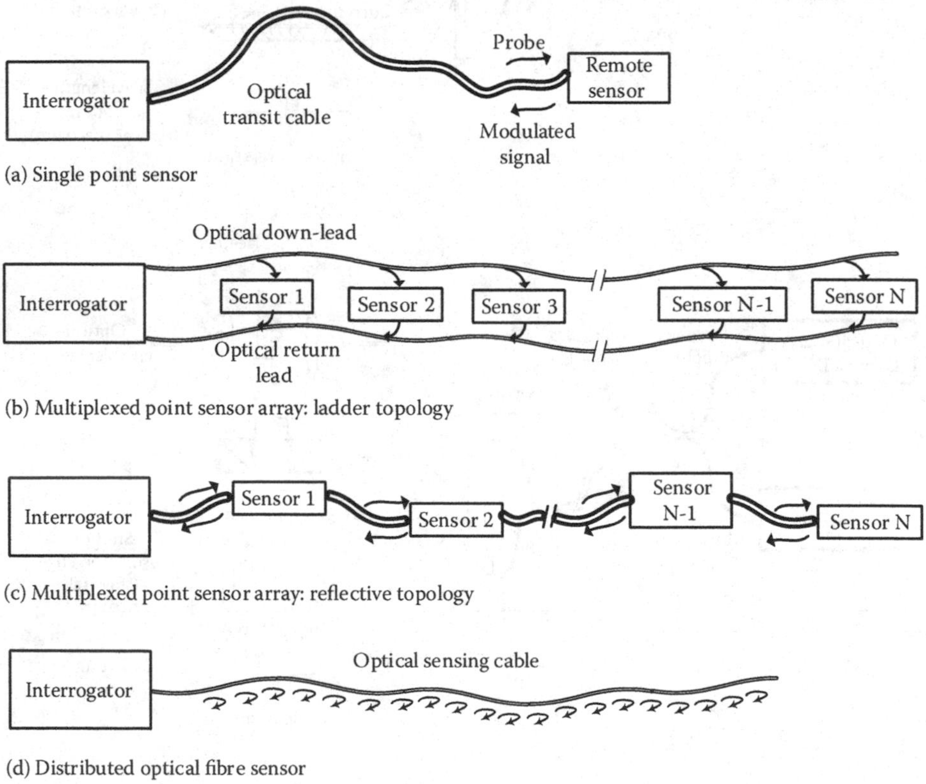

FIGURE 1.4 Topology of single-point (a), multiplexed (b and c) and distributed optical fibre sensors (d).

A number of multiplexing strategies [17] emerged where the sensors are identified through, for example, their allocated slot in the wavelength spectrum of the probe light (wavelength-division multiplexing), their coherence function (coherence-domain multiplexing) or the propagation delay to and from the sensor (time-domain multiplexing). In general, multiplexed sensor systems are more complex and often perform less effectively than single-point sensors. Nonetheless, if they are fit for the intended purpose, they can earn acceptance as a technology.

Multiplexed sensor systems exist in a variety of topologies. In a ladder topology (Figure 1.4b), a down-lead fibre conveys the probe light past each of the sensors in turn, providing a fraction of the probe energy to each sensor. The modulated light from each sensor is collected by a return fibre and brought back to the interrogator. Alternatively, in a reflective topology (Figure 1.4c), the probe light travels to each of the sensors and a portion of its energy is modulated and reflected back towards the interrogator. These multiplexed arrangements are to be contrasted with the single-point sensor (Figure 1.4a), where the sensor is connected directly to the interrogator using one or two fibres.

Multiplexed fibre sensors gained acceptance in applications where an array of sensors in well-defined positions were required, such as hydrophone arrays for naval sonar systems: these systems require repeated measurement points at specified locations, usually on a linear array or two-dimensional grid.

However, in the multiplexed sensing approach, mapping the complete spatial profile of a measurand, without any *a priori* knowledge of the form of the distribution requires an unreasonable number of sensing points, especially where the spatial frequency of the information could be very high. As an example, the current-carrying capacity of a buried energy cable is determined by the temperature of its hottest point, and the width of the hot-spot can be as short as 1–2 m depending on the dimensions of the conductors. So, for a long buried cable of length, say, 10 km, some 10,000 sensing points would be required. Moreover, the temperature distribution is dictated not just by the dissipation of the energy in the cable but also by the ability of the material, known as the *backfill*, surrounding it to conduct the heat away. Because of irregularities in the backfill or the proximity of other heat sources, it is often not possible to predict where the limiting value of the temperature will occur. Therefore, it is insufficient to install just a few sensors in pre-selected places.

Figure 1.5a provides an illustration of the spatial distribution of a measurand, for example, the temperature profile along an energy cable. In Figure 1.5b, the solid curve shows the inferred profile for a sparsely sampled sensor array; although this is based on the output of 11 point sensors (shown at the crosses), the resulting data are clearly unable to map the measurand (shown as the dotted curve) and the two spikes have been completely missed. In Figure 1.5c, with three times the spatial sampling density, a truer map of the profile is obtained, but spikes of similar width could still be partly missed if they were to fall between two sensor positions.

DOFSs solve this problem: they measure a continuous spatial profile of the measurand along the entire length of the sensing fibre (Figure 1.4d). They therefore differ from multiplexed sensors that measure at a finite number of discrete locations. In a distributed sensor, every point in the fibre contributes to the output. Figure 1.4d contrasts the output of a distributed optical fibre sensor with that of multiplexed sensor arrays depicted in Figure 1.4b and c and the output of which is illustrated in Figure 1.5b and c.

1.1.3 Distributed Optical Fibre Sensors

A *distributed optical fibre sensor* is defined as an intrinsic sensor that is able to determine the spatial distribution of one or more measurands at each and every point along a sensing fibre.

DOFSs are unique systems. They bear some resemblance to active sonar and radar in the sense that they resolve the measurand over quite large distances and they apply similar concepts of reflectometry to define the distance of the target from the system. However, radar and sonar systems tend simply to provide an indication of the presence of a target, its size and sometimes velocity. In general, they do not provide

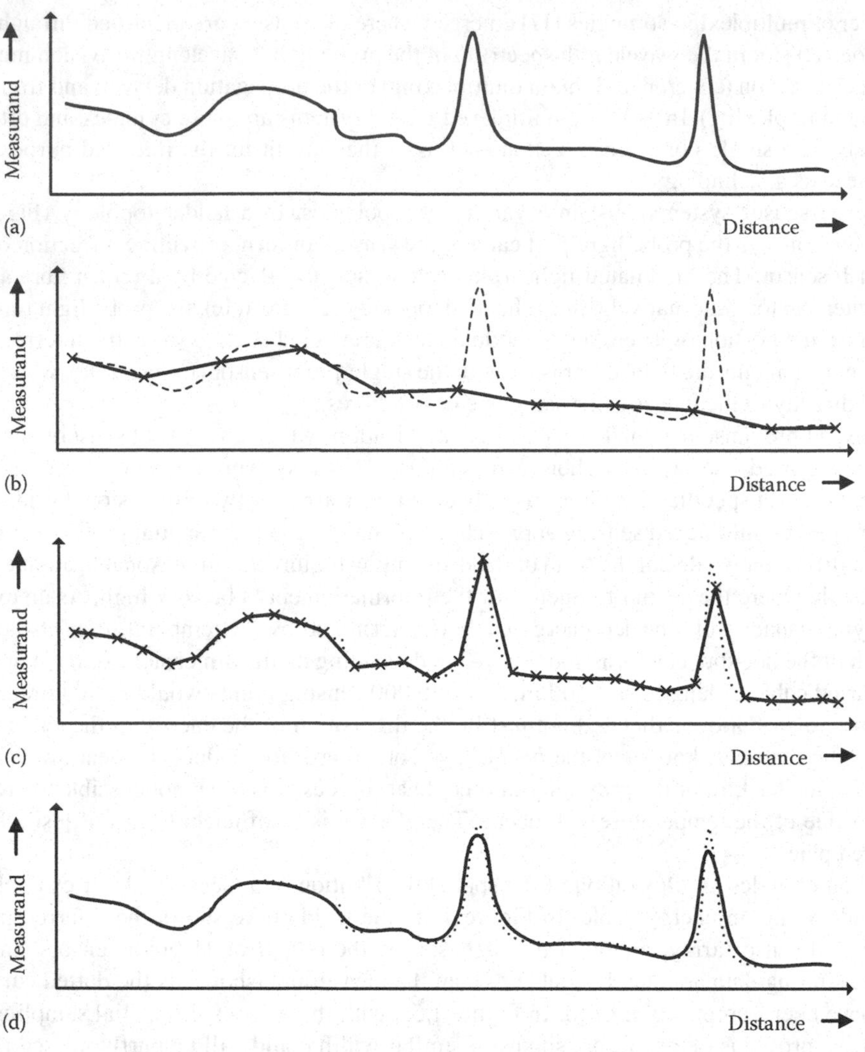

FIGURE 1.5 Spatial distribution of a measurand and how it is sensed. The reference (true) distribution is shown in (a). Multiplexed point sensors sampled sparsely (b) and densely (c). Distributed measurement with marginally adequate spatial resolution (d).

a continuous reading of a parameter of the medium between the system and the target and there is, in general, no control over that medium.

Atmospheric Lidar (Light Radar) systems are closer to DOFSs, in that some variants will provide a profile of temperature and humidity [18], density, wind speed [19,20] or pollutant particle concentration [21] along the beam. Since Lidar systems operate in free space, they can provide a three-dimensional profile of those quantities that they measure by scanning the beam over azimuth and elevation. Lidar systems exploit properties of the gas through which the probe beam travels to analyse one or more parameters of the medium. For example, differential absorption spectroscopy is used to identify certain chemical species. The Raman spectrum of certain molecules provides temperature information as well as composition. The linewidth of molecular transitions is affected by pressure and Doppler shifts can be used for wind speed measurement.

A third analogy to DOFSs is that of time-domain reflectometry (TDR), a pulse-echo technique used to locate discontinuities in electrical cables and waveguides. TDR operates in a one-dimensional medium, just as DOFSs do, but in general, TDR is limited to detecting changes in the characteristic impedance of the propagating medium.

Finally, DOFSs may be seen as an extension of multiplexed sensor arrays in the reflective topology where the sensors have become infinitesimally small and distributed along the length of the fibre. Figure 1.5d illustrates the output of a DOFS measuring the same distribution as the multiplexed arrays in Figure 1.5b and c. In this example, the spatial resolution of the DOFS is not quite sufficient to follow the fastest variations of the measurand as a function of distance. A DOFS with adequate resolution would faithfully reproduce the reference distribution of Figure 1.5a.

1.2 HISTORICAL DEVELOPMENT OF DOFS

1.2.1 Optical TDR: An Enabling Technology

The technological ancestry of DOFSs is to be found in the development of optical TDR (OTDR), which was demonstrated in 1976 and 1977 [22–24]. This technology was developed initially to detect attenuating or reflective faults or imperfections in optical transmission lines, but it was extended to a number of uses, such as determining the distribution of properties of the fibre, e.g. numerical aperture, diameter or mode field diameter.

An OTDR (Figure 1.6) is an instrument that launches a series of pulses (*probe pulses*) into an optical fibre and detects the optical signal returning to the launching end. The optical signal is collected by a photodetector and converted to an electrical current that is amplified and then digitised. The data are averaged over multiple probe pulses and displayed, usually on a plot of optical power (logarithmic scale) versus distance from the start of the fibre, as illustrated in the lower part of Figure 1.6. The instrument

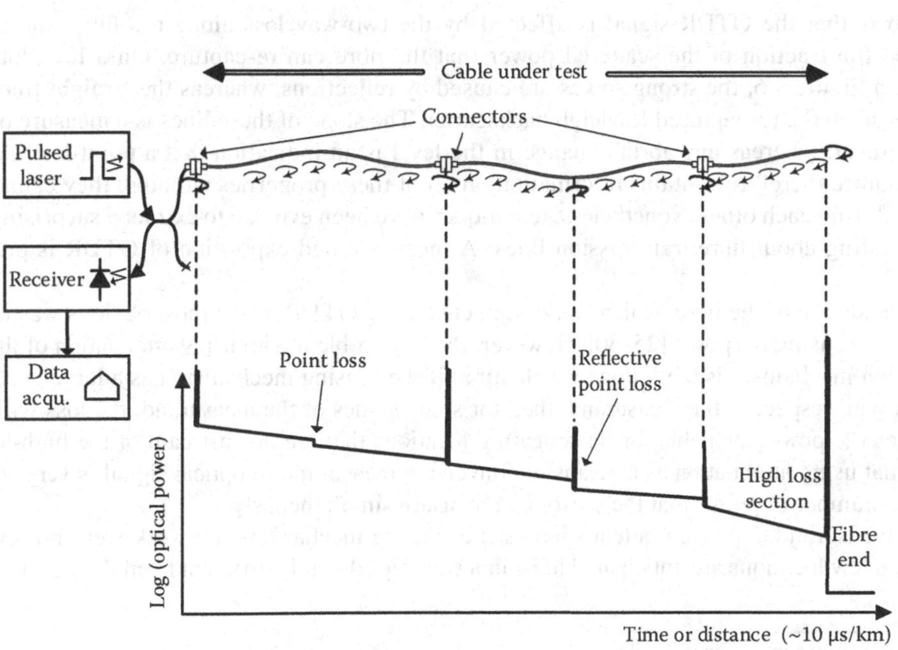

FIGURE 1.6 Schematic diagram of an OTDR and cable under test (upper part) and typical OTDR output (lower part).

actually measures on a timescale defined from the launching of the probe pulses and the distance scale is mapped from the time scale by multiplying by half* the group velocity of light in the fibre (approx. 100 m/μs). At a reflective point, such as a fibre break, a poor splice or a connector, the signal consists of a fraction of the probe pulse power that is reflected by the abrupt transition of the refractive index. This information is useful for locating defects, assessing the length of the fibre or checking the performance of connectors.

However, the major signal of interest in OTDR is a much weaker component that arises from re-captured scattering of the probe pulse along the fibre; this is known as the backscatter signal (or sometimes backscatter signature). In Figure 1.6, the curled arrows represent light returned towards the launching end; the faint ones are used for the continuous re-captured scattering, and the heavier ones, for reflections.

In OTDR, the main contribution to the backscatter signal is Rayleigh backscatter and arises from the following processes:

a. The probe pulse travels along the fibre and is gradually attenuated through a variety of mechanisms, such as absorption, micro-bending† and scattering. The incremental energy lost for each elemental section of fibre that is traversed, expressed as a fraction of the probe energy, is approximately 2.3×10^{-4} m^{-1} for a probe wavelength of 1064 nm and 4.6×10^{-5} m^{-1} at 1550 nm in good-quality fibres.
b. One of the sources of loss is scattering, which re-directs a small fraction of the probe energy in all directions, although not completely uniformly. As mentioned previously, Rayleigh backscatter is the dominant contribution to fibre attenuation in many practical cases; in these cases, most of the light that is removed from the probe pulse whilst it propagates is lost through scattering.
c. OTDRs use the fraction of the scattered light that falls within the acceptance solid angle of the fibre in the return direction. The fraction of the scattered light that is captured depends on the characteristics of the fibre but is typically between 0.1% and 1%.
d. Re-captured light returns to the launching end, suffering further attenuation between the scattering location and the launching end.

It follows that the OTDR signal is affected by the two-way loss along the fibre, the scattering process and the fraction of the scattered power that the fibre can re-capture. On a log plot, such as illustrated in Figure 1.6, the strong spikes are caused by reflections, whereas the straight lines sloping downwards show the re-captured Rayleigh backscatter. The slope of these lines is a measure of the loss of each segment, whereas any abrupt change in the level is an indication that a point-loss exists. The OTDR signature therefore contains information on all of these properties although they cannot always be separated from each other. Nonetheless, techniques have been evolved to extract a surprising amount of understanding about fibre transmission lines. A more detailed exposition of OTDR is provided in Chapter 3.

The attenuation of the fibre, which can be detected using OTDR, is the most obvious way of exploiting OTDR for sensing purposes [25–30]. However, the key problem with using attenuation of the fibre as a transduction mechanism is the following dilemma: if the sensing mechanism has a large coefficient of attenuation with respect to the measurand, then for some values of the measurand, the loss will be high. This reduces the power available for interrogating locations that are downstream of the high-loss point. It follows that using attenuation as a means of converting measurand to optical signal is very limiting in terms of the number of points that the sensor can measure simultaneously.

Using measurand-dependent attenuation as the sensing mechanism can work well, however, when only one or a few locations are anticipated to be in a state significantly different from the normal, low-loss

* To allow for two-way propagation in the fibre.

† Microbending is a distortion of the fibre axis on a small scale, usually caused by pressure applied through a rough surface. It causes light that is normally guided by the fibre no longer to meet the conditions for being propagated and so to leak out of the fibre.

condition. An example of such a use would be the detection and location of a hot spot: in some applications, it is sufficient to determine if any point along the sensing fibre has exceeded some threshold condition. Even in this case, a strongly non-linear response is desirable, so as to discriminate between a small hot-spot that exceeds the threshold and a long region where the temperature is warm but below the selected threshold. Some examples of loss-based sensors are discussed in Chapter 10.

To maintain the ability to measure a very large number of sensing points, it is therefore desirable to exploit a mechanism that does not significantly alter the attenuation of the fibre and yet provides a specific response to a selected measurand. The first step in that direction was the use of polarisation in OTDR (POTDR), proposed by Alan Rogers [31] and demonstrated by the present author and others in 1980 [32]. POTDR is based on the fact that the Rayleigh scattering process largely preserves the state of polarisation (SOP) at the scattering point. Therefore, it provides an indication of the evolution of the SOP of the light as it travels along the fibre. In turn, the SOP is affected by a number of external influences [33], such as magnetic fields [34,35], bending [36] or uniaxial pressure [37].

After a flurry of research activity in the early 1980s, POTDR has seen relatively little use for two main reasons. The first is that it does not provide a complete description of the SOP of the light travelling in the fibre because POTDR signals are influenced by some types of disturbance such as magnetic fields and side pressure, but not the effects of twisting of the fibre, that do however affect the SOP in the forward direction. A second issue with POTDR is that in conventional fibres, the SOP is affected simultaneously by many potential measurands, and it is therefore problematic to apply this technology to sense just one measurand reliably. However, work is continuing at a modest level, with progress made around the year 2000 on the separation of the effects of linear and circular birefringence, under certain assumptions [38] and its use as a distributed pressure sensor [39] with special fibres that are more specific in their response to pressure. POTDR is discussed further in Chapter 10.

1.2.2 Distributed Temperature Sensors

The first distributed sensor with a specific response to one measurand, temperature in this case, was demonstrated in 1982 [40–42]. This sensor used a liquid-core fibre interrogated by OTDR. In this medium, the intensity of the scattering is sensitive to temperature, and it was demonstrated that the device responded locally to temperature. The sensor also exhibited an acceptably small change in attenuation that could in any case be corrected using techniques known in the field of OTDR (see Chapter 3). This work is the first to describe these types of device as *distributed sensors* [42]. Of course, the practical application of a liquid-core fibre was limited by difficulties in deployment and also the restricted temperature range* over which such fibres operate. So the search began for techniques that are specific to well-defined measurands and capable of interrogating many points along the fibre.† The selected fibres were also required to be sufficiently rugged for both practical deployment and measuring over a wide temperature range.

The use of inelastic scattering allowed distributed sensors to become practical devices.

1.2.3 Elastic and Inelastic Scattering

The adjectives 'elastic' and 'inelastic' are used to denote a scattering process in which the energy E_{ph} of the photons in the incident wave is preserved (elastic) or not (inelastic).

Scattering occurs when the medium through which the incident wave travels is not perfectly homogeneous; i.e. its refractive index varies locally. For the three types of scattering discussed here, the

* c. 0°C–100°C.

† The original work on liquid core fibres provided only a few hundred resolvable points at 1 m spatial resolution.

inhomogeneities exist on a distance scale much smaller than the wavelength of the incident light. For these small-scale fluctuations of the refractive index, the elastic scattering is known as Rayleigh scattering. The small-scale fluctuations of the refractive index that cause Rayleigh scattering are frozen in the glass and the elastic process does not involve their motion. The fact that the incident energy is preserved in the scattering process means that the frequency $\nu = E_{ph}/h$ (where h is Planck's constant [~6.626 × 10^{-34} J.s]) of the scattered wave is the same as that of the incident wave.

In contrast, the local variations of the refractive index that are involved in inelastic scattering are caused by thermally driven vibrations carried by phonons. The heat in the material is held in the form of molecular vibrations (stretching, bending or rotation of inter-atomic bonds) or lattice vibrations (longer-scale periodic movement of the material). Molecular vibrations occur at very high frequency (~10 THz), and they cause Raman scattering; in contrast, lower-frequency vibrations in the hypersonic range (10–30 GHz in this case) give rise to Brillouin scattering. In both cases, energy is exchanged between the incident wave and the material; the exchange of energy causes a frequency shift.

When the scattered light emerges at a lower frequency than the incident light, the scattered photon has given up energy, in the form of an additional phonon, to the medium in the interaction; the resulting new spectral features are known as Stokes lines or bands. In the opposite case, where the scattering has increased the frequency, energy has been transferred from the medium to the light. This is known as anti-Stokes scattering, a process in which a phonon is removed and its energy is incorporated in the scattered photon.

The anti-Stokes process is dependent on the population density of the phonons that are used in the process. In Raman scattering, the energy of the phonon is of similar order to the unit thermal energy $k_B T$ (where k_B is Boltzmann's constant [~1.38 × 10^{-23} J/K] and T is the temperature) and so the population density is strongly temperature dependent. The intensity of the anti-Stokes Raman band is therefore temperature sensitive.

However, the Stokes process transfers energy from the probe to the glass, a process that is only slightly temperature sensitive. It follows that the anti-Stokes Raman band is more temperature sensitive than the Stokes band, and the ratio of these two signals is commonly used for estimating temperature.

The Raman (a) and Brillouin (b) spectra for typical fibres are shown in Figure 1.7. The frequency scales are very different for the two cases: the Brillouin linewidth is about six orders of magnitude narrower than that of the Raman spectrum and the frequency shift is three orders smaller.

Figure 1.7 also illustrates how these spectra change with temperature: for Raman, a strong increase in the anti-Stokes intensity and a slight increase for the Stokes band occur, whereas for the Brillouin

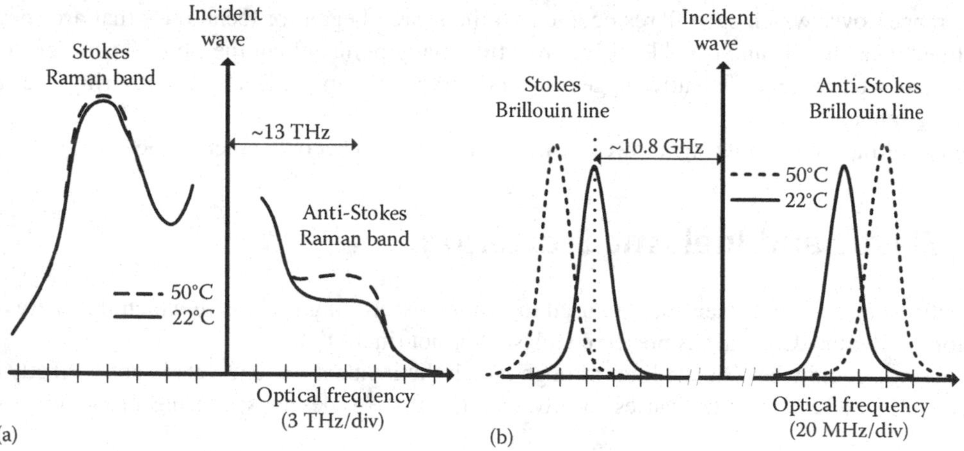

FIGURE 1.7 Illustrative Raman and Brillouin spectra in typical optical fibres; the frequency scale applies to an incident wavelength of 1550 nm in the case of the Brillouin spectrum.

spectrum, the two lines vary essentially in the same way as a function of temperature.* The frequency of the Brillouin lines relative to the probe light also varies with temperature and strain.

For further detail, the reader is referred to Chapters 4 and 5.

A glimpse of a very weak, but linear, temperature response in solid core fibres using OTDR was provided in 1982 [43] when a local change in the backscatter signal of a few percent was demonstrated in response to a temperature change of around 200 K. This effect can be attributed to the presence of additional spectral components in addition to Rayleigh backscatter in the OTDR signature. The results are explained by the contribution of spontaneous Brillouin scattering within the light collected by the acquisition system. Brillouin scattering will be discussed in Chapter 5, but in this case [43], the acceptably strong variation of the Brillouin signal with temperature was largely diluted by a much stronger, temperature-insensitive Rayleigh signal, resulting in a small relative temperature sensitivity.

1.2.4 Raman OTDR

The next technological breakthrough is attributable to John Dakin and co-workers, who proposed [44] and demonstrated [45] the use of the very weak Raman [46,47] components that exist in the scattered light spectrum (see Figure 1.7a).

Rejecting the dominant, temperature-insensitive Rayleigh scattering and selecting the temperature-sensitive Raman bands resulted in a viable distributed temperature sensor (DTS). It should be pointed out that measuring the ratio of the anti-Stokes Raman to the Stokes Raman signals was a well-established method for non-contact temperature measurement in chemistry. The novelty here was applying Raman thermometry to optical fibres and as a distributed measurement. Further details on Raman scattering and its application to distributed temperature sensing may be found in Chapter 4.

Those working in the field of OTDR at the time found that the weakness of the signals available even in conventional OTDR presented considerable difficulties. The Raman OTDR required most (typically 99.8%) of the signal returning in an OTDR to be rejected, and this presented new challenges in the design of DTS systems. Nonetheless, the attraction of Raman ratio thermometry for DTS was that it presented an approach that

a. Used standard fibre;
b. Was specific to temperature and
c. Was suited to the technology of the day.

To provide an historical context, this breakthrough occurred prior to the widespread development of the erbium-doped fibre amplifier or the easy availability of narrowband injection lasers. Large optical powers are required for Raman OTDR, but, at least in glasses, the linewidth of the source of probe light can be quite broad (several nm). Raman OTDR was therefore ideally suited to the technology available in the 1980s. In particular, Raman DTS systems were able to use multimode optical fibres that, at that time, were still being deployed for some types of optical communications systems. Multimode fibres also capture efficiently the emission of broad contact laser diodes that provide high-power, short, optical pulses at moderate cost; the combination of multimode fibre and broad contact lasers was therefore ideally suited for the first commercial generation of DOFS systems. Indeed, Raman DTS systems perform best over most distance ranges on this type of fibre.

Raman OTDR forms the basis of most present commercial DTS systems, and a number of improvements have been made to the optical arrangement, the acquisition electronics and the loss compensation mechanisms since the first generation of interrogators in the mid to late 1980s.

Chapter 4 of this book describes the technology of Raman DTS and Chapter 7 discusses the applications of DTS.

* The temperature sensitivity diverges at cryogenic temperatures.

1.2.5 Brillouin-Based Distributed Sensing

In 1989, Horiguchi and Tateda of NTT Laboratories published a paper [48] that described the use of stimulated Brillouin scattering for measuring the attenuation profile of a single-mode fibre, a technique that they referred to as Brillouin optical time-domain analysis (BOTDA). Stimulated Brillouin scattering is a non-linear optical effect involving the interaction of two counter-propagating optical inputs (i.e. one input launched into each end of the fibre) that are coupled through a mechanical vibration of the glass on a scale comparable to the optical wavelength.

The initial focus of the BOTDA work was to improve the range and accuracy of OTDR by exploiting the optical gain provided by stimulated Brillouin scattering. However, it became rapidly apparent that the technique was also a means of measuring the Brillouin frequency shift. The latter is a function of strain and temperature [49–53], and thus, a new possibility in distributed sensing was conceived.

In common with other DOFS technologies, BOTDA provides position-resolved readings, and the signals that it provides can be strong because of the gain inherent in the measurement technique.

The Brillouin shift, being a function of temperature and strain, has been used as a sensor for both of these measurands; unless this is obvious from the physical context (for example, there is no strain applied to the fibre), it is necessary to distinguish between the two measurands, and a wide variety of techniques have been employed to this end; they are discussed in Chapter 8.

One major disadvantage of BOTDA is that it requires a loop configuration (because light is injected from both ends of the fibre) and if the sensing fibre is broken anywhere in the loop, *all* measurements are lost. This can be a serious issue in practical applications, where the loss of transmission is always a possibility.

An alternative technique, Brillouin OTDR (BOTDR), was demonstrated in 1993 also by a team at NTT [54]. BOTDR is based on spontaneous Brillouin scattering and it uses a single-ended configuration (in the case of BOTDR, a loop configuration allows most of the sensing fibre to remain operational in the event of a single break). BOTDR thus provides a more resilient solution. Although it can be argued that the performance of a BOTDA system is *a priori* superior to that of a BOTDR, the BOTDA range is in effect halved when deployed to monitor a linear object owing to the need for a looped sensing fibre.

SPONTANEOUS AND STIMULATED BRILLOUIN SCATTERING

Both spontaneous Brillouin scattering and stimulated Brillouin scattering are used in distributed sensing. Systems based on spontaneous Brillouin scattering are similar in concept to Raman DTS, with the key difference being that Brillouin scattering causes a much smaller frequency shift. The Brillouin technique therefore requires a finer resolution of the backscattered light spectrum, including the use of sources with narrower linewidths.

The frequency shift in Raman scattering is determined by molecular vibrations and the part of the spectrum that is used for DOFSs is at a shift of about 13 THz. Brillouin scattering arises from the presence of lattice vibrations (hypersonic acoustic waves) and the frequency shift is determined by the condition that the wavelength of the acoustic wave must match the incident optical wavelength. The frequency shift is therefore determined by the ratio of the optical to acoustic velocity in the medium.

Stimulated Brillouin scattering, however, involves the interaction of three waves, namely two optical waves separated by the Brillouin shift and the acoustic wave that couples the two optical waves to each other. The additional complexities involved in stimulated Brillouin scattering and the narrowband nature of the process have led to a host of different approaches for exploiting this physical phenomenon.

BOTDA is the first example of a successful optical fibre sensor using non-linear effects, namely the interaction of two counter-propagating optical signals.

Since the development of these two techniques, a host of improvements have been described in the technical literature, and research is still active; the technology of Brillouin-based distributed sensing is the subject of Chapter 5 of this book.

Comparing Brillouin to Raman technology, the former is far more diverse in the approaches that have been followed and the performance metrics that have been claimed cover a much wider range than is the case for Raman DTS. In addition, for long-distance applications, the narrowband nature of both the probes and the return signals in Brillouin allows them to be amplified effectively using remotely-sited optical amplifiers. This is not the case for Raman DTS because the breadth of the Raman signal is such that the amplification process is far too noisy. On the other hand, the Raman technology is generally simpler to implement and it has the considerable benefit of being insensitive to strain and so avoids the cross-sensitivity that is a serious issue in Brillouin systems. These points will be explored further in Chapters 4, 5 and 8.

1.2.6 Distributed Sensing with Rayleigh Backscatter

Although the first decade of the history of DOFSs involved systems that discarded the Rayleigh component in the backscatter, that same component is central to the next measurand to gain widespread use, namely sensing the distribution of mechanical disturbance along a fibre.

Conventional OTDRs use broadband sources to generate probe pulses. However, when a very narrow linewidth laser is used, new phenomena become apparent. In this case, the entire fibre distance that is occupied by a probe pulse at any time, a *resolution cell*, is illuminated by light that has a consistent phase relationship throughout the section that it occupies. Each scattering centre thus re-radiates light with a fixed phase relationship to all the other scattering centres in the cell. Clearly, owing to the random disposition of the scatterers, the relative phase of their radiation is also random, but it remains fixed as long as the frequency of the laser is stable and the relative optical distance between scattering centres is constant. If the fibre is disturbed, e.g. strained, the relative phase relationships are altered.

The effects of coherence are unwanted in conventional OTDR and considerable care is taken to eliminate them; however, it is precisely these effects that Rayleigh-based distributed vibration sensing (DVS) exploits.

When the backscattered light arrives at the receiver, the electric fields of all of these scatterers are summed on the detector; the photocurrent therefore depends on the vector sum of these electric fields. On average, each unit length of fibre is expected broadly to return the same backscatter intensity. In the case of a low-coherence source, there is very little fluctuation around the average intensity as a function of distance along the fibre. However, when the source is coherent, the relative phase of the light scattered from all of the inhomogeneities within each unit length of fibre is key to the received backscatter intensity. If the random arrangement of the scatterers is such that, by and large, their fields arrive in phase, then their electric fields add constructively, resulting in a strong output signal; conversely, under some combination of laser frequency and disposition of the scatterers, the electric field sum will be close to zero. A coherent Rayleigh OTDR is therefore a distributed multipath interferometer.

This interference behaviour was initially investigated when OTDR was first used in single-mode fibre, especially with coherent sources [55,56]. However, the existence of these interference effects was seen as a noise source; they are coherent between successive pulses and they cannot therefore be removed by simple signal averaging. These effects can be alleviated by using a naturally broadband source; however, where a narrowband source *must* be used, coherence effects are minimised with dithering techniques in which the frequency of the source is varied during an acquisition cycle [57–59].

In the early 1990s, sensing vibration using coherent Rayleigh backscatter was proposed in the literature [60,61] and also demonstrated in the laboratory [62–64]. Temperature fluctuations were also found to affect the Rayleigh signal [64,65], as expected from the change in refractive index (thermo-optic effect) that affects the optical path length between scatterers. The majority of these early papers were based on detecting variations in the intensity of the Rayleigh backscatter signal in response to an external stimulus. However, Dakin and Lamb's patent [60], filed in 1990, is the first to mention the use of the phase of the backscattered light, differentiated over a prescribed fibre interval, for sensing vibration or other changes in environment of the fibre. In 2000, the US Naval Research Laboratory demonstrated a different optical arrangement for estimating the differential phase [66,67], which in turn was a proxy for dynamic strain on the fibre.

Since the mid-2000s, several companies have worked on developing the measurement of coherent Rayleigh backscatter as a distributed vibration sensor, with applications in intrusion detection, borehole seismic acquisition and the detection of noise in hydrocarbon wells for flow determination. Chapter 6 discusses using coherent Rayleigh backscatter for DVS. This chapter also discusses another use of coherent Rayleigh backscatter measured as a function of probe frequency. In this case, the backscatter can provide finely resolved measurements of temperature and strain.

Some applications of DVS are described in Chapter 9.

A number of other distributed sensing techniques have been published in the 30 or so years since DOFSs were first proposed. They are summarised in Chapter 10.

1.3 PERFORMANCE CRITERIA IN DISTRIBUTED FIBRE-OPTIC SENSORS

Distributed sensors bring at least one new dimension to monitoring, namely their ability to determine the spatial distribution of the parameter(s) of interest. In some applications, where conventional technology has involved intermittent measurements, permanently installed distributed sensors also bring the further dimension of measuring as a function of time. This increase in information provided by DOFSs is balanced by a lesser measurement quality for each resolvable point than can be achieved with a single-point sensor. In fact, it can be stated that the measurement performance of any distributed sensor at a single location will be inferior to that of the best available single-point sensor.* There are therefore distinct trade-offs in the specification of distributed sensors. These trade-offs will be discussed in the context of the various DOFS technologies, but some more general points are made in the following section. Even the criteria by which DOFSs are assessed are specific to this technology.

1.3.1 Four Key Criteria

The main aspects of the performance of DOFSs are described by

- The measurand resolution;
- The spatial resolution;
- The range and
- The measurement time required to achieve the stated measurement resolution.

Certain aspects of the performance of a distributed sensor are described in the same way as those of conventional single-point sensors. For example, we need to specify the measurand resolution and the accuracy required of the sensing system, its dynamic range, acceptable operating conditions and the time taken for the measurement.

* This can be shown with a conceptual experiment in which a long, coiled length of optical fibre, used as a distributed sensor, is exposed to a measurand. By averaging the measured value along the coil, better resolution will be obtained than in any one location along that coil.

1.3.1.1 Measurand resolution

The *measurand resolution* is the ability of the sensor to distinguish small changes in the value of the quantity that is measured. This is similar to the resolution of a conventional single-point sensor; however, for DOFSs, each resolvable point is roughly equivalent to one independent sensing point. In general, the measurand resolution is a function of location because the signal is increasingly attenuated with increasing distance along the fibre.

1.3.1.2 Range

The *range* of the sensor is the maximum length of sensing fibre that can be measured. The range also dictates the maximum pulse repetition rate: the interrogator must wait until one pulse has travelled to the far end of the sensing fibre, and its backscatter has returned to the launching end, before it launches the next pulse. A faster pulse repetition rate would cause overlapping backscatter signals in the fibre, resulting in location ambiguity.

The range parameter is, of course, linked to the cumulative loss to the most remote point to be sensed and a system specification can be based on a maximum fibre length with a defined maximum loss per unit length. It is sometimes more appropriate to specify the maximum length and the maximum cumulative loss separately; this is the case when the system includes many splices or connections that may markedly increase the overall loss of the sensing fibre.

1.3.1.3 Spatial resolution

In the case of a distributed sensor, it is also necessary to define its ability to distinguish the value of the measurand at closely spaced locations, i.e. its *spatial resolution*.

Typically, a distributed sensor responds to an abrupt change in the measurand along the fibre with some blurring of that edge in its output. Figure 1.8 illustrates the difference between the actual measurand (solid line) and that detected by the sensor (broken line). The spatial resolution δz is usually defined as the distance over which the sensor output corresponding to an abrupt transition is spread, estimated between 10% and 90% points on the transition (dotted vertical lines in Figure 1.8).

The 10%–90% definition of spatial resolution is appropriate for determining the degree to which a transition can be reproduced in the sensor output. For small features, it is also appropriate to think in terms of the spatial response function of the DOFS as determined by its output when there is a very narrow spike in the measurand, for example, a very small hot-spot in the case of a DTS.

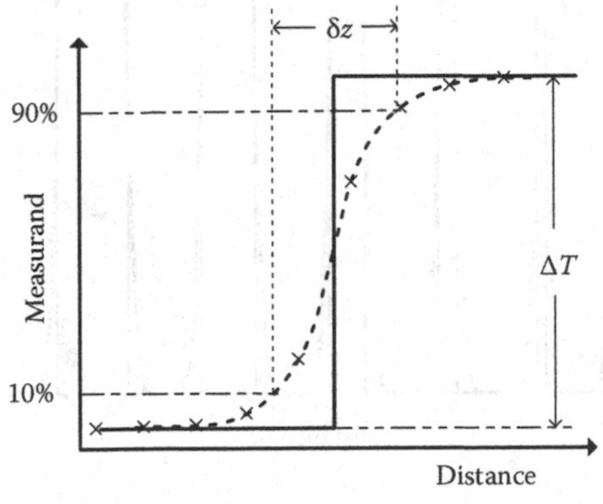

FIGURE 1.8 Definition of spatial resolution.

The measurand distribution, as reported by the DOFS, is the convolution of its spatial response with the true spatial distribution of the measurand. If the spatial features of the measurand resolution are finer than the spatial resolution of the DOFS, the details of the measurand distribution will be attenuated. For simplicity, the following illustrations are based on a DTS measurement.

In the case of a hot-spot smaller than the spatial resolution, the temperature difference measured at its centre will be lower than the actual value. The full temperature difference will not be correctly reported, even if the width of the hot-spot is the same as the spatial resolution (using the 10%–90% definition introduced previously).

Figure 1.9 shows examples of square hot-spots of varying widths (their width relative to the spatial resolution of the DTS is indicated at the top of the figure and ranges from 10% to 400%). The assumed temperature profile, normalised to the departure ΔT from the background temperature, is shown as a solid line. The calculated output of the DTS as a function of distance is shown as a dotted line. As can be seen, the reported temperature difference is less than the true value of ΔT when the spatial resolution is similar to, or smaller than, the width of the hot-spot. For a hot-spot of width equal to the spatial resolution, the peak reported value is only 80% of ΔT. For a hot-spot of half the width of the spatial resolution, the reported value is about 48% of ΔT. For very small hot-spots, the spatial response is dominated by that of the DTS, and for very large hot-spots, the measured temperature profile is close to the true value. The crucial point is the transition region, in which a DOFS is used just beyond its capability; this occurs quite frequently, either for reasons of availability of equipment or because, at the time when the equipment is specified, the users are unaware of the fine features of the measurand.

The peak value reported by the sensor and scaled to ΔT is shown in Figure 1.10 as a function of the hot-spot width (normalised to the DTS spatial resolution). As mentioned previously, the result will depend on the shape of the temperature profile, and Figure 1.10 shows two cases, one a square hot-spot (i.e. a flat-topped thermal profile) and also a Gaussian profile that approximates a hot-spot created locally but where heat can diffuse along the object being measured and also laterally away from it.

The spatial resolution, together with the range, determines the number of independently resolvable points that can be measured by the system.

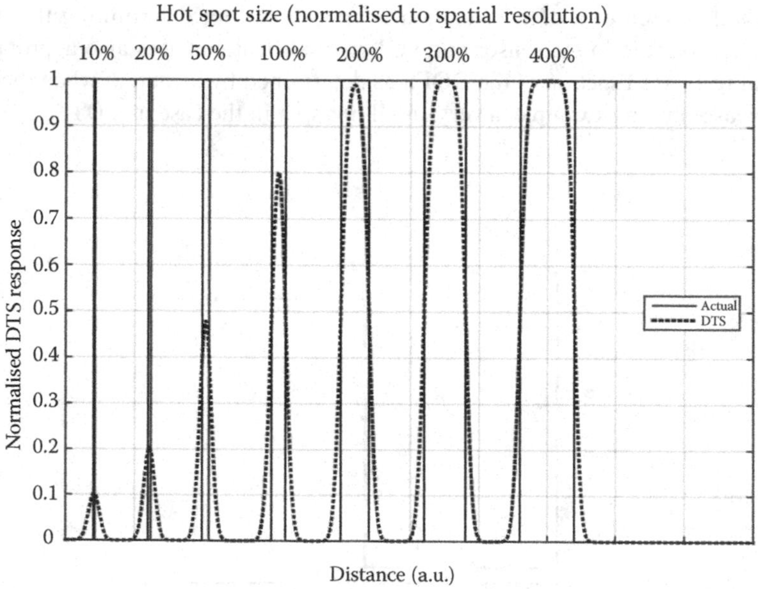

FIGURE 1.9 Normalised response of a distributed sensor to square hot-spots of varying width (relative to spatial resolution).

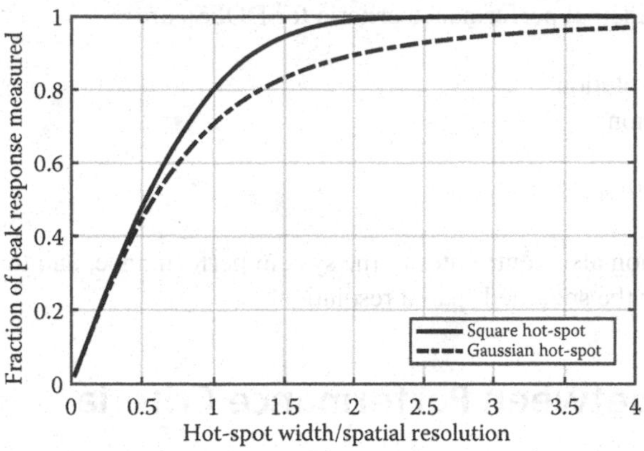

FIGURE 1.10 Peak value reported by a DTS at the centre of a hot-spot, plotted as a function of the ratio of the width of the hot-spot to the spatial resolution (10%–90% definition). The response is normalised to the temperature excursion of the hot-spot relative to adjacent fibre sections. Solid line is for a square (flat-top) spatial thermal profile and the dot-dash curve for a Gaussian thermal profile.

1.3.1.4 Sampling resolution

The output of a DOFS is provided digitally in the form of a series of values of the measurand at discrete locations along the fibre. These locations are related, through the two-way speed of light in the fibre, to the times at which the acquisition system samples the analogue signals provided by the optics. The *sampling resolution* is the fibre distance between these samples of the raw analogue signals.* Ideally, the sampling resolution should be substantially smaller than the spatial resolution in order to reproduce faithfully the information available from the receiver outputs that, in turn, depend on the characteristics of the optics and electronics.

In the example in Figure 1.8, the sample separation (samples illustrated by crosses) is approximately 2.5 times smaller than the spatial resolution, and so the measured edge is represented with reasonable accuracy and, if required, could be interpolated. However, if the sampling resolution were less than the spatial resolution, considerable detail would be lost (as was illustrated in Figure 1.5). As a rule of thumb, it is recommended to ensure that the sampling resolution is at least two times finer than the spatial resolution – this ensures that the profile can be reconstructed and interpolated to an accuracy that is limited only by the spatial resolution.

The sampling and spatial resolutions are two separate quantities. The latter is an analogue measure of the spatial frequency range of the information provided by the sensor. In contrast, the former is merely the inverse of the spatial frequency at which it is sampled. It is perfectly possible to sample the measurand distribution with many times more detail than the spatial resolution. However, once the sampling resolution is about twice the highest spatial frequency present in the analogue output of the distributed sensor [68], little is to be gained by further increasing the spatial sampling density.

1.3.1.5 Measurement time

Finally, the *measurement time* is the time taken by the system to acquire the readings for all points in the sensing fibre to the required measurand resolution.

* The DOFS output can of course be re-sampled in the signal processing. The quantity of interest here is the native sample separation prior to re-processing.

To summarise, the principal performance criteria for DOFSs are

a. The measurand resolution;
b. The spatial resolution;
c. The range and
d. The measurement time.

The sampling resolution also contributes to the system performance, and the spatial dimension must be adequately sampled for the specified spatial resolution.

1.3.2 Interplay between Performance Criteria

A critical point about the four criteria cited is that they are inextricably linked; any specification that does not show the tie between these quantities is misleading.

The four criteria are connected through the signal-to-noise ratio that determines the measurement resolution. In general, DOFSs deal with very small signals, and therefore, their signal-to-noise ratio is low. This limits the resolution of the measurand.

As an example, at room temperature, the temperature sensitivity of the anti-Stokes Raman signal is typically 0.8%/K. Thus, to resolve temperature to 1 K, a signal-to-noise ratio of 125 is required.* If (as is usually the case), the signal-to-noise ratio of this component of the backscatter signal is lower than 125 for a single probe pulse, it is necessary to acquire multiple waveforms with successive probe pulses and to average them to improve the temperature resolution. For a well-designed system operating within the specified range of measurement times, the signal-to-noise ratio should improve in proportion to the square root of the number of traces that are averaged. For a fixed pulse repetition frequency, therefore, the temperature resolution would improve as the inverse square root of the measurement time. This improvement applies as long as the measurand resolution is limited by white noise (for example in the optical receiver). However, as the measurement time increases, eventually, the white noise is reduced to the point where other effects become significant, such as bias stability affecting the offsets in the acquisition circuitry. Drifts in the instrumentation can dominate in the extreme limit of very long-term measurements.

Similar trade-offs apply to the other criteria. For example, as the sensing fibre is extended, the attenuation suffered by the probe light in the outward direction and by backscattered light returning from the most remote point is increased; this attenuation reduces the detected signal and thus degrades the resolution of the measurand. In fact, the degradation of the resolution with increased sensing fibre length is more severe still because the longer the fibre, the lower the maximum permissible pulse repetition frequency; therefore, for a fixed measurement time, fewer backscatter waveforms can be acquired and averaged. A more subtle effect is that, the longer the fibre, the more likely is the build-up of non-linear effects; this limits the probe power, or probe energy, that can be launched into the fibre, further degrading the measurand resolution for a given measurement time.

A finer spatial resolution requires a narrower probe pulse; for fixed probe power, this means lower probe energy and thus lower returned backscatter power. In addition, a finer spatial resolution requires a wider detection bandwidth, which increases the noise appearing with the signal and this also degrades the measurand resolution.

Thus, we see that in DOFSs, there is a deep-seated relationship between four key performance criteria; however, the detail of the interaction between the criteria is complex and depends on the particular measurement type and the specifics of the system design. Nonetheless, the prospective user of these systems must be aware of these fundamental trade-offs.

* This is simplified for the purpose of illustration; we shall see in Chapter 4 that the resolution depends on the signal-to-noise ratio of all the signals used, including the Stokes Raman backscatter.

At a deeper level of detail, it is necessary to understand the correlation between noise contributions. For example, there is a choice of independent variable when defining the resolution in terms of the standard deviation of the measurand. Is it the standard deviation of the measurand at a particular location measured over time? Or is it the standard deviation of the measurand at a particular time, estimated along a section of fibre where the measurand is uniform? The result may not be the same in each case because the statistics of the noise could be quite different in one dimension, compared with those in another. In the context of distributed temperature systems, this has led to the concept of distance-step and time-step resolution [69], i.e. the ability of the system to detect, with a specified degree of confidence, a change in the measurand along the fibre at a given time or a change over time at a given location. Data are acquired at a given resolution, typically the native spatial resolution of the instrument and a relatively fast measurement time (say 30 or 60 s) over a long period. The data can then be further averaged over distance and time to provide an estimate of the time-step or distance-step resolution as a function of averaging in the time and distance axes. The data can be presented in the form of contour lines that typically resemble hyperbolas. Each contour line represents a given confidence level that can be achieved by any combination of time and distance averaging that matches a location on the line [69].

1.3.3 Estimation of Key Performance Parameters

The range and measurement time can be considered independent variables when the performance of a DOFS is assessed, and so, spatial resolution and measurand resolution are usually specified as a function of these variables.

The 10%–90% definition of spatial resolution is conveniently tested by arranging a spatial step-transition in the measurand value. The results can be fitted to the expected function representing the integral of the DOFS spatial response (an erf function for a Gaussian response), and the values at 10% and 90% can then be calculated from the parameters of the fit. This approach is more precise if the sample separation is finer (ideally by a factor of 2 or more) than the spatial resolution.

However, SEAFOM®, a joint industry forum promoting the use of fibre optics in subsea applications, has defined the spatial resolution for a DTS in terms of hot-spots [70]. In fact, their definition uses an arrangement rather similar to that illustrated in Figure 1.9, i.e. consisting of a series of three well-separated fibre coils of varying lengths that are exposed to a differential temperature ΔT above that of the remainder of the test fibre. The longest, at four times the published spatial resolution, must indicate a reading of the temperature excursion equal to the actual ΔT within the claimed temperature resolution. The shortest coil is used to demonstrate that ΔT is *not* measured by the system for this coil length. The third coil is selected such that it shows one measurement point within 10% of ΔT. This definition is therefore also dependent on the sampling resolution. For a very densely sampled DOFS, this definition would be equivalent to ~1.3 times that of the 10%–90% definition.

The simplest way to estimate the measurand resolution is to arrange for a section of fibre to be at a uniform condition and to measure that section repeatedly. The standard deviation of the data points measured along this uniform section of fibre is an estimate of the measurand fluctuation on a fast time scale, assuming that the points are statistically independent. In the SEAFOM specification [70], this performance metric is referred to as 'spatial temperature resolution', and it is related to the previously mentioned 'distance-step resolution', insofar as any variation between traces is ignored; this criterion includes only variations of the measurand occurring at different places at the same time. Generally, the measurement is repeated a number of times in order to estimate the standard deviation more precisely.

However, this method fails to account for the correlation between close sample points. In general, it is desirable for the DOFS to sample the measurand in the spatial dimension at points that are closer than the spatial resolution. As a result, adjacent sample points are correlated and the correlation may extend further depending on the degree of over-sampling. In the case of a DOFS with a Gaussian distribution of the measurement uncertainty and a sample separation equal to half the spatial resolution (a common ratio for DTS systems), the standard deviation estimated along a constant-measurand fibre section would be about 61% of the true value. In other words, unless corrected, this method under-estimates the measurand resolution by a factor of almost two. A correction can be applied to the results if the auto-correlation properties of the measured values are known.

Another estimate of the measurand resolution, termed *repeatability*, is obtained from similar measurements, but now the standard deviation at each point is calculated across a data set. This is closely related to the 'time-step resolution' mentioned earlier. In general, the repeatability varies only slowly along the fibre and so the data are smoothed with a moving-average filter to improve the estimate of the standard deviation.

In a typical set-up, a section of fibre is kept at uniform and constant measurand conditions. A zone is defined where the standard deviation and the mean value of the measurand are estimated for each of a repeated set of measurements. The mean of the standard deviation estimates the measurand resolution and, conversely, the standard deviation of the mean estimates the repeatability. These estimates can, of course, be repeated at different locations along the fibre in order to assess how the performance varies with distance along the fibre.

1.3.4 Sensor Accuracy and Cross-Talk

The accuracy of the sensor is the difference between the measured value and the true value (determined by a trusted reference measurement method).

In the case of DOFSs, the accuracy is frequently a function of position along the fibre, for example, as a result of incorrectly compensated losses in the fibre.

Accuracy is primarily an issue in the measurement of quasi-static quantities (such as temperature and strain). For dynamic quantities (such as vibration), the prime consideration is the time dependence of the signal.

The possibility of cross-talk between sensing positions along the fibre arises if, for example, the acquisition system is not perfectly linear. Clearly, the mechanisms for cross-talk will differ between the various types of sensor, but where accuracy is important, it should be considered and, if necessary, tested.

The accuracy of DOFSs is also affected by cross-sensitivity, i.e. where the reading depends on the measurand of interest as well as another parameter. In general, the problem of cross-sensitivity can be addressed with more measurements that are not linearly related. A considerable body of work, summarised in Chapter 8, has been devoted by the DOFS research and development community to the problem of separating sensitivity to temperature from that to strain.

1.4 HISTORY OF COMMERCIALISATION: SUCCESSES AND FAILURES

A number of DOFS techniques are now well established and commercially successful.

1.4.1 Raman DTS

The Raman-based DTS is the most developed and widely commercialised DOFS technology. DTS systems have been manufactured for almost 30 years, with some of the original market participants still engaged in this business. The original driver of this market was the monitoring of assets, mainly cables and transformers, in energy transmission. Electricity transmission is carried out at high voltage, with high current flows; the monetary value of the energy transported is considerable. Therefore, the failure of a circuit is expensive, particularly since privatised industries are held to performance standards by regulators and are penalised, for example, for being unable to accept a lower-price electricity supply when offered, a situation that can arise if a transmission circuit is out of action or thought to be at maximum capacity. The deregulation of electricity markets in several countries in the 1980s and 1990s separated the generation, transmission and retailing of energy. The beginning of the DTS business was brought about by the need for improved monitoring of the electricity transmission network.

By the late 1980s, several companies were offering DTS systems based on Raman OTDR, including, in the United Kingdom, York Sensors Ltd., and Cossor Electronics (a subsidiary of Raytheon). In Japan, major cable companies, including Hitachi Cable, Sumitomo Electric, Fujikura and Furukawa, all worked on the technology that they saw as a means of monitoring their energy cables. Other companies showed passing activity, at least at the research level, including Brugg Cable and ABB (both Swiss) and Asahi Glass and Toshiba in Japan. Some of these companies had left the market by 1990 either through a merger or through a change in strategic focus. Later, other companies became involved in DTS technology.

The turnover in the DTS market is illustrative of the fact that although the technology is reasonably well understood, the ability to design a DTS system is not, in itself, sufficient for successfully participating in the market. What is required is to provide a complete product from research through to manufacturing with quality systems and industry-specific qualifications in place as well as applications development, sales and customer support. Therefore, a research team in a company or a university can relatively easily put together a first prototype; however, without the strategic focus and investment in manufacturing, in sales and in customer support, there is unlikely to be a successful business. Alternatively, a business that has a sufficient need internally for DTS systems can support their development and manufacture. This is particularly true if the internal requirement is not aligned with the specifications of generally-available interrogators.

Of the entrants to the DTS market, some emerged from the research departments of large energy equipment suppliers and others were optical fibre technology companies. To the author's knowledge several systems were developed by companies in both categories but were never commercialised.

The merging of DTS activity is noticeable in the energy cable business, where several of the large manufacturers at some point (usually in the mid to late 1980s or early 1990s) developed a DTS system for monitoring hot-spots on their cables. Their clients certainly required the availability of such measurements and industrial logic at the time dictated an in-house solution. The energy cable market did not support internal investments because the volume of the business was not sufficient for each cable manufacture to develop and maintain a DTS product line. Many of the energy cable entrants either withdrew from the market and sourced their DTS systems from independent DTS suppliers or merged their cable businesses (including DTS) with competitors, as part of a more general consolidation of the energy cable business.

In the early days of DTS, there was much excitement about measuring the temperature profile of large transformers to detect the formation of hot-spots. In practice, however, the difficulty of installing fibres all along the windings, together with the limited spatial resolution available, has led the industry to instrument just a few transformers of a particular type to validate their thermal models; however, this application never achieved significant volumes for DTS. Instead, manufacturers continue to rely on point optical fibre sensors at selected places in transformers to monitor the temperature at critical points defined by modelling and type-testing.

Other applications that have been evaluated, but abandoned, also seem to be characterised by the unfavourable balance of installation cost compared with the value of the measurement. For example, the use of DTS in chill cabinets in food stores was explored, but the concept was abandoned for reasons of practicality (such as the need to clean, to protect the sensor and to be able to easily move cabinets), as were several other applications in the food industry. Although the cost of DTS systems and optical cable has fallen dramatically since that time, and configuration software would allow much more flexible reconfiguration, other technologies have developed in the intervening time and they are currently viewed as more appropriate than a DTS solution.

The high end of the DTS business, where a long sensing range and a very fine temperature resolution are required, is still technically challenging, and only a few players compete in this market. The driver in this business is now oil and gas production, where DTS measurements provide insight into the performance of injector or producer wells. This market has higher barriers to entry: in addition to providing

DTS systems with cutting edge metrology, the supplier must understand the installation of optical cables into hydrocarbon wells and the reliability of fibre in these hostile conditions. This understanding is a prerequisite to achieve reliable system operation given that intervention on these wells (e.g. to replace a fibre) is expensive. Moreover, the supplier must be able to provide not just temperature profiles but also interpretations that convert the DTS data into the information that the client seeks about the productivity of their wells and reservoirs. The DTS system must also interface with the client's database or reservoir model so that the data can be used jointly with other information such as petro-physical logs and production history. A DTS supplier to this market must ideally master all these aspects or team up with others of complementary competence to provide a complete solution.

In specific markets, intellectual property is still a barrier to entry and several key patents, for example, on fibre installation and data interpretation, are still current.

1.4.2 Commercial Brillouin-Based Distributed Strain and Temperature Sensors

Commercial Brillouin-based distributed strain and temperature sensors (DSTSs), although more recent than DTS, are provided by a number of suppliers. This technology is used primarily for monitoring very long energy cables and other long structures such as pipelines (sometimes as a competing technology to Raman DTS and in other cases complementing Raman DTS) and the sensing ranges achieved can be higher than for Raman systems, partly owing to the use of optical amplification, an option not viable with Raman systems. A key issue is the cross-sensitivity of temperature and strain. This can sometimes be addressed by *a priori* knowledge of the situation. Alternatively, strain-isolated fibres can be provided to measure only temperature, with co-located strain-coupled fibres sensing both quantities. The temperature profile (from the strain-decoupled fibre) is then used to separate the two measurands in the strain-coupled fibre. Some types of Brillouin sensors are able natively to distinguish the two measurands, for example, by measuring intensity and frequency shift. The topic of cross-sensitivity is discussed in further detail in Chapter 8.

The measurement of strain distribution is finding applications in civil engineering, for example, and some of these are described in Chapter 8.

1.4.3 Distributed Vibration Sensing

After a faltering start, DVS (sometimes referred to as DAS, for distributed *acoustic* sensing) is now becoming established, driven by applications in the oil and gas industry. Although this technology is still being trialed, at the time of writing, it does seem that there is sufficient momentum to develop this industry segment. Two main applications are driving this technology.

The first of these applications is vertical seismic profiling, where the fibre is used as a continuous sensor of seismic signals originating at the surface from an active source or possibly from micro-seismic events caused by movement in the geological formations near the well. In this application, DVS replaces an array of electrical sensors that are expensive and time-consuming to install and rig up.

DVS measurements are presently 1–2 orders of magnitude less sensitive than the geophones (velocity sensors) or accelerometers that they replace. However, the fact that they acquire the vibration profile of the entire well simultaneously substantially reduces the acquisition cost by massively reducing the time taken for the measurement. In fact, the cost of a borehole seismic measurement in the exploration phase of the field development is dominated by 'rig time', i.e. the cost relating to the time for which an expensive drilling facility and its crew are required on site. In some cases, the seismic measurement can be performed at the same time as other measurements and so no additional rig time is involved at all and the time required to introduce the seismic tool is completely eliminated, leaving only the provision of a seismic source as the on-cost for acquiring the data.

Equally significantly, the DVS measurement can provide very fine spatial sampling, in contrast to conventional measurements where the signals are often under-sampled on grounds of cost. An adequately sampled signal can be processed more effectively to reduce noise, and this attribute of DVS also partly redresses the balance of performance compared with conventional sensors. Moreover, there is a widely held belief in the industry that measurement performance will improve and close the gap between the existing and new technologies.

DVS is also applied downhole to improve production using noise in the well arising from flow to infer a flow profile; vibration generated by devices such as valves and pumps is also analysed to determine the condition of these assets.

Outside the wellbore, DVS technology is also used in security, e.g. to detect intruders at fences or along pipeline routes and in railways to determine the location of trains.

DVS is now being applied commercially; however, the ultimate ubiquity of the measurement is still to be established.

1.5 OTHER TECHNOLOGIES

Some distributed sensing technologies have yet to catch on properly. The ancestor of them all, POTDR, has been demonstrated as a magnetic field sensor (and by extension, an optical current sensor), but in practice, the demand seems to be to measure well-defined points to rigorous metrology standards (typically 0.1% of full-scale over a very wide range of environmental conditions). Therefore, it is the optical current sensor, a point sensing device, that appears to be the real contender to displace conventional current transformers, rather than POTDR. POTDR has been proposed as a means to sense pressure distributions, for example, in hydrocarbon wells. However, to the author's knowledge, this approach has not been commercialised, and in any case, the promised metrology (1% of full-scale) is not adequate for the application.

Loss induced by micro-bending was developed into a linear fire detection system in the mid-1990s, but the technique was abandoned in favour of Raman DTS. Although measuring the totality of the backscatter signal for determining micro-bending at a particular location is technically easier than selecting the very weak Raman component, the technique requires a special cable that will induce losses abruptly when the temperature exceeds a certain value. This was achieved, notably by Ericsson, but there were difficulties, such as the added cost of the cable and permanent increases in loss as a result of heating, that caused the industry to move away from this approach.

The measurement of static strain using optical frequency domain analysis was developed initially as a tool for measuring optical fibres but has found a niche in certain high-resolution applications, where a strain profile is required on a relatively small body. It is not yet a major business, but it could well become one in the future, for example, in applications where the length scale is modest but extremely good spatial and measurand resolution are required.

1.6 DISTRIBUTED SENSING AND DENSE POINT SENSOR ARRAYS

The concepts of distributed sensors and multiplexed sensor arrays were introduced in this chapter as distinct approaches, one offering a continuous measurement of the measurand profile and the other providing a series of measurements at discrete points. However, in practice, distributed sensors are invariably sampled along the distance coordinate because processing the data captured in the digital domain is much simpler than attempting to do that with analogue electronics. Conversely, one could imagine an array of point sensors with a dense coverage of the distance along the fibre which would faithfully reproduce the measurand distribution. This raises the question of what is the fundamental difference between the two concepts.

A distributed sensor has a defined spatial resolution, and each sampled point is a weighted mean of the value of the measurand in the vicinity of the sample. The value of the measurand within roughly one spatial resolution contributes to the sensed output; however, it is quite possible that every detail of the measurand profile is not fully represented if the spatial resolution is insufficient to track its local variation. Nonetheless, a local variation of the measurand on a scale smaller than the spatial resolution will have *some* influence on the sensor output. As noted previously, the system designer or specifier chooses the spatial resolution as part of a trade-off with other system parameters; it might be varied *a posteriori* by deconvolution or spatial averaging, but these operations merely shift the trade-off.

In contrast, an array of point sensors will faithfully represent the *local* value of the measurand subject to the condition that spatial variations of the measurand are small compared with the size of the point sensor. However, any variation of the measurand between sensing points will be missed by an array of point sensors.

Therefore, in general, in an array of point sensors, the spatial resolution is given by the size of the sensors and its sample separation is larger than its spatial resolution. Conversely, distributed sensors are usually designed with a sampling resolution smaller than the spatial resolution and the sampling resolution is determined by the interrogation electronics rather than by the physical disposition of point devices; it is up to the user to specify a spatial resolution that is appropriate to the expected spatial frequency of the measurand.

It follows that a point sensor array with an element spacing sufficiently small to sample the measurand distribution adequately would be functionally equivalent to a distributed sensor with sufficient spatial resolution for the application. However, the technologies used in each case are completely different.

1.7 ORGANISATION OF THE BOOK

Part 1 covers the technology foundations on which DOFSs have been built. Chapter 2 provides an overview of optical fibres, their construction and light propagation in them; it also introduces briefly the optical fibre components that are used to build the optics of DOFSs that are described in subsequent chapters. Chapter 3 describes OTDR and related techniques that the reader by now will have understood to be the basis of DOFSs.

Part 2 discusses the physics behind the three major types of DOFSs, i.e. Raman DTS (Chapter 4), Brillouin DOFS (Chapter 5) and coherent Rayleigh backscatter techniques (Chapter 6); it also discusses the design of these systems and the variants that exist, together with their performance and limitations.

Part 3 is devoted to how DOFSs are used: the practical issues involved and the applications where they have proven to be valuable. Chapters 7, 8 and 9 describe temperature, strain and vibration measurements, respectively. Part 3 also summarises in Chapter 10 a number of DOFS technologies that are of scientific or technical interest but have not yet found commercial use (or have been abandoned), and Chapter 11 concludes this introduction to the subject of DOFSs with a summary, comments and closing remarks.

REFERENCES

1. Keck, D. B., R. D. Maurer, and P. C. Schultz. 1973. On the ultimate lower limit of attenuation in glass optical waveguides. *Appl. Phys. Lett.* 22: 7: 307–309.
2. Hayashi, I., M. B. Panish, and F. K. Reinhart. 1971. GaAs–$Al_xGa_{1-x}As$ double heterostructure injection lasers. *J. Appl. Phys.* 42: 5: 1929–1941.
3. Vali, V., and R. W. Shorthill. 1976. Fiber ring interferometer. *Appl. Opt.* 15: 5: 1099–1100.

4. Cole, J. H., R. L. Johnson, and P. G. Bhuta. 1977. Fiber-optic detection of sound. *J. Acoust. Soc. Am.* 62: 5: 1136–1138.
5. Bucaro, J. A., H. D. Dardy, and E. F. Carome. 1977. Optical fiber acoustic sensor. *Appl. Opt.* 16: 7: 1761–1762.
6. Johnson, M., and R. Ulrich. 1978. Fibre-optical strain gauge. *Electron. Lett.* 14: 14: 432–433.
7. Norris, J. O. W. 1989. Current status and prospects for the use of optical fibres in chemical analysis. A review. *Analyst* 114: 11: 1359–1372.
8. Wickersheim, K. A. 1978. Optical temperature measurement technique utilizing phosphors. US4075493.
9. Smith, A. M. 1978. Polarization and magnetooptic properties of single-mode optical fiber. *Appl. Opt.* 7: 1: 52–56.
10. Budiansky, B., D. C. Drucker, G. S. Kino, and J. R. Rice. 1979. Pressure sensitivity of a clad optical fiber. *Appl. Opt.* 18: 24: 4085–4088.
11. Quick, W. H., K. A. James, and V. H. Strahan. 1980. Means for sensing and color multiplexing optical data over a compact fiber optic transmission system. US4223216.
12. Bosselmann, T., and R. Ulrich. 1984. High-accuracy position-sensing with fiber-coupled white-light interferometers, 2nd International Conference on Optical Fibre Sensors, Stuttgart, Germany, VDE Verlag: 361–364.
13. Brooks, J., R. Wentworth, R. Youngquist et al. 1985. Coherence multiplexing of fiber-optic interferometric sensors. *J. Lightw. Technol.* 3: 5: 1062–1072.
14. Brooks, J., B. Moslehi, B. Kim, and H. Shaw. 1987. Time-domain addressing of remote fiber-optic interferometric sensor arrays. *J. Lightw. Technol.* 5: 7: 1014–1023.
15. Sakai, I., R. Youngquist, and G. Parry. 1987. Multiplexing of optical fiber sensors using a frequency-modulated source and gated output. *J. Lightw. Technol.* 5: 7: 932–940.
16. Mlodzianowski, J., D. Uttamchandani, and B. Culshaw. 1987. A simple frequency domain multiplexing system for optical point sensors. *J. Lightw. Technol.* 5: 7:1002–1007.
17. Dakin, J. P. 1987. Multiplexed and distributed optical fibre sensor systems. *J. Phys. E Sci. Instrum.* 20: 8: 954.
18. Behrendt, A., and V. Wulfmeyer. 2003. Combining water vapor DIAL and rotational Raman lidar for humidity, temperature, and particle measurements with high resolution and accuracy, Lidar Remote Sensing for Environmental Monitoring IV, San Diego, SPIE 5154:
19. Targ, R., M. J. Kavaya, R. M. Huffaker, and R. L. Bowles. 1991. Coherent lidar airborne windshear sensor: Performance evaluation. *Appl. Opt.* 30: 15: 2013–2026.
20. Diaz, R., S.-C. Chan, and J.-M. Liu. 2006. Lidar detection using a dual-frequency source. *Opt. Lett.* 31: 24: 3600–3602.
21. Fredriksson, K., B. Galle, K. Nyström, and S. Svanberg. 1979. Lidar system applied in atmospheric pollution monitoring. *Appl. Opt.* 18: 17: 2998–3003.
22. Barnoski, M. K., and S. M. Jensen. 1976. Fiber waveguides: A novel technique for investigating attenuation characteristics. *Appl. Opt.* 15: 9: 2112–2115.
23. Barnoski, M. K., M. D. Rourke, S. M. Jensen, and R. T. Melville. 1977. Optical time domain reflectometer. *Appl. Opt.* 16: 9: 2375–2379.
24. Personick, S. D. 1977. Photon probe – An optical time-domain reflectometer. *Bell Syst. Tech. J.* 56: 3: 355.
25. Asawa, C. K., J. W. Austin, M. K. Barnoski et al. 1984. Microbending of optical fibers for remote force measurement. US4477725.
26. Asawa, C. K., and H. F. Taylor. 2000. Propagation of light trapped within a set of lowest-order modes of graded-index multimode fiber undergoing bending. *Appl. Opt.* 39: 13: 2029–2037.
27. Asawa, C. K., and S.-K. Yao. 1983. Microbending of optical fibers for remote force measurement. US4412979.
28. Asawa, C. K., S. K. Yao, R. C. Stearns, N. L. Mota, and J. W. Downs. 1982. High-sensitivity fibre-optic strain sensors for measuring structural distortion. *Electron. Lett.* 18: 362–364.
29. Yao, S. K., and C. K. Asawa. 1983. Microbending Fiber Optic Sensing. 1983 Technical Symposium East, SPIE: 9–13.
30. Griffiths, R. W. 1987. Structural monitoring system using fiber optics. US4654520.
31. Rogers, A. J. 1980. Polarisation optical time-domain reflectometry. *Electron. Lett.* 16: 13: 489–490.
32. Hartog, A. H., D. N. Payne, and A. J. Conduit. 1980. Polarisation optical-time-domain reflectometry: Experimental results and application to loss and birefringence measurements in single-mode optical fibres, 6th European Conference on Optical Communication, York, UK, IEE, 190 (post-deadline): 5–8.
33. Rogers, A. J. 1981. Polarization-optical time domain reflectometry: A technique for the measurement of field distributions. *Appl. Opt.* 20: 6: 1060–1073.
34. Ross, J. N. 1982. Birefringence measurement in optical fibers by polarization optical time-domain reflectometry. *Appl. Opt.* 21: 19: 3489–3495.
35. Kim, B. Y., D. Park, and C. Sang. 1982. Use of polarization-optical time domain reflectometry for observation of the Faraday effect in single-mode fibers. *IEEE J. Quant. Electron.* 18: 4: 455–456.

36. Hartog, A. H., D. N. Payne, and A. J. Conduit. 1981. Polarisation measurements on monomode fibres using optical time-domain reflectometry. *IEEE Proc. Pt. H* 128: 3: 168–170.
37. Namihira, Y. 1985. Opto-elastic constant in single mode optical fibers. *J. Lightw. Technol.* 3: 5: 1078–1083.
38. Rogers, A. 2002. Distributed fibre measurement using backscatter polarimetry, 15th International Conference on Optical Fiber Sensors, Portland, OR, IEEE, 1: 367–370.
39. Rogers, A. J., S. E. Kanellopoulos, and S. V. Shatalin. 2005. Method and apparatus for detecting pressure distribution in fluids. US7940389B2.
40. Hartog, A. H., and D. N. Payne. 1982. A fibre-optic temperature-distribution sensor, IEE Colloquium on Optical Fibre Sensors, London, IEE, Digest 1982/60: 2/1–2/2.
41. Hartog, A. H., and D. N. Payne. 1984. Fibre optic sensing device. GB2122337.
42. Hartog, A. H. 1983. A distributed temperature sensor based on liquid-core optical fibers. *J. Lightw. Technol.* 1: 3: 498–509.
43. Hartog, A. H., and D. N. Payne. 1982. Remote measurement of temperature distribution using an optical fibre, 8th European Conference on Optical Communications, Cannes, France: 215–218.
44. Dakin, J. P. 1984. Temperature measuring arrangement. GB2140554.
45. Dakin, J. P., D. J. Pratt, G. W. Bibby, and J. N. Ross. 1985. Distributed antistokes ratio thermometry. In 3rd International Conference on Optical Fiber Sensors. San Diego, Optical Society of America. PDS3.
46. Raman, C. V. 1929. The Raman effect. Investigation of molecular structure by light scattering. *Trans. Faraday Soc.* 25: 781–792.
47. Long, D. A. 1977. *Raman spectroscopy*. McGraw-Hill, New York, USA.
48. Horiguchi, T., and M. Tateda. 1989. BOTDA – Nondestructive measurement of single-mode optical fiber attenuation characteristics using Brillouin interaction: Theory. *J. Lightw. Technol.* 7: 8: 1170–1176.
49. Horiguchi, T., T. Kurashima, and M. Tateda. 1990. A technique to measure distributed strain in optical fibers. *IEEE Photon. Technol. Lett.* 2: 5: 352–354.
50. Kurashima, T., T. Horiguchi, and M. Tateda. 1990. Distributed-temperature sensing using stimulated Brillouin scattering in optical silica fibers. *Opt. Lett.* 15: 18: 1038–1040.
51. Kurashima, T., T. Horiguchi, and M. Tateda. 1990. Thermal effects on the Brillouin frequency shift in jacketed optical silica fibers. *Appl. Opt.* 29: 15: 2219–2222.
52. Tateda, M., and T. Horiguchi. 1990. Water penetration sensing using wavelength tunable OTDR. IEEE Photon. *Technol. Lett.* 2: 11: 844–846.
53. Tateda, M., T. Horiguchi, T. Kurashima, and K. Ishihara. 1990. First measurement of strain distribution along field-installed optical fibers using Brillouin spectroscopy. *J. Lightw. Technol.* 8: 9: 1269–1272.
54. Shimizu, K., T. Horiguchi, Y. Koyamada, and T. Kurashima. 1993. Coherent self-heterodyne detection of spontaneously Brillouin-scattered light waves in a single-mode fiber. *Opt. Lett.* 18: 3: 185–187.
55. Healey, P. 1984. Fading rates in coherent OTDR. *Electron. Lett.* 20: 11: 443–444.
56. Healey, P. 1984. Fading in heterodyne OTDR. *Electron. Lett.* 20: 1: 30–32.
57. King, J., D. Smith, K. Richards et al. 1987. Development of a coherent OTDR instrument. *J. Lightw. Technol.* 5: 4: 616–624.
58. Izumita, H., S. I. Furukawa, Y. Koyamada, and I. Sankawa. 1992. Fading noise reduction in coherent OTDR. *IEEE Photon. Technol. Lett.* 4: 2: 201–203.
59. Izumita, H., Y. Koyamada, S. Furukawa, and I. Sankawa. 1997. Stochastic amplitude fluctuation in coherent OTDR and a new technique for its reduction by stimulating synchronous optical frequency hopping. *J. Lightw. Technol.* 15: 2: 267–278.
60. Dakin, J. P., and C. Lamb. 1990. Distributed fibre optic sensor system. GB2222247A.
61. Taylor, H. F., and C. E. Lee. 1993. Apparatus and method for fiber optic intrusion sensing. US5194847.
62. Juskaitis, R., A. M. Mamedov, V. T. Potapov, and S. V. Shatalin. 1992. Distributed interferometric fiber sensor system. *Opt. Lett.* 17: 22: 1623–1625.
63. Juskaitis, R., A. M. Mamedov, V. T. Potapov, and S. V. Shatalin. 1994. Interferometry with Rayleigh backscattering in a single-mode optical fiber. *Opt. Lett.* 19: 3: 225.
64. Shatalin, S. V., V. N. Treshikov, and A. J. Rogers. 1998. Interferometric optical time-domain reflectometry for distributed optical-fiber sensing. *Appl. Opt.* 37: 24: 5600–5604.
65. Rathod, R., R. D. Pechstedt, D. A. Jackson, and D. J. Webb. 1994. Distributed temperature-change sensor based on Rayleigh backscattering in an optical fiber. *Opt. Lett.* 19: 8: 593–595.
66. Posey, R. J., G. A. Johnson, and S. T. Vohra. 2000. Strain sensing based on coherent Rayleigh scattering in an optical fibre. *Electron. Lett.* 36: 20: 1688–1689.

67. Posey, R. J., G. A. Johnson, and S. T. Vohra. 2001. Rayleigh Scattering Based Distributed Sensing System for Structural Monitoring, 14th Conference on Optical Fibre Sensors, Venice, Italy, SPIE, 4185: 678–681.
68. Jerri, A. J. 1977. The Shannon sampling theorem – Its various extensions and applications: A tutorial review. *Proc. IEEE* 65: 11: 1565–1596.
69. Hadley, M. R., G. A. Brown, and G. J. Naldrett. 2005. Evaluating permanently installed fiber-optic distributed temperature measurements using temperature-step resolution, SPE International Improved Oil Recovery Conference in Asia Pacific, Kuala Lumpur, Malaysia, Society of Petroleum Engineers, SPE-97677-MS.
70. SEAFOM. 2010. Measurement specification for distributed temperature sensing. SEAFOM-MSP-01-June2010.

Optical Fibre Technology

2

The operation and limitations of distributed optical fibre sensors (DOFSs) are dictated by the design and materials used to manufacture optical fibres, how light propagates in them and how photonic systems are assembled from basic optical components. This chapter provides some basic knowledge needed for understanding the remainder of the book and points the reader to references for more in-depth reading on optical fibre technology.

2.1 PROPAGATION IN OPTICAL FIBRES

An optical fibre is an elongated structure – usually round in cross-section – that guides light. In general, optical fibres are made from high-purity silicate glass and consist of a core surrounded by a cladding. The guidance is achieved by arranging for the core to have a refractive index n_1 higher than that, n_2, of the cladding. The refractive-index contrast causes light arriving within a defined cone of acceptance at the fibre input to be trapped in the waveguide. A further polymer coating layer is added to the outside of this structure to provide mechanical protection.

Assuming an abrupt refractive index change at the core-cladding boundary, the guidance of an optical fibre may be understood by considering the Snell-Descartes law of refraction [1], namely that a plane wave arriving from a medium of refractive index n_1 at angle θ_1 from normal incidence emerges in a second medium of index n_2 at angle θ_2 according to the following relation:

$$n_2 \sin(\theta_2) = n_1 \sin(\theta_1). \tag{2.1}$$

If $n_1 > n_2$, then of course $\theta_2 > \theta_1$. At a particular angle of incidence, the wave is refracted parallel to the interface, i.e. $\theta_2 = \pi/2$. Beyond this *critical angle* of incidence $\theta_c = \sin^{-1}(n_2/n_1)$, all light is reflected back into the first medium; this is known as *total internal reflection*. Therefore, waves travelling at angles less than $\theta_f = \frac{\pi}{2} - \theta_c$ from the fibre axis are guided by the refractive index contrast between the core and the cladding.

Figure 2.1a shows a fibre seen from the side and the ray path shown passes through the centre of the core, which is known as a meridional ray.*

The limiting angle θ_f that ensures guidance leads directly to the concept of the numerical aperture (*NA*), a parameter that defines whether a ray will be guided or not. The construction of angles is shown in Figure 2.1b. It is simple to show with standard trigonometric identities that the largest angle external θ_e at which a wave travelling in air can reach the end face of the fibre and still be guided is given by $\sin(\theta_e) = \sqrt{n_1^2 - n_2^2} \equiv NA$.

* Skew rays following a helical path within the core can also be guided.

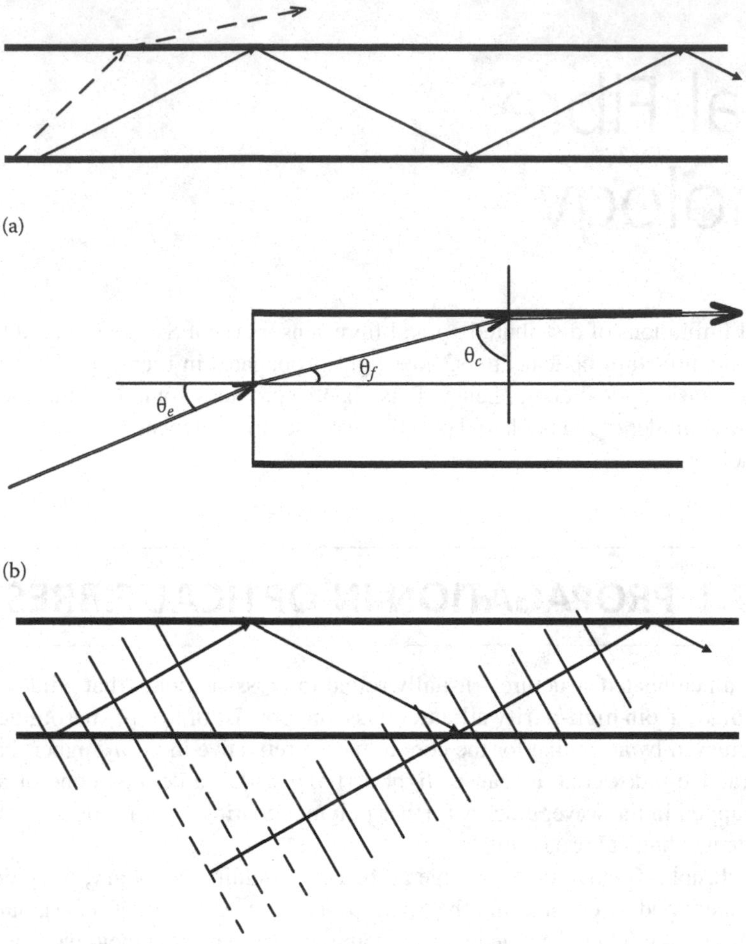

FIGURE 2.1 (a) Ray diagram for total internal reflection. (b) Geometric construction of the critical angle. (c) Plane wave approximation to propagation in an optical fibre.

The *relative index-difference* Δ between core and cladding is defined as

$$\Delta = \frac{NA^2}{2n_1^2} = \frac{n_1^2 - n_2^2}{2n_1^2} \simeq \frac{n_1 - n_2}{n_1}. \tag{2.2}$$

A ray description for light propagation is helpful in visualising various optical phenomena, but it is really an abstraction of the concept of a plane wave,* the ray being normal to the plane of the wave.

In moving from the concept of a ray to one of a plane wave, we associate the wave with a particular wavelength, and we thus recognise that it has a phase and that its phase is spatially periodic; i.e. if the wave is delayed by exactly one wavelength, the phase is unchanged. This leads to modifying the picture of Figure 2.1a to that of Figure 2.1c (which illustrates a ray passing through the core centre) to show

* A plane wave implies of course an infinite spatial extent and it is known that a wave emerging from a small aperture has a curved wavefront caused by diffraction. Nonetheless, the plane-wave approximation remains surprisingly valid down to quite small dimensions and is certainly reasonably accurate [2] when used on multimode fibres, the core diameter of which is typically 50 μm, compared with an optical wavelength of order 0.8–1.6 μm.

associated wavefronts. In order for the wave to be consistent in the waveguide, the phase after two reflections must be delayed by an exact multiple of 2π. Where this is not the case, the wave breaks up into separate waves that *do* match that condition and that, globally, sum to the original wave at the launching point. This demonstrates graphically that only certain ray angles (within the range defined by θ_e) are accepted by the waveguide; these permitted angles correspond to the more proper description of the waveguide in terms of modes that are solutions of the equations for electro-magnetic wave propagation, i.e. Maxwell's equations with the properties of the waveguide as boundary conditions. Each mode has a particular electric field distribution and propagation velocity.

It should also be noted that the guided waves carry a proportion of their power in the cladding in the form of an *evanescent wave* that decays rapidly with increasing distance from the core-cladding interface. The behaviour of the mode is influenced by the properties of the first few microns of the cladding through the evanescent wave.

Readers interested in a more detailed understanding of light propagation in optical waveguides are referred to textbooks such as Refs. [1] and [3].

Whether or not the next mode is guided is dictated by the *normalised frequency*, a parameter that combines the core radius a, the wavelength λ and the NA and is given by

$$V = \frac{2\pi}{\lambda} a \cdot NA. \tag{2.3}$$

As the V value is increased, more modes can be guided by the fibre; the critical value of V at the boundary between a mode being guided or not is known as the *cut-off frequency* for that mode. The number of modes that a fibre can guide depends on the size of the core relative to the wavelength of the light that is guided and also on the index-difference between the core and cladding. A fibre having a small core and a low NA will guide only two modes, each of which has a simple electric field pattern that is similar, but not generally identical, to a Gaussian distribution. The two lowest-order modes differ in that they are polarised orthogonally to one another.

2.1.1 Types of Optical Fibre

Several types of optical fibre are commonly used and they are illustrated in Figure 2.2. For each fibre type, the figure shows a schematic cross-section together with (to the left in each case) an illustrative representation of the refractive index profile along the vertical diameter.

2.1.1.1 Step-index multimode fibres

Figure 2.2a illustrates a multimode, step-index, fibre. Here, the core diameter and numerical aperture are sufficiently large that many modes are guided (the core diameter is typically 50 μm, but frequently much larger, and NA is usually 0.2 or more). A clear boundary is defined between the core and the cladding, where, in the ray picture, total internal reflection takes place.

For a step-index fibre, the number of guided modes is approximately $V^2/2$ [4], which works out at 434 for $a = 25$ μm, $\lambda = 1.064$ μm and $NA = 0.2$. One limitation of step-index fibres is that the modes travel at a wide range of group velocities* and emerge from a fibre of length L over a range of arrival times given by Ref. [4]

$$\delta\tau_{SI} = \frac{L}{c}[1 - 2/V](n_1 - n_2). \tag{2.4}$$

* The *group velocity* is the transit time of the pulse envelope, as opposed to the *phase velocity* n of individual wavelets within the pulse. The group velocity v_g is related to the speed of light in vacuum c through the group index $N_g = \frac{c}{v_g}$ with $N_g = n - \lambda\frac{dn}{d\lambda}$.

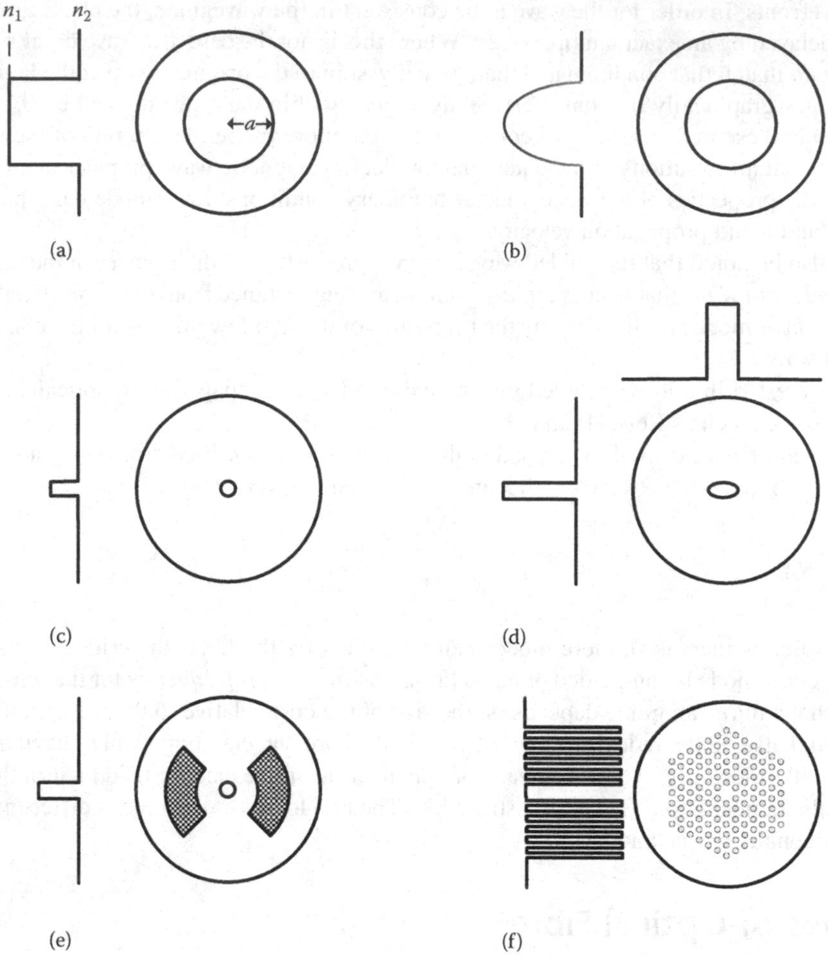

FIGURE 2.2 Schematic cross-section (right) and refractive index profile (left) for a few important types of optical fibre. (a) Step-index multimode fibre, (b) graded-index multimode fibre, (c) single-mode fibre, (d) elliptical core polarisation-maintaining fibre, (e) stress-birefringent polarisation-maintaining fibre and (f) photonic crystal fibre.

For $a = 25$ μm and $NA = 0.2$, $\delta\tau_{SI}$ evaluates to about 43 ns/km. As a result, short pulses launched into a step-index multimode fibre will spread, and this limits the information-carrying capacity of the fibre and, in the case of DOFSs, the spatial resolution.

2.1.1.2 Graded-index multimode fibres

Graded-index multimode fibres were developed to overcome the pulse-broadening of step-index fibres; they are designed so that the transit times are as far as possible the same for all modes.

For graded-index multimode fibres (Figure 2.2b), the refractive index varies smoothly from the core centre to the core-cladding interface. As a result of the gradual refractive index transition, the meridional rays are deflected smoothly, generally before reaching the core-cladding interface.

The transit time equalisation property of a graded-index fibre may be visualised by considering a first ray passing straight down the centre of the core and a second one meandering about the core centre; the first has a shorter path than the second but, on average, it travels in a medium of higher refractive index, and this slows it relative to the second ray. By carefully adjusting the profile, the shorter geometric path length of the first ray can be approximately compensated in the second ray by an average lower velocity path. Of course, this picture does not address the complexities of skew rays or of the dispersive properties of the materials. Nonetheless, it illustrates why graded-index fibres are attractive, where fibre bandwidth is a consideration. In general, the profile function $n^2(r) = n_1^2(1 - 2\Delta f_p(r))$ is close to parabolic. If the shape of the refractive index profile function $f_p(r)$ is optimised, the pulse spreading now becomes [5]

$$\delta\tau_{GI} = \frac{L}{c} n_1 \frac{\Delta^2}{8}. \tag{2.5}$$

For the same parameters as were used in connection with Equation 2.4, $\delta\tau_{GI}$ is calculated to be 60 ps/km for a precisely optimised profile, an improvement by three orders of magnitude over $\delta\tau_{SI}$. Although even small imperfections will increase the actual pulse spreading that is achieved in practice, some commercially available fibres are within a factor 10 of the optimum (a few are within a factor of 3 of the optimum). Multimode fibres are used in certain DOFSs, notably in distributed temperature sensor (DTS) systems, and the fibre bandwidth can limit their spatial resolution, which is why a graded-index profile is almost essential when multimode fibres are used.

It should also be noted that the optimal refractive index profile depends on the operating wavelength [6] owing to the dispersion (wavelength-dependence) of the refractive index of the glasses forming the fibre and how that dispersion varies across the core as the composition changes from that at core centre to that of the cladding. From a practical point of view, this means that the pulse spreading specified at one operating wavelength will usually be different at another wavelength. This is likely to affect the spatial resolution of long DOFSs operating on multimode fibre.

A common standard [7] for multimode, graded-index fibre specifies a core diameter of 50 μm and an *NA* of 0.2 (for multimode fibre, the definition of *NA* is based on the refractive index at the centre of the core).

In the case of multimode, graded-index fibres, the number of modes that are guided is given by $V^2/4$, so a graded-index, multimode fibre having $a = 25$ μm and $NA = 0.2$ will guide about 340, 216 and 100 modes at 850 nm, 1064 nm and at 1550 nm, respectively. However, the cut-offs for the modes tend to be bunched into groups, with several modes of different electric field distributions but similar propagation constant, being admitted as guided modes by the fibre at similar V values. These are known as *mode groups*.

More generally, the electric field relating to a mode may be written as a function of the radial and azimuthal coordinates r and θ, axial coordinate z and time t

$$E(r,\theta,z,t) = E_m(r,\theta)\cdot\exp\left[j(\omega t - \beta_m z)\right], \tag{2.6}$$

where ω is the angular frequency (phase change per unit time) and β_m is the propagation constant (phase change per unit distance along the z axis) specific to mode m.

Each mode can be assigned a number that defines the particular solution of the wave equation that it stems from; given the two-dimensional nature of the field distribution across the fibre core, the mode designation involves an ordered pair of integers. These integers relate to the number of zeros of the electric field in the radial and azimuthal dimensions [4].

Each mode travels independently of the others and has its own particular electric field distribution, phase and group velocity and state of polarisation.

The fraction of the total power launched into a multimode fibre that is carried by each mode is known as the *modal power distribution*. Whereas modes travel independently, power can be coupled from one to another through imperfections in the fibre, such as bends. Therefore, the modal power distribution can vary along the fibre. For example, if the power were launched in only the lowest order mode of a multimode fibre, the effect of random bends in cables would cause power to be transferred in part to other modes. In a sufficiently long fibre, with some mode coupling, it might be envisaged that the power would ultimately be spread equally between modes. However, the highest-order modes generally suffer from fractionally higher losses than lower-order modes do and so the differential attenuation depletes the power in the higher-order modes. The effect of power flowing from low-order modes to higher-order modes, together with a higher loss for the latter, ultimately leads to an equilibrium modal power distribution [8,9], provided that sufficient coupling exists.

In any event, the fact that the properties of modes are different and that power can be transferred between modes means that care must be taken in the interpretation of measured quantities, such as attenuation, that are aggregated values across all modes. Often, the result of the measurement is a function of the modal power distribution and it can vary along the fibre as that distribution settles following the launch. In addition, the power distribution can be radically altered at splices or connectors because of misalignment or if fibres of mismatched dimensions are joined together [10]. The modal power distribution can affect the result of most measurements in multimode fibre. It is therefore important to consider the conditions at the point of launching and how they can be altered by propagation through the link. This applies also to using these fibres for distributed sensing.

2.1.1.3 Single-mode optical fibres

In Figure 2.2c, a single-mode fibre is illustrated. Single-mode fibres generally have a smaller core and lower *NA* than multimode fibres do. By selecting V to be smaller than the cut-off for the second mode, one can ensure that only the lowest-order mode is propagated. For a step-index fibre, the value of V at which the second spatial mode is guided is about 2.405 (the first zero of the J_0 Bessel function), so for $a = 4$ μm and $NA = 0.12$, the wavelength required for $V = 2.405$ is $\lambda_c = 1.25$ μm. Given the definition of V, whether a fibre is single mode or not depends on the operating wavelength, so, for example, a fibre that is single mode at wavelengths longer than, say 1250 nm, will support multiple modes at shorter wavelengths.

The exact shape of the electric field distribution depends, of course, on the refractive index distribution, but for a simple step-index distribution, the mode field patterns are Bessel functions. However, the electric field distribution of the lowest-order mode is very close to Gaussian, to the point that a Gaussian beam of the appropriate spot size can be launched into that mode with some 99.5% efficiency [11–13].

For circularly symmetric fibres, the properties of any two modes of the same field distribution, but of orthogonal polarisation, are the same; this is known as *degeneracy*. In particular, their velocities are the same. If the fibre departs from circular symmetry, the properties of the two modes and in particular their propagation constants differ, an effect known as *birefringence*. Birefringence of the fibre at relatively low levels in general causes the polarisation state of the light to vary as it travels down the fibre; bends, side pressure, twists all influence the fibre birefringence and, hence, the polarisation state of the light.

Most communication systems can accept a drifting state of polarisation of the light, either because their receivers are insensitive to polarisation or because they can compensate for, or control, the state of polarisation. Some systems, notably sensing systems, do require the light to be delivered in a known state. Polarisation-preserving fibres (illustrated in Figure 2.2d and e) achieve this by imposing a strong, built-in birefringence. This causes light travelling in two particular orthogonal states to experience substantially different propagation constants. For power to be coupled between polarisation states, perturbations of the fibre must have sufficient components at a spatial frequency that matches the difference in propagation constants [14]. Coupling between orthogonal states is thus inhibited if there is a marked difference in the propagation constants of the two polarisation modes. In the case of Figure 2.2d, the birefringence is achieved by using a very non-circular core [15]; in Figure 2.2e, the dark regions are stress-applying sectors made from glass having a substantially larger expansion coefficient than the surrounding material. The stress alters the refractive index differentially for light polarised parallel or orthogonally to the line passing through the centre of the stress-applying sectors and thus results in a strong birefringence [16].

Fibres made from other materials, e.g. plastics, are also known. In fact, the very first DTS [17] was made from a fibre consisting of a hollow silica tube (forming the cladding) filled with an ultra-transparent liquid (hexachlorobuta-1,3-diene) that had a higher refractive index than silica and thus forming the core.

In recent years, fibres containing arrays of holes have also been developed; their guidance mechanisms are somewhat different. One example is illustrated in Figure 2.2f. They are known as photonic crystal fibres (PCFs) [18]. PCFs are one example of a class known as *microstructured* fibres.

2.1.2 Loss Mechanisms

2.1.2.1 Absorption

Modern optical fibres are probably the most transparent manufactured objects – at its most transparent wavelength, the best fibre will lose only around 20% of the light that was launched into it after transmission through 5 km. This is one of the reasons why they are such an attractive medium for telecommunications and for sensing technology.

In silicate glasses, the wavelength range in which low-loss transmission occurs is limited at short wavelengths by electronic absorptions that occur around 100 nm; these are the ultraviolet (UV) absorption bands. At longer wavelengths, around 10 μm in the mid-infrared (IR), molecular absorptions dominate the attenuation. The tails of these strong absorption bands stretch into the visible from the UV and into the near-IR from the mid-IR and place a floor under lowest loss that can be achieved for solid core fibres in a given material system. They also shape the dispersion of the glasses, i.e. the variation of their refractive index with wavelength.

Additional absorptions are caused by impurities, such as metallic ions and OH bonds. Modern manufacturing methods have allowed these impurities to be eliminated from good-quality optical fibres. However, in hostile environments, some impurity-related losses can re-appear. In particular, hydrogen becomes mobile at high temperature and penetrates barriers that are hermetic in more benign conditions. Hydrogen can absorb light in its molecular form as an interstitial impurity after diffusing into the glass, and, worse still, it can react with the glass matrix to form OH bonds that, in silica, are strong absorbers at 1383 nm and to a lesser extent at 1240 and 945 nm. A concentration of 1 ppm of OH ions would cause an increase in loss of ~54, 2.3 and 0.83 dB/km at 1383, 1240 and 945 nm, respectively [19]. The presence of additives, such as GeO_2, alters the effect of OH ion impurities by creating additional absorption peaks at slightly different wavelengths.

2.1.2.2 Scattering

Scattering is the process by which small amounts of forward-travelling light is re-directed by interactions with inhomogeneities in the glass from which the fibre is formed. The most prominent of these interactions, Rayleigh scattering, relates to the refractive index being non-uniform on a distance scale much smaller than the wavelength of the incident light. Again, for a given material system, Rayleigh scattering is fundamental in that it cannot be eliminated.

Rayleigh scattering arises from thermodynamically driven fluctuations in the density of the glass. When the fibre is drawn, its material is softened at a very high temperature (c. 2000°), and this results in thermal agitation in the material and thus, in fluctuations in the density of the material, over quite short distances, of order 100 nm. As the fibre is drawn, it is cooled very fast (<1 s) to temperatures well below the softening temperature (indeed, this rapid cooling is required in order to apply a protective polymer coating) and its viscosity increases rapidly as it cools. As a result, the glass becomes kinetically limited against re-ordering the density distribution to a value corresponding to room temperature and reaching the degree of disorder appropriate to that temperature. Instead, the distribution of density is frozen and is considered equivalent to a *fictive temperature* T_f, which, for fibre drawn at 2000°C, is around 1500°C. The density variations are thus fixed in the fibre and form a low-level three-dimensional roughness of the refractive index.

Propagation of light through a dielectric medium involves the electric field polarising the molecules through which it travels, and the molecular polarisation, in turn, results in a secondary field being radiated with a phase delay. In a homogeneous medium (i.e. where the refractive index is uniform), all polarising fields are in phase and cancel in all but the forward direction [20]. The phase shift induced in this process retards the electro-magnetic wave, and this effect accounts for the slower propagation velocity compared with free space.

Now, when the medium is slightly inhomogeneous, the regions of refractive index different from the mean may be regarded as isolated scatterers. These scatterers re-radiate a small fraction of the light in all directions, and because they are anomalies in an otherwise homogenous medium, their polarising fields are not cancelled in all but the forward direction: they may be regarded as individual dipoles that radiate with a toroidal electric field pattern about the dipole axis. It is the accumulation of all such scatters that gives rise to Rayleigh scattering, subject to the scattering centres being much smaller than the wavelength of the incident light.

The Rayleigh scattering coefficient is proportional to the inverse fourth power of wavelength, and in pure silica, its value is about 0.7 dB km^{-1} μm^{-4}. Therefore, at 1 μm, for every metre of fibre that the light travels through, about 0.016% of the incident light is lost.

In nature, Rayleigh scattering from gas molecules in the atmosphere is the mechanism that causes the sky to appear blue because the shorter wavelengths in the solar spectrum are preferentially scattered. Thus, when the sky is viewed in the absence of clouds, the observer sees Rayleigh-scattered light [21].

It should be noted that the blue light from the sky is polarised, if viewed at 90° to the incident sunlight. This effect is used by bees to navigate. Reputedly, Norse seafarers also used polarising crystals to navigate in cases where the sun is obscured, but patches of blue sky are visible, although this interpretation of the myths is disputed [22]. In the same way, the scattered light that can be observed from the side of the fibre contains information on the state of polarisation of the incident light at that point.

In modern telecommunication fibres, Rayleigh scattering dictates the minimum achievable loss. The fundamental absorption and Rayleigh scattering losses modelled for a typical single-mode, step-index index fibre are plotted as a function of wavelength in Figure 2.3. The model is based on approximations available in Refs. [23] and [24] for the material properties. It is also necessary to model the proportion of optical power travelling in the core and cladding; approximations in Refs. [12] and [24] were used for that purpose. The total loss (solid line) shows a minimum near 1550 nm: at this wavelength, the reduction in loss for increasing wavelength due to the UV absorption (dotted line) and Rayleigh scattering (broken line) is cancelled by the sharp increase in IR absorption (dot-dash line). The subset of the data shown as

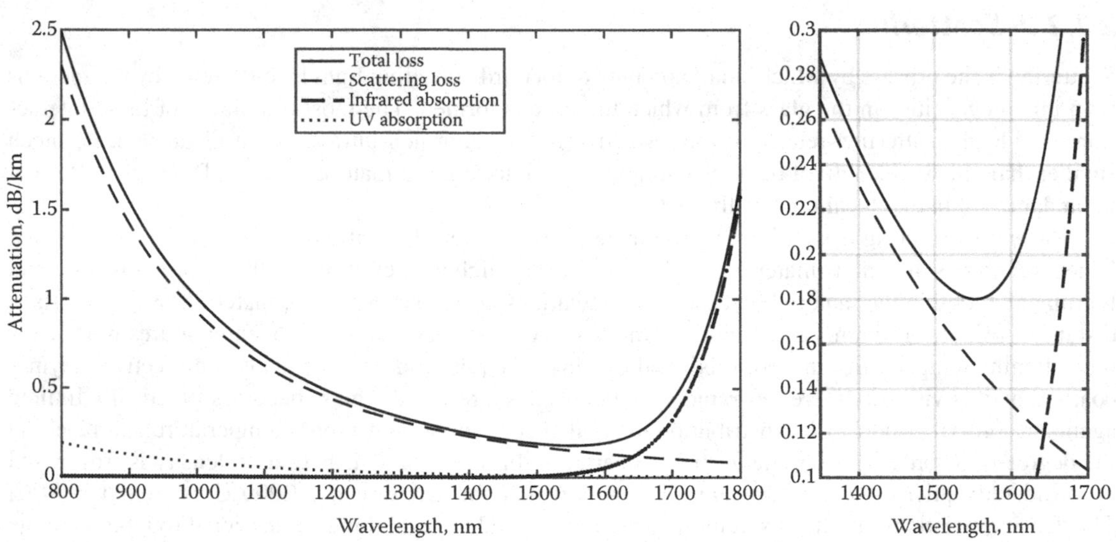

FIGURE 2.3 Plot of the fundamental losses for a typical germano-silicate single-mode fibre (expanded view to the right).

a separate plot to the right on an expanded scale illustrates how narrow the transmission window is when used for long-distance transmission or sensing when the signal-to-noise ratio is marginal. For example, a Raman DTS system having a probe at 1550 nm would generate anti-Stokes and Stokes signals at 1450 nm and 1663 nm, respectively, resulting in an increased loss relative to the probe of 0.035 and 0.11 dB/km, respectively. These seemingly minor differences in attenuation are in fact quite important when sensing over distances of 50–100 km, which is necessary in certain applications.

Although the Rayleigh scattering can be modified slightly by how the fibre is processed, for example, in the drawing conditions [25], it is nonetheless an inescapable effect.

The presence of index-modifying additives (sometimes referred to as dopants by analogy with semiconductor fabrication processes), such as GeO_2, in the glasses has two distinct effects, namely (a) to provide an additional mechanism for creating inhomogeneities, i.e. fluctuations in dopant concentration, and (b) by modifying (usually reducing) the softening temperature of the multi-component glass. For GeO_2, the former effect dominates, and addition of this dopant increases Rayleigh scattering. For P_2O_5, however, the fictive temperature of the glass is substantially reduced, and some success has been achieved in reducing the Rayleigh scattering by incorporation of this additive [26]. Other approaches, based on accelerating the structural relaxation time, for example, by addition of chlorine in the glass, have also shown a slight but useful reduction in the Rayleigh scattering loss [27].

Rayleigh scattering is elastic: to first order, the energy of the scattered photons is the same as that of the incident photons, so the wavelength is preserved in the scattering process.

Rayleigh scattering also, by and large, preserves the polarisation of the incident light: in silica, about 5% of the scattered light intensity emerges in the polarisation orthogonal to that of the incident light, whereas 95% remains in the original polarisation. The depolarisation effect is thought to be due to a small-scale anisotropy of the polarisability of microscopic scattering regions in the glasses that are randomly orientated [28]. Thus, if scatterers are anisotropic and not aligned to the incident electric field, the induced dipoles do not align perfectly to that field. Therefore, in such a case, the scattered light has a component in the orthogonal polarisation to that of the incident light.

However, in order to observe the depolarisation effect in the backscattered light, the fibre must be of the single-mode type: in multimode fibres, the state of polarisation is generally scrambled, unless extreme care is taken to launch only the lowest-order mode and then to prevent mode-coupling.

In forward propagation in conventional single-mode fibres, the *degree* of polarisation is preserved, but not the *state* of polarisation: the lowest-order mode, as previously mentioned, consists of two modes that have similar propagation constants. The optical power can therefore easily be transferred between them. However, in a polarisation-preserving (high-birefringence) fibre, light launched with its electric field aligned to one of the principal axes tends to remain in that state. In this case, the backscattered light reflects the polarisation state at the scattering point and the strong birefringence results in the power being resolved in the two orthogonally polarised modes. Experimental work [29] and also the author's own unpublished work on polarisation-maintaining fibres of a different type from those of Nakazawa et al. [29] showed that in silica-based optical fibres, the depolarisation ratio is c. 5%.

2.1.2.3 Other loss mechanisms

Other elastic scattering mechanisms also exist, such as that caused by inhomogeneities that are not very small compared with the wavelength, an effect known as Mie scattering. In contrast to the case of Rayleigh scattering from molecules, the larger water droplets in clouds cause Mie scattering. Clouds are white or gray because Mie scattering is far less wavelength dependent than Rayleigh scattering.

A rough interface between the core and cladding will cause further scattering. Fortunately, modern fibre production processes have eliminated both Mie-scale impurities and surface roughness* at least in silica-based fibres.

* An exception being PCF fibres, where thermodynamic effects during the draw lead to built-in roughness at the air–glass interfaces [30].

The principal remaining loss mechanism that should be cited is that due to bending, i.e. where the fibre is bent either consistently over significant lengths with constant radius of curvature (macro-bending), or on a small scale, for example, by being pressed against a rough surface (micro-bending). In essence, bending couples light from guided modes to modes that are not guided or are leaky. Cable designs usually avoid micro-bending losses, but it should be noted that some sensor designs are based on inducing a controlled amount of attenuation through bending. Examples will be provided in Chapter 10.

Other scattering mechanisms exist and are exploited extensively in DOFSs. They are Raman scattering and Brillouin scattering; they are inelastic in that the frequency of scattered light is different from that of the incident light. Their use in DOFSs will be discussed in Chapters 4 and 5. However, they contribute only a few percent to the attenuation of the fibre at room temperature.

2.1.3 Bandwidth of Optical Fibres

The bandwidth of an optical fibre is a measure of its information-carrying capacity; it is inversely proportional to the spreading that pulses suffer when propagating along the fibre. In optical communications, short pulses launched into the fibre become broader, and this limits the number of pulses that can be launched in a given time interval and still be distinguished at the fibre output. The bandwidth is largely inversely proportional to length, and so it is usually quantified as a bandwidth × distance product, in units of MHz km.

Bandwidth is a frequency-domain concept. In the time domain, the term *dispersion* is used to describe the pulse broadening; it is measured in ns/km.

In the context of distributed sensors, the dispersion limits the spatial resolution by spreading the probe pulse over a longer time interval and so the pulse occupies a longer section of fibre. Dispersion also influences, sometimes helpfully, the extent of non-linear optical effects, which are distortions of the spectrum of the light guided when the optical intensity is high.

Dispersion is broadly categorised as *intermodal* (i.e. differences of propagation time between modes) and *intramodal* (broadening specific to each mode) [6]. Intermodal dispersion, which applies to multimode fibres only, has already been discussed in the context of the rationale for grading the refractive index distribution in multimode fibres. A carefully controlled refractive index profile is capable of providing a bandwidth of around 10 GHz km for $NA = 0.2$.

Intramodal dispersion, on the other hand, relates to how the group velocity of each mode varies as a function of the wavelength of the propagated light. Intramodal dispersion is also referred to – especially in single-mode fibres – as *chromatic dispersion* (CD), i.e. dispersion relating to the colour of the light.

CD is partly caused by the wavelength-dependence of the refractive index, which is known as material dispersion.

Although material dispersion varies somewhat between glass compositions, broadly, it results in the time of flight of a pulse in a sample of bulk material or in a multimode fibre being minimised between 1270 nm (for pure silica) [31] and 1350 nm (for silica-germania binary glass with a large germania content) [32,33]. The wavelength at which the time of flight is minimised is known as the wavelength of zero material dispersion.

A second cause of intramodal dispersion is the fact that as the wavelength varies, the size of the core varies compared to the wavelength, and this alters the group velocity of the mode, regardless of any material effects. The phase velocity c/n_{eff} of a mode transitions from c/n_2 at cut-off to approach c/n_1 for V values well above cut-off, and of course the effective index n_{eff} transitions from n_2 to n_1. Its phase velocity and, by differentiation, its group velocity, are a function of wavelength through changes in V, completely independently of wavelength dependence of the refractive indices n_1 and n_2. It is this effect that is known as waveguide dispersion.

Waveguide dispersion can be ignored in multimode fibres because, under constant modal power distribution, its effect on the aggregate velocity of all guided modes is largely offset by new modes becoming guided as the V value is increased [34].

However, in single-mode fibres, waveguide dispersion adds to material dispersion to modify the overall variation of pulse transit time as a function of wavelength. Waveguide dispersion can be tailored by designing the refractive index profile to shift the wavelength of zero dispersion to longer wavelength or indeed flatten the dispersion curve over a specified spectral region.

For completeness, it should be mentioned that the CD is further affected by a combination of waveguide and material effects, namely that as the wavelength changes, the fraction of the mode carried in the mode that travels in the core varies, and this means that the average exposure of the mode to the dispersive material properties of the core and the cladding also varies. Furthermore, the relative index difference is also wavelength dependent, an effect known as profile dispersion.

In single-mode fibres, the user focuses on the combination of all these contributions, i.e. on the CD, measured in units of ps nm^{-1} km^{-1}.

2.2 CONSTRUCTION OF OPTICAL FIBRES

The fabrication of optical fibres and the analysis of how light is propagated in them has been described in a number of texts [1,3–5,24]; however, a few basic concepts will be described here to introduce the subject.

Low-loss optical fibres are currently made from high-silica glasses, i.e. a mixture of glasses consisting mainly of fused silica (SiO_2) to which other materials are added to raise or lower the refractive index. Germania (GeO_2), phosphorus pentoxide (P_2O_5) and (less commonly used) alumina (Al_2O_3) raise the refractive index, whereas B_2O_3 and fluorine reduce it. These additives also affect other properties of silica, such as its softening temperature, expansion coefficient, dispersion, stress-optical effect, attenuation and susceptibility to damage from ionising radiation or hydrogen ingress.

Although silica was well known for its low attenuation, the breakthrough in fibre fabrication came with vapour deposition processes, such as chemical vapour deposition, flame hydrolysis and plasma deposition, which all allowed the glasses to be synthesised directly into a starter rod (known as a preform) that could be drawn straight into a fibre with minimal intermediate working. The glasses are created by oxidising starting halide materials (e.g. $SiCl_4$, $GeCl_4$, etc.) that are liquid and can be purified to a high degree through bulk chemical processing. The vapour deposition processes allow the extreme purity of the starting materials and of the oxygen used in the chemical process to be preserved in the glass that is formed. Prior work on purifying glasses in a molten state was simply not up to the task of removing impurities (such as metallic ions or OH ions) to the parts-per-billion level that was required.

The starting halides in liquid form are held in a bubbler. By passing a carrier gas through a fine frit in the bubbler, some of the starting material is carried as a vapour in a gas stream to the reaction. The carrier gas (typically dry nitrogen) is arranged to be saturated with the halide vapour, and the quantity of each material that is brought to the reaction can therefore be controlled through the gas flow rate. By choosing the relative flow rates for each of the constituent starting materials, the ultimate composition of the resultant glass can be accurately controlled.

Most of the processes are designed to deposit successive, radially symmetric, thin layers and thus build up a chosen radial refractive index profile. In the first of these processes [35], outside vapour phase oxidation (OVPO), many layers of unconsolidated (sooty) silicate are created by burning the starting materials in a hydrogen flame (thus the name 'flame hydrolysis'); the solid combustion products are deposited on a mandrel that is removed at the end of the process. The boule of soot is dried in a chlorine-rich atmosphere to remove the water vapour arising from the oxidation process. The boule is then consolidated by slowly heating it to fuse the soot into a clear, bubble-free glass. In a variant of this process, the material is deposited at the end of a boule that is gradually retracted, the so-called vapour-phase axial deposition (VAD) process [36]. Although the flame hydrolysis processes result in a soot that has a high water and OH ion content, the very fine porous structure allows effective drying and OH-free material (down to 1 ppb) was obtained in this way in the 1980s [37].

In the modified chemical vapour deposition process (MCVD) [23], the reaction takes place inside a substrate silica tube that is heated by a hydrogen–oxygen torch from the outside. When all layers are deposited inside the tube, the entire structure is collapsed at a higher temperature to create a solid preform. Although the soot is formed in the flowing materials in the tube, the sooty reaction products are driven to the walls downstream of the reaction by thermophoresis, i.e. they are driven to the colder surface by the temperature gradient.

In the plasma process [38], a plasma is induced electrically from the outside of a tube but created inside the tube, and, similarly to MCVD, thin layers of glass are deposited inside the substrate tube.

Each of these processes has its advantages and drawbacks that can be quantified, for example, through the deposition rates* that can be achieved, the preform size and the type of materials and concentrations of specific dopants† that can be obtained. However, they have all resulted in very low-loss fibres with precisely controlled refractive index profiles.

Using the MCVD technique, a fibre described as having the ultimate low-loss at 1.55 μm was produced in 1979 [39] in which the minimum loss was measured to be 0.2 dB/km, a true breakthrough. However, even this fibre contained a sufficient OH impurity level to cause an absorption peak of ~10 dB/km at 1380–1390 nm. In the intervening years, improved purity in the fabrication process and changes to the fibre design have led to commercially available fibres guaranteed to provide a loss of 0.17 dB/km or less at 1550 nm [40], with typical peaks at 1380–1390 nm of only a few tenths of 1 dB/km. Some commercially available fibres show essentially no discernible absorption due to OH ions at around 1380–1390 nm.

2.3 PRINCIPAL COMPONENTS USED IN DISTRIBUTED SENSING SYSTEMS

The construction of a DOFS requires components that generate, manipulate and detect light.

The purpose of this section is not to explain the details of how these components work but to describe their function and typical characteristics. The symbols used to represent them in the system block diagrams that will be found later in the book are also introduced and shown in Figure 2.4.

2.3.1 Lasers

Lasers are used almost exclusively in DOFSs because they can efficiently launch large power levels into optical fibres. However, they come in many types and the main ones are discussed here in increasing order of coherence length, i.e. in order of reducing spectral width. Generally, a laser consists of a gain medium that amplifies light travelling through it combined with feedback that re-launches the amplified light through the amplifier. So starting from spontaneous emission within the gain medium, a stronger, more coherent wave is produced that depends both on the gain medium and the nature of the feedback. Generally, the feedback provides a resonant structure that sets up one or more modes that individually emit in a narrow spectrum.

2.3.1.1 Broad contact semiconductor lasers

The simplest and earliest form of diode-laser consists of a semiconductor gain region formed by a p–n junction in a direct band-gap material, such as GaAs, with the feedback being provided by reflections

* The mass of material that can be accreted on the preform in a unit time.

† Some dopants, fluorine, for example, are difficult to incorporate in large concentrations with some processes.

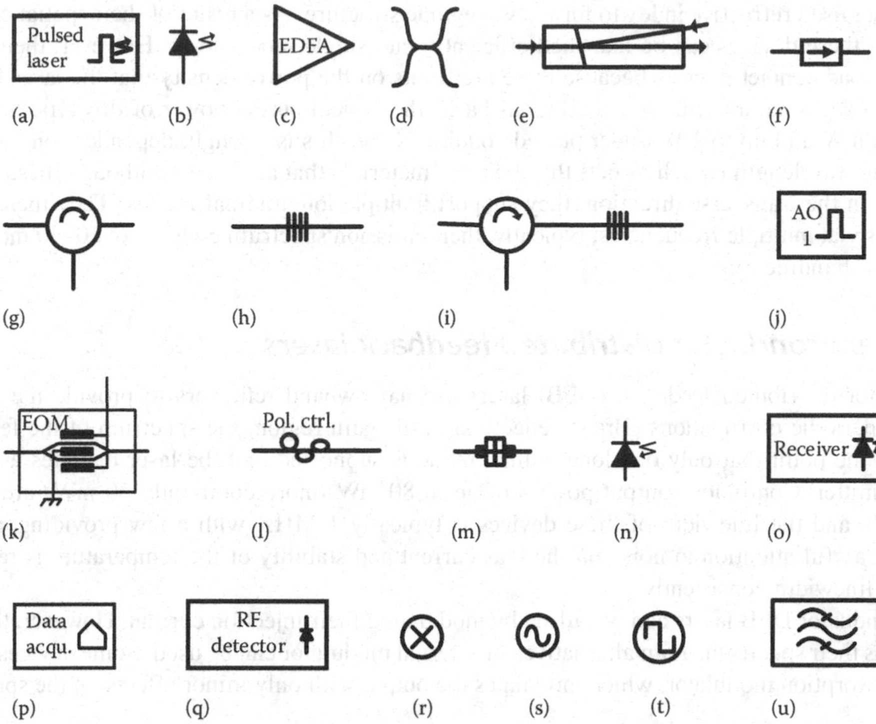

FIGURE 2.4 Symbols for components and functional blocks used in schematic diagrams later in this book. (a) Pulsed laser transmitter, (b) laser diode, (c) erbium-doped fibre amplifier, (d) fused-taper directional coupler, (e) micro-bulk-optic wavelength multiplexer, (f) optical isolator, (g) circulator, (h) FBG, (i) circulator and Bragg grating combined to form a bandpass filter, (j) AOM (used here to generate an optical pulse), (k) EOM, (l) polarisation controller, (m) fibre–fibre connector, (n) PD detector, (o) optical receiver, (p) data acquisition module, (q) RF detector, (r) electrical mixer, (s) sinusoidal voltage generator, (t) square-wave voltage generator, (u) filter (here a bandpass filter) – this symbol is used for both electrical and optical signals.

from cleaved facets of the laser chip; a laser chip is typically a few hundred microns in length. The junction is arranged to confine both the electrical carriers and the light travelling in the amplifying region. Being electrically a diode, these devices have contacts on either side. In the case of broad contact devices, the top contact is several tens of μm wide (typically 75 μm); as a result, the emission occurs in a line of similar width, which is single-mode only in the plane transverse to the junction. In the plane parallel to the junction, the light is emitted in a complex, multimode pattern that can be launched efficiently only into a multimode fibre. However, broad-contact lasers are powerful, and, in pulsed conditions, a 75 μm-contact device can typically deliver more than 10 W of peak power.

Broad contact laser diodes are suited to some types of Raman multimode DTS systems because the demands on the spectral purity are not stringent. They are also applicable to loss-based sensors, for example, those that use controlled micro-bending as the sensing mechanism.

In schematic diagrams to be found in later chapters, a laser diode is represented by a diode with a symbol denoting light emission (Figure 2.4b); the symbol in Figure 2.4a is used when the laser is included in a pulse transmitter.

2.3.1.2 Fabry-Pérot single-mode laser diodes

Fabry-Pérot single-mode laser diodes are designed to emit only in a single transverse mode, i.e. they are spatially coherent, by confining the gain to a narrow stripe. Some designs also confine the light in the gain region with a raised refractive index to form a waveguide structure. As a result of their spatial coherence, the output of these devices can be launched efficiently into single-mode fibres. However, their power is lower than broad contact devices because there are limits on the power density that the laser facets can support, even when passivated; on a continuous basis, the typical output power of this type of laser is a few hundred mW and up to 1 W under pulsed conditions, but this is strongly dependent on the supplier and operating wavelength (which affects the choice of materials that are used). Although these lasers are single mode in the transverse direction, they support multiple longitudinal modes. They therefore emit simultaneously at multiple frequencies; typically, their emission spectrum consists of 10–30 modes separated by a few hundred pm.

2.3.1.3 Semiconductor distributed feedback lasers

Semiconductor distributed feedback (DFB) lasers use narrowband reflectors to provide the feedback: by forming periodic corrugations (Bragg reflectors) in the gain region, the spectrum of the feedback is narrowed to the point that only one longitudinal mode is supported and the laser operates as a single-frequency emitter. Continuous output powers of up to 80 mW (more commonly 50 mW) are available commercially and the linewidth of these devices is typically 1 MHz (with a few providing <500 kHz linewidth). Careful attention to noise on the bias current and stability of the temperature is required to achieve this linewidth consistently.

Semiconductor DFB lasers can be pulsed by modulating their injection current. However, this generally broadens their spectrum; as an alternative, an external modulator can be used. Some devices integrate an electro-absorption modulator, which modulates the output with only minor effects on the spectrum.

2.3.1.4 External cavity feedback semiconductor lasers

The next level of spectral purity is reached, still with Bragg grating reflectors, but extending the cavity by moving the reflector outside the gain chip with a waveguide made from silicon or glass. Open-loop devices provide linewidths below 100 kHz and, with some control circuitry, below 10 kHz with recent products advertised at linewidths around 1.5 kHz. Power levels available without additional amplification are of order 15 mW.

2.3.1.5 Semiconductor DFB lasers with active phase/frequency feedback

Further narrowing of the laser linewidth can be provided by locking the laser to a frequency reference, such as a high-finesse Fabry-Pérot étalon and providing fast feedback to the injection current based on the relative frequency and phase of the laser relative to the frequency reference. Often, the Pound-Drever-Hall technique [41] is used for this purpose, and in the scientific literature, remarkably narrow spectra (of order 80 Hz half-width) have been achieved starting from DFB laser diodes having a natural linewidth of order 700 kHz [42]. Actively narrowed DFB laser diodes are available commercially, although it is not always clear which specific techniques are used to achieve the linewidth.

2.3.1.6 Fibre DFB lasers

Fibre DFB lasers are based on a doped fibre amplifier (see Section 2.3.1) within a cavity defined by fibre Bragg gratings (FBGs) (see Section 2.3.6) that provide a narrowband feedback. The lasers operate continuously and provide output powers of typically 15–40 mW (more with further optical amplification outside the cavity, but within the laser module). They offer the narrowest linewidths available for lasers that are simple to incorporate within transportable equipment. Linewidths of a few hundred Hz are available from several manufacturers.

Fibre DFB lasers are usually designed to provide some adjustment of the output frequency by either stretching the cavity with a piezo-electric transducer, which provides a narrow range (a few hundred MHz to a few GHz, depending on the manufacturer) but fast (a few hundred Hz) adjustment. Wider tunability (typically a few tens of GHz) is often available through thermal adjustment of the main cavity elements.

The main weaknesses of this type of laser are (a) a relative intensity noise (i.e. a fluctuation of the output power) caused by relaxation oscillation in the gain medium that is higher than in some other high-performance laser types and (b) an innate sensitivity to mechanical vibration, although modern designs have improved the resilience to these disturbances to a remarkable extent.

These lasers are sub-systems including embedded control software and, usually, they can be controlled using simple commands.

2.3.1.7 Tunable lasers

Some types of DOFSs require a laser that can be varied (swept) in its wavelength, sometimes over a wide range (e.g. a few tens of nm). Tunable lasers are available commercially, based either on fibre lasers with a fast-scanning tunable filter within the cavity [43] or on tunable semiconductor lasers [44,45] that are variants of the DFB laser diodes, with additional control to allow independent adjustment of the gain, reflector wavelengths and a separate region to adjust the phase in the cavity.

2.3.1.8 Symbols for optical sources

In this book, sources are represented in schematic diagrams by a simple text box, with a description of the device and, in some cases, a symbol representing a pulse and light emission ('z'-like arrow pointing outwards) (Figure 2.4a). Tunability is indicated by an arrow crossing the box and pointing upwards and to the right. In cases where the laser is a diode laser, the electrical symbol for a diode has been added, or used independently (Figure 2.4b).

2.3.2 Optical Amplifiers

Optical amplifiers are used to boost probe signals prior to launching them into the sensing fibres and sometimes also to preamplify the signals returning from the sensor prior to their reaching the detector. Optical amplifiers are characterised, *inter alia*, by the gain they provide; the wavelength range over which the gain (ratio of output to input signal) is available; the amplified spontaneous emission (ASE) that unfortunately, but inevitably, accompanies the gain and, more generally, by a noise figure (ratio of signal-to-noise at the output, relative to that at the input). They are also specified by the power that they can emit, whether in continuous or pulsed mode.

2.3.2.1 Doped fibre amplifiers

Although doped optical fibre amplifiers were demonstrated in 1964 [46], it was the incorporation of rare earths into silica-based optical fibres [47], compatible with transmission fibres and suitable for pumping with laser diodes, that allowed the technology to become convenient and widely available for optical amplification [48–50] and fibre lasers [51]. At its most basic, a doped fibre amplifier consists of an appropriate length of optical fibre incorporating the selected concentration of rare-earth dopant and a pump laser coupled into the amplifier [52,53]. The dopants are chosen for the operating wavelength required, the most common being Nd and Yb, which emit around 1060 nm, and Er, which is used in the 1540–1570 nm region. Laser-diode pumps at c. 980 nm (Yb and Er), 808 nm (Nd) and 1480 nm (Er) are available commercially. In practice, an amplifier also requires a dichroic device to combine, with minimal loss, the pump and the signal arriving from different fibres into the amplifier region. In addition, isolators are used to provide unidirectional gain and so prevent oscillation from stray reflections elsewhere in the system. Doped fibre amplifiers are very versatile in the sense that many configurations exist that are variously optimised for power boosting, for low-noise amplification or for high gain. With small average pump power (tens of mW), a single stage Er-doped fibre amplifier (EDFA) can provide 30 dB small-signal gain and more sophisticated designs achieve gains in excess of 40 dB. At low duty cycles, output powers exceeding 10 W are available from relatively basic configurations (of course specialised designs provide thousands of W continuous output in the field of industrial fibre lasers). Under ideal conditions, the noise figure of EDFAs can approach 3 dB and general-purpose amplifier modules with noise figures below 5 dB that include all the necessary pump, multiplexing and isolation functions as well as control circuitry are available commercially.

2.3.2.2 Semiconductor optical amplifiers

Gain regions similar to those used within semiconductor lasers are sometimes used as optical amplifiers, with the feedback carefully suppressed. Semiconductor optical amplifiers (SOAs) perform less well than fibre lasers in terms of maximum output power, gain and noise figure; however, they have their place in system design in that they provide very fast switching of the gain and so can act to limit the dynamic range of the signals they operate on. Some designs also have non-linear optical properties that allow them to be used for optically controlled switching.

2.3.2.3 Symbol for optical amplifiers

The symbol used in drawings in this book to denote optical amplifiers is the triangle commonly used for electronic amplifiers with the addition of the wording 'EDFA' or 'SOA', as appropriate (Figure 2.4c).

2.3.3 Fused-Taper Couplers

Fused-taper couplers [54] are used for a variety of signal-dividing and combining functions. They are manufactured by heating and simultaneously stretching two or more fibres that are held in contact with each other. The fibres fuse together and the tapering causes the power to spread from the core and to be guided by the medium surrounding the tapered region; in other words, the modes of the cores of the input fibres are converted to modes of the larger structure (i.e. they are guided in the cladding of the original fibres). When the taper reverses (at the exit end of the coupler), the light reverts to the cores, but in a proportion that is dictated by the relative phases of the modes of the cladding. As a result, there is a defined phase relationship between the light of the output fibre. Thus for a 2 inputs × 2 outputs coupler (normally referred to simply as a 2 × 2 coupler), the light waves in the two output ports are in anti-phase. For a perfectly balanced 3 × 3 coupler, the outputs are shifted by 120° with respect to one another. Some sensor designs make use of this property.

Fused-taper devices are available commercially in essentially any desired splitting ratio and the excess loss of modern devices is of order 0.1 dB.

The same technology is used for making wavelength-dependent devices, for example, wavelength splitter or combiners. For example, for pumping optical amplifiers, fused-taper multiplexers are available that combine a signal at 1550 nm with a pump wavelength at 980 nm or 1480 nm. These devices are the equivalent of a dichroic beamsplitter in bulk optics.

Fused-taper devices are also available for polarisation-maintaining fibre, in the form of directional couplers that preserve polarisation. Polarisation-splitting devices that direct the light polarised on one axis out through one port and the light polarised on the orthogonal axis to a second output port (the fibre equivalent of a polarising beamsplitter, such as a Glan-Taylor in bulk optics) have also been demonstrated with the fused-taper approach [55], although most commercial polarisation beamsplitters are manufactured using the approach described in Section 2.3.4.

The symbol for a 2 × 2 coupler used in this book shows the input fibres joining and then separating again (Figure 2.4d). The splitting ratio or wavelength entering and leaving each port is indicated explicitly, where relevant.

2.3.4 Micro-Bulk Optics

Some functions that are required for manipulating light cannot presently be carried out in fused fibre-tapers. This need is fulfilled with miniature optical assemblies consisting of input fibres, collimating lenses, some functional optics and then lenses for focusing the output light back into one or more fibres. The functional optics can be optical filters that are often more precisely controlled than can be achieved in fused-taper technology. Other functions include wavelength-division multiplexers or polarisation multiplexers and also isolators and circulators (discussed in Section 2.3.5).

Typical excess losses of these components are around 0.5 dB. Several technologies are used to align and fix in place the components relative to one another; some of these are limited in their power handling and so this point needs to be checked when specifying a device built with micro-bulk optics.

The symbol for a micro-bulk optic dichroic combiner has been included in Figure 2.4e.

2.3.5 Isolators, Circulators and Faraday Rotation Mirrors

Isolators are devices that transmit light efficiently in one direction but block transmission in the opposite direction; i.e. they are the optical equivalent of a one-way hydraulic valve or an electrical diode. They are used extensively to protect against reflections, which can corrupt some measurements or damage lasers and amplifiers. Typically, in the 1550 nm wavelength region, devices with transmission losses of order 0.5 to 0.7 dB and a backward rejection >40 dB are available, depending on the grade selected. The wavelength range of these devices is limited, but a single device will operate reasonably well over a wavelength range of a few tens of nm. Obtaining similar devices for shorter wavelengths is more problematic for material reasons: the magneto-optic materials used to provide the functionality tend to have significant attenuation, for example, at 1064 nm.

A circulator is a device based on similar technology that accepts light from one port and transfers it with low loss to a second port. However, light entering the second port is transferred to a third port, but not to the first. Three-port devices are the most common; however, four-port devices are available with light entering the third port leaving through the fourth.

Isolators and circulators are normally assembled using micro-bulk-optics. Special components made with polarisation-maintaining fibre are available, some operating on both principal linear polarisations, others just transmitting one polarisation state in one direction and blocking the orthogonal state in both directions.

The Faraday rotation mirror (FRM) [56] is another device made using the same technology. An FRM has only one input/output lead and it reflects the light entering the device, with however, a $\pi/2$ phase between orthogonal polarisation states entering the fibre. This has the effect of reversing the birefringence of the entire fibre from the source to the mirror. FRMs are often used to ensure a high degree of visibility in the interference of the light returning along separate fibres that have been split in an all-fibre Michelson interferometer configuration. It is commonly used in fibre sensors and in DOFSs in the demodulation of the light returning from the sensing fibre.

The symbol for an isolator (Figure 2.4f) is an arrow parallel to the long axis of a rectangle; a circulator is depicted by a circle with a curved arrow indicating the low-loss path around the device, shown in Figure 2.4g.

2.3.6 Fibre Bragg Gratings

FBGs are devices in which the refractive index is modulated periodically along the axis of the fibre; this has the effect of reflecting a narrow wavelength range that matches the pitch of the grating [57]. The first report of an FBG in 1978 [58] was followed by improved understanding of the grating formation process and techniques for designing and creating them were evolved in the following years [57,59,60]. They are now a mainstay of fibre sensing technology [61] because the wavelength at which they reflect is sensitive to temperature and strain. FBGs are also used in telecommunications for narrowband filtering and dispersion compensation. These permanent gratings are formed by creating damage, e.g. by exposure to a spatially periodic pattern of intense UV light.

In the context of this book, it is primarily the filtering properties of FBGs that are of interest. They are versatile tools for manipulating the spectrum of light travelling in a fibre on a very fine scale. The peak reflectivity of an FBG can be controlled from a fraction of 1% to essentially 100% and the spectral width can be as narrow as a few GHz in commercially available (but rather special) devices. FBGs used as sensors typically reflect in a spectral width of 10–30 GHz.

The symbol used for an FBG in this book is a line denoting the fibre crossed by four closely spaced lines perpendicular to the fibre axis to denote the refractive index modulation (Figure 2.4h).

The transmission spectrum of an FBG is that of a notch filter and so can be used to eliminate just one narrow wavelength band; in reflection, only one band is returned. In combination with a circulator, this can be converted into a narrow bandpass filter, as illustrated in Figure 2.4i.

More complex FBGs have been tailored to provide a specific delay/wavelength response and are used for compensating dispersion in optical transmission. Combinations of gratings are also used to interact and form, for example, Fabry-Pérot interferometers that exhibit extremely narrow transmission bands.

2.3.7 Modulators

Modulators are used to alter the intensity, phase and/or frequency of an optical signal travelling through them. In the context of distributed sensors, the modulation speed is usually important, and two technologies dominate, namely acousto-optic modulators (AOMs) and electro-optic modulators (EOMs); both are available coupled to input and output fibres.

2.3.7.1 Acousto-optic modulators

AOMs [62] are based on the interaction between an acoustic wave travelling in the device and the light beam. The sound wave is created by a piezo-electric transducer that converts a radiofrequency (RF) electrical input into a compressional or shear wave. The angle between the acoustic wave and the optical beam is chosen to meet the Bragg condition, and in this case, the optical beam is deflected by an angle that is usually quite small (a few degrees) but sufficient for capturing the undeflected and the deflected beam separately. In sensor applications, it is usually just the deflected beam that is of interest.

The deflected wave is known as the first-order diffracted beam and the undeflected one as the zeroth-order beam. At very high RF powers, the deflected light can appear in additional beams (second and higher orders). Taking the output from the first order results in an output that is controlled by the RF power applied to the piezo-electric transducer. The AOM can therefore be used as an intensity modulator and also as a fast optical switch.

The performance of a typical AOM depends on the switching speed, the design (including the materials selected), the RF frequency and the size of the optical beam that passes through the device. In particular, the acoustic wave travels at a speed that depends on the material (typically of order 5000 m s^{-1} for compressional waves); in this example, the acoustic wavefront crosses an optical beam of diameter 1 mm in about 200 ns. Therefore, in order to produce very fast devices, the optical beam must be focused tightly where it interacts with the acoustic wave, which results in compromises in the design, such as lower deflection efficiency.

Generally, devices using high RF input frequencies exhibit faster switching speeds (or equivalently, wider modulation bandwidths) but also higher transmission losses than those operating at lower RF frequencies. A few RF frequencies, such as 40, 80, 110 and 200 MHz, are commonly used, but there appears to be no standard. Devices operating at 200 MHz designed for a wavelength of 1550 nm typically have switching speeds of 10–15 ns and transmission losses of order 3–4 dB.

A key benefit of AOMs in the context of DFOS is their very high extinction ratio, which is quoted by some suppliers as >50 dB.

Because the optical and acoustic beams meet at an angle that is a few degrees from the perpendicular, there is an in-line component to the acoustic wave that results in a frequency shift of the deflected light. The device can be adjusted by mechanical alignment to shift the frequency to higher (up-shift) or lower (down-shift) frequency than that of the incident light. AOMs are therefore used as frequency shifters as well as for intensity modulation. The frequency shift is exactly that of the RF drive and so it can be controlled precisely.

The symbol used for an AOM is simply a text box with 'AO' or 'AOM' and sometimes a number if more than one device is shown in a single diagram. If the AOM is used as a pulse modulator, a pulse symbol is added in the box (Figure 2.4j) and the frequency shift is added outside the box, if relevant.

2.3.7.2 Electro-optic modulators

EOMs, such as Pockels cells and Kerr effect devices, are very well known in optics [63]. However, in the context of this book, we restrict the term EOM to refer to integrated-optic devices that comprise a waveguide defined in an electro-optic material, typically $LiNbO_3$ or InP, with input and output fibres aligned and attached to the internal waveguide. Electrodes are deposited on the device to apply voltages that modulate the refractive index and are used to vary the phase of the light travelling through the device. Intensity modulators are usually formed by splitting the input into two parallel waveguides and recombining these split inputs to form a Mach-Zehnder interferometer. The output of the interferometer is controlled by modulating (through the voltage applied to the electrodes) the relative phase of the light arriving at the waveguide junction from each arm.

EOMs can operate very fast, well in excess of 20 GHz in some devices; their losses tend to be quite high (of order 3–4 dB for a 20-GHz device) partly due to inefficiencies in coupling light in and out of the device.

The frequency spectrum of light emerging from an EOM depends on the drive and the bias of the device. For a phase modulator, and assuming a constant RF power input, new lines appear in the optical spectrum, spaced from the incident optical spectrum by the RF frequency and its harmonics. The level of each pair of harmonics depends on the RF drive level, i.e. on the modulation index.

In the case of intensity modulators driven by a sinusoidal voltage, a single pair of new lines at the RF can be achieved by carefully selecting the bias point and the RF level. I/Q modulators allow the relative phase of the upper and lower sidebands to be manipulated individually and allow one sideband to be suppressed. In other words, these more complex devices can be used as single-sideband modulators and they act as frequency shifters as well as modulators.

Electro-optic technology is also used for modulating the polarisation of the light; they are essential for very fast polarisation switching or modulation.

The symbol used for an electro-optic intensity modulator is a box marked 'EOM' with a stylised waveguide that splits and recombines (forming a Mach-Zehnder interferometer, MZI) and a pair of electrodes linked to external terminals (Figure 2.4k).

2.3.8 Switches

Fibre switches are used to selectively connect one fibre to one of a set of possible outputs. They are usually mechanical or micro-electro-mechanical (MEMS) devices and achieve switching times of 1–50 ms; they are used in applications where fast switching is not required. Apart from insertion losses, switches are transparent to the optical system: their operation is usually independent of the frequency, the modulation and the type of signal travelling through them. The cross-talk of optical switches is usually extremely low (–50 dB or better), but some designs are more limited in this respect. In addition, some technologies are limited in their power handling and some require the light to be off when the switch position is changed.

The symbol used is the same as is common in electronics.

2.3.9 Polarisation Controllers

Polarisation controllers are used to manipulate the polarisation of light travelling in a fibre, rather like quarter-wave and half-wave plates are used in free-space optics. In laboratory work, they commonly take the form of three single loops of optical fibre each wrapped around a mandrel of carefully selected diameter articulated on a common axis tangent to the mandrel, known as 'Lefèvre loops' after their inventor, Hervé Lefèvre [64].

They are depicted by a fibre with three small ellipses tangential to the fibre (Figure 2.4l).

Other types of polarisation controllers, which are electrically controlled, are available commercially. The internal construction of these devices is not always clear and they are represented simply with a text box worded appropriately.

2.3.10 Connectors and Splices

Large-scale optical fibre systems, such as DOFSs, require the interconnection of multiple sections of fibre. This can be achieved permanently using a splice; alternatively, a connector can be used to allow disconnection and reconnection. Fusion splicers are special machines that align two fibre ends and then melt them together to form permanent splices with very low losses (<0.01 dB for identical fibres) and essentially no back reflection. Fusion splices are the ideal solution, if their use is permissible.

Mechanical splices are used in some cases where the environment is unsuitable for fusion splices (e.g. owing to the presence of an explosive atmosphere). Mechanical splices align the fibres using their external surfaces and hold the fibre in position, using friction or adhesives. Caution must be used in selecting mechanical splices because some exhibit quite strong back-reflections that may not be tolerated by the sensor interrogator; in other cases, index-matching gels are used that may not be appropriate for all power levels or optical wavelengths.

Connectors are used when the connections need to be made and unmade frequently. Connectors are available that, for conventional fibres, allow low loss (c. 0.1 dB) and low back reflection (<–50 dB). Optical fibre connectors mainly rely on tight tolerances of the parts they are made from but also on tightly-specified fibre geometry, in particular core-cladding concentricity, cladding diameter and mode-field or core diameter. Conventional telecommunications fibres and the ferrules used for alignment in fibre connectors are now fabricated extremely precisely, including in their mechanical dimensions. This is not always true of specialty fibres used in DOFSs (e.g. high-temperature fibres).

Optical fibre connectors must be treated with care to avoid degrading their loss and back-reflection performance. This means making sure that they are clean before mating a connector pair and also avoiding transmitting power through them when they are not mated.

The symbol used for connectors is shown in Figure 2.4m.

2.3.11 Detectors

Detectors are devices that convert light to electrical signals. The most common of detector types are semiconductor photodiodes (PDs) that absorb an oncoming photon whose energy goes into creating an electron-hole pair that results in a charge movement from one terminal of the device to the other. A current directly proportional to the rate at which photons are absorbed is therefore passed by the device.

The quantum efficiency Q of the device is the probability that an incoming photon will create an electron–hole pair. On the other hand, the responsivity R_d of the detector is the ratio of the photocurrent to the incident optical power (in units of A/W); it combines the quantum efficiency with the ratio of the electron charge q to the energy of a photon. From the Planck-Einstein relation, the photon energy is given by

$$E_{ph} = h \cdot \nu = h\frac{c}{\lambda}, \tag{2.7}$$

where ν is the frequency of the light and λ its wavelength.

$$R_d = \frac{q}{h \cdot \nu} Q \tag{2.8}$$

So, assuming $Q = 1$, the best values of responsivity that can be expected are

$$E_{ph} = \frac{6.626 \cdot 10^{-34} \cdot 2.998 \cdot 10^{8}}{850 \cdot 10^{-9}} = 2.33 \cdot 10^{-19}\ \mathrm{J}$$

$$\text{so } R_d = \frac{1.602 \cdot 10^{-19}\ \mathrm{C}}{2.33 \cdot 10^{-19}\ \mathrm{J}} = 0.68\ \mathrm{A/W} \text{ for } \lambda = 850\,\mathrm{nm}$$

$$E_{ph} = \frac{h \cdot c}{1300 \cdot 10^{-9}} = 1.528 \cdot 10^{-19}\ \mathrm{J}$$

$$\text{so } R_d = \frac{1.602 \cdot 10^{-19}\ \mathrm{C}}{1.528 \cdot 10^{-19}\ \mathrm{J}} = 1.05\ \mathrm{A/W} \text{ for } \lambda = 1300\,\mathrm{nm}$$

$$E_{ph} = \frac{h \cdot c}{1550 \cdot 10^{-9}} = 1.282 \cdot 10^{-19}\ \mathrm{J}$$

$$\text{so } R_d = \frac{1.602 \cdot 10^{-19}\ \mathrm{C}}{1.282 \cdot 10^{-19}\ \mathrm{J}} = 1.25\ \mathrm{A/W} \text{ for } \lambda = 1550\ \mathrm{nm}$$

In order for the photon to be absorbed, its energy must be larger than the bandgap of the semiconductor material. For silicon, the long-wavelength edge of the absorption band is about 1100 nm (depending on the dopants used in the device and the temperature of the junction), and so only wavelengths shorter than this value can be detected with this material. Other materials, such as germanium or InGaAs, are used at longer wavelengths.

In a further sub-division of the types of PDs used in DOFSs, some devices, such as avalanche PD detectors (APDs), are designed to provide internal gain. APDs have a particular internal structure and are biased with a higher voltage than conventional PDs to create a strong internal electric field. The electric field accelerates the primary photo-generated carriers and they collide with the semiconductor lattice, generating further electron-hole pairs, a process known as avalanche multiplication. The gain available from the avalanche process is varied through the bias voltage and ranges from about ten to several hundreds. In the limit of high bias, the device breaks down and becomes conductive, resulting in effectively infinite gain.

In many cases, the use of APDs improves the sensitivity of the receiver by amplifying the photocurrent before it reaches the preamplifier. So long as the noise added in the internal avalanche multiplication is lower than the noise of the preamplifier, an improvement in the signal-to-noise ratio is achieved. The APD is accompanied by noise because the multiplication experienced by individual photo-generated carrier varies about the mean. This limits the useful range of the APD gain setting. A further limitation arises from technological imperfections in that the breakdown voltage is not uniform across the active area of the device. Therefore, at high gains, some parts of the device are at breakdown even though others are at a more modest gain and so the mean gain that can be achieved is limited.

The subject of detectors for DOFSs, and how they are used, is addressed in more detail in Chapter 3, including a number of alternative devices used in specialised OTDR designs. The symbol used for a detector is the electrical diode symbol with a z-like arrow pointing inwards (Figure 2.4n). In most cases, the detector is used within a receiver as shown in Figure 2.4o.

2.3.12 Other Symbols

Other symbols used in the book are shown in Figure 2.4, including (Figure 2.4u) the symbol for a filter that is used in this book both for optical and electrical signals. The signal type should be clear from the context.

REFERENCES

1. Adams, M. J. 1981. *An introduction to optical waveguides*. Wiley, Chichester and New York.
2. Hartog, A. H., and M. J. Adams. 1977. On the accuracy of the WKB approximation in optical dielectric waveguides. *Opt. Quant. Electron.* 9: 223–232.
3. Snyder, A. W., and J. Love. 1983. *Optical waveguide theory*. Chapman and Hall, London.
4. Gloge, D. 1971. Weakly guiding fibers. *Appl. Opt.* 10: 10: 2252–2258.
5. Gloge, D., and E. A. J. Marcatili. 1973. Multimode theory of graded-core fibers. *Bell Syst. Tech. J.* 52: 9: 1563–1578.
6. Olshansky, R., and D. B. Keck. 1976. Pulse broadening in graded-index optical fibers. *Appl. Opt.* 15: 2: 483.
7. CCITT. 2007. Characteristics of a 50/125 μm multimode graded index optical fibre cable. G.651.
8. Jeunhomme, L., M. Fraise, and J. P. Pocholle. 1976. Propagation model for long step-index optical fibers. *Appl. Opt.* 15: 12: 3040–3048.
9. Olshansky, R. 1975. Mode coupling effects in graded-index optical fibers. *Appl. Opt.* 14: 4: 935–945.
10. Kotov, O. I., L. B. Liokumovich, A. H. Hartog, A. V. Medvedev, and I. O. Kotov. 2008. Transformation of modal power distribution in multimode fiber with abrupt inhomogeneities. *J. Opt. Soc. Am. B* 25: 12: 1998–2009.
11. Gambling, W. A., and H. Matsumura. 1977. Simple characterisation factor for practical single-mode fibres. *Electron. Lett.* 13: 23: 691–693.
12. Marcuse, D. 1977. Loss analysis of single-mode fiber splices. *Bell Syst. Tech. J.* 56: 5: 703–718.
13. Marcuse, D. 1978. Gaussian approximation of the fundamental modes of graded-index fibers. *J. Opt. Soc. Am.* 68: 1: 103–109.
14. Kaminow, I. 1981. Polarization in optical fibers. *IEEE J. Quant. Electron.* 17: 1: 15–22.
15. Dyott, R. B., J. R. Cozens, and D. G. Morris. 1979. Preservation of polarisation in optical-fibre waveguides with elliptical cores. *Electron. Lett.* 15: 13: 380–382.
16. Birch, R., D. N. Payne, and M. Varnham. 1982. Fabrication of polarisation-maintaining fibres using gas-phase etching. *Electron. Lett.* 18: 24: 1036–1038.
17. Hartog, A. H. 1983. A distributed temperature sensor based on liquid-core optical fibers. *J. Lightw. Technol.* 1: 3: 498–509.
18. Knight, J. C. 2003. Photonic crystal fibres. *Nature* 424: 847–851.
19. Kaiser, P., A. R. Tynes, H. W. Astle et al. 1973. Spectral losses of unclad vitreous silica and soda–lime–silicate fibers. *J. Opt. Soc. Am.* 63: 9: 1141–1148.

20. Fabelinskii, I. L. 1968. *Molecular scattering of light.* Plenum Press, New York.
21. Rayleigh, L. 1899. XXXIV. On the transmission of light through an atmosphere containing small particles in suspension, and on the origin of the blue of the sky. *The London, Edinburgh, and Dublin Philosophical Magazine and Journal of Science* 47: 287: 375–384.
22. Roslund, C., and C. Beckman. 1994. Disputing Viking navigation by polarized skylight. *Appl. Opt.* 33: 21: 4754–4755.
23. Nagel, S., J. B. MacChesney, and K. Walker. 1982. An overview of the modified chemical vapor deposition (MCVD) process and performance. *IEEE J. Quant. Electron.* 18: 4: 459–476.
24. Jeunhomme, L. B. 1990. *Single-mode fiber optics. Principles and applications.* Marcel Dekker, New York.
25. Ohashi, M., M. Tateda, K. Shiraki, and K. Tajima. 1993. Imperfection loss reduction in viscosity-matched optical fibers. *IEEE Photon. Technol. Lett.* 5: 7: 812–814.
26. Tajima, K., M. Ohashi, K. Shiraki, M. Tateda, and S. Shibata. 1992. Low Rayleigh scattering P_2O_5-F-SiO_2 glasses. *J. Lightw. Technol.* 10: 11: 1532–1535.
27. Saito, K., M. Yamaguchi, H. Kakiuchida et al. 2003. Limit of the Rayleigh scattering loss in silica fiber. *Appl. Phys. Lett.* 83: 25: 5175–5177.
28. Sosman, R. B. 1927. *Properties of silica.* Chemical Catalog Company, Inc., New York.
29. Nakazawa, M., M. Tokuda, and Y. Negishi. 1983. Measurement of polarization mode coupling along a polarization-maintaining optical fiber using a backscattering technique. *Opt. Lett.* 8: 10: 546–548.
30. Roberts, P. J., F. Couny, H. Sabert et al. 2005. Ultimate low loss of hollow-core photonic crystal fibres. *Opt. Expr.* 13: 1: 236–244.
31. Payne, D. N., and W. A. Gambling. 1975. Zero material dispersion in optical fibres. *Electron. Lett.* 11: 8: 176–178.
32. Payne, D. N., and A. H. Hartog. 1977. Determination of the wavelength of zero material dispersion in optical fibres by pulse-delay measurements. *Electron. Lett.* 13: 21: 627–629.
33. Adams, M. J., D. N. Payne, F. M. E. Sladen, and A. H. Hartog. 1978. Wavelength-dispersive properties of glasses for optical fibres: The Germania enigma. *Electron. Lett.* 14: 22: 703–705.
34. Hartog, A. H. 1979. Influence of waveguide effects on pulse-delay measurements of material dispersion in optical fibres. *Electron. Lett.* 15: 20: 632–634.
35. Schultz, P. 1974. Method of forming a light focusing fiber waveguide. US3826560 A.
36. Inada, K. 1982. Recent progress in fiber fabrication techniques by vapor-phase axial deposition. *IEEE J. Quant. Electron.* 18: 10: 1424–1431.
37. Murata, H. 1986. Recent developments in vapor phase axial deposition. *J. Lightw. Technol.* 4: 8: 1026–1033.
38. Geittner, P., D. Küppers, and H. Lydtin. 1976. Low-loss optical fibers prepared by plasma-activated chemical vapor deposition (CVD). *Appl. Phys. Lett.* 28: 11: 645–646.
39. Miya, T., Y. Terunuma, T. Hosaka, and T. Miyashita. 1979. Ultimate low-loss single-mode fibre at 1.55 μm. *Electron. Lett.* 15: 4: 106–108.
40. Corning. 2014. Corning™ SMF-28 ULL™ Optical Fiber product information sheet. Retrieved from http://www.corning.com/WorkArea/showcontent.aspx?id=63951.
41. Black, E. D. 2001. An introduction to Pound–Drever–Hall laser frequency stabilization. *Am. J. Phys.* 69: 1: 79–87.
42. Nakagawa, K. I., M. Kourogi, and M. Ohtsu. 1992. Frequency noise reduction of a diode laser by using the FM sideband technique. *Opt. Lett.* 17: 13: 934–936.
43. Gloag, A., N. Langford, K. McCallion, and W. Johnstone. 1996. Continuously tunable single-frequency erbium ring fiber laser. *J. Opt. Soc. Am. B* 13: 5: 921–925.
44. Coldren, L. A. 2000. Monolithic tunable diode lasers. *IEEE J. Sel. Topics Quant. Electron.* 6: 6: 988–999.
45. Coldren, L. A., G. A. Fish, Y. Akulova et al. 2004. Tunable semiconductor lasers: A tutorial. *J. Lightw. Technol.* 22: 1: 193–202.
46. Koester, C. J., and E. Snitzer. 1964. Amplification in a fiber laser. *Appl. Opt.* 3: 10: 1182–1186.
47. Poole, S., D. Payne, R. Mears, M. Fermann, and R. Laming. 1986. Fabrication and characterization of low-loss optical fibers containing rare-earth ions. *J. Lightw. Technol.* 4: 7: 870–876.
48. Mears, R. J., L. Reekie, I. M. Jauncey, and D. N. Payne. 1987. Low-noise erbium-doped fibre amplifier operating at 1.54 μm. *Electron. Lett.* 23: 19: 1026–1027.
49. Laming, R. I., L. Reekie, D. N. Payne et al. 1988. Optimal pumping of erbium-doped-fibre optical amplifiers, 14th European Conference on Optical Communication, Brighton, UK IEE 292 Pt II: 25–28.
50. Desurvire, E. 1987. High-gain erbium-doped traveling-wave fiber amplifier. *Opt. Lett.* 12: 11: 888–890.
51. Mears, R. J., L. Reekie, S. B. Poole, and D. N. Payne. 1985. Neodymium-doped silica single-mode fibre lasers. *Electron. Lett.* 21: 17: 738–740.
52. Bjarklev, A. 1993. *Optical fiber amplifiers: Design and system applications.* Artech House, Boston and London.

53. Desurvire, E. 2002. *Erbium-doped fiber amplifiers: Principles and applications*. Wiley-Interscience, Hoboken, NJ.
54. Kawasaki, B. S., K. O. Hill, and R. G. Lamont. 1981. Biconical-taper single-mode fiber coupler. *Opt. Lett.* 6: 7: 327–328.
55. Yataki, M., D. N. Payne, and M. Varnham. 1985. All-fibre polarising beamsplitter. *Electron. Lett.* 21: 6: 249–251.
56. Martinelli, M. 1992. Time reversal for the polarization state in optical systems. *J. Mod. Optic.* 39: 3: 451–455.
57. Kashyap, R. 1999. *Fiber Bragg gratings*. Academic Press, Cambridge, MA.
58. Hill, K. O., Y. Fujii, D. C. Johnson, and B. S. Kawasaki. 1978. Photosensitivity in optical fiber waveguides: Application to reflection filter fabrication. *Appl. Phys. Lett.* 32: 10: 647–649.
59. Erdogan, T. 1997. Fiber grating spectra. *J. Lightw. Technol.* 15: 8: 1277.
60. Hill, K. O., and G. Meltz. 1997. Fiber Bragg grating technology fundamentals and overview. *J. Lightw. Technol.* 15: 8: 1263.
61. Kersey, A. D., M. A. Davis, H. J. Patrick et al. 1997. Fiber grating sensors. *J. Lightw. Technol.* 15: 8: 1442–1463.
62. Dixon, R. W. 1970. Acoustooptic interactions and devices. *IEEE Trans. Electron. Devices* 17: 3: 229–235.
63. Goldstein, R. 1986. Electro-optic devices in review. *Lasers and Applications* April: 67–73.
64. Lefevre, H. C. 1980. Single-mode fibre fractional wave devices and polarisation controllers. *Electron. Lett.* 16: 20: 778–780.

3 Principles of Optical Time-Domain Reflectometry (OTDR) for Distributed Sensing

Optical time-domain reflectometry (OTDR) underpins essentially all of the distributed optical fibre sensor (DOFS) approaches. It is not surprising, therefore, that the principles and techniques that have been applied to enhancing the performance of OTDR are found in DOFS too. This chapter describes some of the basic techniques of OTDR; they are used in various combinations in the DOFS systems to be discussed in the following chapters.

3.1 OPTICAL TIME-DOMAIN REFLECTOMETRY

The concept of the OTDR was demonstrated in 1976 by Barnoski and Jensen [1] and almost simultaneously by Personick [2], who also provided an expression for the signal levels that are received in such systems. The OTDR was initially used in optical telecommunications and fibre quality assurance to measure uniformity and to check the attenuation profile of optical fibres for installation defects or quality control.

In the earliest, and simplest to visualise, implementation of OTDR, a short probe pulse is launched into the fibre under test, illustrated in Figure 3.1. For simplicity, the optics have been shown as a directional coupler that launches the forward probe light into the fibre to be tested and redirects the backscatter return to a receiver.

The probe pulse travels down the fibre, losing light along its way owing to the various attenuation mechanisms, including Rayleigh scattering (see Section 2.1.2.2).

The scattering process re-directs a small fraction of the forward-travelling light into all directions (though not totally uniformly to all solid angles). In particular, a fraction of the scattered light falls within the angle of acceptance of the fibre in the reverse direction; this scattering and re-capture process is illustrated in the inset of Figure 3.1. The recaptured fraction of the scattered light is guided back towards the launching end. The return signal also loses power through attenuation on the return pass.

It should be noted that the re-capture of backscattered light by the waveguide had been considered by Kapron et al. in 1972 [3], but in the context of its possibly deleterious effect on optical communications,* rather than as a diagnostic test for the fibre itself.

* The concern in the work by Kapron et al. related to pulse broadening caused by a double-scattering process whereby light would be scattered, re-captured in the reverse direction, followed by a second similar process leading to parasitic signals propagating in the normal transmission direction.

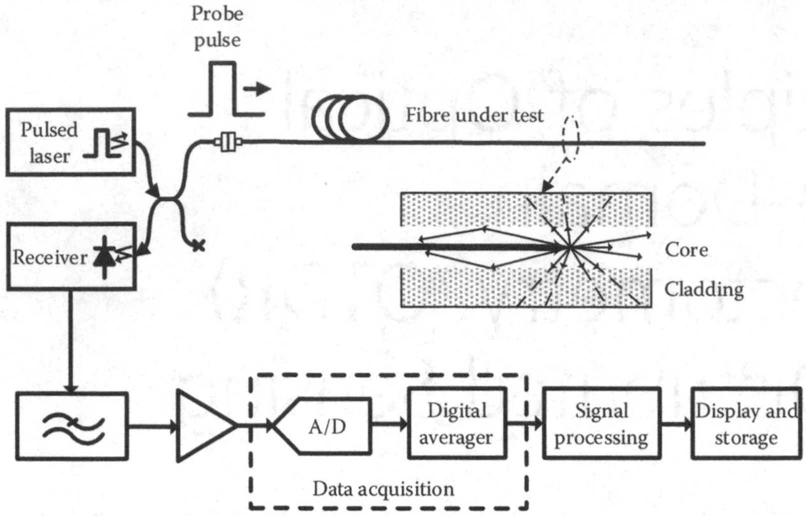

FIGURE 3.1 Schematic diagram of an OTDR.

On returning to the launching end, part of the backscattered light is directed to a receiver. The resulting voltage is filtered, further amplified and digitised, averaged, processed and displayed. The different elements of an OTDR will be discussed in more detail further in this chapter.

Reflections in the optics in the common path to the fibre under test, including the joint to the fibre itself, are to be minimised because the backscatter signal is very weak by comparison with the forward power. Even a very weak reflection is likely to overload the very sensitive receiver designed to detect the low-level backscatter signal. This is why the spare port of the coupler (shown with a cross) must be terminated with low reflectivity, and similarly, low-reflection connectors or fusion splices are preferred for interconnecting fibre sections.

OTDRs provide information on the continuity of a fibre; i.e. they inform the user of the existence and location of any break or high point-loss. The latter might be caused by a kink in the fibre or a poor splice, and it is common practice to check splices with an OTDR before closing splice housings. OTDRs also determine the existence of regions of the fibre where the loss, although distributed, is higher than normal; this might be caused by micro-bending in a cabling process or a poor cable installation. With more sophisticated measurements and interpretation, additional knowledge of the fibre can be gleaned, such as the presence of mismatches in the properties of fibres at joints, axial variation of diameter [4,5], numerical aperture [5], chromatic dispersion [6] or of impurity distribution [7].

3.1.1 Single-Ended Measurement

Following Refs. [2] and [8], we can write a few expressions to establish the relationship between the energy of the probe pulse and backscattered power. Initially, we assume that the pulse launched into the fibre is a Dirac function of energy $E_p(0)$ and of vanishingly narrow width. The implications of a finite pulse duration are discussed at the end of this section.

The forward pulse energy is attenuated at a rate α expressed in nepers/m (1 Np/m ~ 4.34 × 10³ dB/km); its dependence on position z in the fibre is thus

$$E_p(z) = E_p(0) \cdot \exp \int_0^z -\alpha(u)\,du. \tag{3.1}$$

CROSS-SENSITIVITY OF LOCATION TO ENVIRONMENTAL EFFECTS

The implicit assumption in Equation 3.5 is that v_g is constant along the fibre. In most cases, this is true to an acceptable degree. However, the refractive index is sensitive to temperature, strain, isostatic pressure and, to a lesser extent, other external perturbations, and so this assumption must be revisited when the fibre length and the potential range of external disturbances could vary v_g by enough to affect the expected location of a point of interest [9]. As an example, the variation of pulse transit time with temperature was measured as 35.7 ps km^{-1} K^{-1} [10]; so for a 100 K temperature rise, in a 1 km fibre, the two-way transit time to the remote end will have varied by ~7 ns, which will shift the perceived location of the remote end of the fibre by 0.7 m. This situation could easily arise in the monitoring of process plant or in hydrocarbon production wells. The strain imposed by the larger expansion coefficients of tight jacketing materials or steel tubes in which fibres are protected from mechanical damage is much larger (five times larger in the case of a 0.5 mm diameter Nylon jacket [10]). So the relationship between transit time and spatial location must be carefully considered in many cases of practical importance. In general, with knowledge of the conditions (temperature, strain) along the fibre, these effects can be corrected.

The energy scattered whilst the pulse travels a distance element dz situated at z is

$$dE_s(z, z+dz) = E_p(z) \cdot \alpha_{ray}(z) \cdot dz, \tag{3.2}$$

where α_{ray} is the Rayleigh scattering-loss coefficient. The capture fraction $B(z)$ is defined as the proportion of the total energy scattered at z that is recaptured by the fibre in the return direction. The energy dE_{BS} $(z, z + dz)$ returning from the length element dz to the launching end is therefore

$$dE_{BS}(z, z+dz) = E_p(0) \cdot \alpha_{ray}(z) \cdot B(z) \cdot \exp \int_0^z -2\alpha(u)\,du. \tag{3.3}$$

In Equation 3.3, it has been assumed that the attenuation suffered by the backscattered light is the same as that suffered by the probe pulse.* The factor of 2 in the exponential term accounts for the outgoing and return travel up to point z. This energy arrives, spread over a time interval, given by

$$dt = \frac{2}{v_g} dz, \tag{3.4}$$

which is simply the time taken by the impulse to travel the distance dz in both directions. Here, v_g is the group velocity of the probe pulse in the fibre. In the case of a Dirac pulse (which we have assumed), no energy will arrive from any other part of the fibre during this time interval. The power received at the launching end of the fibre at time $t = 2z/v_g$ is thus

$$P_{BS}(t) = \frac{dE_{BS}}{dt} = \frac{v_g}{2} \cdot E_p(0) \cdot \alpha_{ray}(z) \cdot B(z) \cdot \exp \int_0^z -2\alpha(u)\,du \tag{3.5}$$

* This assumption is not always valid if the modal power distribution is not the same in both cases, or if the wavelength of the scattered light differs from that of the probe light, which is the case in certain DOFSs.

We also define the *backscatter factor* $\eta(z)$ as

$$\eta(z) = \frac{v_g}{2} \cdot \alpha_{ray}(z) \cdot B(z). \tag{3.6}$$

$\eta(z)$ is the ratio of the backscatter power returned to the energy travelling in the probe pulse; it has dimensions of s^{-1}, but it is convenient to use W/J to point back to its physical origin.

From Equation 3.5, if the scattering loss, the capture fraction and the total attenuation are independent of the z coordinate, the backscatter power is a pure exponential function of position along the fibre. However, local changes in the scattering loss or the numerical aperture can cause localised departures from that exponential function. Variations of the loss, however, affect the exponent of the decay curve, and as can be seen from Equation 3.5, $\frac{d\left(\log(P_{BS}(t))\right)}{dt}$ is directly proportional to the local attenuation.

A synthetic example of a backscatter curve is shown on a linear scale in the upper left-hand panel of Figure 3.2; the parameters selected are representative of the backscatter power that would be returned from a typical multimode fibre operated at 1064 nm, with a pulse energy of 200 nJ (for example, a peak power of 10 W and a pulse duration of 20 ns); these conditions are relevant to distributed temperature sensors (DTSs). The curve also includes the main noise contributions, although they are not readily apparent on this scale.

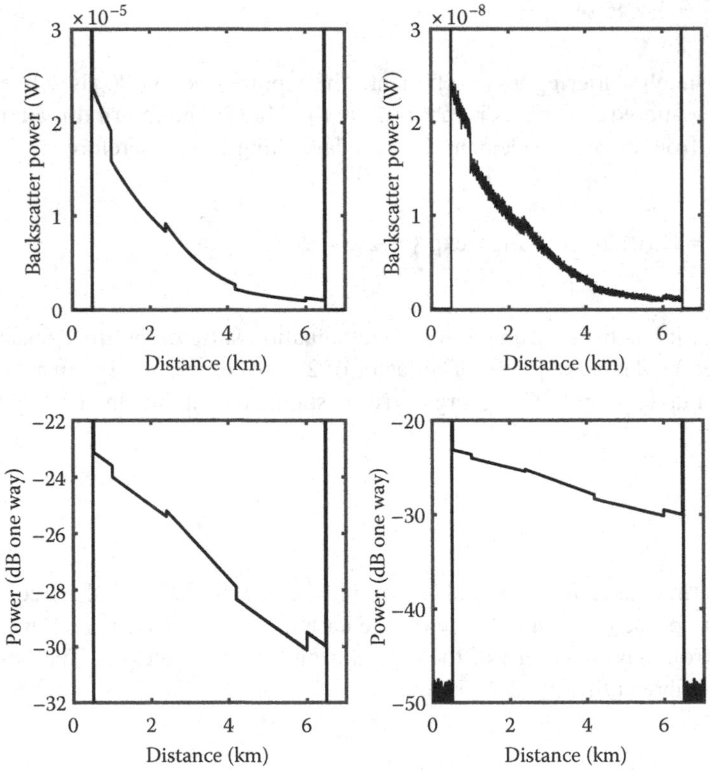

FIGURE 3.2 Synthetic backscatter curves. Upper left-hand panel: typical single shot trace in linear units; upper right-hand panel: data generated as left plot at reduced probe energy; lower left-hand panel: data from panel above plotted in logarithmic (dB) units on the vertical axis and lower right panel: data from lower left on a compressed vertical scale.

In the figure, the overall exponential decay is punctuated by abrupt changes in the signal level at 0.5, 2.4, 4.2 and 5.5 km. At 0.5 km, the signal drops by about 20%. It would be natural to assign this feature to localised loss. However, according to Equation 3.5, it could also be caused by a variation of $\eta(z)$ resulting from either a change in the scattering loss or of the capture fraction (see Equation 3.6). An OTDR trace obtained from a single end of the fibre is therefore ambiguous. For example, at 2.4 km, the backscatter signal increases; this, in a passive fibre, must be caused by an increase in $\eta(z)$. However, by simply viewing this curve, one cannot be certain that the feature at 2.4 km is caused only by a localised variation of $\eta(z)$ or whether there is a larger variation of $\eta(z)$ partially masked by a coincident localised loss. Some variations of $\eta(z)$ exist between batches of fibre, and at a splice, where localised loss is also possible, the single-ended OTDR measurement cannot provide a precise estimate of loss owing to possible variations of $\eta(z)$ on either side of the splice.

The noisy trace in the upper right-hand panel of Figure 3.2 was synthesised with a probe energy of 200 pJ (i.e. 1000 times weaker than that of the upper left plot; note in particular how the noise decays as the signal decreases, which is typical of some of the types of noise found with OTDR signals and that will be discussed further on.)

In long fibres, the weaker signals from the remote end will be below the noise level, and it is usual to apply considerable averaging* to improve the signal-to-noise ratio (SNR).

At the far end of the fibre, a typical reflection has been included; features such as this often occur at the end of the fibre or at reflective interfaces, such as connectors. They are caused by a Fresnel reflection due to the abrupt change in refractive index. Beyond the far-end reflection, the signal drops to essentially zero, since the probe pulse has left the fibre. A reflection is also visible at the start of the fibre.

The lower left plot of Figure 3.2 reproduces the same data from the panel above it but plotted on a logarithmic scale; it shows a feature that was not readily apparent on a linear scale, namely that the slope of the rate of decay is higher between 2.4 and 4.2 km; i.e. the loss here is higher, although, just possibly, there might be a tapering of the one of the fibre properties to induce an equivalent variation of $\eta(z)$.

Finally, the lower right-hand plot in Figure 3.2 shows the same data on a compressed vertical scale and it shows the noise after the end of the fibre. This view is typical of what is seen on the display of an OTDR; the gap between the nearest backscatter signal and the noise level is an indication of the dynamic range of the instrument. The vertical scale in (c) and (d) is in dB (one way); i.e. $5\log_{10}\left(\frac{P_{BS}(z)}{P_{BS}(0)}\right)$, which is half of the backscatter power ratio as measured; it is scaled in this way to be reasonably representative of the loss attributable to a single pass through the fibre. The use of dB (one way) is conventional in OTDRs.

Commercially available OTDRs generally provide a number of functions for setting acquisition parameters (such as maximum fibre length to be measured, probe wavelength, pulse duration), for storing and identifying the results as well as interpretation aids, e.g. automatically locating loss points in the fibre and estimating their severity, finding the fibre end and determining the mean loss over a defined part of the fibre.

In practice, of course, the pulse duration is finite and a pulse duration τ_p (usually defined as the full width at half-maximum of the probe power as a function of time) will result in the spatial resolution of the system being limited to $\delta z = \frac{v_g}{2}\tau_p$. However, the spatial resolution of an OTDR is also limited by the acquisition system: in general, the bandwidth of the receiver and of the analog-digital converter will smear the backscatter signal further. Moreover, in some cases (especially with multimode fibre and where a fine spatial resolution is targeted), the bandwidth of the fibre (see Section 2.1.3) can limit the spatial resolution of the measurement. In general, the spatial resolution of an OTDR is determined by the convolution of the probe pulse shape, the impulse response of the acquisition and that of the fibre, in the two-way transit from launching end to the point of interest and back.

* In this context, signal averaging means making repeated measurements of the backscatter signal and summing the results point-by-point along the fibre. Assuming that the noise is uncorrelated between successive acquisitions, the SNR improves in proportion to the square root of the number of accumulated waveforms.

To summarise, the spatial resolution of an OTDR is determined by the impulse response that the system that, itself, is the convolution of the impulse responses of (a) the probe pulse, (b) the receiver and acquisition electronics and (c) the fibre that is tested.

To a first approximation, the spatial resolution of the system (between 10% and 90% points of the response to a step change in signal) may be modelled as a succession of filters and the impulse response is then given by

$$\delta z = 1.087 \cdot \frac{v_g}{2} \cdot \sqrt{\tau_p^2 + \tau_{rx}^2 + \tau_f^2} \tag{3.7}$$

for Gaussian-like pulse shapes, where τ represents the duration of the impulse response measured at full-width, half-maximum and the subscripts rx and f denote the receiver and the fibre, respectively.

For a rectangular, finite duration, probe pulse of peak power P_p, the pulse energy is

$$E_p(0) = P_p(0) \cdot \tau_p. \tag{3.8}$$

3.1.2 Double-Ended Measurement

Di Vita and Rossi [11] and Matthijsse and de Blok [12] showed that the multiplicative terms can be separated from the exponent in Equation 3.5 if a second measurement of the backscatter signal is performed from the opposite end of the fibre. Thus, for a fibre of length L (after re-stating as a function of position, where z is always defined from End 1), the signals for End 1 and End 2 are given by

$$P_{BS1}(z) = \frac{dE_{BS}}{dt} = \frac{v_g}{2} \cdot E_p(0) \cdot \alpha_{ray}(z) \cdot B(z) \cdot \exp \int_0^z -2\alpha(u)\,du \tag{3.9}$$

and

$$P_{BS2}(z) = \frac{dE_{BS}}{dt} = \frac{v_g}{2} \cdot E_p(0) \cdot \alpha_{ray}(z) \cdot B(z) \cdot \exp \int_L^z 2\alpha(u)\,du. \tag{3.10}$$

It follows that the square root of the product and ratio of Equations 3.9 and 3.10 are given by

$$\sqrt{P_{BS1}(z) \cdot P_{BS2}(z)} = E_p(0) \cdot \alpha_{ray}(z) \cdot B(z) \cdot \frac{v_g}{2} \cdot \exp \int_0^L -\alpha(u)\,du = \eta(z) \cdot E_p(0) \cdot \exp \int_0^L -\alpha(u)\,du \tag{3.11}$$

and

$$\Lambda(z) = \sqrt{\frac{P_{BS1}(z)}{P_{BS2}(z)}} = \exp \int_0^z -2\alpha(u)\,du \cdot \exp \int_0^L \alpha(u)\,du. \tag{3.12}$$

Since the $\exp\int_0^L -\alpha(u)du$ term is independent of location along the fibre, Equation 3.11 provides a measure $\eta(z)$ that is independent of the local attenuation. Likewise, Equation 3.12 provides the integral of the attenuation term in Equation 3.9 and has eliminated the local variation of $\eta(z)$. The double-ended technique therefore allows the backscatter factor and the local attenuation to be determined independently of one another.

The double-ended measurement approach is shown schematically in Figure 3.3. A fibre switch is used to connect the OTDR alternately to one, then the other, end of a fibre to be tested (shown on the spool to the right). Often, lengths of standard fibre (having a known backscatter factor) are placed between the OTDR and the test fibre, to provide a known backscatter level that can be used as a reference.

The separation of backscatter traces is illustrated in Figure 3.4, using the same synthetic example as in Figure 3.2. When the distance scale is defined as starting from End 1, then of course the backscatter signal measured from End 2 appears to grow with increasing z, as shown in Figure 3.4a. The values of $\eta(z)$ and of local attenuation extracted are shown in Figure 3.4b and c, respectively.

In order to evaluate Equations 3.11 and 3.12, it is necessary to time-reverse the data from one end or the other (here, this was carried out on End 2 data) and carefully to align the distance axes, so the features are also aligned. (*NB:* This is a crucial step – failure to properly align the traces gives rise to degraded spatial resolution and spurious features.)

In Figure 3.4b, the backscatter factor is shown, normalised to the mean value in the first and final 500 m sections, that represent standard fibre of known backscatter factor. In this example, the variation of the backscatter factor between the sections has been exaggerated (compared to the spread of values typically found in fibres of the same type) to illustrate the effects.

The local attenuation (lower plot) is calculated from Equation 3.12 by differentiation. Naturally, this has the effect of amplifying the noise. It would be usual to smooth the result or to calculate the difference over a larger step than was used here (1 m).

Figure 3.4 now reveals information that was hidden in the single-ended measurement, namely that the fibre is made up of three sections, in addition to the 500-m sections of standard fibre at each end. The three 'unknown' sections each have a different backscatter factor (0.833, 0.90 and 0.75 relative to the standard value); in addition, the loss of the central section is 1.5 dB/km, compared with 1 dB/km for the other sections.

This rather simplistic example illustrates the process that is used and typical information that can be gleaned from double-ended measurements. We shall see that this concept is central to some types of distributed sensors.

Equations 3.11 and 3.12 are the first step in the interpretation of backscatter traces to understand the origins of non-uniform exponential decay, usually caused by imperfections or changes in fibre properties. However, $\eta(z)$ does not identify the origin of such non-uniformities unambiguously because they could

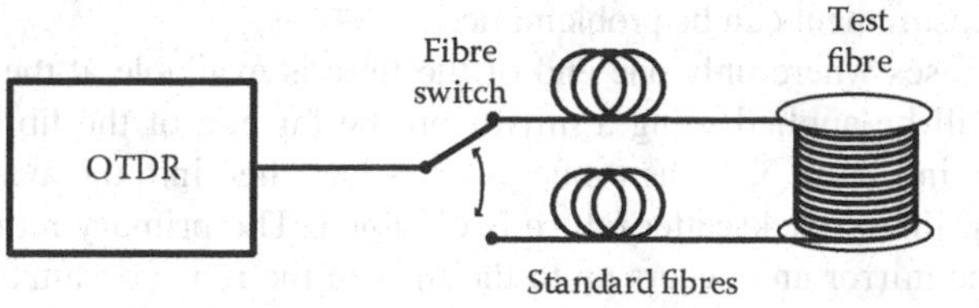

FIGURE 3.3 Double-ended OTDR configuration.

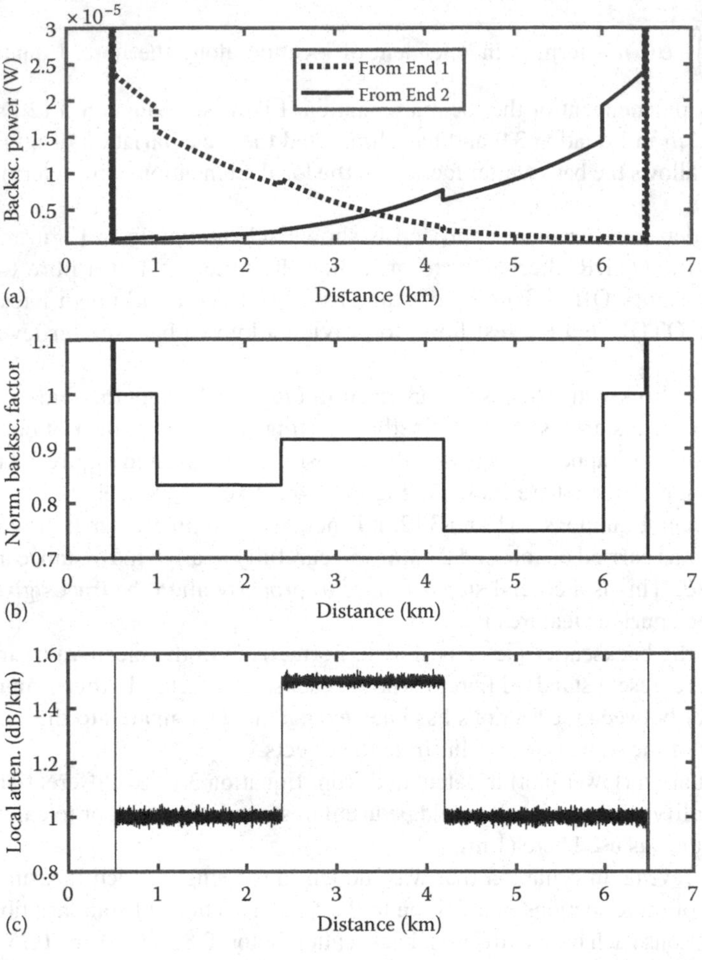

FIGURE 3.4 Separation of double-ended traces. (a) Traces from End 1 and End 2 plotted against the same distance coordinates. (b) Backscatter factor, normalised to first and final sections. (c) Local attenuation.

be caused by changes in the scattering loss or by changes in process parameters that affect the capture fraction.

The loss $\Lambda(z)$ measured in double-ended mode is independent of variations of the backscatter factor, and this is valuable for estimating precisely the quality of splices or assessing the loss component of non-uniform fibre. Although present-day conventional fibres are very uniform in their properties, new types of fibre that may be used for distributed sensing may not be so well controlled and the double-ended technique allows the distribution of their losses to be assessed precisely [13]. Indeed, assembling DOFS often involves fibres of different types connected as a single sensing fibre, and estimating their splicing losses from a single-ended measurement can be problematic.

In those common cases where only one end of the fibre is available at the instrument, the double-ended technique can still be applied using a mirror on the far end of the fibre [14], i.e. a folded-path measurement, as shown in Figure 3.5. The probe pulse is launched into the available end of the fibre in the usual way and the primary backscatter return is obtained. The primary return is the same as would be measured without the mirror and occurs up to the time in the reflectogram at which the probe signal reaches the mirror. At that time, a strong signal, corresponding to the reflection of the probe at the mirror, is received, followed by a secondary, reflected backscatter trace that results from the probe pulse

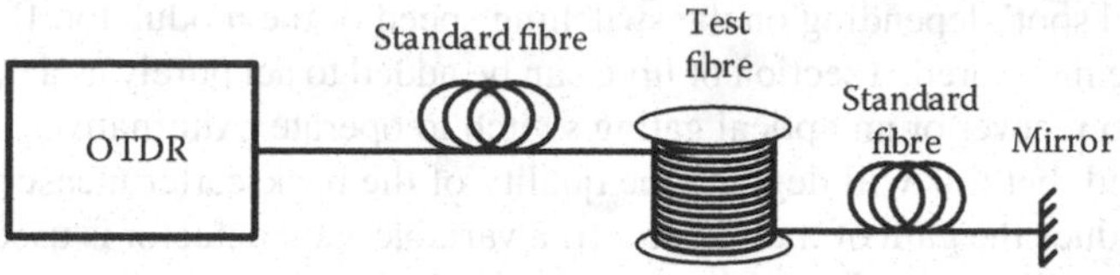

FIGURE 3.5 Folded-path OTDR configuration.

travelling backwards towards the launching end. Its backscatter travels initially away from the instrumentation, towards the mirror, and is then reflected by the mirror and sent towards the instrumentation. The secondary backscatter signal has the same properties as if it had been launched from the remote end, provided that the mirror is exactly perpendicular to the fibre axis (to avoid mode conversion at this point). Of course, the signal in the second backscatter trace has suffered the additional attenuation of a two-way trip through the entire length of the fibre, but as long as the OTDR has sufficient dynamic range, a useful measurement can be obtained that allows losses to be separated from changes in the backscatter factor of the fibre. A typical signal for a folded-path OTDR and synthesised with the same properties and the fibre used for Figure 3.4 is shown in Figure 3.6. This measurement is equivalent to looping the fibre back to the instrument and thus doubles (in logarithmic units) the dynamic range required of the system.

One serious issue arises with the folded-path OTDR approach, namely that the probe pulse reaches the detector with only the attenuation caused by the round-trip to the mirror and back. This reflected probe pulse is a far stronger signal than the backscatter signals that are of primary interest, and it will undoubtedly overload the sensitive receiver required to detect the backscatter signals. It is therefore necessary to prevent this overloading, for example, by temporarily blocking the arrival of the returned light with a fast modulator as was the case in Ref. [14]. A section of fibre will inevitably be obscured to the remote OTDR,

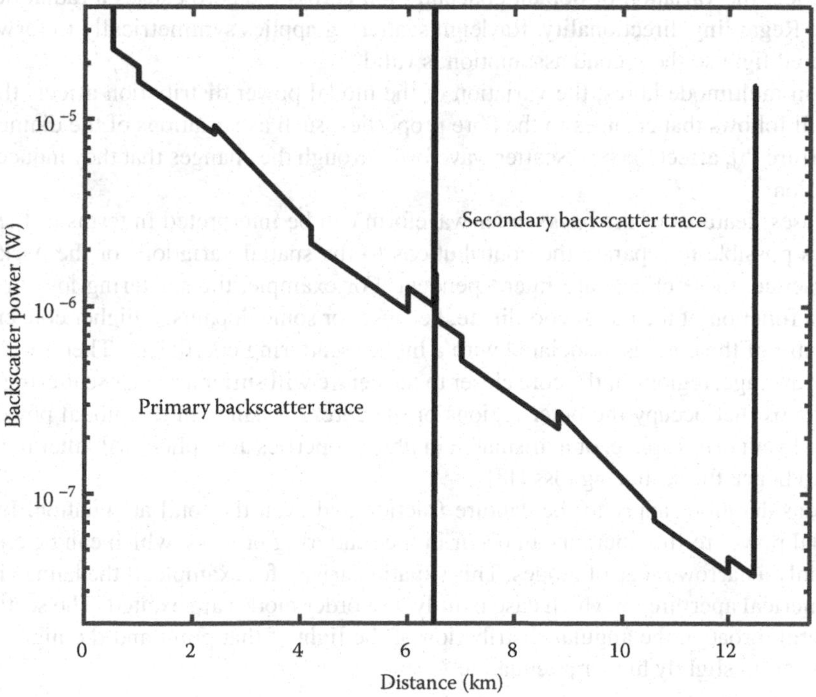

FIGURE 3.6 Folded-path OTDR traces for the assembly of fibres used in Figures 3.2 and 3.4 (Log power scale).

the length of the 'blind spot' depending on the switching speed of the modulator. If this occurs at a point that must absolutely be monitored, a section of fibre can be added to act purely as a delay line and so allow time for the receiver to recover or an optical gating switch to operate. Alternatively, the sensitivity of the receiver can be reduced, but this will degrade the quality of the backscatter measurement. It is also possible temporarily to reduce the gain of the detector (if a variable-gain detector is used); however, restoring the normal operating conditions of the receiver after the reflection has passed will also take a finite time and so lead to a blind spot.

3.1.3 Capture Fraction

The capture fraction for multimode, graded-index fibres is given by Ref. [15]

$$B(z) = \frac{1}{4} \cdot \frac{NA^2}{n_1^2}, \tag{3.13}$$

and for step-index multimode fibres,

$$B(z) = \frac{3}{8} \cdot \frac{NA^2}{n_1^2}. \tag{3.14}$$

Here, *NA* is the numerical aperture of the fibre (see Chapter 2) and n_1 is the refractive index at the centre of the core. However, Equations 3.13 and 3.14 are approximations that, for example, assume that α_{ray} is independent of the radial coordinate at which the scattering occurs and of the direction of propagation. In practice, the variation of dopant concentration across the core causes a radial dependence on scattering loss. Regarding directionality, Rayleigh scattering applies symmetrically to forward-scattered and backscattered light so the second assumption is valid.

However, in multimode fibres, the variation of the modal power distribution affects the backscatter signature [16]. It follows that changes to the fibre properties, such as variations of the diameter [5,17] and numerical aperture [5], affect the backscatter waveform through the changes that they induce in the modal power distribution.

In some cases, features of the backscatter waveform can be interpreted in terms of their root cause. It is not always possible to separate the contributions to the spatial variations of the backscatter factor because, in practice, these effects are interdependent. For example, the scattering loss in graded-index fibres is often a function of the radial coordinate, because for some dopants, a higher concentration (typically at the centre of the core) is associated with a higher scattering coefficient. Therefore, mode groups that occupy, on average, regions of the core closer to the centre will suffer a higher scattering loss than will other mode groups that occupy the outer regions of the core. A change in the modal power distribution caused perhaps by an offset splice, or a mismatch in fibre properties at a splice, will alter the modal power distribution and hence the scattering loss [18].

Similar considerations apply to the capture fraction and even the total attenuation. In particular, a degree of modal power re-distribution can occur in the scattering process, which can be envisaged if the probe excites only a narrow range of modes. This situation arises, for example, if the launch is restricted in radius and numerical aperture, in which case mainly low-order modes are excited. The scattering process would, in general, broaden the angular distribution of the light at that point and the higher order modes generally suffer from slightly higher attenuation.*

* This so-called 'differential mode attenuation' (DMA) results from higher-order modes being closer to their cut-off and so being more susceptible to micro-bending.

It follows that, in a multimode fibre, it cannot be assumed that the attenuation is the same for the probe light and the backscattered light. For detailed OTDR measurements, therefore, it is desirable to control the modal power distribution entering the fibre. For certain mode-filtered excitations, the backscatter signals can provide detailed information on the key properties of the fibre [16], namely diameter fluctuations [4,17], variation of the numerical aperture [5] and even the shape of the refractive index radial distribution [16], particularly when separated into backscatter factor and attenuation using the double-ended technique. It has also been shown that the measurement of splice losses with OTDR in multimode fibre is sensitive to the modal power distribution. Unfortunately, the inversion of this information is not guaranteed: from known variations of the fibre parameters, the backscatter waveforms can be predicted quite accurately; however, it is not always possible to interpret a set of backscatter waveforms unambiguously in terms of the fibre parameters.

For single-mode fibres, the situation is somewhat different: there is no modal dependence to consider, but the backscatter factor is now an inter-related function of the distributions of the mode field across the core (and into the cladding as an evanescent wave) and of the scattering coefficient. However, the effects of attenuation and capture fraction can be separated unambiguously because of the absence of modal power dependence.

The backscatter factor B is the following function of the electric near-field distribution of the mode ψ_N [8],

$$B = \frac{3}{4V^2}\frac{NA^2}{n_1^2}\frac{\int_0^\infty R\cdot\alpha_{ray}(R)\cdot\psi_N^4(R)\,dR}{\int_0^\infty R\cdot\alpha_{ray}(R)\cdot\psi_N^2(R)\,dR\cdot\int_0^\infty R\cdot\psi_N^2(R)\,dR}. \tag{3.15}$$

To compare Equation 3.15 with the equivalent expression for a multimode fibre, Equations 3.13 and 3.14, we can re-write Equation 3.15 in the form

$$B = B'\cdot\frac{NA^2}{n_1^2}, \tag{3.16}$$

where the normalised backscatter capture fraction B' is

$$B' = \frac{3}{4V^2}\frac{\int_0^\infty R\cdot\alpha_{ray}(R)\cdot\psi_N^4(R)\,dR}{\int_0^\infty R\cdot\alpha_{ray}(R)\cdot\psi_N^2(R)\,dR\cdot\int_0^\infty R\cdot\psi_N^2(R)\,dR}. \tag{3.17}$$

B' takes a value between 0.08 and 0.3, but for a fibre having a radially uniform distribution of scattering loss and a V value between 1.6 and 2.2 (which is common for single-mode fibres under normal operating conditions), B' is within the range 0.23 to 0.26 and is thus close to the value for a uniformly excited, graded-index multimode fibre with a uniform scattering loss. It should also be noted that the same approach has been shown [19] also to be valid for calculating the coupling of light scattered from the lowest order mode of a multimode fibre and captured into an arbitrary mode of the same fibre.

In single-mode fibres, backscatter measurements at a range of wavelengths can be used to assess the variation of mode field diameter with wavelength and hence, the chromatic dispersion of the fibre [6].

3.1.4 Coherence Effects

Thus far, we have considered that each elemental section of fibre provides incremental energy to the scattered light signal, independently of all the other parts of the fibre occupied by the probe pulse. Whilst this is true on average, in practice, the detector measures the square of the electrical field impinging on it. That electric field is the summation of all the scattered electric fields captured by the fibre over a distance occupied by the probe pulse at any one time, i.e. the *resolution cell*. If the source occupies a broad spectral width, then its coherence length is short.

The resolution cell may be divided into elemental sections equal to the source coherence length. The power collected in one elemental section is, by definition, not coherently related to the power collected from any other section: they are statistically independent and their powers simply add. However, all the scatterers *within* a coherence length have a fixed, although random, relative phase distribution. As a result, the square of the summation over a coherence length of the electric fields for all scatterers is *not* equivalent to a summation of their power as would be the case for incoherent light: the electric field must be treated as a vector quantity, and the sum of the electric fields therefore includes the effect of their relative phase.

With a coherent (narrowband) source, the phase of the incident light is consistent for all the dipoles within a resolution cell and so too are the phases of the re-emitted (scattered) light from each of them. Because the location of the scattering centres is random, so too are the relative phases of the dipoles; however, their relative phase is fixed as long as the probe frequency is stable and the fibre is undisturbed. The scattered light therefore consists of the summation of this large collection of electric fields that have a stable, but random, phase relationship.

In some sections of the fibre, the phase of the scatterers will happen, by and large, to be the same, and in this case, a strong backscattered signal will result. In other locations, the electric fields will more or less cancel, resulting in a very weak signal. It is common to observe a ratio of 100:1 between the strongest and weakest signal when coherent light is used as a source in OTDR.

This strong modulation that exists in the Rayleigh backscatter with coherent sources is known as *fading*, because in some parts of the fibre, for a given state of the fibre and a particular laser frequency, the signal is near zero. It is a form of multi-path interference, similar to the way in which the strength of a cellular telephone signal can vary over a distance of just a few cm in a cluttered space, such as inside a building. The issue is particularly pronounced in single-mode fibres. In multimode fibres, each mode carries its own fading properties, and so the averaging that occurs over many modes reduces the degree of fading that the total signal is subjected to.

The backscatter resulting from probing with a coherent source was analysed by Healey [20]. The main result is that the coherence overlays a multiplicative modulation on the backscatter signal (as given by Equation 3.5). The electric field of the backscatter signal is the summation of the electric field of all the scatterers within the coherence length of the probe, which, for very coherent sources, is equal to the duration of the pulse. Each of these scatterers re-emits an electric field of amplitude a_e^k that has a fixed phase ϕ_e^k relative to the incident light, but that is random, owing to the random location of the scatterers. The summation is given by

$$E_s(z) = p(t)\sum_k a_e^k \cdot \exp\left(j\phi_e^k\right) = p(t)A_e(z)\cdot \exp(j\Phi_e(z)), \tag{3.18}$$

with $p(z)$ representing the state of polarisation of the backscattered light (that is by and large* the same as that of the incident light) and $[A_e, \Phi_e]$ is a pair of random variables that represent the distribution of the electric field. Φ_e is uniformly distributed over the interval $[-\pi, \pi]$.

* Subject to the depolarisation that occurs in the scattering process. See Section 2.1.2.2.

For coherent OTDRs (C-OTDRs), we are interested in the backscatter power, which is proportional to $|E_s(z)|^2$. The statistical distribution of the power depends on the number of degrees of freedom M_f (the number of statistically independent field components): a single-frequency source in a single-mode fibre in a single polarisation state has just one degree of freedom, and the probability distribution of $P_{BS}(z)$ follows an exponential form given by Ref. [21]

$$\Pr(P_{BS}(z)) = \overline{P_{BS}(z)}^{-1} \exp\left[-\frac{P_{BS}(z)}{\overline{P_{BS}(z)}} \right], \tag{3.19}$$

where $\overline{P_{BS}(z)}$ is the mean of $P_{BS}(z)$.

This applies to situations where the coherence length of the source before pulse modulation is much longer than the pulse duration and when light from only one mode is collected.

A plot of backscatter power vs. distance measured by the author in a single-mode fibre, with a single-frequency probe, is shown in Figure 3.7. In this case, the fibre length is just above 1000 m and the power has been normalised to the mean backscatter signal. A strong reflection can be seen at the remote end (at 1041 m from the start of the trace) that far exceeds the maximum value plotted, followed by a cessation of the backscatter signal. Rather than a smooth exponential decay, the trace takes on a ragged appearance with deep power contrast. In the example shown, about 10 locations are measured at 1/1000 of the mean value and a similar number are five or more times greater than the mean, a contrast in this example of 5000:1. Similar data are shown in Figure 3.8, measured on the same fibre, at the same time, but with 16 different optical carrier frequencies. Although each of these traces has a similar general appearance, each is completely different from the others in the detailed variation of the power vs. distance: these signals are statistically independent.

However, a repeated measure of the same fibre at a single carrier frequency is shown in Figure 3.9; this is an expanded view of a 150-m section of fibre containing data acquired with 10 separate laser pulses and plotted on the same axes. It is clear that although minor differences can be observed, especially at the local extrema, the overall features are well preserved throughout this dataset.

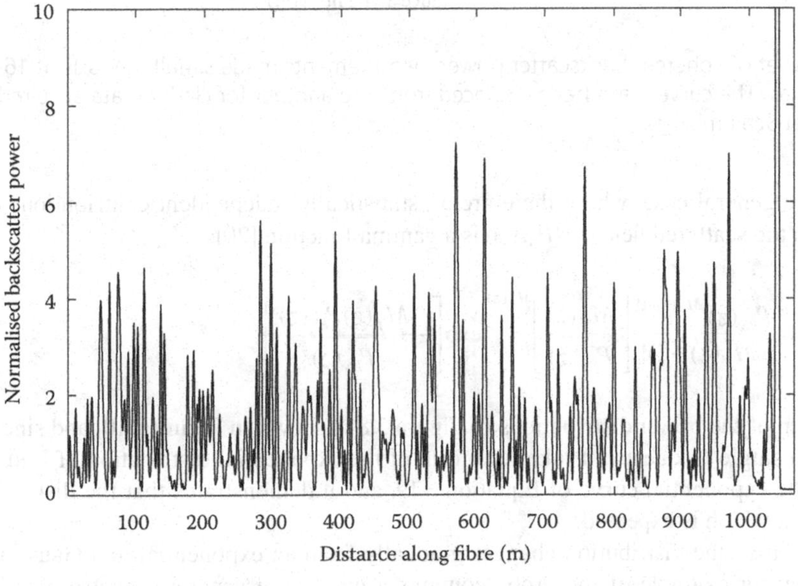

FIGURE 3.7 Coherent backscatter power variation measured along a single-mode fibre.

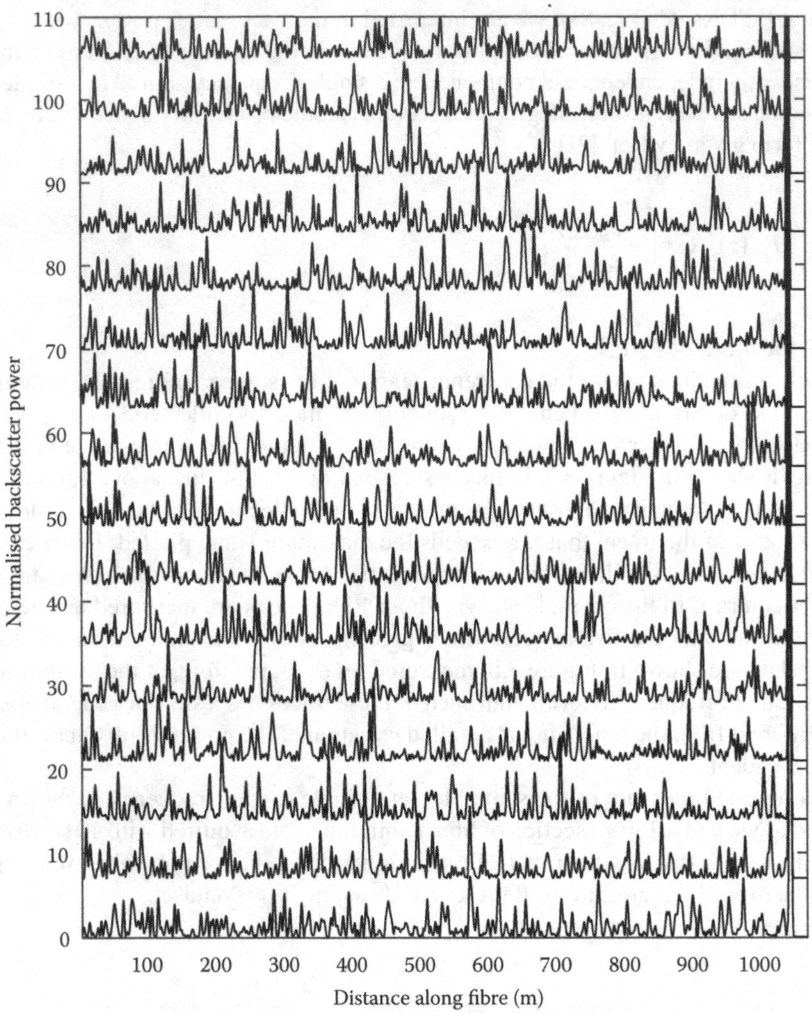

FIGURE 3.8 A set of coherent backscatter power measurements made simultaneously at 16 distinct optical carrier frequencies. The curves have been displaced from one another for clarity. Data acquired by author and Florian Englich at Schlumberger.

In the more general case, where there are M_f statistically independent contributions (equal in their mean power) to the scattered field, $\Pr(P_{BS}(z))$ is a gamma function [20]:

$$\Pr(P_{BS}(z)) = \frac{P_{BS}(z)^{[M_f(z)-1]}}{(M_f(z)-1)!}\left[\frac{M_f(z)}{\overline{P_{BS}(z)}}\right]^{M_f(z)} \exp\left[-\frac{M_f(z)P_{BS}(z)}{\overline{P_{BS}(z)}}\right]. \tag{3.20}$$

A histogram of the power distributions of Figure 3.8 is shown in Figure 3.10, and since each point is recorded with a single frequency, it takes the form of the exponential distribution of Equation 3.19. The dotted line is the exponential curve of Equation 3.19, and it demonstrates that the distribution is indeed very close to that which is expected.

As M_f increases, the distribution changes gradually from an exponential to a Gaussian distribution. This is the case, for example, if the probe contains a number of separate spectral lines – this can be simulated with the dataset shown in Figure 3.8 by summing the backscatter power at each location across

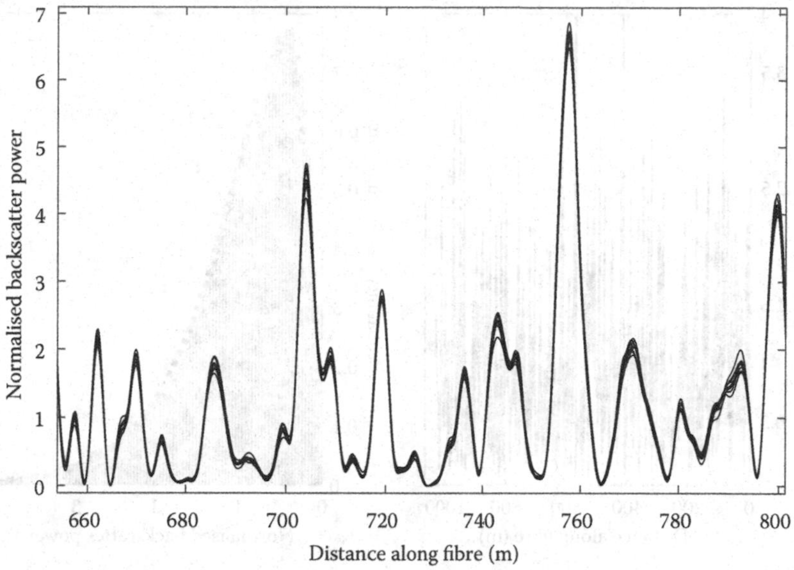

FIGURE 3.9 Ten overlaid backscatter signals recorded with successive probe pulses. (One of the traces is a subset of the data shown in Figure 3.7.)

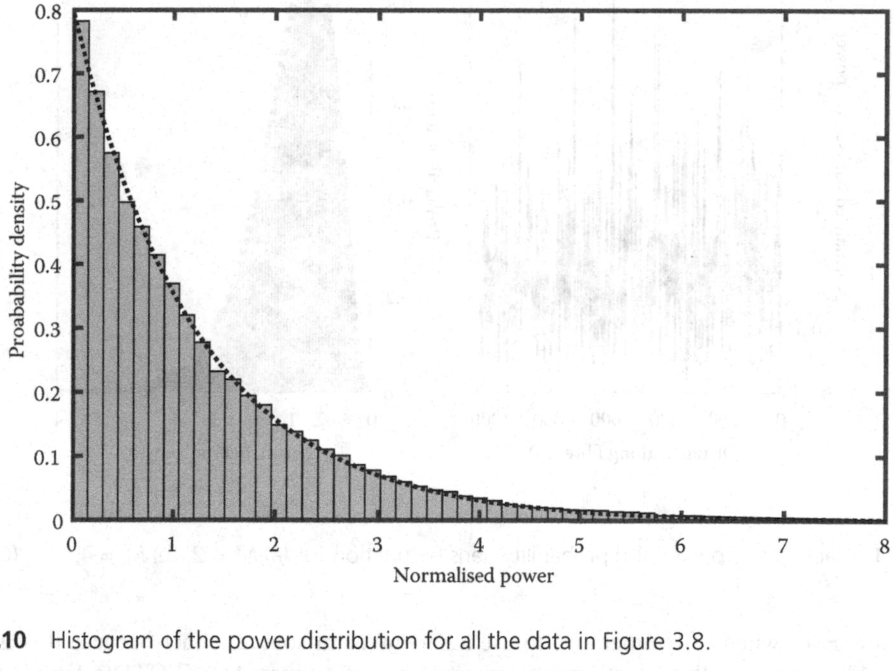

FIGURE 3.10 Histogram of the power distribution for all the data in Figure 3.8.

subsets of the data for the 16 frequencies available. The resulting backscatter power (sum of backscatter power for groups of 2, 4, 8 and 16 frequencies) and the corresponding probability distributions are shown in Figure 3.11a–d, with the curve of Equation 3.20 overlaid over the histogram of the experimental data. As can be seen in the figures, when the number of frequencies included in the power summation increases, the histogram moves from an exponential distribution to one approaching a Gaussian function, as expected based on Equation 3.20. In the limit of large M_f, the fluctuations of the backscatter curve due to the remaining coherence of the source (let alone from truly random noise sources) are indeed Gaussian.

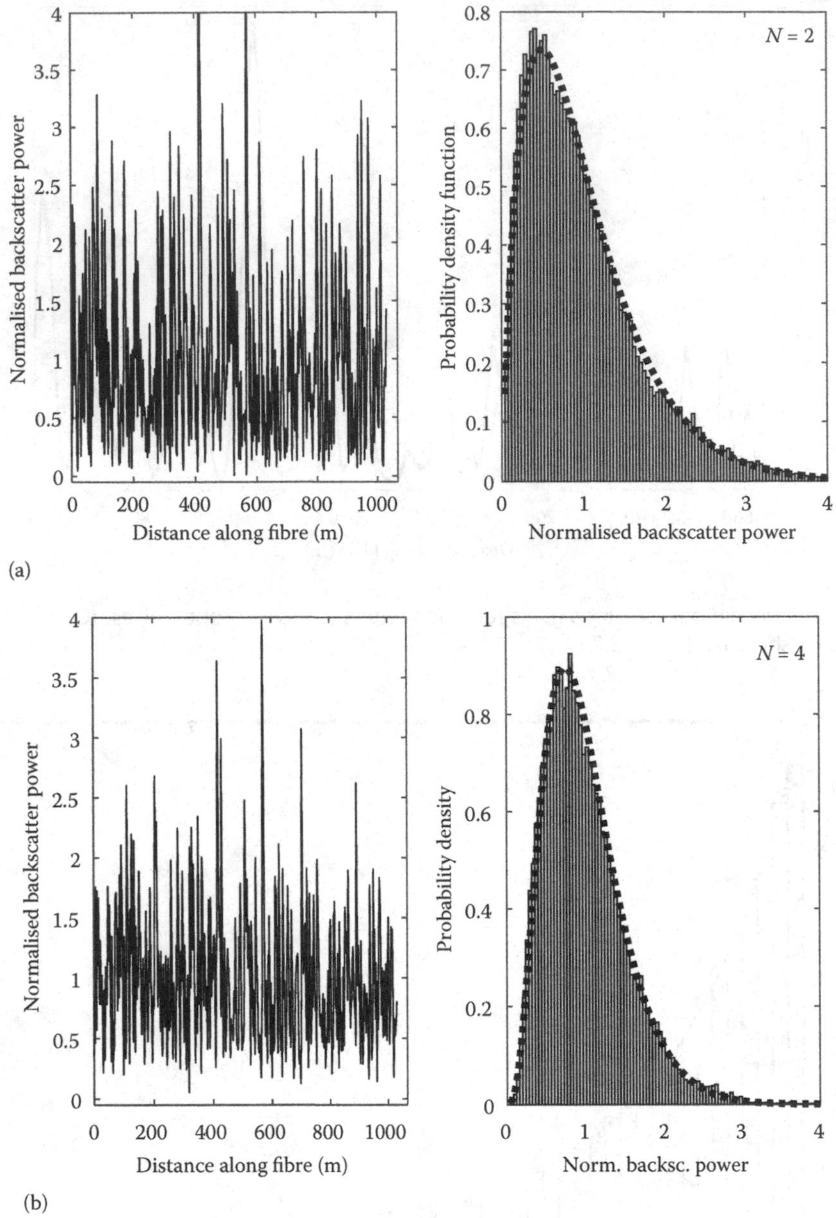

FIGURE 3.11 Backscatter power and probability density function for (a) $M_f = 2$, (b) $M_f = 4$. *(Continued)*

There are cases when it is imperative to use a narrow-band source, either because this is the only source available or because the detection process requires such a source (see C-OTDR, later in this chapter, in Section 3.3.1.1). However, the fading is very specific to the exact frequency of the source and to the precise disposition of the fibre. Should either of these vary during the acquisition, the fading pattern will change, and, by averaging many waveforms acquired under different conditions, the effect of fading on the backscatter signal can be mitigated.

A fibre placed in a buried duct can be very stable over long periods, and there are cases where only negligible change in the fading pattern occurs over long times (the author has observed a stable fading pattern on a fibre in a disused gas well that was stable over more than 30 minutes). This implies that the

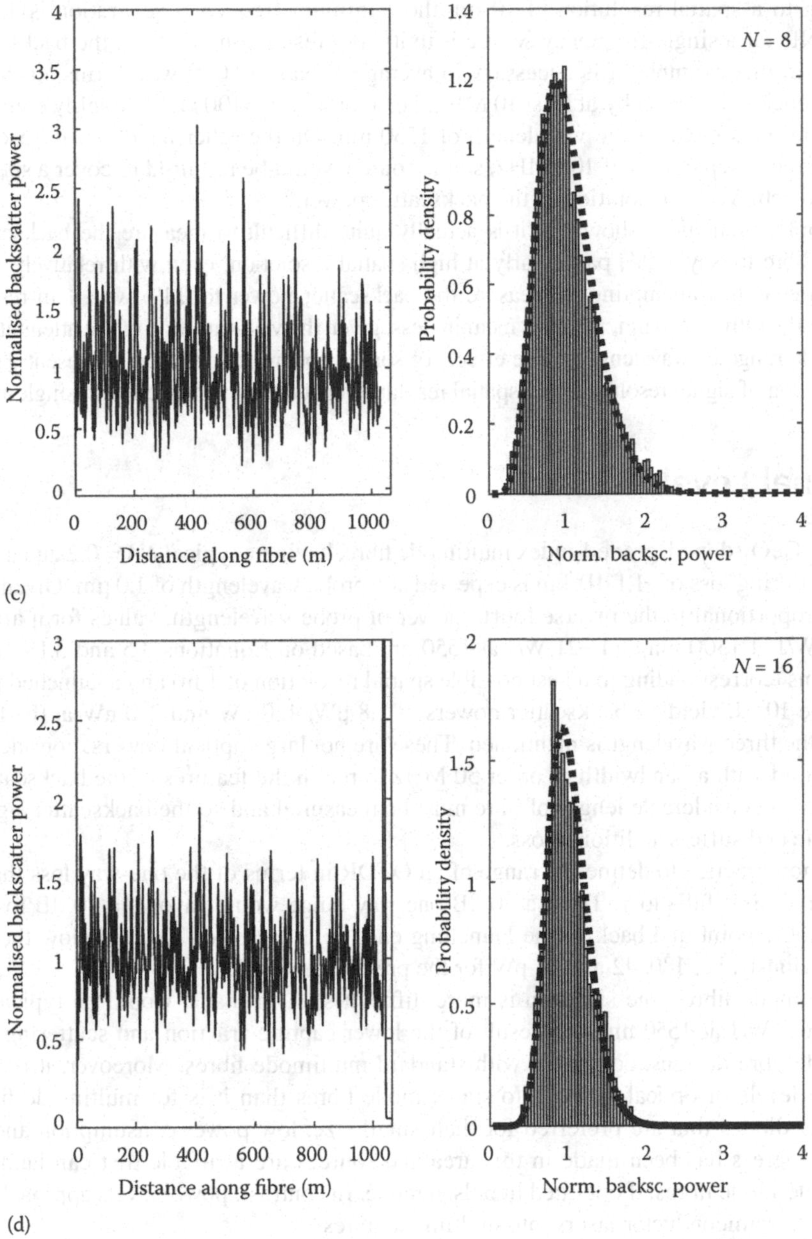

FIGURE 3.11 (CONTINUED) Backscatter power and probability density function for (c) M_f = 8 and (d) M_f = 16.

temperature of the well and the frequency of the laser were both extremely stable over this period (better than 10^{-3} K and 1 MHz, respectively). The variation of the coherent backscatter pattern with external measurands is discussed in Chapter 6.

Given that the user controls the interrogator, but not the fibre, the most reliable way of varying the fading pattern is to modulate the frequency of the laser, so that the fading for each pulse is statistically independent from that of all other pulses in the sequence.

Mermelstein et al. [22] showed that the condition for two probe signals to be statistically independent is that their frequencies differ by at least the inverse pulse duration; for a pulse duration of 100 ns (corresponding to a spatial resolution of 10 m), the minimum frequency separation is 10 MHz. Since the starting SNR for a single-frequency source is unity, if a resolution of 1% on the backscatter signal is required, then in this example, it is necessary to average at least 10,000 waveforms, each generated by pulses at frequencies separated by at least 10 MHz, i.e. a total span >100 GHz, roughly equal to a spectral width of 0.8 nm for a mean probe wavelength of 1550 nm. On the other hand, a 1-m resolution system requires a frequency separation of 100 MHz, so the source would be required to cover a spectral range of at least 8 nm to achieve 1% resolution of the backscatter power.

These simple calculations show that it is actually quite difficult to measure the backscatter power in a single-mode fibre to, say, 0.1%, particularly at high spatial resolution, even with relatively low-coherence sources. As an example, attempting to measure the backscatter power to 0.1% with 1 m resolution would require a spectral width of 800 nm, which is meaningless, given the vast variation of the attenuation coefficient over such a wide range of wavelengths. The effects of source coherence therefore represent a real limitation on the combination of signal resolution and spatial resolution that can be achieved with single-mode fibres.

3.1.5 Signal Levels

In the case of a GeO_2-doped, graded-index multimode fibre, having a typical $NA = 0.2$ and a core diameter of 50 µm, a scattering loss of ~1.1 dB/km is expected at a probe wavelength of 1.0 µm. Given that Rayleigh scattering is proportional to the inverse fourth power of probe wavelength, values for η are ~180 W/J at 900 nm, ~42 W/J at 1300 nm and ~21 W/J at 1550 nm based on Equations 3.5 and 3.13. So, for a pulse duration of 10 ns (corresponding to a best possible spatial resolution of 1 m) and a launched power of 1 W, $Ep(0)$ would be 10 nJ, yielding backscatter powers of 1.8 µW, 420 nW and 210 nW at the launching end of the fibre at the three wavelengths mentioned. These are not large optical powers, considering that they must be measured with a bandwidth of order 50 MHz to retain the features of the backscatter signature. Also, in general, a considerable length of fibre must be measured and so the backscatter signal returning from the remote end suffers additional loss.

It is common practice to define the range of an OTDR in terms of the one-way loss that can be sustained before the SNR falls to 1. Thus, a 20 dB one-way range would involve a 40 dB loss to the most distant measurable point and back to the launching end, i.e. a factor of 10,000 below the power levels mentioned previously, i.e. 180, 42 and 21 pW for the previous examples.

For single-mode fibres, the situation is more difficult still, in that η would be typically 13 W/J at 1300 nm and 6.5 W/J at 1550 nm as a result of the lower capture fraction and scattering loss for typical single-mode fibre designs, compared with standard multimode fibres. Moreover, it is more difficult to launch high levels of optical power into single-mode fibres than it is for multimode fibres, particularly from laser diodes that are preferred for their small size, low power consumption and ruggedness. Nonetheless, progress has been made in this area and sources are available that can launch more than 0.5 W into single-mode fibres, if operated in pulsed mode. In contrast, power levels approaching 10 W can be launched with semiconductor lasers into multimode fibres.

Applying the 0.5 W power criterion to a single-mode OTDR operating at 1550 nm, a 10 ns probe pulse duration would result in a signal of 4.2 pW returning from the remote end of a fibre exhibiting a 20 dB one-way loss. Some OTDRs claim dynamic ranges in excess of 30 dB one-way. At 42 fW, the power received would be far too small for most conventional receivers: it can be calculated that for a time resolution of 10 ns and for a received power of 42 fW, only one photon is received on average for every 300 time slots of duration 10 ns.

The problem of low signal levels in OTDR has been tackled in a number of ways.

Firstly, the measurement is not carried out with a single-laser pulse. In practice, all OTDRs have an averaging function that repeatedly pulses the laser, acquires a backscatter waveform and sums the resulting waveforms point-by-point.

This process was originally carried out with analog boxcar averagers, initially acquiring a single point along the fibre at a time. A little later, it was realised that the pulse-to-pulse power fluctuation that some lasers exhibit could be cancelled by using a pair of boxcar averagers and taking the ratio of their output [7]. It is now performed in digital electronics [23] by digitising the whole backscatter waveform, storing the values in memory and cumulating the point-by-point summations in an accumulator store; this is of course far more efficient.

The SNR improves in proportion to the square root of the number of waveforms that are averaged (at least until some coherent noise effect precludes further improvement). Therefore, it is not uncommon for OTDR measurements to be accumulated over several seconds and even tens of minutes for the most demanding cases.

In general, the power that can be launched into the fibre is limited, and so for the longest distances, it is usual to select wider probe pulse durations. This has two effects: firstly, the probe energy increases in proportion to pulse duration, and secondly, the bandwidth of the acquisition system can be reduced for wider pulses (because the wider probe pulses generate no high-frequency information to be collected in any case), and this reduces the noise presented to the acquisition system. Further methods for extending the range or otherwise improving the performance of OTDRs will be discussed later in this chapter. However, the trade-off between the key parameters of range, measurement time, spatial resolution and resolution of the backscatter power will already be apparent from the foregoing comments.

Although early OTDRs were large instruments, they have now become a basic everyday tool for fibre installation teams, and small, handheld instruments are available that, notwithstanding their compact size, still offer a very useful performance and functionality. Versions also exist that can be added as an extension card to personal computers. Special versions of OTDRs that are designed to be operated on amplified long-distance links also exist, and optical communication systems sometimes include permanently installed OTDRs for link surveillance. In this case, the design of the OTDR goes hand in hand with that of the communication system; thus, means for probing through amplifiers, isolators and filters must be devised whilst ensuring that the OTDR signals do not perturb the transmission of data.

3.2 OTDR SYSTEM DESIGN

3.2.1 Sources and Detectors

OTDRs require high power sources, and, originally, this often meant selecting devices that were not exactly at the wavelength of interest. However, with more development of semiconductor laser technology and optical amplification techniques, sources are by and large available at all the major wavelengths of interest to optical communications. Apart from the major operating windows of 850, 1300 and 1550 nm, there is also interest in 1390 nm (to estimate the OH^- ion content of the fibre) and 1650 nm (beyond the main transmission bands to avoid disrupting active communications links). As noted earlier, sources are available for delivering 10 W or more from broad contact semiconductor laser into multimode fibre.

For single-mode fibres, the source must also emit in a single transverse mode in order to achieve a reasonable launching efficiency. When OTDR technology was being developed, single-mode laser diodes were able to launch at most 10 mW into a single-mode fibre, and this drove a range of innovations in improving receiver sensitivity that will be described later. At the time of writing, however, for most wavelengths of interest for single-mode fibre, pulsed semiconductor lasers able to couple at least 0.5 W into the fibre under test are available. This is adequate for most portable OTDRs used for field testing. In addition, the development of rare-earth-doped fibre amplifiers has allowed, for many operating wavelengths, the probe pulse to be amplified to boost its energy.

OTDRs use photodiodes almost universally as the detector. In general, avalanche photodiode detectors (APDs) are used because the internal gain they provide prior to the photocurrent reaching the preamplifier outweighs the noise added in the process. This is almost always the case at short wavelengths

(<1100 nm) with silicon detectors. A good APD for these purposes absorbs a very large fraction of the light impinging on it (i.e. it will have a high quantum efficiency), exhibits low dark current (i.e. the reverse bias current that exists even when no light is detected) and its multiplication factor is almost the same for all primary photo-electrons generated by the absorption process. A silicon APD offers all of these features in its absorption band. In particular, devices showing quantum efficiency >90%, multiplication gain >100 with excess noise of only a factor of 2 at this gain and a dark current <10 nA at this same gain setting are available.

At longer wavelengths, the situation still favours APDs in general, although the comparison is not as clear-cut as for short wavelengths. Beyond 1100 nm, silicon no longer absorbs the incident light, and other materials, such as InGaAs, are used instead. Compared with silicon, these materials exhibit lower available gain, higher dark current and higher multiplication noise. At very low spatial resolution (and thus low receiver bandwidth), the benefit of APDs is reduced, and in some cases, the excess noise of an APD outweighs its benefits compared to detectors without internal gain.

3.2.2 Common System Designs

The most common OTDR arrangement consists of a pulsed laser delivering a single probe pulse with the light being directly converted to an analog electrical signal that is amplified and digitally sampled. Returning to Figure 3.1, a pulsed laser is launched into a fibre lead that is connected to the fibre under test. It is necessary to separate forward probe light from the returning backscatter. In multimode OTDRs, this function is usually accomplished with a directional coupler, which splits the power arriving in any one port into two parts that appear in a defined ratio on the two opposite ports. Care must be taken to avoid reflections from within the device. Likewise, beyond the directional coupler, low back-reflection connectors are advisable. They can be either convex-polished to provide a so-called 'physical contact' between the cores and thus frustrate the reflection at the interface, or angle-polished connectors that re-redirect any reflection outside the cone of acceptance of the fibre. At best, a passive directional coupler will cause a two-way loss by a factor of 4, i.e. 6 dB(optical)* (50% transmission for each of the outgoing and returning optical signals).

Circulators provide a more efficient solution; they are commonly available for single-mode fibre, but also for multimode fibre. Overall, a two-way loss of ~1.5 dB is achievable in this way; three-port circulators also have excellent directivity; i.e. they isolate backward propagating light very effectively from the probe light. Prior to the advent of circulators, active devices, such as fast optical switches employed as deflectors [24,25], were used to reduce the losses on launching probe light and receiving backscattered light.

The backscatter is directed to a receiver consisting of a photodetector and an electrical preamplifier. The photodetector is generally an APD [26] that provides internal gain by multiplying the primary photo-generated carriers before they exit the device.† In this way, a larger photocurrent is presented to the preamplifier, which generally improves the overall sensitivity of the system. In the case of systems designed exclusively for low-spatial-resolution, ultra-long-range operation, it can be preferable to use detectors without internal gain. The choice of the detector is part of the overall optimisation of the system design.

The output current of a photodiode is proportional to the optical power that is detected, and so a preamplifier of the transimpedance type (current input, voltage output) is preferred in OTDR owing to its low

* The use of dB in optical fibre instrumentation can be a little confusing. The transmission of an optical element quoted in dB is defined as 10 $\log_{10}$ (power out/power in), although sometimes the negative sign is omitted, when there is no ambiguity that the result represents a loss. However, when light reaches a photodiode, the photocurrent is proportional to optical power, and yet in electrical units, the transmission is given by 10 $\log_{10}$ (electrical power out/electrical power in); for the current, such a ratio is of course 20 $\log_{10}$ (electrical current out/electrical current in). So if there is any risk of ambiguity, the ratio is quoted in dB (optical) or dB (electrical).

† The pioneering papers of Barnoski et al. and Personick used rather less ideal detection arrangements, but the APD followed by a transimpedance preamplifier became the prevalent approach within a few years.

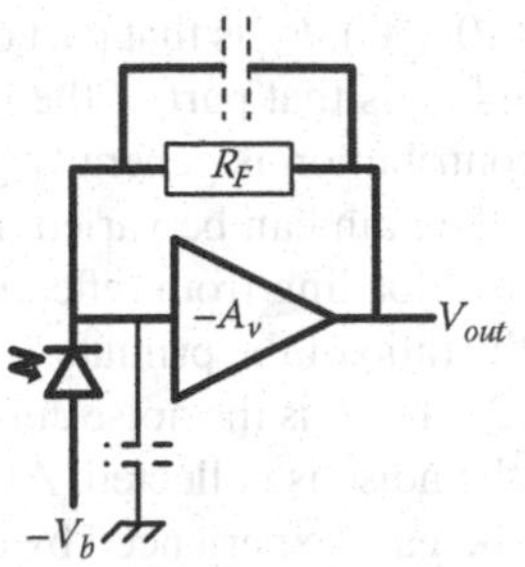

FIGURE 3.12 Schematic diagram of a transimpedance receiver.

input impedance [27]: the photodiode is a current source and the following amplifier should be designed for current input.

Conceptually, such a preamplifier (Figure 3.12) is obtained by connecting a feedback resistor from the output of an inverting voltage amplifier to its input. In the ideal operational amplifier case having very large inverting gain $-A_V$, the transimpedance gain (ratio of voltage out to current in) is simply the resistance R_F of the feedback resistor. This may be visualised by considering a small current removed from the input node by a photo-current flowing through the detector. In order to maintain a zero voltage at that node (which is what the very large inverting gain aims to do), the voltage at the output takes a sufficiently positive value to push an equal current through the feedback resistor. Again, under ideal conditions, the input impedance is simply the value of the feedback resistor divided by the open-loop gain of the amplifier. The low input impedance minimises the change in voltage at the node between the detector and the preamplifier. In practice, the design problem is complicated by the presence of a number of parasitic components, particularly the stray capacitances shown in dotted lines in Figure 3.12, as well as by the fact that the gain and bandwidth of the amplifier are finite and that there is signal delay through the amplifier.

3.2.3 Noise in Optical Receivers

There are three main sources of noise degrading the signal quality, namely (a) shot noise from both the signal and the dark current of the detector, (b) multiplication noise within the detector and (c) the noise of the preamplifier itself. Given that these noise sources are uncorrelated, their effects add as the square root of the sum-of-squares to result in an aggregate noise. It is usual to refer this noise to the input of the preamplifier, i.e. after the detector, and to calculate it in terms of the noise current density measured at that point. The spectral density of most of the noise sources is broadly frequency independent.

The current presented to the preamplifier consists of the optical power detected and multiplied by the APD gain together with the dark current of the detector.

The SNR is usually defined as the signal power/noise power integrated over the bandwidth of the filters in the acquisition system. However, in general, it is the square root of that quantity, $(S/N)_{opt}$, that is of direct interest to the user who usually compares the *optical* signal level with the r.m.s. (root mean square) noise level, for example, after the cessation of the backscatter signal. Therefore, the definition adopted here (and in several parts of the literature) is the ratio of the backscatter signal to the r.m.s. noise level and is given by

$$(S/N)_{opt} = \frac{P_{BS}(z)\cdot R_d \cdot M_G}{\sqrt{\left[2q\left(I_{du}+\left(I_{dm}+P_s(z)\cdot R_d\right)M_G^2F(M_G)\right)+i_n^2\right]B_w}}. \quad (3.21)$$

Here, q is the electron charge (~1.6 × 10^{-19} C), I_{dm} is that part of the dark current of the photodiode that is subject to the multiplication process and I_{du} is that part of the leakage current that is not multiplied. At typical gain settings, the multiplied I_{dm} contribution is generally greater than I_{du}. M_G is the multiplication gain for the APD; although, in principle, the gain can be varied as a function of time in order to increase the dynamic range or alleviate the worst overloading from reflections, we will assume here that it is fixed. R_d is the responsivity of the detector, i.e. the ratio of the primary (before multiplication) photocurrent to the power incident on its surface (see Section 2.3.11). i_n is the noise current spectral density due to the preamplifier, and B_w is the bandwidth over which the noise is collected. Although, on average, the multiplication of a primary carrier has a mean value, M_G, the gain experienced by each individual carrier fluctuates around the mean; this random variation gives rise to additional noise represented by the excess noise factor $F(M_G)$.

$F(M_G)$ depends on the ratio of ionisation rates for electrons and holes and on which type of carrier initiates the multiplication of the carriers; these factors are material dependent, but the design of the semiconductor junction can also help contribute to a low excess noise factor. If only one type of carrier (electron or hole) contributes significantly to the gain, then $F(M_G)$ is relatively low; however, if the ionisation rates are the same (as is the case in germanium APDs, for example), then $F(M_G)$ is close to unity. If k_{he} is the ratio of ionisation rates for holes relative to that for electrons and if the carriers injected into the multiplication region are solely electrons (the general case in modern APDs), then $F(M_G)$ is given by Ref. [28]

$$F(M_G) = k_{he} M_G + \left(2 - \frac{1}{M_G}\right)(1 - k_{he}). \tag{3.22}$$

$F(M_G)$ is sometimes approximated to Ref. [26]

$$F(M_G) = M_G^x. \tag{3.23}$$

x is unity for Ge, typically 0.5 for InGaAs APDs and commonly 0.3 for Si. Therefore, at wavelengths shorter than about 1050 nm (depending on doping levels), where silicon absorbs the incident light, very good detectors are available, given that for x = 0.3 and M_G = 100, $F(M_G)$ evaluates to just 4, degrading the S/N_{opt} in Equation 3.21 by just a factor of 2 for a gain of 100 (before considering the gain-independent contributions).

One of the benefits of using Equation 3.23, in spite of its approximate nature, is that it is simple to select a value of M_G that maximises $(S/N)_{opt}$. At the very least, it provides starting value for empirically optimising $(S/N)_{opt}$ in practical situations.

At longer wavelengths, commercially available InGaAs APDs offer lesser performance but are nonetheless still preferable to non-amplifying detectors in most cases. Nonetheless, research continues, and devices with very low excess multiplication noise have been reported in HgCdTe [29] and InAs [30] and also in AlInAs/InGaAs devices [31]. Recently, compound detectors in which the absorption and multiplication regions are made from different materials have been reported in the literature. In these devices, the multiplication usually takes place in silicon, but the absorption occurs in germanium (effective out to well beyond 1600 nm) [32] or InGaAs [33]. This is another promising technology for detecting weak light beyond the absorption band for silicon.

Unfortunately, $F(M_G)$ is not the only consideration in the noise relating to the detector: the dark current is also important; *both* low dark current *and* low $F(M_G)$ are required.

The term *excess noise* refers to the noise in the multiplied current output of the detector over and above the shot noise. Shot noise arises from the discrete nature of the electric charges forming a current. For electronic devices, this noise follows a Poisson distribution, and so it is proportional to the square root of the current, and thus, the SNR of a current, if limited by shot noise, is proportional to the square root of the signal current. All the currents in Equation 3.21 contribute to the shot noise, i.e. the dark current of the laser and the primary signal current prior to multiplication, namely $P_s(z) \cdot R_d$.

Turning now to the i_n term, it consists of the Johnson noise of the feedback resistor, given in Equation 3.24, and the noise of the front-end amplifying device. It is assumed that the first stage has sufficient gain to ensure that the noise contribution from further amplification stages can be neglected. To avoid the Miller effect* [34], a following stage with low input impedance is preferred [35].

Referred to the input noise, Johnson noise current density is

$$i_{RF} = \sqrt{\frac{4k_B T_A}{R_F}}, \tag{3.24}$$

where T_A is the temperature of the resistor and k_B is Boltzmann's constant. For $R_F = 100$ kΩ, i_{RF} works out at about 0.4 pA Hz$^{-1/2}$. From the sole perspective of this noise source, one would maximise R_F. However, as the bandwidth of the preamplifier is increased, it becomes increasingly difficult to design a circuit using large values of R_F, for reasons that will be discussed next.

The response time of the circuit is affected by the total capacitance C_T at the input node of the preamplifier and is dominated by a time constant τ_R:

$$\tau_R = \frac{C_T \cdot R_F}{A_V}. \tag{3.25}$$

As the frequency of interest increases, τ_R must be reduced, and so R_F/A_V must also be reduced. Moreover, it becomes more difficult to maintain A_V at high frequencies because the signal transit time from input to output must be kept well below the desired impulse response of the receiver in order for the preamplifier to remain stable. The need to keep short transit times through the forward amplifier limits the number of amplifying stages that can be incorporated within the feedback loop of the preamplifier, and in turn, this limits the gain available. At high frequencies, it therefore becomes necessary to reduce the value of R_F.

In any case, at high frequencies, other noise factors dominate, and in general, the noise spectral density of a receiver designed for a wide frequency range is higher than one that can be optimised for low frequencies only. The corollary is that it is difficult to design a receiver for an OTDR that performs well for both high and low spatial resolution, yet this is precisely what is asked of designers because commercial OTDRs frequently offer pulse durations ranging from 10 to 1000 ns and sometimes a wider range still. The design of an OTDR is, like so many engineering problems, the art of the compromise.

For the low-to-moderate bandwidths required in OTDR receivers, a field-effect transistor (FET) is generally the best choice for the front-end amplifying device. The FET amplifier contributes two noise terms, the first of which is one associated with the gate leakage current I_G and can be added to I_{du} in Equation 3.21. The other noise term due to the FET, the channel thermal noise i_{fe}, is given by Refs. [36] and [37]

$$i_{fe}^2 = \frac{4k_B T_C \Gamma C_T^2 (2\pi)^2 B_w^2}{3g_m}, \tag{3.26}$$

where g_m is the transconductance of the FET. Γ is a factor that accounts for the fact that several contributions to i_{fe}^2 are correlated. The excess channel noise factor Γ is about 0.7 for a silicon FET but can take values from 1.4 to more than 2 for other materials and depends also on the design of the device and the preamplifier [37].

Fundamentally, i_{fe}^2 is caused by thermal voltage noise in the channel of the FET that is converted through the front-end capacitance to current noise at the input [38]. Here, the temperature of the FET

* An effect that describes how the effective input capacitance is increased due to gain across the parasitic capacitance of the input amplifying device.

channel, T_C, is used; for the very small devices used in high-performance receivers, T_C can be considerably higher than ambient. Moreover, in devices such as high-electron mobility transistors (HEMT), it is the temperature of the carriers that matters and this can be several times higher (in absolute temperature units) than ambient [39].

Thus, the i_n^2 term that appears in Equation 3.21 is given by

$$i_n^2 = \frac{4k_B T_A}{R_F} + \frac{4k_B T_C \Gamma C_T^2 (2\pi)^2 B_w^2}{3g_m}. \tag{3.27}$$

From Equation 3.26, we see that there is a strong case for minimising C_T and choosing a device with a large g_m. The frequency dependence of i_{fe}^2 means that at low frequencies, effects such as the Johnson noise of the feedback resistor (i_{RF}) or the multiplied dark current dominate. In addition, I_G, which is negligible in silicon FETs, is usually larger (and often not even specified) in devices, such as HEMTs, that are designed for operation at much higher frequencies than apply to OTDRs. Therefore, choosing the FET must also include selecting a component with low I_G.

The process of designing an OTDR receiver must start with defining the finest spatial resolution that is required. This usually limits the value of R_F that can be selected as discussed previously. When the OTDR is used at a lower spatial resolution setting, filtering reduces primarily the channel noise term i_{fe}^2 because of its strongly increasing noise contribution at high frequencies. The choice of R_F imposes a lower limit on the noise that can be achieved for the receiver as a whole, particularly at low frequencies. With a value of R_F defined, then provided that $I_{du} + I_G$ is substantially less than $\frac{4k_B T_A}{qR_F}$, these leakage currents are not significant in the overall noise performance. For $R_F = 1M\Omega$, if the sum $I_{du} + I_G$ is significantly less than 50 nA, the leakage currents will hardly contribute to the receiver noise. At the time of writing, HEMTs in surface-mount packages with input capacitance of order 0.5 pF, $g_m \geq 45$ mS and $I_G < 20$ nA are available commercially. Therefore, for $\Gamma = 1.4$, $T_C = 1800$ K, $g_m = 60$ mS and $C_T = 1.5$ pF, i_{fe} evaluates to 0.8 pA for a bandwidth of 100 MHz (which would be appropriate for systems aimed at a resolution of 0.5 m or longer). Thus, for $R_F =$ 100 kΩ, the overall noise of the preamplifier can still be below 1 pA $\text{Hz}^{-1/2}$ in a bandwidth of 100 MHz.

Two worked examples are provided in the following sections; their purpose is to illustrate the design process and some of the trade-offs involved rather than to provide a solution to a particular problem. The design requirements are likely to be different from those used in the example, as are the components available to a particular engineer. Although the most of the parameters that are used are discussed in the text, they are summarised in Table 3.1.

3.2.3.1 Worked example 1: Multimode, short-wavelength OTDR

As one example of the interplay of signal and noise in an OTDR, let us take the case of an instrument designed for measuring multimode, graded-index fibres at a wavelength of 850 nm at which silicon APDs can be used. We will also assume that a spatial resolution of 1 m is required; for simplicity, the pulse duration and impulse response of the receiver are taken to be equal at around 7 ns and so a bandwidth of 0.44/7 ns $\simeq$ 63 MHz is required throughout the analog signal chain. A probe power launched into the fibre of 3 W has been assumed (a figure that is also taken to include the effect of losses in the optics in the return direction). The probe energy is therefore 21 nJ; for $NA = 0.2$ and $\alpha_s = 2.11$ dB/km, the backscatter factor evaluates to $\eta = 223$ W/J (Equations 3.13 and 3.6) and so the backscatter power at the start of the fibre is about 4.8 μW (Equation 3.5).

It will be assumed that the receiver can be designed with $R_F = 100$ kΩ and that the front-end amplifying device is a HEMT with $g_m = 50$ mS together with a total capacitance at the input node of the preamplifier of $C_T = 4.0$ pF. Further, the sum of the gate leakage current and un-multiplied dark current used in the model is $I_G + I_{du} = 40$ nA, which is achievable with most silicon APDs and with some care in the selection of the HEMT.

TABLE 3.1 Parameters used in the worked examples of Section 3.2.3

PARAMETER	*WORKED EXAMPLE 1 MULTIMODE*	*WORKED EXAMPLE 2 SINGLE-MODE*
Requirements		
Fibre type	Graded index, multimode	Single-mode, step index
Probe wavelength	850 nm	1550 nm
Target spatial resolution	1 m	10 m
(Implied bandwidth)	63 MHz	6.3 MHz
Probe Properties		
Pulse duration	7 ns	70 ns
Effective peak probe power	3 W	0.25 W
Pulse energy	21 nJ	17.5 nJ
Fibre Properties		
Scattering loss	2.11 dB/km	0.165 dB/km
NA	0.2	0.13
Backscatter factor	223 W/J	7.5 W/J
Near-end backscatter	4.8 μW	0.131 μW
Detector		
Type	Si APD	InGaAs APD
k_{he}	0.02	0.3
I_{dm}	0.1 nA	0.3 nA
R_d	0.6 A/W	1 A/W
Receiver		
R_F	100 kΩ	1 MΩ
g_m	50 mS	50 mS
C_T	4 pF	2 pF
Calculated Noise Currents		
i_{RF}	0.4 pA/Hz$^{1/2}$	0.129 pA/Hz$^{1/2}$
i_{fe}	0.8 pA/Hz$^{1/2}$	0.057 pA/Hz$^{1/2}$
i_n	0.89 pA/Hz$^{1/2}$	0.141 pA/Hz$^{1/2}$

The detector is assumed to have an ionisation ratio k_{he} = 0.02 and a multiplied dark current I_{dm} = 0.1 nA.* The shot noise spectral density attributable to the leakage currents $I_G + I_{du}$ is $\sqrt{2q(I_G + I_{du})}$, which evaluates to c. 0.13 pA Hz$^{-1/2}$. This is almost negligible compared with the Johnson noise for the feedback resistor (0.4 pA Hz$^{-1/2}$ based on Equation 3.24; in any case, their combination is calculated to be 0.42 pA Hz$^{-1/2}$). This noise density is independent of signal level and of the bandwidth of the system.

The calculated noise associated with the input HEMT, i_{fe}, is c. 0.8 pA Hz$^{-1/2}$ using Equation 3.27 and so is dominant for the combination of component properties and bandwidth used here. The combined noise spectral density that is attributable solely to the receiver, i.e. i_n, is 0.89 pA Hz$^{-1/2}$.

Figure 3.13a shows plots of the signal and r.m.s. noise as a function of distance for several values of the APD gain. The distance axis, whilst relevant to OTDR design, is also a proxy for signal level. The noise is shown for a single probe pulse, i.e. before any averaging has been applied. The extreme value of 500 for the APD gain in the final panel is well above that recommended for most devices, and in many cases, this value is not achievable. The dashed line shows the multiplied signal photocurrent. The horizontal line shows those contributions to the noise that are independent of signal level and APD gain, i.e. the preamplifier noise and that

* Some of the detector parameters are not specified by some APD suppliers but can be inferred from some of the typical curves that they provide or, in the worst case, from noise measurements on sample devices.

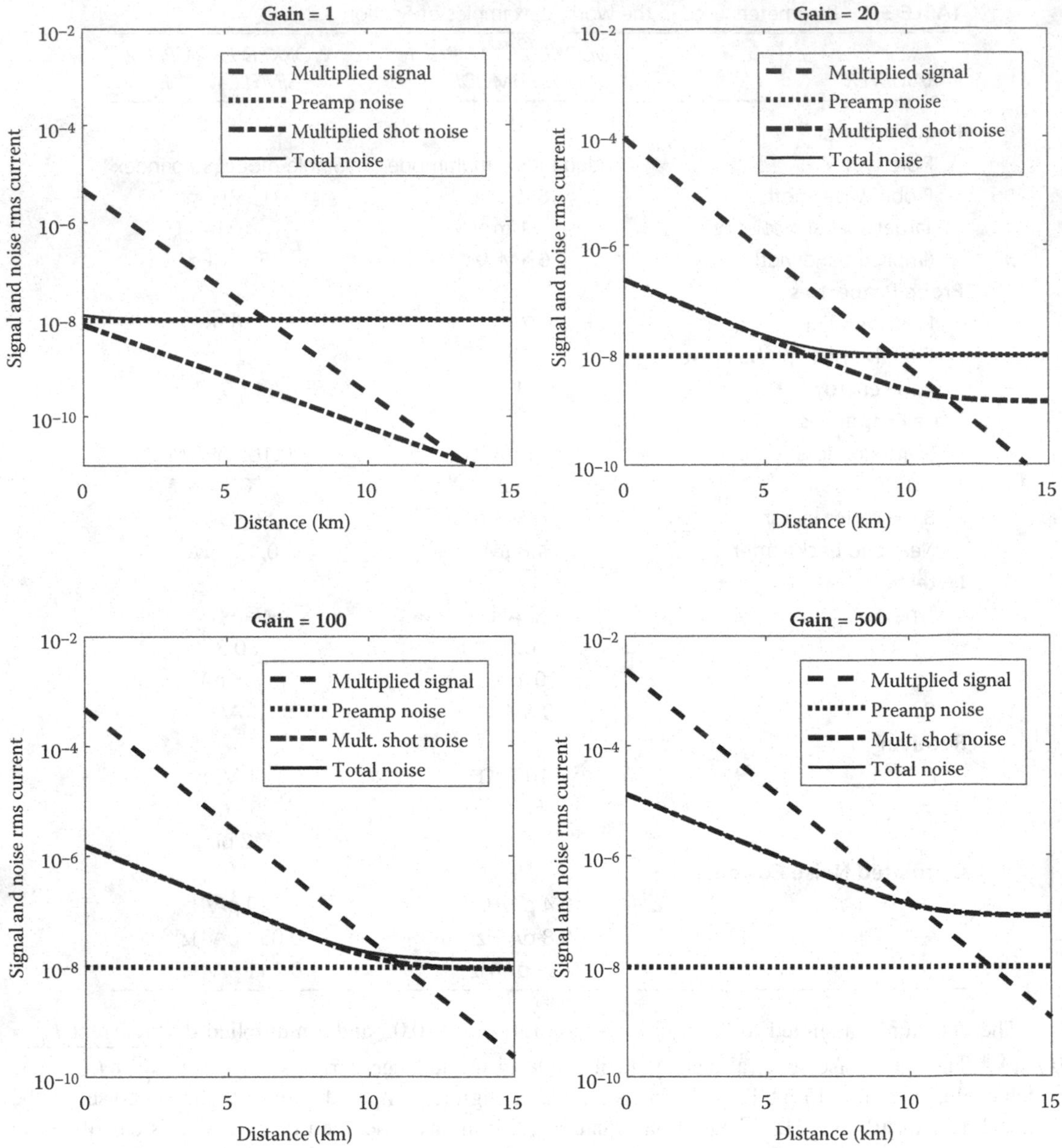

FIGURE 3.13 (a) Signal and noise contributions modelled as function of distance for a multimode, 850 nm OTDR. The APD gain modelled varies between panels and is indicated in the title of each sub-plot. *(Continued)*

related to the dark current. The multiplied shot noise (dot-dash curve) includes the noise from the multiplied dark current as well as from the multiplied signal current signal. The total noise is shown by a thin solid line.

At a gain of 1 (upper left panel) and near the start of the fibre where the signal is strong, the shot noise is just above the receiver noise, which demonstrates that here there is little to be gained from the APD at this signal level; however, beyond about 1 km from the launching end, the decreasing signal results in the receiver noise dominating other noise sources; the signal current falls below the receiver noise at a distance of about 7 km (i.e.

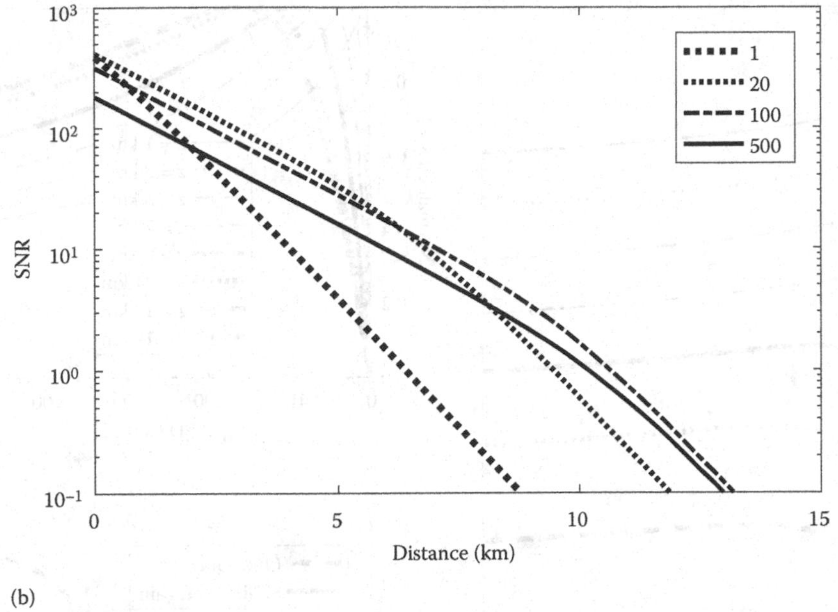

FIGURE 3.13 (CONTINUED) (b) SNR as a function of distance derived from the data in panel (a). *(Continued)*

the SNR is unity at this point). As the gain is increased, see panels for Gain = 20 and Gain = 100, the signal is increased in proportion, but the noise that is subject to avalanche gain (dot-dash curve) increases slightly faster as a result of the multiplication noise.

The total noise follows the multiplied shot noise curve initially until that falls below the preamplifier noise line. In spite of the rise of the multiplied shot noise with increasing APD gain, the point at which the SNR is unity is pushed to towards more distant locations along the fibre (about 9 km for $M_G = 20$ and 11 km for $M_G = 100$): this demonstrates that applying APD gain improves the SNR in this range of parameter values. In the case of $M_G = 20$, at the SNR = 1 point, the noise is still dominated by preamplifier. However, for $M_G = 100$, the noise at the SNR = 1 point is now dominated by shot noise from the amplified signal. It is also noticeable that for $M_G = 100$, the total noise curve always remains above that of the preamplifier, even for distant points where the signal is very weak; this is due to the multiplied dark current of the detector, i.e. the I_{dm} term in Equation 3.21. This effect is more visible still in the final panel ($M_G = 500$), where the I_{dm} term results in the total noise remaining an order of magnitude above that of the preamplifier. As a result of this effect, there is no benefit from the increased gain and in fact it will be shown that this is detrimental.

The SNR is shown in Figure 3.13b as a function of distance for the same four values of APD gain as in Figure 3.13a. Applying a gain to the APD degrades the SNR for the strong signals near the start of the fibre, but the rate of decline of the SNR is reduced for a wide range of signals, resulting in a net improvement for most of the length of the fibre, and of course, it is the most distant points that matter in this context: it is *their SNR* that limits the system performance. We also note that the SNR resumes a steeper rate of degradation (at around 10 km for $M_G = 100$) where the multiplied shot noise curve approaches that of the preamplifier.

To complete the picture, Figure 3.13c shows (left panel) the SNR as a function of APD gain at various locations along the fibre, in steps of 2 km. These curves illustrate a number of points, namely (a) there is an optimum APD gain setting, (b) this setting increases with increasing distance along the fibre (i.e. as the signal is weakened) and (c) the separation between the curves increases with increasing distance along the fibre, demonstrating that the APD is less and less able to overcome the reduction in signal level.

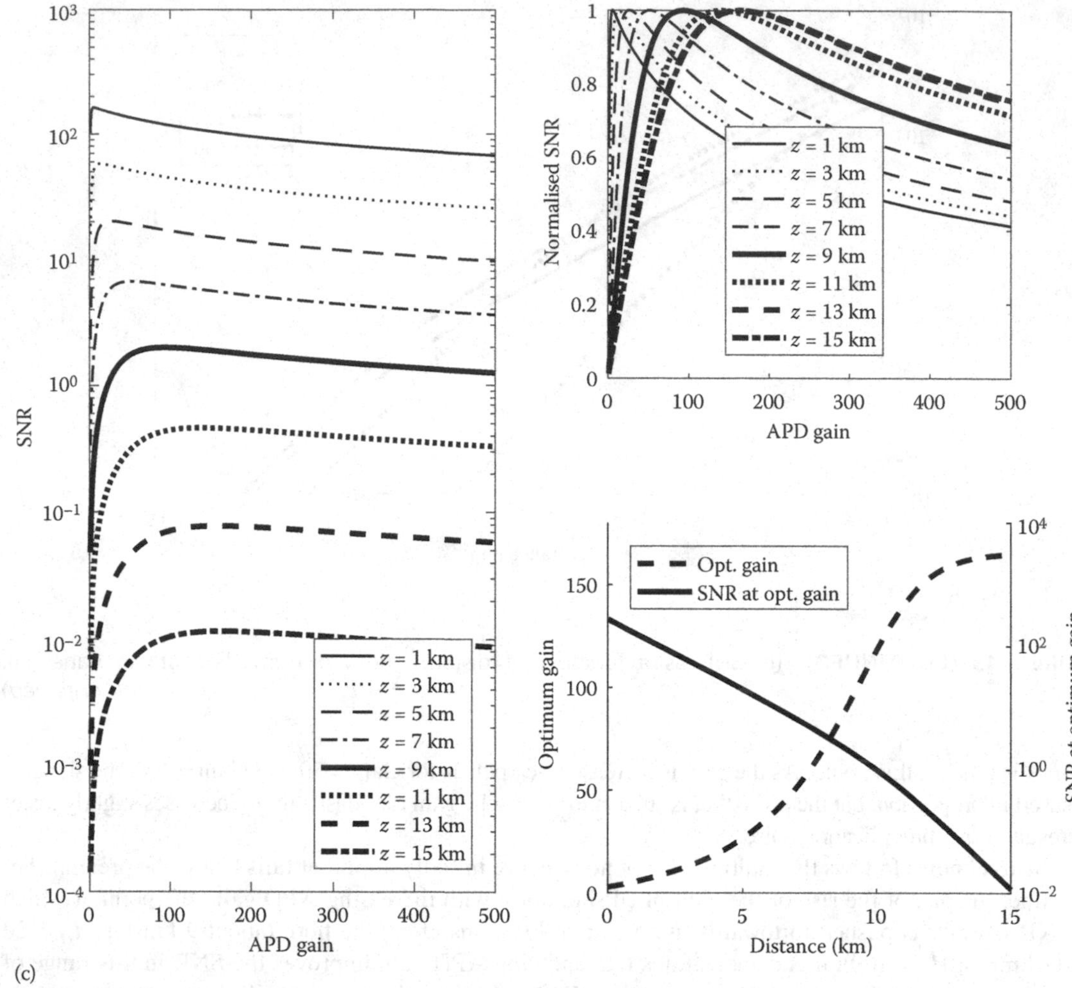

FIGURE 3.13 (CONTINUED) (c) SNR for the 850 nm OTDR as a function of APD gain (left panel). Upper right panel: the data of left panel normalised to the maximum SNR value. Lower right panel: Optimum APD gain as a function of distance along the fibre (broken line, left axis) and corresponding SNR (solid line, right-hand axis).

(*Continued*)

The SNR, as a function of APD gain, normalised to that for the optimum gain, is shown in the upper right-hand panel of Figure 3.13c; it clearly illustrates the existence of an optimum APD gain and that this optimum varies with signal level. For weak signals, the optimum is relatively flat.

The lower right panel shows how the optimum gain varies with distance along the fibre and how, in the limit of the weak signals at the remote end, the optimum reaches a constant value. In this case, the levelling out is caused by the multiplication of I_{dm}. Where the optimum gain levels off, the SNR begins to fall off at a faster rate because the reduction in signal level is no longer partially compensated by the increase in APD gain.

It is worth pointing out that it is extremely unusual for the APD gain to be varied along the fibre, particularly during the course of a measurement. Therefore, the design must be optimised for the most remote signal (on the basis that the SNR elsewhere, although not optimum, will still be better than that for the remote end), supported by considerations illustrated in Figure 3.13a–c.

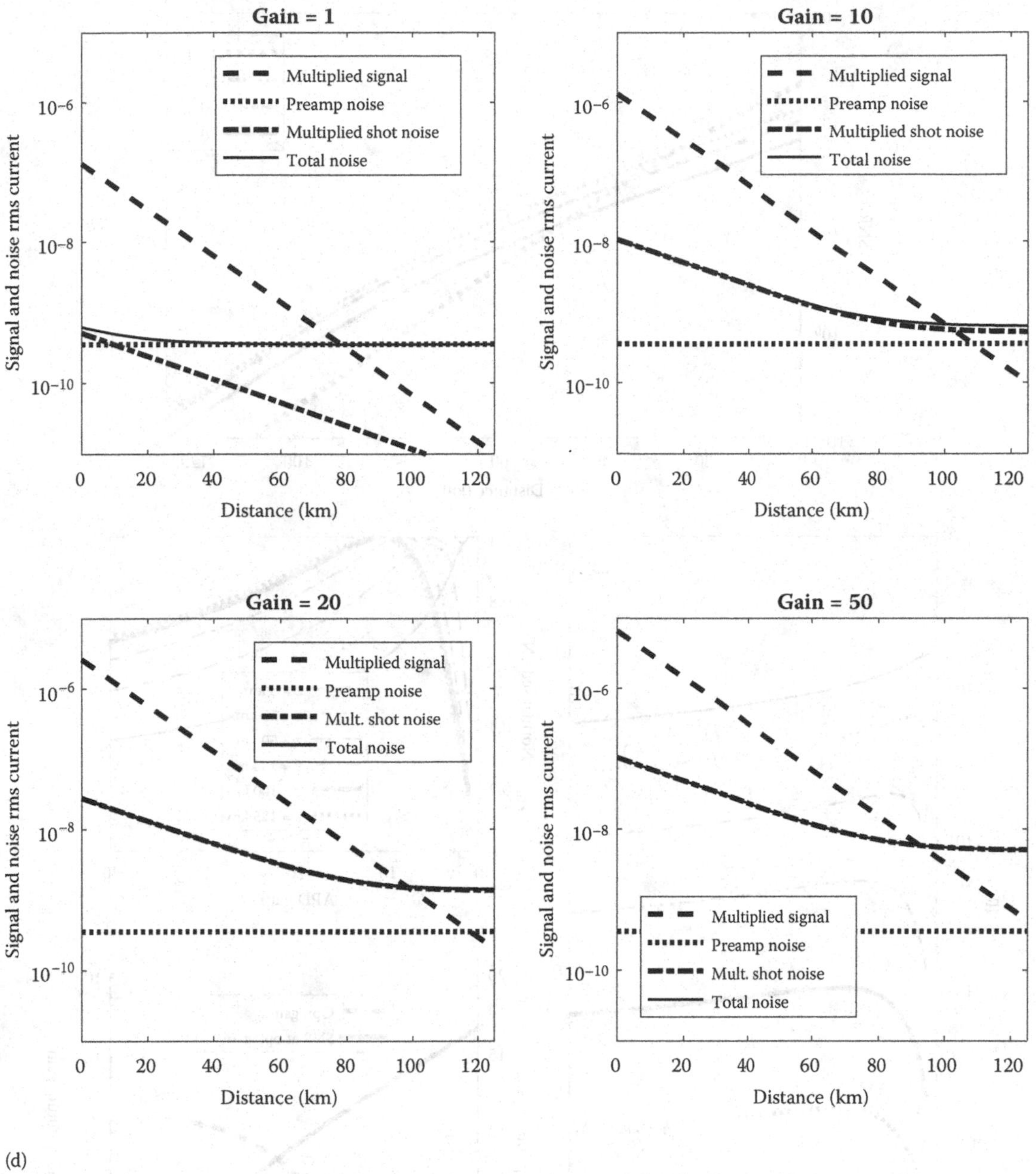

FIGURE 3.13 (CONTINUED) (d) Signal and noise contributions modelled as function of distance for a single-mode, 1550 nm OTDR. The APD gain modelled varies between panels and is indicated in the title of each sub-plot. *(Continued)*

We have also seen that the APD optimisation can be limited by the multiplied dark current. This parameter will vary between devices of the same type as well as between devices of different designs. Some thought must be given to how the operating point will be chosen in the setting up of the instrument.

The detector dark current is a strong function of temperature (doubling for every 5 K increase in temperature in the case of silicon devices). In the case of laboratory instruments, the detector can be temperature controlled, for example, with thermo-electric (TE) devices that can provide a modest degree of

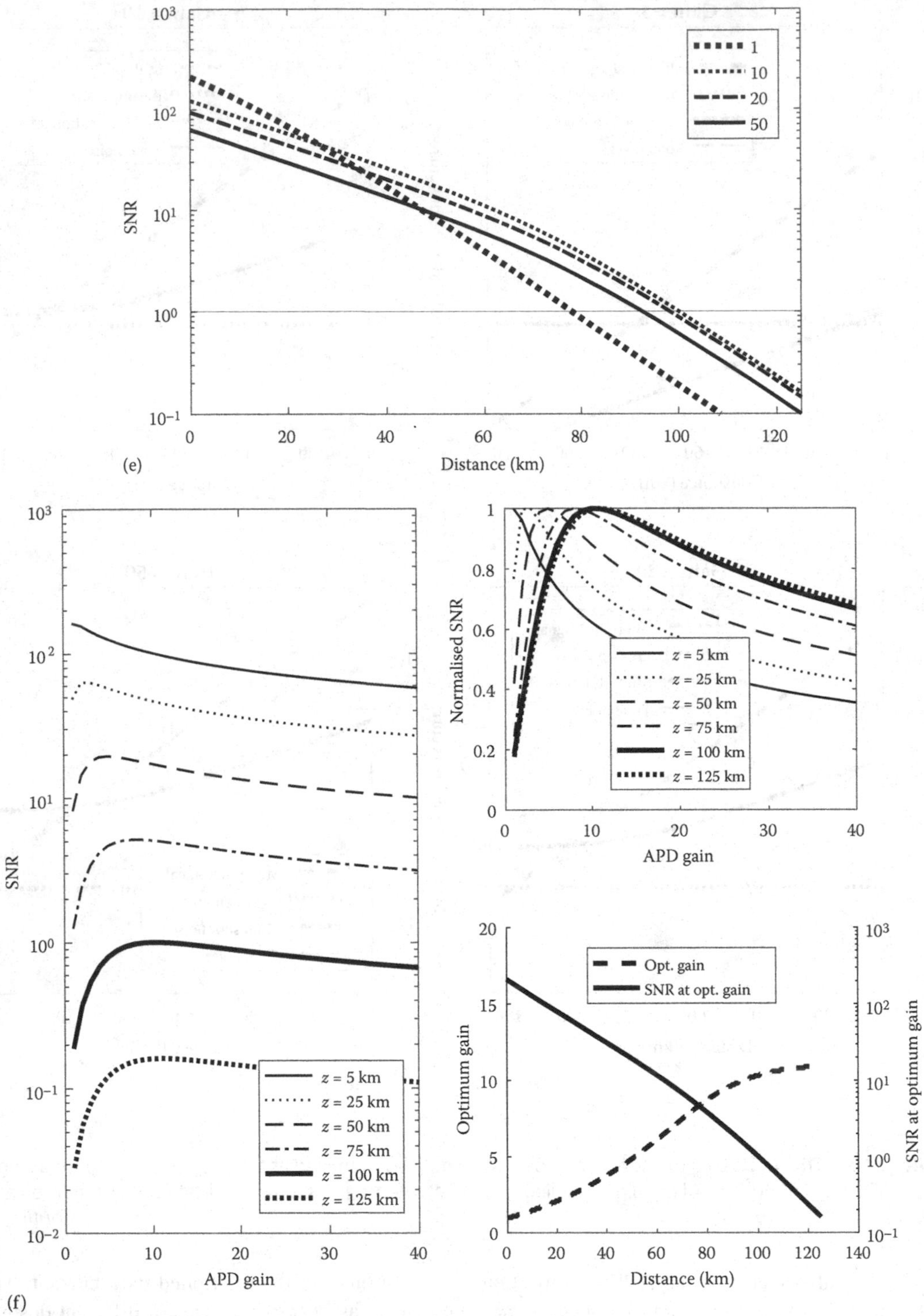

FIGURE 3.13 (CONTINUED) (e) SNR as a function of distance derived from the data in panel (d). (f) SNR for the 1550 nm single-mode OTDR example as a function of APD gain (left panel). Upper right panel: the data of left panel normalised to the maximum SNR value. Lower right panel: Optimum APD gain as a function of distance along the fibre (broken line, left axis) and corresponding SNR (solid line, right-hand axis).

cooling. In the case of battery-powered field instruments, this is unlikely to be an option because of the power consumption of TE devices and the wider range of operating temperatures that are encountered. Therefore, the design must cope with varying dark currents and the fact that, at a particular bias voltage, the APD gain is a function of temperature.

3.2.3.2 Worked example 2: Single-mode, long-wavelength OTDR

Similar sets of curves are shown in Figures 3.13d–f for an example of a single-mode OTDR operating at 1550 nm in a fibre having $NA = 0.13$. For this second worked example, a target of 10 m spatial resolution has been selected, which allows the transimpedance to be increased to $R_F = 1\ \mathrm{M\Omega}$. At 1550 nm, a long-wavelength APD, typically based on InGaAs, is the only realistic option at present, and so $k_{he} = 0.3$ and $I_{dm} = 0.3$ nA were chosen as being reasonably representative of this type of device.* The lower capacitance of these types of detector permits the assumption of $C_T = 2.0$ pF. The value of the fibre loss included in the model is 0.2 dB/km. Other parameters may be found in Table 3.1.

As can be seen in Figure 3.13d, the initial backscatter signal is about two orders of magnitude lower than for the multimode, 850-nm case (a reduction by a further factor of 10 in signal level would have occurred if the spatial resolution of the first example had been maintained). However, the receiver noise is somewhat lower than in the multimode case, thanks to the larger feedback resistor and the lower bandwidth and input capacitance (the latter two parameters both affect the noise from the FET, based on Equation 3.27). Of course, the attenuation coefficient is lower, which allows single-mode, long-wavelength designs to aim for long ranges – the figures are calculated for up to a range of 125 km. A similar behaviour of the noise as a function of distance is shown with, however, a more pronounced effect of the noise increasing relative to that of the preamplifier for weak signals and as the APD gain is set higher. It should be noted that the highest value of APD gain modelled is far lower than was the case for the silicon APD.

The SNR, plotted vs. distance (Figure 3.13e), is similar in appearance to the equivalent plot for the multimode case, after allowing for the change in distance scale. At the near-end of the fibre, the SNR is only a factor of 2 lower than in the multimode case at unity gain in spite of the substantial reduction of the signal, a result of the lower receiver noise. The dynamic range of the backscatter signal (i.e. the ratio of the near-end to far-end signal) is a factor of 2 lower in logarithmic units than in the multimode case, in spite of the longer lengths, simply because the attenuation is about 10 times higher at 850 nm for a multimode fibre than at 1550 nm in a single-mode fibre at 1550 nm.

The final set of plots, shown in Figure 3.13f, is to be contrasted with Figure 3.13c, and they demonstrate a much lower range of useful APD gain: there is little point in increasing the gain much beyond 10.

3.2.4 Overload Considerations

Unfortunately, noise is not the only consideration in the receiver design. The dynamic range of signal strength ranges from the weak backscatter at the lower end to a reflection from a cleaved fibre end or from a connector. The strong reflections will usually saturate a receiver that is designed for the weak backscatter signal. We have seen that the backscatter signal is several orders of magnitude lower than the probe pulse power. Of course, this ratio depends on the spatial resolution, because the backscatter power is proportional to pulse energy (see Equation 3.8), whereas a reflection varies in proportion to probe power P_p. For a multimode OTDR operating at 1300 nm on a typical graded-index fibre, the backscatter factor is about 50 W/J. Thus, a 10 nJ (1 W × 10 ns) pulse would return a backscatter signal of about 0.5 μW from a fibre section near the launching end. In contrast, a 4% reflection would return 40 mW, i.e. nearly

* There is a wide variety of InGaAs devices offering quite different values of ionisation ratio, multiplied dark current and maximum gain. In addition, the values of these parameters (or proxies, such as the noise figure) are sometimes quoted in ways that cannot be related directly to the gain at the conditions tested. Consequently, considerable interpretation and, often, complementary measurements are required.

5 orders of magnitude higher. Admittedly, a 4% reflection requires a perfect cleave on the fibre end, and a pulse duration of 10 ns (corresponding to a spatial resolution of order 1 m) is at the lower end of settings typically available. A receiver optimised for detecting backscatter signals will overload under these conditions.

A transimpedance preamplifier, as previously discussed, changes its output voltage to draw a current through R_F of equal magnitude to the input current, but of an opposite sign. The net current flowing into the input node is thus zero and the voltage at that node is constant and nominally 0.* An overload occurs when the input current is greater than the maximum voltage output (voltage swing) V_{sw} of the preamplifier divided by R_F. V_{sw} is limited by issues such as the voltage supply available or the maximum operating voltage of the output amplifying device. When the output voltage of the preamplifier reaches its limit and the input current continues to rise, charge is stored on the input node. After the input current returns to within the normal operating range of the preamplifier, the charge is drained away through the feedback resistor. However, for severe overloads, this can take a considerable time. For example, if the overload is due to a reflection, then the duration of the excess current fed to the amplifier is equal to the pulse duration, but the time taken to clear the overload is given by the pulse duration multiplied by the overload factor (the ratio of actual input current divided by the maximum permissible current for correct operation of the circuit). This could be many times the spatial resolution, and the OTDR is 'blind' for the time it takes for the receiver to come out of saturation. In addition, the receiver must then settle following the large transient after it comes out of overload. Until the settling time has elapsed, the output of the receiver will be distorted.

Of course, avoiding reflections would solve this problem, but this is not always feasible, especially in the event of a fault. Alternatively, the designer can choose a value of R_F well below that which minimises the receiver noise and they can therefore balance the ability of the receiver to recover from overload against achieving the best possible sensitivity. Modifications to the front end, such as adding a reverse-biased Schottky diode to limit the departure of this node from zero, can help accelerate the overload recovery, but they do not eliminate the problem and they add some noise to the receiver, for example, by increasing the input capacitance and leakage current at the input node.

Using an optical feedback path [40] could, in principle, alleviate this problem. In this case, the feedback resistor is replaced with the combination of a light-emitting diode (LED) or laser diode driven by the output of the receiver and a photodiode that is illuminated by the LED; in turn, the photodiode causes the current to flow into the input node. This combination fulfils the same role as the feedback resistor, i.e. to flow current into the input node in exact proportion to the current drawn by the optical signal; however, the circuit can be designed with better resilience to overload (it is not limited by the voltage swing of the amplifier) and it eliminates the Johnson noise from the feedback resistor. However, the additional detector increases the capacitance and the leakage current at the input node and there is a delay around the feedback loop that limits the bandwidth of this type circuit. To the author's knowledge, it has not been used in OTDR (although this was proposed in Ref. [40]) or DOFS; it would be particularly appropriate for long-range systems, where the bandwidth requirements are limited but the dynamic range needed is large.

3.2.5 Acquisition and Further Processing

The acquisition of the OTDR signal is now made relatively straightforward by the availability of high-resolution (e.g. 12-bit) A/D converters sampling at very high rates (>1 G sample/s, if required). Also, advances in digital field-programmable gate array logic allow the acquisition and real-time averaging of many waveforms to be carried out in a single device. Alternative approaches, such as the fast and low-power processors now available, make the design of this part of the system very manageable. This is not to minimise the issues faced by designers of the digital part of an OTDR who must handle constraints of power consumption, heat dissipation, ruggedness, limited space and cost.

* Other than the input-voltage offset of the device.

3.2.6 Limitations to Power Launched

In OTDRs, the power that can be launched is not limited solely by source availability but also by non-linear optical effects, eye safety and explosion hazard considerations. High power launched into an optical fibre can result in a number of non-linear optical effects that transfer probe energy to other wavelengths. One of the lowest threshold effects is stimulated Brillouin scattering (SBS); it can cause a strong conversion from forward to backward energy (with a frequency shift), but fortunately, in conventional OTDRs, it can be eliminated by using a source with a broad spectrum, which is desirable in any case to limit coherence effects.* Self-phase modulation (SPM) is caused by the optical Kerr effect, a change in the refractive index in proportion to optical intensity. With pulsed light, it results in a broadening of the optical spectrum, particularly at the edges of a pulse. Again, because OTDRs generally deal with broadband sources, SPM is not a serious limitation on their performance.

Nonetheless, OTDR systems based on coherent detection employ sources with very narrow instantaneous linewidth. These systems are limited by SPM and SBS (depending on the details of the design); this point is discussed later in this chapter.

Stimulated Raman scattering (SRS) is a more serious limitation because it occurs even with broad sources and also because it transfers optical energy to substantially different wavelengths (a frequency shift of ~13 THz, equivalent to about 100 nm for a source wavelength of 1550 nm). If significant SRS occurs, the precision of the measurement is compromised because the probe and return wavelengths become uncertain and position dependent. However, SRS has been used to create probe pulses [41,42] for OTDR at wavelengths where powerful lasers were unavailable or for tunable† OTDR.

It has also been proposed to enhance the range of OTDRs by launching probe pulses at relatively short wavelength and at levels well beyond the threshold for SRS and allowing the conversion process to take place along the fibre [43]. The technique fell out of favour once suitable sources at the wavelengths of interest became available, because the range of wavelengths present at each location along the fibre is somewhat uncertain when using SRS in this way. Nonetheless, the use of a separate Raman pump to amplify both the probe and the backscatter has been employed in long-range OTDR systems [44] and also in long-range distributed sensors [45]. The limitation is, however, that using a continuous Raman pump amplifies the probe pulse near the start of the fibre and can lead to it exceeding the maximum allowable power for linear operation, which led to techniques using modulated Raman pumps [46].

Eye safety considerations [47] can limit the power that is launched into the fibre where the fibre might be accessible to the public or where the OTDR is functioning continuously, e.g. when it is used to monitor the quality of a network.

The possibility that the power carried by an optical fibre could cause an explosion should also be considered in OTDRs and especially in the context of DOFS where, in some applications, the sensing fibre is situated in a potentially explosive environment. The optical power is normally confined to the core, that itself is inside the cladding, a coating and a protective cable structure. Under normal operating conditions, therefore, there is no hazard. However, in the event of a rupture of the cable, the optical power will reach the potentially explosive atmosphere.

A study of the potential for power guided by an optical fibre for initiating explosions [48] showed that the principal concern is the absorption of the light by small particles on or very near to the end-face of the fibre. If the power delivered by the fibre is sufficient, the particle can be heated to a point where it ignites the flammable or explosive gas mixture in which it is immersed. In the context of an optical fibre, the main result [48] is that the potential for initiation of explosions has been demonstrated for power levels of 35 mW. Designers must be aware of this risk and consider the likelihood of this power being exceeded, including under fault conditions.

* Brillouin scattering is, however, a key sensing technology and will be described in more detail in Chapter 5.

† A tunable OTDR measures the attenuation as a function of distance and wavelength by varying the source wavelength and making backscatter measurements at a set of probe wavelengths.

Depending on the hazardous zone classification, the sensing systems are required to remain demonstrably safe in the event of one or more faults. Most installations in hazardous environments require certification by an authorised test laboratory. This process usually involves a review of the optical sources and their control electronics, including their behaviour in the event of one or more faults developing (such as the failure of an active device, or short circuits developing between nodes of the control circuity).

3.3 ENHANCING THE OTDR RANGE

3.3.1 Improved Detection

As described previously, the performance of a classical, single-pulse OTDR is bounded, on the one hand, by the probe energy that can be launched and, on the other, by the sensitivity of the receiver. Researchers have tackled this problem by looking for ways of improving the receiver sensitivity, using photon counting or coherent receivers, or by increasing the probe energy using spread-spectrum techniques. The reader will recognise all of these techniques applied, as appropriate, to DOFS in later parts of this book.

3.3.1.1 Coherent detection systems

Using a very narrow band source allows coherent detection techniques to be employed; in the literature, such systems are often referred to as coherent OTDR (C-OTDR). This concept is illustrated in Figure 3.14. The source is a continuous-wave laser, the output of which is split into two paths. The first path is modulated into the probe pulse that is launched into the fibre to be tested. A second path takes another fraction of the laser output, known as the *optical local oscillator* (OLO), and directs it to a directional coupler,

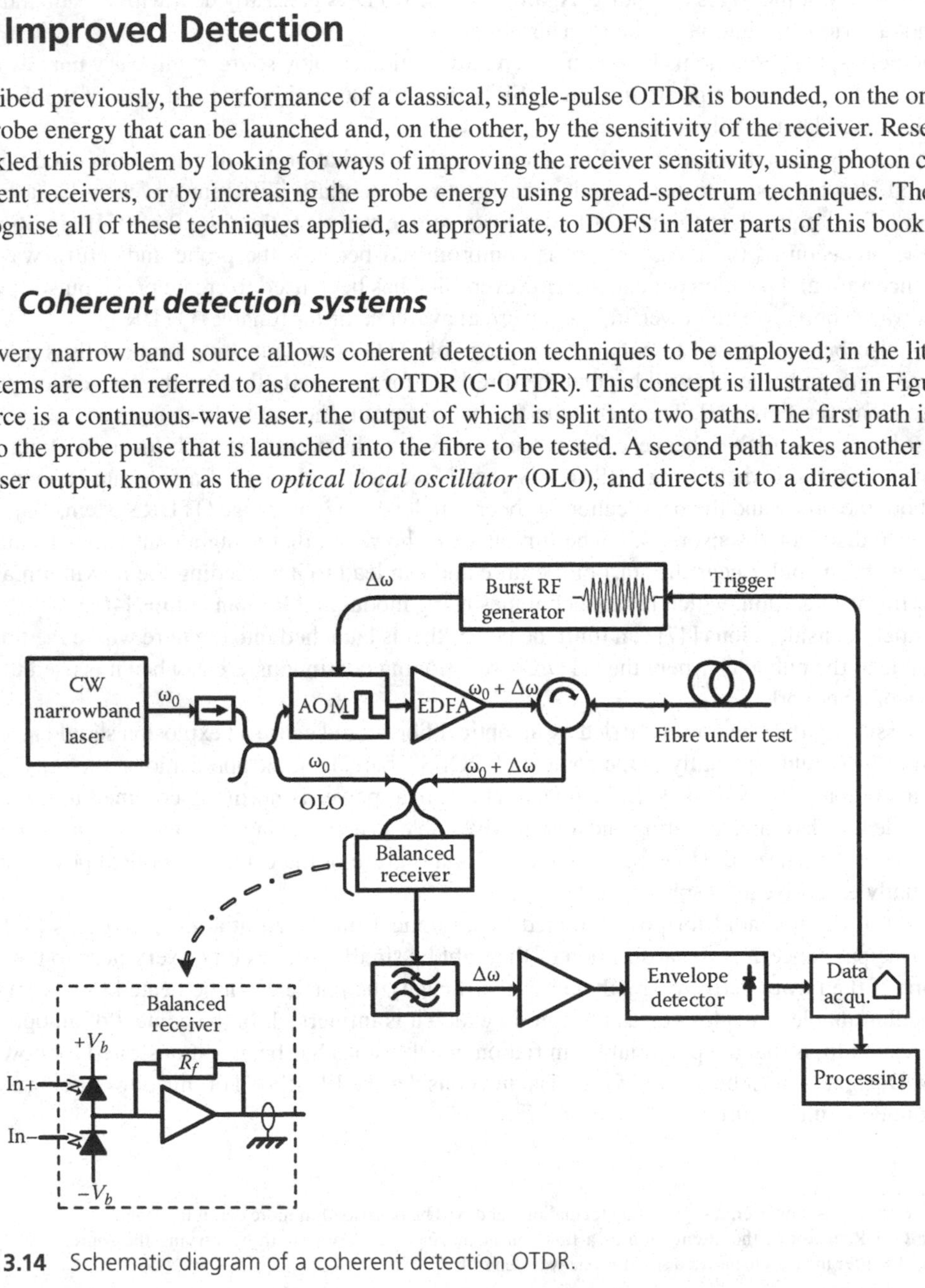

FIGURE 3.14 Schematic diagram of a coherent detection OTDR.

where it is combined with the backscatter returning from the fibre under test. The combined signal is fed to a receiver, or preferably (as illustrated) a balanced receiver. There are two basic implementations of the C-OTDR concept, namely heterodyne and homodyne detection.

In the first, heterodyne detection, the probe pulse is frequency shifted in the modulation process, typically using an acousto-optic modulator (AOM) [49] (see Chapter 2 for a description of AOMs).

The fact that the frequencies of the backscatter and the local oscillator are different means that the signal at the receiver contains a mixing product at the frequency difference, known as the *intermediate frequency* (IF). The IF is selected from the output of the preamplifier using a bandpass filter. Crucially, a non-linear operation is applied at this point, such as envelope detection, rectification or squaring. This operation extracts only the amplitude information (eliminating the phase information). Prior to the non-linear operation, the signal is quasi-oscillatory, and because the phase of the IF drifts, a straight average of the IF signal would return zero. The signal can be sampled after envelope detection and averaged as in conventional OTDRs. Alternatively, the output of the receiver can be sampled and rectified in the digital domain prior to averaging.

A commercial C-OTDR system was developed by STC Ltd. in 1987 [50].

A laser emitting at angular frequency ω_0 has a time-dependent electric field of the form $E_{LO} \cdot \exp(j\omega_0 t)$ and the backscatter electric field, shifted to $\omega_0 + \Delta\omega$ by the AOM*, relative to ω_0 can be written as

$$E_b(t) = E_b(t) \cdot \exp\left\{ j\left[(\omega_0 + \Delta\omega)t + \Phi(t) \right] \right\}, \tag{3.28}$$

where $E_b(t)$ is the position-dependent amplitude of the backscatter signal, i.e. it is proportional to the square root of the backscatter power returning from location $z = t/(2v_g)$, and $\Phi(t)$ is the local phase of the backscatter signal. $\Phi(t)$ is not important in C-OTDR, but we shall see later that it is crucial to distributed vibration sensors. The electric field falling on the detector is then given by

$$E_{tot} = E_{LO} \cdot \exp(j\omega_0 t) + E_b(t) \cdot \exp\left\{ j\left[(\omega_0 + \Delta\omega)t + \Phi(t) \right] \right\}. \tag{3.29}$$

Since the photodetector provides a current in response to optical power arriving at its surface, the photocurrent is given by

$$i_{phot} = R_d \left\{ E_{LO}^2 + E_b^2 + 2E_{LO}E_b \exp\left[j\left(\Delta\omega t + \Phi(t) \right) \right] \right\}. \tag{3.30}$$

By filtering the receiver output to select the IF at $\Delta\omega$ only, i.e. the third term in Equation 3.30, a signal is obtained that is the product of the electric field of the OLO and of the backscatter signal. Subject to damage limitations for the detector, the OLO can be made sufficiently large that the IF signal dominates the noise of the receiver; under these conditions, the receiver noise no longer impacts the sensitivity of the system.

In general, a balanced receiver (illustrated in the inset of Figure 3.14) is used by connecting both outputs of the coupler to two separate photodiodes that are connected in series and reverse-biased between voltages $+V_b$ and $-V_b$. As mentioned in Chapter 2, the phases at the outputs of a 2 × 2 coupler differ by 180°, and so inputs from Port 1 or Port 2 emerge anti-phase and the balanced receiver amplifies the difference between these inputs. The input from the OLO, for example, is split equally into the two output ports, and so the difference between its contributions to the positive and negative inputs of the balanced receiver is negligible.

However, the signal resulting from a combination of the signals arriving simultaneously at Port 1 and Port 2 depends on their relative phase. A balanced receiver therefore largely eliminates the DC component at ω_0 and at $\omega_0 + \Delta\omega$ (the latter being insignificant in any case if sufficient OLO power is applied) by

* See Section 2.3.7 on the frequency-shifting properties of AOMs.

amplifying only a combination of signals arriving at the OLO and signal inputs of the coupler, but rejecting those signals arriving at just one of the input ports. The balanced receiver therefore further rejects the quasi-DC signals and also makes more efficient use of the available signal (in a conventional, single-detector receiver, only one half of the signal photons are detected). The heterodyne term

$$i_{het} = R_d \left\{ 2E_{LO}E_b \exp\left[j\left(\Delta\omega t + \Phi(t) \right) \right] \right\} \tag{3.31}$$

therefore dominates the output of the receiver and is the only optical contribution that is used. i_{het} is rectified (either in analog circuitry or in digital form). By squaring this signal, a waveform that is directly comparable to the conventional backscatter signals is obtained. Without this squaring operation, the decay rate of the measured signal is that of the electric field, not backscatter power and, in logarithmic units, its slope is half of that of a conventional OTDR.

The S/N_{opt} after detection is given by

$$S/N_{opt} = \sqrt{\frac{2P_{LO} \cdot P_{BS} \cdot R_d^2}{\left[2q \cdot I_{du} + i_n^2 + 2q \cdot R_d(P_{LO} + P_{BS}) \right] . B_w}}. \tag{3.32}$$

The first two terms in the denominator, $2 \cdot q \cdot I_{du} + i_n^2$, represent the noise spectral density due to the dark current of the detector(s) and the pre-amplifier, respectively. The final term is the shot noise relating to the photo-electrons produced in the detectors. It can be seen that if $P_{LO} \gg P_B$, then clearly, only the local oscillator's contribution is important.

Typically, the receiver noise current density is a few pA/Hz$^{1/2}$; for a 5 nA dark current, the shot noise attributable to the dark current is 40 pA/Hz$^{1/2}$, so the pre-amplifier current noise is the main concern. The local oscillator power required to match the receiver noise is given by

$$P_{LO} = i_n^2 \cdot \frac{R_d}{2q}. \tag{3.33}$$

For i_n = 8 pA/Hz$^{1/2}$, a local oscillator power of 0.2 mW is required for the two noise contributions to match. Obviously, because designers have control over P_{LO}, they would generally ensure that it is well above this value; the shot noise due to the local oscillator is then the only significant contribution to the denominator of Equation 3.32. Under these conditions, the SNR ratio becomes

$$S/N_{opt} = \sqrt{\frac{P_{BS} \cdot R_d}{q \cdot B_w}}. \tag{3.34}$$

Equation 3.34 simply means that the SNR is proportional to the square root of the number of detected photons. Thus, the C-OTDR technique allows the sensitivity to approach limits set fundamentally by the number of photons that are detected, eliminating any other effects. However, there are a few *caveats*.

Firstly, the heterodyne process results from the superposition of electric fields at the surface of the detector. So far, we have ignored the polarisation of the two electric fields but, in practice, the heterodyne process selects from the backscatter signal only that part that is in the same state of polarisation as the local oscillator. This can be alleviated by using two detector systems and demodulating orthogonal polarisation states individually, an approach known as *polarisation diversity*. Alternatively, the polarisation of one of the electric fields (OLO or backscatter) that are interacting can be varied much faster than the measurement time, so that the polarisation effects average out. This method is sometimes referred to as *polarisation scrambling*; it reduces the average detected optical power by a factor of 2.

Secondly, the heterodyne technique, being frequency shifted, also has an image band at symmetric, negative frequencies. If information in that image band is not used, its noise passes through the filters, but no additional signal is contributed. This degrades the SNR by a factor of $\sqrt{2}$ [51].

Thirdly, the C-OTDR technique relies on using a narrowband source. However, we have discussed the fading effects that the use of such sources gives rise to. The solution is to vary the frequency of the source over the duration of the measurement. Therefore, a source with very high coherence in the short-term but capable of being modulated in frequency or that naturally hops in frequency is required. In Ref. [50], the authors used the elegant solution of allowing the backscatter from a subsidiary fibre to feed back into the semiconductor laser that they were using as a source. The Rayleigh scattering spectrum as a function of frequency has a number of spiky features (see Chapter 6), one of which, near the centre of the laser's emission spectrum, injection-locks the laser [52]. When the condition of the subsidiaray fibre drifts, the reflection spectrum changes; the laser then hops in frequency and locks to another reflection feature. In this way, a source with a broad long-term spectrum, but highly coherent on a short-term basis, was obtained. To complement this approach, the temperature of the laser can be slewed during the acquisition [53], which shifts the centre of its gain curve and stimulates more frequency hopping. It is, however, preferable for the frequency shift to occur between backscatter signals, so that the local oscillator is consistent with the backscatter signal and the laser frequency is changed only after the cessation of the backscatter signal and before the next probe pulse is issued. This can be achieved by current injection [54], complemented by temperature sweeping. Theses by Mark [55] and Izumita [56] provide between them an excellent knowledge base on the subject of C-OTDR.

It is also possible to perform a C-OTDR measurement using the same frequency for the probe and the local oscillator, an approach known as homodyne detection [57]. In telecommunications systems, homodyne detection relies on the signal being at a known phase that allows the signal (and associated noise) to be collected only from positive or negative frequencies; in contrast, in heterodyne, noise is collected from upper *and* lower sidebands. Homodyne detection therefore has a 3 dB sensitivity advantage over heterodyne detection; this does not apply to OTDR because the phase of the signal is fundamentally random. Nonetheless, there are claimed benefits in terms of system simplicity. In practice, most of the work on C-OTDR has involved heterodyne, rather than homodyne, detection. This is possibly for practical reasons, such as the fact that the acousto-optical modulators that are commonly used for cutting the probe pulse from the source laser, owing to their excellent extinction ratio, naturally shift the frequency of the modulated pulse and this simplifies the detection process.

3.3.2 Photon Counting

In photon counting, a special detector is used that, when it detects the arrival of a single photon, generates an output pulse that is sufficiently large that it can be processed by digital circuits and therefore no further degradation of the signal quality occurs in the subsequent handling of the signal by the electronics.

Photon counting detectors [58] are a well-known technology, and the fastest devices can resolve times of the order of tens of picoseconds [59]; this corresponds to a spatial resolution for OTDR measurements of the order of a few mm. Recent OTDR publications, e.g. Ref. [60], on photon counting have indeed reported achieving a resolution of 2 cm at a range of 20 km.

Photon counting has been used extensively not only for analysing fast fluorescence decay in chemistry [61,62] but also for time-of-flight ranging [63] and 3-D imaging [64]. Recent advances in quantum encryption have largely relied on photon counting [65].

As early as 1980, Healey demonstrated an OTDR using single-photon counting [66] and, in the following year, a multi-channel photon counting OTDR [67]; in both cases, the aim was to overcome the limitations caused by the very weak signals obtained from single-mode fibres with the low-power single-mode laser diodes then available. In Healey's case, the objective was to achieve a sufficient range on single-mode fibre to be able to span the then typical distance between repeaters in terrestrial optical telecommunications networks (30 km at that time), and a poor spatial resolution (100 m) was an acceptable compromise. However, recent developments in the technology of photon counting detectors [60] keep this technology competitive with conventional approaches, in spite of the gain in performance of sources and detectors in conventional OTDRs.

Photon counting OTDRs are able to measure far lower signals than conventional systems can. However, they tend to suffer more from large changes in dynamic range, such as those that occur at reflections or abrupt losses. In addition, their best performance is achieved only with strongly cooled detectors, which limits field use. Nonetheless, at least one supplier provides a range of instruments based on this technology [68].

As far as DOFSs are concerned, the approach has been directed primarily at Raman DTS systems, and a further description of the implementations will be provided in Chapter 4.

3.3.3 Spread-Spectrum Techniques

The previously discussed approaches to stretching the range of OTDRs were based on improving the receiver sensitivity. An alternative set of approaches are based on maximising the probe energy launched without breaching the limits set by non-linear effects and, in particular, SRS. Although it would be possible simply to increase the pulse duration, this would have, of course, the effect of degrading the spatial resolution.

Spread-spectrum techniques encode information about the location of the scattered signal whilst maintaining either a continuous or a very-long-duration probe signal. They involve transmitting long-duration probe signals that have much wider bandwidth than would be implied by the inverse of the duration of the transmission. The additional bandwidth is used to encode information on the probe signal that allows the backscatter signal to be reconstituted with a resolution comparable to the inverse of the bandwidth of the transmitted signal, rather than duration of the probe.

Thus, in a spread-spectrum technique, the probe energy is determined by the probe duration, but the resolution is determined by the bandwidth of the probe signal. In single-pulse systems, these quantities are bound together because the probe bandwidth is essentially the inverse of the pulse duration. Spectrum-techniques break that connection by allowing the probe bandwidth to be much wider than the inverse probe duration.

3.3.3.1 OFDR

In optical frequency-domain reflectometry (OFDR), the source is swept in frequency, but the amplitude is maintained essentially constant. The technique is also sometimes referred to as frequency-modulated, continuous wave (FMCW). There are two types of OFDR, namely coherent and incoherent OFDR.

3.3.3.2 Coherent OFDR

In coherent OFDR (C-OFDR), the frequency of the source itself is frequency-modulated [69]; the source frequency is swept and a heterodyne detection technique is used to beat the frequency of the light leaving the laser with that which has been emitted a little earlier and was scattered back from some distance down the fibre. The system is relatively simple to understand if the sweep is linear. In this case, a particular beat frequency is associated with location down the fibre because the frequency difference between the reference (local oscillator) and the scattered light is proportional to time delay and hence to distance for a linear sweep.

The distance axis of a C-OFDR is mapped to frequency according to Ref. [70]

$$z = \frac{v_g}{2\gamma} f_b, \tag{3.35}$$

where γ is the sweep rate (rate of change of frequency) and f_b is a particular frequency in the spectrum that identifies z.

For a linear sweep of duration T_r, γ is related to the frequency range Δf by

$$\gamma = \frac{\Delta f}{T_r}, \qquad (3.36)$$

and so,

$$z = \frac{v_g}{2}\frac{T_r}{\Delta f} f_b. \qquad (3.37)$$

The spatial resolution is given by

$$\delta z = \frac{v_g}{2 \cdot \Delta f}. \qquad (3.38)$$

So, in C-OFDR, the spatial resolution is given by the range of the frequencies in the probe sweep. Note that, although the source is required to span a wide range for a high spatial resolution, if the sweep rate is slow, the frequencies that need to be measured are quite modest. For a fibre of length L, the maximum beat frequency $f_{b_{max}}$ that needs to be recorded is given by

$$f_{b_{max}} = \frac{\Delta f}{L} \frac{2}{v_g T_r}. \qquad (3.39)$$

For example, for a fibre length of 1 km, if T_r is the same as the two-way pulse transit time, the sweep duration would be 10 μs, and for a spatial resolution of 0.1 m, Δf would be 1 GHz, and so $f_{b_{max}}$ would also be 1 GHz. However, if the source were swept more slowly, say in 10 ms, then $f_{b_{max}}$ would be 1 MHz, which is far less challenging technologically. In this case, the system is averaging the information, not by sweeping repeatedly as would an OTDR but by acquiring the information over a longer time.

This reduction in the acquisition bandwidth allows systems with very high spatial resolution (where the *source* is swept over a wide range) to use moderate speed electronic acquisition systems. So the C-OFDR technique allows fine resolution to be achieved with relatively slow electronics provided that the source can be swept reliably over a wide range.

Although the technique was demonstrated some 30 years ago in short optical fibres [71,72], achieving a long range requires a very well controlled source, which must sweep over a wide range and yet provide a highly predictable phase over long times. So in practice, initially, the technique was used mainly for short-distance applications, such as analysing connected fibre sections for small reflections and imperfections or studying integrated optical circuits [73].

However, the development of well-controlled single-frequency fibre lasers has opened up the possibility of using C-OFDR in long fibres, and Geng et al. [74] showed backscatter measurements over more than 95 km. However, even these very sophisticated lasers had insufficient coherence to avoid distorting the backscatter amplitude measurement. As a result, it has been necessary to estimate the actual frequency sweep of the laser with the aid of an auxiliary interferometer and use that information to correct the OFDR signal [75]. The implementation of this approach has led to the development of long-range systems with very high spatial resolution, such as 5 cm over a fibre distance of 40 km [76], which is well beyond what has been reported in the time domain. Similar techniques have been used extensively over short distances in medical applications for optical coherence tomography, e.g. for diagnosing macular degeneration and detached retinas in patients' eyes [77–79]. Here, the aim is extremely fast scanning with a very fine spatial resolution, requiring very broad and fast sweeping.

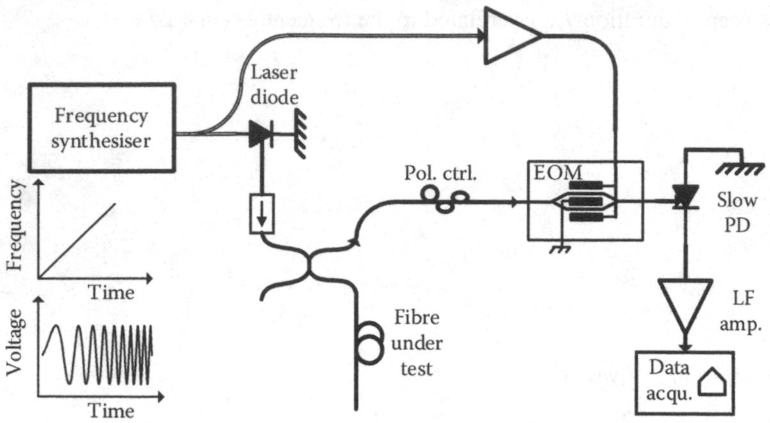

FIGURE 3.15 Schematic diagram of an incoherent OFDR system. (Adapted from Venkatesh, S., Dolfi, D. W., *Appl. Opt.* 29, 1323–1326, 1990.)

3.3.3.3 Incoherent OFDR

In incoherent OFDR, a sinusoidal amplitude modulation of the source intensity is applied, and it is the frequency of that modulation that is swept. MacDonald [80] provided an early demonstration of this approach, but only for measuring discrete reflections, rather than the backscatter signal.

Probably the most conclusive work in the area of incoherent OFDR is that of Venkatesh and Dolfi [81]. Here (Figure 3.15, after Ref. [81]), an electrical frequency synthesiser provides a frequency-swept voltage drive to a laser diode, resulting in an amplitude modulation at a swept frequency of the laser output power. The laser output is launched into the fibre that is tested and the backscatter or reflected return is first modulated again by an electro-optic modulator and then collected by a detector. The detector output is low pass filtered and digitised. In effect, the backscatter signal is multiplied by the same modulation function as the laser, but because of the two-way propagation time in the test fibre, there is a delay between the modulation frequency applied to the source of the probe light and that applied at the detection, and therefore a Fourier transform of the digitised signal relates a particular frequency to a location in the fibre. This technique separates the spatial resolution (dictated by the span of the frequency sweep) from the frequencies present in the detector output, which are determined by the sweep rate (i.e. the derivative of modulation frequency with respect to time) and the length of the fibre tested. In addition, the re-modulation of the returning light performs a mixing operation essentially in the optical domain and it alleviates the need for high-frequency detectors and fast acquisition, an issue that was far more important when this work was reported than it is today.

Thus, similarly to the C-OFDR approach, incoherent OFDR allows the frequency range required of the acquisition to be scaled down by using a sweep time that is much longer than the transit time in the fibre.

To the author's knowledge, incoherent OFDR is not used at present, probably because the technological progress on controlling the frequency of optical sources has progressed so much that C-OFDR has fulfilled the needs in this area, in addition to progress in the spatial resolution of OTDR systems. Nonetheless, the work on incoherent OFDR is a pillar of techniques used in DOFS and in Raman DTSs in particular (see Chapter 4).

3.3.3.4 Pulse compression coding

Similar to OFDR in its basic premise, pulse compression coding involves launching trains of coded ('1' or '0') pulses into the fibre. Each '1' results in a backscatter trace. In the absence of any distinguishing feature between the pulses (see next section, 'Frequency-diverse C-OTDR'), the backscatter energy simply adds and results in larger backscatter signal that, however, is smeared over the time of the pulse train.

Like OFDR, pulse compression coding has its roots in radar [82], where it has been used for a long time to circumvent the fact that pulsed sources have limited peak power and that increasing the pulse duration degrades the distance resolution.

The basic concept of pulse compression coding is to transmit a sequence of pulses that has particular auto-correlation properties; these pulses result in overlapped return signals that are garbled compared with that from a single pulse. By correlating the return signal with the original sequence, the backscatter signal can be restored to a waveform that is equivalent to a single pulse, but of much higher energy. In this way, one aims to improve the SNR of the measurement.

A major difference exists between radar signals and OTDR, namely that – at least in incoherent OTDR – what is measured is optical power that is proportional to the square of the electric field amplitude. So, in this case, the optical signals received can only be positive or zero, not negative. Therefore, the codes that have been developed for conventional radars that use coherent detection are not appropriate for OTDR.

C-OTDR would, in principle, allow a number of coding schemes that are available in radar to be applied to the optical domain because, in C-OTDR, the backscatter is measured as an electric field, which therefore has a phase and can be negative. However, this approach is also limited in OTDR because unlike a radar echo, the phase of the effective reflection that the OTDR returns is not consistent between sections of fibre and therefore the correlation operation that is used to recover the fibre backscatter response does not result in a narrow impulse response for the correlated signals. Therefore, even in C-OTDR, it is necessary to estimate the absolute value of the electric field (or the power) of the backscatter before applying correlation operators.

A pulse compression code must have an auto-correlation peak equal to the number of '1' chips for zero shift and, at worst, only '1' and '0' values for non-zero shifts. Ideally, its auto-correlation function would be non-zero only for a zero shift; nonetheless, with a sufficiently long code, a trail of '1' and '0' for non-zero shifts could be acceptable.

m-sequences are continuously repeating sequences of pseudo-random bits and can be created relatively simply with a shift register and some feedback that performs a logical Exclusive OR operation with re-circulating values held in the shift register [83]. m-sequences repeat after a fixed number of bits have been provided; the repeat interval is $m = 2^n - 1$, where n is the length of the shift register. Therefore, very long sequences can be created using just a few logic gates and Ref. [83] provides the design for sequences up to $2^{40} - 1$; i.e. sequences are known that repeat only after $>10^{12}$ chip periods, but shorter sequences are known for any integer power of 2 below that as well as for some combinations of prime numbers.

m-sequences have the property that their auto-correlation function consists of just $-1/m$ except at zero shift, or after a delay corresponding to a complete sequence duration, where it is 1. Thus, they provide an essentially perfect coding scheme. This approach was used to modulate the probe signal and then correlate the received backscatter signal with the same sequence [84]. Although a backscatter trace with correctly reconstituted spatial resolution was obtained, the improvement due to the coding was not clear.

More generally, the weakness of using continuous sequences such as m-sequences is that their output chip stream is also continuous. For spread-spectrum techniques, this means that the weak signals from the remote end arrive at the same time as strong signals from the near end. When the fibre is long, the signal will have a large dynamic range and so the signals returning from the remote end are orders of magnitude smaller than the near-end signal. In this situation, the strong near-end signals generate a large level of shot noise at the detector that is superimposed on the weak signal from the remote end. Therefore, beyond a certain dynamic range, the benefits of pulse-compression coding are negated because the remote signal is swamped by noise accompanying the strong near-end signals.

What is really required is a code having the good auto-correlation properties already discussed and a finite code length. No such codes exist for unipolar signals (such as optical pulse trains). Healey was probably the first to realise in this context [85] that a similar function could be achieved by using multiple codes that, *individually*, had unsatisfactory autocorrelation functions but when used *in combination* would provide the desired autocorrelation property. Healey used a pair of relatively short codes that had complementary autocorrelation functions that when summed gave the perfect response equal to N times the signal corresponding to a '1' pulse for zero shift and '0' only for non-zero shifts. It should be noted that this work was carried out with photon counting, and this is the only example of such a combination, to the author's knowledge.

In the field of non-coherent radar, where a similar problem exists of signals being unipolar only, it was recently shown that by encoding a bipolar sequence with a Manchester code, it is possible to achieve an autocorrelation function with significant response only near zero-shift [86]. However, in this work, the main, desired, autocorrelation peak is surrounded by negative signals, which makes it unsuitable for OTDR and DOFS applications.

A schematic arrangement of a pulse-compression coding OTDR is shown in Figure 3.16, which is very similar to a conventional OTDR, except that rather than using a pulsed laser, the laser is now a transmitter able to generate strings of '0' and '1' pulses. These pulses must be highly consistent – each '1' having the same energy and delay as all other '1's' in the sequence and a high extinction for the '0' bits. The laser transmitter is driven by a code generator, which (in the case of m-sequences) generates a continuous sequence. In more common cases, the code generator selects one finite-length code at a time from a code list and sends this to the laser. In many cases, the same code will be sent repeatedly and the resulting backscatter averaged before the code is changed. The signal processing is somewhat more complex because it must deal with a plurality of codes (and it must separate and process the results of their backscatter individually).

The breakthrough in pulse-compression coding was made by Nazarathy et al. [87], who showed how the family of codes of arbitrary length (provided they are integer powers of 2) developed by Golay [88] could be used to provide a set of four codes that, when used in combination, provide the perfect autocorrelation function.

Golay described, and provided methods for creating, sequence pairs $[A,B]$ of length M such that the sum of their auto-correlation functions is

$$A_k * A_k + B_k * B_k = \begin{cases} 2M & \text{for } k = 0 \\ 0 & \text{for } k \neq 0 \end{cases}, \tag{3.40}$$

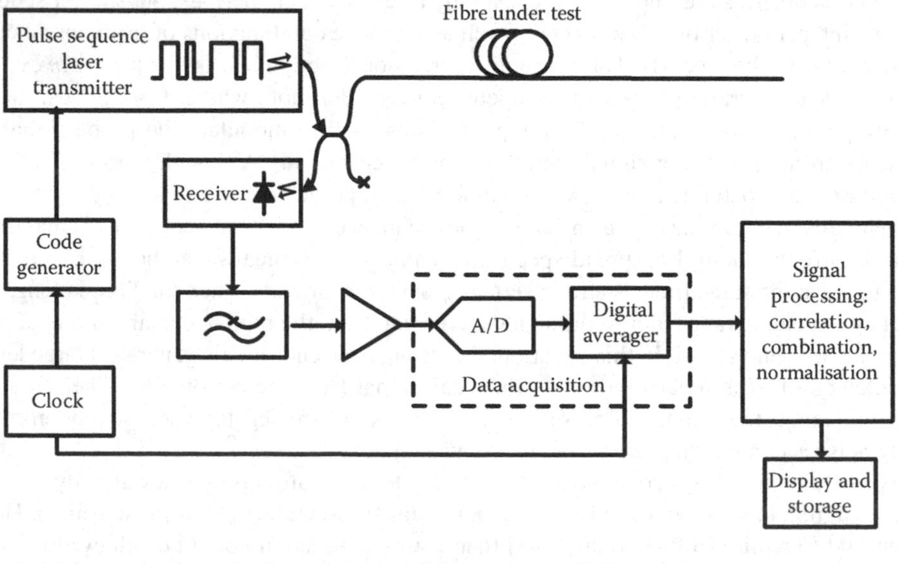

FIGURE 3.16 Schematic diagram of a pulse compression coding OTDR.

where * denotes the discrete-time correlation operator. Thus, pairs of codes can be created that, when used together, provide a perfect autocorrelation function. The autocorrelation functions of the *A* and *B* sequences, as illustrated in Figure 3.17, individually show significant sidelobes (with spikes >10 for a 64-chip code); however, the sidelobes for $A_k * A_k$ and $B_k * B_k$ complement each other perfectly, and therefore, the summation in Equation (3.40) results in a perfect autocorrelation function, with non-zero value only at zero shift, as shown by the dotted line in Figure 3.17 that has been shifted upwards for clarity.

However, the problem with using Golay sequences in the context of OTDR is that they are bipolar; i.e. the value of a chip is either −1 or 1. In incoherent optical systems, we cannot have 'negative light', so the Golay approach cannot be used directly.

The elegant solution devised in Ref. [87] is to simulate a bipolar code by launching each code, then its complement on top of a bias signal β, so a total of four codes derived from the original code pair are used:

$$\begin{aligned}
&\text{a)} \quad u_k^A = \beta(1 + A_k) \\
&\text{b)} \quad u_k^{\bar{A}} = \beta(1 - A_k) \\
&\text{c)} \quad u_k^B = \beta(1 + B_k) \\
&\text{d)} \quad u_k^{\bar{B}} = \beta(1 - B_k)
\end{aligned} \tag{3.41}$$

The backscatter returns from u_k^A and $u_k^{\bar{A}}$ are subtracted from each other before being correlated with the sequence. Likewise, the difference between the returns from probing with u_k^B and $u_k^{\bar{B}}$ is correlated with the *B* sequence and the sum of these correlations forms the output of the processing.

It turns out [87] that the value of S/N_{opt} resulting from the four-sequence coding and correlation is $\sqrt{4M}$ greater than that which would be achieved with a single pulse, for equal number of averages. However, since

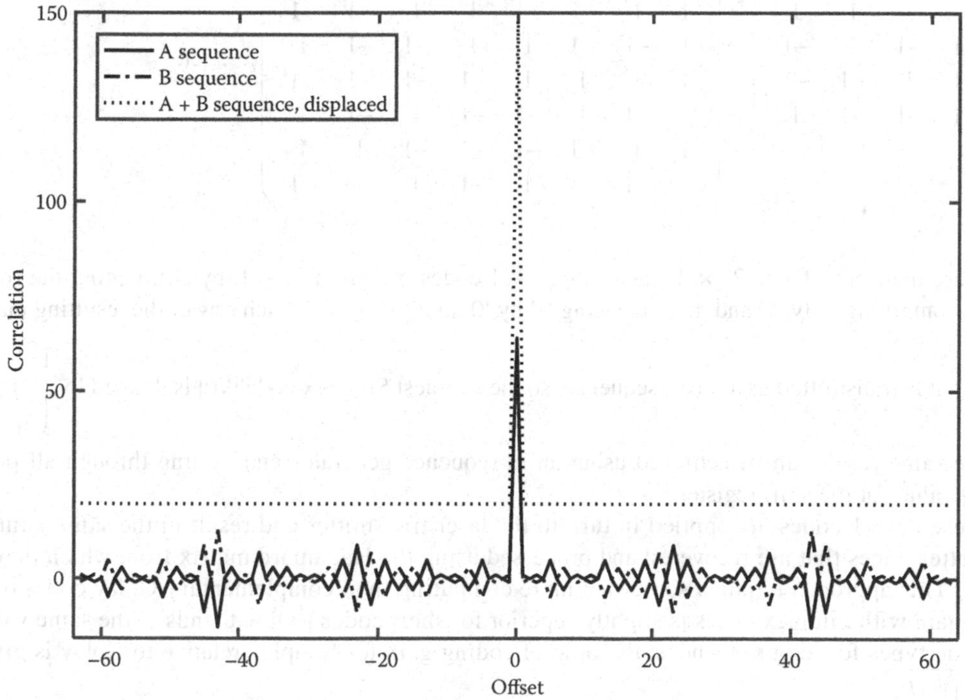

FIGURE 3.17 Autocorrelation functions of two complementary Golay sequences and their sum.

four codes must be used and their returns combined, four individual pulses could be launched and their backscatter returns acquired in the same time in a conventional OTDR. It follows that the gain using Golay coding is just $\sqrt{M}$. Although this result may seem disappointing, it follows from the fact that, although the signal collected is M times larger than that received for a single pulse, noise is also collected for each of the time slots within the sequence. The accumulated noise, assumed to be uncorrelated from slot to slot, increases as $\sqrt{M}$, and thus the overall gain in SNR from the coding is also $\sqrt{M}$. The coding gain has been demonstrated experimentally [87], both by demonstrating similar measurement quality achieved in *1/M* of the acquisition time with coding than without and also by showing substantial improvements with coding in the same measurement time compared with the reference single pulse results.

There must be an optimum code length in pulse compression. A single pulse has no coding gain. At the other extreme of a continuous sequence, such as an m-sequence, the gain will reduce as the backscatter from the near end of the code sequence dominates that at the far end. From a pure SNR point of view, the shot noise from the largest part of the return could exceed the signal from the most remote parts of the code, so the limitation relates more to the cumulative fibre loss over the section of fibre occupied by the code, rather than the number of bits in the code, *per se*. In practice, none of the reports of pulse compression coding OTDR appear to have been limited by this effect because the combination of fibre loss, code length and spatial resolution quoted only cover a dynamic range of only a few dB.

An alternative to Golay coding has been demonstrated in recent years, often referred to as Simplex codes, which are derived from Hadamard matrices [89]. The most basic Hadamard matrix is given by $\begin{pmatrix} 1 & 1 \\ 1 & -1 \end{pmatrix}$ and one way of generating larger Hadamard matrices is to multiply this matrix repeatedly by itself, viz.

$$\begin{pmatrix} 1 & 1 & 1 & 1 \\ 1 & -1 & 1 & -1 \\ 1 & 1 & -1 & -1 \\ 1 & -1 & -1 & 1 \end{pmatrix}, \begin{pmatrix} 1 & 1 & 1 & 1 & 1 & 1 & 1 & 1 \\ 1 & -1 & 1 & -1 & 1 & -1 & 1 & -1 \\ 1 & 1 & -1 & -1 & 1 & 1 & -1 & -1 \\ 1 & -1 & -1 & 1 & 1 & -1 & -1 & 1 \\ 1 & 1 & 1 & 1 & -1 & -1 & -1 & -1 \\ 1 & -1 & 1 & -1 & -1 & 1 & -1 & 1 \\ 1 & 1 & -1 & -1 & -1 & -1 & 1 & 1 \\ 1 & -1 & -1 & 1 & -1 & 1 & 1 & -1 \end{pmatrix}, \ldots \text{etc.}$$

These matrices of size $2^N \times 2^N$, generate $2^N - 1$ codes of length $2^N - 1$, by eliminating the row and column containing only '1' and then replacing '1' by '0' and '–1' by '1'. Each row of the resulting matrix is a code that is transmitted as a probe sequence, so the simplest Simplex code [90] is defined by $\begin{pmatrix} 1 & 0 & 1 \\ 0 & 1 & 1 \\ 1 & 1 & 0 \end{pmatrix}$.

The same result can be achieved using an m-sequence generator and cycling through all possible starting values of the shift register.

These $2^N - 1$ codes are applied in turn to the laser transmitter and result in the same number of backscatter traces that are recovered and processed using the Hadamard matrix from which they were created. This approach requires somewhat more computing than complementary Golay codes do. The coding gain with Simplex codes is slightly superior for short codes [90] but tends to the same value for both code types for long sequences; the ratio of coding gain for Simplex relative to Golay is given by $\sqrt{(M+1)/M}$.

For the correlation operation to return exactly the expected response, i.e. the same shape as a single probe pulse, but M times larger, there must be perfect linearity in every link of the chain, from transmitter to A/D conversion. Imperfections in the coding arise from the fact that directly modulated laser diodes can experience slightly different turn-on delay and pulse energy depending on the bit pattern; moreover, the photodiode and preamplifier, and any further amplifiers and conditioning circuits, must also exhibit perfect linearity. This is of course never fully the case and there will therefore be small artefacts in the recovered response; these can be particularly marked after a sharp transition, such as at a splice, where some distortion is most likely to be noticed.

The Simplex approach is roughly two times more tolerant than the complementary Golay codes of non-linearity in the transmitter or receiver. This is observed empirically, but the reason is probably that a wider combination of bit sequences is present, and this may cause distortion to be averaged to some extent over the entire code set. It was also shown that in the case of Simplex coding, certain types of imperfections can be corrected, if known [90].

In pulse-compression OTDRs, the most difficult issue is that of overloads caused by reflections. By definition, when the receiver is overloaded, it is not responding linearly to the input, and this causes serious distortions that can last for the entire duration of the code sequence. This may be why correlation OTDRs have not seen very much commercial success.

However, in certain DOFSs, the reflections can be much reduced because the most troublesome reflections are at a different wavelength from the signal of interest (and so can be rejected by spectral filtering) and pulse compression coding has been used in a number of sensing applications.

3.3.3.5 Frequency-diverse C-OTDR

Frequency-diverse OTDR [91] combines the receiver sensitivity enhancement of coherent detection with spreading the spectrum of the probe pulse to make, in effect, multiple measurements simultaneously. The technique consists of launching a train of pulses, which are launched at distinct times, each having a slightly different frequency, such that each pulse generates a distinct IF.

This approach is illustrated in Figure 3.18, where in the particular example,* an electro-optic phase modulator generates a set of distinct frequencies ω_1, ω_2, ω_3,… ω_n offset from that of source laser ω_0. An AOM selects probe pulses at these frequencies and a multi-frequency composite probe pulse is launched into the fibre under test. The time separation between the probe pulses of differing frequency allows the power to be set to a far higher level than if the frequencies were generated at precisely the same time. However, the backscatter traces at each of the frequencies are delayed with respect to one another and this delay must be reversed in the subsequent processing.

The combined backscatter, containing signals at each of the probe frequencies, is converted to a set of IFs that is passed to a signal-processing function that initially was implemented in hardware but in more recent work is carried out in the digital domain after A/D conversion, as illustrated in Figure 3.18.

Each IF band is detected separately; therefore, each IF provides an independent measurement of the backscatter. Provided that the IFs are sufficiently distinct, then the fading of each backscatter signal will be statistically uncorrelated from that of the others in the pulse train, and so the fading for each of the signals will occur at different locations in the fibre.

To avoid cross-talk between the backscatter signals that were launched at separate times, the spectral width of each pulse must be narrower than the separation between IFs. In principle, this is easy to arrange: in Ref. [91], the spatial resolution sought was only 1 km, resulting in a fundamental requirement on frequency separation of only 100 kHz, although in practice, a somewhat wider separation, such as 200 kHz, would be used to ensure adequate separation between IFs for the purposes of filtering the signals without distortion. This condition also ensures statistical independence between probe pulses for the purposes of eliminating fading. However, in the first work that described the technique [91], the source itself had a linewidth of 30 MHz, so the IF bands were separated by 250 MHz, with a total of four frequency-diverse pulses used simultaneously at 500, 750, 1000 and 1250 MHz from the OLO. The authors of Ref. [91]

* The original work of Sumida [91] modulated the frequency of the source laser for a short time during the formation of the probe pulses. The arrangement shown is broadly based on Ref. [92].

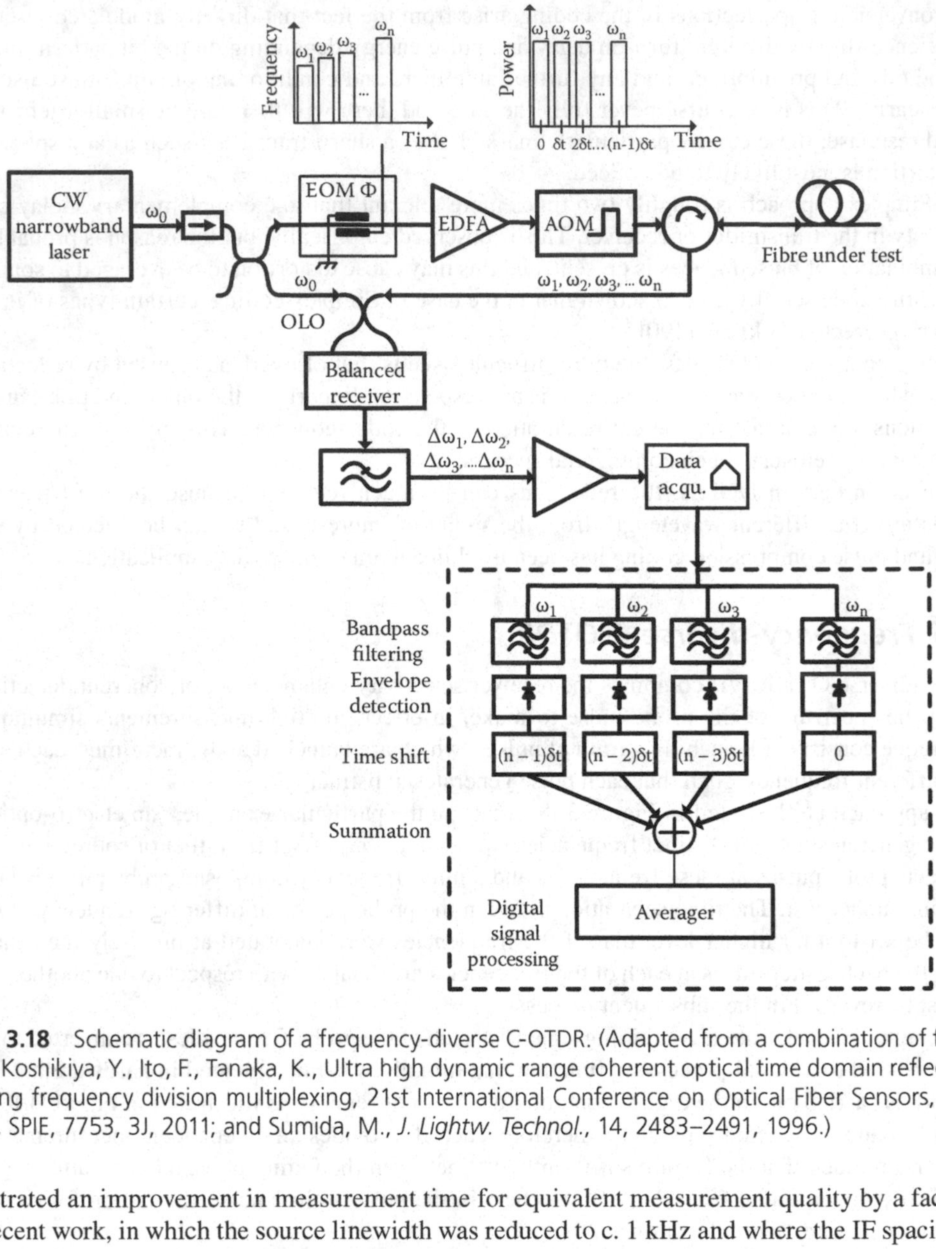

FIGURE 3.18 Schematic diagram of a frequency-diverse C-OTDR. (Adapted from a combination of figures in Iida, H., Koshikiya, Y., Ito, F., Tanaka, K., Ultra high dynamic range coherent optical time domain reflectometry employing frequency division multiplexing, 21st International Conference on Optical Fiber Sensors, Ottawa, Canada, SPIE, 7753, 3J, 2011; and Sumida, M., *J. Lightw. Technol.*, 14, 2483–2491, 1996.)

demonstrated an improvement in measurement time for equivalent measurement quality by a factor of 4. More recent work, in which the source linewidth was reduced to c. 1 kHz and where the IF spacing could therefore be reduced to the fundamental values dictated by pulse duration, demonstrated the use of 10 frequency-diverse probe pulses separated by 200 kHz, still for a target spatial resolution of 1 km. A very impressive 100 dB dynamic range was reported [93].

In most reports, successive pulses of slightly different carrier frequencies are launched sequentially; as a result, non-linear effects cause only limited interaction between the pulses because they are not present in a given section of fibre at any one time. This allows the power in each probe pulse to be increased to their individual maximum level and so to maximise the probe energy launched, and this was the case in the examples cited previously.

Experimentally, it may be simpler to launch a frequency-diverse probe consisting of a single pulse containing simultaneously many frequencies [94]. However, in this case, a number of non-linear effects disrupt the measurement, so the power must be reduced and there is no net gain in system performance in spite of the increased complexity. The non-linear effects that affect the system performance include SPM and cross-phase modulation that are caused by the Kerr effect, i.e. a change in refractive index caused by the presence of optical energy in the fibre; at the edges of a pulse, where the optical intensity is varying, the frequency of the probe is slightly chirped (i.e. varies in time), and this broadens the spectrum of the probe light.

The next non-linear effect to consider is SBS, which converts probe energy to a frequency-shifted backward propagating signal. The effect of SBS depends on probe energy, so for long pulses, it can be the limiting factor. If the pulses are separated in frequency by the Brillouin linewidth (a few tens of MHz) or more, then the SBS will affect each pulse individually and the energy of each can be brought up to the maximum energy for SBS. However, in the case of long-range systems, where pulse durations are long, the minimum spacing required for frequency diversity is quite small (~200 kHz for 1 km spatial resolution); as a result, the SBS could operate over a large part of the pulse train as if it were a single pulse. In practice, the systems that have been reported to date [92,95] were intended for monitoring very-long-distance communication links where the power required to reach the SBS threshold is not available.

In frequency-diverse C-OTDR, it is important to ensure that the spectra of the backscatter signals from all of the IFs do not overlap. This can be achieved by shaping the pulses to reduce the sidelobes in the frequency domain (also known as *apodising* the pulses). However, especially in ultra-long-range systems, the variation of probe power during the composite pulse brings back the SPM. The SPM in the propagation of the optical pulses can be alleviated by launching simultaneously *guard* light at a different, but close frequency [96]. The guard light provides a complementary intensity to the probe pulse of interest – its sole function is to ensure a uniform intensity before, during and after the probe pulse; the guard contribution to the backscatter signal is discarded by a filter. The guard light technique was demonstrated, together with very effective apodisation, in recent reports on frequency-diverse C-OTDR [95,97].

The frequency-diverse C-OTDR approach is particularly appropriate when the spatial resolution requirement is modest (as in long-distance submarine systems, for example). However, it has been extended, in the context of DOFS to pulse durations of a few tens of ns, equivalent to a spatial resolution of a few m in an OTDR.

3.3.4 Amplified OTDR Systems

Optical amplifiers are now commonly used in telecommunications systems, and not surprisingly, they have naturally found their way into OTDR and DOFS designs. However, attempts to use optical amplification were made well before the common availability of the components required for this task. In the earliest report [98] of this approach, the authors used a pulsed semiconductor laser as a probe source and, after the pulse was emitted, switched the laser to a state below the lasing threshold, but where it still provided some gain, i.e. the same laser was used as a source and as an optical preamplifier. Although elegant in concept, this approach suffered from noise in the optical amplifier, which was designed primarily as a laser. Nonetheless, an improvement by 4.5 dB one way was claimed, although much of that was attributable to eliminating the directional coupler.

Boosting the probe pulse with an optical amplifier is now relatively straightforward at wavelengths where rare-earth-doped fibre gain is available (1530–1580 nm [the C-band of optical communications and the short-end of the L-band], 1040–1080 nm and, to a lesser extent, at other wavelengths, such as 1300 nm). This was demonstrated in Ref. [99], where an erbium-doped fibre amplifier was placed in series between a 1536-nm OTDR and the fibre to be tested; an extension of the dynamic range of the measurement by 10 dB one-way was reported. The use of Raman amplification in the fibre under test has also been mentioned [44,100].

Using amplification optimally to enhance the performance of an OTDR requires the separation of the source and the receiver amplifiers, so that the power of the pulse can be adjusted based on avoiding non-linear effects and the optical amplifier prior to the receiver is optimised to provide low-noise gain for the weak backscatter signals.

3.4 TRENDS IN OTDR

Progress continues in OTDR, even though this is a mature technology. The key objective of reaching the distance between repeaters for terrestrial systems was met in the 1980s. The challenge then moved to monitoring submarine systems incorporating optical amplifiers; this was overcome by a combination of allowing the OTDR to use optically amplified signals (which requires some adaptation in the instrument) and telecommunications systems designed with OTDR functionality in mind, allowing, for example, repeaters to be bypassed if required and dedicated channels for OTDR probing.

In the 1990s, field OTDRs of small size (similar in size to a laptop computer or smaller) became part of the standard tool box of optical fibre installation and repair teams. These OTDRs are simple to use, run on batteries and are capable of storing many waveforms for future analysis. In particular, it is common practice to record and archive the OTDR signature of installed communication links for later analysis and to detect evidence of system degradation. In these OTDRs, the emphasis shifted from raw performance to questions of user interface, automatic fault determination and location as well as automated analysis of losses in designated fibre sections or at splices and of back-reflections.

By 2000, OTDR was part of modern optical telecommunication system design, with built-in OTDR capability, i.e. dedicated OTDR modules that are permanently installed in terminal and intermediate equipment that constantly probe the optical layer of the communications system to detect faults and thus allow down-time to be minimised. In the case of optical communications systems including optical amplifiers, special care must be taken to probe the link with roughly constant power to avoid disrupting the gain of the intermediate amplifiers or interfering signal data-carrying wavelengths through non-linear optical effects. One solution to this problem is to apply a continuous low-level signal (that the authors refer to a 'dummy pulse'), which constrains the gain of the amplifiers and then to shift the frequency to launch somewhat stronger probe pulses. In this way, the build-up of population inversion in the amplifiers between probe pulses is limited and the gain of the optical amplifiers is mostly constant over the pulse repetition interval. The backscatter from these pulses can be distinguished in the frequency domain [101]. A separate issue arises when the optical link includes regenerators in which the optical signals are detected, error-corrected and re-transmitted. Arrangements need to be made for OTDR signals to bypass these devices [92].

Specialised OTDRs with short range, very high spatial resolution, are available for applications such as building networks or aboard aircraft or ships. In crowded spaces, finding the location of a break or severe bend loss means determining on which side of a bulkhead or partition the fault lays. This implies a very fine spatial resolution, far better than is required for long-rang optical telecommunication fault-finding.

Optical fibre systems serving buildings and communities are often branched, and this has prompted research into OTDRs able to determine the performance of each branch separately.

3.5 APPLYING FIBRE REFLECTOMETRY TO DISTRIBUTED SENSING

Returning to Equation 3.5, we can see that the backscatter signal depends directly on the local scatter loss and the local capture fraction and that its derivative is determined by the local attenuation. This provides much scope for applying the OTDR technique to distributed sensing. Thus, according to Equation 3.5, a distributed sensor can be realised by modulating any term that affects $P_{BS}(t)$, i.e. the loss, the capture fraction or the scattering loss. In addition, in the measurement of $P_{BS}(t)$ (e.g. Figure 3.1), information on the phase and the state of polarisation of the backscattered light has been discarded but can be used for sensing.

Regarding the total attenuation, it is relatively simple to induce excess loss in an optical fibre, for example, by applying controlled bends [102]. Many physical inputs can be converted to bending. Side pressure and, thus, mechanical movement are obvious approaches, but swellable materials have been used for temperature detection and chemical sensing [103].

Modulation of the numerical aperture of multimode fibres can be used for sensing, provided that there is an angular filter prior to the receiver.

The modulation of the scattering loss has a central role in distributed sensing. By selecting different spectral components of the backscatter signal, temperature, strain and dynamic strain can be sensed. Many other measurands (force, presence of certain chemicals, electric and magnetic fields, flow) can be converted to these quantities by design of the fibre coating or cable or inferred from the primary measurements. How these spectral components are detected and used for sensing forms the bulk of the remainder of this book.

One of the issues that this versatile technology is constantly faced with is the fact that distributed sensors, being sensitive to many parameters, can lack specificity. Cross-sensitivity, whilst often overlooked in scientific papers, is a serious issue for practical application of the technology. This can force the design of cable types that cancel or enhance particular measurands. Alternatively, deploying several different types of sensors with distinct and separable response to relevant measurands can be the only way to address the problem. The separation of temperature and strain is discussed in further detail in Chapter 8.

In Part 2 of this book, it is shown how the reflectometry techniques described in this chapter are used to measure the spatial distribution of Raman (Chapter 4), Brillouin (Chapter 5) and Rayleigh (Chapter 6) backscattered light. These measurements, implemented in a wide variety of optical and signal processing arrangements, have allowed precise and well-resolved profiles of temperature, strain and vibration to be acquired and put to many practical purposes that are summarised in Part 3.

REFERENCES

1. Barnoski, M. K., and S. M. Jensen. 1976. Fiber waveguides: A novel technique for investigating attenuation characteristics. *Appl. Opt.* 15: 9: 2112–2115.
2. Personick, S. D. 1977. Photon probe – An optical time-domain reflectometer. *Bell Syst. Tech. J.* 56: 3: 355.
3. Kapron, F. P., R. D. Maurer, and M. P. Teter. 1972. Theory of backscattering effects in waveguides. *Appl. Opt.* 11: 6: 1352–1356.
4. Conduit, A. J., A. H. Hartog, M. R. Hadley et al. 1981. High-resolution measurement of diameter variations in optical fibres by the backscatter method. *Electron. Lett.* 17: 20: 742–744.
5. Eriksrud, M., and A. Mickelson. 1982. Application of the backscattering technique to the determination of parameter fluctuations in multimode optical fibers. *IEEE J. Quant. Electron.* 18: 10: 1478–1483.
6. Murdoch, S. G., and D. A. Svendsen. 2006. Distributed measurement of the chromatic dispersion of an optical fiber using a wavelength-tunable OTDR. *J. Lightw. Technol.* 24: 4: 1681–1688.
7. Conduit, A. J., A. H. Hartog, and D. N. Payne. 1980. Spectral- and length-dependent losses in optical fibres investigated by a two-channel backscatter technique. *Electron. Lett.* 16: 3: 77–78.
8. Hartog, A. H., and M. P. Gold. 1984. On the theory of backscattering in single-mode optical fibers. *J. Lightw. Technol.* 2: 2: 76–82.
9. Bisyarin, M. A., O. I. Kotov, L. B. Liokumovich, and N. A. Ushakov. 2013. Relationship between temporal and spatial reflectogram scales in fiber optical time domain reflectometry systems. *OM&NN(IO)* 22: 2: 104–111.
10. Hartog, A. H., A. J. Conduit, and D. N. Payne. 1979. Variation of pulse delay with stress and temperature in jacketed and unjacketed optical fibres. *Opt. Quant. Electron.* 11: 265–273.
11. Di Vita, P., and U. Rossi. 1979. Backscattering measurements in optical fibres: Separation of power decay from imperfection contribution. *Electron. Lett.* 15: 15: 467–469.
12. Matthijsse, P., and C. M. de Blok. 1979. Field measurement of splice loss applying the backscattering method. *Electron. Lett.* 15: 24: 795–797.

13. Poole, S., D. Payne, R. Mears, M. Fermann, and R. Laming. 1986. Fabrication and characterization of low-loss optical fibers containing rare-earth ions. *J. Lightw. Technol.* 4: 7: 870–876.
14. Gold, M. P., A. H. Hartog, and D. N. Payne. 1984. A new approach to splice-loss monitoring using long-range OTDR. *Electron. Lett.* 20: 8: 338–340.
15. Neumann, E.-G. 1980. Theory of the backscattering method for testing optical fibre cables. *Arch. Elektr. Übertragungstechnik* 34: 157–160.
16. Mickelson, A. R., and M. Eriksrud. 1982. Theory of the backscattering process in multimode optical fibers. *Appl. Opt.* 21: 11: 1898–1909.
17. Conduit, A. J., D. N. Payne, A. H. Hartog, and M. P. Gold. 1981. Optical fibre diameter variations and their effect on backscatter loss measurements. *Electron. Lett.* 17: 8: 307–308.
18. Kotov, O. I., L. B. Liokumovich, A. H. Hartog, A. V. Medvedev, and I. O. Kotov. 2008. Transformation of modal power distribution in multimode fiber with abrupt inhomogeneities. *J. Opt. Soc. Am. B* 25: 12: 1998–2009.
19. Bisyarin, M. A., O. I. Kotov, A. H. Hartog, L. B. Liokumovich, and N. A. Ushakov. 2016. Rayleigh backscattering from the fundamental mode in multimode optical fibers. *Appl. Opt.* 55: 19: 5041–5051.
20. Healey, P. 1985. Statistics of Rayleigh backscatter from a single-mode optical fibre. *Electron. Lett.* 21: 6: 226.
21. Healey, P. 1987. Statistics of Rayleigh backscatter from a single-mode fiber. *IEEE Trans. Commun.* 35: 2: 210–214.
22. Mermelstein, M. D., R. J. Posey, G. A. Johnson, and S. T. Vohra. 2001. Rayleigh scattering optical frequency correlation in a single-mode optical fiber. *Opt. Lett.* 26: 2: 58–60.
23. Jeffery, R. D., and J. L. Hullett. 1980. *N*-point processing of optical fibre backscatter signals. *Electron. Lett.* 16: 21: 822–823.
24. Nakazawa, M., T. Tanifuji, M. Tokuda, and N. Uchida. 1981. Photon probe fault locator for single-mode optical fiber using an acoustooptical light deflector. *IEEE J. Quant. Electron.* 17: 7: 1264–1269.
25. Gold, M. P., and A. H. Hartog. 1984. Improved dynamic-range single-mode OTDR at 1.3 μm. *Electron. Lett.* 20: 7: 285–287.
26. Melchior, H., M. B. Fisher, and F. R. Arams. 1970. Photodetectors for optical communication systems. *Proc. IEEE* 58: 10: 1466–1486.
27. Alexander, S. B. 1997. *Optical communication receiver design*. Institute of Electrical Engineers, Stevenage, Herts, UK.
28. McIntyre, R. J. 1966. Multiplication noise in uniform avalanche diodes. *IEEE Trans. Electron. Devices* ED-13: 1: 164–168.
29. Williams, G. M., M. A. Compton, and A. S. Huntington. 2009. High-speed photon counting with linear-mode APD receivers, Advanced Photon Counting Techniques III, SPIE, 7320: 732012.
30. Marshall, A. R. J., C. H. Tan, M. J. Steer, and J. P. R. David. 2009. Extremely low excess noise in InAs electron avalanche photodiodes. *IEEE Photon. Technol. Lett.* 21: 13: 866–868.
31. Campbell, J. C., S. Demiguel, F. Ma et al. 2004. Recent advances in avalanche photodiodes. *IEEE J. Sel. Top. Quant. Electron.* 10: 4: 777–787.
32. Kang, Y., H.-D. Liu, M. Morse et al. 2008. Monolithic germanium/silicon avalanche photodiodes with 340 GHz gain-bandwidth product. *Nat. Photon.* 247: 1–5.
33. Kang, Y., P. Mages, A. R. Clawson et al. 2002. Fused InGaAs–Si avalanche photodiodes with low-noise performances. *IEEE Photon. Technol. Lett.* 14: 11: 1593–1595.
34. Miller, J. M. 1920. Dependence of the input impedance of a three-electrode vacuum tube upon the load in the plate circuit. *Sci. P. Bur. Std.* 15: 351: 367–385.
35. Hullett, J. L., and T. Muoi. 1976. A feedback receive amplifier for optical transmission systems. *IEEE Trans. Commun.* 24: 10: 1180–1185.
36. Gold, M. 1985. Design of a long-range single-mode OTDR. *J. Lightw. Technol.* 3: 1: 39–46.
37. Ogawa, K. 1981. Noise caused by GaAs MESFETs in optical receivers. *Bell Syst. Tech. J.* 60: 6: 923–928.
38. van der Ziel, A. 1962. Thermal noise in field-effect transistors. *Proc. IRE* 50: 8: 1808–1812.
39. Schneider, M. 1991. Reduction of spectral noise density in p-i-n-HEMT lightwave receivers. *J. Lightw. Technol.* 9: 7: 887–892.
40. Kasper, B. L., A. R. McCormick, C. A. Burrus, Jr., and J. R. Talman. 1988. An optical-feedback transimpedance receiver for high sensitivity and wide dynamic range at low bit rates. *J. Lightw. Technol.* 6: 2: 329–338.
41. Nakazawa, M., and M. Tokuda. 1983. Measurement of the fiber loss spectrum using fiber Raman optical-time-domain reflectometry. *Appl. Opt.* 22: 12: 1910–1914.
42. Cisternino, F., B. Costa, M. Rao, and B. Sordo. 1986. Tunable backscattering for single-mode optical fibers in 1.1–1.6-μm spectral region. *J. Lightw. Technol.* 4: 7: 884–888.
43. Noguchi, K. 1984. A 100-km-long single-mode optical-fiber fault location. *J. Lightw. Technol.* 2: 1: 1–6.

44. Spirit, D. M., and L. C. Blank. 1989. Raman-assisted long-distance optical time domain reflectometry. *Electron. Lett.* 25: 25: 1687–1689.
45. Alahbabi, M. N., Y.-T. Cho, and T. P. Newson. 2005. 150-km-range distributed temperature sensor based on coherent detection of spontaneous Brillouin backscatter and in-line Raman amplification. *J. Opt. Soc. Am. B* 22: 6: 1321–1324.
46. Kee, H. H., G. P. Lees, and T. P. Newson. 1998. Extended-range optical time domain reflectometry system at 1.65 μm based on delayed Raman amplification. *Opt. Lett.* 23: 5: 349–351.
47. IEC. 2004. Safety of laser products – Part 2: Safety of optical fibre communication systems (OFCS). IEC 60825-2.
48. McGeehin, P. Optical techniques in industrial measurement: Safety in hazardous environments. 1994. European Commission Directorate-General XIII. EUR 16011EN.
49. Healey, P., and D. J. Malyon. 1982. OTDR in single mode fibre at 1.55 μm using heterodyne detection. *Electron. Lett.* 28: 30: 862–863.
50. King, J., D. Smith, K. Richards et al. 1987. Development of a coherent OTDR instrument. *J. Lightw. Technol.* 5: 4: 616–624.
51. Personick, S. D. 1971. An image band interpretation of optical heterodyne noise. *Bell Syst. Tech. J.* 50: 1: 213–216.
52. Lang, R. 1982. Injection locking properties of a semiconductor laser. *IEEE J. Quant. Electron.* 18: 6: 976–983.
53. Izumita, H., S. I. Furukawa, Y. Koyamada, and I. Sankawa. 1992. Fading noise reduction in coherent OTDR. *IEEE Photon. Technol. Lett.* 4: 2: 201–203.
54. Izumita, H., Y. Koyamada, S. Furukawa, and I. Sankawa. 1997. Stochastic amplitude fluctuation in coherent OTDR and a new technique for its reduction by stimulating synchronous optical frequency hopping. *J. Lightw. Technol.* 15: 2: 267–278.
55. Mark, J. 1987. Coherent measuring technique and Rayleigh backscatter from single-mode fibers. PhD Technical University of Denmark.
56. Izumita, H. 2008. Highly developed coherent detection OTDR technology and its applications to optical fiber networks monitoring. PhD Waseda University, Japan.
57. Healey, P., R. C. Booth, B. E. Daymond-John, and B. K. Nayar. 1984. OTDR in single-mode fibre at 1.55 μm using homodyne detection. *Electron. Lett.* 20: 9: 360–361.
58. Cova, S. D., and M. Ghioni. 2011. Single-photon counting detectors. *IEEE Photon. J.* 3: 2: 274–277.
59. Cova, S., A. Lacaita, M. Ghioni, G. Ripamonti, and T. A. Louis. 1989. 20-ps timing resolution with single-photon avalanche diodes. *Rev. Sci. Instrum.* 60: 6: 1104–1110.
60. Legré, M., R. Thew, H. Zbinden, and N. Gisin. 2007. High resolution optical time domain reflectometer based on 1.55 μm up-conversion photon-counting module. *Opt. Expr.* 15: 13: 8237–8242.
61. Louis, T. A., G. Ripamonti, and A. Lacaita. 1990. Photoluminescence lifetime microscope spectrometer based on time-correlated single-photon counting with an avalanche diode detector. *Rev. Sci. Instrum.* 61: 1: 11–22.
62. Wahl, M., I. Gregor, M. Patting, and J. Enderlein. 2003. Fast calculation of fluorescence correlation data with asynchronous time-correlated single-photon counting. *Opt. Expr.* 11: 26: 3583–3591.
63. Pellegrini, S., G. S. Buller, J. M. Smith, A. M. Wallace, and S. Cova. 2000. Laser-based distance measurement using picosecond resolution time-correlated single-photon counting. *Meas. Sci. Technol.* 11: 6: 712.
64. Itzler, M. A., M. Entwistle, M. Owens et al. 2010. Geiger-mode avalanche photodiode focal plane arrays for three-dimensional imaging LADAR, SPIE Optical Engineering + Applications SPIE, 7808: 0C.
65. Gisin, N., G. Ribordy, W. Tittel, and H. Zbinden. 2002. Quantum cryptography. *Rev. Mod. Phys.* 74: 145–195.
66. Healey, P., and P. Hensel. 1980. Optical time domain reflectometry by photon counting. *Electron. Lett.* 16: 16: 631–633.
67. Healey, P. 1981. Multichannel photon-counting backscatter measurements on monomode fibre. *Electron. Lett.* 17: 20: 751–752.
68. Luciol Instruments. 2015. v-OTDR. Retrieved 20 September 2015, http://www.luciol.com/v-OTDR.html.
69. Eickhoff, W., and R. Ulrich. 1981. Optical frequency domain reflectometry in single-mode fiber. *Appl. Phys. Lett.* 39: 9: 693–695.
70. Tsuji, K., K. Shimizu, T. Horiguchi, and Y. Koyamada. 1997. Coherent optical frequency domain reflectometry using phase-decorrelated reflected and reference lightwaves. *J. Lightw. Technol.* 15: 7: 1102–1109.
71. Uttam, D., and B. Culshaw. 1985. Precision time domain reflectometry in optical fiber systems using a frequency modulated continuous wave ranging technique. *J. Lightw. Technol.* 3: 5: 971–977.
72. Venkatesh, S., and W. V. Sorin. 1993. Phase noise considerations in coherent optical FMCW reflectometry. *J. Lightw. Technol.* 11: 10: 1694–1700.
73. Glombitza, U., and E. Brinkmeyer. 1993. Coherent frequency-domain reflectometry for characterization of single-mode integrated-optical waveguides. *J. Lightw. Technol.* 11: 8: 1377–1384.

74. Geng, J., C. Spiegelberg, and J. Shibin. 2005. Narrow linewidth fiber laser for 100-km optical frequency domain reflectometry. *IEEE Photon. Technol. Lett.* 17: 9: 1827–1829.
75. Ahn, T.-J., J. Y. Lee, and D. Y. Kim. 2005. Suppression of nonlinear frequency sweep in an optical frequency-domain reflectometer by use of Hilbert transformation. *Appl. Opt.* 44: 35: 7630–7634.
76. Ito, F., X. Fan, and Y. Koshikiya. 2012. Long-range coherent OFDR with light source phase noise compensation. *J. Lightw. Technol.* 30: 8: 1015–1024.
77. Yun, S., G. Tearney, B. Bouma, B. Park, and J. de Boer. 2003. High-speed spectral-domain optical coherence tomography at 1.3 μm wavelength. *Opt. Expr.* 11: 26: 3598–3604.
78. Leitgeb, R., W. Drexler, A. Unterhuber et al. 2004. Ultrahigh resolution Fourier domain optical coherence tomography. *Opt. Expr.* 12: 10: 2156–2165.
79. Bachmann, A., R. Leitgeb, and T. Lasser. 2006. Heterodyne Fourier domain optical coherence tomography for full range probing with high axial resolution. *Opt. Expr.* 14: 4: 1487–1496.
80. MacDonald, R. I. 1981. Frequency domain optical reflectometer. *Appl. Opt.* 20: 10: 1840–1844.
81. Venkatesh, S., and D. W. Dolfi. 1990. Incoherent frequency modulated CW optical reflectometry with centimeter resolution. *Appl. Opt.* 29: 9: 1323–1326.
82. Barton, D. K. 1978. *Radars: Volume 3: Pulse compression.* Artech House, Dedham, Mass, USA.
83. MacWilliams, F. J., and N. J. Sloane. 1976. Pseudo-random sequences and arrays. *Proc. IEEE* 64: 12: 1715–1729.
84. Okada, K., K. Hashimoto, T. Shibata, and Y. Nagaki. 1980. Optical cable fault location using correlation technique. *Electron. Lett.* 16: 16: 629–630.
85. Healey, P. 1981. Pulse compression coding in optical time domain reflectometry, 7th European Conference on Optical Communication Copenhagen, Denmark, P. Peregrinus Ltd: 5.2-1.
86. Levanon, N. 2006. Noncoherent pulse compression. *IEEE Trans. Aerospace Electron. Syst.* 42: 2: 756–765.
87. Nazarathy, M., S. A. Newton, R. P. Giffard et al. 1989. Real-time long range complementary correlation optical time domain reflectometer. *J. Lightw. Technol.* 7: 1: 24–38.
88. Golay, M. T. E. 1961. Complementary series. *IRE Trans. Inform. Th.* IT-7: 82–87.
89. Lee, D., H. Yoon, N. Y. Kim, H. Lee, and N. Park. 2004. Analysis and experimental demonstration of simplex coding technique for SNR enhancement of OTDR, IEEE LTIMC 2004 – Lightwave Technologies in Instrumentation & Measurement Conference, Palisades, NY, USA, IEEE: 118–122.
90. Jones, M. D. 1993. Using simplex codes to improve OTDR sensitivity. *IEEE Photon. Technol. Lett.* 5: 7: 822–824.
91. Sumida, M. 1996. Optical time domain reflectometry using an M-ary FSK probe and coherent detection. *J. Lightw. Technol.* 14: 11: 2483–2491.
92. Iida, H., Y. Koshikiya, F. Ito, and K. Tanaka. 2012. High-sensitivity coherent optical time domain reflectometry employing frequency-division multiplexing. *J. Lightw. Technol.* 30: 8: 1121–1126.
93. Iida, H., Y. Koshikiya, F. Ito, and K. Tanaka. 2011. Ultra high dynamic range coherent optical time domain reflectometry employing frequency division multiplexing, 21st International Conference on Optical Fiber Sensors, Ottawa, SPIE, 7753: 3J.
94. Lu, L., Y. Song, F. Zhu, and X. Zhang. 2012. Performance limit of a multi-frequency probe based coherent optical time domain reflectometry caused by nonlinear effects. *Chin. Opt. Lett.* 10: 4: 040604.
95. Iida, H., K. Toge, and F. Ito. 2013. 200-subchannel ultra-high-density frequency division multiplexed coherent OTDR with nonlinear effect suppression. In Optical Fiber Communication Conference and Exposition and the National Fiber Optic Engineers Conference (OFC/NFOEC), 2013. 1–3.
96. Hartog, A., and P. Wait. 2007. Optical pulse propagation. WO2007141464.
97. Iida, H., K. Toge, and F. Ito. 2014. Pulse waveform manipulation in FDM-OTDR for suppressing inter-channel crosstalk. *J. Lightw. Technol.* 32: 14: 2569–2576.
98. Suzuki, K., T. Horiguchi, and S. Seikai. 1984. Optical time-domain reflectometer with a semiconductor laser amplifier. *Electron. Lett.* 20: 18: 714–716.
99. Blank, L. C., and D. M. Spirit. 1989. OTDR performance enhancement through erbium fibre amplification. *Electron. Lett.* 25: 25: 1693–1694.
100. Spirit, D. M., and L. C. Blank. 1994. OTDR using distributed optical amplification in optical waveguide under test. US5298965A.
101. Sumida, M., S. I. Furukawa, K. Tanaka, and M. Aiki. 1996. High-accurate fault location technology using FSK-ASK probe backscattering reflectometry in optical amplifier submarine transmission systems. *J. Lightw. Technol.* 14: 10: 2108–2116.
102. Fields, J. N., C. K. Asawa, O. G. Ramer, and M. K. Barnoski. 1980. Fiber optic pressure sensor. *J. Acoust. Soc. Am.* 67: 3: 816–818.
103. Michie, W. C., B. Culshaw, M. Konstantaki et al. 1995. Distributed pH and water detection using fiber-optic sensors and hydrogels. *J. Lightw. Technol.* 13: 7: 1415–1420.

PART TWO

Distributed Sensing Technology

Raman-Based Distributed Temperature Sensors (DTSs)

4

In this chapter, we discuss the technology and performance of Raman distributed temperature sensor (DTS) systems. The system description is initially based on Raman optical time-domain reflectometry (OTDR) using pulsed lasers and conventional analogue receivers. However, alternative designs based on optical frequency-domain reflectometry (OFDR), pulse compression and photon counting are also described.

4.1 PRINCIPLES OF RAMAN OTDR

4.1.1 Spontaneous Raman Scattering

Raman scattering is a form of inelastic light scattering, i.e. a scattering process where the energy of the incident photon is *not* conserved. Of course, this means that the scattered photon has lost or gained energy in the interaction with matter and its wavelength is therefore shifted relative to that of the incident photons.

In the case of Raman scattering, the light interacts with molecular vibrations, for example, stretching modes between atoms [1] within the molecule. In suitable molecules (so-called Raman-active ones), the molecular vibration causes a fluctuation of the polarisability at the vibrational frequency which is then periodically different from that of other parts of the medium, resulting in scattering. In this process, energy is transferred between the incident photon and the vibration states of the molecule. In the most frequent case, the incident photon excites a molecular vibration and sheds energy in the process. As a result, the scattered photon carries a reduced amount of energy and its frequency is therefore also reduced. Its wavelength is thus increased in the process, which is referred to as *Stokes* Raman scattering. In the opposite case, where vibrational energy is transferred to the scattered photon, the process is known as *anti-Stokes* Raman scattering and the resulting scattered light appears at a shorter wavelength.

Of course, molecules other than the simplest diatomic ones have multiple vibration modes, and therefore, several new lines can appear in the spectrum of the scattered light, associated with the various possible extensional, bending and rotational modes of the molecule. In the case of glasses, which may be viewed as micro-crystalline on the scale of a few molecules, the environment of each molecular bond is different, and this leads to a slight change in the vibrational modes that take part in the Raman-scattering process. It follows that instead of well-defined lines, the Raman spectrum in glasses is more akin to one or more bands associated with major types of vibration in the molecule, because the fine spectral detail that is visible in Raman scattering in gases is smeared in the spectrum of the scattered light in glasses.

In our previous overview of scattering (Chapter 2, Section 2.1.2.2), we commented that scattering of interest for DOFS arises when the medium is not homogeneous on a scale smaller than a wavelength. In the case of Raman scattering, this inhomogeneity arises because in some molecular bonds, the polarisability of the molecule varies during the vibration – so it fluctuates on a time scale consistent with the period of the vibration, typically a few THz to tens of THz. The process occurs on a dimensional scale much smaller (<1 nm) than typical wavelengths (800–1600 nm) used in optical fibres.

Raman scattering is quite different from fluorescence: although both processes convert some of the incident light into a new wavelength band, fluorescence is caused by the absorption of the photon, the photon energy being used to move an electron from a low-energy state (usually the ground state) to a meta-stable, higher-energy state within the electronic structure of an atom or a molecule. Eventually, the electron decays to a lower energy state, releasing some of the energy in the form of fluorescence. Thus, fluorescence relies upon the presence of an absorption band (corresponding to the energy required for the electron to transition from low to high energy state), and this absorption band is therefore fixed in the spectrum. By contrast, in Raman scattering, the emission band tracks the incident light by a fixed frequency shift that is independent of the wavelength of the incident light. Also, the time scales are quite different: fluorescence decay times can be quite slow in glasses (in the μs to ms range); in contrast, the Raman response is quasi-instantaneous (in the femtosecond regime). Confusion sometimes arises between the two effects because Raman scattering is sometimes described by reference to a virtual state to which the molecule transitions from a vibrational state before immediately releasing the energy by moving back to another vibrational state. However, this virtual state does not relate in general* to a real electronic transition of an atom, as would be the case for fluorescence. Raman and fluorescence are therefore quite distinct effects.

The fact that the wavelength of Raman-scattered light is different from that of the incident light implies some form of non-linearity. However, in most of the distributed sensors based on Raman scattering, the Raman process is linear in the sense that the scattered power is proportional to the probe power. Non-linear (stimulated) Raman scattering also exists, but it is not commonly used as a sensing mechanism [2].† In fact, the onset of stimulated Raman scattering (SRS) often limits the amount of power that can be launched as a probe, and this effect ultimately limits the performance of Raman-based distributed temperature sensors.

Figure 4.1 (after Ref. [3]) shows the Raman intensity in various types of glass as a function of the frequency shift. In fact, a mixture of these glass components forms the majority of the materials used in optical fibres. It illustrates many aspects of Raman scattering relevant to distributed sensing. The horizontal axis is expressed in wavenumbers (cm^{-1}), a unit used in spectroscopy that is proportional to frequency and that, for historical reasons, uses the centimeter-gram-second (CGS) unit system. A frequency shift ω expressed in radian/s is equivalent to wavenumber ν through $\nu = \omega/(2\pi c)$, where c is the speed of light in vacuum expressed in cm s^{-1}.

The intensities in Figure 4.1 are normalised to the maximum for the SiO_2 band, which occurs at around 440 cm^{-1}, but it is worth noting that this band is very broad; it corresponds to longitudinal optical and transverse optical vibrations of the glass molecules. It is also worth noting that other glasses that are frequently used as additives to silica in optical fibres have substantially larger Raman intensities than silica does. Since, at least for germano-silicate compositions, the Raman cross-section is linearly dependent on the relative amount of each component [4], the data in the figure are very useful for estimating the Raman scattering for a given composition. It is also interesting to note that there is a large overlap between the GeO_2 and SiO_2 bands around 400–500 cm^{-1}.

* The exception being *resonance Raman scattering*, a technique used in chemistry to enhance the Raman signal from very low concentration of the target species [1].

† Although the reference cited was based on stimulated Raman scattering, it was used to estimate the relative polarisation between two counter-propagating signals in the fibre; it was therefore more a variant of polarisation OTDR than an implementation of Raman OTDR.

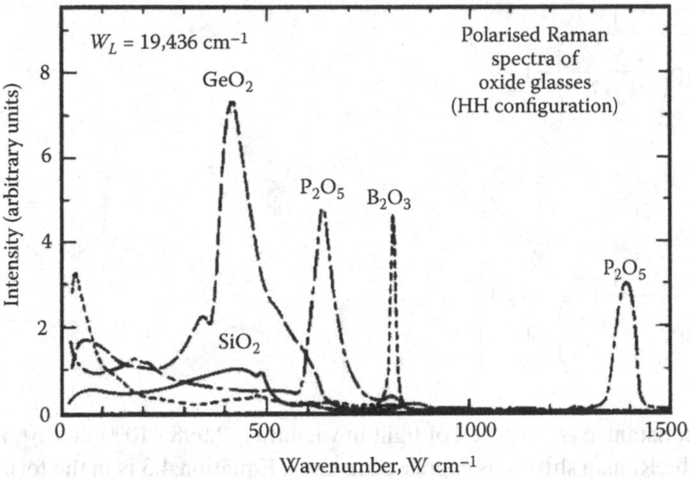

FIGURE 4.1 Raman spectra (normalised to the peak intensity for silica) for SiO_2, GeO_2, B_2O_3 and P_2O_5 glasses. (Reprinted with permission from Galeener, F.L., Mikkelsen, J.C., Geils, R.H., Mosby, W.J., *Appl. Phys. Lett.* 32, 34–36, 1978. Copyright 1978, American Institute of Physics.)

For a system using this Raman shift, the spontaneous scattering level scales as

$$I_{total} = \left(1 - C_{GeO_2}\right) I_{SiO_2} + C_{GeO_2} \cdot I_{GeO_2}, \tag{4.1}$$

where C_{GeO2} is the molar percentage of germania in the glass and I_{SiO2} and I_{GeO2} are the cross-sections for each glass component. Naturally, this expression must be weighted over the bandpass spectrum of whatever filter is used to select the Raman scattered light.

The Raman process results in a frequency shift that is independent of the *incident wavelength* and, therefore, the resulting wavelengths of the anti-Stokes (λ_{as}) and Stokes (λ_s) Raman scattered light dependent on the probe wavelength. Their relation to the incident light at wavelength λ_0 is as follows for wavelengths expressed in m and frequency shift ν_R expressed in cm^{-1}:

$$\lambda_{as} = \frac{0.01}{\dfrac{0.01}{\lambda_0} + \nu_R}, \quad \lambda_s = \frac{0.01}{\dfrac{0.01}{\lambda_0} - \nu_R}. \tag{4.2}$$

As an example, the wavelength shifts correspond to ν_R = 440 cm^{-1} work out at about 35, 50 and 100 nm for probe wavelengths of 904, 1064 and 1550 nm, respectively.

4.1.2 Temperature Sensitivity of the Raman Scattering

The availability of molecular bonds in excited states clearly influences the likelihood of a transition exchanging energy with an incident photon.

For a single line, the intensity of the anti-Stokes and Stokes scatter is given by Long [5] as a function of temperature *T*, where subscripts *as* and *s* refer to the anti-Stokes and Stokes Raman bands, respectively:

$$I_{as}(T) = \frac{K_{as}}{\lambda_{as}^4} \cdot \frac{1}{\exp\left(\frac{h \cdot \nu_R \cdot c}{k_B \cdot T}\right) - 1} \tag{4.3}$$

and

$$I_s(T) = \frac{K_s}{\lambda_s^4} \cdot \left\{ \frac{1}{\exp\left(\frac{h \cdot \nu_R \cdot c}{k_B \cdot T}\right) - 1} + 1 \right\}, \tag{4.4}$$

where h is Planck's constant, c is the speed of light in vacuum (~$2.998 \cdot 10^{10}$ cm s^{-1})*, k_B is the Boltzmann constant and where the Raman shift ν_R is expressed in cm^{-1}. Equation 4.3 is in the form of a Bose-Einstein distribution. The coefficients $K_{as(s)}$ are determined primarily by the Raman cross-section for the anti-Stokes (Stokes) bands, averaged across the wavelength band defined by the filters used to select the Raman components.

It is common in Raman OTDR DTS systems to use the ratio $R(T) = I_{as}(T)/I_s(T)$, which evaluates to

$$R(T) = \left(\frac{K_{as}}{K_s}\right)\left(\frac{\lambda_s}{\lambda_{as}}\right)^4 \exp\left(-\frac{S_{LE}}{T}\right). \tag{4.5}$$

The coefficient S_{LE} that defines the temperature sensitivity of the system is defined as

$$S_{LE} = h \cdot \nu_R \cdot c / k_B, \tag{4.6}$$

and for $\nu_R = 440$ cm^{-1}, a typical value, it evaluates to 633.1 K.

The $I_{as}(T)$, $I_s(T)$ and $R(T)$ functions (normalised to $K_{as,s}/\lambda_{as,s}^4 = 1$) are shown in Figure 4.2 for $\nu_R = 440$ cm^{-1}; we note that at room temperature, $I_{as}(T)$ is roughly one tenth of $I_s(T)$ and that at high temperature, the ratio tends to a constant value. Thus, the sensitivity of the Raman ratio is reduced at high temperature. It is also instructive to plot (Figure 4.3) the relative sensitivity curves, $\varepsilon_{as} = \frac{1}{I_{as}(T)}\frac{dI_{as}(T)}{dT}$, $\varepsilon_s = \frac{1}{I_s(T)}\frac{dI_s(T)}{dT}$ and $\varepsilon_R = \frac{1}{R(T)}\frac{dR(T)}{dT}$. Figure 4.3 illustrates clearly that the sensitivities of the anti-Stokes and Stokes Raman signals are c. 0.83% K^{-1} and 0.096% K^{-1}, respectively, at room temperature (293 K) for a typical silica-based glass. The sensitivity of their ratio at around this temperature is about 0.74%/K. At 593 K, however, the sensitivity of the Raman components are 0.27%/K (anti-Stokes) and 0.094%/K (Stokes), and the temperature sensitivity of the ratio has fallen to 0.18%/K.

It is not surprising that the sensitivity of the Raman ratio falls at high temperatures because (by inspection of Equations 4.3 and 4.4) the anti-Stokes and Stokes signals converge to the same value at very elevated temperatures. In practice, the ability of presently available silica-based optical fibres to survive at high temperature is limited to about 900 K (and even then with many restrictions as to coatings, glass composition and conditions in which the sensing fibre is held). Therefore, the variation of the anti-Stokes/Stokes ratio is not a key limitation in practical applications. It can be overcome by using the anti-Stokes signal rather than the Raman ratio and finding a different means of correcting for the many factors that the Raman ratio eliminates, which are described next.

* Exceptionally, the speed of light is expressed here in CGS units for consistency with the use of wavenumbers.

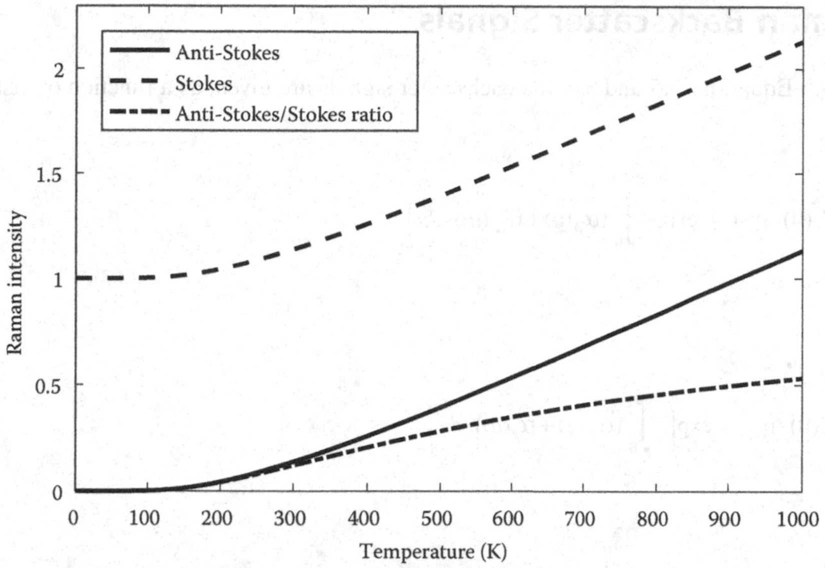

FIGURE 4.2 Normalised intensity of the Raman anti-Stokes and Stokes and their ratio in a typical multimode optical fibre.

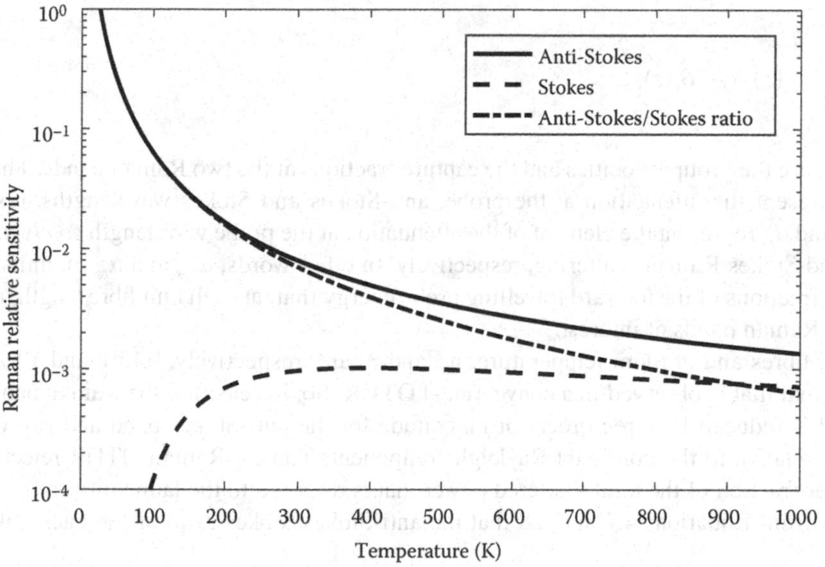

FIGURE 4.3 Relative temperature sensitivity of the Raman anti-Stokes, Stokes intensities and their ratio in a typical multimode optical fibre.

4.1.3 Raman Backscatter Signals

By analogy with Equations 3.5 and 3.6, the backscatter signals are given, as a function of distance z along the fibre, by

$$P_{as}(z) = E_p(0) \cdot \eta_{as}(z) \cdot \exp\left(-\int_0^z (\alpha_p(u) + \alpha_{as}(u)) \cdot du\right) \tag{4.7}$$

and

$$P_s(z) = E_p(0) \cdot \eta_s(z) \cdot \exp\left(-\int_0^z (\alpha_p(u) + \alpha_s(u)) \cdot du\right) \tag{4.8}$$

with

$$\eta_{as}(z) = \frac{1}{2}\alpha_{ras}(z) \cdot v_{g_{as}} \cdot B_{as}(z) \tag{4.9}$$

and

$$\eta_s(z) = \frac{1}{2}\alpha_{rs}(z) \cdot v_{g_s} \cdot B_s(z). \tag{4.10}$$

$v_{g_{as,as}}$ and $B_{as,s}$ are the group velocities and the capture fractions at the two Raman bands. The coefficients α_p, α_{as}, α_s represent the attenuation at the probe, anti-Stokes and Stokes wavelengths, respectively. In contrast, α_{ras} and α_{rs} represent the element of the attenuation at the probe wavelength *solely attributable* to anti-Stokes and Stokes Raman scattering, respectively. In other words, α_{ras} and α_{rs} (in units of dB/km or Np/m) are the fractions of the forward travelling probe energy that, at each unit fibre length, are converted to each of the Raman bands of interest.

In typical fibres and at room temperature, η_{as} and η_s are, respectively, 0.15% and 1.5% of the total backscatter signal that is observed in a conventional OTDR. So, in selecting the Raman bands, the signal to be detected is reduced by three orders of magnitude for the anti-Stokes band and two orders for the Stokes signal, relative to the dominant Rayleigh component. Thus, a Raman OTDR rejects most of the already meager fraction of the total scattered power that is returned to the launching end.

It follows from Equations 4.7 and 4.8 that the anti-Stokes/Stokes ratio of the backscatter signals is given by

$$R(T(z)) = \frac{K_{as}}{K_S} \cdot \left(\frac{\lambda_s}{\lambda_{as}}\right)^4 \cdot \exp\left(-\frac{S_{LE}}{T(z)}\right) \cdot \exp\left(-\int_0^z (\alpha_{as}(u) - \alpha_s(u)) \cdot du\right) \tag{4.11}$$

provided that $v_{g_{as}} = v_{g_s}$ and $B_{as}(z) = B_s(z)$. All the dependencies on pulse energy, capture fraction and attenuation at the probe wavelength that were present in Equations 4.7 and 4.8 have been cancelled by the ratiometric approach.

Regarding the group velocities, their variation is sufficiently small [6] to be ignored in the intensity-ratio calculation, particularly as this difference is normally cancelled in the system design by use of a reference coil, as will be seen later. As to the capture fraction terms, they are proportional to the square of the numerical aperture of the fibre, and this is dispersive, an effect known as profile dispersion [7]. However, since the profile dispersion is independent of dopant concentration [8,9] in typical fibre materials, even a variation in the numerical aperture would not invalidate the assumption; in addition, their effect is largely cancelled by use of a reference coil.

The main weakness of the $B_{as}(z) = B_s(z)$ assumption is, for multi-mode fibre, the possibility that the system is sampling a different modal power distribution for the two signals measured. If such a difference exists, then any mode selectivity in the separate anti-Stokes and Stokes channels could be converted to a change of the Raman ratio and could thus bias the result.

If we examine Equation 4.11, there are two main terms that are functions of position, namely the temperature-dependent $\exp\left(-\frac{S_{LE}}{T(z)}\right)$ term and the differential attenuation term $\exp\left(-\int_0^z (\alpha_{as}(u) - \alpha_s(u)) \cdot du\right)$ that represents the difference in the loss suffered by the backscatter signals on their return from the scattering location z to the launching end of the fibre. Thus, taking the anti-Stokes/Stokes ratio has also eliminated most of the loss terms present in Equations 4.7 and 4.8. However, this is not sufficient; an inaccurate estimate of the differential loss will seriously bias the temperature measurement of the DTS system.

Unfortunately, without further information, it is not possible to separate the effect of temperature from that of differential loss. This is a fundamental issue with Raman OTDR systems. Many systems simply assume that the loss is known, for example, from a single-point calibration of the temperature at the far-end. With the temperature known at a remote point, it is possible to find a value for $\alpha_{as} - \alpha_s$ that forces a match between the temperature reported at the remote calibration and the DTS reading at that point. The differential loss is then assumed (a) to be uniform along the fibre and (b) to be constant in time. These assumptions can be valid under benign conditions, where the fibre is reasonably straight, has no major splicing losses and where it is deployed in conditions where its attenuation is not degrading over time. However, in many practical applications, such as in high-temperature corrosive environments, this is not true. In this case, as the fibre degrades, for example, as a result of hydrogen ingress, the differential loss varies and the temperature profile becomes gradually more incorrect with increasing distance from the launching end and as the fibre ageing process continues. Some DTS suppliers have provided improved techniques to measure the additional information required. These will be discussed further on, after a description and discussion of basic Raman DTS systems.

4.2 RAMAN OTDR DESIGN

A typical arrangement of a Raman OTDR is shown in Figure 4.4.

A pulsed light source is used to provide probe pulses at wavelength λ_p. The probe pulses are launched into the sensing fibre through a directional coupler that separates the backscatter returned by the sensing fibre from the forward-travelling probe light. In Raman ratio systems, the directional coupler can be implemented as a dichroic splitter, which allows highly efficient transfer in each direction because the wavelengths are different. In the example of Figure 4.4, the dichroic coupler is designed to transmit the probe power efficiently to the sensing fibre, but in the return direction, to deflect each of the Raman signal bands onto separate detectors. It is a form of wavelength-division multiplexer, and devices of this type have been developed for telecommunications systems [10], although not necessarily for the specific

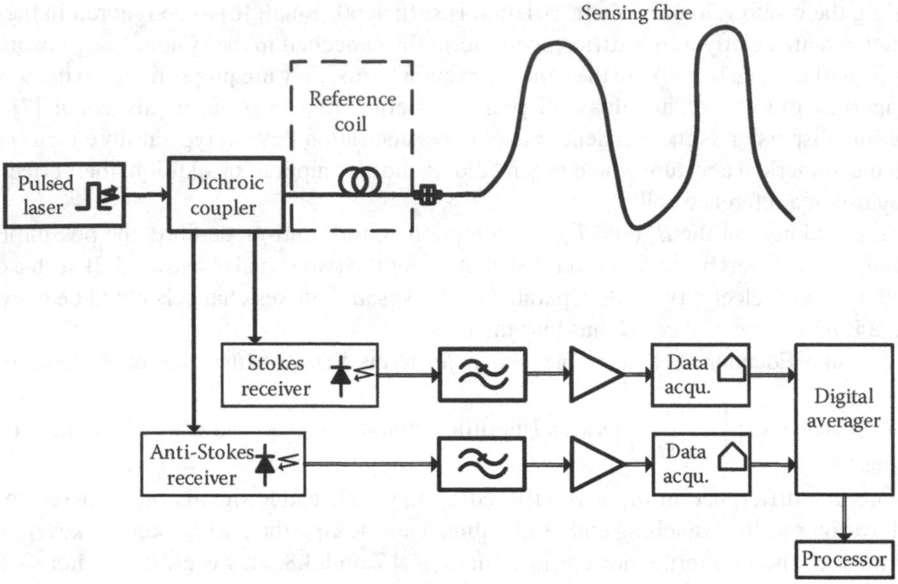

FIGURE 4.4 Schematic diagram of a Raman ratio distributed temperature sensor.

wavelengths and bandwidths required by the DTS system designers, who will therefore, in general, need to design or specify their own devices.

An optical connector is generally used to link the sensing fibre to the optics within the Raman OTDR.

In a Raman ratio measurement, the directional coupler has separate outputs for the anti-Stokes and Stokes Raman bands. These optical signals are fed to receivers, i.e. an optical detector that converts optical power to electrical current, followed by an electrical preamplifier. The resulting electrical signal is usually further amplified and filtered prior to being digitised. The signals in Raman OTDR are very low (<1 nW), so it is usual to average the backscatter from many probe pulses in order to provide an adequate signal-to-noise ratio (SNR) to the signals that are sent to the processor.

In general, the Raman outputs are generated in separate physical channels, each with optical filters, detectors and acquisition. However, alternative approaches using a single receiver channel in which the filter is switched mechanically from anti-Stokes to Stokes bands have also been developed, either on grounds of cost or for perceived reasons of improved reproducibility between the two measurements.

The role of the processor is to collect the data from the averager and possibly further condition it, to carry out the calculations needed to convert the data into the form of temperature information as well as to monitor the health of each of the elements of the system. In some systems, further processing is carried out, such as providing a statistical output for designated fibre sections (for example, the mean or maximum temperature in a zone) or providing alerts based on the measured values or the state of the sensing system. In the case of systems used for critical applications, the output (possibly in the form of a relay closure) directly actuates a separate system, such as initiating an extinguishing operation in a fire protection application.

It is common practice to add a section of fibre, the *reference coil*, between the directional coupler and the sensing fibre. The reference coil is at a known temperature (either stabilised or simply monitored), and this allows the influence of a number of potentially variable parameters of the system to be cancelled out.

In some systems, an optical fibre switch is used to multiplex a number of sensing channels to the same acquisition system, either after or before the reference coil. In the latter case, a separate reference coil is provided for each sensing fibre. This multiplexing approach allows the cost of the acquisition system to be shared between more sensing points, and it is appropriate if there is no requirement for continuous monitoring of each sensing location.

With the knowledge of the temperature T_{ref} in the reference coil, and by normalising $R(T(z))$ to its average value in the reference coil, $R(T(\mathrm{rc}))$, based on Equation 4.11, we obtain a normalised Raman ratio $\bar{R}(T(z))$ as a function of temperature at distance z,

$$\bar{R}(T(z)) = \frac{R(T(z))}{R(T(\mathrm{rc}))} = \exp\left(S_{LE} \cdot \left(\frac{1}{T_{ref}} - \frac{1}{T(z)} \right) \right) \cdot \exp\left(-\int_{z_{ref}}^{z} (\alpha_{as}(u) - \alpha_s(u)) \cdot du \right). \quad (4.12)$$

As can be seen from Equation 4.12, all the terms dependent on system design (embedded in K_{as}, K_S), such as the relative transmission of the dichroic couplers, filters, sensitivity of the detectors etc., have been cancelled. Here, z_{ref} is the end of the reference coil, i.e. the start of the sensing fibre. The normalised Raman ratio is now dependent only on the temperature where the scattering occurs and the differential loss in the return path along the sensing fibre. However, it should be noted that some of the nominally cancelled loss terms could be mode dependent. In the event of some form of mode-filtering within the optics, their values could change as a function of location and the cancellation illustrated by Equation 4.12 may then not be perfectly valid [11–13].

Starting from a measured value of the Raman ratio as a function of z, with the knowledge of T_{ref} and with an estimate of the distribution of the differential attenuation along the fibre, the temperature distribution may be calculated using Equation 4.13, where $T(z)$ and T_{ref} are expressed in K [14].

$$\frac{1}{T(z)} = \frac{1}{T_{ref}} - \frac{1}{S_{LE}} \left(\ln\left(\bar{R}(z)\right) + \int_{z_{ref}}^{z} (\alpha_{as}(u) - \alpha_s(u)) \cdot du \right) \quad (4.13)$$

We now discuss the various elements of a Raman system.

4.2.1 Wavelength, Laser and Fibre Selection

Although Raman scattering can be collected for any source wavelength, there are nonetheless constraints in the designer's selection of an operating wavelength. For example, single-mode fibres must be operated wavelengths longer than the cut-off of their second transverse mode, so unless specialty fibres are selected, the wavelength of the probe and the related anti-Stokes Raman signal must be greater than about 1300 nm, this being the shortest operating wavelength for telecommunications single-mode fibres.

Multimode fibres are not restricted to operating beyond a mode cut-off. Nonetheless, this type of fibre is designed for a specific wavelength, especially as regards to its bandwidth, owing to the *profile dispersion* effect, i.e. the fact that the refractive index profile that minimises intermodal pulse broadening is a function of wavelength [15]. This issue is particularly important because commercially available multimode fibres are generally optimised in the region 850–1300 nm and their properties are not guaranteed at the wavelength of minimum attenuation, 1550 nm.

The bandwidth of the sensing fibre can limit the spatial resolution of the system by broadening the probe pulse and also spreading the backscatter from each point over time. If this pulse broadening approaches the design spatial resolution for the entire system, it will degrade the overall system spatial resolution.

Multimode fibres, as discussed in Chapter 2 (Section 2.1.1.2), suffer from intermodal dispersion, which usually limits their bandwidth. In modern optical fibres, we can usually assume that the pulse broadening scales roughly with distance, because mode mixing is relatively low. Let us take the example of a good-quality graded-index fibre for which the bandwidth × distance product might typically* be of order 1 GHz km around a probe wavelength of 1064 nm which, for a Gaussian impulse response, corresponds to an approximate pulse broadening of 0.44 ns km^{-1}. At a distance of 10 km, therefore, the probe pulse will have spread by about 4.4 ns. The same will apply to the returning backscatter, with the *caveat* that the backscattering process may well re-distribute the power between modes, so the dispersion in the forward and backward directions combines approximately as the square root of the sum of squares. As a result, we can expect a two-way dispersion of about 6.2 ns for light scattered at the remote end of a 10 km fibre. This value will limit the attainable spatial resolution to 0.62 m or longer and will affect the bandwidth budget of other parts of the system. Conversely, a system having a resolution of 1 m at short distances will be degraded in its resolution to approximately $\sqrt{1^2+0.62^2}=1.17\,\text{m}$ at a distance of 10 km, assuming the impulse responses of the fibre and the system are at least approximately Gaussian. Fibres of lesser bandwidth could seriously affect the resolution of DTS systems at ranges beyond a few km.

First-generation DTS systems, designed in the 1980s, used a variety of fibre designs, some of which are no longer available. For example, the 85/125 fibre† was ideal for the sources available at the time. However, that fibre design was abandoned by users and manufacturers who then concentrated on the 50/125 (0.2 NA) design [16] and, to a lesser extent, on the 62.5/125 (0.27 NA) design. A loss of launching efficiency results for certain source types. This illustrates the fact that the distributed temperature sensing market is very small compared to telecommunications, and the fibres and components that are available to DTS designers are dictated by a separate business. This situation is likely to prevail until the DTS business grows sufficiently to support its own fibre designs. Some fibres designed for high-temperature DTS measurements are available commercially, but so far, the focus has been on their reliability in harsh environments, rather than the process control on the refractive index profile that is required to maintain the spatial resolution over long distances.

For single-mode fibres, the pulse broadening is due to chromatic dispersion (CD; discussed in Chapter 2, Section 2.1.3). For a common step-index fibre design, CD at 1450 nm (the anti-Stokes central wavelength for a probe pulse at 1550 nm) is about 10 ps nm^{-1} km^{-1}. So for a spectral width of the anti-Stokes filter of 15 nm and a range of 30 km, the broadening of the anti-Stokes component would be about 4.5 ns, equivalent to a resolution limitation of 0.45 m. The spectral linewidth of the probe pulse can be chosen to be substantially narrower than that of the anti-Stokes backscatter passed by the filter, so its dispersion can be neglected. It should also be noted that the CD for the Stokes and anti-Stokes bands can well differ, given that they propagate at materially different wavelengths.

All other things being equal, it is desirable to maximise the signal received for the longest design range of the system. In modern optical fibres, the attenuation is limited by Rayleigh scattering between 800 and 1500 nm except in the vicinity of the absorption peak at 1390 nm (and in lesser quality fibres also at 945 and 1240 nm) due to OH^- ions being present in the glass as impurities. Rayleigh scattering falls off with the inverse fourth power of wavelength, from around 1 dB/km at 1000 nm (for a typical multimode fibre) to under 0.2 dB/km at 1500 nm. Thus, as the wavelength is increased, the loss per unit length is reduced, and so too is the cumulative loss to the most distant point of interest. On this basis, maximising the wavelength within the region bounded by electronic and molecular absorption would appear to be the best approach. However, the scattering loss and, thus, also the intensity of the signal are also reduced. It follows that for each operating range, there is a roughly optimal operating wavelength, all else being equal.

Figure 4.5 shows the product of the scattering intensity and the cumulative loss (average of the losses of probe and anti-Stokes wavelengths) as a function of distance along the fibre. These curves represent the relative anti-Stokes power available at the receiver for a number of different systems. A selection of wavelengths has been chosen. For the first two of these, 850 nm and 904 nm, powerful semiconductor

* At the time of writing, the highest bandwidth on multimode fibre offered by a leading supplier was 4.7 GHz km at 850 nm, but only 500 MHz km at 1300 nm. The bandwidth at 1064 nm is seldom specified even though, as we shall see further, this is an important operating wavelength for Raman OTDR systems.

† A graded-index fibre having a core diameter of 85 μm surrounded by a cladding of 125 μm with a numerical aperture of 0.27 that was used for fibre-to-the home applications, notably in France, in the 1980s.

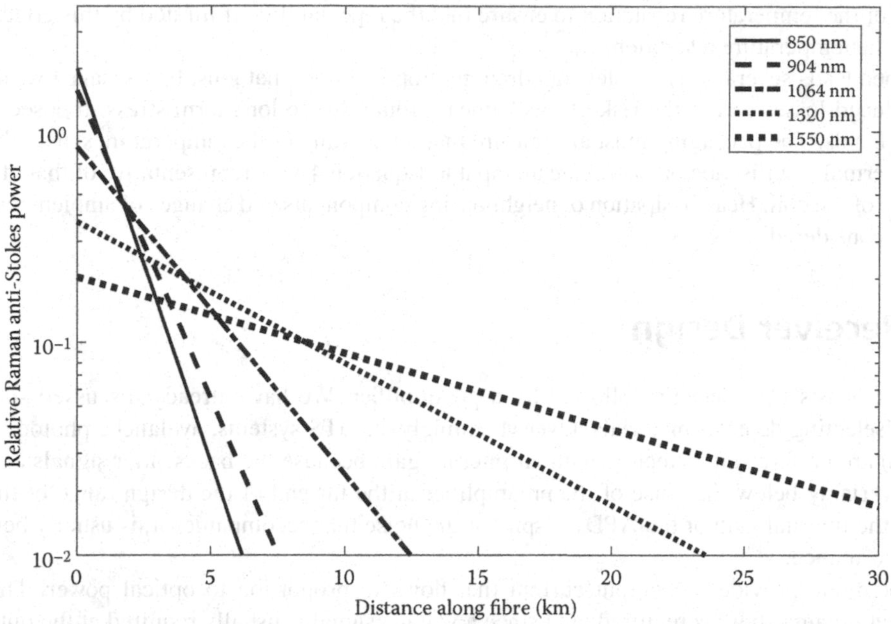

FIGURE 4.5 Anti-Stokes power available at the receiver as a function of distance for a typical graded-index multimode fibre at various operating wavelengths.

multimode pulsed lasers are available; at 1064 nm, a number of options for pulsed lasers also exist; the other two wavelengths, 1300 nm and 1550 nm, are common operating wavelengths for telecommunications, the former corresponding to the wavelength of zero material dispersion for typical germanosilicate glasses and the second to the wavelength of minimum loss for silica-based optical fibres. The figure suggests that for fibres shorter than 2.2 km, the preferred wavelength is for 850 nm or 904 nm; these options are overtaken by 1064 nm up to 4.2 km and the cross-over between 1300 and 1550 nm occurs at 8.6 km. However, this picture is over-simplified. It ignores the following effects:

a. Availability of powerful sources at each wavelength
b. The relative performance of detectors at each wavelength and their responsivity as a function of wavelength
c. The maximum power that can be launched as a function of wavelength and fibre length
d. The actual losses of the fibre must be taken into account, not just the scattering loss, although they are frequently closely related.

In general, factors b–d lead to somewhat longer cross-over points; in addition, systems can be operated beyond the cross-over points; they are simply no longer optimal from the point of view of SNR.

4.2.2 Reference Coil Design

The reference coil is used to normalise signals returning from the sensing fibre to those returned by the reference coil itself. It is therefore important that it does not add significantly to the noise, particularly to the repeatability (defined in Chapter 1). Assuming that the noise in the reference coils is uncorrelated between locations separated by one spatial resolution, then the noise contributed by the reference coil is equal to that in the near-end of the sensing fibre divided by $\sqrt{z_{ref}/\delta z}$. This term (and hence the length of the reference coil) is then chosen to ensure that the noise contribution of the coil is a certain fraction

(0.2 or 0.1) of the temperature resolution to ensure that the repeatability, if limited by this effect, is much lower than the temperature resolution.

This then leads several tens or a few hundred m of optical fibre that must be packaged with attention to bend-induced losses and to the risk of mechanical failure due to long-term stress imposed by excessively tight bends. The packaging must also ensure that the reading of the temperature sensor that is used either for thermal stabilisation or to provide an input to Equation 4.13 is representative of that of the mean temperature of the coil. Heat dissipation of neighbouring components and changes of ambient temperature must all be considered.

4.2.3 Receiver Design

A receiver consists of a detector followed by a preamplifier. We have already discussed some of the criteria for selecting detector materials. Overwhelmingly, in DTS systems, avalanche photodiode detectors (APDs) are preferred to detectors without internal gain because the backscatter signals are so weak as to be invariably below the noise of the preamplifier at the far end of the design range of the system. Therefore, the internal gain of the APD, in spite of the noise that accompanies it, is usually beneficial to system performance.

A photodiode provides an output current that flows in proportion to optical power. Therefore, a current-input preamplifier* is required, and since a voltage signal is usually required at the output of the receiver, a transimpedance amplifier is generally selected. The noise, impulse response and overload handling of the preamplifier are critical to overall system performance.

In general, the front-end amplifying device is a field-effect transistor (FET) owing to their low current-input noise. The noise mechanisms in these receivers were discussed in Chapter 3.

4.2.4 Principal Types of Raman OTDR Systems

In practice, Raman OTDR systems fall into three broad categories that relate maximum range to certain technological choices. The exact boundaries between these categories depend slightly on the preferences of manufacturers and the constraints that they operate under. In general, the performance of DTS systems is limited by the power that can practically be launched: this is a matter of trade-offs between cost and performance.

4.2.4.1 Short-range systems

Firstly, short-range systems (up to about 4–5 km in range) operate around 850–910 nm. In this wavelength range, powerful pulsed broad-contact laser diodes are available at relatively low cost. In addition, standard silicon APDs absorb the backscattered light efficiently and provide a first level of low-noise amplification. These systems tend to be limited by how much optical power can be launched into the sensing fibre, from a combination of the peak output of available lasers and the moderate launching efficiency that can be achieved from these sources. Although lasers with single-mode outputs can be launched efficiently into single-mode or multimode fibre, broad-contact lasers emit in a wide stripe (typically 75 μm in length) and, in the orthogonal plane, emit over a wide range of angles, well beyond the acceptance angle of 50/125 fibre. Therefore, unless complex anamorphic optical systems are used to image the laser differently in each axis, the launching efficiency is usually less than 30%.

---

* A preamplifier designed with low input impedance.

4.2.4.2 High-performance systems operating at around 1064 nm

A separate category of systems operates at around 1064 nm; 1064 nm is an emission wavelength for the Nd and Yb ions, which can be used in optical sources. Options for these sources include microchip Nd:YAG Q-switched lasers [17], Q-switched fibre lasers [18,19] or a master oscillator, power amplifier (MOPA) source [20]. Because these sources are fundamentally single-mode, they can be launched efficiently into the sensing fibre (which typically is multimode).

In contrast with systems driven by pulsed semiconductor laser sources, these systems are limited by non-linear optical effects, primarily SRS, rather than the inability to launch more power from available sources. This is a fundamental limitation that is discussed in Section 4.3.3. For a 1064-nm probe, the Raman bands, particularly the Stokes band, are on the edge of the absorption band of silicon APDs, but extended silicon devices usually perform better, even at reduced quantum efficiency, than longer wavelength detectors. The 1064 nm category of systems dominates the high-performance DTS market over distances of up to about 15 km. Coincidentally, but very conveniently, this wavelength region is least affected by damage that afflicts optical fibres subjected to hydrogen ingress at high temperature.

It should be noted that high-power pulsed laser diodes are also available at 1064 nm; however, systems based on these sources usually perform worse than those based on rare-earth-doped fibre sources, certainly beyond a range of a few km.

4.2.4.3 Long-range systems

A third operating wavelength is in the long-wavelength region for optical fibres, around 1550 nm. The loss of a pristine fibre is minimised in this wavelength region and powerful sources are available, whether Q-switched fibre lasers [19] or MOPA sources [21]. Detector technology must shift to long-wavelength devices, such as InGaAs APDs, which show lower gain and higher noise than silicon does. Long-wavelength systems are preferred at ranges longer than about 10 km on multimode fibre or 15 km on single-mode fibre. Preferably, the system should be designed for the specific fibre type since the optical efficiency and noise performance can be optimised. Nonetheless, some systems designed for one fibre type are occasionally operated on sensing fibre of the other type with significant penalties in metrological performance. Long-wavelength systems operating on multimode fibre generally perform worse than 1064-nm systems at ranges below 15 km, owing to the reduced signal level for a given probe energy, a slightly reduced maximum launch power and worse detector performance. However, beyond about 15 km, they are the only viable solution, given the cumulative losses suffered at shorter operating wavelengths.

A system operating at 1400 nm has also been discussed [22]. In this case, the source was a solid-state pulse source (Nd:YLF diode-pumped Q-switched laser) emitting at 1320 nm that was shifted to 1400 nm using SRS [23]. The resulting anti-Stokes and Stokes backscatter were reported to be 1320 nm and 1500 nm, respectively. This approach has seen limited adoption, possibly because the probe wavelength is close to the absorption peak at 1390 nm that is caused by inclusion on OH^- ions in the glass. Although modern fibres exhibit extremely low levels of OH^- content, their presence can dominate the fibre transmission at these wavelengths, especially after some ageing at high temperature.

4.3 PERFORMANCE OF RAMAN OTDR DTS SYSTEMS

4.3.1 Spatial Resolution

The spatial resolution of a Raman OTDR is, like any other OTDR, determined by the impulse response of the system consisting of the response of the laser, the receiver and the fibre, as discussed in Chapter 3, i.e.

$$\delta z = 1.087 \cdot \frac{v_g}{2} \cdot \sqrt{\tau_p^2 + \tau_{rx}^2 + \tau_f^2}\ . \qquad (4.14)$$

Part of the system design involves allocating the allowable spatial resolution between the three contributions mentioned. In multimode systems, the pulse broadening caused by propagation in the fibre often limits the spatial resolution at long range. Although it would be possible to optimise the bandwidth of multimode fibres to achieve a resolution of 1 m at 10 km or more, in practice, the design of optical fibres is driven by the needs of telecommunications applications, and these do not always coincide with those of DTS applications.

Regarding the allocation of the impulse response between probe pulse and receiver response, the optimum depends on the design of the receiver and how its noise varies with bandwidth. It also depends on how the peak power that can be launched varies with pulse duration, a point that we will return to in Section 4.3.3. As a rough rule of thumb, setting the pulse duration equal to the impulse response of the receiver and acquisition is generally a reasonable compromise.

4.3.2 Temperature Resolution

The temperature resolution of a Raman OTDR system is directly related to the uncertainty on the Raman ratio. In general, this is dominated by the anti-Stokes signal, being the weaker of the two signals, but in some cases, the contribution of the Stokes signal is significant. This is the case, for example, in systems operating at a probe wavelength of 1064 nm, where the detectors are silicon photodiodes: the absorption of silicon falls off beyond 1000 nm and so the detection efficiency for a given detector type can be far lower for the Stokes than the anti-Stokes band. In any event, the resolution is given by

$$\delta T(z) = \frac{\delta R(z)}{R(z) \cdot \varepsilon_R}. \qquad (4.15)$$

Clearly, $\delta R(z)$, the uncertainty on the estimate of the Raman ratio, will be dictated by the SNR on each of its constituents; other DTS systems where the temperature estimate is derived from a different combination of signals (such as those from both ends or Rayleigh backscatter as a reference) require their own error analysis.

At room temperature, ε_R is approximately 0.7%/K, and so for a 1 K temperature resolution, the combined signal-to-noise $\delta R(z)/R(z)$ must be at least 142. This relationship is a function of temperature: as the temperature increases, the Raman ratio becomes less sensitive to temperature, and hence, $R(z)$ must be measured with better SNR in order to achieve a given measurement resolution. The converse applies as the temperature is reduced, with the additional issue that at very low temperatures, the anti-Stokes signal may fall so rapidly that it becomes dominated by receiver noise and thus is more difficult to measure.

4.3.3 Fundamental Limitations

The SNR is limited on the one hand by the energy that can be launched into the fibre and on the other hand by noise associated with the limited photon arrival rate.

In a basic pulsed Raman OTDR system, the pulse duration is determined by the target spatial resolution and therefore the energy is limited by the peak power that can be delivered into the fibre. In some systems, particularly those used in cost-sensitive applications, the probe power is limited by the output power of the source laser and the efficiency with which it can be launched into the fibre.

However, even if unlimited source power were available, the power that can be launched is limited by non-linear effects. In Raman OTDR systems, the launched power is usually limited by the SRS [23,24]. Although other non-linear effects, such as stimulated Brillouin scattering or self-phase modulation, can take place (sometimes at lower power than SRS), the other effects, being dependent on a narrow linewidth, can be avoided by choosing a broad-spectrum source. A model of the evolution of the stimulated Raman emission along a fibre is provided in Refs. [25] and [26], but the effects are described briefly as follows.

SRS affects Raman OTDR systems in two ways. The first effect of excessive launch power is that the measurement becomes biased: SRS converts power from the probe pulse to the Stokes band and results in the build-up of a secondary pulse at the Stokes wavelength that travels with the probe pulse. The secondary pulse returns, through Rayleigh backscatter, an additional signal at the Stokes wavelength of the probe pulse. It therefore adds a spurious contribution to the Stokes signal and so artificially reduces the Raman ratio; in turn, the measured value of the temperature is also reduced. In single-mode fibres at 1550 nm, the maximum launched power is of order 1 W, and for multimode 50/125 fibres at 1064 nm, the power limitation is about 15 W, but this value depends strongly on the modal power distribution and, therefore, on the optical design of the system.

Some approaches to Raman DTS avoid using the Stokes signal or at least generate this signal at lower probe power. Even in this case, SRS limits the power that can be launched, albeit at a higher power level, because the conversion probe to Stokes power reduces the power in the probe pulse and therefore also the power in the anti-Stokes signal.

In the case of short fibres and wide pulses, the Raman interaction is dominated by the probe pulse intensity; i.e. the probe power divided by the area occupied by the mode or modes in which the probe is travelling. As the fibre becomes longer, the effect of attenuation suffered by probe and Stokes pulses must be considered and the interaction length can be approximated [24] to an effective length $L_{eff} = 1/\alpha_p$ (here, the attenuation must be expressed in the natural logarithmic unit, i.e. Np/m). Here, α_p has been included as the limiting factor; this is appropriate for most DTS systems where the attenuation is higher at the wavelength of the pump than at that of the Stokes wavelength. This is not the case, however, in long wavelength systems, where the pump can be at the wavelength of minimum attenuation; it is also not true when the fibre is degraded. In the case of a probe pulse at 1064 nm, the presence of hydrogen diffused as a gas into the core will increase the attenuation at the Stokes wavelength (c. 1115 nm). A more complex model [25,26] (that includes the attenuation of each optical wave) is then required to estimate L_{eff}.

The foregoing discussed broad pulses, where SRS limits the *peak* power of the probe pulse. However, as the pulse duration is reduced, the permissible probe power *increases* owing to the walk-off (defined below) between the probe and the secondary Stokes pulse. SRS is caused by the interaction between the probe power and the Stokes light, which initially is caused by spontaneous emission. However, once some power exists at the Stokes wavelength, it causes further power to be transferred to itself. In any elemental part of the fibre, the amount of probe power transferred from probe to Stokes is proportional to the intensity in the probe pulse and to that in the Stokes pulse. The optical power at the two interacting wavelengths must be present at a location in the fibre simultaneously. For short pulses, CD will generally cause the probe and Stokes pulses to travel at slightly different group velocities and, therefore, to drift apart. So as the Stokes power grows due to SRS, it moves either ahead or behind the probe pulse (depending on the sign of the CD). This effect is known as pulse *walk-off* [27], and it reduces the length of the fibre over which pulses can interact.

The walk-off effect is illustrated in Figure 4.6, where the power transferred to the Stokes band is plotted as a function of distance in a conventional (9 μm core diameter, *NA* = 0.13) step-index, single-mode fibre and also as a function of the time axis on the scale of the pulse duration. The simulated pulses are all Gaussian and their wavelength is 1550 nm; their duration is 2 ns and 50 ns full width at half-height in the left- and right-hand panels, respectively. These plots show some interesting features. Firstly, in comparing the two upper panels (where the probe power was simulated at 10 W peak), the peak value of the Stokes power is ~2.5 W for the wide pulse but only ~10 μW for the narrow pulse: the walk-off has prevented the large-scale energy transfer from pump to Stokes in the case of the narrow pulse. It will also be observed that for the wide pulses, the peak

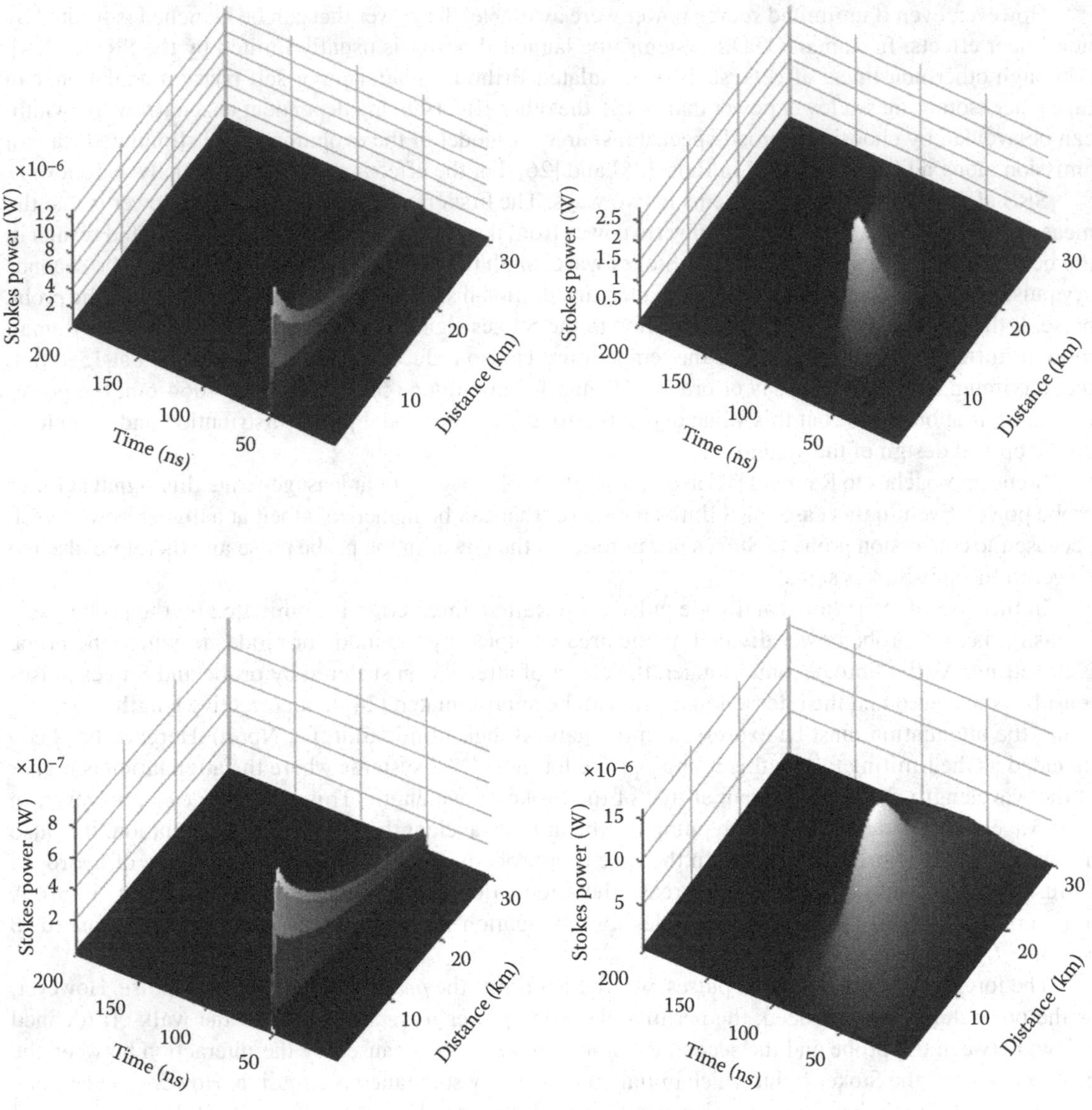

FIGURE 4.6 Stokes power build-up for short pulses in a standard single-mode fibre (see text for fibre parameters). The pulse duration is 2 ns for the left panels and 50 ns for the right-hand panels. The upper row is for a peak power of 10 W, and the lower row is for 2 W peak power.

value of the Stokes power occurs at around 15 km from the launching end: if SRS has not occurred on any significant scale at that distance, the losses will prevent it taking place further along the fibre.

It should also be noted that the Stokes power in the case of the narrow pulse exhibits a decay that is roughly exponential, further demonstrating that SRS plays a minor role even for the 10 W pulse, which is well into the non-linear regime for the wider pulse. In the lower panels, the peak power has been simulated at 2 W and in this case neither wide nor narrow pulse experiences significant conversion due to SRS. However, the faster decay in the case of the 2 ns, 10 W pulse (compared with lower power 2 ns pulse) demonstrates that some power is transferred non-linearly to the Stokes wavelength but that the contribution is weak and that it occurs only near the launching end.

As a result of the walk-off effects, the permissible peak power is not strictly constant: at the constant extreme, pulse durations are long (coarse spatial resolution) and/or the CD is small, and therefore, the walk-off effect does not lead to a significant increase in permissible launch power. However, for systems intended for fine spatial resolution, where probe pulses are short, and where these systems are operated in a wavelength region where CD is large, then the walk-off effect will allow the probe power to be increased, and in this situation, the system is limited more by pulse energy than by pulse power. This is important for the trade-off characteristics of fine spatial resolution DTS systems.

In estimating the effects of walk-off, the CD as well as the losses at all relevant wavelengths (including at multiple Stokes orders) must be taken into account.

4.3.4 Signal-to-Noise Ratio

The SNR and, therefore, the temperature resolution of a Raman DTS system, are determined on the one hand by the signal returning from the point of interest and on the other from the noise at that same point. One is usually concerned with the SNR at the far end of the range, given that closer locations will perform better. In single-pulse DTS systems, the SNR is usually limited at the near end by noise-in-signal, a combination of the shot noise arising from the backscatter and the multiplication noise of the detector; this falls off as the signal itself decays, leading to a sub-linear (close to square root) relationship between optical power and SNR. At the remote end, the noise sources within the preamplifier (Johnson noise of the feedback resistor and the noise associated with leakage currents in the detector and front-end amplifying device) tend to dominate and they are independent of the signal level.

The DTS designer can only increase the probe energy to the non-linear limit and then design the most efficient optical system in the return direction together with the lowest-noise receiver that they can conceive of. At that point, the practical limits to DTS performance have been reached. It is for these reasons that DTS designs using spread-spectrum approaches (OFDR and pulse compression coding [PCC]) have been employed; interestingly, in commercial systems, the single-pulse approach still offers better performance, possibly owing to practical issues in making full use of the energy that can be launched with spread-spectrum techniques.

4.3.5 Performance Comparisons: A Figure of Merit and Typical Specifications

The diversity of systems available and the range of comparison criteria can lead to confusion as to the relative performance of the various systems available. Some aspects of system performance are non-negotiable in each application. For example, a hot-spot detection system might require a resolution below the expected hot-spot size, or the system might be required to cover the entire length of a cable. Nonetheless, in most systems, some aspects of performance can be traded off for others.

The key performance parameters that affect distributed sensing systems are (a) spatial resolution, (b) length of the sensing fibre, (c) resolution of the measurand σ and (d) the measurement time required to achieve that resolution. Items (a) and (b) can be aggregated in a comparison by considering the number of independently sensed points, N_s, i.e. the ratio of sensing fibre length to the spatial resolution. Furthermore, in most DTS systems, the temperature resolution scales as the inverse of the square of the data acquisition time t_{acq}, so we can introduce a figure of merit M_{DTS} [28]

$$M_{DTS} = N_s \cdot \sigma^{-1} \cdot t_{acq}^{-1/2}. \tag{4.16}$$

M_{DTS} represents the number of sensed points that can be acquired to a resolution of 1 K in an acquisition time of 1 s. It allows meaningful comparisons to be made between systems specified in different ways, provided that all parameters discussed in (a) to (d) are provided consistently.

Table 4.1 provides some examples of performance achieved. In the top of the table, typical values of early (pre-1990), second-generation (roughly years 1990–2000) and recent commercial systems are provided.

TABLE 4.1 Examples of performance achieved in commercial Raman OTDR systems and in high-spatial resolution systems reported in the scientific literature

REFERENCE	*FIBRE TYPE*	*PROBE WAVELENGTH (nm)*	*SPATIAL RESOLUTION (m)*	*TEMPERATURE RESOLUTION σ (°C)*	*UPDATE TIME T_{ACQ} (s)*	*FIBRE LENGTH (m)*	*SENSED POINTS N_S*	*M_{DTS} ($K^{-1}\,s^{-1/2}$)*
Commercial Single Pulse								
First gen	MM	904	7.5	0.4	60	2,000	266	86
Second gen	MM	1,064	1	2.0	30	8,000	8,000	730
Second gen	SM	1,550	5	1	600	15,000	3,000	122
Second gen	SM	1,550	10	1.5	600	30,000	3,000	82
Third gen	MM	1,064	1	0.3	10	8,000	8,000	8,430
Third gen	MM	1,064	1	0.2	12	4,000	4,000	5,770
Commercial Pulse Compression								
	MM	1,064	1.5	0.65	600	8,000	5,333	335
	MM	1,064	1.5	0.15	600	4,000	2,666	726
Commercial – OFDR								
	MM	980	1.0	0.80	200	4,000	4,000	353
	MM	980	1.5	0.29	158	4,000	2,666	731
	MM	980	3.0	0.88	27	4,000	1,333	257
	MM	1,480	3.0	3.0	60	10,000	3,333	143
Research Papers								
High Resolution, Direct Detection								
[101]	SM	1,550	0.24	2.5	2,400	135	562	4.6
[28]	MM	1,064	0.37	0.56	1	1,100	2,972	5,307

(Continued)

TABLE 4.1 (CONTINUED) Examples of performance achieved in commercial Raman OTDR systems and in high-spatial resolution systems reported in the scientific literature

REFERENCE	*FIBRE TYPE*	*PROBE WAVELENGTH (nm)*	*SPATIAL RESOLUTION (m)*	*TEMPERATURE RESOLUTION σ (°C)*	*UPDATE TIME T_{ACQ} (s)*	*FIBRE LENGTH (m)*	*SENSED POINTS N_S*	*M_{DTS} ($K^{-1}\ s^{-1/2}$)*
[28]	MM	1,064	0.39	0.09	10	1,100	2,972	10,442
[28]	MM	1,064	0.39	0.37	10	3,100	7,848	6,707
PCC								
[41]	MM	1,064	1	0.3	600	4,000	4,000	544
[41]	MM	1,064	1	3		8,000	8,000	109
[39]	SM	1,550	17	3	100*	50,000	2,941	98
[45]	SM	1,550	1	3	30	26,000	26,000	1,582
High Spatial Resolution Using Photon Counting								
[46]	MM	840	0.3	4	180	40	133	2.4
[46]	MM	820	1.3	3	90	350	269	9.6
[62]	SM	1,550	0.012	3	60	1.2	100	4.3
[63]	MM	635	0.1	2	60	20	200	12.9
[51]	MM	850	0.4	0.8	2,700	900	2,250	54.1
[47]	SM	532	0.1	5	360	0.9	9	0.1
[64]	MM	780	1.0	1.5	600	2,500	2,500	68

* The paper provides only a number of probe pulses or probe pulse sequences, not the acquisition time that was required. The value of 100 s is derived from the pulse number (180,000) and the fibre length (50 km) that limits the pulse repetition frequency to 2 kHz at most. This leads to a fastest possible update time of 90 s and the value of 100 s, in the table allows for some overhead in the instrumentation.

The values provided by suppliers may vary in aspects of how they are specified, and so it has been necessary to apply some interpretation to available data, and this section of the table should be taken as an indication only. In particular, the extent to which contingencies are included to allow for losses of the fibre itself or at connectors may vary as can indeed the safety margin between typical and guaranteed performance.

As an example of the difference between the best possible performance in practical use and that which is achieved in the laboratory, it is possible to select the lowest-loss fibre from a batch and splice it with essentially zero loss. However, in practice, the user is often faced with fibre that is already installed, or that has suffered excess loss in cabling or installation or indeed has been selected for price rather than ultimate transmission. The same applies to the bandwidth of multimode fibres (as mentioned, these are often not even specified for the operating wavelength of DTS systems) and CD for single-mode fibres.

In addition, operational considerations frequently dictate the use of connectors that after a few mating/de-mating cycles may have degraded in their loss or become more reflective through incorrect maintenance during operation; even when installation conditions allow fusion splicing (by far the best option if available), poor working conditions (cramped, dusty or fibre leads that are too short for many attempts at a good splice), rushed timescales or indeed mismatches in fibre dimensions may cause losses well above those that are theoretically achievable. The incremental losses from imperfect connections and fibre quality below the very best that is available often degrades the performance that is achieved in practice with DTS systems.

The remainder of the table shows published performance values of commercial systems based on other approaches (OFDR and PCC) as well as results from research papers focused on specific goals, such as high spatial resolution.

4.4 ALTERNATIVE APPROACHES TO RAMAN OTDR

In pulsed Raman DTS systems, we have seen that the performance is limited fundamentally by the probe pulse energy that can be launched, which in turn is usually limited by the peak power. In order to overcome this limitation, spread-spectrum designs have been implemented by a number of researchers and system manufacturers.

Some of the techniques adapted from radar technology and used in OTDR were described in Chapter 3; in these cases, only the application of these techniques to Raman DTS is described here, and the reader is referred back to Chapter 3 (Section 3.3.2) for more detail on the techniques themselves.

Spread-spectrum techniques use some form of coding on the probe and, critically, increase its duration. An increase in duration of the probe would normally imply a degraded spatial resolution. However, in spread-spectrum techniques, the coding allows the backscatter signals to be processed in such a way that the resolution of the system is recovered.

4.4.1 Optical Frequency-Domain Reflectometry (OFDR)

DTS systems based on frequency-modulated, continuous-wave (FMCW) modulation were disclosed by Glombitza [29] and were developed by the German cable company Felten & Guilleaume, a business now operating under the Lios brand name and that is, at the time of writing, under the ownership of Danish company NKT.

In coherent FMCW, as used for example in radar or in Rayleigh backscatter systems [30–32], it is the optical carrier that is modulated and the signal returned is compared with that emitted by the source by mixing the two signals on a photodetector. However, owing to the phase-incoherence and broadband nature of the Raman signal,* for Raman DTS applications, the only available option is to apply the

* The linewidth of the Raman signal, on the order of 6 THz, far exceeds the bandwidth of available detector/preamplifier circuits.

modulation as a sub-carrier, i.e. the incoherent OFDR approach described in Section 3.3.2.3. Thus, the intensity of a continuous source is modulated by a sinusoidal waveform, the frequency of which is further modulated to encode the backscatter signal and allow discrimination between light returning from each section of fibre. In the incoherent approach, the detailed behaviour of the optical carrier frequency is of little importance.

4.4.1.1 Data acquisition in Raman OFDR

Incoherent OFDR can be implemented in two distinct modes. In the first, described in Chapter 3, the frequency sweep is fast compared with the round-trip time of the fibre, in the sense that light from more than one frequency is present in the fibre at any one time. In this case, the receiver output is digitised at a rate sufficient to capture the fine detail of the highest frequency present in the backscatter response. The duration of the sweep can be much longer than the round-trip time in the fibre, but the key is that the frequency changes by a measurable amount (more than one frequency data point) over the round-trip time.

An alternative, which appears to be used in commercial systems, is to set the frequency at a rate slow enough that the light travelling in the fibre is at a single frequency at any one time or indeed to adjust the modulation frequency in discrete steps: in this case, the data acquisition can measure the amplitude and phase response of the fibre for that one frequency in a narrow bandwidth (of order of the inverse time between steps of the frequency modulation function). Therefore, one complex data point is obtained for each input frequency; a complete data set, acquired from a number of separate modulation frequencies, provides the amplitude and phase response of the fibre in the anti-Stokes or Stokes wavelength band. This data set can simply be converted to the spatial domain (distance along the fibre) through a discrete Fourier transform. The information relating to distance is now encoded in the relationship between the amplitudes and phases for all points in the data set.

A typical arrangement is shown in Figure 4.7. In this case, the source is a laser diode that is directly modulated through its injection current. The modulation signal is a sinusoidal signal, the amplitude of

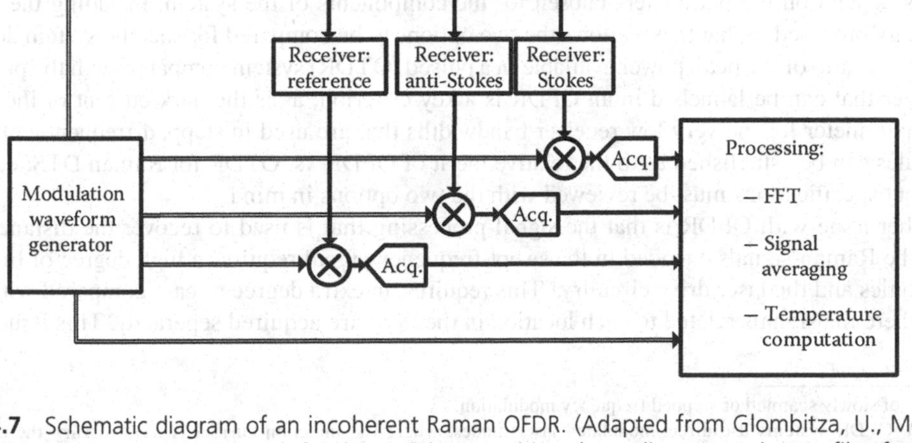

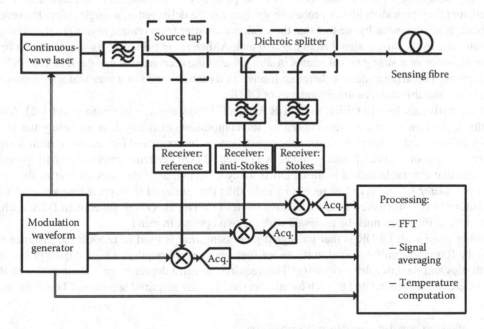

FIGURE 4.7 Schematic diagram of an incoherent Raman OFDR. (Adapted from Glombitza, U., Method for evaluating backscattered optical signals for determining a position-depending measuring profile of a backscattering medium, EP0692705B1, 1998.)

which is nominally constant, but the frequency of which is varied. It is known that laser diodes can be driven in this manner up to frequencies well above 1 GHz [33,34], and this translates to a theoretical spatial resolution below 10 cm. The authors of Ref. [29], from which Figure 4.7 is adapted, found it necessary to filter the output of the laser to eliminate any emission that might overlap with the Raman bands. A small sample of the source output is acquired in a reference channel that is used to normalise the Raman signals and thus cancel any variations in the spectrum that are related to distortion in the modulation actually achieved with the source, as opposed to the intended modulation imposed via the injection current.

The remainder of the modulated light is launched into the sensing fibre via a dichroic device; as in Raman OTDR, the Raman backscatter signals in each of the bands are directed after optical filtering to separate detectors and preamplifiers. The signal generator provides a sinusoidal output with the required frequency modulation. The output of the signal generator is mixed electrically with the outputs of the backscatter receivers and the reference channel. The mixing process shifts the received signal to a constant frequency* that can be filtered into a narrow bandwidth.

4.4.1.2 Signal recovery

In the case of a slowly stepped frequency modulation scheme, a number of probe frequencies are launched in sequence, and the signal from each of the three receivers is acquired after mixing with the modulation function to bring the result into a narrow, constant frequency range.

A discrete Fourier transform of the data (normalised by the reference channel data) converts the frequency-domain backscatter directly into the time (i.e. distance) domain. The Fourier transformed data can then process exactly as if it had been acquired in the time domain with a pulsed source.

4.4.1.3 Comparison with OTDR methods

Although OFDR is directly equivalent to the OTDR approach, there are important differences that affect the design and optimisation of the systems in different ways.

The key benefit of OFDR is that sources of low peak power, but able to emit continuously, can be employed, and they provide a higher probe energy than can be delivered in a single pulse. However, this benefit is negated somewhat by the fact that the weak backscatter from remote regions of the fibre arrives at the same time as the strong signal from the near end. Although, in principle, these signals can be separated, the presence of a strong signal results in shot noise at the receiver that can degrade the SNR in the remote regions. The arrangement is therefore limited in dynamic range in a way that single-pulse interrogation is not and this reduces the advantages of OFDR.

A comparative analysis of OFDR vs. OTDR Raman DTS systems can be found in Ref. [35]. Although the results depend on the parameters chosen for the components of the system, including the fibre, the methodology provided in that thesis allows the two options to be compared for specific system designs. In particular, the ratio of the peak power available in a pulsed (OTDR) system compared with the permissible mean power that can be launched in an OFDR is a key criterion, as is the dark current of the detector, a critical parameter for the very low receiver bandwidths that are used in stepped-frequency OFDR. No general rules can be established as to the relative merit of OFDR vs. OTDR for Raman DTS: each set of requirement specifications must be reviewed with the two options in mind.

Another issue with OFDR is that the signal processing that is used to recover the distance dependence of the Raman signals encoded in the swept-frequency signal requires a high degree of linearity of the electronics and the laser drive circuitry.† This requires an extra degree of care compared with Raman OTDR, where the signals related to each location in the fibre are acquired separately. This issue is offset

* In the case of slowly scanned or stepped frequency modulation.

† In general, OFDR systems use direct-modulated laser diodes, where the injection current is used to vary the output. The same requirements on linearity would apply in the case of an external modulator being used to encode the probe signal with a sweep-frequency.

to some extent by the fact that the signals are converted to low, constant frequency, and high-resolution data converters and precision amplifiers can be used. However, the receivers and mixers are required to handle broadband signals and their frequency response must be precisely the same for the Raman ratio compensation to work to be fully effective. This is not easy to arrange in the electronics.

As will be noted in Figure 4.7, a third channel has been used to acquire the signature of the modulated source, which allows distortions in the probe signal to be corrected in principle, subject to the limitation on channel matching just discussed.

The maximum power that can be launched is dictated in both OTDR and OFDR by SRS because other effects, such as stimulated Brillouin scattering, can be suppressed, for example, by choosing a suitably broad source linewidth. In general, the peak optical power is expected to be lower in OFDR because of the way the probe energy is spread in time. This suggests that the Raman limit on pulse energy will be reached in OTDR before it is reached in OFDR. This is undoubtedly the case: an OFDR will tolerate a higher probe energy than will an OTDR DTS system.

However, because the probe signal is pulsed in OTDR, it interacts with the backscatter only over the short section of fibre occupied by the pulse. In OFDR, in contrast, both signals are broadly continuous, although modulated on a short time scale. In the forward direction, the SRS, which transfers power from the probe wavelength to the Stokes wavelength, will not see the same walk-off effects that help high resolution OTDR systems maintain performance for high-power, short-duration pulses.

In OFDR, the stimulated scattering limit will depend on the modulation frequency, so SRS has the potential to distort the spectrum of the probe energy as well as to add a background contribution mainly to the Stokes signal. In fact, the walk-off effects lead to a frequency-dependent amplitude modulation of the backscatter signal that becomes periodic above approximately 40 kHz [35], and its effect is substantial even at the modest mean power level of 100 mW in single-mode fibre (in contrast, Raman OTDR systems in single-mode fibre can be operated at peak powers approaching 1 W, depending on the types of signal that are used). Although the distortion caused by SRS can be corrected, the correction is based on *a priori* knowledge of the distortion and so may not be reliable in practical applications.

Backward SRS hardly affects Raman OTDR systems because there is essentially no overlap between the spontaneous Raman scattering and the forward probe pulse. In Raman OFDR, however, the counter-propagating backscatter and the probe co-exist over the entire length of the fibre, resulting in some level of stimulated transfer from the anti-Stokes to the probe waves and from the probe to the Stokes, both effects resulting in a skew of the Raman ratio in favour of the Stokes wavelength.

In the case of backward SRS, the impact is not very specific to any location in the fibre because the shift of the Raman ratio takes place over a substantial distance in the fibre. Rather, it will appear as a non-linear contribution to the differential loss and lead to a progressively worse under-reading of the temperature as the distance from the launching end increases.

To summarise, the OFDR method allows substantially more probe energy to be launched than is the case for OTDR. However, the peak allowable power is substantially lower for OFDR than OTDR. This tends to reduce the apparent advantage of the OFDR approach.

The lower power limitations, together with the effects of shot noise from the near end impacting weak signals from the far end, seriously degrade any benefit of the frequency-domain approach. Published results do not support any clear superiority of OFDR over OTDR, and, as is often the case, the preference as to the approach adopted may be more a matter of the particular expertise of the designer.

Table 4.1 includes the performance for a number of Raman OFDR systems; as can be seen, their performance (based on the figure of merit) is certainly no better than single-pulse Raman OTDR.

4.4.2 Pulse Compression Coding (PCC)

PCC is a related technique to OFDR in that it also spreads the probe energy over a longer duration and thus increases the probe energy for a given peak power. In the case of PCC, however, a sequence of pulses is launched into the fibre and the backscatter from each pulse is super-imposed on the backscatter returns from the other pulses in the code.

PCC has antecedents in the field of radar [36], for example. The single-pulse response is recovered by correlating the measured backscatter waveform with the coding that has been applied to the source. Ideally, the response of such a correlated signal would be a single pulse, free of any spurious ripples or pulses. In practice, one usually settles for a maximum ripple of $\pm 1/M$, where M is the length of the code.

Of course, radar systems are different from incoherent optical systems in that in radar, the signal can take a positive or negative sign, which simplifies the design of codes for PCC. However, in a system such as a Raman OTDR, the light, being incoherent, cannot be coded in phase: it can only be coded in intensity from zero to one. In this case, where the intensity of the source is modulated, the intensity cannot be negative, and therefore, some of the codes known in other fields are not directly applicable. Nonetheless, recurrent pseudo-random sequences are known that, when correlated, result in a maximum value of the residual ripple modulation artefacts of $1/M$. However, these sequences are continuous, which brings about the same issues that were raised in connection with OFDR, namely of strong light from the near-end swamping weaker light from the remote end. What is really needed is a finite length code that can occupy only a fraction of the fibre length and yet bring some coding benefit to the system.

There are no known single optical codes that fulfil this requirement; however, approaches involving combinations of codes have been used successfully for compressing Raman DTS signals. Complementary Golay codes were developed at Hewlett-Packard for OTDRs, and it is believed, based on Ref. [37], that this concept has been transferred to Raman DTS systems. In this approach [38], a total of four codes are launched into the fibre and a backscatter signal is collected from probe pulse sequence. The performance of Simplex code-enhanced systems has been extensively discussed in recent publications [39–44]. These concepts of PCC were discussed in the context of Rayleigh backscatter in Chapter 3.

The relative performance of commercial Raman OTDR systems using PCC is also shown in Table 4.1. Generally, they are similar to Raman OFDR and fall short of recent single-pulse Raman OTDR systems, at least in the range 1–10 km, with the notable exception of recent work on single-mode fibre [45], where substantial progress was reported, namely a spatial resolution of 1 m at 26 km, resolving in 3 K in a measurement time of 30 s.

4.4.3 Systems for Very High Spatial Resolution

Equations 4.7 and 4.8 demonstrate that the Raman signal is proportional to probe pulse energy and, therefore, for a fixed peak power, to the pulse duration. Thus, as the spatial resolution is made finer, the pulse energy must be reduced and so the backscatter signals become weaker. Compounding this problem is the fact that a finer spatial resolution also requires a broader bandwidth in the receiver and subsequent analog processing chain. Both of these effects conspire to degrade the SNR by a factor proportional to the power 1.5 (for shot-noise limited systems) to 2.5 (for systems limited by noise unrelated to the signal power) of the inverse impulse response of the system.

As a result, an approach that is suited to low-light conditions has been investigated for very-high-resolution systems, namely photon counting, the first report on this subject dating back to 1987 [46]. Photon counting, as implied by the name, is a set of techniques in which the detectors are designed to respond with a clearly detectable pulse when a single photon is detected. The output, being already in quasi digital form, suffers little further degradation in the subsequent parts of the signal chain, and very small power levels (in the aW regime) can be detected. However, these techniques are limited in the maximum signal that they can detect before overloading.

The photon-counting approach has been implemented in two broad approaches, namely time-correlated single-photon counting (TC-SPC) and multi-channel photon counting (MC-PC). They are described in turn after a brief summary of the available detector types.

Photomultiplier tubes (PMT), single-photon avalanche detector (SPAD) and superconducting nanowire single-photon detectors (SNSPDs) have all been used for single-photon counting DTS experiments. The most common detectors in the published works are PMTs [46–48], which have the benefit of a very large amplification of the photo-electrons (10^5 being typical), fast response and low dark counts. The very large internal gain allows the signal arising from a single photon to dominate the noise of a suitable following amplifier. Devices with response times below 100 ps are available commercially, which would allow for spatial resolutions in DTS of order 1 cm. However, PMTs are relatively large vacuum tube devices requiring very large bias voltages (in the kV range) and screening from magnetic fields. The fact that their quantum efficiency is relatively poor is not too much of an issue given that the photon flux available in most DTS experiments can usually be arranged to overcome the fact that only 1% or 10% (depending on the wavelength and photocathode material) of the photons reaching the device will give rise to a photoelectron.

SPADs, on the other hand, are a variant of solid-state avalanche detectors that are generally operated above their breakdown voltage, such that the detection of a single photon can switch the device from a non-conductive state into a low-resistance, high-current state. As soon as the detection occurs, a passive [49] or an active [50] circuit temporarily reduces the bias of the detector below the breakdown voltage, allowing the device to return to its normal, high-resistance state, whilst in the low-resistance state, a large current can pass through the device, providing a signal sufficient for a following preamplifier to provide a clear pulse, an unambiguous signal that a photon has been counted. With active quenching, the time during which the device is in the low-resistance state and unable to detect another photon can be reduced to about 40 ns. SPADs were used in the only published MC-PC work [51] and are thought also to have been applied in two commercial systems manufactured for a whilst by Cossor Electronics [52] and Tyree Optech Pty. Ltd. [53]. DTS work published with SPADs was based at near infrared wavelengths (<900 nm) using silicon devices. The probability of detection with silicon falls off rapidly beyond about 950 nm. Long-wavelength APDs (initially based on germanium, then InGaAs), in general, have higher multiplication noise and dark current than silicon does and are thus less suited to photon-counting applications, but some success was achieved by cooling the devices [54] and later by gating them [55–60].

SNSPDs [61] are a relatively new type of detector; they are based on the absorption of a single photon on the surface of an extremely thin serpentine wire that is held just below its critical temperature and through which a DC bias current is maintained below the critical value at which superconductivity is lost. If designed correctly, the heat resulting from the absorption of a single photon can cause the device to switch locally from superconductive to a resistive state, which gives rise to a measurable electrical signal. The biasing circuitry is arranged (rather like passive quenching in SPADs) to return the device to its previous superconducting state. Whilst the device is recovering, it is insensitive to the arrival of another photon and so a dead-time also exists.

Of the three detector types, PMTs are the longest established, and, at their optimum wavelength (in the visible spectrum), they are probably the best-performing detector in terms of dark count and maximum counting rate. In addition, they do not require cooling, and although they are a vacuum-tube technology, they can be packaged for adequate field reliability in moderate environments, such as in a control room. However, their detection efficiency falls off dramatically at wavelengths longer than about 650 nm. Silicon SPADs offer sensitive detection up to about 1000 nm, although the wavelength of peak sensitivity for most SPADs is 900 nm or shorter. SNSPDs can be operated at long wavelength (e.g. 1550 nm) and are relatively wavelength-insensitive because their fundamental operating principle is the heating caused by the absorption of a single photon.

Photon counting for Raman DTS has been adopted by several groups, spread over a 25-year period [46,47,51,62–64], justified by the very small signal levels available, especially as the pulse duration is reduced. Key results are included in Table 4.1.

Whereas photon counting avoids additive noise originating in a preamplifier, it does nonetheless still suffer from other forms of noise, including jitter in the timing circuits and also the fundamental noise arising from the Poisson statistics of the signal that is measured. The main limitation of the photon-counting approach is that it is limited to very low light levels: only one photon must arrive at the detector in a time interval determined by the requirement for the detector to recover from the previous detection event and to return to a state where it can, once again, detect a new photon. This interval is known as the *dead time*, and any further photons arriving during that time will not be counted, which biases the estimate of the signal strength. Although algorithms exist to correct for dead-time bias, their range of effectiveness is limited and the photon arrival rate is still limited to about 0.5 events per dead-time interval.

In addition, in photon counting, care must be taken to avoid limitations caused by after-pulsing, i.e. the tendency of the detector to exhibit a higher (or sometimes lower) probability of detection shortly after a photon is detected [49,50,65]. These effects may be characterised by the auto-correlation function of the detector, and ideally, the device should return to its mean detection probability immediately after the end of the dead time. Of course, no detector is perfect, and one of the publications [51] on the subject shows an undershoot (i.e. the signal, after a falling edge drops below the subsequent steady value) immediately beyond a heated zone along the fibre, followed by a region with an artificially high temperature reading as a result of after-pulsing.

4.4.3.1 Time-correlated single-photon counting

Most of the single-photon counting experiments that have been reported used TC-SPC. In this technique, the photon arrival rate is reduced to well below one event per observation interval by reducing the pulse duration (which is required in any case for high resolution) as well as the peak power. In the case of DTS, the observation interval is the two-way transit time in the fibre. A basic schematic diagram is shown in Figure 4.8 (for simplicity, only one acquisition channel is shown in full; some elements of the second channel are outlined in broken lines). The basic principle of TC-SPC is that a laser pulse is fired into the sensing fibre and the time to the arrival of the first scattered photon passed by the filter is detected using a time-to-digital converter (TDC). In most cases, the time measurement has been implemented as shown in Figure 4.8 using a time-to-amplitude converter (TAC), i.e. a circuit that provides a voltage uniquely related to the time between

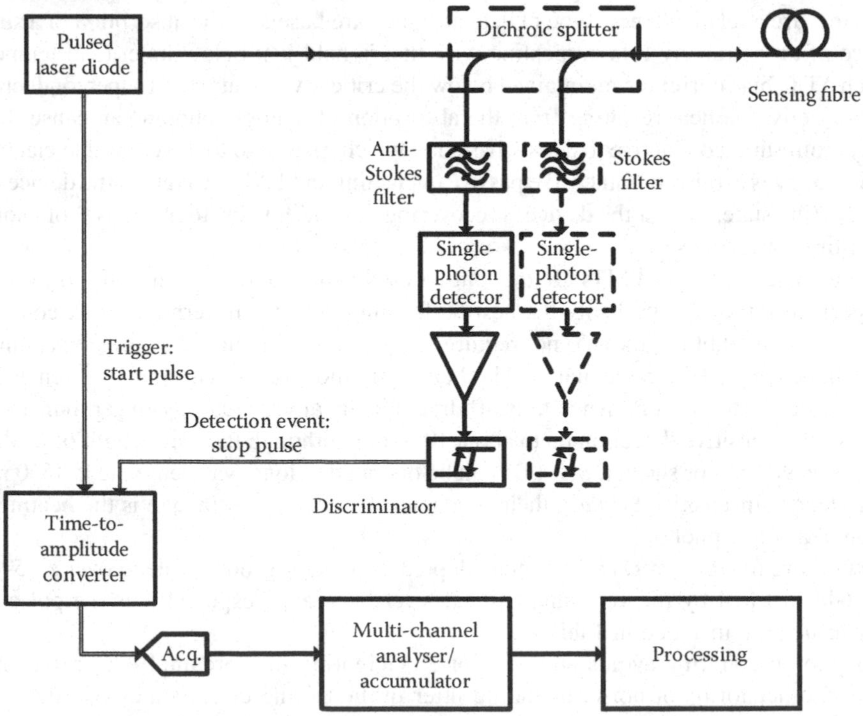

FIGURE 4.8 Schematic diagram of a time-correlated single-photon counting DTS.

start and stop pulses (for example a linear voltage ramp initiated by the start pulse). On arrival of the stop pulse, the TAC voltage is digitised by an analog-to-digital converter. Thus, the fibre is divided into time slots (or 'bins'), each slot being a resolvable point along the fibre, so the digital value of the time difference relates to an address in a bank of registers. After the conversion of the photon arrival time into the digital domain, the relevant memory location is incremented; this is the function of the multi-channel analyser/accumulator.

Systems requiring a longer time interval without sacrificing time resolution can complement the TAC approach with a digital counter, the counter being used for a coarse measurement and the TAC to interpolate between cycles of the clock [66]. In this way, a delay resolution of 10 ps has been achieved with a range limited solely by the number of bits in the counter. A number of other techniques, such as interpolating Vernier delay lines [67] or dual-slope interpolators [68], have been used for constructing high-resolution TDCs. For practical purposes, with modern electronics, the range is limited by factors other than the TDC.

By firing the laser many times, a histogram is built up of the distribution of photon arrival times and, thus, of time dependence of the backscatter intensity.

The main issue with TC-SPC is that the first photon stops the measurement, so too high an intensity at the near-end of the fibre will prevent any events being detected further downstream. Even with correction algorithms, this limits the photon arrival rate to less than 0.1 photon/pulse for the entire fibre [48]. Therefore, the more resolvable points are targeted, the lower the photon detection rate per point will be. This explains why experimental TC-SPC systems have usually been demonstrated only on short fibres.

Assuming that there is no limitation on the photon arrival rate that the acquisition electronics can handle, then the time t_{acq} required to measure the temperature to resolution δT is given by Ref. [46]

$$t_{acq} = \frac{2 \cdot L^2}{\delta z_{PC} \cdot \delta T^2 \cdot \varepsilon_{AS}^2 \cdot P_d \cdot v_g} \tag{4.17}$$

where P_d is the probability that one photon will be detected from the backscatter return for the entire sensing fibre of length L and ε_{AS} is the relative sensitivity of the anti-Stokes Raman signal to temperature.* Here, δz_{PC} is the length of fibre corresponding to each time slot (equivalent to the sampling resolution in an analog system). In Equation 4.17, the ratio $L/\delta z_{PC}$ is the number of points along the fibre for which the photon arrivals are separately tallied, whilst $2 \cdot L/v_g$ is the time taken for the pulse to travel along the fibre in both directions and therefore it is also the minimum time between probe pulses (assumed here also to be equal to the actual pulse repetition period). The remaining terms in the equation, namely $\frac{1}{\delta T^2 \cdot \varepsilon_{as}^2 \cdot P_d}$, represent the number of probe pulses per time bin that are required to accumulate sufficient photons to estimate the temperature to an adequate resolution. t_{acq} is the minimum possible time that this measurement will take because in practice, the pulse repetition period will need to be extended by the recovery time for the detector and its bias circuit. In addition, the source may not be able to deliver exactly the optimum pulse repetition period and the digital circuitry will need some time to reset after detecting a photon.

Equation 4.17 also assumes that only one Raman component is measured. This was the case in Ref. [46], where no loss compensation of any sort was provided. In fact, most of the experiments described in the TC-SPC literature use the anti-Stokes Raman component only in proof-of-principle demonstrations. In practical applications, it is likely that either a double-ended measurement or two signals will be required, which would extend the measurement further, even if the system incorporates dual-detector/counting circuits, simply because the noise from each channel affects the processed result. For a pair of anti-Stokes and Stokes channels operating at their respective optimum photon arrival rate, the measurement time required for a given resolution will be doubled.

Finally, Equation 4.17 assumes uniform backscatter intensity, i.e. a fibre so short that the attenuation is minimal over its length and a relatively uniform temperature distribution. Any departure from this

* In this work, only the anti-Stokes Raman signal was used.

condition will extend the measurement time because the peak intensity is dictated by the strongest signal whereas the noise is determined by the weakest signal. This factor explains why the TC-SPC technique is particularly suited to short distances, and in practice, the systems that have been described have generally been demonstrated over very short distances. For example, Dyer et al. [69] describe a TC-SPC system that uses a SNSPD as a detector. A spatial resolution of 1 cm is reported, but the length of the sensing fibre is shorter than 3 m.

4.4.3.2 Multichannel photon counting

So far, only one group has reported using the multi-photon counting technique [51] in spite of the clear benefit over TC-SPC, namely that it can, in principle, detect more than one photon per laser pulse. The basic principle of MC-PC is that a fast counter is started on the launching of the probe pulse and, on each receipt of a detection pulse, the counter output is latched in a register and rapidly used to increment a memory location defined by the count value. In Ref. [51], the clock operated at 250 MHz, resulting in a timing resolution of 4 ns, equivalent to a spatial resolution of ~0.4 m. Electronics has progressed rapidly in the time since that publication, and it is clear that far higher resolution would be achievable nowadays using similar techniques.

Although MC-PC overcomes the TC-SPC limitation of ≪1 photon detected per probe pulse, it can still only detect one photon at a time, and the system must be designed in such a way the probability of another photon arriving during the dead time caused by a previous detection is small. Typically, this requires a mean arrival rate of <0.1 photon per dead time interval. For dead time of 50 ns, therefore, the photon count rate must be kept below 1 photon per 500 ns (i.e. per 50 m of fibre). It follows that, for fibre lengths in excess of 50 m, MC-PC can improve system performance substantially.

The result reported in Ref. [51] was a spatial resolution of 0.4 m (limited by the speed of the digital counting circuitry) and a temperature resolution of 0.8 K for a measurement time of 2700 s over a fibre length of 900 m. This work stands out amongst the photon-counting results because it provided proper loss correction (through use of the Raman ratio and doubled-ended measurement – the latter will be described later in this chapter) and also because it demonstrated a measurement range of practical interest. However, the spatial resolution demonstrated in Ref. [51] was not as fine as in some of the TC-SPC results.

4.4.3.3 Correction for dead time and other detector imperfections

Techniques for dead-time correction are used in, for example, nuclear physics. However, as discussed in Ref. [70], their accuracy is limited and there are no adequate correction models for single-photon avalanche detectors. Nonetheless, they have been attempted, based on photon arrival statistics: if a given count rate is measured, an estimate can be made of those counts that have been missed and thus apply a correction. Although this approach adds noise (by multiplying the measured count by an estimate of the correction that itself is noisy), it is unbiased if the fact that a photon is missed is due purely to the arrival of previous photons and if these arrivals have a well-defined impact on the circuit (e.g. reset the dead time). Unfortunately, this is not the case because other effects such as dark counts and after pulsing distort the simple behaviour that is expected when only signal photons are detected [71]. Nonetheless, if the behaviour of each detector is well characterised, corrections can be applied that yield useful increase in allowable count rates [71].

4.4.3.4 Recent improvements in photon counting

Technology for photon counting has progressed in recent years driven in part by quantum cryptography. The need to detect single photons efficiently has led to the use of gating approaches where the internal gain of the detector is modulated. Originally, the modulation was applied synchronously with some expected signal [72] or to measure a particular part of the backscatter curve. However, very-high-speed (of order 1 GHz) square wave or sinusoidal modulation is now commonly applied [59,60,73,74] to the detector bias voltage. Although the device is inactive during the time at which the bias is lowered, this process also has the effect of avoiding (or at least reducing) after-pulsing, and the result is a substantial improvement in a combination of quantum efficiency and dead-time. It is likely that this approach will be applied to DTS at some time, but no reports on this topic have appeared in the literature, to the author's knowledge.

Another technological improvement in the field of photon counting is the use of multi-element detectors. In this approach, the active area of a detector is divided into separate SPAD pixels, each of which is passively quenched. Although each pixel has a dead-time, if the light is spread over a large number of pixels, then each pixel can respond separately to the light falling on it. It follows that the photon arrival rate required to overwhelm the entire array of pixels is increased, relative to a single detector, in proportion to the number of pixels. It is even possible for multiple photons that arrive simultaneously on different pixels each to be detected. This is one way in which *photon number resolving detectors* are being constructed, although other approaches where the avalanche charge is carefully measured are also used [58,75]. With such detectors, the power level that a photon-counting Raman DTS can receive would be increased significantly, and this solves one of the major issues with photon counting, namely the need to reduce the received power level to avoid detector saturation.

At the time of writing, the main applications for multi-element photon counting are in biology, at short wavelengths, and detectors optimised for those applications not suited for Raman DTS over distances of more than a few hundred metres. However, there is no fundamental reason why detector arrays should not be built for wavelengths suited to Raman DTS.

4.4.3.5 Comparison with analog direct detection

Until recently, the resolution of commercial, and indeed experimental, Raman DTS systems using analog detection has been limited to about 1 m. However, technology has advanced, and it has been demonstrated that analog direct detection can outperform photon counting [28], certainly down to a spatial resolution of 0.4 m. In this approach, rather than limiting the photon arrival rate and counting the few returning photons, the signal is increased as far as possible, close to the non-linear limits, and other aspects of the system are optimised for transmission efficiency and receiver sensitivity. In this work, short-duration, high-power pulses (~2 ns, 60 W) were obtained from a MOPA source operating at 1064 nm. The anti-Stokes and Stokes Raman signals were detected with Si-APDs and linear preamplifiers, the bandwidth of which had been adapted to the higher frequency signals that the improved spatial resolution entails.

The authors showed that the performance of such a system is far superior to photon-counting systems in this resolution range. This is primarily due to the fact that photon counters used to date are extremely limited in their count rate: the analog approach allows a far higher backscatter power to be received, and this increase in the signal is more significant than the increased noise due the receiver that accompanies it in analog systems. A summary of the performance achieved in Ref. [28] is shown in Table 4.1.

4.5 PRACTICAL ISSUES AND SOLUTIONS

4.5.1 Loss Compensation

The differential attenuation between the anti-Stokes and Stokes Raman bands is a fundamental issue in Raman DTS systems. An error on the ratio of the transmission at these wavelengths of only 0.7% (0.03 dB) accumulated from the scattering position to the launching end of the fibre is sufficient to generate a measurement error of 1 K. Therefore, great care must be taken in compensating for the losses at the two scattered wavelengths. Most of the commercially available DTS systems use a loss correction factor that is set for a given installed fibre, by checking the temperature at the remote end of the fibre. This approach is adequate, where the precision of the measurement is not critical or where there is confidence that the loss of the fibre is stable over time and invariant along its length.

Where a long sensing fibre is made up of sections of fibre from different batches, it is quite likely that the attenuation will vary between sections; this effect would require calibration of each fibre section separately. Although achievable in principle, this is cumbersome in the factory because careful tracking of inventory is required post-calibration and it is difficult to guard against errors in the order of installation of the fibres that are spliced in the field. Once the cables are installed, a calibration can be carried out in the field, but this is time-consuming and the installed fibre is not always accessible. Furthermore, the act of installing the fibre can result in subtle changes in its loss; this implies that a post-installation calibration is mandatory.

4.5.1.1 Fibre degradation

The attenuation of optical fibres can also vary over time. Insufficient buffering in optical cables can transfer mechanical strain to the fibre and result in micro-bending loss. Stresses induced in the cabling and installation process could relax over time and invalidate a post-installation calibration.

A more pernicious effect, however, is that of hydrogen ingress. Although glass is regarded as hermetic, it is known [76] that, at high temperature, hydrogen can defuse into silica glasses and remain in solution [77,78]; this gives rise to a number of absorption bands relating to the gas in solution in the glass. Although hydrogen as a gas at low pressure is transparent in the visible and near-infrared owing to its symmetric diatomic structure, at high pressure or in solution within a glass, the symmetry is not quite preserved and absorption lines appear at wavelengths of interest to DOFS [79]. Newly installed submarine optical telecommunications systems in the early 1980s exhibited increases of attenuation associated with hydrogen diffusion even at the moderate temperatures (4°C–30°C) that these cables are exposed to [80–82]. It turns out that even at room temperature, hydrogen, if present, will diffuse into the core of an optical fibre over about 1000 hours; this is a short time compared with the design life of subsea communication systems (10^5 hours or more). Whilst the diffusion rate is strongly dependent on temperature, the key point is that the partial pressure of the gas can reach equilibrium in the glass in less than 800 hours at 20°C and about 8000 hours at −20°C [81]. In solution in glass, new absorption lines appear at 1080, 1130, 1170, 1190, 1240 and 1590 nm. At 1240 nm, for example, a partial pressure of 2 atm will lead to an increase in absorption of 15 dB/km [80]. In practice, a relatively broad absorption band appears between 1070 and 1250 nm (incorporating the peaks listed within that range) that overlaps with the Stokes band of a DTS operating with a probe at 1064 nm. Since the gas is in solution, the new attenuation peaks persist until the gas diffuses back out of the glass.

However, with persistent presence of hydrogen in the glass, and at somewhat elevated temperature, the hydrogen can react with defects in the glass to form OH^- bonds that have a much stronger absorption per molecule (by a factor of ~40). The OH^--related losses are irreversible. Some discussion of these effects is provided in Chapter 2 (Section 2.1.2.1). A detailed study of the growth of attenuation at elevated temperature may be found in Ref. [83].

An example of the losses created by diffusion of hydrogen at high temperature is shown in Figure 4.9, where the ordinates represent the increase in loss relative to that of the pristine fibre. Although the fibre has a nominally hermetic coating (graphitic carbon in this case), at 250°C, hydrogen rapidly permeates through this coating and into the glass, and within a day, new absorption features appear in the loss spectrum. In this case, hydrogen is present in solution (spectral features identified by the 'H2 peaks' label); even after one day (second curve in the lower panel), some reaction with the glass has taken place, leading to irreversible losses. The growth of the permanent damage at 1385 and 1400 nm is related to silicon–hydroxyl and germanium–hydroxyl bonds, respectively; this is the first overtone of the fundamental OH^- line. A long-wavelength edge also results from fundamental absorption of the OH^- bond at around 2700 nm that spreads into the wavelength range of interest. A smooth increase of the attenuation at short wavelength (the so-called short-wavelength edge [SWE]) is also apparent. It is thought that this additional absorption is caused by the creation of Ge-H bonds at pre-existing defect

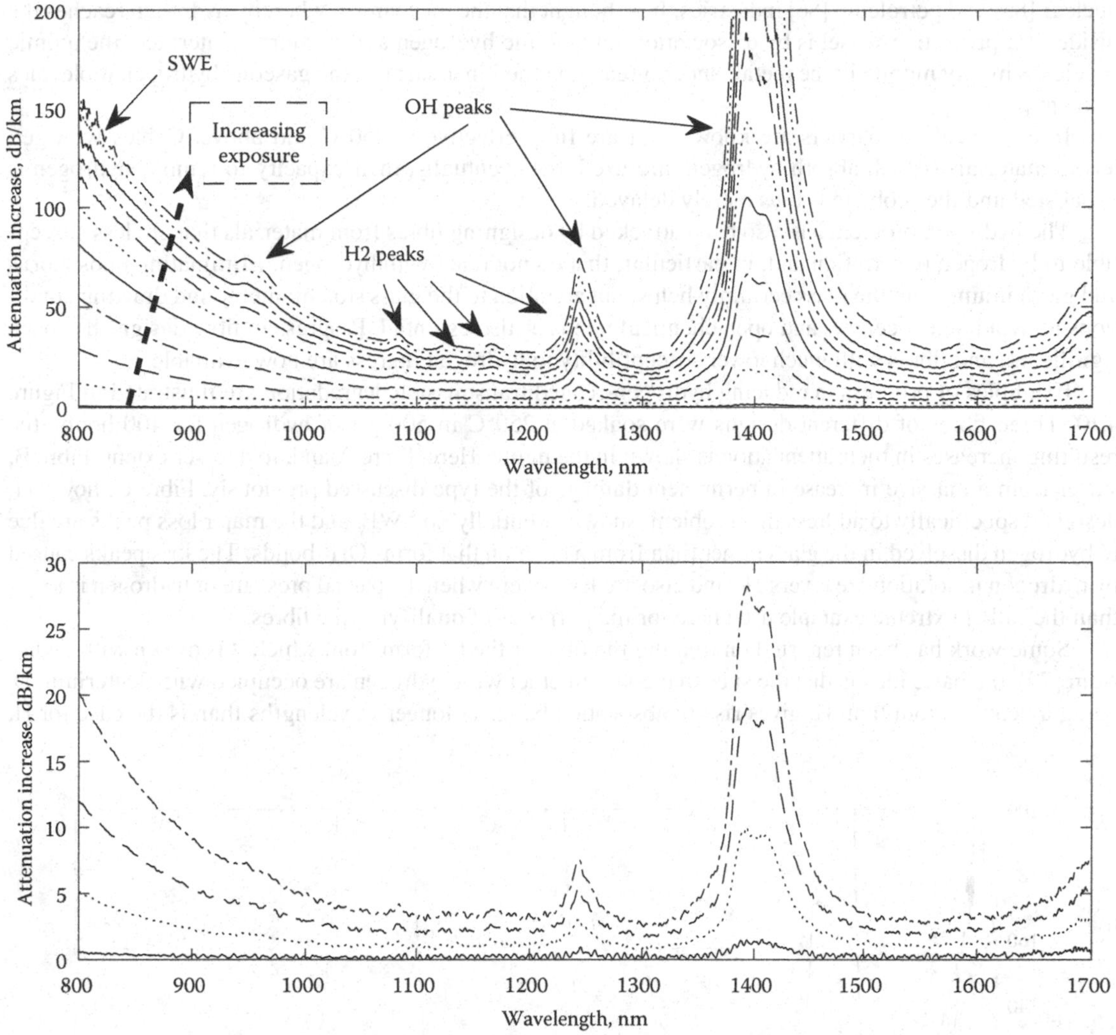

FIGURE 4.9 Growth of attenuation in multimode, graded index fibre (GeO_2 doped) with a hermetic coating at 250°C and in 500 psi (3.4 MPa) of H2 partial pressure. Upper panel: time scale covers 400 hours. Lower panel: initial stages covering 72 hours. Data recorded by Will Hawthorne at Schlumberger.

centres. These bonds result in a gradual increase in loss at short wavelength as well as fluorescence at around 650 nm. An additional loss peak at 1590 nm is caused by hydrogen in solution; there is also an overall rise in the base loss of the fibre, caused by the merging of the tails of the various new absorption bands.

Hydrogen barriers are commonly used to protect optical fibres in harsh environments, including so-called 'hermetic' coatings, such as carbon deposited directly onto the glass surface when the fibre is drawn before any polymer coating is applied. These coatings are effective at moderate temperatures, up to about 150°C [84]. However, at higher temperatures, they become increasingly porous to hydrogen, and at 250°C, the hydrogen penetration is delayed by only a few days. Metal tubes, laser welded around the fibres, have also been developed to prevent hydrogen reaching the fibre. The structures were designed originally to prevent the failure of trans-oceanic communications systems and they are well suited to this role. Again, at high temperature, these barriers have only limited effectiveness, and eventually, hydrogen penetrates through them. Studies of hydrogen penetration through metals have been conducted for the nuclear [85] and petroleum [86] industries. It is thought that the mechanism whereby hydrogen reaches the inside of a protective vessel is by dissociation into atomic hydrogen at the external interface; the atomic species is highly mobile in the metal, and, on reaching the inner surface, the gaseous hydrogen molecules re-form [87].

In any event, no barriers are known that are fully effective at 250°C and above. Cables with gettering materials (which absorb hydrogen) are used, but eventually, their capacity to remove hydrogen is exhausted and the problem is thus merely delayed.

The hydrogen problem has also been attacked by designing fibres from materials that are less susceptible to hydrogen penetration and, in particular, that do not react with hydrogen. Eliminating phosphorus and germanium from the glass certainly helps, but attention to the glass stoichiometry and drawing conditions to avoid defect centres and open chemical bonds is also essential. Progress in fibre designs has been significant, and fibres with much lower susceptibility to hydrogen ingress are now available.

Examples of progress in reducing hydrogen-related transmission degradation are illustrated in Figure 4.10. Three fibres of different designs were soaked at 250°C in 500 psi of hydrogen for 400 hours; the resulting increases in their attenuation is shown in the figure. Here, Fibre A and, to a lesser extent, Fibre B, suffer from a massive increase in permanent damage of the type discussed previously. Fibre C, however, designed specifically to address this problem, shows essentially no SWE, and the major loss peaks are due to hydrogen dissolved in the glass rather than from a reaction that forms OH^- bonds. The loss peaks caused by hydrogen in solution are reversible and also are less severe when the partial pressure of hydrogen is lower than the rather extreme example used here for the purposes of qualifying the fibres.

Some work has been reported on treating the fibre or the preform from which it is drawn with deuterium [79]; the basic idea is that the sites that could interact with hydrogen are occupied with deuterium. D, being a heavier atom than H, gives rise to absorption bands at longer wavelengths than is the case for H,

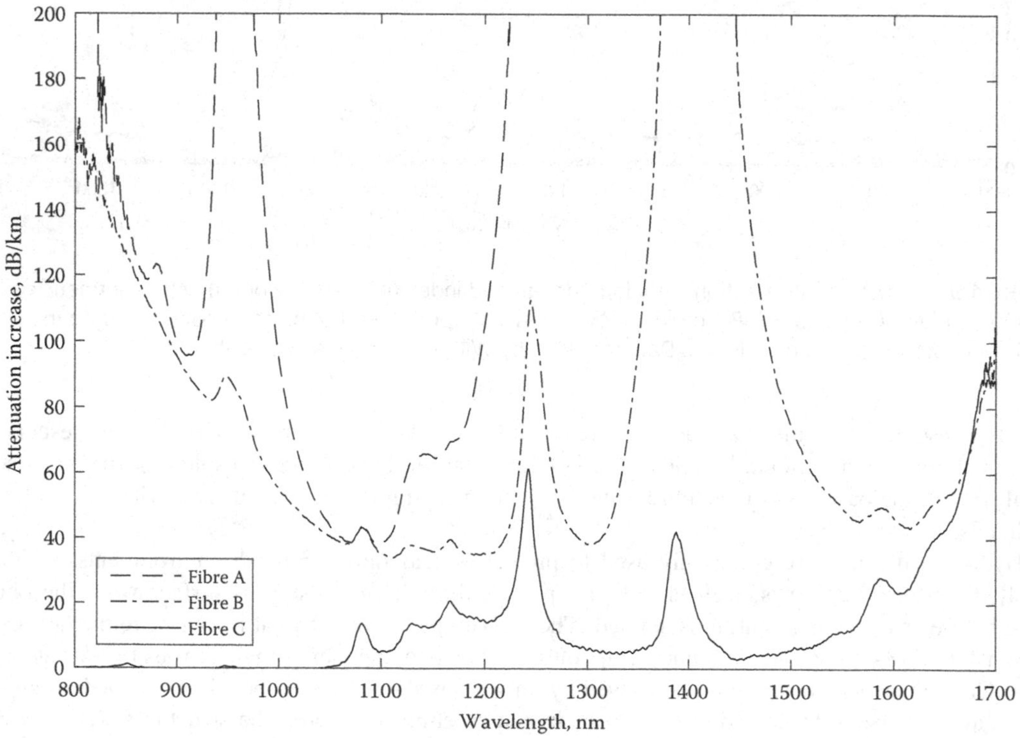

FIGURE 4.10 Attenuation increase resulting from exposure to hydrogen at 250°C for three fibre samples. Data recorded by Will Hawthorne at Schlumberger.

thus shifting the problem to a region that is not of concern. Although convincing results were produced, this approach does not seem to be in common use, possibly owing to the cost of deuterium.

Hydrogen penetration and reaction induce changes in the differential attenuation of the fibre; this process defeats the calibration of single-ended Raman DTS systems. In many practical cases, changes in the differential attenuation take the system outside of its specified accuracy before the reduction in signal (due to additional attenuation) results in an excessive standard deviation of the measured values. In other words, the *accuracy* of the system degrades before its *resolution* becomes unacceptable.

An example of the effect of fibre degradation is shown in Figure 4.11. A conventional fibre having a conventional (GeO_2 doped) core was installed in a SAGD well (for details of this process, please see Chapter 7) and the time stated is measured from when steam was injected to heat the reservoir. To the left, the Stokes Raman backscatter signal is shown sampled on four different days. These curves demonstrate the dramatic increase in attenuation in the heated fibre section that starts at a distance of about 150 m from the launching end; the result is a fall in the power received from the far end by a factor 10 in just a few days. The effect on the temperature measurement is shown in the right-hand panel; it can be seen that the measured temperature falls off dramatically, in spite of the temperature at the start of the heated section continuing to rise as a result of continuous injection of steam. This demonstrates that, for this type of degradation, the attenuation at the anti-Stokes wavelength is even more severe than for the Stokes signal. The Stokes signal is shown because it is relatively insensitive to temperature and so it is a less ambiguous measure of fibre loss than would be the case for the anti-Stokes signal. Note also that the noise on the inferred temperature trace is not apparent, demonstrating the point made earlier that the first issue arising from fibre degradation in this application is a skewing of the measurement rather than a loss of resolution.

The problem of the loss of calibration caused by fibre degradation has prompted the search for more robust measurement approaches than the basic single-ended Raman ratio DTS. Several approaches (double-ended and J-fibre) use fibre placement, illustrated in Figure 4.12; in these cases, additional

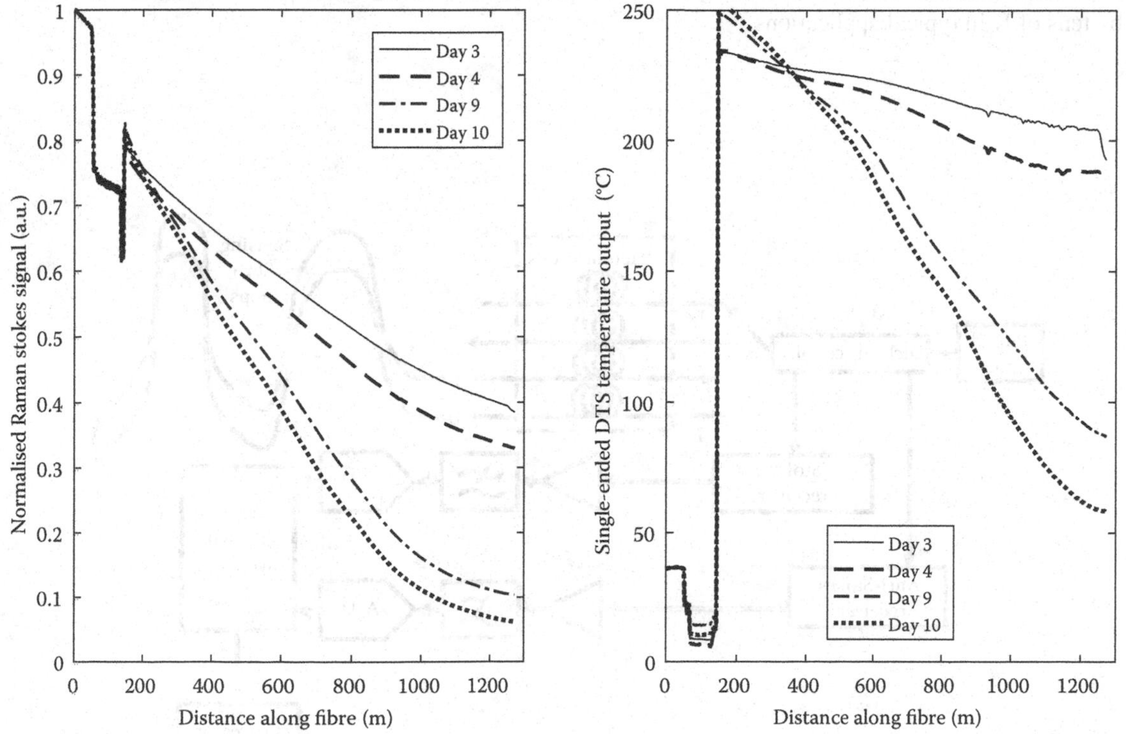

FIGURE 4.11 Example of the degradation of an optical fibre in a high-temperature oil well. Left: Raman Stokes power as a function of distance; right: single-ended DTS inferred temperature.

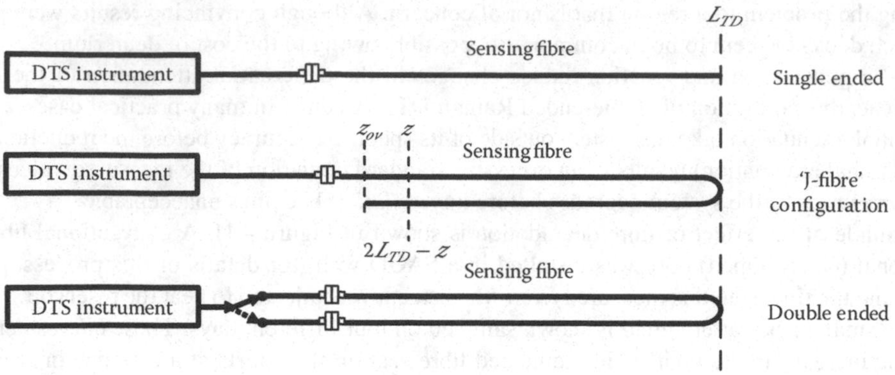

FIGURE 4.12 Alternative approaches to DTS fibre deployment.

information is gained from the fibre layout over and above that provided by the conventional single-ended deployment. Other approaches introduce additional functionality in the instrument to gain the information required to determine the differential loss profile independently from the temperature and correct for it.

An alternative of using an independent sensor at some location, or locations, along the fibre preferably at the remote end has been proposed [88,89]. The information provided by the additional point sensor(s) can be used to constrain the temperature calculation at their location. In general, this is helpful, but not sufficient, because it is often the case that the differential attenuation, when affected by fibre degradation, is also strongly non-uniform. In the case of non-uniform differential attenuation corrected at a single remote point, the DTS output will still be incorrect at other locations; in this situation, a single-point correction is better than no correction at all, but the temperature output can still depart from the correct value by tens of K in typical applications.

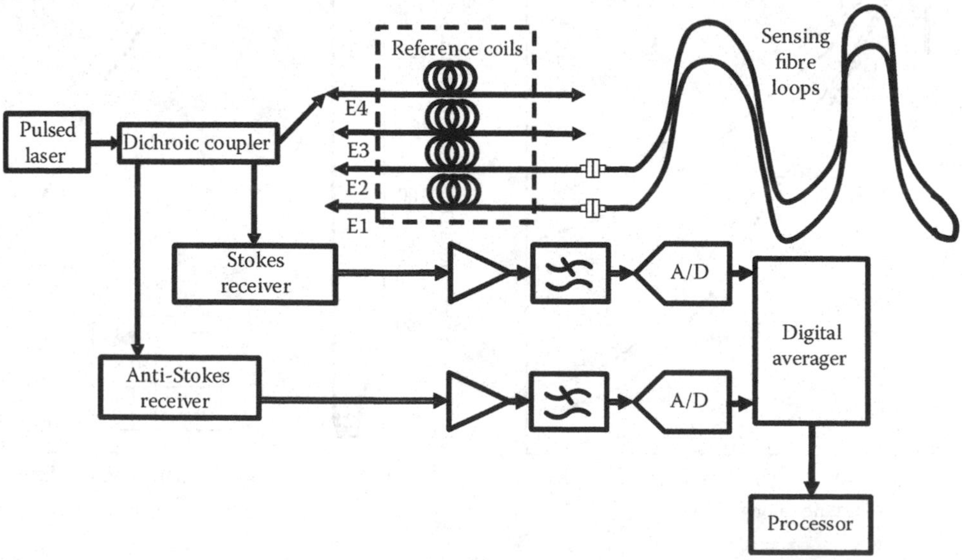

FIGURE 4.13 Double-ended, multi-loop DTS system diagram.

4.5.1.2 Doubled-ended measurements

Doubled-ended DTS measurements have been employed since the first commercial DTS systems supplied by one manufacturer, namely York Sensors Ltd. The basic principles are similar to those described in Chapter 3 but have been extended to the Raman ratio measurement for DTS. A typical arrangement is shown in Figure 4.13, where the basic configuration of Figure 4.4 has been extended by replicating the reference coils and adding a fibre switch to select each reference coil in turn. The sensing fibre is looped back so that each end is connected to a separate channel of the interrogating system. The loop is seen to be connected to reference coils marked 'E1' and 'E2'. Given that a switch is provided within the interrogator, it is often the case that further reference coils and switch positions are provided, which allows the measurement resources of the interrogator to be shared between more than one sensor loop, thus further reducing the cost per measurement point of the system.

Used with a Raman anti-Stokes/Stokes DTS system, the double-ended measurement successively probes the loop from one end and then the other. In this way, a Raman ratio can be calculated for each fibre end.

By analogy with the OTDR example (see Chapter 3, Section 3.1.2),

$$R_{E1}(T(z)) = \frac{K_{as}}{K_S}\cdot\left(\frac{\lambda_s}{\lambda_{as}}\right)^4\cdot\exp\left(-\frac{S_{LE}}{T(z)}\right)\cdot\exp\left[-\int_0^z(\alpha_{as}(u)-\alpha_s(u))\cdot du\right]$$
$$R_{E2}(T(z)) = \frac{K_{as}}{K_S}\cdot\left(\frac{\lambda_s}{\lambda_{as}}\right)^4\cdot\exp\left(-\frac{S_{LE}}{T(z)}\right)\cdot\exp\left[\int_L^z(\alpha_{as}(u)-\alpha_s(u))\cdot du\right]. \tag{4.18}$$

The geometric mean of $R_{E1}(T(z))$ and $R_{E2}(T(z))$ is then

$$\sqrt{R_{E1}(T(z))R_{E2}(T(z))} = \frac{K_{as}}{K_S}\cdot\left(\frac{\lambda_s}{\lambda_{as}}\right)^4\cdot\exp\left(-\frac{S_{LE}}{T(z)}\right)\cdot\exp\left[-\int_0^L(\alpha_{as}(u)-\alpha_s(u))\cdot du\right]. \tag{4.19}$$

Similarly to the double-ended technique in conventional OTDR, the position dependence of differential attenuation has been eliminated, and so when this equation is normalised by its value in the reference coils, the loss is completely eliminated from the temperature computation.

$$\sqrt{\bar{R}_{E1}(T(z))\bar{R}_{E2}(T(z))} = \sqrt{\frac{R_{E1}(T(z))}{R_{E1}(T(z_{ref}))}\frac{R_{E2}(T(z))}{R_{E2}(T(z_{ref}))}} = \exp\left(S_{LE}\cdot\left(\frac{1}{T_{ref}}-\frac{1}{T(z)}\right)\right), \tag{4.20}$$

and so the temperature distributed can be extracted from $\sqrt{\bar{R}_{E1}(T(z))\bar{R}_{E2}(T(z))}$ as shown in Equation 4.21

$$\frac{1}{T(z)} = \frac{1}{T_{ref}} - \frac{1}{S_{LE}}\left\{\ln\left[\sqrt{(\bar{R}_{E1}(z))(\bar{R}_{E2}(z))}\right]\right\}. \tag{4.21}$$

The double-ended approach thus provides a measure of the local temperature independent of the distribution of differential attenuation. This is true if the differential attenuation is independent of direction of propagation; whereas in the case of conventional OTDR, the modal power distribution can invalidate this statement to some extent, the referencing of the anti-Stokes/Stokes ratio is far less sensitive to such effects and the double-ended approach is found in practice to be a robust means of correcting for changes in the differential loss.

In the figure, the path of the fibre is seen to be folded with a short loop connecting the fibre connected to each of the channels. This may be a convenient way of arranging for a looped measurement, for example, when a single linear object is measured, such as a borehole: in this case, the function of the loop is purely to solve the differential attenuation problem. However, there is no requirement in the double-ended method for the fibre to follow the same path or for the fibre connected to E1, say, at a given distance from the interrogator to be at the same temperature, or suffer the same attenuation as the fibre connected to E2.

In some applications, such as in boreholes, the fibres forming the two sides of the loop are naturally co-located. In other applications, however, the measurement capability in the return path may be fully utilised. For example, in high-voltage energy cables, the fibre on the outward path may monitor one phase and the return fibre monitors another phase, so in this case, the fibres are fully utilised each to measure different temperature profiles. The path of the cable in the loop is arbitrary and could, for example, follow the perimeter of a facility.

Using the double-ended method on a single linear object requires twice as much fibre for sensing as the single-ended configuration, and it is therefore more expensive. However, much of the cost is often in the cable and deployment rather than the fibre itself and so the financial impact for an additional fibre can be relatively small. In all cases, however, the interrogation system is required to probe twice the length of an equivalent single-ended system. The double-ended technique can be used with signals other than the conventional anti-Stokes/Stokes ratio [90] such as just the anti-Stokes signals measured from each end or anti-Stokes Raman signals normalised to Rayleigh backscatter [90].

In common with other doubled-ended techniques, rather than looping the fibre back, a mirror can be placed at the remote end [90].

4.5.1.3 'J-fibre' configuration

When the object to be monitored is linear, the sensing fibre can also be deployed to the far end of a linear structure and brought back over part of the return path. The fibre is still measured as a single-ended sensor, but there is some duplication of the measurement that can be utilised to alleviate the differential loss problem. The approach is used primarily in borehole (e.g. geothermal or hydrocarbon wells) where the issues of fibre degradation are serious, but the user does not always have access to the better solution of full double-ended acquisition.

Denoting the remote end of the object to be measured as L_{TD}, the Raman ratios at positions symmetric with respect to L_{TD} are

$$R(z) = C \cdot \exp\left(-\frac{S_{LE}}{T(z)}\right) \cdot \exp\left(-\int_0^z \Delta\alpha(u) \cdot du\right) \quad (4.22)$$

and

$$R(2L_{TD} - z) = C \cdot \exp\left(-\frac{S_{LE}}{T(2L_{TD} - z)}\right) \cdot \exp\left(-\int_0^{2L_{TD}-z} \Delta\alpha(u) \cdot du\right), \quad (4.23)$$

where $C = \frac{K_{as}}{K_S} \cdot \left(\frac{\lambda_s}{\lambda_{as}}\right)^4$ and $\Delta\alpha = \alpha_{as} - \alpha_s$.

Given that z and $2L_{TD} - z$ are co-located, Equations 4.22 and 4.23 can be combined to give

$$\ln\left(\frac{R(z)}{R(2L_{TD} - z)}\right) = \int_z^{2L_{TD}-z} \Delta\alpha(u) \cdot du. \quad (4.24)$$

Likewise, the geometric mean $\sqrt{R(z)\cdot R(2L_{TD}-z)}$ of the near-end and far-end signal for co-located sections of fibre can be used to eliminate the differential attenuation between the Raman wavelengths, but only over that part of the fibre where the fibre is looped back (i.e. between z_{ov} and $2L_{TD}-z_{ov}$). The section between the launching end and z_{ov} is not loss-compensated.

Using Equation 4.24, the mean differential attenuation over the section from z to $2L_{TD}-z$ may be deduced and used to estimate a temperature correction profile, or a corrected temperature obtained directly from the geometric mean. The temperature resolution will, of course, be far worse at $2L_{TD}-z$ than at z, and therefore, in the J-fibre approach, the overall temperature resolution is degraded. This effect can be mitigated to some extent by filtering the derived $\int_z^{2L_{TD}-z}\Delta\alpha(u)\cdot du$ over much longer times than the temperature measurement update period, on the basis that fibre degradation is a slow process. They can usually also be filtered over distance along the fibre because changes in differential attenuation (at least those caused by hydrogen darkening) are reasonably even over short fibre distances given that the processes involved are temperature driven and the temperature usually varies smoothly with depth in boreholes.

4.5.1.4 Multi-wavelength measurements

The ambiguity caused by uncertain differential attenuation arises primarily because, in a conventional Raman DTS, the losses in the return direction are not known. Several approaches to overcoming this limitation have been devised based on measuring the additional information required.

Conceptually, the simplest approach is to measure the attenuation at the anti-Stokes and Stokes wavelengths with additional OTDR wavelengths and to use that information to calculate a differential attenuation correction that is updated periodically [91]. This method requires either additional OTDR instruments or, if built into the DTS, at least two additional lasers.

Several techniques have been proposed that require only one laser in addition to that used in a conventional Raman DTS. Collectively, they are referred to here as 'loss-compensated single-ended' measurements. They are illustrated in Figure 4.14 by reference to the wavelengths employed in each case. For reference, the usual anti-Stokes/Stokes Raman ratio approach is illustrated at the top of the figure. Although the prime motivation for developing these configurations is to solve the differential attenuation problem, they also can help with optical non-linearity, i.e. the effects that occur when the probe power exceeds the threshold for SRS in the case of Raman DTS systems. The solid arrows identify processes that occur and are used in a conventional Raman DTS system. Arrows with broken lines denote new sources and scattering processes. Also, it has been necessary to use new symbols for the wavelengths involved, such as λ_0, λ_{-1} and λ_1, because although the wavelengths may be *similar* to those used in conventional systems, the potential slight differences must be tracked through the modelling in order to identify potential sources of imperfect loss correction. For the sake of conciseness, similar subscripts (0, −1, 1...) have been applied in the following sections to the probe energy for sources at the various wavelengths, so E_0 represents the energy of the source at λ_0, etc.).

4.5.1.5 Raman ratio with separate sources

An elegant approach, disclosed by Bibby in 1985 [92], uses a pair of pulsed sources separated by one Stokes shift ν_R. In a first measurement, Laser 1 at λ_p launches probe pulses of energy E_p, and the resulting anti-Stokes backscatter $P_{as}(z)$ at λ_{as} is acquired.

$$P_{as}(z)=E_{p2}\cdot\eta_{as}(z)\cdot\exp\left(-\int_0^z(\alpha_p(u)+\alpha_{as}(u))\cdot du\right)\tag{4.25}$$

This is identical to the anti-Stokes backscatter signal in a conventional DTS, assuming the probe wavelength is the same. However, for the Stokes measurement, a separate probe pulse at a shorter wavelength, λ_{p2} with loss α_{p2} and separated in frequency from λ_p by ν_R (i.e. nominally at λ_{as}), is launched, and

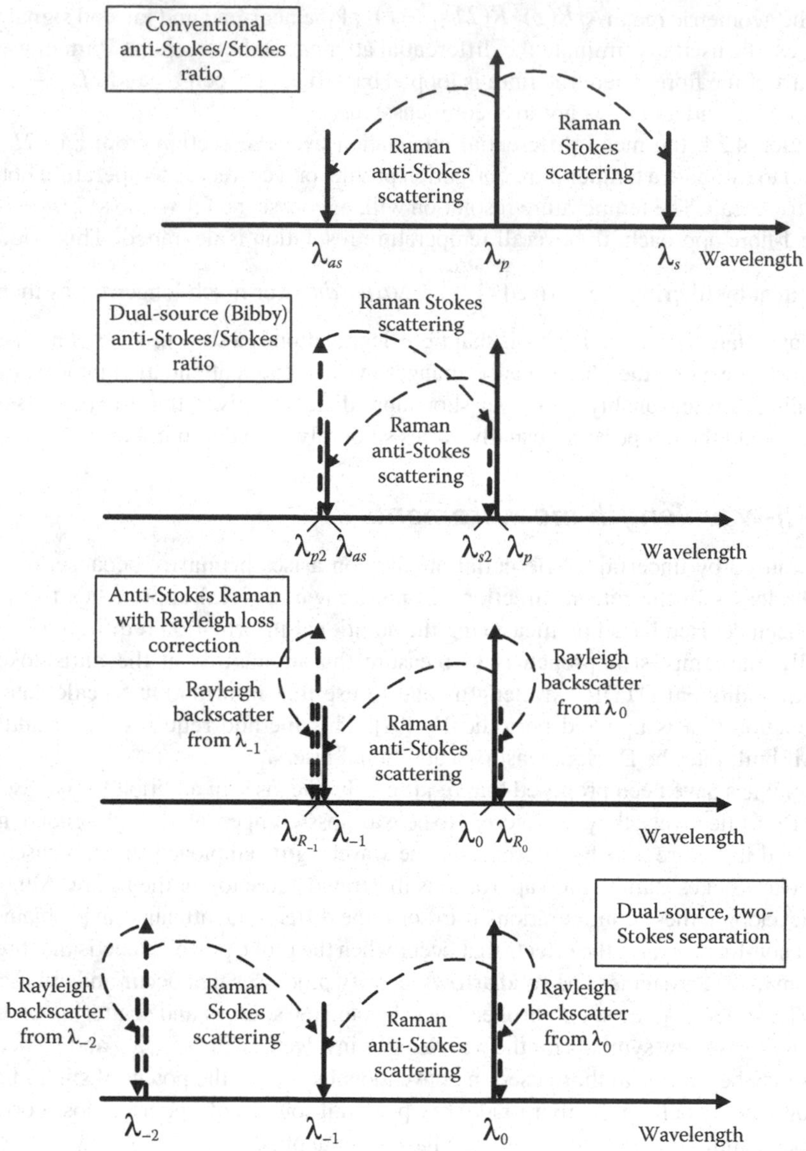

FIGURE 4.14 Loss-compensated single-ended Raman DTS configurations.

the backscatter at its Stokes wavelength λ_{s2} with loss α_{s2} is collected as a function of distance along the fibre, leading to a second signal $P_s(z)$:

$$P_s(z) = E_p \cdot \eta_s(z) \cdot \exp\left(-\int_0^z (\alpha_{p2}(u) + \alpha_{s2}(u)) \cdot du\right). \tag{4.26}$$

The Raman ratio is then given by

$$R(z) = \frac{P_{as}(z)}{P_s(z)} = \frac{E_p}{E_{p2}} \frac{\eta_{as}(z)}{\eta_s(z)} \exp\left(-\int_0^z \left(\alpha_p(u) + \alpha_{as}(u) - \alpha_{p2}(u) - \alpha_{s2}(u)\right)\right). \tag{4.27}$$

λ_{p2} is set such that $\lambda_{p2} = \lambda_{as}$, and therefore, $\lambda_p = \lambda_{s2}$. It follows that, in principle, $\alpha_{p2} = \alpha_{as}$ and $\alpha_p = \alpha_{s2}$; to first order, the integral term in Equation 4.27 should therefore vanish. The accuracy of this cancellation is dependent on how closely the centre wavelengths of the lasers match the wavelengths of the Raman bands at the same nominal wavelengths. In addition, the spectral of widths of the sources are unlikely to match the filtered Raman spectra, and if the attenuation is strongly wavelength dependent, unmatched loss spectra could skew the correction.

The ratio of probe pulse energies can be normalised by use of reference coils. The Bibby technique requires additional measurement time because the Stokes and anti-Stokes signals cannot be acquired at the same time. Fortunately, the number of pulses used for the Stokes measurement can often be less than that for the anti-Stokes measurement because the Stokes signal is stronger.

Although Bibby did not publish any experimental results using his technique, it was used in the 1980s by Cossor Electronics [52] and more recently by Sensortran [93].

4.5.1.6 Anti-Stokes Raman with Rayleigh loss correction

An alternative approach to loss-compensated single-ended measurement was disclosed in 1995 [94]*; it also involves two sources separated by one Stokes shift. However, in this case, three signals are acquired. Thus, from a first source, Laser A, at λ_0, an anti-Stokes backscatter signal $P_{as}(z)$ at wavelength λ_{-1} and a Rayleigh backscatter signal $P_{R0}(z)$ at λ_0 are obtained:

$$P_{as}(z) = E_0 \cdot \eta_{as}(z) \cdot \exp\left(-\int_0^z (\alpha_0(u) + \alpha_{-1}(u)) \cdot du\right) \tag{4.28}$$

and

$$P_{R_0}(z) = E_0 \cdot \eta_{R_0}(z) \cdot \exp\left(-\int_0^z 2\alpha_0(u) \cdot du\right), \tag{4.29}$$

where α_0 and α_{-1} are the fibre losses at λ_0 and λ_{-1}, respectively. A second source, Laser B, centred at λ_{-1}, is used to collect an additional Rayleigh backscatter trace, i.e.

$$P_{R_{-1}}(z) = E_{-1} \cdot \eta_{R_{-1}}(z) \cdot \exp\left(-\int_0^z 2\alpha_{-1}(u) \cdot du\right). \tag{4.30}$$

The geometric mean

$$\sqrt{P_{R_{-1}}(z) P_{R_0}(z)} = \sqrt{E_{0(\lambda_0)} E_{-1(\lambda_{-1})}} \sqrt{\eta_{R_0} \eta_{R_{-1}}} \exp\left[-\int_0^z 2(\alpha_0(u) + \alpha_{-1}(u))\, du\right] \tag{4.31}$$

is a synthetic trace that, nominally, has the same loss as the anti-Stokes Raman backscatter signal. η_{R_0} and $\eta_{R_{-1}}$ are the Rayleigh backscatter factors for the λ_0 and λ_{-1}, respectively, and $\sqrt{E_{0(\lambda_0)} E_{-1(\lambda_{-1})}}$ is the geometric mean of the probe energies for the two sources. Therefore, the anti-Stokes Raman trace normalised to the new synthetic trace is

* The grant of the patent cited occurred in 2003; 1995 refers to the filing date.

$$\frac{P_{as}(z)}{\sqrt{P_{R_{-1}}(z)P_{R_0}(z)}} = \sqrt{\frac{E_0}{E_{-1}}} \cdot \frac{\eta_{as}(z)}{\sqrt{\eta_{R_0}(z)\eta_{R_{-1}}(z)}}, \tag{4.32}$$

with the usual proviso that the attenuation spectrum of the second source is representative of that seen by the anti-Stokes backscatter from the first. However, because the outgoing pulse is the same for two of the signals, the outgoing losses are rigorously compensated. $\sqrt{\eta_{R_0}(z)\eta_{R_{-1}}(z)}$ is essentially temperature independent. The temperature sensitivity of the combination in Equation 4.32 is therefore that of the anti-Stokes signal alone, an increase of about 15% at room temperature relative to the usual Raman ratio. The benefit of using the anti-Stokes alone for temperature sensing increases with increasing temperature.

A potential drawback of this technique, however, is that the Rayleigh signals allow reflections (e.g. from connectors) through to the detectors; this can cause overloads that distort the temperature measurement for some fibre length beyond the reflections themselves. This issue is alleviated by use of low-reflectivity connectors and there are a number of applications where the use of connectors can be avoided. A separate issue relating to spontaneous Brillouin scattering included with the Rayleigh signals is discussed in Section 4.5.7.

4.5.1.7 Dual sources separated by two Stokes shifts

Another approach [95,96] uses two sources, A and B, separated by two Stokes shifts, i.e. at λ_0 and λ_{-2}. The references cited do not disclose exactly how the data obtained from this configuration are used. However, it is inferred that four traces are collected, namely anti-Stokes Raman $P_{as}(z)$ and Rayleigh $P_{RA}(z)$ backscatter arising from probing with Laser A and Stokes Raman and Rayleigh $P_{RB}(z)$ backscatter resulting from probing with Laser B.

Using a similar notation as in the previous approach, $P_{as}(z)$ is given by Equation 4.28 and the Rayleigh backscatter from Laser A by Equation 4.29. However, the Stokes measurement obtained with Laser B is

$$P_s(z) = E_{-2} \cdot \eta_s(z) \cdot \exp\left(-\int_0^z (\alpha_{-2}(u) + \alpha_{-1}(u)) \cdot du\right). \tag{4.33}$$

The second Rayleigh measurement at λ_{-2} may be written as

$$P_{RB}(z) = E_{-2} \cdot \eta_{R_{-2}}(z) \cdot \exp\left(-\int_0^z 2\alpha_{-2}(u) \cdot du\right). \tag{4.34}$$

The following combination of these four traces yields a profile that is nominally differential-loss independent. The term under the square root symbol is temperature-independent and serves only for loss correction

$$\frac{P_{as}(z)}{P_s(z)}\sqrt{\frac{P_{RB}(z)}{P_{RA}(z)}} = \frac{E_0 \cdot \eta_{as}(z)}{E_{-2} \cdot \eta_s(z)}\sqrt{\frac{E_{-2} \cdot \eta_{R_{-2}}(z)}{E_0 \cdot \eta_{R_0}(z)}} \tag{4.35}$$

In practice, it seems as if Laser A is used primarily with a stored anti-Stokes/Rayleigh loss correction function. Equation 4.35 is used periodically to estimate the error that may have resulted from changes in the differential loss and the loss correction function is updated appropriately.

4.5.1.8 Summary of the loss compensation methods

Of the different approaches, the single-ended anti-Stokes/Stokes Raman is by far the *least* reliable – its accuracy depends entirely on the uniformity and stability of the differential loss. The loss-compensated single-ended methods are more reliable and can ensure continued accuracy even when the fibre is degrading, but they rely on the additional measurements being an accurate representation of the transmission that they seek to calibrate and so they fail if the spectrum of the additional sources is a poor match to that of the Raman signals. The J-fibre suffers from the increased loss at the remote end, and it requires that the loss suffered in both fibre legs and their temperature distribution are the same.

The double-ended technique is the most accurate and does not rely on the losses being the same in the two fibres forming the up and down path. Indeed, it does not require the two fibre ends to follow the same path. Even the double-ended measurement, however, can suffer from non-reciprocal losses caused, for example, by the combination of poor splices and particular modal power distributions. In single-mode fibres, however, it should offer perfect correction (although it is not used in single-mode fibre to the author's knowledge).

4.5.2 Calibration

Equation 4.13 relates the measured signals to the actual temperature in each point along the fibre. The underlying model assumes that the signals that are measured are purely due to Raman backscatter and that the system collects anti-Stokes and Stokes Raman backscatter signals. This assumption holds provided that certain conditions are met, such as sufficient rejection of out-of-band light by the filters; there is also an implicit assumption that the Raman band is narrow relative to the frequency shift so that S_{LE} is a single number representative of the entire scattering process. It also implies that the effective Raman shift is the same for each of the bands (Stokes, anti-Stokes) used; although this is true fundamentally, in practice, the Raman bands are broad, and so the effective frequency shift applicable to the light reaching the detectors may not be the same for the two bands if the filters that select them are not symmetric relative to the laser frequency, for example, because of fabrication tolerances on the filters. The integral loss term in Equation 4.13 is handled by the double-ended measurement technique, but it requires additional calibration in the case of single-ended systems.

So in the case of a double-ended system, calibration consists of mapping $\frac{R(T(z))}{R(T(\text{rc}))}$ as a function of temperature. For well-designed systems, the exponential relationship between the Raman ratio in the sensing fibre to that in the reference coils and temperature (see Equation 4.20) is accurate over a wide temperature range, and only a few calibration points (in principle, just three) are sufficient to evaluate S_{LE} and T_{ref}. Although T_{ref} can be measured directly, its function in Equation 4.12 is also to eliminate certain properties of each realisation of the optical system (such as the transmission of the filters and the losses in the optics), and so in practice, it is best to evaluate it from calibration measurements rather than use a direct measurement of the reference coil temperature.

The temperature calibration points can be realised by varying the temperature of a section of fibre and measuring the system output as well as the calibration fibre section. However, from a practical point of view, it is preferable to install stable temperature zones that are held continuously at a stable temperature, for example, in oil baths, because the time required for stabilising a single section of fibre precisely (to say 0.01 K) can be prohibitive. A multi-temperature calibration set up can also be used for checking other aspects of system performance, such as temperature and spatial resolution. Moreover, a fixed, multi-temperature calibration set-up can allow the fibre to pass through the same temperature regions several times and so it can be verified that the calibration applies at several sections of the fibre and, thus, that any position-dependent dependence of the calibration is within acceptable bounds.

One possible calibration process is to expose separate sections of fibre to a known temperature profile and adjusting the parameters of Equation 4.13 until the system output is equal to the known temperature profile within acceptable limits. It is usual then to run a calibration check to confirm that the system calibration is effective.

The calibration intervals that are required for Raman DTS systems depend primarily on the stability of (a) the filters, (b) the laser spectrum (or spectra) and (c) the temperature sensor in the reference coils.

Each of these components is subject to ageing, and so the designer needs to consider their possible drift based upon the conditions that the equipment will be subjected to. In particular, variations of temperature and humidity can affect the central wavelength of thin-film dielectric filters that are frequently used for separating the wavelength components.

In the case of single-laser, single-ended systems, the correction for differential loss becomes critical to the calibration. In practice, there is no rigorous means of ensuring that these systems are calibrated at every point along the fibre, so the best that can be done is to measure the temperature near the sensing fibre at one or a few key locations and deriving a loss correction profile from those measurements. The weakness of this approach is that a change in the differential loss cannot be detected or corrected without *in situ* re-calibration. This rather defeats the purpose of a DTS system, and in a number of practical applications, the calibration has been found to drift seriously with time. In some cases, the single-ended DTS measurement is rendered useless in just a few weeks. More generally, there can be no confidence in the long-term that the calibration of the DTS system as a whole will be stable. Unfortunately, this potential calibration drift cannot be fixed by re-calibrating the interrogation system: it is the sensing fibre itself that must be calibrated, and this is usually inaccessible after installation. The best palliative for single-ended, single-source DTS systems is to check their calibration periodically with a system that is robust to drifts in fibre losses, such as a double-ended DTS (if a return fibre is installed) or one of the multi-wavelength systems.

In the case of multi-laser single-ended systems, the information is in principle available for an accurate measurement. There are, however, additional drift mechanisms in that the frequency separation of the lasers is now critical to the calibration.

It should also be noted that the details of the Raman scattering spectrum can vary between fibres of nominally the same type fabricated by different suppliers. At the time that DTS technology was first developed, batch-to-batch variations were observed even in fibres provided by the same supplier. This was particularly noticeable when using fibres designed for high-temperature operation; these fibres have a different glass composition from conventional telecommunications fibres. In these cases, it is usually necessary to calibrate the DTS system with the fibre type intended for use in the target application.

Finally, the parameters in the calibration model have so far been assumed to be independent of temperature, i.e. that the exponential relationship between these parameters and the temperature reading is accurate. In some systems, this assumption breaks down, and this must be guarded against by a calibration over the full temperature range of interest, and, if necessary, a more complex model in which the effective shift used is itself a function of temperature must be employed [97]. This correction is small and can be estimated by iteration, starting with a temperature-independent value of ν_R.

4.5.3 Probe Power Limitations

As previously mentioned, the performance of Raman DTS systems is limited by SRS. The first effect of SRS is to skew the measured value by increasing the Stokes signal artificially. This problem is eliminated to first order in the last two approaches in Figure 4.14, which involve measuring the anti-Stokes Raman signals with a loss correction provided by Rayleigh backscatter. However, what is described here is the fact that these compensation methods are not valid when the power level is too high and means of alleviating that problem are illustrated.

As the probe power increases, some power transfers to the Stokes wavelength, meaning that, beyond a certain probe power, the remaining backscatter signals are no longer precisely proportional to probe energy. To first order, this does not affect the anti-Stokes/Rayleigh ratio because the forward energy is the same for both. Equation 4.32, when written out to include the loss terms separated into the forward and backward losses, is as follows:

$$\frac{P_{as}(z)}{\sqrt{P_{R_{-1}}(z)P_{R0}(z)}} = \sqrt{\frac{E_0}{E_{-1}}} \cdot \frac{\eta_{as}(z)}{\sqrt{\eta_{R0}(z)\eta_{R_{-1}}(z)}} \cdot \frac{\exp\left[-\int_0^z (\alpha_{0f}(u)+\alpha_{-1b}(u))\cdot du\right]}{\exp\left[-\frac{1}{2}\left(\int_0^z (\alpha_{0f}(u)+\alpha_{0b}(u))\cdot du + \int_0^z \left((\alpha_{-1f}(u)+(\alpha_{-1b}(u))\right)\cdot du\right)\right]}, \tag{4.36}$$

where the f and b suffices refer to forward and backward propagation directions, respectively. In the case of Laser B, which is used for a Rayleigh measurement only, a low energy probe pulse is usually sufficient because the Rayleigh backscatter is so much stronger than the anti-Stokes Raman signal for a given probe energy. Therefore, we can arrange for $\alpha_{-1f}(z) = \alpha_{-1b}(z)$ at least as far as optical non-linear effects are concerned by arranging for Laser B to emit a sufficiently low power without impacting the system performance to any significant extent. However, as we attempt to maximise the crucial anti-Stokes Raman signal, an element of non-linear loss $\alpha_{0fnl}(z)$ will be added to the forward loss term, giving $\alpha_{0fh}(z) = \alpha_{0f}(z) + \alpha_{0fnl}(z)$. Taking $\alpha_{0b}(z) = \alpha_{0f}(z)$, for high energy probe pulses of energy E_{0h}, Equation 4.36 becomes

$$\frac{P_{as}(z)}{\sqrt{P_{R_{-1}}(z)P_{R0}(z)}} = \sqrt{\frac{E_{0h}}{E_{-1}}} \cdot \frac{\eta_{as}(z)}{\sqrt{\eta_{R0}(z)\eta_{R-1}(z)}} \exp\left[-\frac{1}{2}\left(\int_0^z \alpha_{0fnl}(u)\cdot du\right)\right], \tag{4.37}$$

and so most losses are still cancelled, but the $\frac{P_{as}(z)}{\sqrt{P_{R_{-1}}(z)P_{R0}(z)}}$ ratio shows that half of the non-linear loss is *not* compensated for.

4.5.3.1 Addressing the linear power limitation

The problem may be addressed with an additional measurement that requires no additional hardware [98], provided that the power of Laser A can be varied under software control. By repeating the measurement of $P_{RA}(z)$ at a sufficiently low power that it is linear with pulses of energy E_{OL}, an additional signal is obtained:

$$P_{ROL}(z) = E_{0L} \cdot \eta_{R0}(z) \cdot \exp\left[-\int_0^z (\alpha_{0f}(u)+\alpha_0\,\mathrm{b}(u))\cdot du\right]. \tag{4.38}$$

By replacing $P_{R_{-1}}(z)$ with $\frac{P_{R_{-1}}(z)^2}{P_{ROL}(z)}$ in Equation 4.36, it is found that the loss terms, including the non-linear one, cancel precisely.

There is a small measurement time penalty for acquiring $P_{R_{-1}}(z)$; however, the correction that it adds to Equation 4.36, i.e. $\exp\left[-\frac{1}{2}\left(\int_0^z \alpha_{0fnl}(u)\cdot du\right)\right]$, is a smooth function of distance along the fibre. In more sophisticated approaches, therefore, it can be estimated separately and filtered in that dimension; to all intents and purposes, this eliminates any degradation of the system performance due to the additional measurement that is acquired.

In practice, therefore, the distortion of the Raman ratio by non-linear effects can be eliminated, and the probe power is then limited by the fact that any further increases in launched power actually reduce the available probe power at the most remote point of interest due to the increased transfer of power to the (unused) Stokes wavelength.

4.5.3.2 The role of wavelength selection in extending linearity

The non-linear limit can be pushed back still further by careful wavelength optimisation: a cursory examination of Equations 4.28 would suggest that $P_{as}(z)$ is maximised at the remote end by selecting λ_0 such that $\int_0^L (\alpha_{0f}(u) + \alpha_{-1b}(u)) \cdot du$ is minimised, and, at low probe power, this implies selecting λ_0 and λ_{-1} to have equal losses, i.e. roughly symmetrically about the wavelength of lowest loss. However, the maximum probe power is also a function of wavelength [99]. The energy transferred by the SRS process is proportional to the product of the intensity at the probe and the Stokes wavelengths. Therefore, if the attenuation at the Stokes wavelength is very large, the growth of the power at the Stokes wavelength is held back. If the wavelength has been selected for equal loss at the anti-Stokes and probe wavelengths, the probe will be at a longer wavelength than the minimum loss, and in general, the loss at the Stokes is higher than at the probe. In the 1550 nm region, the loss increases more steeply on the long-wavelength side of the minimum than the short-wavelength side. So, shifting the probe wavelength slightly to longer wavelength extends the permissible launch power for linear operation because, as the wavelength increases, the loss for the Stokes signal increases faster than that of the probe. It follows that shifting the probe to a somewhat longer wavelength allows more power to be transmitted to the remote end, even though the losses at low power are slightly increased.

The concept of suppressing the growth of the Stokes power can be extended by, in single-mode fibres, deliberately increasing the loss at the Stokes wavelength by controlled bending of the fibre because bending loss is strongly wavelength dependent in these fibres or by selecting the cut-off to provide little guidance of light at the Stokes wavelength [99].

An active approach in which a so-called guard pulse is propagated at the same time as the probe pulse has also been considered [99]. In this approach, the guard pulse is shifted to longer wavelengths by two Stokes shifts relative to the probe pulse. As the probe propagates, it starts to generate amplified spontaneous emission at the Stokes wavelength. In the presence of a strong guard pulse, the Stokes light is further converted to the guard wavelength; this has the effect of suppressing the build-up of energy at the Stokes wavelength. The effectiveness of the technique is limited by the guard pulse itself suffering losses (through the normal loss mechanism, usually higher than the probe with typical wavelength selections, and through SRS conversion to the third Stokes). Once the guard pulse is depleted, the conversion of power from probe to first Stokes can resume; nonetheless, in specific applications, this approach can increase the threshold for non-linear effects by almost an order of magnitude.

The guard pulse is not used in the measurement itself and so it can be of quite long duration to avoid walk-off effects between the built-up Stokes pulse and the guard pulse.

4.5.4 Insufficient Filter Rejection

Given that the Raman signals are substantially weaker than Rayleigh backscatter, the rejection of the filters used to select the chosen Raman light is critical. At room temperature, the anti-Stokes Raman light is three orders of magnitude weaker than the Rayleigh backscatter, and so the anti-Stokes filters must reject at least 50 dB at the probe wavelength in order for the Raman signal to contain less than 1% of Rayleigh light. This can be achieved with good-quality optical interference filters, but their design must be specified with care. It may be that multiple filters must be employed to achieve the desired combined rejection.

The question of filter rejection is particularly important if the intention is to measure low temperatures. For example, in liquefied natural gas storage applications, the temperature to be monitored could be as low as 110 K. At this temperature, the anti-Stokes signal will have fallen to about 2% of its room-temperature value, so a filter with even 50 dB rejection will pass almost as much Rayleigh backscatter as anti-Stokes Raman. DTS systems designed for cryogenic applications therefore need yet higher filter rejection specifications than would be required for operation at temperatures higher than say, –50°C.

The very weak Raman signal does not preclude using DTS at these temperatures because the increased sensitivity of the anti-Stokes signal roughly compensates for the reduced signal level, at least for shot-noise limited systems. However, for cryogenic measurements, the filter rejection is even more important than for applications at room temperature and above.

4.5.5 Forward Raman Scattering

The scattering process is symmetric in the sense that the same fraction of the scattered power is collected in the forward and backward directions. So even under low power conditions, where spontaneous emission dominates, two pulses having power $P_{fas,fs}(z)$ build up and travel with the probe pulse. The portion of power $dP_{fas,fs}$ added to these pulses at each incremental position along the fibre is

$$dP_{fas,fs}(z) = P_0 S(z)\alpha_{ras,rs}(z) \cdot \exp \int_0^z -\alpha_p(u) \cdot du - P_{fas,fs}(z) \cdot \alpha_{as,s}. \tag{4.39}$$

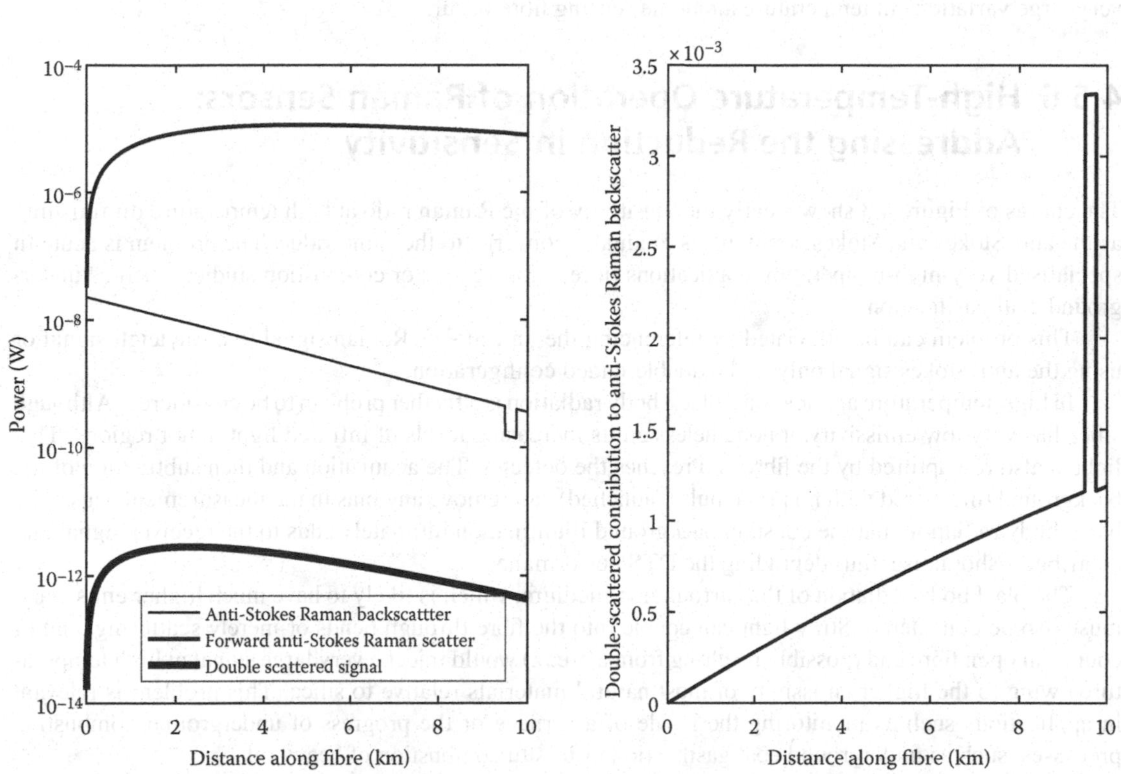

FIGURE 4.15 Modelled forward Raman scatter and Rayleigh backscatter double scattering. Left: Accumulated forward Raman scattering (upper curve) and resulting double-scattered Raman/Rayleigh backscatter (lower curve) together with the usual anti-Stokes Raman backscatter (centre curve); right: additional double-scattered contribution to the anti-Stokes Raman backscatter. In both plots, the fibre is at 300 K, other than a cold-spot near the far end of the fibre.

Here, the subscripts *fas* and *fs* refer, respectively, to the forward anti-Stokes and Stokes Raman power. The relevance of the forward scattered pulse is that it is itself subject to Rayleigh backscatter; the resulting, double-scattered signal will pass through the filters for the Raman signals in the interrogator. $\alpha_{ras,rs}$ refers to the Rayleigh scattering coefficient at the anti-Stokes and Stokes wavelengths.

As an example, Figure 4.15 (left plot) shows $P_{as}(z)$ (anti-Stokes Raman backscatter), $P_{fas}(z)$ and the double-scattered Rayleigh $P_{dsas}(z)$ resulting from $P_{fas}(z)$. In this case, it is assumed that a multimode, graded-index ($NA = 0.2$) sensing fibre is probed at 1064 nm with a spatial resolution of 1 m. Typical losses in a pristine fibre of this design have been assumed for each of the relevant wavelengths. The probe power of 10 W results in a maximum forward anti-Stokes pulse power approaching 10 μW. The ratio of $P_{dsas}(z)/P_{as}(z)$ is also shown (right plot), and it grows slightly sub-linearly reaching about 0.12% at 10 km, for a sensing fibre held at 300 K in this example. This is a relatively trivial distortion of the DTS accuracy, and, if necessary, it could be corrected from an initial measurement of the temperature profile. A more serious issue arises if a long, warm section of fibre is followed by a cold section. For example, a room-temperature fibre with a short section at 109 K (liquid natural gas) shows a local error on the anti-Stokes signal in the cold section of about 6%, as illustrated in the figure. This results in a local temperature measurement error of more than 1 K, which is still not highly significant, given that the main interest in an application such as this is leak detection and the actual accuracy is probably not critical.

We can therefore conclude that in most practical cases, the influence of double-scattered light is only of marginal significance. Nonetheless, this point should be borne in mind for unusual applications where very large variations in temperature along the sensing fibre occur.

4.5.6 High-Temperature Operation of Raman Sensors: Addressing the Reduction in Sensitivity

The curves of Figure 4.3 show clearly the sensitivity of the Raman ratio at high temperature diminishing as the anti-Stokes and Stokes components gradually converge to the same value. The problem is acute in specialised very-high-temperature applications in reaction vessels or combustion studies, such as underground coal gasification.

This problem can be alleviated by referencing the anti-Stokes Raman signal to a Rayleigh signal or using the anti-Stokes signal only in the double-ended configuration.

In high-temperature applications, black-body radiation is a further problem to be considered. Although silica has very low emissivity, it nonetheless emits increasing levels of infrared light in hot regions. This light is also re-captured by the fibre and reaches the detector. The acquisition and then subtraction of the background (measured with no probe pulse launched) can remove any bias in the measurement caused by black-body radiation, but the constant background illumination ultimately adds to the receiver signal and contributes shot noise, thus degrading the DTS performance.

The blackbody radiation of the surrounding medium, which is likely to have much higher emissivity, must also be considered. Stray light can couple into the fibre through bends or merely scattering, and of course an open fibre end (possibly resulting from a break) would inject a very large signal at high temperature owing to the higher emissivity of most natural materials relative to silica. This problem is relevant in applications such as monitoring the inside of a furnace or the progress of underground combustion processes, such as underground coal gasification or in-situ combustion of heavy oil.

4.5.7 Brillouin Backscatter

Raman systems use relatively broadband sources. This means that the wavelength bands normally thought of as Rayleigh scattering also contain Brillouin backscatter spectral components that are very close to the incident frequency. Given the breadth of the sources, the Rayleigh backscatter is mixed with Brillouin backscatter. This is not important in pure Raman ratio systems, but where the Rayleigh backscatter is used as a reference signal, or to infer differential loss, it can be important. Brillouin backscatter will be examined more closely in Chapter 5, but it is sufficient to state here that it adds a small component to the Rayleigh signal that is temperature sensitive, whereas the model for Rayleigh backscatter signals is that they are temperature *insensitive*. A small temperature sensitivity of an unfiltered backscatter trace was reported in Ref. [100] and can be attributed to spontaneous Brillouin backscatter. The effect of Brillouin backscatter is a distortion of the calibration of systems using Rayleigh backscatter that becomes significant at high temperature. For example, a system calibrated over the range [0°C–150°C] might show a departure from the modelled response of about 3°C at 250°C. This can be eradicated by improving the internal model that calculates temperature from the measured signals. The exact form of the model will of course depend on the precise measurement configuration.

4.6 SUMMARY

DTS technology using Raman backscattering in optical fibres is well developed and proven. Several approaches have seen practical implementation in commercial systems. Nonetheless, the limitations must be understood by designers and specifiers alike: although the basic principles are relatively simple, it should be remembered that, fundamentally, this is an intensity ratiometric measurement and quite small measurement errors can induce serious departures from a correct calibration. However, if care is exercised in the choice of fibre, in the design or specification of the interrogator and in the installation of the sensing cable, the technology provides a valuable, low-cost means of measuring the temperature of large objects.

REFERENCES

1. Long, D. A. 2002. *The Raman effect: A unified treatment of the theory of Raman scattering by molecules*. John Wiley & Sons, Chichester.
2. Farries, M. C., and A. J. Rogers. 1984. Distributed sensing using stimulated Raman interaction in a monomode optical fibre. 2nd International Conference on Optical Fibre Sensors. Stuttgart, Germany, SPIE/VDE Verlag. 121–132.
3. Galeener, F. L., J. C. Mikkelsen, R. H. Geils, and W. J. Mosby. 1978. The relative Raman cross sections of vitreous SiO_2, GeO_2, B_2O_3 and P_2O_5. *Appl. Phys. Lett.* 32: 1: 34–36.
4. Davey, S. T., D. L. Williams, B. J. Ainslie, W. J. M. Rothwell, and B. Wakefield. 1989. Optical gain spectrum of GeO2–SiO2 Raman fibre amplifiers. *IEE Proc. J.* 136: 6: 301–306.
5. Long, D. A. 1977. *Raman spectroscopy*. McGraw-Hill, New York, USA.
6. Payne, D. N., and A. H. Hartog. 1977. Determination of the wavelength of zero material dispersion in optical fibres by pulse-delay measurements. *Electron. Lett.* 13: 21: 627–629.
7. Sladen, F. M. E., D. N. Payne, and M. J. Adams. 1979. Definitive profile-dispersion data for germania-doped silica fibres over an extended wavelength range. *Electron. Lett.* 15: 15: 469–470.
8. Sladen, F. M. E., D. N. Payne, and M. J. Adams. 1978. Profile dispersion measurements for optical fibres over the wavelength range 350 to 1900 nm. 4th European Conference on Optical Communication, Genoa, Italy. 48–57.
9. Adams, M. J., D. N. Payne, F. M. E. Sladen, and A. H. Hartog. 1978. Wavelength-dispersive properties of glasses for optical fibres: The germania enigma. *Electron. Lett.* 14: 22: 703–705.
10. Ishio, H., J. Minowa, and K. Nosu. 1984. Review and status of wavelength-division-multiplexing technology and its application. *J. Lightw. Technol.* 2: 4:448–463.
11. Eriksrud, M., and A. Mickelson. 1982. Application of the backscattering technique to the determination of parameter fluctuations in multimode optical fibers. *IEEE J. Quant. Electron.* 18: 10: 1478–1483.

12. Mickelson, A. R., and M. Eriksrud. 1982. Theory of the backscattering process in multimode optical fibers. *Appl. Opt.* 21: 11: 1898–1909.
13. Kotov, O. I., L. B. Liokumovich, A. H. Hartog, A. V. Medvedev, and I. O. Kotov. 2008. Transformation of modal power distribution in multimode fiber with abrupt inhomogeneities. *J. Opt. Soc. Am. B* 25: 12: 1998–2009.
14. Dakin, J. P., D. J. Pratt, G. W. Bibby, and J. N. Ross. 1985. Distributed antistokes ratio thermometry. 3rd International Conference on Optical Fiber Sensors. San Diego, CA, USA. Optical Society of America. PDS3.
15. Olshansky, R., and D. B. Keck. 1976. Pulse broadening in graded-index optical fibers. *Appl. Opt.* 15: 2: 483.
16. International Telecommunications Union. 2007. Characteristics of a 50/125 μm multimode graded index optical fibre cable for the optical access network. G.651.1.
17. Zayhowski, J. J., and C. Dill. 1994. Diode-pumped passively Q-switched picosecondmicrochip lasers. *Opt. Lett.* 19: 18: 1427–1429.
18. Morkel, P. R., K. P. Jedrzejewski, E. R. Taylor, and D. N. Payne. 1992. Short-pulse, high-power Q-switched fiber laser. *IEEE Photon. Technol. Lett.* 4: 6: 545–547.
19. Lees, G. P., P. C. Wait, A. H. Hartog, and T. P. Newson. 1996. 1.64 μm pulsed source for a distributed optical fibre Raman temperature sensor. *Electron. Lett.* 32: 19: 1809–1810.
20. Lees, G. P., A. H. Hartog, and P. C. Wait. 2011. Optical pulse generator for distributed temperature sensing operating at a characteristic wavelength in a range between 1050 nm and 1090 nm. US7881566B2.
21. Wakami, T., and S. Tanaka. 1994. 1.55-μm long-span fiber optic distributed temperature sensor, 10th International Conference Optical Fibre Sensors, Glasgow, United Kingdom, SPIE, 2360: 134–137.
22. Shiota, T., and F. Wada. 1992. Distributed temperature sensors for single-mode fibers, Distributed and Multiplexed Fiber Optic Sensors, SPIE, 1586: 13–18.
23. Agrawal, G. P. 2007. *Nonlinear fiber optics*. Academic Press. New-York, USA.
24. Smith, R. G. 1972. Optical power handling capacity of low loss optical fibers as determined by stimulated Raman and Brillouin scattering. *Appl. Opt.* 11: 11: 2489–2494.
25. Urquhart, P., and P. J. R. Laybourn. 1986. Stimulated Raman scattering in optical fibers with nonconstant losses: A two-wavelength model. *Appl. Opt.* 25: 11: 1746–1753.
26. Urquhart, W. P., and P. J. R. Laybourn. 1986. Stimulated Raman scattering in optical fibers with nonconstant losses: A multiwavelength model. *Appl. Opt.* 25: 15: 2592–2599.
27. Stolen, R., and A. Johnson. 1986. The effect of pulse walkoff on stimulated Raman scattering in fibers. *IEEE J. Quant. Electron.* 22: 11: 2154–2160.
28. Chen, Y., A. H. Hartog, R. J. Marsh et al. 2014. A fast, high-spatial-resolution Raman distributed temperature sensor, 23rd International conference on optical fibre sensors, Santander, Spain, SPIE, 9157: 52.
29. Glombitza, U. 1998. Method for evaluating backscattered optical signals for determining a position-depending measuring profile of a backscattering medium. EP0692705B1.
30. MacDonald, R. I. 1981. Frequency domain optical reflectometer. *Appl. Opt.* 20: 10: 1840–1844.
31. Ghafoori-Shiraz, H., and T. Okoshi. 1986. Fault location in optical fibers using optical frequency domain reflectometry. *J. Lightw. Technol.* 4: 3: 316–322.
32. Shadaram, M. 1986. Analysis of power ratio of Rayleigh and Fresnel reflections in optical fibers using OFDR. *Appl. Opt.* 25: 2: 174–175.
33. Boisrobert, C. Y., J. Debeau, and A. H. Hartog. 1976. Sinusoidal modulation of a CW GaAs laser from 9 MHz to 1.1 GHz. *Opt. Commun.* 19: 2: 305–307.
34. Kobayashi, S., Y. Yamamoto, M. Ito, and T. Kimura. 1982. Direct frequency modulation in AlGaAs semiconductor lasers. *IEEE J. Quant. Electron.* 18: 4: 582–595.
35. Karamehmedovic, E. 2006. Incoherent optical frequency domain reflectometry for distributed thermal sensing. PhD Technical University of Denmark.
36. Barton, D. K. 1978. *Radars: Volume 3: Pulse compression.* Artech House, Dedham, MA, USA.
37. Beller, J. 2010. Optical time domain reflectometry for determining distributed physical properties EP 1 807 686 B1.
38. Nazarathy, M., S. A. Newton, R. P. Giffard et al. 1989. Real-time long range complementary correlation optical time domain reflectometer. *J. Lightw. Technol.* 7: 1: 24–38.
39. Park, J., G. Bolognini, L. Duckey et al. 2006. Raman-based distributed temperature sensor with simplex coding and link optimization. *IEEE Photon. Technol. Lett.* 18: 17: 1879–1881.
40. Bolognini, G., J. Park, M. A. Soto, N. Park, and F. D. Pasquale. 2007. Analysis of distributed temperature sensing based on Raman scattering using OTDR coding and discrete Raman amplification. *Meas. Sci. Technol.* 18: 10: 3211–3218.
41. Soto, M. A., P. K. Sahu, S. Faralli et al. 2007. High performance and highly reliable Raman-based distributed temperature sensors based on correlation-coded OTDR and multimode graded-index fibers, 3rd European Workshop on Optical Fiber Sensors, Porto, Portugal, SPIE, 6619: 3B.

42. Baronti, F., A. Lazzeri, R. Roncella et al. 2010. SNR enhancement of Raman-based long-range distributed temperature sensors using cyclic Simplex codes. *Electron. Lett.* 46: 17: 1221–1223.
43. Soto, M. A., T. Nannipieri, A. Signorini et al. 2011. Advanced cyclic coding technique for long-range Raman DTS systems with meter-scale spatial resolution over standard SMF. *IEEE Sens. J.* 878–881.
44. Bolognini, G., and A. H. Hartog. 2013. Raman-based fibre sensors: Trends and applications. *Opt. Fiber Technol.* 19: 6B: 678–688.
45. Soto, M. A., T. Nannipieri, A. Signorini et al. 2011. Raman-based distributed temperature sensor with 1 m spatial resolution over 26 km SMF using low-repetition-rate cyclic pulse coding. *Opt. Lett.* 36: 13: 2557–2559.
46. Stierlin, R., J. Ricka, B. Zysset et al. 1987. Distributed fiber-optic temperature sensor using single photon counting detection. *Appl. Opt.* 26: 8: 1368–1370.
47. Thorncraft, D. A., M. G. Sceats, and S. B. Poole. 1991. An ultra high resolution distributed temperature sensor, 8th International Conference on Optical Fiber Sensors, Monterey, CA, USA, IEEE: 258–261.
48. Feced, R., M. Farhadiroushan, V. A. Handerek, and A. J. Rogers. 1997. Advances in high resolution distributed temperature sensing using the time-correlated single photon counting technique. *IEE Proc. J.* 144: 3: 183–188.
49. Brown, R. G. W., K. D. Ridley, and J. G. Rarity. 1986. Characterization of silicon avalanche photodiodes for photon correlation measurements. 1: Passive quenching. *Appl. Opt.* 25: 22: 4122–4126.
50. Brown, R. G. W., R. Jones, J. G. Rarity, and K. D. Ridley. 1987. Characterization of silicon avalanche photodiodes for photon correlation measurements. 2: Active quenching. *Appl. Opt.* 26: 12: 2383–2389.
51. Höbel, M., J. Ricka, M. Wuethrich, and T. Binkert. 1995. High-resolution distributed temperature sensing with the multiphoton-timing technique. *Appl. Opt.* 34: 16: 2955–2967.
52. Basnett, T., and S. J. Barber. 1989. Distributed temperature measurement with optical fibres, Fibre Optics '89, London, United Kingdom, SPIE, 1120: 291–297.
53. Peck, D., and P. Seebacher. 2000. Distributed temperature sensing using fibre-optics (DTS systems), Electricity Engineers' Association Annual Conference, Auckland, New Zealand: 1–8.
54. Healey, P. 1981. Multichannel photon-counting backscatter measurements on monomode fibre. *Electron. Lett.* 17: 20: 751–752.
55. Yuan, Z. L., B. E. Kardynal, A. W. Sharpe, and A. J. Shields. 2007. High speed single photon detection in the near infrared. *Appl. Phys. Lett.* 91: 4: 041114.
56. Zhang, J., R. Thew, C. Barreiro, and H. Zbinden. 2009. Practical fast gate rate InGaAs/InP single-photon avalanche photodiodes. *Appl. Phys. Lett.* 95: 9: 091103.
57. Campbell, J. C., W. Sun, Z. Lu, M. A. Itzler, and X. Jiang. 2012. Common-mode cancellation in sinusoidal gating with balanced InGaAs/InP single photon avalanche diodes. *IEEE J Sel Topics in Quant Electron* 48: 12: 1505–1511.
58. Yan, L., R. Min, E. Wu et al. 2012. High-speed photon-number resolving with sinusoidally gated multipixel photon counters. *IEEE Photon. Technol. Lett.* 24: 20: 1852–1855.
59. Restelli, A., J. C. Bienfang, and A. L. Migdall. 2013. Single-photon detection efficiency up to 50% at 1310 nm with an InGaAs/InP avalanche diode gated at 1.25 GHz. *Appl. Phys. Lett.* 102: 14: 141104.
60. Zhang, Y., X. Zhang, and S. Wang. 2013. Gaussian pulse gated InGaAs/InP avalanche photodiode for single photon detection. *Opt. Lett.* 38: 5: 606–608.
61. Natarajan, C. M., M. G. Tanner, and R. H. Hadfield. 2012. Superconducting nanowire single-photon detectors: Physics and applications. *Supercond. Sci. Technol.* 25: 6: 063001.
62. Tanner, M. G., S. D. Dyer, B. Baek, R. H. Hadfield, and S. Woo Nam. 2011. High-resolution single-mode fiber-optic distributed Raman sensor for absolute temperature measurement using superconducting nano-wire single-photon detectors. *Appl. Phys. Lett.* 99: 201110–201113.
63. Feced, R., M. Farhadiroushan, V. A. Handerek, and A. J. Rogers. 1997. A high spatial resolution distributed optical fiber sensor for high-temperature measurements. *Rev. Sci. Instrum.* 68: 10: 3772–3776.
64. Stoddart, P. R. et al. 2005. Fibre optic distributed temperature sensor with an integrated background correction function. *Meas. Sci. Technol.* 16: 6: 1299.
65. Brown, R. G. W., and M. Daniels. 1989. Characterization of silicon avalanche photodiodes for photon correlation measurements. 3: Sub-Geiger operation. *Appl. Opt.* 28: 21: 4616–4621.
66. Maata, K., and J. Kostamovaara. 1998. A high-precision time-to-digital converter for pulsed time-of-flight laser radar applications. *IEEE Trans. Instrum. Meas.* 47: 2:521–536.
67. Dudek, P., S. Szezepanski, and J. V. Hatfield. 2000. A high-resolution CMOS time-to-digital converter utilizing a Vernier delay line. *IEEE Trans. Solid-State Circuits* 35: 2: 240–247.
68. Raisanen-Ruatsaliainen, E., T. Rahkonen, and J. Kostamovaara. 2000. An integrated time-to-digital converter with 30-ps single-shot precision. *IEEE J. Solid-State Circuits* 35: 10: 1507–1510.
69. Dyer, S. D., M. G. Tanner, B. Baek, R. H. Hadfield, and S. W. Nam. 2012. Analysis of a distributed fiber-optic temperature sensor using single-photon detectors. *Opt. Expr.* 20: 4: 3456–3466.

70. Cova, S., M. Ghioni, A. Lacaita, C. Samori, and F. Zappa. 1996. Avalanche photodiodes and quenching circuits for single-photon detection. *Appl. Opt.* 35: 12: 1956–1976.
71. Höbel, M., and J. Ricka. 1994. Dead-time and afterpulsing correction in multiphoton timing with nonideal detectors. *Rev. Sci. Instrum.* 65: 7: 2326–2336.
72. Bethea, C. G., S. Cova, G. Ripamonti, and B. F. Levine. 1988. High-resolution and high-sensitivity optical-time-domain reflectometer. *Opt. Lett.* 13: 3: 233–235.
73. Comandar, L. C., B. Fröhlich, M. Lucamarini et al. 2014. Room temperature single-photon detectors for high bit rate quantum key distribution. *Appl. Phys. Lett.* 104: 2: 1101.
74. Patel, K. A., J. F. Dynes, A. W. Sharpe et al. 2012. Gigacount/second photon detection with InGaAs avalanche photodiodes. *Electron. Lett.* 48: 2: 111–113.
75. Jian, Y., E. Wu, X. Chen, G. Wu, and H. Zeng. 2011. Time-dependent photon number discrimination of InGaAs/InP avalanche photodiode single-photon detector. *Appl. Opt.* 50: 1: 61–65.
76. Lee, R. W., R. C. Frank, and D. E. Swets. 1962. Diffusion of hydrogen and deuterium in fused quartz. *J. Chem. Phys.* 36: 4: 1062–1071.
77. Shackleford, J. F., P. L. Studt, and R. M. Fulrath. 1972. Solubility of gases in glass. II. He, Ne, and H2 in fused silica. *J. Appl. Phys.* 43: 4: 1619–1626.
78. Shelby, J. E. 1977. Molecular diffusion and solubility of hydrogen isotopes in vitreous silica. *J. Appl. Phys.* 48: 8: 3387–3394.
79. Stone, J. 1987. Interactions of hydrogen and deuterium with silica optical fibers: A review. *J. Lightw. Technol.* 5: 5: 712–733.
80. Namihira, Y., K. Mochizuki, M. Kuwazuru, and Y. Iwamoto. 1983. Effects of hydrogen diffusion on optical fibre loss increase. *Electron. Lett.* 19: 24: 1034–1035.
81. Mochizuki, K., Y. Namihira, M. Kuwazuru, and M. Nunokawa. 1984. Influence of hydrogen on optical fiber loss in submarine cables. *J. Lightw. Technol.* 2: 6: 802–807.
82. Mochizuki, K., Y. Namihira, M. Kuwazura, and Y. Iwamoto. 1984. Behavior of hydrogen molecules adsorbed on silica in optical fibers. *IEEE J. Quant. Electron.* 20: 7: 694–697.
83. Ramos, R. T., and W. D. Hawthorne. 2008. Survivability of optical fiber for harsh environments, SPE Annual Technical Conference and Exhibition, 21–24 September 2008, Denver, CO, USA, SPE: paper SPE 116075-MS.
84. Lemaire, P., K. L. Walker, K. Kranz, R. Huff, and F. DiMarcello. 1989. Diffusion of hydrogen through hermetic carbon films on silica fibers, MRS Proceedings, Cambridge University Press, 172: 85.
85. Kishimoto, N., T. Tanabe, T. Suzuki, and H. Yoshida. 1985. Hydrogen diffusion and solution at high temperatures in 316L stainless steel and nickel-base heat-resistant alloys. *J. Nucl. Mater.* 127: 1: 1–9.
86. Shanabarger, M. R. 1994. Comparison of the high temperature hydrogen transport parameters for the alloys Incoloy 909, Haynes 188, and Mo-47.5 Re, Hydrogen Effects in Materials: 5th International Conference on the Effects of Hydrogen on the Behavior of Materials, Moran, WY, USA, TMS: 243–250.
87. Duval, S., R. Antano, F. Ropital, and M. Jerome. 2004. Hydrogen permeation through ARMCO iron membranes in sour media, CORROSION 2004, New Orleans, USA, NACE International: 04740.
88. Yamate, T., R. J. Schroeder, R. T. Ramos, and O. C. Mullens. 2007. Method for measuring using optical fiber distributed sensor. US7215416B2.
89. Pruett, P. E. 2004. Methods and apparatus for calibrating a distributed temperature sensing system. US6807324B2.
90. Hartog, A. H., M. P. Gold, and A. P. Leach. 1987. Optical time domain reflectometry. EP0213872B2.
91. Yamate, T., R. J. Schroeder, R. T. Ramos, and O. C. Mullens. 2006. Method for measuring and calibrating measurements using optical fiber distributed sensor. US 7 126 680B2.
92. Bibby, G. W. 1985. Temperature measurement. EP0225023B1.
93. Suh, K., and C. E. Lee. 2008. Auto-correction method for differential attenuation in a fiber-optic distributed-temperature sensor. *Opt. Lett.* 33: 16: 1845–1847.
94. Hartog, A. H. 2003. Optical time domain reflectometry. EP 1 338 877 B1.
95. Lee, C. E., K. Kalar, and M. E. Sanders. 2007. Methods and apparatus for dual source calibration for distributed temperature sensors. US7628531B2.
96. Suh, K., C. Lee, M. Sanders, and K. Kalar. 2008. Active plug & play distributed Raman temperature sensing, 19th International Conference on Optical Fibre Sensors, Perth, Australia, SPIE, 7004: 35.
97. Fukuzawa, T., H. Shida, K. Oishi, N. Takeuchi, and S. Adachi. 2013. Performance improvements in Raman distributed temperature sensor. *Photonic Sensors* 3: 4: 314–319.

98. Hartog, A. H. 2007. Distributed optical fibre measurements. US7284903B2.
99. Hartog, A. H. 2003. Optical time domain reflectometry EP0636868B1.
100. Hartog, A. H., and D. N. Payne, 1982, Remote measurement of temperature distribution using an optical fibre, 8th European Conference on Optical Communications Cannes, France: 215–218.
101. Belal, M., Y. T. Cho, M. Ibsen, and T. P. Newson. 2010. A temperature-compensated high spatial resolution distributed strain sensor. *Meas. Sci. Technol.* 21: 015204.

5 Brillouin-Based Distributed Temperature and Strain Sensors

Brillouin scattering is an inelastic scattering process that occurs in optical fibres. It results in two additional frequencies on either side of the incident light. It differs from Raman scattering in that the frequency shift is much smaller (~10 GHz vs. ~13 THz for Raman scattering) and its natural linewidth is also far narrower (~30 MHz vs. ~6 THz).

These quantitative differences have made Brillouin scattering far more versatile than Raman scattering for distributed sensing for the following reasons.

Firstly, Brillouin scattering in glasses is sensitive to strain and temperature in its frequency and intensity, whereas the Raman *strain-dependent* frequency shift is only readily detected in crystalline [1] or reasonably ordered solid materials such as polymer fibres with aligned molecular chains [2]. In glasses, the breadth of the Raman line precludes its use in the measurement of the frequency shift and so it deprives us of the information that it yields in crystals. Thus, the ability to measure precisely the frequency shift of the Brillouin scattering has allowed distributed sensing technology to provide strain as well as temperature data. Some interrogation approaches allow the strain and temperature to be determined independently [3].

Secondly, the very narrow linewidth of the Brillouin process allows optical amplification [4] and heterodyne [3,4] techniques to be employed for enhancing the signal quality; this has enabled Brillouin measurements to be carried out over extremely long distances [5]. In contrast, optical amplification techniques are ineffective for very broad signals such as Raman scattering and heterodyne detection is completely ruled out given that the receiver would need a bandwidth of several THz to capture the signal.

Thirdly, stimulated Brillouin scattering (SBS) can readily be used in sensing; indeed, the first published Brillouin sensing approaches were based on SBS [6,7]. The signal obtained in this case is a measure of the gain (or loss) in the interaction of two counter-propagating waves. This has led to some very interesting configurations where the signals are engineered such that they can interact only over very short distances, and, as a result, spatial resolutions in the millimetre range [8] have been demonstrated. These configurations have no equivalent in Raman distributed temperature sensor (DTS).

Finally, the spontaneous Brillouin backscatter signal is substantially stronger (by about one order of magnitude) than the Raman anti-Stokes backscatter, although that advantage is mitigated by the weaker sensitivity of its intensity to temperature.

Although the physics is different in the two cases, there are many similarities between spontaneous Brillouin scattering (SpBS) and spontaneous Raman scattering. The values of the parameters involved, notably the frequency shift, sensitivity and linewidth, are different, but the basic principles are familiar to those who have worked in Raman DTSs. The term *Brillouin optical time-domain reflectometer* (BOTDR) is used for systems based on time-domain interrogation of SpBS.

SBS, however, brings quite different concepts in that, in the stimulated case, it is the gain profile that is measured rather than the backscatter intensity spectrum (although of course these two are closely related) and, in the process of measuring the gain spectrum, there is a large amount of flexibility in the implementation. This has led to many schemes for exploiting SBS, including Brillouin optical time-domain analysis (BOTDA), Brillouin optical frequency-domain analysis (BOFDA), Brillouin optical coherence domain analysis (BOCDA) and Brillouin optical coherence domain reflectometry (BOCDR) and techniques based on dynamic Bragg gratings (DBGs). The field further expanded into methods for expanding the length of the sensing fibre that can be interrogated and for accelerating the speed at which the measurement is updated.

5.1 SPONTANEOUS BRILLOUIN SCATTERING

5.1.1 The Effect

As mentioned in Chapter 1, SpBS is an inelastic process that results from the interaction between an incident wave (the probe wave in our case) and thermally driven material-density fluctuations that travel at the speed of sound. In the thermal agitation, acoustic phonons of all frequencies are constantly being created and re-absorbed. During the lifetime of these phonons, they modulate the refractive index through the stress-optical effect* and produce a travelling fluctuation of the refractive index caused by the acoustic wave of the phonon [9,10]. Those phonons, whose acoustic wavelength happens to match the optical wavelength of the incident light, have a finite probability of interacting with the probe by either adding or subtracting energy to an incident photon.

In contrast with Raman scattering, where the interaction is with molecular vibrations, i.e. *optical* phonons, here the interactions occur with *acoustic* phonons, i.e. vibrations of the lattice. The process has some similarity to the acoustic diffraction of light in an acousto-optic modulator (AOM) [11], wherein an incident optical beam interacts with an ultrasonic acoustic wave to diffract some of the light.

In materials that support shear stress, such as glasses, Brillouin scattering exists with both longitudinal acoustic (LA, pressure waves) and transverse acoustic (TA, shear waves) [12], the TA waves being substantially slower than the LA waves. In practice, the LA waves are used exclusively in distributed sensing, primarily because we are considering only backscattered light, whereas scattering from TA phonons – in any case weaker than that from LA phonons – falls off in intensity in the angles captured by optical fibres in the backward direction.

5.1.2 Frequency Shift

The frequency ν_B of the Brillouin LA lines is dictated by the condition of the acoustic and optical wavefronts matching [12], i.e.

$$\nu_B = 2n_1\nu_0\frac{V_A}{c}\sin\left(\frac{1}{2}\theta\right) \tag{5.1}$$

for an incident optical frequency ν_0, an acoustic velocity V_A and an angle θ between the incident and scattered light; c is the speed of light in vacuum. In the case of backscatter where $\theta \simeq \pi$, of course, the sine

* The presence of an acoustic wave induces a periodic deformation of the glass, which in turn alters its refractive index.

term is close to unity.* Equation 5.1 essentially states that, for the backscatter case, the acoustic wavelength must match the optical wavelength for the scattering and re-capturing process to occur. Typically, values of $V_A = 5960$ m/s and $n_1 = 1.45$ yield about 11 GHz for ν_B at an optical frequency of 193.4 THz (1550 nm). The Brillouin frequency is proportional to probe frequency so at the wavelength of 633 nm (where the present author carried out initial experiments in the early 1980s), the Brillouin shift is around 27 GHz. The precise value of the frequency shift is a function of the material composition and the design of the fibre [14]. For example, the addition of GeO_2 to the core reduces the frequency shift. Some of the modern single-mode fibre designs that tailor the dispersion and bending-loss characteristics exhibit complex Brillouin spectra with multiple peaks.

5.1.3 Spectral Width

The spectral width of the spontaneous Brillouin backscatter is dictated by the lifetime of the thermal phonons. The lifetime of the hypersonic phonons is a function of frequency and temperature, and it is known that the attenuation of the hypersonic wave (which determines the phonon lifetime) is a function of its frequency. In silica, there is a resonance of the phonons at energies corresponding to temperatures around 130 K for hypersonic frequencies of 35 GHz [15], i.e. corresponding to a probe wavelength of 488 nm. As the wavelength increases, the corresponding hypersonic frequency that is probed is reduced and the phonon resonance occurs in the region of 70 K (depending on the fibre type) for probe wavelengths in the 1300–1550 nm region [16]. Above the temperature at which that resonance occurs, the absorption coefficient for hypersonic phonons is proportional to the square of the frequency [17], i.e. inversely proportional to the square of the probe wavelength. The Brillouin linewidth is therefore also inversely proportional to the square of the probe wavelength.

For a probe wavelength of 1550 nm and for those photons scattered at exactly 180° (i.e. on axis) in silica, the natural linewidth is about 17 MHz [14], corresponding to a phonon lifetime of around 10 ns. However, an optical fibre is not a single material, and variations in the glass composition lead to broadening of the Brillouin spectrum.

In a radially non-uniform fibre, the Brillouin contribution of each part of the core and cladding will generally be at a different shift, and this effect broadens the Brillouin linewidth.

A further contribution to the broadening of the Brillouin spectrum is the geometric effect shown in Equation 5.1. In essence, as the numerical aperture increases, so the range of angles that is captured by the waveguide broadens. Since the Brillouin frequency shift reduces for decreasing angles from the pure 180° (on axis) backscatter, the Brillouin spectrum is therefore broadened towards lower frequencies relative to the case of a very-low-*NA* fibre. This effect was discussed in Ref. [18], where it was shown that the resulting linewidth $\Delta\nu_B$ is given by

$$\Delta\nu_B(NA) = \sqrt{\Delta\nu_B(0)^2 + \nu_B^2 \frac{NA^4}{4n_1^4}}, \tag{5.2}$$

where ν_B specifically refers to the Brillouin shift on axis. The geometrical effect contributes materially to the overall linewidth even at low *NA*, such as 0.1. A conventional single-mode fibre designed for long

* It is worth noting that Brillouin scattered light, measured perpendicular to the axis of the fibre, was studied well before its use in sensing in order to understand how optical fibres perform in terms of scattering loss, power distribution between core and cladding and the actual material composition. For further details, the reader is referred to Ref. [13]. An example of the use of forward stimulated Brillouin scattering is discussed in Chapter 8 (Section 8.2.3).

distance applications has an *NA* around 0.12–0.14, although modern fibre designs are seldom of the simple step-index type.

So far, the spectrum has been discussed only in terms of the optical contributions. However, Brillouin scattering is the interaction between at least one optical signal (in the case of spontaneous scattering) and acoustic waves resulting in a third, scattered wave. The acoustic waves themselves are in general guided by the acoustic waveguide formed by the contrast in acoustic velocity in the different parts of the fibre. Most of the core dopants, used to raise the refractive index, also raise the acoustic index (i.e. they increase the *slowness* of the acoustic wave) [19], the major exception being Al_2O_3, which has the opposite effect.

The effect of acoustic waveguides on Brillouin scattering is discussed in the literature only in relation to the stimulated process. Although SpBS does not involve the regenerative feedback between probe light and acoustic phonons through electrostriction that results in the stimulated process, the interaction nonetheless takes place over a substantial fibre length dictated by the acoustic lifetime, and this length represents several hundred acoustic wavelengths and about 10 core radii.

It follows that the acoustic modes shape the Brillouin scattering process, in terms of the overlap integral that dictates the efficiency of the process. Much of the work on Brillouin scattering is carried out in single-mode fibres, by which it is meant that the fibre is operated at optical frequencies below the cut-off of the second mode. However, the waveguide is not necessarily single mode for *acoustic* waves, and it is frequently the case that more than one acoustic mode is guided and shapes the Brillouin spectrum [20,21].

5.1.4 Sensitivity to Temperature and Strain

The frequency shift and the intensity of the spontaneous Brillouin backscatter are sensitive to temperature and strain. These relationships can be described with a 2 × 2 matrix, the elements of which will be discussed next.

5.1.4.1 Sensitivity coefficients

The coefficients of frequency shift, $C_{\nu_B\varepsilon}$, C_{ν_BT}, and intensity, $C_{I_B\varepsilon}$, C_{I_BT}, due to changes in strain $\Delta\varepsilon$ and temperature ΔT can be determined experimentally for a given fibre type and measurement wavelength, and so a given change in the frequency and intensity may be written as Refs. [3] and [22]

$$\begin{bmatrix} \Delta\nu_B \\ \Delta I_B \end{bmatrix} = \begin{bmatrix} C_{\nu_B\varepsilon} & C_{\nu_BT} \\ C_{I_B\varepsilon} & C_{I_BT} \end{bmatrix} \begin{bmatrix} \Delta\varepsilon \\ \Delta T \end{bmatrix}. \tag{5.3}$$

Provided that the determinant of this matrix is non-zero, it can be inverted so that measured changes in frequency and intensity may be used to infer the temperature and strain relative to some baseline as follows [3,23]:

$$\begin{bmatrix} \Delta\varepsilon \\ \Delta T \end{bmatrix} = \frac{1}{\left|C_{\nu_B\varepsilon}C_{I_BT} - C_{I_B\varepsilon}C_{\nu_BT}\right|} \begin{bmatrix} C_{I_BT} & -C_{\nu_BT} \\ -C_{I_B\varepsilon} & C_{\nu_B\varepsilon} \end{bmatrix} \begin{bmatrix} \Delta\nu_B \\ \Delta I_B \end{bmatrix}. \tag{5.4}$$

The uncertainties in calculated strain and temperature, for errors $\delta\nu_B$ and δI_B, on the frequency and intensity are then given by Refs. [3] and [23]

$$|\delta\varepsilon| = \frac{|C_{I_BT}||\delta\nu_B| + |C_{\nu_BT}||\delta I_B|}{|C_{\nu_B\varepsilon}C_{I_BT} - C_{I_B\varepsilon}C_{\nu_BT}|} \tag{5.5}$$

and $$|\delta T| = \frac{|C_{I_B\varepsilon}||\delta\nu_B| + |C_{\tilde{\nu}_B\varepsilon}||\delta I_B|}{|C_{\tilde{\nu}_B\varepsilon}C_{I_BT} - C_{I_B\varepsilon}C_{\tilde{\nu}_BT}|}. \tag{5.6}$$

Typical values of the sensitivity coefficients may be found in Refs. [3] and [24] and are summarised, for typical step index single-mode fibres, as

$$\begin{bmatrix} C_{\nu_B\varepsilon} & C_{\nu_BT} \\ C_{I_B\varepsilon} & C_{I_BT} \end{bmatrix} = \begin{bmatrix} 0.046\ \text{MHz}/\mu\varepsilon & 1.07\ \text{MHz/K} \\ -8\cdot10^{-4}\%/\mu\varepsilon & 0.36\%/\text{K} \end{bmatrix}.$$

It will be noted that the terms forming the determinant are of opposite sign, and this helps slightly in improving the conditioning of Equation 5.3. Interestingly, if we select $\delta\nu_B = 1.1$ MHz and $\delta I_B = 0.36\%$, i.e. each equivalent to a 1 K uncertainty if affecting the temperature measurement only, then from Equations 5.5 and 5.6, the overall temperature uncertainty is still 1 K, but the strain uncertainty is far worse than that dictated by frequency or intensity uncertainty alone.

The sensitivity of intensity to temperature is considerably weaker than is the case with Raman scattering (c. 0.36%/K vs. 0.8%/K for the anti-Stokes line at room temperature), and this presents a challenge for precision measurements.

5.1.4.2 Sensitivity of the intensity to temperature

The same equations relate the intensity of the Brillouin scattering to temperature as applied to Raman scattering, namely [15]

$$I_{B_{as}}(T) = \frac{K_B}{\lambda_{as}^4}\cdot\left(\frac{1}{\exp\left(\frac{h\cdot\nu_B}{k_B\cdot T}\right)-1}\right) \quad \text{anti-Stokes}$$

$$I_{B_s}(T) = \frac{K_B}{\lambda_s^4}\cdot\left(\frac{1}{\exp\left(\frac{h\cdot\nu_B}{k_B\cdot T}\right)-1}+1\right) \quad \text{Stokes.} \tag{5.7}$$

The intensity coefficient K_B is related to the Rayleigh scattering coefficient through the Landau-Placzek ratio (LPR) (see Section 5.1.4.4). In this case, the Brillouin shift is in the region of 10 GHz (for incident light at 1550 nm, rising to 27 GHz at 633 nm), so the energy in the numerator of the exponents in Equation 5.7 (representing the energy of the phonon responsible for the scattering) is about three orders of magnitude smaller than the denominator (thermal energy for each degree of freedom). It follows that, at temperatures above a few K, the intensity of Brillouin scattered light is nearly the same for the Stokes and the anti-Stokes lines and so the anti-Stokes/Stokes ratio cannot be used in the same way as it is for Raman scattering. At room temperature, Brillouin scattering is essentially proportional to absolute temperature and the relative sensitivity, a more useful measure for distributed sensing, is roughly $1/T$. However, at cryogenic temperatures, where the phonon energy is closer to k_BT, the difference between the Stokes and anti-Stokes lines is re-asserted and measurable differences in their intensities can be observed [15].

5.1.4.3 Sensitivity of intensity to strain

The coefficient K_B in Equation 5.7 is also a function of strain. At temperatures well above $k_B/h \cdot \nu_B$, the Brillouin intensity in just one of the two lines, normalised to the Rayleigh intensity, was shown to be the following function of Young's modulus E_Y, Poisson's ratio ν_p, fictive temperature T_f and the isothermal compressibility B_T [25]:

$$\frac{I_B}{I_R}(T,\varepsilon) = \frac{T}{T_f}\frac{1}{E_\gamma \cdot B_T \cdot \dfrac{(1-\nu_p)}{(1+\nu_p)\cdot(1-2\nu_p)} \cdot (1+5.75\varepsilon) - 1}. \tag{5.8}$$

5.1.4.4 Loss compensation for intensity measurements

In the case of Brillouin intensity measurements, the problem of referencing the backscatter signal against an equivalent temperature-insensitive signal with similar attenuation is solved far more effectively than in Raman DTS systems by use of the Rayleigh line. The latter is so close in its frequency to the Brillouin lines that there is no significant difference in its attenuation compared to that of the Brillouin signals. In principle, therefore, we have almost perfect compensation for effects other than those of interest, i.e. the ratio of Rayleigh to Brillouin lines in the scattered light spectrum, known as the LPR. In a liquid, this ratio relates purely to the properties (mainly the compressibility) of the material and is not temperature dependent because that term is the same for the two signals forming the LPR. However, in a glass, the temperature applicable to Rayleigh scattering is the fictive temperature (the temperature at which inhomogeneities are fixed during the cooling process), whereas that applying to the Brillouin scattering relates to the present, ambient temperature. So in glasses, the LPR is a function of temperature [9].

For a single-component glass, the LPR is given by half the inverse of Equation 5.8:

$$LPR(T) = \frac{I_R}{2I_B(T)} = \frac{T_f}{T}\frac{\rho V_A^2 B_T - 1}{2} \tag{5.9}$$

(parameters are defined in Equation 5.8). It should be noted that some sources, notably Fabelinskii [10], use the definition in Equation 5.9, which is the ratio of the Rayleigh backscatter to the sum of the energy in both Brillouin lines, whereas others omit the factor of 2 in the denominator; i.e. they use the energy in a single Brillouin line. The choice of definition is not particularly important provided the ratio is used consistently.

Although the Rayleigh scattered signal provides an effective reference signal, using the LPR is still challenging, partly because of the noise relating to coherence in the Rayleigh backscatter. Thus, a broad spectrum is required to average the effects of coherence and so the same source cannot be used for the Brillouin interrogation as for Rayleigh backscatter acquisition, unless the source is widely tunable. As an example, with a spatial resolution of 10 m, the frequency separation required for each statistically independent Rayleigh measurement is 10 MHz (see Chapter 6, Section 6.5.1), and so for a 1 K resolution (using a sensitivity of

0.3%/K), 10^5 independent measurements are required, thus spanning ~1 THz (equivalent to about 8 nm at an operating wavelength of 1550 nm). These are modest values of spatial and temperature resolution and yet the tunability required is a substantial fraction of the range of typical tunable sources; this limits the scope for substantial improvements in the spatial or temperature resolution. Furthermore, as the spectral width of the source used for the Rayleigh measurement is increased, the assumption that the attenuation that is measured is representative of that suffered by the narrowband Brillouin signal or signals becomes less reliable.

The use of the LPR for a truly independent measurement of strain and temperature is therefore not straightforward, although there are a number of reports of its successful implementation either intrinsically in a single apparatus [3,26–29] or by multiplexing a BOTDR with a conventional OTDR using an optical switch [30].

Even when the temperature/strain discrimination based on the LPR amplifies the uncertainty and the most reliable output is frequency derived, the availability of intensity, as well as frequency-shift information, often helps interpret the frequency measurement. For example, if it is not clear whether a feature on the frequency-shift profile is due to temperature or strain, access to intensity data can resolve the ambiguity on a qualitative basis at least, and in many applications of practical importance, this is sufficient for the interpretation of the results.

5.1.4.5 Sensitivity of the Brillouin frequency shift

The Brillouin frequency shift is also a function of strain and temperature.

Some insight into the sensitivities of frequency shift to temperature and strain [31] may be gained by differentiating Equation 5.1 and considering its connection to the basic physical properties of the materials forming the fibre. The coefficient relating the Brillouin frequency shift to strain and temperature is given by Ref. [31]

$$C_{\nu_B\varepsilon,T} = \frac{1}{n_{eff}}\frac{\partial n_{eff}}{\partial(\varepsilon,T)} + \frac{1}{V_A}\frac{\partial V_A}{\partial(\varepsilon,T)}. \tag{5.10}$$

An effective refractive index n_{eff} that allows for the fact that the wave propagates in the cladding as well as the core (and for any radial variation of the core index) must be used. The acoustic velocity is related to the Young's modulus, Poisson's ratio and density ρ (c. 2.21 × 10^3 kg/m^3) through

$$V_A = \sqrt{\frac{E_Y(1-\nu_p)}{(1+\nu_p)(1-2\nu_p)\rho}}. \tag{5.11}$$

So its strain sensitivity is related to those of the three parameters in Equation 5.11 as follows:

$$\frac{1}{V_A}\frac{\partial V_A}{\partial\varepsilon} = \frac{1}{2E_Y}\frac{\partial E_Y}{\partial\varepsilon} + \frac{\nu_p(2-\nu_p)}{\left(1-\nu_p^2\right)(1-2\nu_p)}\frac{\partial\nu_p}{\partial\varepsilon} - \frac{1}{2\rho}\frac{\partial\rho}{\partial\varepsilon}. \tag{5.12}$$

The variation of the density with strain follows from the Poisson ratio:

$$\frac{1}{\rho}\frac{\partial\rho}{\partial\varepsilon} = -(1-2\nu_p), \tag{5.13}$$

whereas the other terms have been determined experimentally as $\frac{1}{E_Y}\frac{\partial E_Y}{\partial \varepsilon} = 5.75$ and $\frac{\partial \nu_p}{\partial \varepsilon} = 3.07$ [32].

The sensitivity of the refractive index to strain is determined by the photo-elastic constants p_{11} and p_{12}:

$$\frac{1}{n_{eff}}\frac{\partial n_{eff}}{\partial \varepsilon} = -\frac{n_{eff}^2}{2}\left[p_{12} - \nu_p(p_{11} + p_{12})\right]. \tag{5.14}$$

If the photo-elastic coefficients vary across the core, their intensity-weighted value would be used. The generally accepted values of E_Y and ν_p for fused silica are, respectively, 73.3 GPa and 0.17 [33]. The photo-elastic constants for bulk fused silica were reported in the form of stress-optical coefficients [34,35], and the latter results were p_{11} = 0.126 and p_{12} = 0.26 at 633 nm. However, in practice, optical fibres are not made from pure silica, nor are they in the same annealed state as the samples used for bulk measurements. So experimental values obtained on similar fibres would be somewhat more relevant, although they are usually not available. In optical fibres, somewhat slightly lower results were reported at the same wavelength [36] and the photo-elastic constants are known to be dispersive [37,38].

Given the wide variety of optical fibre types and the photo-elastic dispersion properties, it is far preferable to determine $C_{\nu_B\varepsilon}$ directly by straining a sample section of the selected fibre as will be used for sensing and measuring the resulting change in ν_B. Nonetheless, the results can be checked for plausibility using the basic relationships given previously with the elastic and photo-elastic constants; some extrapolation for different wavelengths can be made using dispersion models [38].

Using similar expressions to Equations 5.10 and 5.12, C_{ν_BT} can be estimated from published values [39] of the temperature sensitivity of Young's modulus $\frac{\partial E_Y}{\partial T} = 1.02$ GPa/K and shear modulus $\frac{\partial G}{\partial T} = 0.31$ GPa/K and using the relation $\nu_p = E_Y/2G$. The calculation of C_{ν_BT} also requires the thermal coefficients of expansion (~0.5×10^{-6}/K at room temperature* [33]) and refractive index (0.95×10^{-5}/K [33] for silica).

5.1.4.6 Coating and coupling effects

The foregoing discussion focused on the fibre itself and the measurands that it experiences. In practice, this is highly simplistic, particularly with regards to strain; the temperature influence can, under quasi-static assumptions, usually be assumed to transfer efficiently from the outside of a cable to the core of the fibre.† However, the various coating layers that are applied to the fibre will usually modify the temperature and strain response. Although the coatings applied to the fibre usually have a lower Young's modulus than silica does, their substantial cross-sectional area and far higher coefficient of thermal expansion will often radically alter the temperature sensitivity under most circumstances.

The attachment of the sensing cable to a structure leads to further potential for error, and so, using distributed strain measurement for monitoring structures requires careful consideration of the strain coupling through the cable to the sensing fibre.

* Note, however, that the expansion coefficient of silica varies strongly and non-monotonically with temperature [33].

† The major exception being actively heated cables, where the heat transfer from a resistive element to the outside of the cable is the property that is being measured through the temperature sensed by the fibre.

Before distributed strain sensing techniques were known, the average strain along a fibre was measured by determining the time required for a pulse to travel through a section of fibre of known length (under zero strain) as the fibre is stretched. When the fibre is strained, the transit time changes partly because the fibre length varies but also because its refractive index is altered (the strain-optical effect). In any event, having calibrated the variation of transit time as a function of strain, measurements of the pulse transit time are a proxy for fibre strain. This technique was used to understand the effect of strain applied by coatings on the fibre.

Experiments using pulse transit times on fibres coated to a diameter of 0.5 mm with nylon [40] showed an increase in stress due to thermal expansion by about 34 με/K; the effect was analysed in more detail as a function of coating thickness [41]. Tests using Brillouin scattering on more modern fibres [42] (0.9 mm jacket diameter) yielded a value of 44 με/K. These are substantial effects that cannot be ignored even in a suspended, loosely coupled cable.

5.2 STIMULATED BRILLOUIN SCATTERING

Associated with the SpBS process is a stimulated process that results in gain at the Stokes wavelength. At sufficient probe power [43], essentially all the power can return to the launching end, shifted to the Brillouin Stokes frequency. This effect was demonstrated in 1972 [44] in short lengths (5–20 m) of single-mode fibre at short wavelength (535.5 nm) with moderate power (1–2.5 W). The use of short wavelengths enhances the gain in a number of ways, partly through the higher intensity resulting from a small core and also because the gain is inversely proportional to wavelength squared.

SBS was demonstrated at thresholds of ~300 mW at the yet shorter wavelength of 514 nm in similar lengths of fibre [45], and at higher powers, it was shown that multiple orders of SBS occur (with 14 orders observed) [46]. An early experimental result on long (13 km), low-loss single-mode fibres at 1320 nm showed that the SBS reaches threshold at around 5 mW of launched power [47]. The Brillouin return rises sharply for further increases of launched power until, at around 15 mW, essentially all of the launched power is returned as SBS (allowing, of course, for losses in the propagation).

A theory of SBS in optical fibres showed the pivotal role of the acoustic modes of the waveguide [48]: the fibre is as much an acoustic as an optical waveguide, and so the presence of guided acoustic waves shapes the Brillouin spectrum.

Returning to the case of SpBS, we have seen that it gives rise to scattered waves travelling in the opposite direction at a frequency close to that of the probe. The counter-propagating optical waves are similar to standing waves in that the interference between them creates regions where the electric fields associated with these waves add and other regions where they subtract, giving rise to a modulation of the mean electric field. In the case of Brillouin scattering and in contrast to a standing wave, this beat pattern travels along the fibre at the acoustic velocity, as illustrated in Figure 5.1. This picture applies to SpBS as well as to SBS. The travelling modulation of the electric field interacts with the material through electrostriction, a property of dielectric materials that causes them to change shape when they are subjected to an electric field. Thus, the presence of a periodic electric field strains the fibre very slightly in a periodic manner. In turn, the strain-optical effect converts this periodic strain into minute variations of the refractive index, and so a moving grating is created. At low optical powers (well below the threshold for stimulated scattering), the grating created through electrostriction is so weak as to be negligible. However, as the incident power increases, so too does the backscatter, and because the index modulation is proportional to the intensity of both counter-propagating waves, a point is reached where the moving acoustic wave produces a sufficient index modulation to strengthen the scattering process, which has the effect of increasing the backscattered wave and so to enhance the strength of the moving grating through increased electrostriction (and thus strain modulation

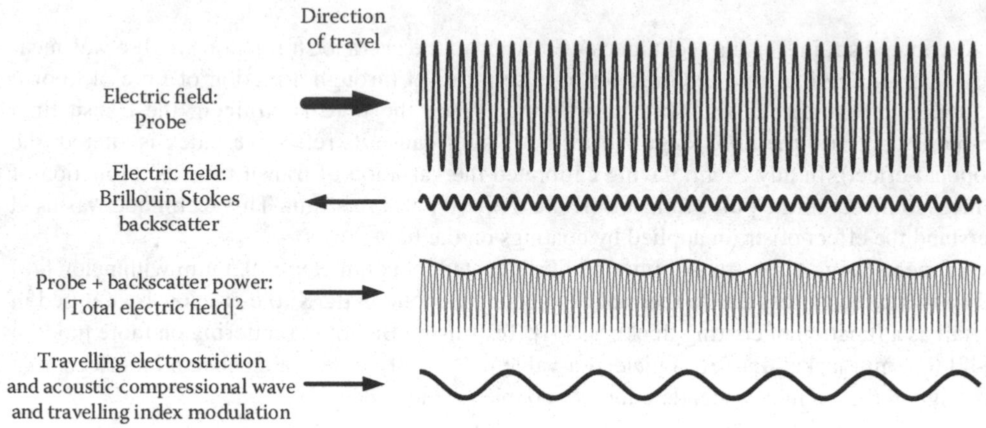

FIGURE 5.1 Illustration of the fields involved in spontaneous Brillouin backscatter.

and so index modulation). Therefore, as the incident power is increased, these processes (scattering, electrostriction, index modulation and further scattering) begin to reinforce one another to an extent sufficient to materially alter the intensity of the scattered light. This is when the process becomes stimulated.

The discussion has so far involved only a single incident beam interacting with acoustic waves to produce a stimulated scattered wave. The same mechanism exists where two counter-propagating waves separated by the Brillouin frequency shift meet: in this case, the same physical mechanisms give rise to amplification of the light at the lower frequency and therefore to transfer of energy from the higher frequency waves.

The relationship between the pump field E_p, the Stokes field E_s of the scattered or amplified wave and the acoustic wave E_a is given by Ref. [49]

$$\begin{aligned} &\frac{\partial E_p}{\partial z} - \frac{n_1}{c}\frac{\partial E_p}{\partial t} = E_a E_s && \text{a)} \\ &\frac{\partial E_s}{\partial z} + \frac{n_1}{c}\frac{\partial E_s}{\partial t} = E_a^* E_p && \text{b)} \\ &\frac{\partial E_a}{\partial z} + \Gamma_d E_a = \frac{1}{2}\Gamma g_B E_p E_s^* && \text{c),} \end{aligned} \tag{5.15}$$

where the damping factor $\Gamma_d = \Gamma_B + i2\pi\delta\nu_B$ is related to the phonon lifetime $1/\Gamma_B$ and the detuning factor $\delta\nu_B$, i.e. the difference between the Brillouin shift and the actual value of the frequency difference between Stokes and pump wave.

The gain at line centre, for a pump power P_p, wavelength l and an effective core area A_{eff}, is given by Ref. [47]

$$g_B = \frac{2\pi n_1^7 p_{12}^2 K_p}{c\lambda^2 \rho V_A \Delta\nu_B} \cdot \frac{P_p}{A_{eff}}. \tag{5.16}$$

K_p is a factor that depends on the relative alignment of the polarisation states. For a long, low-birefringence fibre, these states are randomly aligned and, on average, K_p is 0.5. However, locally, the value can vary from 1 to near 0. In fibres where the polarisation states are maintained, K_p can be 1.

The gain at angular frequency ω_s, for an incident (pump) at ω_p is a function of frequency in close proximity to the line centre. It is a complex function of frequency:

$$g_B(\omega_s) = \frac{g_B(o)}{1 - j2(\omega_p - \omega_s - 2\pi\Delta\nu_B)}. \tag{5.17}$$

The real part denotes the Brillouin signal gain, whereas the imaginary part describes a change in the refractive index. These two quantities are related by the Kramers-Kronig relation; the imaginary part is used in so-called 'slow light' to provide a controllable delay to optical signals [50].

Following the notation of Urricelqui et al. [51], the transfer function of the Brillouin gain process is given as a function of the detuning $\delta f = \left(\frac{\omega_p - \omega_s}{2\pi}\right)$ by

$$H_{SBS}(\delta f) = \exp\left(g_B(\delta f)\right) = \exp\left(G_{SBS}(\delta f) + j\phi_{SBS}(\delta f)\right). \tag{5.18}$$

Most sensors that use SBS exploit only the gain part, i.e.

$$G_{SBS}(\delta f) = \frac{g_B(0) \cdot \Delta\nu_B^2}{\Delta\nu_B^2 + 4\delta f^2} \tag{5.19}$$

However, we shall see (in Section 5.5.8) that recent developments have demonstrated the use of the phase term.

5.3 SYSTEMS BASED ON SpBS

The measurement of SpBS requires narrow linewidth sources: the linewidth must, at the very least, be substantially narrower than the frequency shift in order to ensure that the Brillouin lines are distinct from the dominant Rayleigh scattering in the spectrum of the scattered light. This would, in principle, allow the intensity of the Brillouin lines to be measured independently from that of the Rayleigh backscatter. In addition, a means of separating the components of the backscattered light onto separate signal channels is also required. No practical instrument has an infinitely abrupt frequency cut-off, and therefore, the filter transmission will always exhibit some residual response well beyond the cut-off frequency.

The foregoing considerations address only the intensity measurement; the main attraction of Brillouin backscatter is its ability to measure the frequency shift as well as the intensity. A means of measuring the frequency shift, such as a discrimination function that converts frequency to amplitude, is therefore also required of the interrogation.

5.3.1 Optical Separation

Spontaneous Brillouin backscatter can be acquired using optical filtering techniques that are analogous to the methods used in Raman DTS, the main difference being that the frequency resolution is required to be far finer for Brillouin scattering, given the small frequency shift involved.

A suitable arrangement is shown in Figure 5.2, where a narrowband laser is modulated to provide a series of probe pulses that are amplified prior to being launched into the sensing fibre. In the return path, a

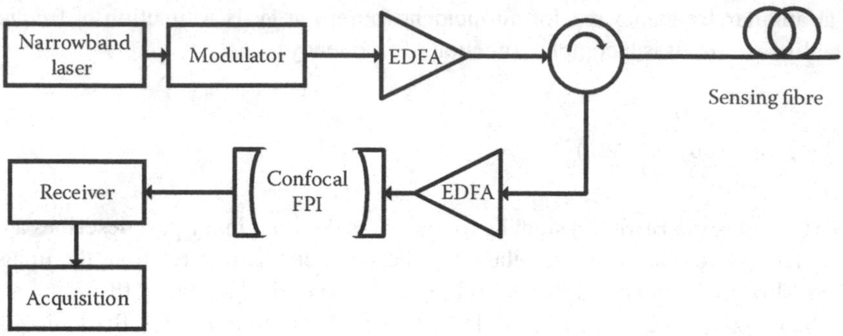

FIGURE 5.2 Acquisition arrangement for the detection of Spontaneous Brillouin backscatter with a scanning optical filter.

scanning narrowband filter (here a confocal Fabry-Pérot interferometer [FPI]) is placed before the optical receiver and functions as an optical spectrum analyser. Scanned FPIs are essentially narrow optical filters that are created using a pair of mirrors. They are commonly used for fine measurements of optical spectra, including in relation to measurements of Brillouin scattering in bulk media [52]. With high reflectivity mirrors, the transmission spectrum of an FPI is a series of spectral lines separated by a *free spectral range* (FSR) given by $FSR = c/(2nl)$, where n and l are the refractive index of the medium separating the mirrors and their separation, respectively. Thus, an FPI with mirrors spaced by 100 mm of air or vacuum would have an FSR of 1.50 GHz. The frequency resolution (full-width at half-maximum) δF is related to the FSR to the *finesse* $\mathcal{F} = \frac{FSR}{\delta F}$. $\mathcal{F}$ is given by

$$\mathcal{F} = \frac{\pi(\mathcal{R}_1\mathcal{R}_2)^{1/4}}{1-(\mathcal{R}_1\mathcal{R}_2)^{1/2}}, \tag{5.20}$$

where $\mathcal{R}_1$ and $\mathcal{R}_2$ are the reflectivity of the mirrors, and so an FPI with 100 mm mirror spacing and reflectivities of 99% would resolve about 5 MHz.

In practice, for distributed sensing, the problem is not so simple. Firstly, the previous expression for $\mathcal{F}$ assumes no excess losses, and at high finesse, this is difficult to achieve; for large mirror separations, it involves carefully aligned confocal mirrors (as illustrated in Figure 5.2). Nonetheless, $\mathcal{F}$ values in excess of 1000 are available. As an alternative to the free-space F–P interferometer, a fibre interferometer (where high-reflectivity mirrors are deposited directly onto the ends of the fibre) may be used. This approach and the confocal arrangement both avoid the limitations of flat FPIs, where losses from beam-spreading degrade the finesse.

In addition, keeping the interferometer stable relative to the laser is extremely challenging, and in practice, the interferometer is scanned over the entire range covering the two Brillouin peaks and the Rayleigh peak and is re-centred based on the Rayleigh peak, in the inevitable occurrence of drift. The finesse dictates a trade-off between spectral range and frequency resolution, but this can be overcome by operating the FPI with an FSR close to the Brillouin shift, so that different orders appear close to each other in the FPI output; this means that the length of the cavity is selected so that the number of half-wavelengths between the mirrors differs by about, (but not exactly), 1 for the Brillouin anti-Stokes, Rayleigh and Brillouin Stokes lines. Whilst the FPI is scanned, OTDR measurements are made of its transmitted signal. There are only two publications where this approach has been used successfully [26,53], and its use appears to have been abandoned.

A more successful optical approach is based on all-optical-fibre interferometers that are selected to provide exactly the filtering function required [54,55]. It was used initially to measure only the LPR but was later extended also to measuring the frequency shift and, thus, a full spontaneous Brillouin measurement.

An unbalanced Mach-Zehnder interferometer (MZI), illustrated in Figure 5.3a, can provide a frequency separation function. It consists of a wave splitter (here a 2 × 2 port coupler with exactly 50%/50% splitting ratio), followed by two independent paths (or arms), which are also the input ports of a second coupler. Light entering at Port 1.1 is divided evenly to flow to Ports 1.2 and 1.3, with, however, a 180° phase shift (see Section 2.3.3) between the two arms. The power transferred to each port of the second coupler depends on the relative phase of the light at each of its inputs. It therefore depends critically on the difference in the lengths of the two arms. As the length of one of the arms is made longer relative to the other (this is what is meant by an unbalanced MZI), the power switches periodically between the two output ports of the second coupler as a function of the relative phase of the signals entering the device. Thus, the transmission efficiency from Port 1.1 to Port 2.2 is a sinusoidal function of the path imbalance.

In an MZI with arms deliberately mismatched by ΔL, the power exiting each of the output ports (2.2. and 2.3) also oscillates as a function of the frequency of the light entering Port 1.1. The frequency change required to go from maximum transmission from Port 1.1 to Port 2.2 to a minimum and back, i.e. the FSR for the Mach-Zehnder configuration, is given by

$$FSR = \frac{c}{\Delta L n_1}. \tag{5.21}$$

Of course, the frequency change required to alter the transmission from the condition where all the power from Port 1.1 exits at Port 2.2 to that where it all arrives at Port 2.3 is half this value. The output of the unbalanced MZI is thus a periodic function of the input frequency. It is therefore clear that an

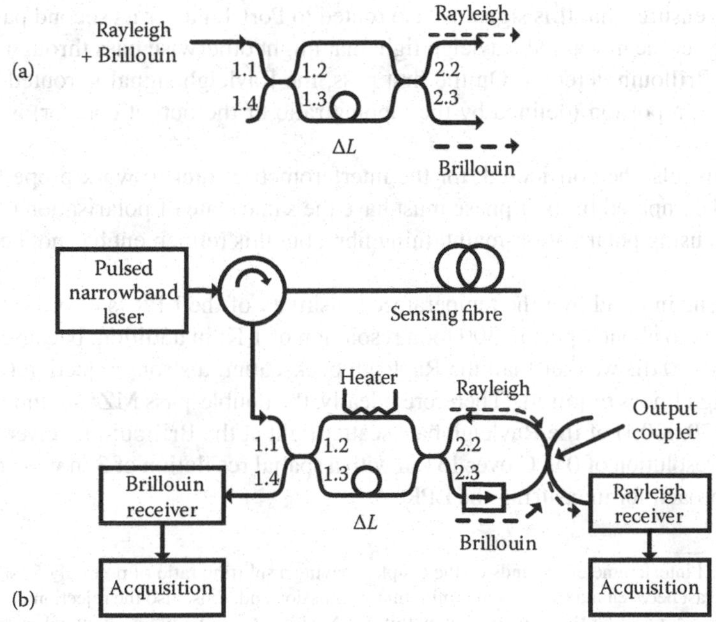

FIGURE 5.3 Spontaneous Brillouin OTDR using a Mach-Zehnder fibre interferometer. (a) An MZI used as a frequency discriminator; (b) schematic arrangement of the Brillouin OTDR using the Mach-Zehnder in a double-pass configuration. (Adapted from De Souza, K., Fibre-optic distributed sensing based on spontaneous Brillouin scattering, PhD, University of Southampton, 1999.)

unbalanced MZI can be used as a frequency-selective filter and output the Rayleigh backscatter at Port 2.2 and the Brillouin lines at Port 2.3, for example, provided that

$$\Delta L = \left(\frac{1}{2} + m\right)\frac{c}{\nu_B n_1}, \text{ where } m \text{ is an integer.} \tag{5.22}$$

So, for example, at an interrogation wavelength of 1550 nm, the Brillouin shift is about 11 GHz, and so inserting that value into Equation 5.22 leads to a path imbalance of about 9.4 mm for $m = 0$.

Although the MZI filter has an elegant simplicity, several practical issues need to be addressed in applying it.

Firstly, the path imbalance must be chosen to match the frequency shift, and so this must be known beforehand and the path imbalance must then be inserted into one of the arms of the MZI with high precision (better than 1 mm). This is challenging, but can be realised, especially with modern fusion splicers and reliable fibre cleavers.

A second issue is that the frequency of the source and the path imbalance of the interferometer can drift independently. The design relies on one signal exiting consistently through Port 2.2 and the other through Port 2.3. This problem was overcome by tuning the path imbalance of the MZI to match the frequency of the source [56]. A small heater was attached to one arm so as to add a small variable phase shift selected to maximise the Rayleigh signal through the intended port. This process can be implemented in a simple control loop by adding some dithering on the source frequency or the interferometer path imbalance.

A third issue is that the rejection of a single MZI* is not sufficient in general – in order to boost that aspect, a double-pass configuration [56] was developed that passes the Brillouin backscatter through the MZI, as illustrated in Figure 5.3b. A pulse laser provides probe pulses to a sensing fibre; the entire backscatter is directed via a circulator to Port 1.1 of the double-pass MZI. The light emerging from the first pass through MZI at Port 2.3 is at the Brillouin frequency, and it is fed back into the MZI at Port 2.2. The path imbalance ensures that this signal is also routed to Port 1.4 after its second pass in the MZI. An isolator is used to reject the unwanted Rayleigh light that might otherwise pass through the MZI a second time and reach the Brillouin detector. On the first pass, the Rayleigh signal is routed to Port 2.2 of the MZI, and from there, a portion (defined by the tapping ratio of the output coupler) is transferred to the Rayleigh receiver.

Polarisation must also be considered: for the interferometric filter to work properly, the beams that are recombined and compared in their phase must have the same state of polarisation (SOP). In principle, this can be achieved using polarisation-maintaining fibre, but this refinement has not been reported so far in the literature.

It should be borne in mind that the temperature sensitivity of the LPR is of order 0.3%/K and so the LPR must be resolved to about 1 part in 300 for a resolution of 1 K; in addition, because the two Brillouin lines together are 20–30 dB weaker than the Rayleigh backscatter, a strong rejection (ideally better than 0.01%) of the Rayleigh line is required.† Therefore, ideally, the double-pass MZI should achieve a rejection in excess of 45 dB (30,000:1) of the Rayleigh backscatter light at the Brillouin receiver.

A temperature resolution of 0.9°C over 16 km with a spatial resolution of 3 m was achieved [57] using the double MZI approach for measuring the LPR.

* The rejection of an MZI interferometer depends on the couplers having a splitting ratio of precisely 50%. Departures from this value degrade the contrast between maximum and minimum transmission and, thus, also the rejection of unwanted frequencies. The losses of the splices connecting the two couplers within the MZI must also be precisely equal in order to achieve a high rejection ratio.

† BOTDR systems can tolerate slightly worse rejection, provided that the calibration takes into account the presence of unwanted Rayleigh signal within the Brillouin backscatter. However, the Rayleigh backscatter usually includes coherent Rayleigh noise (CRN), which is not removed by averaging. In addition, the presence of a temperature-independent signal mixed with the temperature dependent one degrades the resolution by contributing shot noise and reducing the overall sensitivity.

A further issue with the MZI approach as illustrated so far is that of coherent Rayleigh noise (CRN) discussed in Chapter 3, that adds an unwanted modulation onto the Rayleigh signal. Any residual Rayleigh scatter within the Brillouin backscatter will add CRN onto that signal as well. It is particularly difficult to eliminate CRN at fine spatial resolution, and for that reason, a separate source for each of the signals is beneficial – a narrowband source is used for measuring the spontaneous Brillouin signal and a broadband source provides the Rayleigh signal with minimal coherent noise.

An alternative arrangement, demonstrated in Ref. [58], is illustrated in Figure 5.4. Probe pulses are generated by a narrowband source consisting of a distributed feedback (DFB) laser, modulated by an AOM, AO1, amplified optically in an Er-doped fibre amplifier (EDFA) and further gated by modulator AO2. The probe pulses are launched into the sensing fibre through a four-port circulator. The reflection peak of grating FBG1 matches the narrowband laser emission, and FBG1 thus behaves as a narrowband reflection filter that serves to eliminate amplified spontaneous emission (ASE) from the EDFA and passes the filtered probe pulses to the sensing fibre through the third port of the circulator. The backscattered light is returned through Port 4 of the circulator, where it encounters a notch filter, consisting of a pair of Bragg gratings (FBG2 and FBG3) that reflect the same frequency as FBG1, i.e. the laser (and Rayleigh backscatter) frequency. FBG2 and FBG3 are separated by an isolator to avoid resonance effects. In this way, a filter with a rejection better than 34 dB at the Rayleigh frequency (and yet providing low-loss transmission at the two Brillouin lines) was achieved. The Rayleigh signal is obtained using a separate broadband source launched into Port 2 of the circulator; the spectrum of broadband source spills around the grating filters and reaches the same detector. The narrowband and broadband sources are operated at different times, so that the Brillouin and Rayleigh signals can be collected separately on the same receiver. The system of Wait and Hartog [58] achieved a spatial resolution of 2 m at a range of 25 km and a temperature resolution at that distance of 7°C with an integration time of 10 minutes (or 1.7°C in 3 hours).

It can therefore be seen that optical filtering techniques can be used to measure the temperature distribution using the LPR, assuming an unstrained fibre.

In practice, it is of interest to measure the strain as well as the temperature; this requires measurement of the frequency shift as well as of the intensity of the backscatter. In this case, a much finer frequency resolution is needed. In addition, even if only the temperature distribution is of interest, it cannot always be guaranteed that the strain state of the fibre is known, and in Brillouin measurements, this leads to an

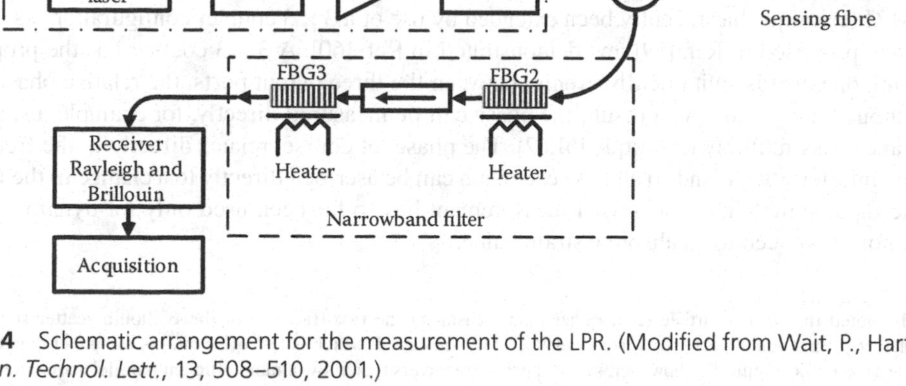

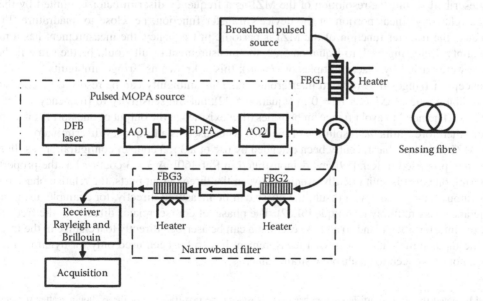

FIGURE 5.4 Schematic arrangement for the measurement of the LPR. (Modified from Wait, P., Hartog, A.H., *IEEE Photon. Technol. Lett.*, 13, 508–510, 2001.)

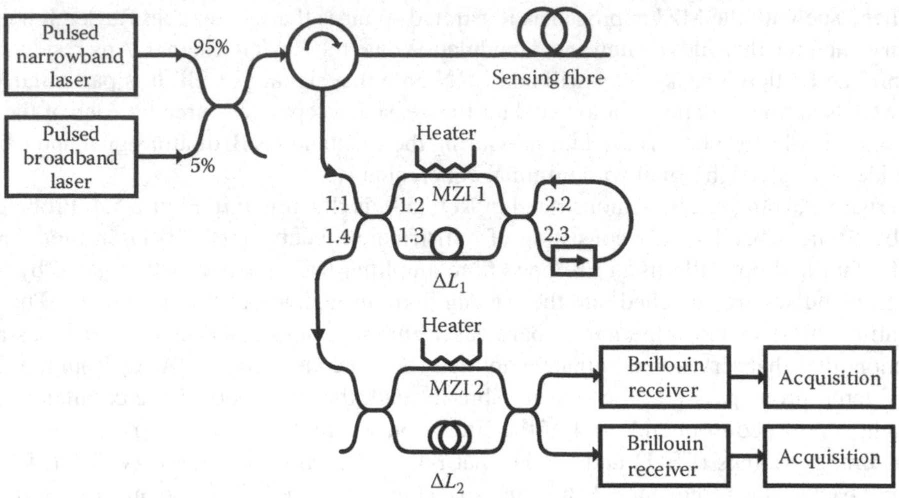

FIGURE 5.5 All-fibre filter arrangement for measuring the frequency shift of the Brillouin backscatter and the Landau-Placzek ratio. (Modified from Kee, H.H., Lees, G.P., Newson, T.P., *Opt. Lett.*, 25, 695–697, 2000.)

ambiguity in what is measured, requiring both the intensity and the frequency of the Brillouin backscatter to be determined.

A refinement to the MZI approach [27] allows the frequency distribution of the Brillouin backscatter, as well as its intensity, to be measured, as illustrated in Figure 5.5. The arrangement is essentially the same as Figure 5.2 with the addition of a second, single-pass MZI, MZI 2, with a much smaller FSR (7 GHz, in Ref. [27]). With proper biasing, MZI 2 provides a discrimination function; i.e. the proportion of the Brillouin light exiting the two ports is a direct, almost linear, function of its frequency.* The LPR is obtained by firstly summing the Brillouin outputs obtained from each receiver, to give a measure of total Brillouin intensity, and secondly by activating the second source, a broadband pulsed laser (and turning off the narrowband source) to obtain a Rayleigh backscatter measurement.

As described so far, the resolution of the MZI as a frequency discriminator is limited by the need to use the relatively linear portion of the device's transfer function, i.e. close to quadrature. In addition, because the transfer function of the MZI is periodic in frequency, the measurement has a modulo 2π uncertainty. Referring back to Equation 5.22, the measurement result would be the same if the path imbalance were varied by exact multiples of $c/(\nu_B n_1)$; this is known as 'fringe ambiguity' – a reference to the concept of fringes in two-beam interferometers. This ambiguity can be resolved by choosing the lowest possible value of ΔL (i.e. m = 0 in Equation 5.22), but the sensitivity to frequency variations is then reduced. The need to avoid fringe ambiguities (that exist when the output of one port is taken past an extremum) therefore limits the resolution of the MZI when used as a frequency discriminator.

The MZI technique has recently been extended by use of a 3 × 3 coupler configuration, as illustrated in Figure 5.6, proposed in Ref. [59] and demonstrated in Ref. [60]. A 3 × 3 coupler has the property that light entering one port is split (ideally evenly) between the three output ports, the relative phase between any two outputs being 120°. As a result, the phase can be measured directly, for example, using the differentiate and cross-multiply technique [61,62]. The phase, of course, relates directly to the free-spectral range of the interferometer, and so a phase excursion can be ascribed directly to a change in the frequency shift of the input signal. The 3 × 3 MZI measurement has so far been used only for dynamic measurements, i.e. not referenced to an absolute strain value.

* It should be noted that in this article (27), rather than measuring the two fractions of the Brillouin scatter simultaneously with two detectors, a single detector was used and MZI 2 was re-tuned, which is less efficient, but the paper nonetheless fully demonstrated the point. Figure 5.5, however, shows a pair of receivers that allow a more efficient parallel acquisition.

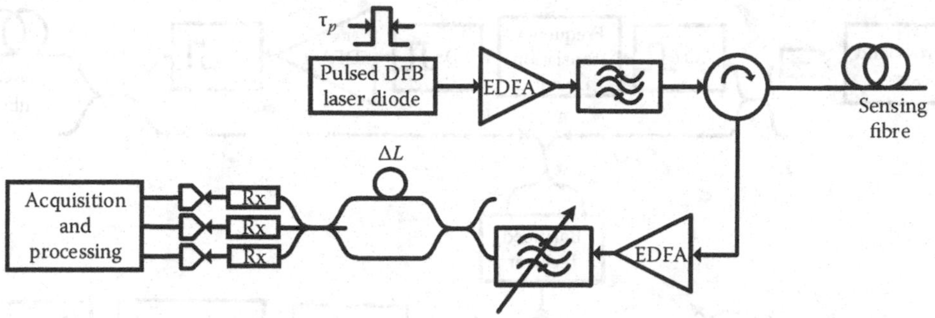

FIGURE 5.6 MZI and 3 × 3 coupler-based frequency discrimination arrangement for Brillouin frequency-shift measurement. (Modified from Masoudi, A., Belal, M., Newson, T.P., *Opt. Lett.*, 38, 3312–3315, 2013.)

This configuration is versatile and has also been used in connection with distributed vibration sensing. One of its benefits over the arrangement shown in Figure 5.5 is that it eliminates the inherent dynamic range limitation that restricts the measurement to occur over only approximately 1/3 of the FSR of the interferometer: thus, for large anticipated signals, it is necessary to reduce the path's imbalance in MZI 2 in order to extend the range, but this simultaneously degrades the sensitivity. The arrangement of Figure 5.6, which allows the phase to be tracked and unwrapped over many FSRs, is limited in the rate of change of the frequency shift, not its overall range. Thus, in applications where a large, but slow, change in the frequency shift is anticipated, a far higher sensitivity can be achieved. It should be noted that the discriminating interferometer is prone to drift relative to the laser frequency, and some means of stabilising their relative frequency would be required for reliable measurements of quasi-static quantities.

To summarise the work of several teams, it is clear that optical filtering techniques can be used to measure not only the Brillouin intensity but also its frequency shift to a precision of order 1 MHz, i.e. 1 part in 2×10^8. This is a quite remarkable relative resolution. However, electrical separation has many advantages over the optical filtering and is now more commonly used for BOTDR.

5.3.2 Electrical Separation

An alternative approach that has gained pre-eminence amongst measurements of SpBS involves using a heterodyne technique to select the Brillouin signal in the electrical, rather than the optical, domain. A heterodyne technique was already used in 1986 to determine the spontaneous Brillouin spectrum of various fibres [14]; however, this work did not resolve the spectrum as a function of distance along the fibre, so it was an average measurement, not a distributed one. A distributed fibre-optic sensor using heterodyne detection of spontaneous Brillouin backscatter was demonstrated by Shimizu et al. in 1993 [63,64]. The basic principle, similar to coherent OTDR (C-OTDR) (see Chapter 3), is that the backscattered light is mixed with a strong signal (the optical local oscillator, OLO) taken from the same laser that provides the probe pulse. As in C-OTDR, the detector output contains beat frequencies, including the two bands at the Brillouin frequency. At the time when Ref. [63] was written, no optical receivers were available with sufficient bandwidth and noise performance to be able to mix the probe frequency directly with the backscatter.

The first implementations of what is now termed BOTDR (Brillouin OTDR) were therefore effected with a probe pulse that was frequency-shifted from the local oscillator in order to bring the local oscillator and *one* of the Brillouin lines closer together and within the bandwidth of the optical receivers that were available.

This approach, illustrated in Figure 5.7a, opened the road to a wide use of BOTDR. Figure 5.7a is considerably simplified from Ref. [63], but in essence, it is a C-OTDR arrangement with two major

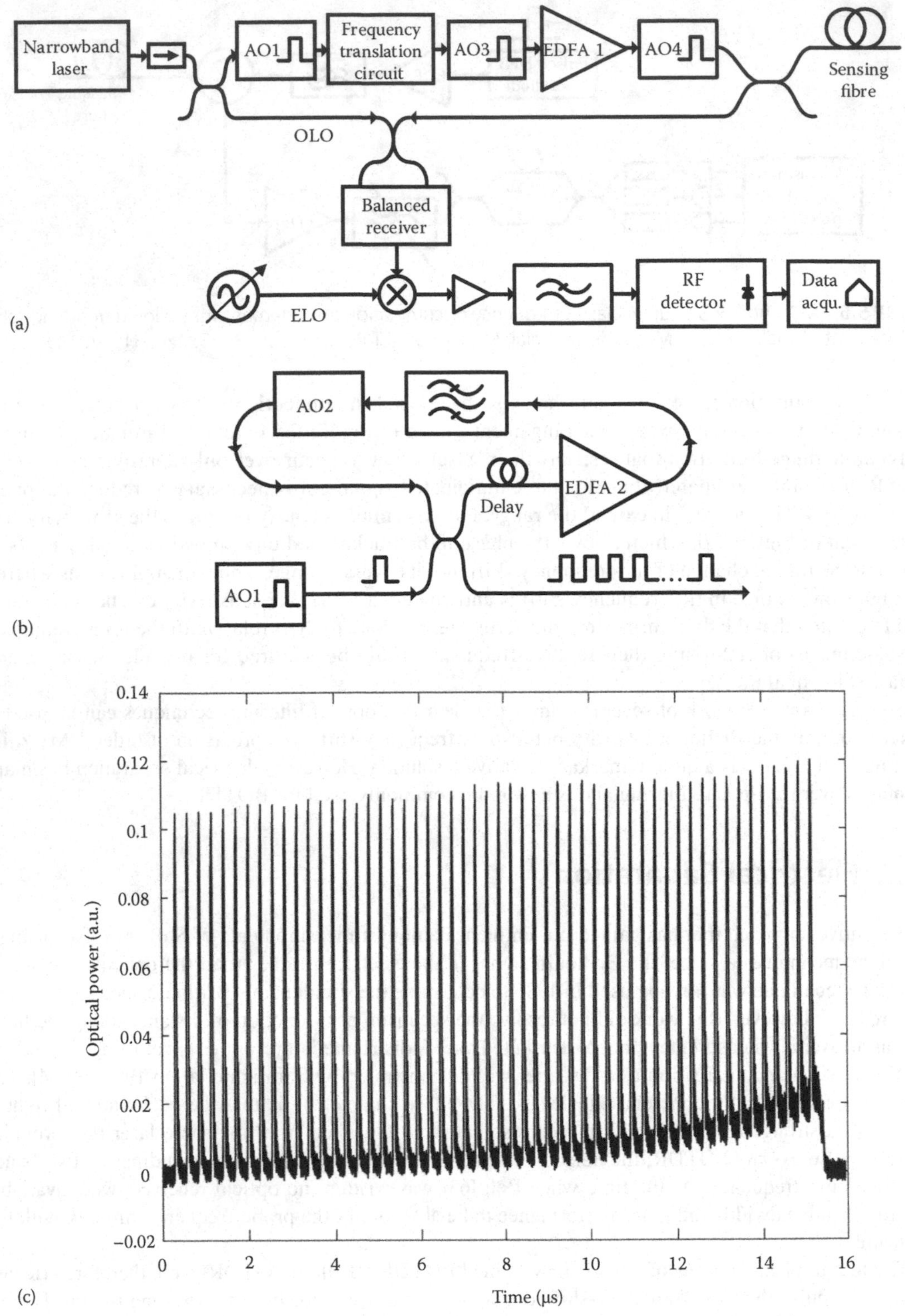

FIGURE 5.7 (a) Initial arrangement for the heterodyne BOTDR measurement. (b) Frequency translation circuit concept. (c) Example of a pulse train emerging from a re-circulating frequency translation circuit, such as that shown in panel (b).

differences. First is the addition of a frequency translation circuit (detailed in Figure 5.7b) that shifts the pulse extracted by AOM AO1 by an amount close to the value of the Brillouin shift.

A train of pulses is generated by the translation circuit, each shifted relative to the previous pulse by a frequency dictated by the radiofrequency (RF) drive to a second AOM (AO2) (Figure 5.7b). One of these pulses is selected (by turning on the third AOM, AO3, at a time coinciding with the arrival of the pulse that is required), further amplified by EDFA 1 and launched as a probe pulse into the sensing fibre; a fourth AOM, AO4, can be used for rejecting most of the ASE from the final optical amplifier. Note the use of a fused taper coupler to launch the probe into the sensing fibre: circulators were not readily available when Ref. [63] was published. For simplicity, many of the functions used to set and monitor the system that were described in Ref. [63] were omitted from Figure 5.7a and b.

When the backscattered light is returned, one of the Brillouin lines (depending on the direction of the shift in the translation circuit) will be close to the frequency of the OLO when mixed with the backscatter on a balanced receiver. Only the portion of the backscattered light spectrum that is close to the OLO is detected in the receiver and passed, at the difference frequency between OLO and the closest Brillouin line, to an electrical mixer that combines the receiver output with a tunable electrical local oscillator (ELO) that further shifts the mixer output to close to DC. This signal is amplified, filtered, rectified or squared [63], averaged and acquired.

By varying the frequency of the probe pulse on successive acquisitions, the Brillouin intensity spectrum can be explored. From that information, the frequency shift and Brillouin intensity can be determined as a function of distance along the fibre, for example, by fitting the data at each location and extracting the parameters of a model describing the shape of the spectrum. For a simple Lorentzian distribution, the parameters in question might specify the position of the peak, its intensity and spectral width.

The electrical separation method has a number of advantages, including the use of heterodyne detection, which allows the effect of thermal noise in the receiver to be overcome. In addition, the filter is now defined in the electrical domain at frequencies where the achievement of a given spectral resolution is far easier than implementing a filter of similar width at optical frequencies. Furthermore, the tuning of the filter across the frequency range of interest is achieved electrically, without moving parts, and this allows faster, more precise setting of the location of the passband.

An additional benefit of electrical-domain separation is that the method measures directly the frequency difference between a sample of the laser light and one or both of the Brillouin lines. As a result, a drift in the frequency of the laser is only of minor consequence. In contrast, in the optical techniques, the filter must be made to track the frequency of the optical source; this can be quite challenging, and it certainly requires a control loop that itself requires error signals to feed it, which adds further complexity to the arrangement.

The frequency shifting circuit that was used in Ref. [63] is based on a re-circulating ring, using an AOM, illustrated in Figure 5.7b. Here, the pulse created by the first AOM is re-circulated in a ring arrangement, the gain of the fibre amplifier (EDFA 2) being just sufficient to cancel the attenuation of other components in the ring. The delay line at the start of the ring is included to avoid any overlap of successive circulating pulses (i.e. it increases the travel time around the ring to a value greater than the pulse duration). The key component is the second AOM (AO2), which shifts the frequency by 80 MHz, in the case of Ref. [63]. The shift for each pulse in the train exiting the ring is thus an increasing integer multiple of 80 MHz. The pulse selected by AO3 is chosen for having a frequency close to that of the Brillouin shift. In this arrangement, rather than tuning the electrical oscillator, the frequency shift of AO2 can be adjusted slightly to vary the frequency of the selected pulse. This was the approach adopted in the original work [63]; these small changes are amplified by the large number of re-circulations in the translation circuit (138 in the original work), and so the RF drive remains within the bandwidth of AO2. Of course, coarser adjustments can be made by selecting a different pulse from the sequence generated by the translation circuit.

It should be noted that the ring-frequency shifter was originally conceived in a different context by Hodgkinson and Coppin [65,66].

There are several limitations to the use of the re-circulating ring frequency translation circuit in the context of BOTDR [67,68]. Firstly, with more than 100 circulations, build-up of ASE occurs; this has the effect of adding noise to the probe pulse and of depleting the population inversion in the amplifier. A bandpass filter is included in the ring, and subsequent time-domain gating of the output to select just the pulse of interest after the ring both help limit the intensity of the ASE in the output. Nonetheless, a build-up of noise that overlaps in time and in the spectral domain with the pulse of interest still occurs. Another issue is controlling the amplification in the ring: the gain must precisely compensate the losses in the ring; however, the gain of EDFA 2 is not constant owing to the build-up of ASE, which drains increasing amounts of energy from the optical amplifier. Furthermore, relaxation oscillations occur in the pulse train, whereby, if the power increases (for example) in successive pulses, this depletes the population inversion in the amplifier, thus decreasing the gain, which eventually recovers as pumping continues at a level consistent with a higher gain. Whilst the time dependence of the round-trip loss can be controlled to some extent by altering the RF power to the AOM, this is intrinsically rather difficult to control. Nonetheless, it was achieved by Shimizu and colleagues, and at least one commercial product was based on this technique.

An example, taken from the author's own work in a different context, is shown in Figure 5.7c: here, a train of 53 pulses emerges from a ring similar to that of Figure 5.7b. In this case, the ring filter is rather broad (5 nm) and the build-up of ASE is clearly visible as a pedestal to the pulses, particularly after ~10 μs.

Of course, the frequency shifting can be carried out in other ways, and in Ref. [69], the sidebands of an integrated electro-optic modulator (EOM) were used to provide frequency-shifted probe pulses for BOTDR.

The purpose of the frequency translation circuit is to reduce the frequency difference between the OLO and a selected Brillouin line. The same can be achieved by using a second laser as the OLO and controlling the second laser to be at a specified offset from the primary laser. This was described in Ref. [70], and it avoids the issues already discussed that arise with a ring frequency translation circuit (of noise and stability), at the price of introducing a second source and the requirement of controlling the frequency separation of the two sources.

It was recognised when BOTDR was first proposed [71] that directly detecting the Brillouin signal in the electrical domain was simpler than the optical shifting techniques that were employed initially as a matter of necessity. As detectors and receivers with wider bandwidth and lower noise became available, the use of a local oscillator at the same frequency as the probe pulse was demonstrated [3,72]. This simplifies the optics but transfers the problem to the electrical (microwave) domain.

A typical arrangement of the microwave coherent detection approach to BOTDR is shown in Figure 5.8. A continuous wave (CW) source is split into two paths, one of which is modulated into probe pulses, amplified, usually filtered to reduce ASE from the amplifier and launched into the sensing fibre. The filtering, as represented in the figure, is commonly performed in the time domain to limit the noise to

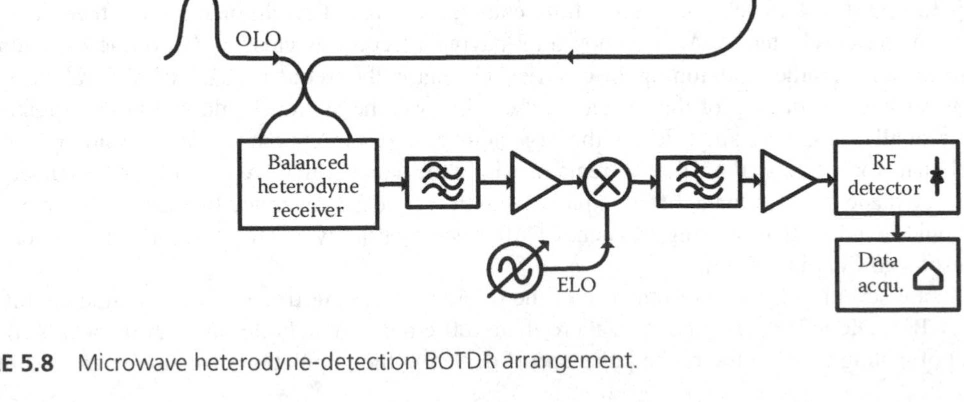

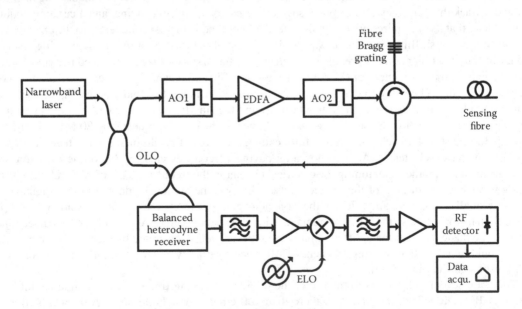

FIGURE 5.8 Microwave heterodyne-detection BOTDR arrangement.

times only coinciding with the probe using a second modulator (AO2) and preferably also with spectral filtering (the reflective Bragg grating filter attached to Port 2 of the circulator performs this function in Figure 5.8) that narrows the bandwidth of the noise. By choosing AO1 and AO2 to shift the frequency in opposite directions but by equal amounts, the probe pulse emerges at the same frequency as the source laser.

The other, usually smaller, fraction of the laser output forms the optical local oscillator and is taken to a directional coupler that mixes it with the backscatter returning from the fibre onto a detector or preferably a balanced receiver (as shown).

In some publications, e.g. Ref. [73], the backscatter signal is pre-amplified optically prior to the mixing process. Adding an optical pre-amplifier should not be necessary with coherent detection, but it has the benefit of making the system more resilient to optical losses prior to detection and it also reduces the local oscillator power required to overcome the noise in the receiver.

The output of the receiver contains a complete spectrum of the light from the Rayleigh backscatter (not shifted from the laser and thus at zero frequency) up to the bandwidth of the receiver, which is selected to include the Brillouin lines. It should be noted that this arrangement 'folds' the downshifted spectrum symmetrically about the Rayleigh scatter line and so the energy in both Brillouin lines is collected. In contrast, the arrangement of Figure 5.7a, in which a frequency-translation circuit is used, collects light only from one of the Brillouin lines at a time.

The frequency of the Brillouin shift is expected to cover a range of 1 GHz for most applications and usually far less, so its location in the spectrum can be reasonably predicted and a bandpass filter is used to select the range of frequencies of interest. The filtered signal, still at the Brillouin shift (i.e. about 10 GHz for a probe at 1550 nm), is then usually mixed with an ELO to a frequency where it can be processed more conveniently. This intermediate frequency is selected to be low enough to simplify the design of the electronics required to amplify and further condition the signal. However, it is also desirable that the relative bandwidth (ratio of upper to lower frequency of interest) is moderate because this simplifies the design of the filter. Therefore, the intermediate frequency is likely to be chosen in the range of 1–2 GHz.

A signal, down-converted in two stages from the optical to the ultra-high frequency (UHF) region, is thus available. The most common way of extracting its features (intensity and central frequency as a function of distance along the fibre) is to pass the intermediate frequency band through a narrow filter whilst stepping the ELO over a range of frequencies. The output of the filter is amplified, rectified (i.e. detected in the same way as happens in analogue radio receivers) and digitised. The average of a chosen number of backscatter traces for each ELO frequency is acquired.

Scanning the ELO and applying a narrowband filter to the output of the electrical mixer has the effect of scanning a filter of the same width across the optical spectrum itself: the heterodyne process has the useful property of projecting a filter back into the original domain with a precision that could not be attained with optical filtering. As an example, a typical bandpass frequency would be 5–10 MHz. The length of a hypothetical fibre Bragg grating having this linewidth would be tens of metres long, and precise control of the relative position of every single cycle of the refractive index modulation would be required (its spectral characteristics would be ruined by any temperature or strain non-uniformity) – this is not achievable at present and, if ever accomplished, is likely to result in a very cumbersome device.

A 3-D dataset containing the intensity of the Brillouin backscatter as a function of distance along the fibre and Brillouin frequency is thus built up. At each location along the fibre, the data of intensity vs. Brillouin frequency shift is processed to extract the main features, namely intensity, central frequency and linewidth. It should be noted that some fibres exhibit multiple peaks in the Brillouin spectra, and algorithms that take this into account are then required.

Figure 5.9 shows an example of a plot of Brillouin intensity as a function of frequency and distance along the fibre. Here, the frequency axis is offset to 10.4 GHz and the frequency spans a range of 500 MHz above that value. The majority of the fibre is on reels at nominally similar conditions, but small changes in winding tension and fibre properties result in four sections (100, 200, 250 and 300 m in length) that are clearly distinguishable. In this case, the second section (200 m in length) is heated by about 35°C relative to the other fibres. The short piece of fibre at the beginning of the plot is of a different type, and its frequency shift and backscatter factor are quite different from the remainder of the fibre sections.

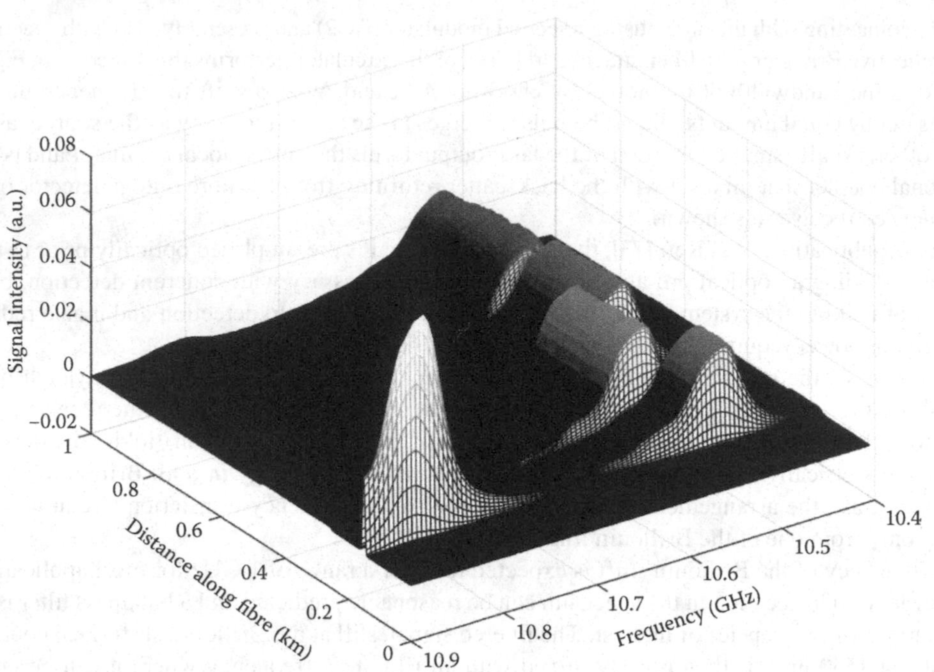

FIGURE 5.9 Pseudo-3-D plot of Brillouin backscatter intensity vs. frequency and distance.

The acquisition process is relatively slow because each frequency must be measured in sequence. Nonetheless, a resolution of the frequency to better than 1 MHz over a range of 25 km with a spatial resolution of 5 m can be achieved in less than 600 s.*

If there is no frequency shift in the pulse generation, the two Brillouin lines overlap in the electrical output spectrum of the receiver. This means that twice the optical power is collected compared with arrangements previously discussed where, owing to the large shift of the OLO relative to the probe, energy from only one Brillouin line is collected in the heterodyne process. Unless the measurement is at cryogenic temperatures, the intensity of the two lines is essentially equal and no further information about the measurand is gained by measuring them separately.

Nonetheless, it may be useful to measure the Brillouin lines separately in order to determine whether the system is approaching the onset of SBS; this allows the power that is launched to be optimised. In fact, with both lines measured separately, the probe power can be increased somewhat and the initial stages of the SBS can be corrected by considering the anti-Stokes/Stokes ratio. This was discussed in the context of optical filtering [74]. The same result can be achieved [75] with the heterodyne set-up of Figure 5.8, for example, by arranging for the two AO modulators to shift the light in the same direction (or using a single-AO modulator) and thus imposing a small frequency shift between the probe light and the OLO. The heterodyne process folds the frequency axis about the LO, and so if the anti-Stokes and Stokes are not symmetric with respect to the LO, they will appear at different frequencies in the heterodyne spectrum.

In Figure 5.10, separation of Stokes and anti-Stokes Brillouin lines was achieved by adding a further AOM in the LO path that shifts the LO frequency down by 110 MHz, resulting in a Brillouin shift that is lower for the Stokes signal and higher for the anti-Stokes line. The spectra shown in Figure 5.10 therefore show the Brillouin lines separated by 220 MHz. In Figure 5.10, the measurement is shown for two probe power levels; at the lower launched power, the two peaks are of essentially the same value (the Stokes line being the stronger by ~4%), whereas when the probe power is doubled, a clear distinction (26%) between the peak values for the two lines is apparent.

* Unpublished work within Schlumberger.

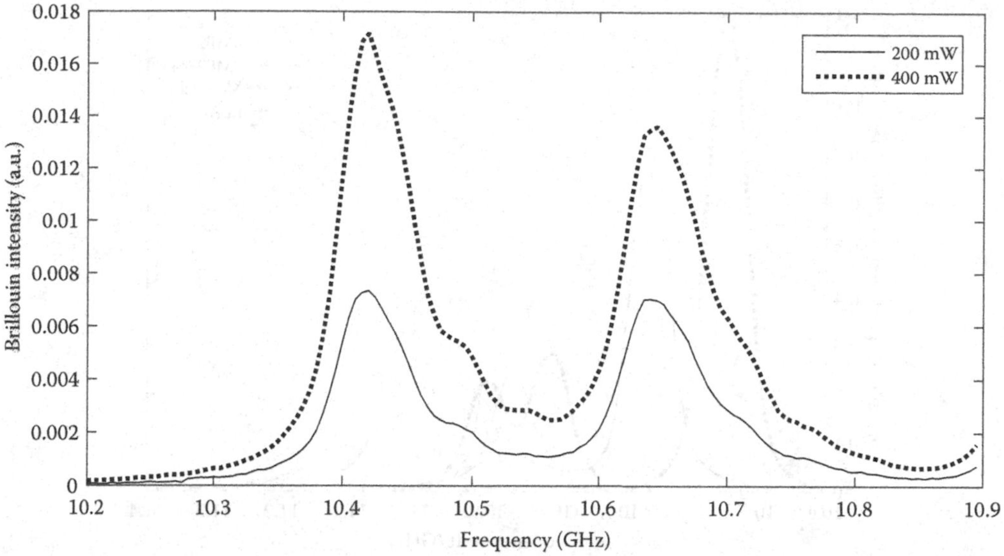

FIGURE 5.10 Brillouin intensity spectrum measured with offset Stokes and anti-Stokes lines (the Stokes line appears at the lower frequency). (Data originally shown in Hartog, A.H., Lees, G.P., Distributed sensing in an optical fibre using Brillouin scattering, US7504618B2, 2009.)

The measurement of the two Brillouin lines separately is particularly useful when the intensity is used: it allows the probe to be increased close to the SBS limits and yet still to measure the spontaneous signal accurately.

5.3.3 Estimation of Frequency, Gain and Linewidth

A single Brillouin peak has a Lorentzian spectral distribution (see Equation 5.19) and the Levenberg-Marquart non-linear least squares algorithm is commonly used [3] to fit a model function to the data. This allows the peak frequency, peak gain and spectral width to be estimated at each location in the fibre. However, in practice, there are several complications to this ideal scenario. Firstly, abrupt transitions of temperature or strain occurring faster than the spatial resolution of the interrogation system result in a local overlap of multiple spectra, with relative Brillouin shifts resulting in different gain peaks that may or may not be distinct.

In contrast, smooth changes in the Brillouin distribution broaden the measured spectrum [76], but the result depends on many factors such as the axial distribution function (whether linear, parabolic etc.) of the Brillouin shift and its width relative to the spatial resolution of the system. Although the effect can be modelled in specific conditions, the fact is that, in general, there is little *a priori* knowledge of the actual distribution. This option is therefore useful only in specific circumstances, such as where the sensing fibre is attached to a structure with known deformation profiles, and it is the extent of that deformation that is to be measured.

Another issue arises with some fibre designs that are optimised for their application in telecommunications or local area networks, for example, to improve their bend resistance or tune the dispersion properties. Some of these fibre designs have complex Brillouin spectra extending over sometimes more than 100 MHz with multiple peaks. The Brillouin spectra of four commonly used optical fibres are shown in Figure 5.11. It is clear that their frequency shift, linewidth and intensity are different; moreover, two of the fibre types exhibit more than one prominent peak.

This makes the fitting more complex; in addition, the fitting needs to be adapted to the specific spectrum or spectra that are present, and this means that a 'universal' fit is unlikely to be possible. Therefore, where the sensing fibre consists of multiple sections each of a different fibre type, a different fitting routine may be required for each section.

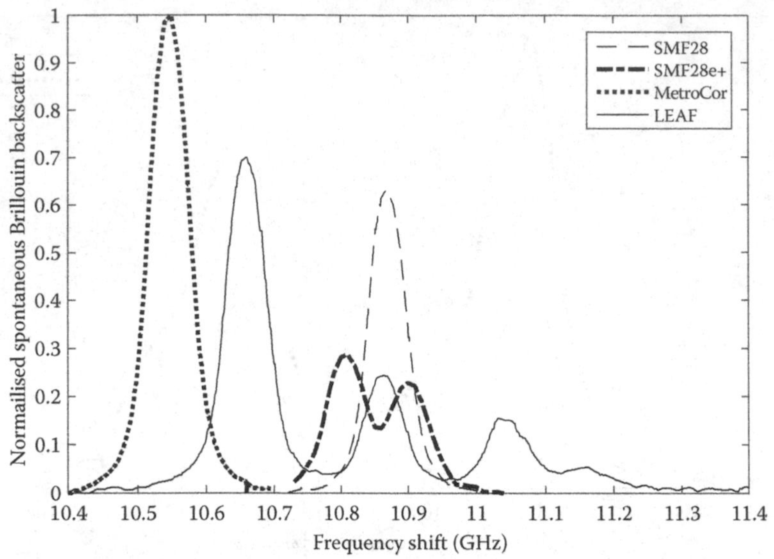

FIGURE 5.11 Spontaneous Brillouin spectra for four types of fibre, measured at a probe wavelength of 1550.92 nm. The trade names for each of the fibres are indicated in the legend and they are all trademarks of Corning Glass Works, Inc. Data recorded by Peter Wait and Graeme Hilton at Schlumberger.

The fitting routines can also be more sensitive to noise at poor signal-to-noise ratios. As an alternative to fitting to a theoretical line shape, methods based on comparing the measured spectrum to a reference line shape have been demonstrated, such as cross-correlating the measured spectra with a reference spectrum [77] or using a similarity method [78], which chooses a frequency shift that minimises some measure* of the distance between the measured and reference traces. Another benefit of these spectrum-matching methods is that the entire shape of the spectrum is used. Some improvement (a few dB) in the fit has been demonstrated over the curve-fitting approaches.

5.3.4 Limitations and Benefits

BOTDR has two main benefits over most other Brillouin measurements. Firstly, it is a single-ended measurement: the complete measurement is carried out from one end of the fibre; this contrasts with other approaches, to be discussed, that require coordinated signals to be launched from both ends. Being single ended, the method also has the advantage of simplicity of installation, a similar argument to that which is frequently made against double-ended Raman OTDR measurements. In effect, the range is doubled compared with other Brillouin systems of equal loop range. As a result of its single-ended architecture, BOTDR is unique amongst the Brillouin measurements in that, in the event of a fibre break, the measurement can continue unhindered up to the location of the break. Other Brillouin systems, such as BOTDA, lose all capability in these circumstances. Furthermore, if system resilience is crucial, arrangements where a cable is interrogated from each end with separate acquisition systems, for example, on different fibres [79], can be implemented. In the event of a rupture of the cable, the redundancy is removed, but no loss of capability occurs, other than in the immediate vicinity of the break.

A second benefit is that a measurement of intensity, as well as of the frequency shift, is provided by BOTDR. This allows a natural separation of strain and temperature. This advantage should not be overstated, however, because it is generally easier to achieve a given resolution of one of these measurands

* The measure used could be a sum of squares (Euclidian measure) or a sum of absolute values (Manhattan measure) of the difference between points of the same frequency value on the two curves.

using the frequency rather than the intensity. In other words, the intensity is usually the 'weakest link', and it is usually its contribution to the measurement uncertainty that limits the overall system performance. This is particularly the issue in long-distance applications, where in addition to a poor signal-to-noise ratio, it often proves challenging to acquire the signal with sufficient linearity over a large range of signal levels. As a result, is it difficult to avoid distortion that ultimately works through to errors on the estimated temperature or strain.

5.3.4.1 Launched power

In SpBS, the limitation on probe power is rather different from that which exists in Raman DTS because, in SpBS, it is mandatory to use a narrow linewidth. Therefore, several non-linear effects that can be avoided in Raman DTS become pre-eminent in the present context.

With a narrow linewidth, SBS re-emerges as a key limitation on probe power. The limitation due to SBS relates in fact to probe energy, and so for low-resolution systems (long probe pulse), it dominates the power that can be launched, and the threshold is given, for a fibre having a loss α_p at the probe wavelength, by Refs. [43] and [44]

$$P_{crit,B} \approx 21\frac{A_{eff}}{g_B \cdot L_{eff}}, \tag{5.23}$$

with the effective length L_{eff} taking the value $(1 - \exp(-\alpha_p L))/\alpha_p$ for long pulses or continuous incident light and $L_{eff} = \tau_p \cdot v_g/2$ for pulses of duration τ_p where this is much shorter than $1/\alpha_p$. A_{eff} is the effective area occupied by the mode [43]. It should be noted that the critical power in Ref. [43] is defined as that at which the ASE is equal to the input power, so at this level, the process is far into a non-linear regime. However, given the exponential nature of the relationship between launched power and ASE, only a modest reduction in power is required to be at the level where the non-linearity begins to be noticeable, rather than where it is totally dominant.

Unlike stimulated Raman scattering, SBS occurs near the beginning of the sensing fibre and so does not become worse with increasing distance from the launching end.

Two further non-linear effects also limit the power that can be launched. The first of these is modulation instability (MI), which results from parametric gain on either side of the probe frequency if the sign of the chromatic dispersion is negative. In the absence of any launched signal in the spectral region where the gain occurs, power builds up from noise and results in new sidebands at frequency offsets of 20–60 GHz for probe power in the range of 0.1–0.4 W for a typical single-mode fibre at an operating wavelength of 1550 nm [80,81].

The initial effect of MI is to build up power at frequencies offset from the probe frequency that overlap with the Brillouin lines. This results in an incorrect measurement of the Brillouin intensity. For example, where the intensity is used to estimate temperature, an error of more than 50 K at a distance of 12 km for a probe power of 0.4 W was reported in Ref. [81]. If only the Brillouin frequency is used, MI still pollutes the measurement at somewhat higher probe powers, by masking the Brillouin peaks themselves.

If the designer of the sensing system has a free choice over the selection of the operating wavelength or the fibre type, then MI can be avoided altogether by operating at a combination of wavelength and fibre design where the chromatic dispersion has a positive sign [81]. In this case, the probe power can be increased considerably, and it was shown that for a combination of fibre design and wavelength where the sign of the chromatic dispersion is negative, the effect of MI is negligible at 0.4 W and at the maximum distance reported, namely 18 km. It was anticipated that 900 mW could be launched in 25 km sensing fibre sections without deleterious results from MI. However, the choice of the fibre type is often imposed by other considerations,* and in this case, MI can seriously limit the probe power that can be launched.

* It is often a requirement to use fibres that have already been installed for other purposes, such as telecommunications or telemetry. The choice of these fibres is dictated initially by the primary need, and their later use for sensing is not usually considered.

Self-phase modulation (SPM), on the other hand, is an effect caused by the non-linearity of the refractive index that responds slightly to the square of the electric field, i.e. in proportion to optical power. This would be of no consequence if the probe power were constant; however, it is modulated in some way, usually as a pulse. This means that the refractive index is slightly higher in the centre of the pulses than outside it. The phase velocity of the pulse is therefore reduced slightly as the intensity is increased; therefore, the leading edge of the pulse experiences a small positive frequency shift, and the trailing edge, a negative frequency shift. These effects are cumulative along the fibre, and it is the *derivative* of the optical power across the pulse that causes the phase modulation. As a result, the effect of SPM is exacerbated as the spatial resolution is made finer, for a given peak power. The reader is referred to the Section 3.3.3.5 of Chapter 3 on frequency-diverse C-OTDR for a discussion of means of alleviating the effect of SPM.

5.3.4.2 Spatial resolution

The spatial resolution of an SpBS sensor is, as for other distributed sensors, limited by the probe pulse duration, the receiver bandwidth and chromatic dispersion in the fibre. The latter can generally be neglected in single-mode fibre for the distances and resolutions that are relevant to distributed sensing. When using the intensity of the SpBS for the measurement, achieving a very fine spatial resolution can be achieved by ensuring that the pulse duration is short and the receiver bandwidth is sufficiently wide. A resolution of 5 cm has been demonstrated in this way with a resolution of the intensity that was equivalent to about 2 K over a relatively modest distance of 18 m [82]. Clearly, the same trade-offs as apply in other OTDR methods are valid here too, namely a reduction in the pulse energy that can be transmitted and an increase in the noise of the receiver as the spatial resolution is improved. On the figure of merit criterion introduced in Chapter 4, this result (at $M = 23$) is an improvement on all the single-photon counting results reported with Raman DTS (it should be noted that the multi-channel photon counting results of Höbel et al. achieved a higher M figure, but at a resolution of 40 cm).

However, the frequency shift is most often used as part of the measurement process; in this case, the fact that the spectrum is broadened as the probe pulse duration is decreased then degrades the ability to locate the centre frequency of the Brillouin line. Thus, when the probe pulse duration is reduced beyond the value at which it broadens the backscatter spectrum, the resolution of the Brillouin frequency is degraded [83].

A synthetic spectrum vs. distance is illustrated in Figure 5.12, where an abrupt transition in the Brillouin frequency is modelled, but it is smeared by the limited spatial resolution (a pseudo-3-D picture is shown on the left, with a few individual spectra plotted in the right-hand panel). In this situation, the Brillouin frequency does not change from one value to another passing through intermediate values; rather, one frequency fades as the other appears, and in the transition region, both frequencies (10.79 and 10.81 GHz, in this example) are present simultaneously. The signal processing routines used to extract the frequency must be set up to recognise and deal with this situation and to process the spectra differently in a transition region; some enhancement may be brought to the spatial resolution in this way. In effect, the software needs some built-in awareness of the 3-D nature of the data that it is processing rather than trying to deal with it as a succession of unrelated spectra. This approach was carried out successfully in Refs. [84] and [85]. An instrumental resolution of 1 m was improved to 0.5 m in the case of Ref. [84]. In essence, the spectrum that the fitting routine seeks to match is a compound one of two Lorentzian peaks with unknown proportions and frequency separation.

This dual spectral feature approach seems to work quite well for the case where the spectrum consists of two distinct frequency shift values. It applies to a number of practical applications, but it is not a universal solution that can handle arbitrary Brillouin shift distributions.

Similar concepts were applied in Ref. [86], where again, a signal model is built up on the assumption of a dual peak and the algorithm attempts to separate the contributions from each. In this paper, a specific criterion was applied to define regions where either a single peak or a dual peak is present. Again, the underlying assumption is that the signal is accurately modelled by one or other of these categories. It would not apply, for example, to a gradual transition between two states that occurs over a distance of the order of the spatial resolution.

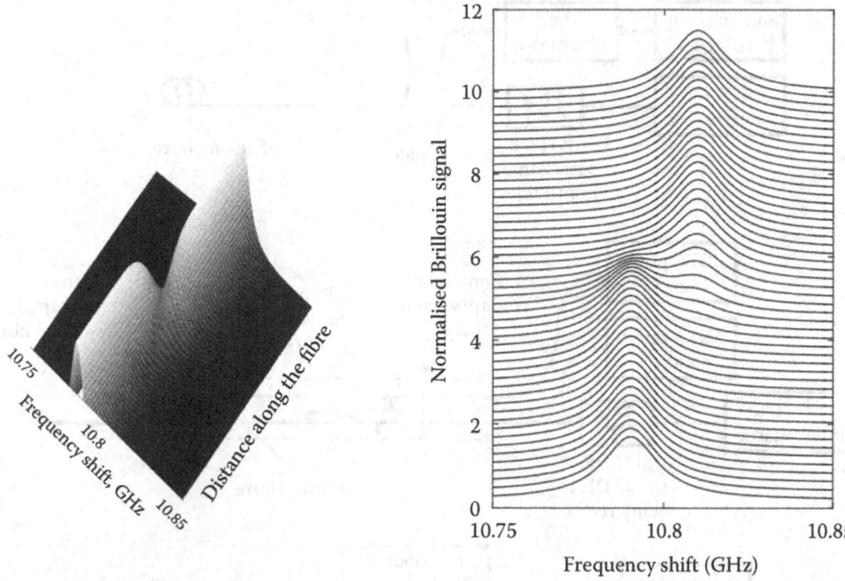

FIGURE 5.12 Synthetic Brillouin spectra at a transition of the measurand with limited spatial resolution.

5.4 RECENT ADVANCES IN DISTRIBUTED STRAIN SENSING BASED ON SpBS

The basic BOTDR methods, just discussed, have been enhanced to improve various aspects of the system performance, such as the range, the speed of measurement and the spatial resolution.

5.4.1 Remote Amplification for Range Extension

The narrowband nature of the Brillouin scattering and the requirement for the sources also to be narrowband allow the probe pulses and the backscatter signals to be amplified optically using doped fibre amplifiers and/or Raman amplification [87]. Although the optical amplification contributes noise, it has nonetheless been demonstrated to extend the range of distributed sensing systems to approximately 100 km from the instrumentation. It should be borne in mind that when the range is extended, the pulse repetition frequency (PRF) must be reduced to avoid the occurrence of overlapping signals from different parts of the fibre, and so even if the amplification process were noiseless, the performance of an extended-range distributed sensing system would still be degraded relative to a shorter one. Nonetheless, remote amplification has proven benefits in long-range monitoring applications such as transmission pipelines [88] and energy cables.

Optical amplification can be used as a pre-amplification method prior to conversion of the signal to the electrical domain. This is particularly appropriate in direct detection with optical filtering, where the receiver is likely to degrade the signal-to-noise ratio by far more than the best achievable noise figure of an optical amplifier (~4 dB). In a well-designed heterodyne system, optical pre-amplification immediately prior to the detector is likely to yield substantial benefits only if there is significant optical loss in the return path between the circulator and the receiver.

However, the major attraction of optical amplification in SpBS systems is remote amplification. This has two benefits. Firstly, in amplifying the probe pulse some distance from the interrogation system,

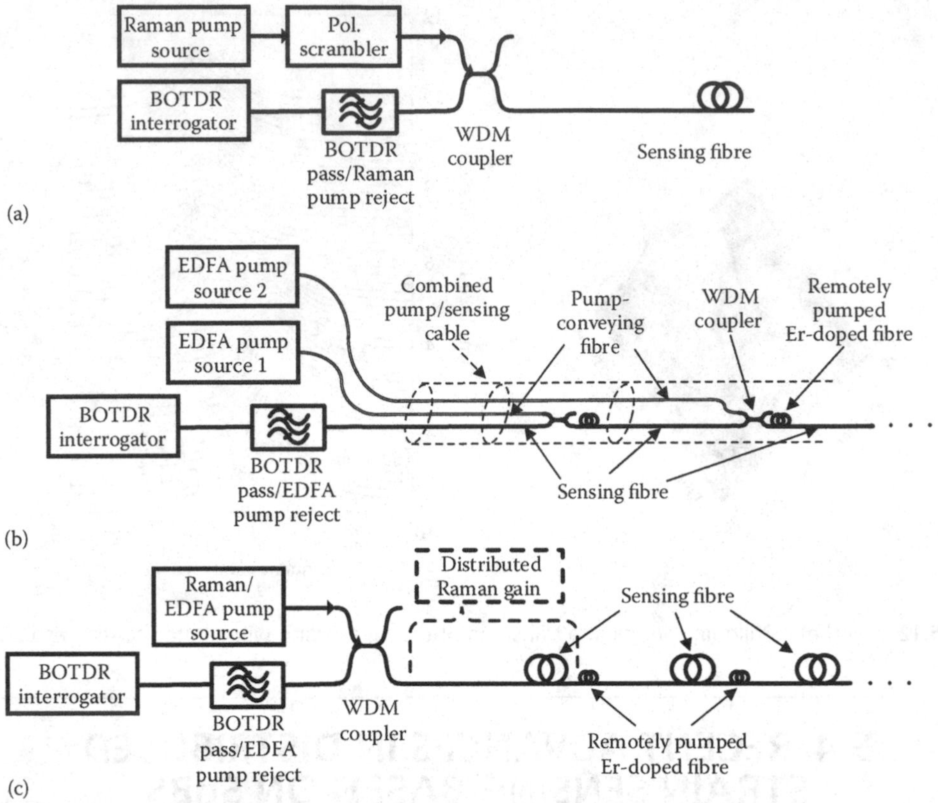

FIGURE 5.13 (a) Distributed Raman amplification in long-range BOTDR. (b) BOTDR sensing system, range-enhanced by means of remotely pumped EDFAs. (c) Range-extended BOTDR sensing system with hybrid distributed Raman amplification and discrete EDFAs.

the *probe* power can be maintained at a level reasonably close to the maximum value allowable by nonlinear effects throughout the length of the sensing fibre. Secondly, by amplifying the *backscatter* signal remotely from the interrogation system, the amplification acts on a still relatively powerful signal, before the transmission losses are applied to the signal returning to the interrogator. The noise contribution of the amplification process is therefore substantially reduced relative to amplifying the backscatter signal after it has returned to the interrogator.

Several configurations for remote amplification are illustrated in Figure 5.13.

5.4.1.1 Distributed Raman amplification

Remote amplification can be achieved with distributed Raman amplification, by launching pump power along the sensing fibre, i.e. co-propagating the probe and the Raman pump, as illustrated in Figure 5.13a. For a probe at 1550 nm, a Raman pump wavelength of c. 1450 nm is appropriate, although the Raman process is tolerant of quite wide ranges of pump wavelength and the use of 1480 nm is convenient in that it can provide distributed Raman gain and also pump remote EDFAs. As shown in the figure, providing Raman remote amplification requires multiplexing the pump and probe waves onto the sensing fibre (achieved here with a wavelength-division multiplexing [WDM] coupler). Owing to the polarisation sensitivity of the Raman gain process, it is also necessary to scramble the SOP of the pump, so that, on average, all sections of the fibre benefit from equal gain; in some cases, two Raman pumps are multiplexed with orthogonal polarisation states to eliminate the polarisation sensitivity of the gain. Finally, the residual pump (in

RAMAN GAIN

Although optical non-linear processes are often described in terms of the incremental power accruing to a signal I_s, what is important in this case is the fractional increase in signal power, i.e. the signal gain that is driven by pump intensity $I_{pump} = P_{pump}/A_{eff}$ given by

$$\frac{1}{I_s}\frac{dI_s}{dz} = \gamma_R I_{pump} \tag{5.24}$$

The Raman parameter γ_R has a typical value of 3×10^{-14} m W^{-1} at 1550 nm assuming an unpolarised pump [89] for a conventional single-mode fibre. So a fibre having effective area A_{eff} = 80 μm^2 would achieve a gain of about 1.5 dB km^{-1} W^{-1} at the peak of the gain curve, a frequency shift of about 13–14 THz (from 1450 nm to 1550 nm). The gain falls off by about one third if the pump is at 1480 nm for the same signal frequency (i.e. a shift of about 9 THz). In practice, Raman amplification in distributed sensors usually aims to cancel or at least reduce the loss of the sensing fibre, and a pump power of a few hundred mW is generally sufficient for this purpose.

the form of Rayleigh-backscattered power returning to the interrogator) must be removed, thus avoid adding unnecessary shot noise in the detection process; this is achieved with the bandpass filter shown. In other respects, the BOTDR interrogation equipment could be the same as that shown, for example, in Figure 5.8. In extreme-range demonstrations, additional pump power is also launched at the remote end or in both directions from a mid-point along the sensing, which, for practical applications, is possible only if electrical power is available to drive the pumps.

The Raman gain process in the sensing fibre amplifies both the probe and the backscatter signal. However, the amplification needs can be quite different for these two waves: in general, the probe is limited in the maximum power that can be carried, and so it is not desirable for the probe to be amplified strongly, in particular near the launching end. In contrast, in general, the backscatter signal can usually benefit from the maximum gain that can be achieved at the remote end of the fibre.

The first reported work [90] on Raman gain in Brillouin sensing used a pulsed pump that was designed specifically to amplify the probe. Here, a pump pulse (120 mW at a wavelength of 1450 nm) was launched into the sensing fibre slightly later than the probe pulse: the probe pulse thus propagated initially without gain until the difference in group velocity for the pump and probe pulses caused them to overlap. The walk-off in this case is about 1 ns/km, and so the process of one pulse catching up with the other is a gradual process, with complete overlap being achieved at 30 km along the sensing fibre, where the probe had already been attenuated.

With a pulsed pump, co-propagating with the probe, the amplification is primarily experienced by the probe. Whereas some small gain also applies to the backscatter, the interaction length is limited to the pump pulse duration.

In contrast with a CW pump, both probe and backscatter are amplified. This approach was demonstrated in Ref. [91], where at the mid-point of a 100 km sensing fibre, Raman pump light was inserted in the backward direction (420 mW) for the first section (closest to the interrogation) and in the forward direction (660 mW) for the second section. Net gains of 13 dB for the first section and 24.5 dB for the second section were demonstrated, although not simultaneously.

Raman amplification is very simple to implement in principle; however, it does require very careful filtering of the received signals owing to the substantial ASE that accompanies the gain. It also requires significant pump power (of order 0.5 W in the example discussed). Clearly, using certain fibre types, e.g. those with small mode field diameter and/or higher GeO_2 content, lowers the pump power that is required; however, many of the applications in long-distance sensing require the use of conventional fibres, often already installed for telecommunications or telemetry, and the designer of the sensing system often has no control over fibre type.

The large pump powers required usually involve multiplexing several pump laser diodes or the use of a dedicated Raman pump laser. Both technologies are commercially available, but in either case, they add to system complexity.

At these substantial pump powers, the power rating of the optical components and the dissipation of any waste heat from optical losses must be considered. Often, this pump source will fall in the Class 4 laser safety classification, which requires special measures to avoid harm to personnel who might service the system. Another safety consideration relates to explosion risk [92] in the event that an open fibre end is exposed to a potentially flammable or explosive atmosphere.

In addition, Raman gain is highly polarisation dependent: the gain for orthogonal pump and signal is roughly one tenth of that for the case where these waves are in the same polarisation state [93]. This problem can be resolved by using more than one pump source launched with different (e.g. orthogonal) polarisation states. Again, commercially available components are available for that purpose. Polarisation scrambling, if sufficiently fast, is another means of circumventing the polarisation dependence of Raman gain, as illustrated in Figure 5.13a.

Another consideration with Raman amplification is that it is desirable to control the gain of the probe and that of the backscatter separately. For example, in several publications, e.g. Ref. [4,91], it was found necessary to reduce the initial probe power to ensure that the Raman gain (that was nominally the same for the probe and the backscatter) would not cause the probe power to exceed non-linear limits.

Adjusting the time dependence of the pump power [94] allows the gain experienced by the pump and the backscatter to be controlled independently. In this way, the pump power can be reduced when it co-propagates with the probe so as to (ideally) just cancel the fibre loss and still allow optimum gain for the backscatter signals. The pump power that can be launched is limited by stimulated Raman scattering provided that the pump has sufficient linewidth to avoid other effects.

Whilst it is desirable to launch high pump powers, it is also necessary to ensure that the power is transferred as far as possible along the sensing fibre, rather than converted to the Stokes wavelength near the launching end. This places an upper limit on the pump power well below the accepted threshold for SRS. In turn, this constrains the energy that is available to transfer from the pump to the probe; the distance over which Raman amplification can be used to cancel the effects of fibre loss is therefore limited.

These limitations are alleviated somewhat by pulse walk-off, which brings new energy from different parts of the pump time distribution to reach the pump as it travels along the fibre.

This process can be enhanced by passive devices that extract the pump, delay it and re-inject it into the sensing fibre. This palliative can provide further extension of the probe amplification range. Ultimately, however, the range extension to be expected from Raman amplification is limited by the attenuation of the pump itself; nonetheless, these methods can be used to extend the range of a BOTDR system by at least 50 km.

5.4.1.2 Remotely pumped rare-earth-doped fibre amplifiers

Remotely pumped discrete doped fibre amplifiers provide another method of boosting both probe and backscatter signals. EDFAs are ideal for this purpose in BOTDR systems operating around 1550 nm. This option is illustrated in Figure 5.13b. In this example, a separate fibre is used to convey the pump power to each of the EDFAs that provide optical gain.

Remotely amplified sensing systems do however require care in their design: the gain of each amplifier must be reasonably matched to the losses, and in practical cases, these may not be fully predictable. The monitoring of long assets such as pipelines or energy cables often involves fibre sections with many splices and sometimes with poorly specified fibre that result in higher losses than planned. It is therefore sometimes necessary to carry the pump power required for one or more of the remote amplifiers on dedicated fibres, possibly in the same cable, as illustrated in Figure 5.13b.

REMOTE PUMP POWER REQUIREMENTS

When using EDFAs, the fibre span between amplifiers is typically relatively short (25–40 km), and so only modest gain (5–10 dB) is required to restore the signal to that at the start of the span. The average power to be delivered is also modest. It can be calculated as follows.

Here, we will consider a remotely amplified sensing fibre of total length 100 km divided into spans of 25 km between boosters. Thus, there are three remotely powered amplifiers, and, of course, the most challenging is the most remote of these. Let us further assume a probe power of 0.25 W, a pulse duration of 80 ns and a PRF of 1 kHz. The final amplifier is required to boost the probe back to the original power and so to deliver an average power of

$$P_{pr} = 0.25\ \text{W} \cdot 80\ \text{ns} \cdot 1\ \text{kHz} = 20\ \mu\text{W}.$$

Concerning the amplification of the backscatter, this is of course dominated by the Rayleigh backscatter. So using the initial energy (0.25 W · 80 ns = 20 nJ) and a backscatter factor $\eta = 5$ W/J, the initial backscatter is 100 nW. To estimate the energy returning to the remote amplifier from the section downstream, we integrate this power over the 25 km (i.e. 0.25 ms), assuming a decay rate corresponding to the fibre loss. So for a loss of 0.2 dB/km, the decay rate is 4.0×10^{-2} dB/μs, and so the integrated energy returning to the preamplifier is ~9.8 pJ/pulse, i.e. an average power of 9.8 nW (given a PRF of 1 kHz). Allowing for a gain of ×10, the amplifier is required to deliver 95 nW to amplify the backscatter from the final sensing fibre section.

These estimates show that the optical power to be delivered by the most remote amplifier is dominated by amplifying the probe pulses and is in the tens of μW. In practice, such amplifiers can be powered by delivering less than 10 mW of pump power in order to achieve the gain of 5 to 10 dB that is required in typical systems [95].

Temperature variations of the remote amplifier must also be considered [96,97]: the gain of a doped fibre amplifier is temperature sensitive, particularly when pumped around 1480 nm, and diurnal and seasonal variations of more than 50 K can easily be experienced; depending on the doped fibre type, such a temperature variation could lead to a change in the gain of at least 1 dB per amplifier for typical values of gain. This consideration contributes to the need to control the gain of remote amplifiers from the interrogation system. Temperature variations also affect the central wavelength of filters used remotely to reject unwanted parts of the spectrum. Controlling the gain of each amplifier requires a separate pump for each device, carried on an individual fibre. In the case of new terrestrial applications, the additional cost of more fibres in a sensing cable is usually not material relative to that of the entire sensing system, and this approach provides considerable flexibility in the design and maintenance of the system.

A further consideration is the build-up of ASE: the peak gain of EDFAs is at somewhat shorter wavelength than the wavelength of lowest loss for the sensing fibre. Therefore, when the system is operated at the wavelength of lowest loss, considerable ASE can appear in spectrum of the light travelling in the fibre and this will extract more power from the remote amplifiers than is used for amplifying the signal of interest. Therefore, careful filtering, certainly in the interrogator, but also in the remote amplifiers, is required in some systems to avoid a loss of system performance. This complicates the design of remote amplifiers that could otherwise consist simply of a carefully chosen length of doped fibre, as illustrated in Figure 5.13b. An optimised remote amplifier would then include a filter to pass the spectral region close to the probe wavelength and to reject most of the ASE [87]. This in turn probably requires the pump to bypass the filter so additional wavelength selective components would then be required to extract the pump before the filter and re-inject it after the filter. It can also be beneficial to separate forward and backward optical paths using circulators, in order to optimise separately the gain of each amplifier. The remote amplifiers can therefore vary from a simple piece of Er-doped fibre spliced in series with the sensing fibre and pumped through the sensing fibre through to a relatively complex, but passive, optical sub-system.

5.4.1.3 Hybrid Raman/EDFA amplification

The third remotely pumped amplifier configuration, illustrated in Figure 5.13c, is a hybrid Raman/EDFA arrangement, where the pump power is co-propagated with the probe in the sensing fibre [4]. It initially provides Raman gain, but the residual pump power is then used to power discrete EDFAs – the power requirements for which are far lower than for Raman distributed amplification. This is the configuration selected in Ref. [98], where 700 mW of Raman pump at 1470 nm was co-propagated with the probe pulse over the first 50 km section of the sensing fibre. At 50 km, an Er-doped fibre remotely pumped by the residual Raman pump power provided further gain for the remaining 38 km of sensing fibre. Thus, a rather simple enhancement of the interrogation system is able to massively extend the sensing range. The same team showed [4] that in this configuration, sufficient pump power can be delivered to a further EDFA, and in this way, the range was extended to 100 km.

5.4.2 Fast Measurement Techniques

The speed of Brillouin systems that measure frequency (as opposed to intensity) was initially limited by the requirement to scan the filtering across the spectrum of possible values of the peak to estimate its location. The process of shifting the frequency of the filter can itself be time-consuming: for example, some of the microwave tunable oscillators used in the conversion from the Brillouin frequency to the detection frequency are tuned with micro-heaters, and they exhibit a finite response time. More fundamentally, the frequency-scanning approach wastes the signal energy within the backscatter spectrum that falls outside the filter passband for any particular frequency setting. Several techniques, to be described next, have therefore been devised to improve the measurement speed.

5.4.2.1 Frequency discriminators

One approach that alleviates the fundamental inefficiency is the use of a discriminator [27], i.e. a device that converts a frequency change into a variation of the intensity using a transfer function that is frequency dependent. In the case of Refs. [27] and [60], this was an unbalanced MZI.

Whereas Kee et al. [27] demonstrated a purely optical discrimination function, Watley et al. [59] also showed the same function in the electrical domain where similar techniques can be used in the heterodyne approach after an initial optical-to-electrical down-conversion. The electrical approach was demonstrated using an electrical bandpass filter detuned from the Brillouin line so that the expected signal was somewhere on the slope of its transmission function [99].

In these approaches to frequency discrimination, the acquisition bandwidth must be sufficiently wide to include the entire span of frequencies at which the Brillouin peak could occur, given the temperature and strain range to be measured. The noise bandwidth that the system has to contend with is thus far wider than when a narrowband filter is scanned across the same frequency band. The gain in measurement time achieved through the use of a discriminator is therefore offset, at least in part, by the higher noise resulting from a wider signal bandwidth. It can therefore be argued that there is no net gain in using a discriminator other than one of convenience and a reduction in the minimum time for a measurement by eliminating the need repeatedly to change the local oscillator frequency.

5.4.2.2 Parallel multi-frequency acquisition

A multi-channel acquisition approach where a number of narrow frequency bands are acquired in parallel [100] allows each of these frequencies to be captured with a noise bandwidth that relates to the linewidth of the signal that is detected. In the fitting process, only those bands that contribute to the signal power need to be used, and the system noise bandwidth is thus reduced whilst retaining the speed of simultaneously acquiring all the data required for determining the Brillouin frequency. The multi-band approach can be implemented in the digital domain by digitising the down-converted Brillouin backscatter signal at a sampling rate sufficient to analyse its frequency content. So, for example, if the Brillouin backscatter has been converted to a possible signal range of say 100–800 MHz (which would still cover a temperature range of 700 K), sampling this electrical signal at 1800 MSPS allows the complete frequency spectrum to be determined with some margin for anti-alias filtering. One process that meets this requirement is

a. To decompose the time series into short time windows and apply a windowing function to control the frequency response (i.e. apodise it);
b. To calculate the Fourier transform for each window;
c. To calculate the power spectral density in each window;
d. To average the power spectral density function for each window over successive probe pulses and
e. To fit the resulting averaged spectra in the usual way to estimate the Brillouin shift and other parameters of interest (such as linewidth or intensity).

Of course, in this process, there is a trade-off between the frequency resolution (which improves with the number of points in the fast Fourier transform [FFT] window) and the spatial resolution. It was recently pointed out that more sophisticated time-frequency analysis yields better results than the short-window Fourier transform does [101].

It should be noted that since the SpBS originates from thermally generated phonons, there is no phase correlation between successive backscatter signals, and so it is necessary to average the power spectral density, rather than the time series. The need to extract the distance-dependent spectrum for each backscatter trace increases the computational workload, but this is tractable and similar to that required for some types of distributed vibration sensing systems, to be described in Chapter 6.

Alternatively, the RF signal may be divided into a number of discrete frequency channels [102]. The output of each channel is then rectified and digitised. Although this could involve a large number of A/D converters, they are required to sample only at a rate consistent with the spatial resolution, and so their cost and power consumption is quite reasonable (c. 40 mW per channel).

5.4.2.3 Signal tracking

There are special situations where the Brillouin frequency distribution varies along the fibre but is expected to be reasonably static with fast variations that are limited in excursion. An example of this might be detecting fast vibration signals on a structure where the temperature is non-uniform along the object to a significant extent but reasonably stable over time. In this case, the problem is to detect fast changes in the Brillouin frequency that occupy a limited frequency span and also to track slow variations of the total frequency distribution.*

In applications such as these, the optical or electronic local oscillator used to heterodyne the backscatter from the Brillouin frequency to a UHF band can be varied in time using a fast tunable oscillator [103]: once a pattern of the frequency distribution in time is established as a function of distance along the fibre, this is fed to the local oscillator and the acquisition is required only to measure small departures from a nominal value, which can be accomplished with just a few digitised channels or a limited-range discriminator. However, the local oscillator is required to be able to vary its output frequency sufficiently fast to follow the spatial distribution of the measurand, and so a tuning bandwidth of 10 MHz would be required for a spatial resolution of order 5 m.

* As an example, the risers that connect subsea wellheads to floating production storage and offloading vessels will see marked changes in temperature as the sea water warms towards the surface and the oil cools. Against the rather large, but quasi-static, change in Brillouin frequency that this causes, there is also a need to detect and measure fast variations in strain caused by vortex-induced vibration.

5.4.3 Pulse Compression Coding

Pulse compression coding in BOTDR follows the same principles as were described in Chapters 3 and 4. Both Simplex [104] and Golay [105] coding have been explored. However, in BOTDR, coding has only limited scope for improvement if the system has otherwise been optimised for launched power. It will be recalled that the SBS in OTDR systems is dictated by the probe energy, and so if the power of a single pulse is already close to the non-linear limit, then increasing the number of pulses to form a probe code is analogous to increasing the pulse duration, and this lowers the threshold for launched power [104].

Nonetheless, high-spatial-resolution BOTDR sensors, where the probe power is limited by the stimulated Raman scattering threshold, can benefit from pulse-compression coding.

5.4.4 Simultaneous Multi-Frequency Measurement

Given that the energy that can be launched in a single-probe pulse is limited and that pulse compression coding is not effective in BOTDR, the simultaneous use of multiple, independent probe pulses launched at different carrier frequencies has been explored.

As illustrated in Figure 5.14 [106], the output of a laser was split into three pairs of side bands separated by 2 GHz by an EOM, EOM1, prior to being pulse modulated (by EOM2) and launched into the sensing fibre.

To complete the BOTDR configuration, the local oscillator was frequency-shifted (by EOM3) to provide an LO signal for each Stokes and anti-Stokes Brillouin signal. This results in three times the probe power being launched, but six signals with their LO being collected. This does cause additional noise, and the overall improvement was just short of a factor of 2 on the frequency resolution. One of the reasons that the gain in performance was modest was the fact that the photodiodes were limited in the power that they could detect, and so the system may not have been fully optimised. Nonetheless, this work shows that there are opportunities for using the available spectrum to launch more probe energy and so improve somewhat the overall performance.

Clearly, the technique could be extended by increasing the number of frequencies launched. It should be noted that in this arrangement, the probe pulses travel simultaneously. Their difference in frequency (2 GHz in the experiment that was reported) avoids any risk of SBS being caused by the higher power. However, other non-linear effects such as stimulated Raman scattering or SPM would ultimately limit the total power that can be launched unless measures are taken to spread the probe power over time, an approach that has been tested in the context of BOTDA and that will be described later in this chapter.

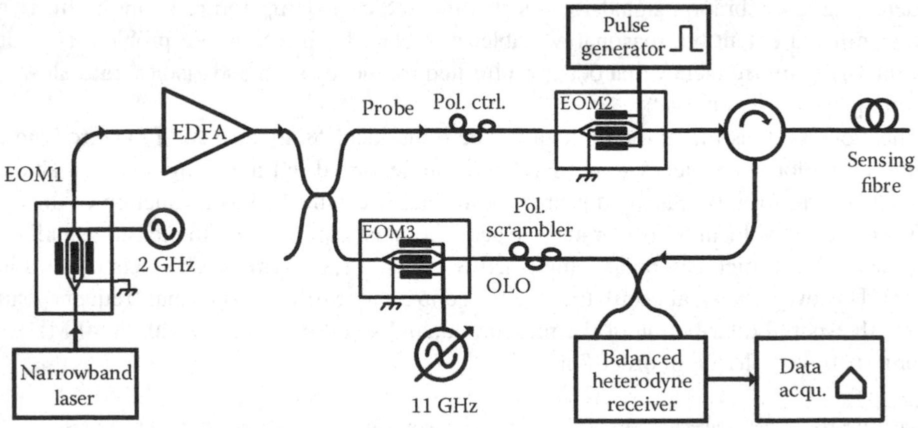

FIGURE 5.14 Multi-frequency implementation of BOTDR. (Modified from Li, C., Lu, Y., Zhang, X., Wang, F., *Electron. Lett.*, 48, 1139–1141, 2012.)

5.4.5 Dual-Pulse Method in BOTDR

A host of techniques have been proposed for improving the spatial resolution of Brillouin sensing systems. Many of them share the concept that one wave will set up a stimulated acoustic wave that is interrogated by a second pulse. In general, these techniques have been developed in the context of BOTDA and will be described later in this chapter.

In BOTDR, there seems only to be one such concept [107,108] involving a pair of closely spaced probe pulses and aimed at improving the spatial resolution. It is worth discussing in some detail because this approach introduces some of the ideas in dual-pulse methods that will be found in connection with stimulated Brillouin systems.

We recall that SpBS is the result of the interaction of a probe pulse with thermally generated acoustic phonons that have their own velocity and lifetime. Two probe pulses that are closely spaced (i.e. launched within the decay time of the acoustic phonons) are in fact scattered by largely the same phonons and their scattered light waves are therefore mutually coherent, assuming that the probe pulses themselves are mutually coherent. As a result of their mutual coherence, the scattered light from the two probe pulses can be made to interfere with one another by mixing the detected backscattered signal with a delayed version of itself, the delay being equal to the time separation between the probe pulses.

Generally, spontaneous scattering is not coherent; the dual-pulse BOTDR technique exploits the fact that the phonons involved in the scattering have a lifetime that is sufficiently long to be observed on the scale of a sub-metre spatial resolution.

The experimental arrangement, illustrated in Figure 5.15, is identical in concept to that of Figure 5.8 with the exception that the AOMs are replaced by a faster EOM that is pulsed twice in rapid succession. The EDFA and circulator were included in later publications [109,110] by the same team. The key difference between Figure 5.8 and Figure 5.15 is the inclusion of a matched filter circuit that splits the signal into two paths and then adds the signal arriving at input A of the adder to a delayed copy of itself arriving at input B – this provides the interference function that the dual-pulse technique exploits (it is equivalent

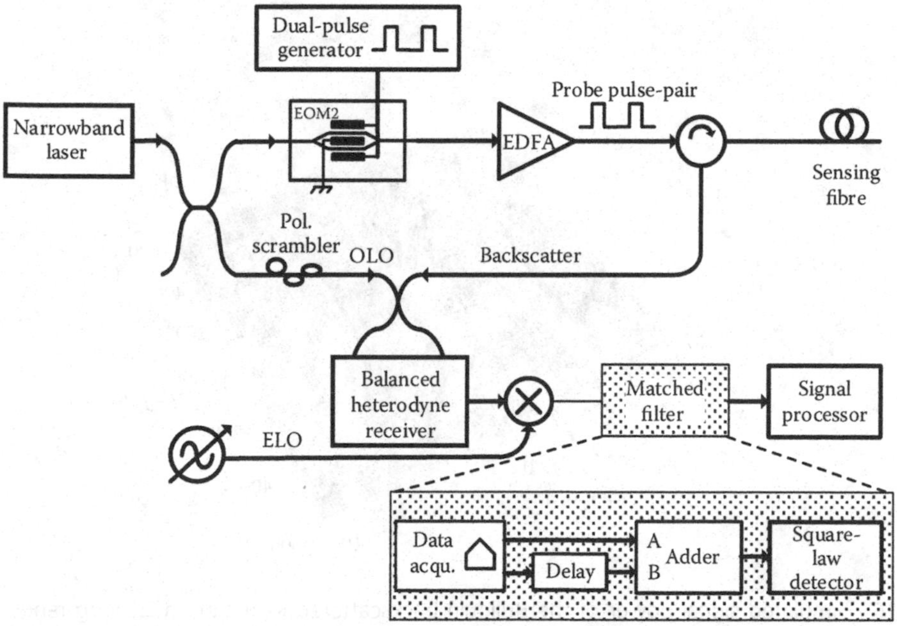

FIGURE 5.15 Dual-pulse BOTDR. (Combined from several drawings in Koyamada, Y., Sakairi, Y., Takeuchi, N., Adachi, S., *IEEE Photon. Technol. Lett.*, 19, 1910–1912, 2007; Koyamada, Y., Apparatus for measuring the characteristics of an optical fiber, US7873273B2, 2011.)

to an MZI). The matched filter can be provided in the electrical domain or in the digital domain after data acquisition, which is what the authors of Ref. [107] actually did; the inset of Figure 5.15 shows one of the digital-domain approaches proposed in Ref. [108] for the matched-filtering function.

The mixed Brillouin backscatter signal exhibits an interference that can be observed as an oscillatory behaviour in the spectral domain. A synthetic signal generated as a function of time from the launching of the pulse-pair and as a function of the frequency offset from the Brillouin shift is shown in Figure 5.16. Here, the pulses have a duration of 2 ns and are separated by 5 ns. The broad peaks at −5 ns and +5 ns are the usual BOTDR signals (spectrally broad owing to the short duration of the pulses). They correspond, respectively, to the time at which the signal for the first pulse arrives at Adder Input A and when the signal from the second pulse arrives at Adder Input B.

The interesting feature, however, is the oscillatory signal at 0 ns that results from the interference between the signal and its delayed copy. The interference occurs over a narrow time duration, equal to about one pulse duration, and its oscillatory nature provides narrow peaks in the frequency domain that allow the centre of the Brillouin shift to be located precisely. The interference signal is highly local to the region between the pulses, and the spatial resolution was demonstrated to be better than 20 cm and, by some definitions, the results reported suggest an even finer resolution approaching 10 cm.

In a subsequent paper [109], the same team showed, as might have been expected, how increasing the separation between the probe pulses reduces not only the frequency of the modulation on the backscatter signal but also the amplitude of this modulation. The first effect relates simply to the frequency change required to induce a 2π phase shift between the Brillouin scatter originating in the two separated regions. The second effect is caused by the phonon decay: the likelihood of a given phonon still existing when the second probe pulse arrives is reduced as the delay between the pulses is increased.

The method was recently extended to making the same measurement in multimode fibre where a spatial resolution of order 40 cm was demonstrated [110], albeit over a relatively short (120 m) sensing fibre, with a frequency resolution of 340 kHz (i.e. equivalent to ~0.34 K or 7 με).

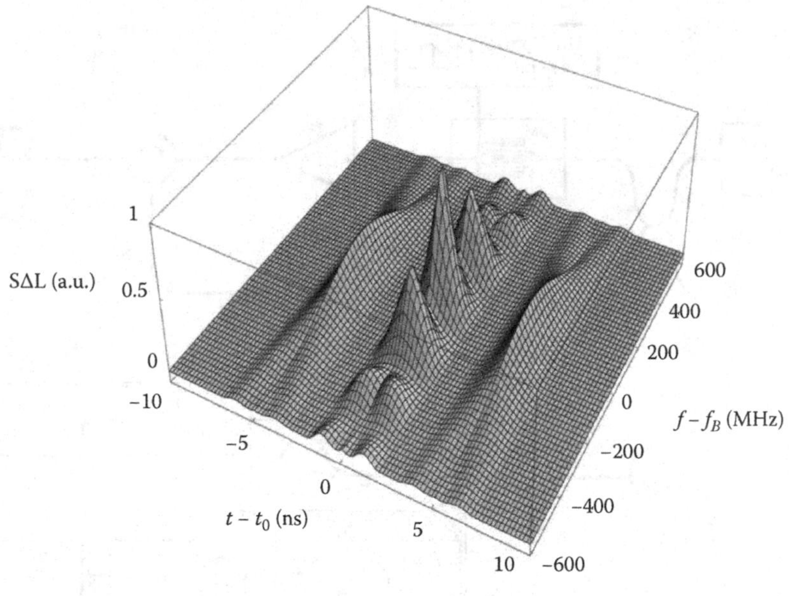

FIGURE 5.16 Simulated signal power of the Brillouin backscattered light from a ΔL-long range along the fibre as a function of the frequency f of the microwave local oscillator and time t. f_B represents the Brillouin frequency shift of the fibre. (Reprinted with permission from Koyamada, Y., Sakairi, Y., Takeuchi, N., Adachi, S. Novel technique to improve spatial resolution in Brillouin optical time-domain reflectometry. *IEEE Photon. Technol. Lett.*, 19, 1910–1912. © 2007 IEEE.)

5.4.6 Spectral Efficiency

In conventional OTDR, the pulse duration is the key parameter defining the spatial resolution, and the trade-off is primarily one of spatial resolution vs. signal-to-noise ratio. Some simple measure of the pulse duration, such as its root mean square (r.m.s.) value, is usually sufficient to characterise it.

In BOTDR, however, we are interested in the spectral response to the probe and so the details of the spectrum of the probe pulse itself are important when one is operating close to transform-limited conditions, i.e. when the pulse duration is close to the phonon lifetime. In this case, there is scope for optimising the pulse shape to provide the narrowest Brillouin spectrum. The situation was analysed [111], and it turns out that some apodising is helpful and that for a constant pulse duration (e.g. measured at half-height), the spectrum of the Brillouin backscatter broadens as the pulse edges are sharpened.

Furthermore, a triangular shape minimises the error on the estimate of the peak frequency of the Brillouin distribution. This was confirmed experimentally [112], and it was shown that for a given pulse duration, a triangular pulse shape maximises the peak power in the Brillouin spectrum. There are therefore clearly benefits in controlling the pulse shape in BOTDR systems in order to jointly optimise the spectrum and the time-domain response.

5.5 SYSTEM BASED ON SBS

5.5.1 BOTDA

BOTDA, initially proposed in 1989 [6,113], is a method that uses SBS to probe the gain spectrum as a function of distance along the fibre. BOTDA involves launching light from both ends of a single fibre, and, in the original configuration, one of the optical signals is continuous and the other is pulsed.

In BOTDA, it has become the common terminology to refer to the pulsed and CW waves as 'pump' and 'probe', respectively, regardless of which of these waves experiences loss or gain. This is of course different from other OTDR methods, including BOTDR, where the probe is a pulsed wave; it also differs from the usage in optical amplifiers, where the pump always provides the energy and the signal is amplified. The reason for this change in usage is probably that the CW signal, on emerging from the fibre, is the wave that carries the signal imparted by the SBS process. However, it can be confusing, and in this chapter, we have attempted as far as possible to refer explicitly to pulsed and CW waves, sometimes in addition to the use of the pump and probe terminology to try to limit the ambiguity.

The basic arrangement is shown in Figure 5.17a. As originally intended, BOTDA was a method for measuring attenuation in the fibre, i.e. a means of making a better OTDR. The two signals were set at a frequency separation equal to the Brillouin shift, and the result of their interaction is observed at the circulator port corresponding to transmission of the CW wave. In the original intent, the transmitted light is simply amplified (for $f_{cw} = f_0 - \nu_B$) or attenuated (for $f_{cw} = f_0 + \nu_B$).

The power that is detected at the receiver consists of a DC component unmodified by interaction with the probe and independent of position along the fibre, together with a position-dependent element that reflects the interaction between pulsed and CW signal. The variable element is given by Ref. [6]

$$P_b^{\pm}(z) = \frac{g_B}{A_{eff}}\frac{v_g}{2}E_p(0)\exp(-\alpha_p z)P_{cw}(L)\exp(-\alpha_{cw}L)\cdot$$
$$\exp\left[\mp\left(\frac{g_B}{A_{eff}}\right)P_{cw}(L)\exp(-\alpha_{cw}L)\left\{\frac{\exp(\alpha_{cw}z)-1}{\alpha_{cw}}\right\}\right]. \tag{5.25}$$

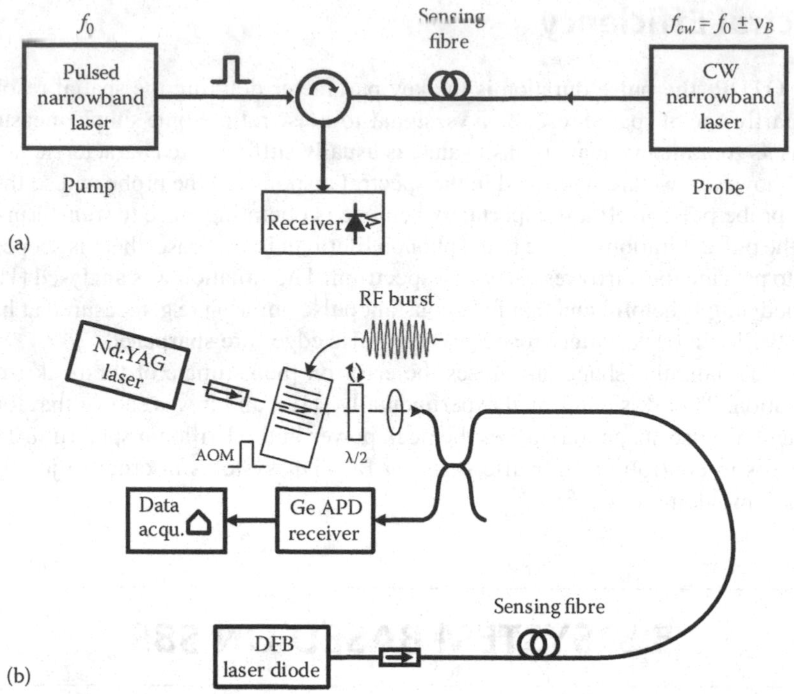

FIGURE 5.17 (a) Basic BOTDA arrangement. (b) First practical realisation of BOTDA. (Modified from Horiguchi, T., Tateda, M., *Opt. Lett.*, 14, 408–410, 1989.)

The DC signal is the residual power in the CW light, i.e.

$$P_{dc} = P_{cw}(L)\exp(-\alpha_{cw}L). \tag{5.26}$$

Here, the sign in ± refers to $f_{cw} = f_0 + \nu_B$ for the + case and $f_{cw} = f_0 - \nu_B$ for the – case.

The multiplicative term in the second line of Equation 5.25 refers to changes (gain or loss) affecting the pulse and can be neglected at moderate powers, leading to the following expression in both cases (gain or loss).

$$P_b(z) = \frac{g_B}{A_{eff}}\frac{v_g}{2}E_p(0)\exp(-\alpha_p z)P_{cw}(L)\exp(-\alpha_{cw}L). \tag{5.27}$$

Even where the powers are sufficient for the second line of Equation 5.25 to depart substantially from unity (i.e. the pulse is depleted or accretes substantial additional energy in the Brillouin interaction), an accurate measurement can still be achieved by measuring in both conditions (gain and loss) and forming the geometric mean of the two results [6], i.e.

$$P_b(z) = \sqrt{P_b^+(z)P_b^-(z)} = \frac{g_B}{A_{eff}}\frac{v_g}{2}E_p(0)\exp(-\alpha_p z)P_{cw}(L)\exp(-\alpha_{cw}L). \tag{5.28}$$

Here, L is the length of the fibre; α_p and α_{cw} are the attenuation of the pulse and CW light, respectively (in practice the parameters take the same value); A_{eff} is the effective cross-sectional area of the fibre over which the light is propagated; E_p represents the probe energy and g_B is the gain factor given in Equation 5.16.

It should be noted that the contribution of the CW light is integrated over the entire length of the fibre and so the local variation of $P_b(z)$ is due only to changes in the pulse energy, not those of the counter-propagating wave, and so the form of $P_b(z)$ is similar to that of backscattered pulse. However, the critical difference between OTDR and BOTDA are that

a. The interaction with the CW light substantially increases the signal level;
b. The gain is dependent on the match between the frequency difference of the sources and the local value of the frequency shift ν_B;
c. The decay rate is exp $(-\alpha_p z)$, not exp $(-2\alpha_p z)$, as would be the case in a conventional direct-detection OTDR and
d. The gain or loss is affected by the relative polarisation states of the probe and CW light.

In practice, the latter effect can be overcome by repeated measurement whilst scrambling the polarisation states to randomise their interaction; the original work [114] used a pair of orthogonal polarisation launch states and averaged the results of these two sets of measurements. The second effect is what allows BOTDA to be used as a distributed sensor, because ν_B is a function of strain [115] and temperature [7].

The gain that can be achieved is limited by depletion of the CW or the pulsed light. Initially, these types of depletion result in a distortion of the attenuation curve. As mentioned, the distortion in its initial stages can be overcome by using the system both in the gain and attenuation modes (i.e. probing successively with $f_{cw} = f_0 - \nu_B$ and $f_{cw} = f_0 + \nu_B$) and taking the geometric mean of the two signals obtained in these ways [6], although ultimately, the lack of signal in one of the cases will degrade the signal-to-noise ratio.

The gain increases the signal level by a factor of a few hundred before limiting effects become significant; although the DC power contributes shot noise that does not exist in a conventional OTDR, an improvement in the SNR by a factor of more than 100 was predicted in Ref. [6]. However, it was the variation of the Brillouin frequency along the fibre that made BOTDA the widely used distributed sensing technique that it is today.

As in BOTDR, the BOTDA measurement requires the scanning of the frequency over a span covering the range of possible values of ν_B, the range being dictated by the measurand and fibre type, with a sufficient margin to cover frequencies on either side of the possible peak values to allow a fitting process to be undertaken.

5.5.2 Optical Arrangements for BOTDA

As originally proposed and demonstrated [116], the BOTDA technique involved two independent lasers. The linewidth requirement is not particularly onerous and, initially, the probe (CW) light was provided by a DFB diode laser having a 10 MHz linewidth whilst the pulse (pump) laser was a solid-state Nd:YAG laser, both sources operating around 1320 nm. The arrangement used is shown in Figure 5.17b. In particular it is worth noting that the pulsed source is made up of bulk optic components given that in these early years, pigtailed AOMs were not available. The Brillouin interaction is polarisation dependent, and so a rotatable half-wave plate is used to adjust the polarisation of the (pulsed) pump. The measurement is repeated with two orthogonal input states and the results are averaged; this was shown to largely cancel the polarisation dependence of the measurement. Polarisation control is still present in most current BOTDA arrangements.

The two independent lasers must be controlled in their relative output frequency to a precision substantially better than the intended measurement accuracy and resolution, i.e. much better than 1 MHz for a 1 K temperature error, i.e. 1 part in 10^8. This was achieved initially by mixing the outputs of the two lasers onto a fast photodiode and monitoring the beat frequency on an electrical spectrum analyser [114]. Very small changes in bias current (typically 170 MHz/mA) or laser chip temperature are sufficient to shift the relative frequency well outside the required limits. A practical system must be able to lock the lasers together without human intervention.

One way of ensuring that the two sources track each other is to lock them with a feedback loop, as explained in Refs. [117] and [118] and illustrated in Figure 5.18 (which shows just the laser control part of the BOTDA arrangement). Here, the outputs of the two lasers are combined onto a mixing coupler with output arms of unbalanced length and detected on two separate photodiodes. The detector outputs are at the same frequency, but they are phase-shifted by the path imbalance δL, which results in a minimum in the mixer output when the two photodiode signals are in anti-phase, a condition that is dictated by the beat frequency of the two lasers. The frequency difference can then be tuned by varying δL. There are more sophisticated means of controlling the relative frequency of two lasers, but this one has the merit of simplicity and is adequate for its purpose. Although there appear to be no reports of this in the literature, in principle, this method of locking two lasers could be used for lasers positioned at either end of a long link. In other words, there is no requirement for the sources to be co-located because the control electronics can simply force the laser close to it to track the frequency of the remote laser.

The requirement for tracking pump and probe signals can be achieved using a single laser and a frequency translation circuit. In principle, the frequency shifting could be realised using the ring arrangement of Figure 5.7b. However, the first realisation [31,119] of BOTDA with just one laser used an EOM, able to shift the frequency by the amount required in a single pass. All frequency-shifted approaches are intrinsically robust against a drift in the laser frequency because the corresponding change in the Brillouin frequency is reduced by a factor of order 20,000 relative to the drift of the laser. Several such configurations have been published.

In the earliest configuration, both probe and pump were pulsed from either end of the fibre, causing the interaction to happen only where the pulses meet; this location can be adjusted by their relative timing; the configuration is shown in Figure 5.19. The output of the laser is pulsed twice by the EOM. Initially, this is done through the microwave input to the EOM that causes a pair of side bands to be created and these are adjusted in frequency to match the intended shift for the Brillouin interaction. (Only the lower sideband was used in this work and the upper sideband was rejected by the optical filter placed before the receiver.) This first pulse is launched (at time T_0) into the sensing fibre and allowed to travel to the far end, where it is reflected from a cleaved end. Sometime later, at time T_1, the EOM is driven in pulsed mode, allowing a second pulse (not frequency shifted) also to be launched. These two pulses meet in the sensing fibre and interact at that location and only there. So this arrangement probes only one location in the fibre at a time, which is slow (since it is also necessary to scan the frequency to determine the Brillouin shift). In practice, measuring a 4 km fibre at 4-m intervals required 20 minutes. However, this has the benefit of requiring a very modest bandwidth of the receiver and acquisition; a boxcar signal averaging system was used for this purpose. In this arrangement, the spatial resolution is defined purely in the optics by the distance over which the pulses interact and is given by $(\tau_p + \tau_s) \cdot v_g/2$, where τ_p, τ_s are the durations of the two pulses. The use of a reflection from the remote end of the fibre allows the measurement to take place

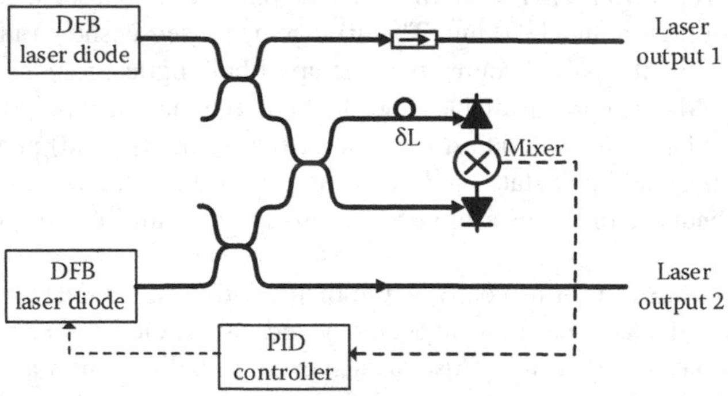

FIGURE 5.18 Locking arrangement for two sources used in BOTDA. (Adapted and simplified from drawings in Bao, X., Ponomarev, E., Li, Y. et al., Distributed Brillouin sensor system based on DFB lasers using offset locking, US7499151B2, 2009.)

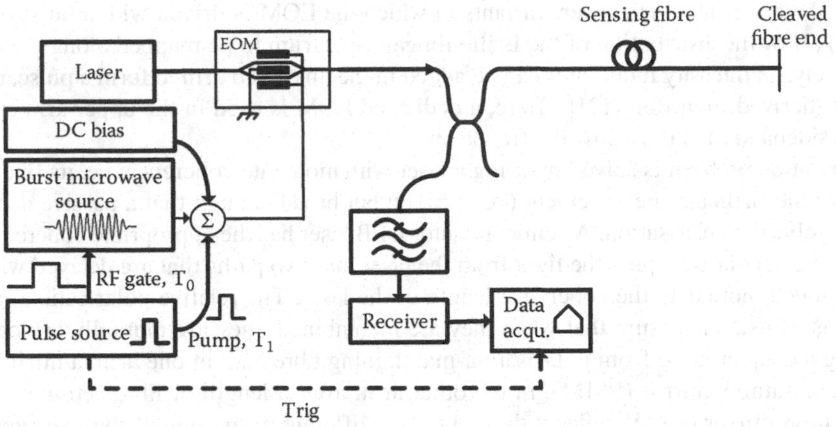

FIGURE 5.19 Single-laser, single-ended, BOTDA arrangement. (Adapted from Nikles, M., La diffusion Brillouin dans les fibres optiques: Étude et application aux capteurs distribués, PhD Ecole Polytechnique Fédérale de Lausanne, 1997.)

when there is access to only one fibre end, provided that the remote end is protected to ensure the presence of the reflection and this was a primary aim of this work [31].

By re-arranging the timing of the drive signals to the EOM [119], a quasi-continuous frequency-shifted probe signal can be generated ahead of an unshifted pump pulse. This emulates more closely the original BOTDA arrangement; of course, it then requires faster acquisition electronics, but the measurement is speeded up.

An arrangement that is more time efficient and closer to the original BOTDA was evolved by the same team and is shown in Figure 5.20. Here, the fibre is looped and attached to the two output ports of a directional coupler. In the upper arm, the light from the laser is launched (unshifted) continuously into End A of the sensing fibre. In the lower arm, the light first encounters an isolator (to prevent light travelling through the sensing fibre from A to B reaching the detector) and then an EOM. The latter generates an optical pulse travelling from B to A that contains side bands that are frequency shifted with respect to the light travelling from A to B in the sensing fibre. Finally, a polarisation controller is used to adjust the relative polarisation states of the light travelling in both directions in the sensing fibre. This arrangement, which is functionally equivalent to that of Figure 5.17b, was used initially [120] with a continuous drive signal applied to the EOM in order to characterise the Brillouin gain spectrum in the sensing fibre,

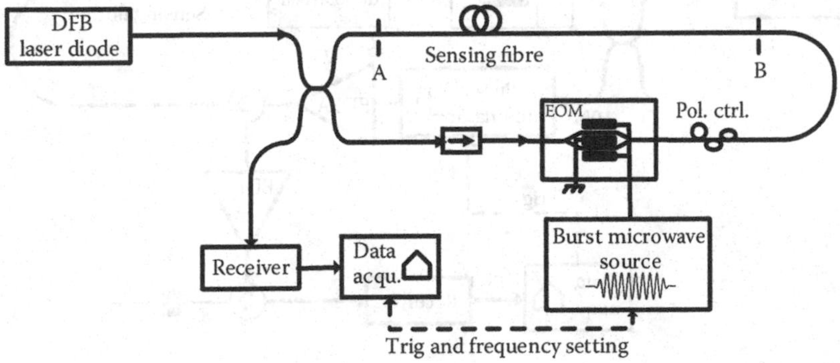

FIGURE 5.20 Looped single-laser BOTDA arrangement. (Modified from Nikles, M., Thevenaz, L., Robert, P.A., *J. Lightw. Technol.*, 15, 1842–1851, 1997.)

averaged over the entire fibre. However, variants in which the EOM is driven with a burst of microwave (as illustrated) allow the distribution of the Brillouin gain spectrum to be mapped along the fibre.

Alternatively, an intensity modulator can be added in the unshifted arm to form a pulse, as illustrated in Figure 5.21 (derived from Ref. [121]). Here, a dedicated EOM is used in the upper arm of the coupler only to create sidebands and thus shift the frequency.

The polarisation problem is solved by using a laser with moderate coherence; i.e. its linewidth is narrow enough for the Brillouin measurement (<c. 1 MHz) but broad enough that a passive depolariser can be used to scramble the polarisation. A semiconductor DFB laser has the appropriate coherence length for that purpose. The depolariser splits the light from the laser into two paths that are delayed with respect to each other by much more than the coherence length of the laser. The relative polarisation of the light in the two paths is adjusted to ensure that when they are recombined, they are mutually orthogonal; this is achieved using a coupler made from polarisation-maintaining fibre and, in one arm, a mirror made from polarisation-maintaining mirror (PMM); in the other arm, after a length of fibre acting as a delay line, a Faraday rotation mirror (FRM) reflects the light. The different properties of the two types of mirror ensure that when the light recombines, the polarisation of the light in the two arms is mutually orthogonal. It is also necessary to ensure that the power returning from the two mirrors is equal by matching the losses in the two arms. This arrangement, applied to the CW light in the Brillouin process, ensures that the gain is independent of polarisation.

The lower arm of the coupler includes another modulator, whose function is purely to cut a pulse from the light arriving from the laser. Although shown as an EOM in Ref. [121], in a separate publication [122], a semiconductor optical amplifier was used for this purpose. Other refinements shown in Figure 5.21 include the addition of optical amplifiers for the pulse and the stimulated scattering shown in two ports of the circulator. A narrowband filter (implemented as a reflective fibre Bragg grating) is used to select one or other of the sidebands and thus select the Brillouin gain or loss measurement. It suffices to tune the laser wavelength (e.g. with temperature) to move from one measurement configuration to the other.

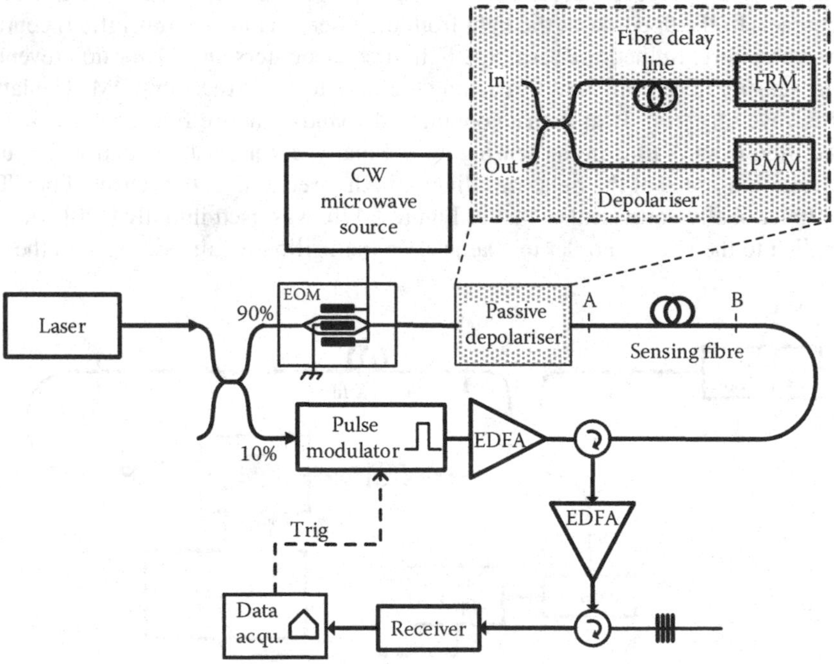

FIGURE 5.21 Single-laser BOTDA arrangement with separate pulse modulation and depolariser. (After Foaleng Mafang, S., Brillouin echoes for advanced distributed sensing in optical fibres. PhD Ecole Polytechnique Fédérale de Lausanne, 2011.)

5.5.3 Sensitivity and Spatial Resolution

For a simple Lorentzian peak, the resolution of the peak frequency is generally quoted as Ref. [123]

$$\delta\nu_B = \frac{\Delta\nu_B + \Delta\nu_L}{2^{1/2} \cdot SNR_e^{1/4}}. \tag{5.29}$$

Here, $\Delta\nu_L$ is the laser linewidth; it can usually be neglected in relation to $\Delta\nu_B$. SNR_e is the ratio of the electrical signal power to noise power. The complex spectra mentioned in the previous section of course complicate the picture, and in general, they degrade the frequency resolution.

However, more recent work [124] based on an analysis of the fitting of a parabolic curve to the near-centre of the distribution has derived the following dependency on the signal-to-noise ratio

$$\delta\nu_B(z) = \frac{1}{SNR(z)}\sqrt{\frac{3 \cdot \delta_{fs} \cdot \Delta\nu_B}{8 \cdot \sqrt{2} \cdot (1 - \delta_{fsu}/\delta_{fs})^{3/2}}}, \tag{5.30}$$

where δ_{fs} and δ_{fsu} are, respectively, the number of frequency steps used to sample the Brillouin spectrum and the subset of those that are used in the fit.

Even taking into account a probable discrepancy between optical and electrical definitions of SNR (Equation 5.30), this still differs by a square-root dependency on the SNR from Equation 5.29. In fact, in Ref. [124], SNR is defined as the deviation of the signal normalised to the Brillouin gain peak, whereas in Ref. [123], it refers to the electrical SNR of the backscatter signals.

The authors of Ref. [124] supported their expression with experimental data in good agreement with their prediction. The absence of the laser linewidth term in Equation 5.30 is probably due to it being neglected, which is overwhelmingly appropriate with modern lasers but not necessarily those available at the time that Ref. [123] was written.

5.5.4 Trade-Offs in BOTDA

A similar trade-off exists in BOTDA as does in BOTDR, namely that as the pulse duration is reduced, the fibre length over which the Brillouin interaction (gain or loss) takes place is also reduced and so the signal to be measured also falls, all other parameters being equal. In addition, the gain spectrum is the convolution of the Brillouin linewidth and the spectrum of the pulse and so, as the pulse duration is reduced, the gain spectrum is broadened. In accordance with Equation 5.29 or 5.30, the ability to resolve the measurand is then degraded, as was demonstrated in Ref. [125]. So in the basic BOTDA arrangement (one pulse interacting with a counter-propagating CW), the limitation is just below 1 m, at an operating wavelength of 1550 nm in conventional single-mode fibre.

Below the limit dictated by the natural linewidth, some detection of change is possible [126], but not a full distributed measurement of the frequency shift. For very short pulses, the broadening of the spectrum and the residual slowly varying signal from adjacent parts of the fibre preclude a reliable estimation of the local Brillouin frequency shift.

5.5.5 Composite Pulse Methods in BOTDA

5.5.5.1 Single shaped pump interrogation

The picture, just described, of worsening spatial and frequency resolution as the pulses are shortened, was called into question by observations [127] of a reversal of this trend: the authors showed that after broadening as the pulse duration was reduced from 40 to about 9 ns, a rapid decrease in the spectral width was observed for further shortening of the pulse duration. An explanation for this counterintuitive result [128] was proposed using a model that showed that the spectral narrowing could be attributed to a poor extinction ratio of the modulator used in the initial experimental results [127]. This is illustrated in Figure 5.22 as the difference between the intended pulse (a) and the pump actually launched (b) in the case of Ref. [127]. This interpretation was then confirmed experimentally [129]. The key point is that the leakage of the modulator first introduces into the sensing fibre a low-level optical field and prior to the launching of the main pulse at the same frequency. The leakage field interacts with the counter-propagating wave to set up a stimulated acoustic wave that is *later* probed by the main pulse. This process separates the interrogation of the acoustic field from its creation or decay: in this case, a fast variation in the optical field can interact with the counter-propagating light faster than the acoustic wave can respond.

Several configurations have been demonstrated to exploit this effect, all based on a first signal that sets up the acoustic wave, with duration sufficient to avoid broadening the Brillouin spectrum from its natural value. The first signal is then followed by a short disruption of the established acoustic wave. Most of these have been summarised in Figure 5.22.

The work reported in Ref. [127], however accidental in its discovery, may be regarded as the first of these, i.e. a low-level seed signal followed by a short pulse (Figure 5.22b). A similar technique [130,131] was described in which a broad pre-pulse of low level seeds the acoustic field and is followed by a narrow, higher-intensity pulse that interrogates the acoustic field already set-up (Figure 5.22c). Kishida and Li [130] demonstrated a resolution of 10 cm and a commercial instrument is available from Neubrex Ltd.

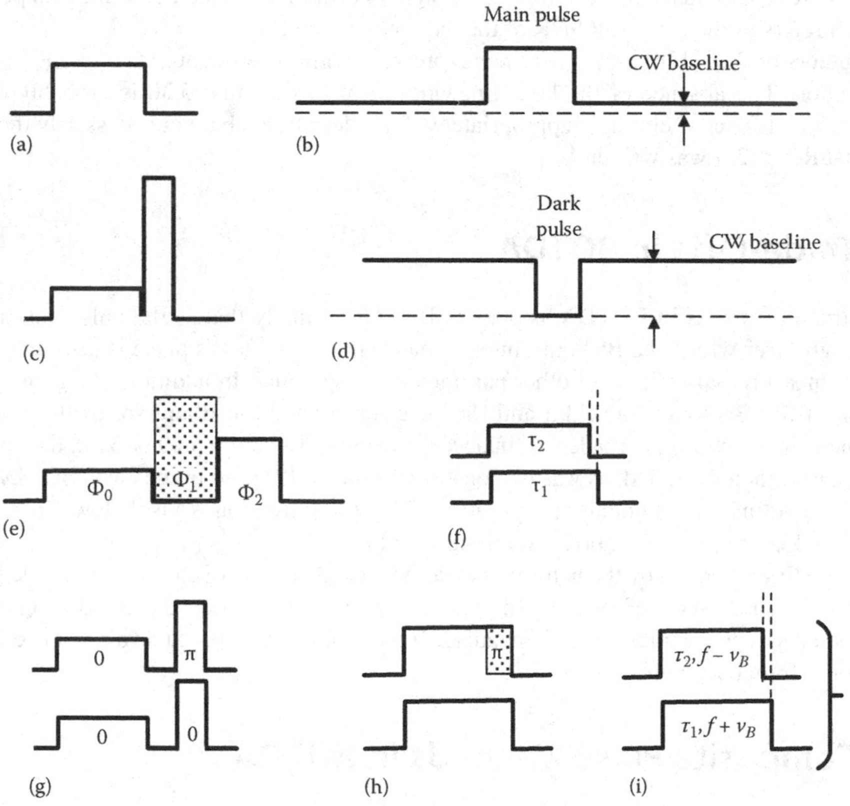

FIGURE 5.22 Composite pulse methods in BOTDA. (a) Single pulse; (b) single pulse with pedestal; (c) pump pre-pulse; (d) dark pulse; (e) generalised amplitude and phase coded pulse; (f) DPP-BOTDA; (g) negative Brillouin gain; (h) Brillouin echoes; (i) ODPA-BOTDA.

(Hyogo, Japan) that achieves this resolution over a distance of 1 km. More recent results from a different group [132] achieved 30 cm resolution over 3.6 km.

A so-called 'dark-pulse' approach (Figure 5.22d) was introduced in which continuous counter-propagating optical waves are launched into the sensing fibre [133,134]; a stimulated acoustic field was set up in those locations where the Brillouin frequency matches the frequency difference between the optical waves. The pump wave is of course partially depleted as a result. A brief interruption of one of these waves (a 'dark pulse') is then created, and this momentarily ceases the Brillouin interaction, thereby reducing the depletion of the pump and so resulting in a temporary increase in the pump signal that is detected. Again, if the dark pulse is much shorter than the acoustic lifetime, the acoustic wave will be scarcely affected, so the observed effect applies mainly to the optical waves. A spatial resolution of around 20 mm was reported – a real breakthrough.

The pre-pulse technique was generalised by considering a composite pulse of three sections [121,135], each section having arbitrary amplitude and phase (Figure 5.22e) and considering in particular the case where the amplitude of the middle section is 0 or equal to that of the other sections, but the phase is flipped by π, i.e. $\Phi_0 = \Phi_2 = 0$; $\Phi_1 = \pi$.

The phase shift has the effect of reversing the gain during its short duration and could be thought as a 'negative' pulse. An analysis of a composite pulse that encompasses all three techniques (pre-pulse, dark pulse and phase shifted) is provided in Ref. [121] and unifies all three concepts. It is shown that the phase-shifting approach doubles the signal of interest, relative to the other two approaches.

In the initial set-up, the phase shift was applied briefly during a continuous signal (so the interaction was between two CW counter-propagating waves). A spatial resolution of 5 cm was demonstrated. However, this approach gives rise to a number of issues, such as the build-up of noise from SpBS and depletion of the pump wave by the continuous interaction. It is also likely to be dependent on the degree of uniformity of the frequency shift in the sensing fibre. Therefore, the timing of Figure 5.22e is more useful than a brief modification of two continuous waves.

5.5.5.2 Differential pump interrogation

A separate, but related, branch in this quest for finer spatial resolution involves repeating the measurement with two slightly different pump pulses and subtracting the results from the two independent acquisitions. The first of these methods [136], Figure 5.22f, termed differential pulse-width pair BOTDA (DPP-BOTDA) involved launching a single probe pulse followed, in a second acquisition of the Brillouin signal, by another of slightly different duration and forming the difference in the digital domain. Although this *did* show spatial resolution consistent with the differences between the two pulses, the resulting difference-trace also showed a wide pedestal that is detrimental to an accurate reading of the Brillouin frequency profile, even though it allows small hot-spots to be detected.

In spite of the earlier comment about these techniques acting 'faster than the acoustic wave can respond', a small response is initiated as soon as the wave changes and after the pump has resumed its normal state, the signal gradually returns to its previous level, resulting in a distortion of the sensor response – this is not an issue for hot-spot detection, but it does preclude an accurate measurement of the frequency distribution. The distance over which this distortion of the response occurs is representative of the distance travelled by the optical waves during the phonon life (a few metres at most). This applies to techniques where the acoustic field is set up for long times before the arrival of the short altered part of the pulse that provides the spatial resolution.

This problem can be addressed by deconvolution, but in the solution adopted (so-called differential gain approach, DGA [137] or 'Brillouin echoes', Figure 5.22h, or negative Brillouin gain [138], Figure 5.22g), three changes were made to the DPP-BOTDA concept, i.e. (a) to create a finite duration pump by adding an intensity modulator to the phase modulator path, (b) to truncate the signal after the phase-reversed section and (c) to compare measurements made with pulses of either uniform phase or including the differential phase section and to subtract the results, i.e. to compare the results of two different interrogations with different pump pulse combinations. In the case of Ref. [137], a resolution of 5 cm over a sensing fibre length of 5 km was achieved. Both deconvolution and the subtraction of similar signals will of course degrade the signal-to-noise ratio, but in the case of the DGA [137], a standard deviation of the frequency measured of 3.0 MHz (at 2σ) was reported, equivalent to a temperature resolution of 3.0 K (2σ). The earlier concept [138] is very similar, but fewer details were provided as to performance.

In the negative Brillouin gain [138] (Figure 5.22g), again, a composite pulse is launched in which a first, low level pre-pulse excites the phonon population, followed by a second pulse having the same phase as the pre-pulse. This process is repeated, but in the second interrogation, the phase of the second pulse is reversed relative to the first pulse, as illustrated in Figure 5.22g. The phase reversed pulse fails to meet the conditions for stimulated scattering, and in fact, it interferes with the pre-existing SBS. The DGA technique is a refinement of the earlier approach (DPP-BOTDA) of numerically subtracting the response resulting from two probe pulses of slightly different durations [136]; it provides roughly double the response and a lesser degree of distortion. The numerical subtraction of two successive waveforms gives rise to questions about the additive contribution of noise to the small difference between the two signals that are subtracted. In addition, the resulting frequency measurement still showed some variation across a region where it should have been uniform, indicating an imperfect correction for the finite reaction time of the acoustic phonons.

5.5.5.3 Simultaneous gain/loss interrogation

The issue of the flatness of the measured frequency in the region of uniform Brillouin frequency was addressed [139] (Figure 5.22i) by making a simultaneous measurement using a pair of pulses of slightly different durations that are launched *at the same time*, but one of which contributes to Brillouin gain (at the anti-Stokes frequency relative to the counter-propagating CW wave), and the other, loss (at the Stokes frequency). This technique is termed optical differential parametric amplification BOTDA (ODPA-BOTDA). If care is taken to balance the powers of these two pulses, the net result is just the subtraction of pulses of dissimilar duration; of course, this condition is truly met only if the frequencies of the two pulses are symmetric about the centre of the Brillouin shift. This means that the frequency separation between the two pulses must vary according the local value of the Brillouin shift. In the experimental demonstration of the technique [139], the two frequencies were generated from a single laser by using the sidebands of an EOM and then separating them into two paths to provide the pulse modulation (of different width) in each path. Careful matching of the path lengths would also be required to ensure that the two pulses overlap as intended in the sensing fibre. Finally, in order to ensure that the Brillouin interaction is balanced for the gain and loss pump pulses, it was necessary to use polarisation-maintaining fibre, which is restrictive in practical applications. The results were reported over a short sensing fibre only; they demonstrated that the linewidth was largely preserved (compared with the broadening that occurs with narrow pulses). An important feature of the approach is that the differentiation, similar to that of DPP-BOTDA, occurs in the optical rather than the digital domain.

5.5.6 Pulse Compression Coding

In Section 5.4.3, we noted that pulse compression coding has limited benefits in BOTDR. However, in BOTDA, the pump (pulsed) wave is generally weaker than the probe pulse in BOTDR and so some performance gain can be achieved through pulse compression coding.

5.5.6.1 Pulse compression coding for range extension

Initial work on On-Off keyed Simplex coding showed distortion resulting from the fact that the pre-existing stimulated acoustic wave that is encountered by each pulse in the sequence depends on the previous values in the code word [140]. Therefore, the SBS response of each pulse in the code is not the same, and this violates the conditions for decoding the overall backscatter in a single-pulse response.

This problem was overcome using return-to-zero coding and spacing the pulses (whether 1 or 0) by an amount sufficient for the acoustic wave to decay between pulses. A spacing of order 100 ns, i.e. several times the typical phonon lifetime, is sufficient to ensure that the effect of previously excited phonons has decayed to an insignificant level.

It will be recalled from Chapter 3 (Section 3.3.3.4) that a key issue in pulse compression coding in OTDR is that it is necessary to work around the absence of a negative optical signal, for example, with complementary Golay codes. However, BOTDA offers further possibilities in that the interaction can be switched from loss to gain by arranging for the modulation to create an up-shift or down-shift relative to the probe. In the small-gain or loss regime, this amounts to providing bipolar coding. Up/down frequency shifting was demonstrated in recent work [141] where the pump wave was switched from gain to loss according to a binary Golay code. In effect, in this approach, the Brillouin process itself is accomplishing the correlation process required to decode the pseudo-random modulation that has been applied. The results showed the expected coding gain (15 dB for 1024 chip code); in addition, the complementary effect of successive Brillouin gain and loss within the code word reduces the effect of pump depletion. A spatial resolution of 2 m was demonstrated at a range of 100 km with a resolution of order 0.8 MHz; these results were achieved with 2000 traces per frequency acquired in the scan (the total time required will then depend on the efficiency of the acquisition and processing, as well as the frequency span that is required).

5.5.6.2 Pulse compression coding as an extension of composite pulse techniques

Coded pulse sequences have been primarily used as a means of distance extension rather than an enhancement of the pre-pulse methods. The exception here is the work of Dzulkefly Bin Zan, Tsimuraya and Horiguchi [142–144], where a resolution of 10 cm was demonstrated with relatively short codes (4–16 bit). The basic approach in this work is to use a long pre-pulse followed by a short sequence of phase-coded sections within the same composite pulse. The short sequence takes place within the phonon lifetime. A further composite pulse is launched after an interval sufficient to allow the photons to decay, so the approach is really a two-level coding scheme.

Several conclusions emerge from this work, namely (a) that Walsh (or Simplex) codes are more robust to distortion from the Brillouin gain of prior bits than are Golay codes, particularly when the entire duration of the code sequence is of order of the phonon lifetime, (b) that when a code exceeds the phonon lifetime, the coding benefit declines, (c) however, multiple code sets can be launched provided that they are separated in time by more than the phonon lifetime.

5.5.7 Slow Light Correction

BOTDA is associated with gain or loss of the probe. Given its very narrow gain spectrum, the gain or loss is associated with a change in the group index through the Kramers-Kronig relations, an effect referred to as *slow light.* The subject of slow light has received considerable attention in the last decade, partly as a means of varying pulse delays for telecommunications [145]; it has also been claimed* to enhance the sensitivity of certain interferometric measurements [146,147]. Delays (for a gain) or advancements (for a Brillouin induced loss) of order 1 ns/dB have been reported [148].

In the context of BOTDA and other Brillouin sensing configurations, the presence of gain or loss modifies the expected location along the fibre of the sensed frequency shift. This effect was analysed in Ref. [149], where it was found that for the Brillouin gain configuration, the time delay due to slow light is a few ns. However, for the Brillouin loss configuration, a delay of up to 130 ns was predicted in the worst case of a fibre having completely uniform loss spectrum along its length.

* There is some dispute of these claims, particularly in the context of rotation sensors.

This delay can be corrected by re-mapping the distance axis to take into account the slow-light-induced change in group delay. However, the slow light phenomenon also affects the spatial resolution of the distributed sensor. In order to overcome this limitation on the spatial resolution (that could be significant), it is necessary to re-map the distance axis as a function of the offset frequency; this could add substantial complexity to the algorithms used to convert from the measured spectra to the frequency shift/distance function.

5.5.8 BOFDA

The probe–pump interaction that is involved in BOTDA can also be implemented in the frequency domain, by replacing the pulsed pump with an amplitude-modulated CW. The technique has been demonstrated [150–154] using a stepped frequency approach, in which a selected modulation frequency is applied, the global response captured and the frequency approach then set to the next in the sequence. A network analyser performs these functions automatically. The usual gain/distance output can be obtained by Fourier transformation. Although the anticipated spatial resolution has been demonstrated, claims of improved resolution of the Brillouin shift for a given measurement time have not really been supported with published experimental evidence comparing its performance directly with BOTDA. Nonetheless, this approach continues to be used and developed, in particular for applications in soil stability monitoring [155].

5.5.9 Performance and Limitations

The basic BOTDA arrangement provides long distance measurement with moderate spatial resolution and update times. In addition to the work just discussed on composite pulses (for enhanced spatial resolution) and pulse-compression coding, a number of enhancements have been devised to improve the measurement time using somewhat more complex arrangements of the probe and pump waveforms and frequency spectra.

5.5.9.1 Parallel (multi-tone) acquisition approaches

As discussed in the context of BOTDR, in general, Brillouin measurements require a scan of the two-dimensional parameter space consisting of location along the fibre and frequency offset between the probe pulse and the Brillouin signal (whether spontaneous, in BOTDR, or stimulated, in BOTDA). This is necessarily time consuming: the fibre dimension can be covered by acquiring a single long backscatter trace, but the classical technique, involving stepping the offset between two sources, requires repeated measurements of the axial profile. Whereas in BOTDR, the need for repetition comes (as a minimum) from a need for improving the SNR, in BOTDA, it also follows from the requirement to scramble the polarisation.

It is therefore of interest to make a complete measurement of the Brillouin spectrum for each probe pulse (even if the launching of the probe pulse must be repeated for signal quality reasons). However, most of the approaches that are available in BOTDR are not appropriate in BOTDA because of the non-linear interaction that is central to the stimulated process and sensing mechanism. Fast Brillouin measurements can be achieved by measuring more than one frequency of the Brillouin spectrum at a time, by simultaneously interrogating with multiple frequencies (or tones).

A first step in this direction was reported by Chaube et al. [156], who replaced the Brillouin CW pump* with a frequency comb (the frequencies in the comb being separated in their case by 20 MHz). The principle of this measurement is that each of the frequencies in the comb interacts separately with the probe pulse, and therefore, information on multiple frequencies is collected simultaneously. The frequency plan is illustrated in Figure 5.23a. As a result, it is not necessary to scan the pump frequency relative to that of the probe, and so all the information is collected in parallel, without scanning the frequency difference.

* In Ref. (156), the term 'pump' is used for the CW waves, whereas in most of the literature, this wave is known as 'probe'.

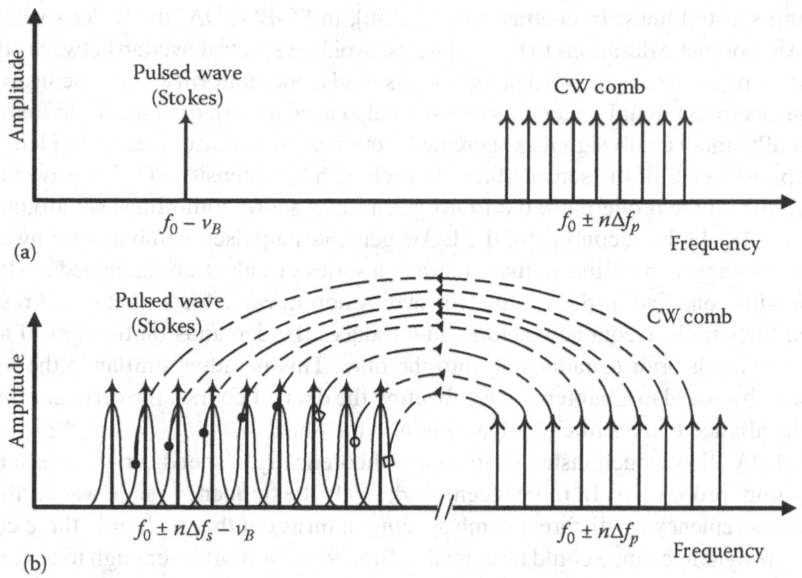

FIGURE 5.23 Frequency maps for multi-tone BOTDA interrogation. (a) Single comb used to frequency-multiplex the pump. (After Chaube, P., Colpitts, B.G., Jagannathan, D., Brown, A.W., *IEEE Sens. J.*, 8, 1067–1072, 2008.) (b) Dual-comb parallel multi-frequency acquisition arrangement. (After Voskoboinik, A., Wang, J., Shamee, B., Nuccio, S.R., Zhang, L., Chitgarha, M., Willner, A.E., Tur, M., *J. Lightw. Technol.*, 29, 1729–1735, 2011.)

This technique is limited in that the probe pulses must be of sufficiently long duration to avoid broadening their spectrum and so interact to any significant degree with only one of the frequencies in the comb. On the other hand, the spectral lines must be sufficiently close that they enable the reconstruction of the gain spectrum. In practice, in Ref. [156], some compromise was made in this respect, in that the measurement based on a comb with a 20 MHz spacing was repeated after shifting the entire comb by 10 MHz, i.e. interleaving two measurements to provide denser sampling in the frequency domain without degrading inordinately the spatial resolution which was set to 12 m. Further repetition of the measurement was conducted with a different relative polarisation state between probe and pump to overcome polarisation sensitivity of the measurement.

In spite of the need to repeat the measurement four times, the time required to acquire all the information on a short (120 m) fibre was 256 μs, leading to a theoretical update rate of ~3.9 kHz. In practice, the measurement was much slower (~1 s update time) because of the time required to transfer and process the data, steps that can in principle be made efficient through the use of dedicated processing power and sufficient data bandwidth. The signal processing involves separating the frequency corresponding to each line in the CW comb, extracting the intensity of the Brillouin signal at each frequency (by squaring and low-pass filtering) and then fitting the intensity at each location to a model Brillouin gain curve in the usual way. Although this demonstration showed a very substantial speed improvement, it suffered from the connection between spatial resolution and separation of the comb lines through the probe pulse duration.

This dilemma was overcome by manipulating the spectra of both the probe and pump such that each consists of widely spaced combs, but of a slightly different pitch [157,158], a technique referred to as sweep-free BOTDA (SF-BOTDA). The wide comb spacing allows more flexibility in the choice of the pulse duration, whilst the slight difference in the comb pitch provides a fine frequency resolution of Brillouin spectrum.

The basic concept is illustrated in the frequency in Figure 5.23b. Each of the comb lines for the probe has an associated line in the pump comb, so a series of probe–pump combinations is created. Within each combination, the probe and pump interact with each other. However, because of the slight difference in the comb spacing, $\Delta f_p \neq \Delta f_s$, each probe wave interacts with a different part of the Brillouin spectrum of its own pump; therefore, each probe/pump combination addresses one point on the Brillouin spectrum. Δf_p and Δf_s can be selected to be quite wide (the work referenced used ~100 MHz) so that there is no interaction between adjacent Brillouin spectral lines. In contrast to Ref. [156], in SF-BOTDA, the wider spacing between the comb lines allows short pulse durations to be used whilst avoiding spectral overlap between adjacent lines. In addition, in the first paper in the series [159], the authors used a non-uniform comb spacing to avoid spurious signals from the electronic signal generators (it would also have the effect of avoiding four-wave mixing).

Experimentally, these comb signals are created from a common laser that is divided into two paths, launched into opposite ends of the sensing fibre. In each path, an intensity EOM is driven by an arbitrary waveform generator. In the probe path, the EOM generates a set of comb lines set around the Brillouin shift with spacing Δf_p. In the second path, the EOM generates a pulsed comb at spacing Δf_s. It was also found preferable to stagger each line so that in effect, a series of pulses are launched at slightly different times, each providing one line in the comb. This avoids non-linear effects (that could result from interactions between lines in the frequency comb) and it reduces the demands on the optical amplifiers used to boost the power levels prior to launching into the fibre. This is rather similar to the multi-frequency C-OTDR systems discussed in Chapter 3. On collecting the data, of course, the different frequencies must be re-timed to be aligned to the same physical origin in the fibre.

The SF-BOTDA allows much faster acquisition – theoretically, it speeds up the measurement time by the number of pump–probe pairs. In more recent work [160], the frequency range was further extended by repeating the measurement with different comb spacing; it turns out that with only three combinations of comb spacings, the dynamic range could be extended from 90 to 900 MHz, enough to cover most practical ranges of temperature and strain.

5.5.9.2 Slope-assisted fast acquisition

If one has an approximate knowledge of the Brillouin shift as a function of location along the fibre, then the pump and probe waves can be arranged, at each location, to be sited at any desired offset from that frequency on a slope of the Brillouin gain peak, for example, at the point of inflection, hence the nomenclature 'slope-assisted BOTDA (SA-BOTDA)'. As a result, a change in the frequency shift is converted to a variation of the Brillouin interaction. This was demonstrated by colliding a probe-pump pulse pair at a single point of interest in a sensing fibre using an otherwise relatively standard BOTDA arrangement [161]. In this particular experiment, having located the frequency difference between probe and pump on to the slope of the Brillouin spectrum, one of these was modulated in frequency and the variation of the Brillouin signal at the modulation frequency allows dynamic variations of the Brillouin shift to be detected with both speed (200 Hz sampling rate) and reasonable resolution (95 με, i.e. about 5 MHz). However, this is a single-point measurement, and if another point is to be interrogated, it must be measured separately.

A variant of SA-BOTDA has recently been demonstrated [162,163] that allows the entire sensing fibre to be mapped dynamically based on the slope of the Brillouin spectrum. Again, the technique relies on prior knowledge of the static distribution of the Brillouin frequency along the fibre. Rather than using a pair of pulse–probe pulses as in Ref. [161], here, a pump pulse interacts with a long probe optical wave that has a time/frequency function designed to map the Brillouin shift along the fibre. As in BOTDA, the two optical waves are injected into opposite ends of the fibre, and as the pump pulse travels down the fibre, the frequency of the probe wave that it encounters is at, or close to, one of the half-height points on the Brillouin spectrum at each location. This can be seen as the BOTDA equivalent of the tracking technique [103], which has already been discussed in the context of BOTDR.

The SA-BOTDA approach has demonstrated remarkable performance on moderate-length (100 m) fibres, namely sampling rates of 8.3 kHz and a resolution of the Brillouin frequency of 0.25 MHz with a pump pulse duration of 13 ns (corresponding to a theoretical spatial resolution of 1.3 m).

5.5.9.3 Distance enhancements in BOTDA through remote amplification

The use of optical amplification to extend the sensing range has also been applied to BOTDA; the problem is rather more complex than is the case for BOTDR because of the need for maintaining appropriate power levels for both pump and probe. In addition, in BOTDA, the average power can be higher, given a continuous probe wave. One recent example of this approach may be found in Ref. [164]. Here, a genuine range of 325 km was reached, i.e. a loop length of 650 km. In this case, the probe was amplified at intervals of 130 km to reach the far end of the sensing fibre; however, this half of the loop was not used for sensing, only for conveying the probe to the remote end of the sensing fibre. The sensing part of the loop was amplified at intervals of 62 km and the pump pulse was carried on a separate fibre, also amplified at 65-km intervals. The probe pulse was inserted at each of the remote amplifiers. This system, whilst a remarkable demonstration, required a total of 10 remote amplifiers and three fibres (one each to convey the probe and pump and one for sensing); only the return fibre (i.e. from the launching end for the probe) was monitored. A measurement time of 104 minutes was required to achieve a resolution of 2 K with a spatial resolution of 3 m. In this system, the optical amplifiers were powered locally, not remotely over optical fibre.

5.5.9.4 Power limitations due to SPM

Although there is a desire to extend the measurement range of BOTDA, limitations similar to those applying to BOTDR are encountered due to the power launched; SPM is one of these effects. It was analysed theoretically and experimentally [165], and as in the BOTDR context, it broadens the spectrum progressively as the pulse travels in the fibre. SPM is an effect that responds to the time derivative of intensity, and so it is particularly acute for short pulses; it was found that at 30 ns, the linewidth began to increase even around 100 mW. At twice that value, the spectrum of a 30 ns pulse had broadened by a factor of 2 over 25 km. So even in this relatively moderate power regime, on distances that are well within the range of interest for many applications, the performance is degraded.

One way of overcoming the power limitations is through frequency-diversity, i.e. by launching multiple interrogating signals at well-separated frequencies. Of course in BOTDA, diversity is required of both the probe and the pump. In one experiment [166], the output of a laser was converted to three spectral lines separated by 17 GHz, the result of driving an intensity modulator at a RF signal level chosen to provide equal power in each of the first pair of sidebands as that which remains in the incident light. These signals were launched into a usual BOTDA arrangement, with a further EOM providing a frequency-shifting function for the pump and an intensity modulator generating the pump pulses (that occur simultaneously at three frequencies offset by 17 GHz). Of course, this only helps reduce some of the non-linear effects and to fully suppress them it is necessary to time-shift the pulses with respect to one another and then re-align the timing of the resulting BOTDA signals. This was achieved using a network of fibre Bragg gratings that reflect each frequency after a delay defined by the length of fibre between them, a concept already known in conventional OTDR for similar reasons [167]. A complementary network of FBGs cancels the time offset between BOTDA signals on their return from the fibre, as in Ref. [167]. Experimental results showed an improvement of the frequency resolution proportional to the number of probe/pump frequency sets.

5.5.10 Phase Measurement in BOTDA

In the discussion on BOTDA, so far, the measurement has involved estimating of the Brillouin shift using the real part of the Brillouin gain. However, there is a phase shift associated with the gain, described by the imaginary term in Equation 5.18. The full complex representation of the Brillouin gain is then

$$H_{SBS}(\delta f) = \exp\left(\frac{g_B(0)\Delta\nu_B^2}{\Delta\nu_B^2 + 4\delta f^2}\right)\exp\left(-j\frac{2g_B(0)\delta f\Delta\nu_B}{\Delta\nu_B^2 + 4\delta f^2}\right). \tag{5.31}$$

In a BOTDA measurement, Equation 5.31 states that the probe wave emerging from the sensing fibre is modulated by the Brillouin interaction in its phase, as well as in its amplitude. An illustration for a modest gain (×1.27 at the peak) and a natural linewidth of 20 MHz is shown in Figure 5.24. Whereas the magnitude of the gain takes the form of a peak, the phase is proportional to its derivative w.r.t. frequency,

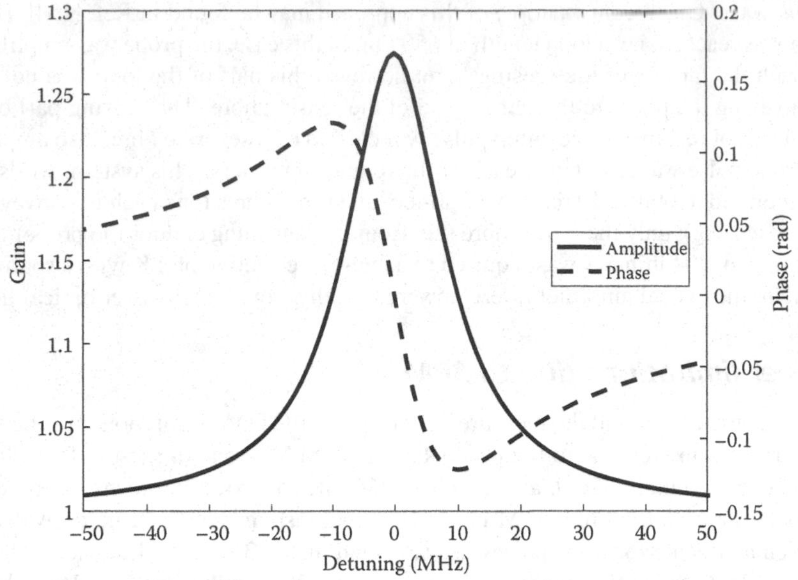

FIGURE 5.24 Amplitude and phase of the complex Brillouin gain function, plotted as a function of the detuning frequency.

resulting in a function that, at the centre of the gain, is approximately linear and crosses zero. Under some circumstances, this makes for faster and more precise measurements of the Brillouin frequency shift.

The benefit of operating on a well-defined slope of the transfer function is similar to that of the SA technique, although the physical principles leading to the transfer function are different between the phase measurement and SA-BOTDA. If the range of the Brillouin frequency shift is confined to the monotonic part of the phase response, no scanning is required and Brillouin frequency can be measured very fast: in Ref. [51], only 64 averages were required to achieve a frequency resolution of 1 MHz and the authors achieved an update rate of 1660 measurements per second, sufficient to track the dynamic strain of a vibrating cantilever beam.

The plots in Figure 5.24 are based on a probe duration that is sufficiently long that the natural line-width and line-shape are not affected. For shorter pulses, the Brillouin spectrum is broadened and the line shape is no longer Lorentzian, and Equation 5.31 would need to be modified in this case [51]. Urricelqui et al. [51] demonstrated how the monotonic measurement range can be extended by narrowing the pump pulse duration (which of course broadens the measured Brillouin gain spectrum).

The phase of the probe signal emerging from the sensing fibre has been measured in a number of ways [51,168–170]. In one approach, the probe is modulated by adding two sidebands that then beat with the carrier tone of the probe and this beat signal is demodulated to yield the magnitude and phase of the gain. The modulation frequency is selected (~800 MHz) to be sufficiently high that it does not interact with the Brillouin gain process and yet is not so high as to complicate the design of the demodulation electronics.

In Ref. [170], a vector network approach was used to extract the magnitude and phase of the Brillouin gain but the basic concept of the measurement is the same. This paper shows examples of the magnitude and the phase of the Brillouin shift in response to a temperature and a good agreement between the temperature dependence of the phase shift and amplitude response.

Lopez-Gil et al. [171] compares the merits of using the phase information and the amplitude of the gain. Under conditions of small Brillouin frequency excursion, it was concluded that the phase approach provides the better estimate of the Brillouin shift. This is primarily a result of the more sensitive nature of a linear fit (for the phase) vs. a parabolic fit (for the amplitude) in the vicinity of the Brillouin shift. In particular, fewer frequency samples are required for the phase estimate. However, this analysis applies only to a narrow range of Brillouin frequencies and to a particular fitting strategy. The more general cases have yet to be assessed in terms of the relative performance of the two techniques.

5.6 BRILLOUIN SENSORS BASED ON COHERENCE-DOMAIN REFLECTOMETRY: BOCDA AND BOCDR

BOCDA and later BOCDR, although still based on the concept of Brillouin scattering, employ a quite different interrogation approach. These ideas were developed originally by Hotate and Hasegawa [172]. The time/frequency modulation used in BOCDA/BOCDR localises the Brillouin interaction to a very narrow region of the fibre, and this provides an extremely fine spatial resolution (of the order of 1 cm) and also allows very fast (<1 s) interrogation of single points. The location of the point that is sensed is 'addressable' and is selected electronically, and so in principle, the system can hop between the points of interest, if desired, ignoring any locations that are not relevant and so the measurement can be very efficient as well as fast and finely resolved spatially. On the other hand, because the fibre is measured point by point, scanning the entire fibre can be slow, and it should also be pointed out that these methods have been demonstrated primarily on relatively short fibres, up to about 1 km with range extensions, compared with the >100 km ranges demonstrated with pulse techniques. The correlation techniques should therefore be seen as complementary to, rather than in competition with, the time and frequency-domain approaches.

5.6.1 BOCDA

The basic arrangement of BOCDA is shown in Figure 5.25. A laser diode is modulated in its output frequency at f_m with modulation depth m_d. This light is launched in both directions into a fibre loop and the resulting waves meet in the sensing fibre. Before discussing the other components in the system, we should consider the relative frequency of the optical signals travelling in clockwise $f_{cw}(t, z)$ and counter-clockwise $f_{ccw}(t, z)$ directions, ignoring the effect of any other devices.

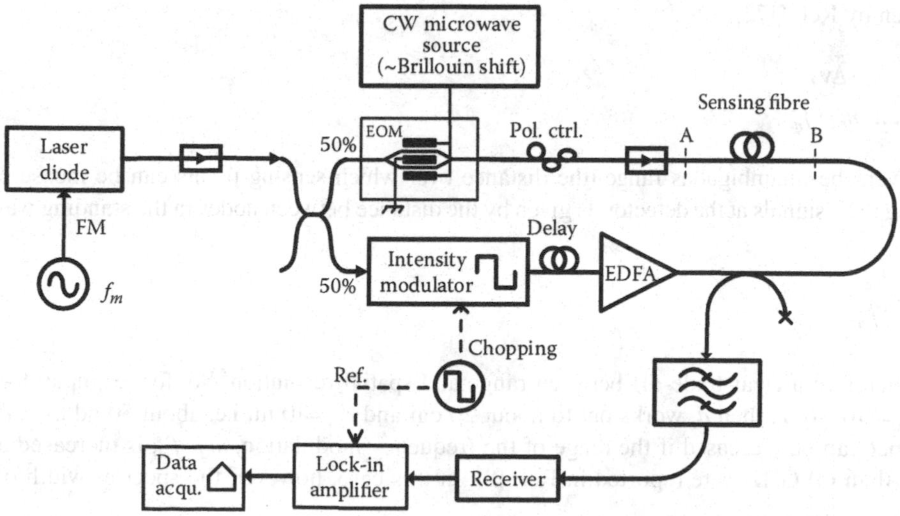

FIGURE 5.25 Schematic arrangement of the BOCDA technique. (Adapted from Hotate, K., Hasegawa, T., *IEICE Trans Electron*, 83, 405–412, 2000.)

The two waves that are launched will have frequency offsets from their nominal value f_0, represented by

$$f_{cw}(t,z) = m_d f_m \sin\left(2\pi f_m \cdot \left(t - z/v_g\right)\right) \tag{5.32}$$

and

$$f_{ccw}(t,z) = m_d f_m \sin\left[2\pi f_m \cdot \left(t - (\mathrm{L}_{tot} - z)/v_g\right)\right], \tag{5.33}$$

and so their difference (using standard trigonometric identities) is

$$\Delta f(t,z) = 2m_d f_m \cos\left[\pi f_m \cdot \left(2t - \mathrm{L}_{tot}/v_g\right)\right] \cdot \sin\left[\pi f_m \cdot (\mathrm{L}_{tot} - 2z)/v_g\right]. \tag{5.34}$$

As in double-ended DTS measurements, the distance axis is measured relative to End A for both waves and L_{tot} is the total length of fibre in the loop. Equation 5.34 shows that the frequency difference is a sinusoidal function of distance along the sensing fibre; it behaves like a standing wave and so there are nodes where the frequency difference between the two waves is constant (and in fact zero until further modified). At these locations, the counter-propagating waves are mutually coherent; i.e. their relative frequency and phase are predictable.

In the upper arm (clockwise direction) of the loop in Figure 5.25, an EOM generates sidebands (in later publications, this is a single-sideband modulator, generating just one frequency-shifted optical signal). This means that the counter-propagating waves are now at a frequency difference close to that required for Brillouin interaction, but this still happens only at the nodes of the standing wave, which give rise to so-called 'correlation peaks'. It is this process of localised coherence that provides the distance discrimination in BOCDA.

The location of the point that is sampled at any one time can be varied through the modulation frequency.* The function of the lock-in amplifier is to detect a change in the signal resulting from Brillouin gain as the microwave frequency applied to the EOM is swept over the range at which the interaction is expected.

The spatial resolution is no longer limited by the spectral broadening that occurs as the pulse duration is reduced. Instead, it is limited by the distance over which the light is coherent. The higher the modulation frequency and the greater the modulation depth, the sharper are the correlation peaks, and the resolution is given by Ref. [172]

$$\delta z = \frac{v_g \cdot \Delta\nu_B}{2\pi \cdot m_d \cdot f_m^2}. \tag{5.35}$$

However, the unambiguous range (the distance over which sensing points can be measured without overlapping their signals at the detector) is given by the distance between nodes in the standing wave pattern:

$$d_m = \frac{v_g}{2f_m}. \tag{5.36}$$

There is therefore a clear trade-off between range and spatial resolution. So, for example, for $m_d \cdot f_m =$ 5 GHz, f_m = 10 MHz, then d_z works out to about 20 cm and d_m =10 m, i.e. about 50 addressable points. This number can be increased if the range of the frequency modulation, $m_d \cdot f_m$, is increased and values of greater than 60 GHz were reported in Ref. [8]; in this case, however, the spectral width of the filter

* Alternatively, the delay or phase of the wave in one direction could be varied. This was carried out in Ref. [173] by modulating the light entering each end of the sensing fibre with separate modulators that impart the same waveform but with a controlled delay between them.

must be increased and many of the artefacts that the filtering is designed to remove can then re-emerge. Nonetheless, features that were only 3 mm long were fully resolved and the results were consistent with a theoretical value of the resolution 1.6 mm.

Equation 5.35 is true only up to $f_m < \Delta\nu_b/2$ [172]. Beyond that value, only some of the spectrum generated by the modulation falls within the gain spectrum and the following expression (which describes just the resolution dependence of the lowest line in the modulated source spectrum) applies:

$$\delta z = \frac{1.52 \cdot v_g}{2\pi \cdot m_d \cdot f_m} \quad (f_m > \Delta\nu_b/2). \tag{5.37}$$

The range/resolution trade-off was addressed by further modulation, which gates out all but one period of the correlation function [174]: in this way, each point of one period of the coherence function can be addressed, and, by re-arranging the time of the modulation, other periods can be selected.

5.6.1.1 Simplified BOCDA approach

In order to reduce the component cost of a BOCDA system, Song and Hotate [173] and Hotate and Yamauchi [175] demonstrated an approach that replaces the fast EOM devices with a careful control of the frequency of a distributed-feedback laser diode. It will be recalled that the emission frequency of DFB-LDs is a function of their injection current, and so it is possible to modulate the frequency by changing the DC-injection current in addition to any sidebands caused by the sinusoidal FM that is central to the BOCDA technique.

However, the emission frequency resulting from a current modulation of a DFB-LD, particularly with a square-wave input, is anything but linear [176,177]: there are many separate time constants in a laser diode, from the carrier injection (very fast) to progressively slower thermal processes in the chip, the submount and the thermal control loop. So in Ref. [175], it was necessary to carefully calibrate the frequency response of the LD frequency to a step in injection current and devise current waveforms that compensate for the deviations from the intended frequency to an accuracy of order 0.1%. As a result, the authors were able to generate a laser output that jumps between two values (separated by a Brillouin shift) that can be used for the probe and pump waves, a fibre delay line being used in one path and an intensity modulator in the other to ensure that only the signals of interest travel when required.

A further benefit of this approach is that the phase of the FM sinusoidal modulation that is applied to both half-cycles of the square wave (i.e. for the pump and probe waves) can be adjusted by the control electronics to sweep the position of the correlation regions. This was carried out at a rate of 40 Hz by applying a ramp to the signal generator [178], and this allowed the location along the fibre that is interrogated to be swept; this overcomes one of the major objections to BOCDA, i.e. that it interrogates one point at a time. These modifications make the system able fully to address any resolvable point along the fibre by the application of a simple control voltage. In contrast, the conventional BOCDA approach requires a change in the modulation frequency, which is less linear and involves a more detailed consideration of the order of the correlation function that is used. The authors of Ref. [178] achieved a spatial resolution of 16 cm using a set-up that requires relatively simple optics, but much expertise in the control electronics and calibration of the equipment.

The further simplification of Ref. [175], relative to Ref. [173], allowed one modulator to be replaced by a delay-line; importantly, this work demonstrated sampling rates of 1 kHz.

5.6.1.2 Range extension

The fundamental trade-off in BOCDA between spatial resolution and range relates to the requirement for a high modulation frequency for good spatial resolution that, however, brings adjacent correlation peaks closer together. This dilemma is overcome by additional modulation at multiples of the original frequency. For a multiple $m \cdot f_m$ of the modulation frequency, only every mth correlation peak remains [179]; a range of 1.5 km with 53 cm spatial resolution was demonstrated.

In another variant of BOCDA, a reflection from the remote end of the fibre was used, so allowing the sensing fibre to be deployed in a straight line [180].

5.6.1.3 Background noise suppression

A technique that synchronously varies the amplitude of the laser diode together with its frequency excursion was found to shape the spectrum and remove a spurious response that distorts the output [181], particularly when the majority of the fibre is at one Brillouin frequency and just short sections are at another; it improves the spatial resolution by avoiding these small regions being missed by the signal analysis.

5.6.1.4 BOCDA using phase coding

The approach developed in Prof. Hotate's laboratory based on the generation of correlation peaks using colliding frequency-modulated waves is just one example of achieving the goal of ensuring that the Brillouin interaction occurs only in a restricted section of the fibre. More recently, it has been realised that there are other ways of achieving similar results.

By launching counter-propagating waves that are phase coded (and frequency shifted to match the Brillouin requirement), it can be arranged for the stimulated Brillouin process to occur on average at only one location in the fibre [182]. In principle, this could be achieved by including two independent phase modulators at each end of the fibre: then by delaying one code with respect to the other, the location of the point that provides the response could be controlled electronically. In Ref. [182], however, an approach to tuning the location, similar to that used in the original BOCDA, was used, namely to delay one arm substantially and operate on a multiple-order correlation peak. Similarly to Ref. [172], the pseudo-random code used repeats after a number ($2^{15} - 1$ in this case) of bit-periods; thus, by operating at a distance that is a multiple (7 in this case) of the distance over which the code repeats, the location of the responsive point can be tuned with small adjustments of the clock that is used to generate the coding sequence.

Whereas in conventional BOCDA, the spatial resolution is directly tied to the length of the fibre that can be scanned, this connection is broken in phase-coded BOCDA: the code can, in principle, be made arbitrarily long, whereas the spatial resolution is dictated by the duration of each bit in the code, i.e. the frequency of the clock driving the pseudo-random code generator.

The first demonstration of phase-coded BOCDA achieved a resolution of 1.2 cm over a length of 40 m. Although this technique can, in principle, be extended arbitrarily, it suffers from two limitations that also exist in some ways in the original BOCDA.

Firstly, these are point-by-point measurements: the BOCDA method measures one point at a time and with a fast measurement. Scanning a few hundred points is not too onerous, whereas for tens of thousands of points, which is the ultimate ambition of such schemes, the time taken to acquire the profile of the entire fibre can be prohibitive.

This problem in phase-coded BOCDA was addressed, just as it had been in the original BOCDA technique, by using additional levels of modulation. In order to add time encoding to the interrogation, it is necessary to use short codes, and a key enabler for this purpose is the availability of perfect finite duration codes, i.e. codes whose auto-correlation function is unity for zero shift and zero elsewhere. Whereas such codes do not exist for optical on/off modulation, in the case of phase modulation, a phase reversal is equivalent to '–1' and so bipolar codes are available in this context and finite duration codes with the ideal auto-correlation function are known. Thus, in Ref. [183], short (127-chip) codes were used in combination with time-domain analysis: the phase-coding provides the very high spatial resolution, and the time-domain encoding is used to repeat the measurement at intervals of one code length along the fibre. Whereas the pure BOCDA technique measures only one point at a time, in this case, it measures just one point in each of the code lengths, i.e. one point in 127; however, independent measurements are made simultaneously at each of the time intervals of one code length along the fibre. In this work, a 400 m length of fibre was measured with 2 cm resolution.

Further variants of dual-level time and correlation coding are described in Refs. [184] and [185].

5.6.1.5 Low coherence BOCDA

BOCDA, having been extended by the use of phase-coding schemes, has been pushed to the ultimate random coding, namely that produced by natural fluctuations of the optical field in a broadband source [186]. The same principles apply, of causing the interaction of counter propagating waves (one shifted by the approximate Brillouin shift and the other pulsed), with a controllable delay between them.

There are a number of practical issues with this approach, one of which being that the delay, which in other schemes can be conducted in the electronic or software domains, must be carried by mechanically scanning a delay line. This also limits the maximum range that can be achieved, 5 cm in this case, with a 4-mm spatial resolution and a 1.5 MHz resolution of the Brillouin shift. In addition, the random nature of the source results in a modulation of the amplitude as well as the frequency, the latter being the element of interest.

The signal-to-noise ratio is also quite poor, owing to the existence of a strong signal, on top of which the contribution of the Brillouin gain is relatively weak.

There may be practical applications of this work – the authors suggest it might lie in the investigation of photonic circuits. Scientifically, however, its value is in confirming the principles underlying BOCDA in the broadest sense.

5.6.2 BOCDR

BOCDR applies similar principles as discussed in BOCDA to the detection of SpBS [187–189]. Like BOTDR, it utilises SpBS, but it uses the same concept of correlation peaks created by sinusoidal frequency modulation of the laser to resolve the distance axis in the fibre. In this case, the correlation peaks are created by the coincidence of the modulation on the scattered signal (with delay encoded as distance along the sensing fibre) with a similar frequency modulation waveform that is used as a reference wave in a heterodyne detection set-up, as illustrated in Figure 5.26. Thus, light scattered from a location z on the sensing fibre has a time–frequency relation that is characteristic of that location due to the round-trip propagation from the start of the fibre and back to the launching end. Although the Brillouin backscatter is weaker than in BOTDA and frequency-shifted, it retains an imprint of the frequency modulation. The scattered and reference waves move together at only one location in each spatial period d_m. This location specifies the spatial origin of the signal that is detected. As in BOCDA, the point that is interrogated can be moved by altering f_m. BOCDR generally provides weaker signals than BOCDA does owing to the absence of gain at the point of scattering; however, it has the advantage of requiring access to one end only of the sensing fibre, a key advantage in some

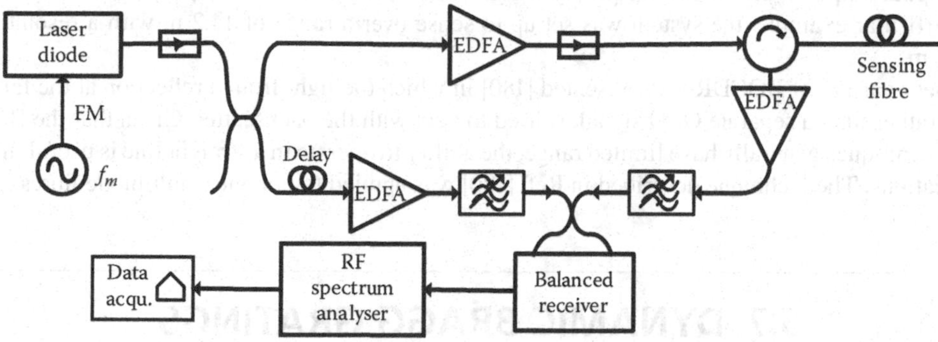

FIGURE 5.26 Schematic arrangement of the BOCDR technique. (Adapted from Mizuno, Y., Proposal and performance improvement of Brillouin optical correlation-domain reflectometry. PhD The University of Tokyo, 2009, with permission.)

applications. Although a single-ended BOCDA has been demonstrated using a mirror at the far-end of the sensing fibre [190], this does not address the concern about a potential break in that fibre, which applies also to BOTDA.

As shown in Figure 5.26, the optical aspect of the BOCDR arrangement is essentially identical to a BOTDR system, with however additional optical amplifiers and tunable filters. The optical power that contributes to the detected signal is a very small fraction of the light returned from the sensing fibre and this proportion becomes smaller as the spatial resolution is made finer; this explains the need for optimising the SNR by optical amplification in the launch, reference and backscattered paths.

The key difference compared with Figure 5.8, for example, is that in BOCDR the pulse modulation is completely missing and it has been replaced with a frequency modulation of the source common to the reference and probe wave. This sets up the conditions for the Brillouin and the reference wave to move together only for specific locations along the fibre. As in BOCDA, the modulation frequency range and the sensing fibre length are arranged so that only one location possesses this property.

The spatial resolution and measurement range are determined by the same equations (Equations 5.35 and 5.36) as in BOCDA. f_m is used to adjust the location of the sensed point, and it also determines the spatial resolution. In order to ensure that the spatial resolution is reasonably constant over the length of the sensing fibre, a high-order correlation peak is selected by using a long delay line (several km) shown in the figure. In this way, a small adjustment of f_m can move the Nth order correlation peak and cover the entire span of the sensing fibre with a relative change of just $1/N$. The power launched can be quite high (28 dBm in Ref. [187]) because the frequency modulation suppresses the SBS. The tunable filters used in each branch of the coupler leading to the receiver are intended to reduce as far as possible the ASE from the optical amplifiers.

As examples of the numerical values of some of the parameters, in a first case [187], f_m was set to ~458 kHz, giving an unambiguous measurement range of 228 m with $m_d \cdot f_m = 5.4$ GHz, which resulted in a spatial resolution of ~40 cm. In a second case, f_m was increased to ~13.46 MHz with the same value of $m_d \cdot f_m$, resulting in a spatial resolution of 13 mm but a range of only 7.6 m. In these cases, a single point could be sampled at 50 Hz.

The same concepts for extending the range, i.e. providing additional levels of modulation, as have been discussed in the context of BOCDA can be applied to BOCDR.

It has recently been demonstrated that the BOCDR technique can be optimised to scan the Brillouin spectrum extremely fast (up to 100 kHz) [191]. Although some averaging is necessary, the technique is still able to track a mechanical wave travelling along a beam, resolving 100 spatial points at an update rate of 100 samples/s (after averaging each point 10 times). The key to achieving this rapid update rate is, in the case of Ref. [191], to mix the output of the balanced receiver (as shown in Figure 5.26) with a signal that is rapidly, and linearly, scanned in frequency (using a voltage-controlled oscillator). Selecting a specific frequency component from the mixed output then converts the output to the time domain. In practice, the value of signal output at the chosen frequency was measured using, in effect, a phase-sensitive amplifier. In this particular example, the system was set up to sense over a range of 14.2 m with a resolution just below 0.4 m.

A linear variant of BOCDR was presented [180] in which the light from a reflection at the far end of the fibre, rather than a separate OLO signal, is used to beat with the backscatter. Given that the BOCDR/BOCDA techniques generally have limited range, the ability to operate in a straight line is useful in practical applications. The technique described in Ref. [180] was applied to polymer multimode fibres.

5.7 DYNAMIC BRAGG GRATINGS

Conventional Bragg gratings (see brief introduction in Section 2.3.6) in optical fibres are permanent devices in which a periodic modulation of the refractive index has been inscribed along the axis.

DBGs are transient devices that fade rapidly. An early example of this class of device [192] was created by a standing wave in a rare-earth doped fibre, of the type that is used for optical amplification but, when unpumped, absorbs the incident wavelength in certain bands. At the correct level of power in the standing wave, the absorption of the doped fibre is saturated in those regions where the light is most intense, i.e. away from the nodes of the standing wave. This creates a periodic region of high absorption that, through Kramers-Kronig relations, will also modify the refractive index periodically, thus forming an index grating. The grating fades at a rate dictated by the fluorescence decay time (0.1–10 ms depending on the rare-earth species and the host material). This process thus creates a Bragg grating that is temporary; these gratings are known as DBGs.

The SBS process also creates a DBG, by the interaction of the counter-propagating wave with stimulated acoustic vibrations. The lifetime of these gratings is of course much shorter than for rare-earth DBGs. In addition, because they already reflect at frequencies that are involved in the Brillouin process, their utility would seem to be limited.

However, in a high-birefringence fibre, there are two orthogonal polarisation modes that have similar mode-field distributions but substantially different propagation constants. It follows that a DBG written by a pair of waves in one polarisation state (say along the x axis) causes a reflection at a different wavelength in the y axis because the propagation constants of the two modes are different [193]. The stimulated acoustic wave is common to both polarisations and results in a spatially periodic modulation of the refractive index, which is of course the same for both polarisations because, physically, there is only one pitch of this grating. However, the corresponding optical frequency that is reflected differs for the two polarisation states because the birefringence converts a common *spatial* frequency to a different *optical* frequency for the two axes. Given refractive indices n_x and n_y for the two principal axes, the reflected wavelengths are related to each other by

$$\frac{2n_x V_a}{\lambda_x} = \frac{2n_y V_a}{\lambda_y}. \tag{5.38}$$

For a birefringence Δn_{x-y}, the difference in the frequency of the reflections on the two axes is

$$\Delta \nu_{x-y} = \frac{\Delta n_{x-y}}{n} \nu. \tag{5.39}$$

Δn_{x-y} values of a few times 10^{-4} are available commercially and even higher values are found in some photonic crystal fibres, so separation between the reflection wavelengths of approaching 1 nm can be achieved.

The main attraction of the Brillouin DBG technique is that it provides two measurements, namely the distribution of the Brillouin shift and of the birefringence. These data can be used to estimate strain and temperature independently through frequency measurements only. The separation of these two measurands (which is discussed in Chapter 8) has been a long-sought goal in distributed sensing, and Brillouin DBG is one of the very few techniques that achieve it.

The basic arrangement that was used to demonstrate the existence of Brillouin DBGs [193] is shown in Figure 5.27 (the items within the shaded area were not present in the initial experiments). The small circles with opposing arrows drawn across a diameter illustrate the state of polarisation of the light travelling in the fibre at these points. The output of a laser diode is split into two pump waves, one of which is shifted by one Brillouin shift using a single-sideband EOM. The pump waves are launched into opposite ends, but into the same polarisation mode of a high-birefringence fibre (x axis in this case). The pump beams interact and set up a moving grating as in all Brillouin measurements. In DBG measurements, however, a separate wave (probe) generated by a tunable laser is used to interrogate the DBG on the orthogonal axis. Tuning the probe laser allows the reflection wavelength of the DBG in the y axis to be measured. When the probe wavelength matches the DBG, the signal measured at Receiver 2 will peak. Without the components in the shaded area, only the mean Brillouin and DBG frequencies can be determined; more precisely, these values must be uniform along the fibre for a reasonable reading to be obtained.

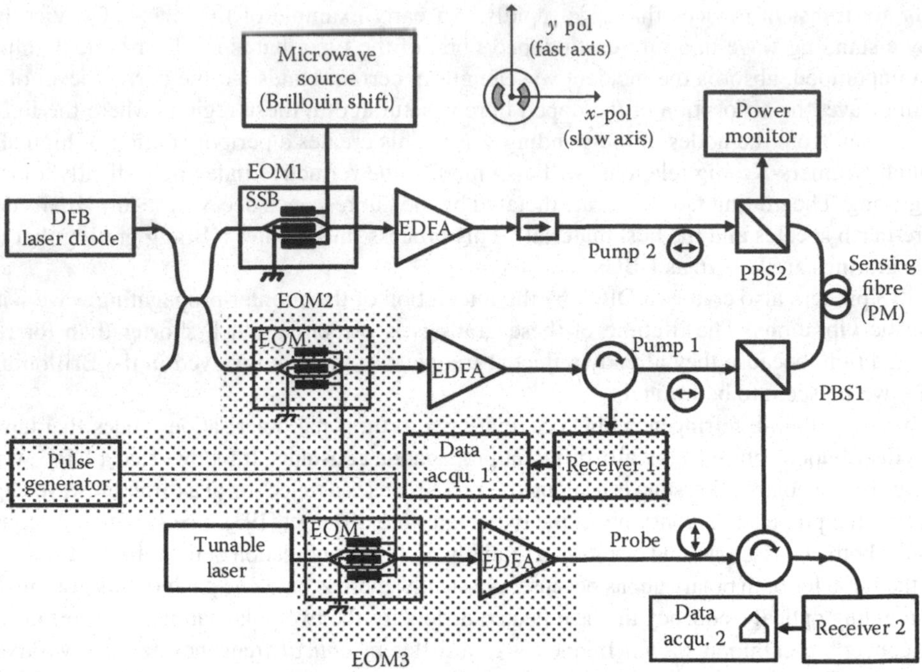

FIGURE 5.27 Basic arrangement for demonstrating Brillouin dynamic Bragg gratings. (Adapted from Song, K.Y., Zou, W., He, Z., Hotate, K., *Opt. Lett.*, 33, 926–928, 2008; Song, K.Y., Zou, W., He, Z., Hotate, K., *Opt. Lett.*, 34, 1381–1383, 2009.)

However, with all the components in Figure 5.27 present as discussed in Ref. [194], then a full, spatially resolved, BODTA measurement is obtained at Receiver 1, and at Receiver 2, the spatial distribution of the DBG central wavelength can be obtained.

The function of the components in the shaded area, introduced in later publications, and particularly modulators EOM2 and EOM3, is to convert both Pump 1 and the probe into pulsed signals to provide positional resolution. This provides a measure of the Brillouin shift and of the birefringence. However, in the general case of a distributed measurement of unknown parameters, it would be necessary to establish the Brillouin frequency distribution, by sweeping the relative frequencies of EOM1 and EOM2 and then to sweep the probe frequency whilst making an OTDR measurement with a pulse tunable light from EOM3 and detected by Receiver 2 to establish the local birefringence. This measurement might therefore be a little time consuming, and in fact, in Ref. [194], it was reported that the measurement of a 100-m fibre took about 7 minutes with a spatial resolution of 2 m for both measurements.

5.8 POLARISATION EFFECTS

The efficiency of the SBS process is strongly dependent on the relative polarisation states of the interacting light waves. Although the efficiency of SpBS does not depend on polarisation, those BOTDR or BOCDR systems that use coherent detection implicitly select one polarisation state through the mixing on the photodetector, the efficiency of which depends on the relative polarisation of the signal and reference waves. Polarisation is therefore an issue in essentially all types of Brillouin distributed sensors. The problem has been tackled in a number of ways that will be discussed here to cover the generality of Brillouin schemes.

Polarisation scrambling involves adding at least one polarisation modulator so as to vary the SOP of the light entering the sensing fibre. Provided that the polarisation modulation function is *not* harmonically related to the PRF, this has the effect of randomising the distortion of the measurement caused by polarisation sensitivity of the measurement. Although this works well in applications where signal averaging will take place, fast measurement set-ups (such as BOCDA) require the polarisation sensitivity to be removed in as few measurements as possible. In these cases, it is preferable to use *polarisation diversity*, where the polarisation is flipped precisely between two orthogonal states on alternate cycles of the measurement [195] and the results of two successive measurements are aggregated by taking the square root of the sum of their squares. It is important to ensure that the polarisation states transmitted for successive interrogation cycles are precisely orthogonal and of equal power, although probe power errors could be compensated in the aggregation.

The concept of polarisation diversity can also be applied passively in the case of BOTDR or BOCDR, by dividing the returned light with a polarisation beamsplitter and mixing each of the polarisation states with a reference wave that is correctly aligned in polarisation with the portion of the backscatter that it is detecting. This approach is commonly used in telecommunications; it requires duplicated receivers and a dual acquisition system, but it is entirely passive and avoids the need to repeat the measurement with orthogonal launch states.

Brillouin measurements do not require extremely narrow sources; it suffices that the source linewidth is much narrower than $\Delta\nu_B$ and a value of 1 MHz is quite acceptable. This allows the use of a third approach (Figure 5.21), *depolarisation*, where one portion of the light is delayed by much more than one coherence length of the source and recombined in orthogonal states [122]. It should be noted that the portions of the light that are recombined must be mutually incoherent; it is also important that the power in each of the combined beams should be as close as possible to each other and precisely orthogonal.

Finally, it should be pointed out that the polarisation sensitivity of the various Brillouin measurement techniques can be used to estimate the birefringence of the sensing fibre [196,197].

Apart from a modulation of the signal level measured at a particular frequency offset in a Brillouin measurement, polarisation dependency also causes a measurement error [198] in the detected central frequency of the Brillouin spectrum and in its shape. Substantial birefringence is required for this problem to be significant; it would not be evident in telecommunications fibres, but it could easily be present in fibres designed for high birefringence that are commonly used in sensing applications.

5.9 SUMMARY

Figure 5.28 summarises the main Brillouin measurement techniques used in distributed sensing. Leaving DBGs aside, either the spontaneous Brillouin backscatter is measured or the Brillouin gain resulting

	Interrogation domain		
	Time domain	Frequency domain	Coherence domain
What is measured			
Spontaneous Brillouin backscatter spectrum	BOTDR		BOCDR
Brillouin gain distribution	BOTDA	BOFDA	BOCDA
Dynamic Bragg gratings	DBG		

FIGURE 5.28 Summary of the techniques reported for distributed Brillouin-based sensing.

from the interaction of two-counter propagating waves. However, the reader will appreciate that these two categories have split further into a very wide variety of interrogation techniques, split firstly into time-, frequency- and coherent-domain approaches. Although it is a possible approach in principle, there are no reports, to the author's knowledge, of the use of OFDR for measuring the spontaneous Brillouin backscatter.

The coherence domain approaches excel at providing very fine (<1 cm) spatial resolution, but on short sensing fibres.

The spontaneous Brillouin methods have the distinct benefit of requiring access from only one end, which can be critical in certain applications.

However, the BOTDA technique, because it involves three-wave interaction (two optical, one acoustic), allows a wealth of variants that can be used to tune the performance, in particular, using composite-pulse techniques for refining the spatial resolution. New refinements of the various Brillouin techniques are still being discovered and tested.

DBGs offer the possibility of separating temperature from strain, but this technique is still in its infancy.

REFERENCES

1. Krishnan, R. S. 1946. Temperature variations of the Raman frequencies in diamond. *Proc. Indian Acad. Sci. (Math. Sci.)* 24: 1: 45–57.
2. Robinson, I. M., M. Zakikhani, R. J. Day, R. J. Young, and C. Galiotis. 1987. Strain dependence of the Raman frequencies for different types of carbon fibres. *J. Mater. Sci. Lett.* 6: 10: 1212–1214.
3. Maughan, S. M., H. H. Kee, and T. P. Newson. 2001. Simultaneous distributed fibre temperature and strain sensor using microwave coherent detection of spontaneous Brillouin backscatter. *Meas. Sci. Technol.* 12: 7: 834–842.
4. Cho, Y.-T., G. P. Lees, G. Hilton et al. 2006. 100 km Distributed fiber optic sensor based on the coherent detection of Brillouin backscatter, with a spatial resolution of 10 m, enhanced using two stages of remotely pumped Erbium-doped fiber combined with Raman amplification, 18th International Conference on Optical Fiber Sensors, Cancun, Mexico, Optical Society of America, 1-55752-817-9: ThC4.
5. Alahbabi, M. N., Y.-T. Cho, and T. P. Newson. 2005. 150-km-range distributed temperature sensor based on coherent detection of spontaneous Brillouin backscatter and in-line Raman amplification. *J. Opt. Soc. Am. B* 22: 6: 1321–1324.
6. Horiguchi, T., and M. Tateda. 1989. BOTDA – Nondestructive measurement of single-mode optical fiber attenuation characteristics using Brillouin interaction: Theory. *J. Lightw. Technol.* 7: 8: 1170–1176.
7. Kurashima, T., T. Horiguchi, and M. Tateda. 1990. Distributed-temperature sensing using stimulated Brillouin scattering in optical silica fibers. *Opt. Lett.* 15: 18: 1038–1040.
8. Song, K. Y., Z. He, and K. Hotate. 2006. Distributed strain measurement with millimeter-order spatial resolution based on Brillouin optical correlation domain analysis. *Opt. Lett.* 31: 17: 2526–2528.
9. Schroeder, J., R. Mohr, P. Macedo, and C. Montrose. 1973. Rayleigh and Brillouin scattering in K_2O–SiO_2 glasses. *J. Am. Ceram. Soc.* 56: 10: 510–514.
10. Fabelinskii, I. L. 1968. *Molecular scattering of light.* Plenum Press, New York.
11. Dixon, R. W. 1970. Acoustooptic interactions and devices. *IEEE Trans. Electron. Devices* 17: 3: 229–235.
12. Heiman, D., D. S. Hamilton, and R. W. Hellwarth. 1979. Brillouin scattering measurements on optical glasses. *Phys. Rev. B* 19: 12: 6583.
13. Rich, T. C., and D. A. Pinnow. 1974. Evaluation of fiber optical waveguides using Brillouin spectroscopy. *Appl. Opt.* 13: 6: 1378–1380.
14. Tkach, R. W., A. R. Chraplyvy, and R. M. Derosier. 1986. Spontaneous Brillouin scattering for single-mode optical-fibre characterisation. *Electron. Lett.* 22: 19: 1011–1013.
15. Vacher, R., and J. Pelous. 1976. Behavior of thermal phonons in amorphous media from 4 to 300 K. *Phys. Rev. B* 14: 2: 823.
16. Fellay, A. 2003. Extreme temperature sensing using Brillouin scattering in optical fibers. PhD Ecole Polytechnique Fédérale de Lausanne.

17. Bömmel, H. E., and K. Dransfeld. 1960. Excitation and attenuation of hypersonic waves in quartz. *Phys. Rev.* 117: 5: 1245–1252.
18. Kovalev, V. I., and R. Harrison. 2002. Waveguide-induced inhomogeneous spectral broadening of stimulated Brillouin scattering in optical fiber. *Opt. Lett.* 27: 22: 2022–2024.
19. Jen, C. K., J. E. B. Oliveira, N. Goto, and K. Abe. 1988. Role of guided acoustic wave properties in single-mode optical fibre design. *Electron. Lett.* 24: 23: 1419–1420.
20. Shibata, N., Y. Azuma, T. Horiguch, and M. Tateda. 1988. Identification of longitudinal acoustic modes guided in the core region of a single-mode optical fiber by Brillouin gain spectra measurements. *Opt. Lett.* 13: 7: 595–597.
21. Shibata, N., K. Okamoto, and Y. Azuma. 1989. Longitudinal acoustic modes and Brillouin-gain spectra for GeO_2-doped-core single-mode fibers. *J. Opt. Soc. Am. B* 6: 6: 1167–1174.
22. Ohno, H., Y. Uchiyama, and T. Kurashima. 1999. Reduction of the effect of temperature in a fiber optic distributed sensor used for strain measurements in civil structures. 1999 Symposium on Smart Structures and Materials. Newport Beach, CA, USA, SPIE 3670: 486–496.
23. Jones, J. D. 1997. Review of fibre sensor techniques for temperature-strain discrimination, 12th International Conference on Optical Fiber Sensors, Washington, DC, Optical Society of America, 16 OTuC1.
24. Parker, T. R., M. Farhadiroushan, V. A. Handerek, and A. J. Rogers. 1997. Temperature and strain dependence of the power level and frequency of spontaneous Brillouin scattering in optical fibers. *Opt. Lett.* 22: 11: 787–789.
25. DeSouza, K., P. C. Wait, and T. P. Newson. 1997. Characterisation of strain dependence of the Landau-Placzek ratio for distributed sensing. *Electron. Lett.* 33: 7: 615–616.
26. Parker, T., M. Farhadiroushan, R. Feced, V. A. Handerek, and A. J. Rogers. 1998. Simultaneous distributed measurement of strain and temperature from noise-initiated Brillouin scattering in optical fibers. *IEEE J. Quant. Electron.* 34: 4: 645–659.
27. Kee, H. H., G. P. Lees, and T. P. Newson. 2000. All-fiber system for simultaneous interrogation of distributed strain and temperature sensing by spontaneous Brillouin scattering. *Opt. Lett.* 25: 10: 695–697.
28. Belal, M., and T. P. Newson. 2011. Evaluation of a high spatial resolution temperature compensated distributed strain sensor using a temperature controlled strain rig. 21st International Conference on Optical Fibre Sensors. Ottawa, Canada, SPIE 7753: 27–24.
29. Belal, M., and T. P. Newson. 2012. Experimental examination of the variation of the spontaneous Brillouin power and frequency coefficients under the combined influence of temperature and strain. *J. Lightw. Technol.* 30: 8: 1250–1255.
30. Sakairi, Y., H. Uchiyama, Z. X. Li, and S. Adachi. 2002. A system for measuring temperature and strain separately by BOTDR and OTDR. Advanced Sensor Systems and Applications, Shanghai, China, SPIE, 4920: 274–284.
31. Nikles, M. 1997. La diffusion Brillouin dans les fibres optiques: étude et application aux capteurs distribués. PhD Ecole Polytechnique Fédérale de Lausanne.
32. Mallinder, F., and B. Proctor. 1964. Elastic constants of fused silica as a function of large tensile strain. *Phys. Chem. Glasses* 5: 4: 91–103.
33. Brückner, R. 1970. Properties and structure of vitreous silica. I. *J. Non-Cryst. Solids* 5: 2: 123–175.
34. Primak, W., and D. Post. 1959. Photoelastic constants of vitreous silica and its elastic coefficient of refractive index. *J. Appl. Phys.* 30: 5: 779–788.
35. Borrelli, N. F., and R. A. Miller. 1968. Determination of the individual strain-optic coefficients of glass by an ultrasonic technique. *Appl. Opt.* 7: 5: 745–750.
36. Bertholds, A., and R. Dandliker. 1988. Determination of the individual strain-optic coefficients in single-mode optical fibres. *J. Lightw. Technol.* 6: 1: 17–20.
37. Namihira, Y. 1985. Opto-elastic constant in single mode optical fibers. *J. Lightw. Technol.* 3: 5: 1078–1083.
38. Barlow, A., and D. Payne. 1983. The stress-optic effect in optical fibers. *IEEE J. Quant. Electron.* 19: 5: 834–839.
39. Spinner, S. A. M. 1956. Elastic moduli of glasses at elevated temperatures by a dynamic method. *J. Am. Ceram. Soc.* 39: 3: 113–118.
40. Hartog, A. H., A. J. Conduit, and D. N. Payne. 1979. Variation of pulse delay with stress and temperature in jacketed and unjacketed optical fibres. *Opt. Quant. Electron.* 11: 265–273.
41. Namihira, Y., K. Mochizuki, and K. Tatekura. 1981. Effects of thermal stress on group delay in jacketed single mode fibres. *Electron. Lett.* 17: 21: 813–815.
42. Kurashima, T., T. Horiguchi, and M. Tateda. 1990. Thermal effects on the Brillouin frequency shift in jacketed optical silica fibers. *Appl. Opt.* 29: 15: 2219–2222.
43. Smith, R. G. 1972. Optical power handling capacity of low loss optical fibers as determined by stimulated Raman and Brillouin scattering. *Appl. Opt.* 11: 11: 2489–2494.

44. Ippen, E. P., and R. H. Stolen. 1972. Stimulated Brillouin scattering in optical fibers. *Appl. Phys. Lett.* 21: 11: 539–541.
45. Hill, K. O., B. S. Kawasaki, and D. C. Johnson. 1976. cw Brillouin laser. *Appl. Phys. Lett.* 28: 10: 608–609.
46. Hill, K. O., D. C. Johnson, and B. S. Kawasaki. 1976. cw generation of multiple Stokes and anti-Stokes Brillouin-shifted frequencies. *Appl. Phys. Lett.* 29: 3: 185–187.
47. Cotter, D. 1982. Observation of stimulated Brillouin scattering in low-loss silica fibre at 1.3 μm. *Electron. Lett.* 18: 12: 495–496.
48. Thomas, P., N. Rowell, H. Van Driel, and G. Stegeman. 1979. Normal acoustic modes and Brillouin scattering in single-mode optical fibers. *Phys. Rev. B* 19: 10: 4986.
49. Galindez-Jamioy, C. A., and J.-M. Lopez-Higuera. 2012. Brillouin distributed fiber sensors: An overview and applications. *J. Sensors* 2012. Article ID 204121.
50. Zadok, A., A. Eyal, and M. Tur. 2011. Stimulated Brillouin scattering slow light in optical fibers [invited]. *Appl. Opt.* 50: 25: E38–E49.
51. Urricelqui, J., A. Zornoza, M. Sagues, and A. Loayssa. 2012. Dynamic BOTDA measurements based on Brillouin phase-shift and RF demodulation. *Opt. Expr.* 20: 24: 26942–26949.
52. Pinnow, D. A., S. J. Candau, J. T. LaMacchia, and T. A. Litovitz. 1968. Brillouin scattering: Viscoelastic measurements in liquids. *J. Acoust. Soc. Am.* 43: 1: 131–142.
53. Wait, P. C., and T. P. Newson. 1996. Landau Placzek ratio applied to distributed fibre sensing. *Opt. Commun.* 122: 4–6: 141–146.
54. DeSouza, K., G. P. Lees, P. C. Wait, and T. P. Newson. 1996. Diode-pumped Landau-Placzek based distributed temperature sensor utilising an all-fibre Mach-Zehnder interferometer. *Electron. Lett.* 32: 23: 2174–2175.
55. De Souza, K. 1999. Fibre-optic distributed sensing based on spontaneous Brillouin scattering. PhD University of Southampton.
56. De Souza, K., P. C. Wait, and T. P. Newson. 1997. Double-pass configured fibre Mach-Zehnder interferometric optical filter for distributed fibre sensing. *Electron. Lett.* 33: 25: 2148–2149.
57. Lees, G. P., P. C. Wait, A. H. Hartog, and T. P. Newson. 1998. Recent advances in distributed optical fibre temperature sensing using the Landau-Placzek ratio. Distributed and Multiplexed Fiber Optic Sensors VII. Boston, MA, USA, SPIE 3541: 292–296.
58. Wait, P., and A. H. Hartog. 2001. Spontaneous Brillouin-based distributed temperature sensor utilizing a fiber Bragg grating notch filter for the separation of the Brillouin signal. *IEEE Photon. Technol. Lett.* 13: 5: 508–510.
59. Watley, D., M. Farhadiroushan, and B. J. Shaw. 2005. Direct measurement of Brillouin frequency in distributed optical sensing systems. WO2005/106396 A2.
60. Masoudi, A., M. Belal, and T. P. Newson. 2013. Distributed dynamic large strain optical fiber sensor based on the detection of spontaneous Brillouin scattering. *Opt. Lett.* 38: 17: 3312–3315.
61. Todd, M. D., G. A. Johnson, and C. C. Chang. 1999. Passive, light intensity-independent interferometric method for fibre Bragg grating interrogation. *Electron. Lett.* 35: 22: 1970–1971.
62. Posey, R. J., G. A. Johnson, and S. T. Vohra. 2000. Strain sensing based on coherent Rayleigh scattering in an optical fibre. *Electron. Lett.* 36: 20: 1688–1689.
63. Shimizu, K., T. Horiguchi, Y. Koyamada, and T. Kurashima. 1993. Coherent self-heterodyne detection of spontaneously Brillouin-scattered light waves in a single-mode fiber. *Opt. Lett.* 18: 3: 185–187.
64. Lee, T., N. Takeuchi, K. Shimizu, and T. Horiguchi. 1995. Light-frequency control apparatus. US5438578.
65. Hodgkinson, T. G., and P. Coppin. 1990. Pulsed operation of an optical feedback frequency synthesizer. *Electron. Lett.* 26: 15: 1155–1157.
66. Smith, D. W., P. Healey, I. W. Stanley, P. Coppin, and T. G. Hodgkinson. 1990. Apparatus for and method of generating a comb of optical teeth of different wavelengths. EP0385 697B1.
67. Shimizu, K., T. Horiguch, and Y. Koyamada. 1993. Frequency translation of light waves by propagation around an optical ring circuit containing a frequency shifter: I. Experiment. *Appl. Opt.* 32: 33: 6718–6726.
68. Shimizu, K., T. Horiguch, and Y. Koyamada. 1994. Frequency translation of light waves by propagation around an optical ring circuit containing a frequency shifter: II. Theoretical analysis. *Appl. Opt.* 33: 15: 3209–3219.
69. Izumita, H., T. Sato, M. Tateda, and Y. Koyamada. 1996. Brillouin OTDR employing optical frequency shifter using side-band generation technique with high-speed LN phase-modulator. *IEEE Photon. Technol. Lett.* 8: 12: 1674–1676.
70. Uchiyama, H., and Z. Lee. 2000. Apparatus for measuring characteristics of optical fiber. US6055044.
71. Shimizu, K., T. Kurashima, T. Horiguch, and Y. Koyamada. 1992. A new technique to shift lightwave frequency for distributed fiber-optic sensing. Distributed and multiplexed fiber optic sensors II, Boston, MA, USA, SPIE, 1797: 18–27.
72. Uchiyama, H., and T. Kurashima. 2002. Optical-fiber characteristics measuring apparatus. US6335788B1.

73. Alahbabi, M. N., N. P. Lawrence, Y.-T. Cho, and T. P. Newson. 2004. High spatial resolution microwave detection system for Brillouin-based distributed temperature and strain sensors. *Meas. Sci. Technol.* 15: 8: 1539–1543.
74. Feced, R., T. R. Parker, M. Farhadiroushan, V. A. Handerek, and A. J. Rogers. 1998. Power measurement of noise-initiated Brillouin scattering in optical fibers for sensing applications. *Opt. Lett.* 23: 1: 79–81.
75. Hartog, A. H., and G. P. Lees. 2009. Distributed sensing in an optical fiber using Brillouin scattering. US7504618B2.
76. Naruse, H., M. Tateda, and A. Shimada. 2003. Deformation of the Brillouin gain spectrum caused by parabolic strain distribution and resulting measurement error in BOTDR strain measurement system. *IEICE Trans. Electron.* 86: 10: 2111–2121.
77. Farahani, M. A., E. Castillo-Guerra, and B. G. Colpitts. 2011. Accurate estimation of Brillouin frequency shift in Brillouin optical time domain analysis sensors using cross correlation. *Opt. Lett.* 36: 21: 4275–4277.
78. Feng, W., Z. Weiwei, L. Yuangang, Y. Zhijun, and Z. Xuping. 2015. Determining the change of Brillouin frequency shift by using the similarity matching method. *J. Lightw. Technol.* 33: 19: 4101–4108.
79. Strong, A. P., and A. H. Hartog. 2015. Fault-tolerant distributed fiber optic intrusion detection. US8947232B2.
80. Agrawal, G. P. 2007. *Nonlinear fiber optics*. Academic Press, Boston, MA.
81. Alahbabi, M. N., Y.-T. Cho, T. P. Newson, P. C. Wait, and A. H. Hartog. 2004. Influence of modulation instability on distributed optical fiber sensors based on spontaneous Brillouin scattering. *J. Opt. Soc. Am. B* 21: 6: 1156–1160.
82. Belal, M., and T. P. Newson. 2011. A 5 cm spatial resolution temperature compensated distributed strain sensor evaluated using a temperature controlled strain rig. *Opt. Lett.* 36: 24: 4728–4730.
83. Naruse, H., and M. Tateda. 1999. Trade-off between the spatial and the frequency resolutions in measuring the power spectrum of the Brillouin backscattered light in an optical fiber. *Appl. Opt.* 38: 31: 6516–6521.
84. Nitta, N., M. Tateda, and T. Omatsu. 2002. Spatial resolution enhancement in BOTDR by spectrum separation method. *Opt. Rev.* 9: 2: 49–53.
85. Brown, A. W., M. D. DeMerchant, X. Bao, and T. W. Bremner. 1999. Spatial resolution enhancement of a Brillouin-distributed sensor using a novel signal processing method. *J. Lightw. Technol.* 17: 7: 1179–1183.
86. Ravet, F., X. Bao, Y. Li et al. 2007. Signal processing technique for distributed Brillouin sensing at centimeter spatial resolution. *J. Lightw. Technol.* 25: 11: 3610–3618.
87. Hartog, A. H., and P. C. Wait. 2009. Optical time domain reflectometry for two segment fiber optic systems having an optical amplifier therebetween. US7595865B2; GB2416588B.
88. Strong, A. P., N. Sanderson, G. P. Lees et al. 2008. A comprehensive distributed pipeline condition monitoring system and its field trial. 7th International Pipeline Conference, 26th September–3rd October 2008, Calgary, Alberta, Canada: IPC2008-64549.
89. Stolen, R. H., and E. P. Ippen. 1973. Raman gain in glass optical waveguides. *Appl. Phys. Lett.* 22: 6: 276–278.
90. Cho, Y.-T., M. N. Alahbabi, M. J. Gunning, and T. P. Newson. 2003. 50-km single-ended spontaneous-Brillouin-based distributed-temperature sensor exploiting pulsed Raman amplification. *Opt. Lett.* 28: 18: 1651–1653.
91. Cho, Y.-T., M. N. Alahbabi, M. J. Gunning, and T. P. Newson. 2004. Enhanced performance of long range Brillouin intensity based temperature sensors using remote Raman amplification. *Meas. Sci. Technol.* 15: 8: 1548–1552.
92. McGeehin, P. 1994. Optical techniques in industrial measurement: Safety in hazardous environments. European Commission Directorate-General. XIII. EUR 16011EN.
93. Bromage, J. 2004. Raman amplification for fiber communications systems. *J. Lightw. Technol.* 22: 1: 79–93.
94. Hartog, A. H. 2007. Raman amplification in distributed optical fiber sensing systems. GB2441016B.
95. Pedersen, B., A. Bjarklev, J. H. Povlsen, K. Dybdal, and C. C. Larsen. 1991. The design of erbium-doped fiber amplifiers. *J. Lightw. Technol.* 9: 9: 1105–1112.
96. Kagi, N., A. Oyobe, and K. Nakamura. 1991. Temperature dependence of the gain in erbium-doped fibers. *J. Lightw. Technol.* 9: 2: 261–265.
97. Suyama, M., R. Laming, and D. Payne. 1990. Temperature dependent gain and noise characteristics of a 1480 nm-pumped erbium-doped fibre amplifier. *Electron. Lett.* 26: 21: 1756–1758.
98. Cho, Y.-T., M. N. Alahbabi, G. Brambilla, and T. P. Newson. 2005. Distributed Raman amplification combined with a remotely pumped EDFA utilized to enhance the performance of spontaneous Brillouin-based distributed temperature sensors. *IEEE Photon. Technol. Lett.* 17: 6: 1256–1258.
99. Sun, A., B. Chen, J. Chen et al. 2007. Detection of Brillouin scattering temperature signal in Brillouin optical time-domain reflectometer sensing system based on instantaneous frequency measurement technology. *Opt. Eng.* 46: 12: 124401.
100. Hartog, A. H. 2008. Measuring Brillouin backscatter from an optical fibre using digitisation. GB2440952B.

101. Yuguo, Y., L. Yuangang, Z. Xuping, W. Feng, and W. Rugang. 2012. Reducing trade-off between spatial resolution and frequency accuracy in BOTDR using Cohen's class signal processing method. *IEEE Photon. Technol. Lett.* 24: 15: 1337–1339.
102. Hartog, A. H. 2011. Measuring Brillouin backscatter from an optical fibre using channelisation. US8013986B2.
103. Hartog, A. H. 2012. Measuring Brillouin backscatter from an optical fibre using a tracking signal. US8134696B2.
104. Soto, M. A., G. Bolognini, and F. Di Pasquale. 2008. Analysis of optical pulse coding in spontaneous Brillouin-based distributed temperature sensors. *Opt. Expr.* 16: 23: 19097–19111.
105. Shengpeng, W., X. Yuhua, and H. Xingdao. 2014. The theoretical analysis and design of coding BOTDR system with APD detector. *IEEE Sens. J.* 14: 8: 2626–2632.
106. Li, C., Y. Lu, X. Zhang, and F. Wang. 2012. SNR enhancement in Brillouin optical time domain reflectometer using multi-wavelength coherent detection. *Electron. Lett.* 48: 18: 1139–1141.
107. Koyamada, Y., Y. Sakairi, N. Takeuchi, and S. Adachi. 2007. Novel technique to improve spatial resolution in Brillouin optical time-domain reflectometry. *IEEE Photon. Technol. Lett.* 19: 23: 1910–1912.
108. Koyamada, Y. 2011. Apparatus for measuring the characteristics of an optical fiber. US7873273B2.
109. Sakairi, Y., S. Matsuura, S. Adachi, and Y. Koyamada. 2008. Prototype double-pulse BOTDR for measuring distributed strain with 20-cm spatial resolution. Society of Instrument and Control Engineers Annual Conference, Chofu City, Japan. IEEE. 1106–1109.
110. Matsuura, S., M. Kumoda, Y. Anzai, and Y. Koyamada. 2013. Distributed strain measurement in GI fiber with submeter spatial resolution by DP-BOTDR. 18th OptoElectronics and Communications Conference held jointly with 2013 International Conference on Photonics in Switching. Kyoto, Japan, Optical Society of America. MS2_2.
111. Naruse, H., and M. Tateda. 2000. Launched pulse-shape dependence of the power spectrum of the spontaneous Brillouin backscattered light in an optical fiber. *Appl. Opt.* 39: 34: 6376–6384.
112. Hao, Y., Q. Ye, Z. Pan, H. Cai, and R. Qu. 2012. Improvement of pulse shape on Brillouin optical time domain reflectometry. 22nd International Conference on Optical Fiber Sensors. Beijing, China, SPIE. 8421: 9E.
113. Horiguchi, T., and M. Tateda. 1988. Optical fiber evaluation methods and system using Brillouin amplification. EP0348235B1.
114. Horiguchi, T., T. Kurashima, and M. Tateda. 1990. A technique to measure distributed strain in optical fibers. *IEEE Photon. Technol. Lett.* 2: 5: 352–354.
115. Horiguchi, T., T. Kurashima, and M. Tateda. 1989. Tensile strain dependence of Brillouin frequency shift in silica optical fibers. *IEEE Photon. Technol. Lett.* 1: 5: 107–108.
116. Horiguchi, T., and M. Tateda. 1989. Optical-fiber-attenuation investigation using stimulated Brillouin scattering between a pulse and a continuous wave. *Opt. Lett.* 14: 8: 408–410.
117. Li, Y., X. Bao, F. Ravet, and E. Ponomarev. 2008. Distributed Brillouin sensor system based on offset locking of two distributed feedback lasers. *Appl. Opt.* 42: 2: 99–102.
118. Bao, X., E. Ponomarev, Y. Li et al. 2009. Distributed Brillouin sensor system based on DFB lasers using offset locking. US7499151B2.
119. Nikles, M., L. Thévenaz, and P. A. Robert. 1996. Simple distributed fiber sensor based on Brillouin gain spectrum analysis. *Opt. Lett.* 21: 10: 758–760.
120. Nikles, M., L. Thevenaz, and P. Robert, A. 1997. Brillouin gain spectrum characterization in single-mode optical fibers. *J. Lightw. Technol.* 15: 10: 1842–1851.
121. Foaleng Mafang, S. 2011. Brillouin echoes for advanced distributed sensing in optical fibres. PhD Ecole Polytechnique Fédérale de Lausanne.
122. Diaz, S., S. Mafang-Foaleng, M. Lopez-Amo, and L. Thévenaz. 2008. A high-performance optical time-domain brillouin distributed fiber sensor. *IEEE Sens. J.* 8: 7: 1268–1272.
123. Horiguchi, T., K. Shimizu, T. Kurashima, M. Tateda, and Y. Koyamada. 1995. Development of a distributed sensing technique using Brillouin scattering. *J. Lightw. Technol.* 13: 7: 1296.
124. Soto, M. A., and L. Thévenaz. 2013. Modeling and evaluating the performance of Brillouin distributed optical fiber sensors. *Opt. Expr.* 21: 25: 31347–31366.
125. Fellay, A., L. Thévenaz, M. Facchini, M. Niklès, and P. Robert. 1997. Distributed sensing using stimulated Brillouin scattering: Towards ultimate resolution. 12th International Conference on Optical Fiber Sensors, Williamsburg, Virginia, Optical Society of America: OWD3.
126. Kalosha, V., E. Ponomarev, L. Chen, and X. Bao. 2006. How to obtain high spectral resolution of SBS-based distributed sensing by using nanosecond pulses. *Opt. Expr.* 14: 6: 2071–2078.
127. Bao, X., A. W. Brown, M. D. DeMerchant, and J. Smith. 1999. Characterization of the Brillouin-loss spectrum of single-mode fibers by use of very short (10-ns) pulses. *Opt. Lett.* 24: 8: 510.
128. Lecoeuche, V., D. J. Webb, C. N. Pannell, and D. A. Jackson. 2000. Transient response in high-resolution Brillouin-based distributed sensing using probe pulses shorter than the acoustic relaxation time. *Opt. Lett.* 25: 3: 156–158.

129. Afshar, S., G. A. Ferrier, X. Bao, and L. Chen. 2003. Effect of the finite extinction ratio of an electro-optic modulator on the performance of distributed probe–pump Brillouin sensor systems. *Opt. Lett.* 28: 16: 1418–1420.
130. Kishida, K., and C. Li. 2006. Pulse pre-pump-BOTDA technology for new generation of distributed strain measuring system. *Struct. Health Monit. Intell. Infrastructure* 5: 471–477.
131. Kishida, K., T. Li, and S. Lin. 2010. Distributed optical fiber sensor. US7719666B2.
132. Galindez Jamioy, C. A., and J. M. Lopez-Higuera. 2011. Decimeter spatial resolution by using differential preexcitation BOTDA pulse technique. *IEEE Sens. J.* 11: 10: 2344–2348.
133. Brown, A. W., B. G. Colpitts, and K. Brown. 2005. Distributed sensor based on dark-pulse Brillouin scattering. *IEEE Photon. Technol. Lett.* 17: 7: 1501–1503.
134. Brown, A. W., B. G. Colpitts, and K. Brown. 2007. Dark-pulse Brillouin optical time-domain sensor with 20-mm spatial resolution. *J. Lightw. Technol.* 25: 1: 381–386.
135. Foaleng-Mafang, S., J.-C. Beugnot, and L. Thévenaz. 2009. Optimised configuration for high resolution distributed sensing using Brillouin echoes. 20th International Conference on Optical Fibre Sensors, Edinburgh, United Kingdom, SPIE, 7503: 2C.
136. Li, W., X. Bao, Y. Li, and L. Chen. 2008. Differential pulse-width pair BOTDA for high spatial resolution sensing. *Opt. Expr.* 16: 26: 21616–21625.
137. Foaleng, S. M., M. Tur, J.-C. Beugnot, and L. Thévenaz. 2010. High spatial and spectral resolution long-range sensing using Brillouin echoes. *J. Lightw. Technol.* 28: 20: 2993–3003.
138. Horiguchi, T., R. Muroi, A. Iwasaka, K. Wakao, and Y. Miyamoto. 2007. Negative Brillouin gain and its application to distributed fiber sensing. 33rd European Conference and Exhibition on Optical Communication (ECOC). Berlin, Germany, VDE: 1–2.
139. Li, Y., X. Bao, Y. Dong, and L. Chen. 2010. A novel distributed Brillouin sensor based on optical differential parametric amplification. *J. Lightw. Technol.* 28: 18: 2621–2626.
140. Soto, M. A., G. Bolognini, and F. Di Pasquale. 2010. Analysis of pulse modulation format in coded BOTDA sensors. *Opt. Expr.* 18: 14: 14878–14892.
141. Soto, M. A., S. Le Floch, and L. Thévenaz. 2013. Bipolar optical pulse coding for performance enhancement in BOTDA sensors. *Opt. Expr.* 21: 14: 16390.
142. Dzulkefly Bin Zan, M. S., T. Tsumuraya, and T. Horiguchi. 2013. The use of Walsh code in modulating the pump light of high spatial resolution phase-shift-pulse Brillouin optical time domain analysis with non-return-to-zero pulses. *Meas. Sci. Technol.* 24: 9: 094025.
143. Dzulkefly Bin Zan, M. S. 2014. A study on the coding techniques for Phase-Shift pulse Brillouin optical time domain analysis (PSP-BOTDA) fiber optic sensor. D. Eng Shibaura Institute of Technology, Japan.
144. Dzulkefly Bin Zan, M. S., and T. Horiguchi. 2012. A dual Golay complementary pair of sequences for improving the performance of phase-shift pulse BOTDA fiber sensor. *J. Lightw. Technol.* 30: 21: 3338–3356.
145. Boyd, R. W., D. J. Gauthier, and A. L. Gaeta. 2006. Applications of slow light in telecommunications. *Opt. Photonics News* 17: 4: 18–23.
146. Shahriar, M., G. Pati, R. Tripathi et al. 2007. Ultrahigh enhancement in absolute and relative rotation sensing using fast and slow light. *Phys. Rev. A* 75: 5: 053807.
147. Shi, Z., R. Boyd, R. Camacho, P. Vudyasetu, and J. Howell. 2007. Slow-light Fourier transform interferometer. *Phys. Rev. Lett.* 99: 24: 240801.
148. Gonzalez-Herraez, M., K.-Y. Song, and L. Thevenaz. 2005. Optically controlled slow and fast light in optical fibers using stimulated Brillouin scattering. *Appl. Phys. Lett.* 87: 8: 081113–081113.
149. Thévenaz, L., K. Y. Song, and M. González-Herráez. 2006. Time biasing due to the slow-light effect in distributed fiber-optic Brillouin sensors. *Opt. Lett.* 31: 6: 715–717.
150. Garus, D., T. Gogolla, K. Krebber, and F. Schliep. 1997. Brillouin optical-fiber frequency-domain analysis for distributed temperature and strain measurements. *J. Lightw. Technol.* 15: 4: 654–662.
151. Garus, D., K. Krebber, F. Schliep, and T. Gogolla. 1996. Distributed sensing technique based on Brillouin optical-fiber frequency-domain analysis. *Opt. Lett.* 21: 17: 1402.
152. Bernini, R., L. Crocco, A. Minardo, F. Soldovieri, and L. Zeni. 2002. Frequency-domain approach to distributed fiber-optic Brillouin sensing. *Opt. Lett.* 27: 5: 288–290.
153. Bernini, R., L. Crocco, A. Minardo, F. Soldovieri, and L. Zeni. 2003. All frequency domain distributed fiber-optic Brillouin sensing. *IEEE Sens. J.* 3: 1: 36–43.
154. Bernini, R., A. Minardo, and L. Zeni. 2004. Stimulated Brillouin scattering frequency-domain analysis in a single-mode optical fiber for distributed sensing. *Opt. Lett.* 29: 7: 1977–1979.
155. Wosniok, A., N. Nöther, and K. Krebber. 2009. Distributed fibre optic sensor system for temperature and strain monitoring based on Brillouin optical-fibre frequency-domain analysis. *Procedia Chemistry* 1: 1: 397–400.
156. Chaube, P., B. G. Colpitts, D. Jagannathan, and A. W. Brown. 2008. Distributed fiber-optic sensor for dynamic strain measurement. *IEEE Sens. J.* 8: 7: 1067–1072.

157. Voskoboinik, A., D. Rogawski, H. Huang et al. 2012. Frequency-domain analysis of dynamically applied strain using sweep-free Brillouin time-domain analyzer and sloped-assisted FBG sensing. *Opt. Expr.* 20: 26: B581-F586.
158. Voskoboinik, A., W. Jian, B. Shamee et al. 2011. SBS-based fiber optical sensing using frequency-domain simultaneous tone interrogation. *J. Lightw. Technol.* 29: 11: 1729–1735.
159. Voskoboinik, A., O. F. Yilmaz, A. W. Willner, and M. Tur. 2011. Sweep-free distributed Brillouin time-domain analyzer (SF-BOTDA). *Opt. Expr.* 19: 26: B842–B847.
160. Voskoboinik, A., A. Willner, and M. Tur. 2015. Extending the dynamic range of sweep-free Brillouin optical time-domain analyzer. *J. Lightw. Technol.* 33: 14: 2978–2985.
161. Bernini, R., A. Minardo, and L. Zeni. 2009. Dynamic strain measurement in optical fibers by stimulated Brillouin scattering. *Opt. Lett.* 34: 17: 2613–2615.
162. Peled, Y., A. Motil, L. Yaron, and M. Tur. 2011. Slope-assisted fast distributed sensing in optical fibers with arbitrary Brillouin profile. *Opt. Expr.* 19: 21: 19845–19854.
163. Peled, Y., A. Motil, and M. Tur. 2012. Fast Brillouin optical time domain analysis for dynamic sensing. *Opt. Expr.* 20: 8: 8584–8591.
164. Gyger, F., E. Rochat, S. Chin, M. Niklès, and L. Thévenaz. 2014. Extending the sensing range of Brillouin optical time-domain analysis up to 325 km combining four optical repeaters, OFS2014 23rd International Conference on Optical Fiber Sensors, Santander, Spain, SPIE, 9157: 6Q1–4.
165. Foaleng, S. M., F. Rodríguez-Barrios, S. Martin-Lopez, M. González-Herráez, and L. Thévenaz. 2011. Detrimental effect of self-phase modulation on the performance of Brillouin distributed fiber sensors. *Opt. Lett.* 36: 2: 97–99.
166. Soto, M. A., A. L. Ricchiuti, L. Zhang et al. 2014. Time and frequency pump-probe multiplexing to enhance the signal response of Brillouin optical time-domain analyzers. *Opt. Expr.* 22: 23: 28584–28595.
167. Hartog, A. H. 2003. Optical time domain reflectometry method and apparatus. US6542228B1.
168. Zornoza, A., A. Loayssa, and M. S. Garcia. 2014. Device and method for measuring the distribution of physical quantities in an optical fiber. US2014/0306101A1.
169. Zornoza, A., M. Sagues, and A. Loayssa. 2012. Self-heterodyne detection for SNR improvement and distributed phase-shift measurements in BOTDA. *J. Lightw. Technol.* 30: 8: 1066–1072.
170. Xiaobo, T., S. Qiao, C. Wei, C. Mo, and M. Zhou. 2014. Vector Brillouin optical time-domain analysis with heterodyne detection and IQ demodulation algorithm. *IEEE Photon. J.* 6: 2: 1–8.
171. Lopez-Gil, A., X. Angulo-Vinuesa, M. A. Soto et al. 2016. Gain vs phase in BOTDA setups. Sixth European Workshop on Optical Fibre Sensors (EWOFS'2016), Limerick, Ireland, SPIE, 9916: 31.
172. Hotate, K., and T. Hasegawa. 2000. Measurement of Brillouin gain spectrum distribution along an optical fiber using a correlation-based technique – Proposal, experiment and simulation. *IEICE Trans. Electron.* 83: 3: 405–412.
173. Song, K. Y., and K. Hotate. 2007. Distributed fiber strain sensor with 1-kHz sampling rate based on Brillouin optical correlation domain analysis. *IEEE Photon. Technol. Lett.* 19: 23: 1928–1930.
174. Song, K.-Y., and K. Hotate. 2006. Enlargement of measurement range in a Brillouin optical correlation domain analysis system using double lock-in amplifiers and a single-sideband modulator. *IEEE Photon. Technol. Lett.* 18: 3: 499–501.
175. Hotate, K., and T. Yamauchi. 2005. Fiber-optic distributed strain sensing system by Brillouin optical correlation domain analysis with a simple and accurate time-division pump-probe generation scheme. *Jap. J. App. Phys.* 44: 7L: L1030.
176. Economou, G., R. Youngquist, and D. Davies. 1986. Limitations and noise in interferometric systems using frequency ramped single-mode diode lasers. *J. Lightw. Technol.* 4: 11: 1601–1608.
177. Shalom, H., A. Zadok, M. Tur et al. 1998. On the various time constants of wavelength changes of a DFB laser under direct modulation. *IEEE J. Quant. Electron.* 34: 10: 1816–1822.
178. Song, K. Y., M. Kishi, Z. He, and K. Hotate. 2011. High-repetition-rate distributed Brillouin sensor based on optical correlation-domain analysis with differential frequency modulation. *Opt. Lett.* 36: 11: 2062–2064.
179. Mizuno, Y., Z. He, and K. Hotate. 2010. Measurement range enlargement in Brillouin optical correlation-domain reflectometry based on double-modulation scheme. *Opt. Expr.* 18: 6: 5926–5933.
180. Hayashi, N., Y. Mizuno, and K. Nakamura. 2014. Simplified configuration of Brillouin optical correlation-domain reflectometry. *IEEE Photon. J.* 6: 5: 1–7.
181. Song, K. Y., Z. He, and K. Hotate. 2006. Optimization of Brillouin optical correlation domain analysis system based on intensity modulation scheme. *Opt. Expr.* 14: 10: 4256–4263.
182. Antman, Y., N. Primerov, L. Thévenaz, and A. Zadok. 2012. High-resolution Brillouin fiber sensing using random phase coding of the pump and probe waves. 22nd International Conference on Optical Fiber Sensors, Beijing, China, SPIE, 8421: 842116.

183. Elooz, D., Y. Antman, N. Levanon, and A. Zadok. 2014. High-resolution long-reach distributed Brillouin sensing based on combined time-domain and correlation-domain analysis. *Opt. Expr.* 22: 6: 6453–6463.
184. London, Y., Y. Antman, R. Cohen et al. 2014. High-resolution long-range distributed Brillouin analysis using dual-layer phase and amplitude coding. *Opt. Expr.* 22: 22: 27144–27158.
185. Denisov, A., M. A. Soto, and L. Thévenaz. 2013. Time gated phase-correlation distributed Brillouin fibre sensor. Fifth European Workshop on Optical Fibre Sensors, Krakow, Poland, SPIE, 8794: 3I.
186. Cohen, R., Y. London, Y. Antman, and A. Zadok. 2014. Brillouin optical correlation domain analysis with 4 millimeter resolution based on amplified spontaneous emission. *Opt. Expr.* 22: 10: 12070–12078.
187. Mizuno, Y. 2009. Proposal and performance improvement of Brillouin optical correlation-domain reflectometry. PhD The University of Tokyo.
188. Mizuno, Y., W. Zou, Z. He, and K. Hotate. 2008. Proposal of Brillouin optical correlation-domain reflectometry (BOCDR). *Opt. Expr.* 16: 16: 12148–12153.
189. Mizuno, Y., Z. He, and K. Hotate. 2009. One-end-access high-speed distributed strain measurement with 13-mm spatial resolution based on Brillouin optical correlation-domain reflectometry. *IEEE Photon. Technol. Lett.* 21: 7: 474–476.
190. Song, K.-Y., and K. Hotate. 2008. Brillouin optical correlation domain analysis in linear configuration. *IEEE Photon. Technol. Lett.* 20: 24: 2150–2152.
191. Mizuno, Y., N. Hayashi, H. Fukuda, K. Y. Song, and K. Nakamura. 2017. Ultrahigh-speed distributed Brillouin reflectometry. *Light: Science & Applications* 6: e16186.
192. Cheng, Y., J. T. Kringlebotn, W. H. Loh, R. I. Laming, and D. N. Payne. 1995. Stable single-frequency traveling-wave fiber loop laser with integral saturable-absorber-based tracking narrow-band filter. *Opt. Lett.* 20: 8: 875–877.
193. Song, K. Y., W. Zou, Z. He, and K. Hotate. 2008. All-optical dynamic grating generation based on Brillouin scattering in polarisation-maintaining fiber. *Opt. Lett.* 33: 9: 926–928.
194. Song, K. Y., W. Zou, Z. He, and K. Hotate. 2009. Optical time-domain measurement of Brillouin dynamic grating spectrum in a polarisation-maintaining fiber. *Opt. Lett.* 34: 9: 1381–1383.
195. Hotate, K., K. Abe, and K. Y. Song. 2006. Suppression of signal fluctuation in Brillouin optical correlation domain analysis system using polarisation diversity scheme. *IEEE Photon. Technol. Lett.* 18: 24: 2653–2655.
196. Thévenaz, L., M. Facchini, A. Fellay, M. Nikles, and P. Robert. 1997. Evaluation of local birefringence along fibres using Brillouin analysis. Optical Fibre Measurement Conference OFMC'97, Teddington, UK, National Physical Laboratory: 82–85.
197. Gogolla, T., and K. Krebber. 2000. Distributed beat length measurement in single-mode optical fibers using stimulated Brillouin-scattering and frequency-domain analysis. *J. Lightw. Technol.* 18: 3: 320–328.
198. Williams, D. 2014. Theoretical investigation of stimulated Brillouin scattering in optical fibers and their applications. PhD University of Ottawa.

Rayleigh Backscatter *Distributed Vibration Sensors and Static Measurements** 6

In the discussion of distributed sensing technologies in previous chapters, Rayleigh (elastic) scattering, if used at all, was seen as a measurand-insensitive signal that could provide loss compensation. This remains true whilst it is a broadband probe wave that is scattered.

However, when the probe signal is coherent, the character of Rayleigh backscatter changes and it becomes extremely sensitive to temperature and strain.

This has led to two types of distributed optical fibre sensors (DOFSs) based on Rayleigh scattering, namely dynamic strain on the one hand and static strain and temperature sensing on the other hand. By dynamic sensing, we mean that the DOFS is able to measure very fast changes of the measurand, but not necessarily relate the measurand presently acquired to an older data set. In other words, a dynamic sensor cannot measure down to zero frequency, and for example, if the equipment is shut off and the user returns at some later time, in all likelihood, the value measured will be unrelated to the earlier measurement even if the sensing fibre has remained undisturbed in the meantime. In contrast, a static (or quasi-static) measurement requires that the value presently acquired can be compared in a traceable manner to an earlier measurement, even if the equipment has been switched off or disconnected in the interim.

In the first case, dynamic variations of strain are detected as a function of position along the fibre, and this provides a distributed sensor of mechanical vibration. In this technology, the frequency of the source is (in most cases) essentially constant even though, sometimes, a few different frequencies are used to alleviate some of the problems caused by the coherence of the source. In the case of static measurements, the design of the Rayleigh-based DOFS is quite different in that the frequency of the probe is varied over a wide range and data are collected at fine intervals of the source frequency. It is the variation of the backscatter signal as a function of probe frequency that is interpreted in terms of the quasi-static value of temperature and strain.

The main distinction between vibration and quasi-static measurements can be summarised by the fact that, in the vibration measurement, the backscatter signature is analysed as a function of time at constant source frequency, whereas for the static measurements, the backscatter signature is measured and processed as a function of probe frequency.

In this chapter, these two technologies are addressed separately starting with vibration sensing.

* This chapter is the subject of work that was, at the time of writing, active amongst direct colleagues at the Schlumberger Fibre-Optic Technology Centre (SFTC) in Romsey, United Kingdom, and I am grateful to the many colleagues there and wider afield with whom I have worked closely on the subject of distributed vibration sensing. I would like to thank especially (in alphabetical order) Florian English, Graeme Hilton, Ian Hilton, Kamal Kader, Gareth Lees, Richard Marsh and Adam Stanbridge at SFTC and, in St Petersburg (Russia), Mikhail Bisyarin, Oleg Kotov, Leonid Liokumovich and Nikolai Ushakov.

6.1 DISTRIBUTED VIBRATION SENSORS

Acoustic signals or mechanical vibrations were amongst the first measurands to be investigated with optical fibre sensors, driven by interest in marine sonar applications. Although the pressure sensitivity of bare [1,2], then coated [3–5] optical fibres was pursued initially, the focus of research moved to wound devices, e.g. hydrophones [6–11] and accelerometers [12–14], which were more sensitive and more specific to a particular measurand and angle of arrival of the acoustic wave. The multiplexing of these sensors was developed in the 1980s using the time [15], frequency [16] and coherence [17] domains to distinguish the different sensor elements positioned along a fibre. Some of the techniques investigated then are to be found in present-day distributed vibration sensors (DVSs).

Alongside the development of acoustic sensor arrays, optical time-domain reflectometry (OTDR) moved from multimode to single-mode fibre around 1980 and 1981 [18,19], and this led to the discovery of the coherence effects [20–22] in these fibres. In multimode fibre, these effects had been masked by mode diversity and the relatively low coherence of the sources that were generally used. However, early measurements on single-mode fibre showed that the source coherence could profoundly degrade the OTDR measurement, and this was regarded as a nuisance, with a flurry of papers appearing in the literature to describe and quantify the problem [23–28].

Dakin and Lamb [29] and then Taylor and Lee [30] realised that coherent Rayleigh backscatter could be used to measure the spatial distribution of disturbances in optical fibres, and a practical demonstration of the sensitivity of the backscatter intensity in coherent OTDR was published in 1994 [31].

Distributed acoustic or vibration sensing has seen a rapid increase in research and commercial interest since about 2005, and the technology and its applications are progressing rapidly, driven particularly by applications in the oil and gas industry. Although the terms *distributed acoustic sensor* (DAS) and *distributed vibration sensor* (DVS) are both used by different authors, the term *acoustic* is etymologically related to pressure waves, whereas what is usually measured is the strain on the fibre caused by a mechanical wave that is not necessarily a pressure wave: shear waves and surface waves commonly exert a stronger influence on the fibre than pressure waves do, and so, in the present author's view, the term *distributed vibration sensor* is to be preferred as a more general description of this class of distributed sensors.

6.1.1 Recall of Coherent Rayleigh Backscatter

Recapitulating briefly the description in Chapter 3, Rayleigh scattering is an elastic process caused by localised inhomogeneities of the refractive index of the medium through which an electro-magnetic wave is travelling. These inhomogeneities cause re-radiation of a small fraction of the probe light with a dipole pattern and have a fixed phase relation to the incident light. The backscattered light that arrives at the detector at any one time is collected from a section of the fibre that was occupied by the probe pulse at the time of scattering. It therefore includes the contribution of a myriad of dipoles, and it is the summation of their electric fields (as recaptured by the fibre and propagated back to the detector) that provides the signal.

With a coherent (narrowband) source, the phase of the incident light is consistent for all the dipoles within a resolution cell, and so too are the phases of the re-emitted (scattered) light from each of them. The scattered light therefore consists of the summation of this large collection of electric fields that have a stable, although random, phase relationship.

6.1.2 Timescales in Distributed Vibration Sensing

DVS uses the time axis both to discriminate locations along the fibre (as is the case in other forms of OTDR) and to represent the measurand in time. The time scales involved are quite different: the connection between time and distance along the fibre is determined by the speed of light c (modified by two-way propagation and the group refractive index N_g of the material). The tracking of vibration signals at specific locations of the fibre occurs on a much slower timescale, appropriate to the monitoring of mechanical vibration. For obvious reasons, we refer to these timescales, respectively, as 'fast time' and 'slow time'. Their mapping is illustrated in Figure 6.1. The rate at which the signal is digitised in fast time is referred to as f_d and relates to the sampling rate of the A/D converter used to acquire the data, whereas the frequency of the readings in slow time is dictated by the probe repetition frequency f_R.

The fast time scale is marked out by sample number i with $\tau_i = i/f_d$. Of course, the maximum duration of the fast time axis is limited by f_R, so that given f_R, the longest possible acquisition duration in fast time is given by $\tau_{max} = 1/f_R$. In turn, the highest frequency that can be detected in the measurand is dictated by the Nyquist frequency $f_R/2$. f_R is limited by the need to avoid multiple probe pulses being present in the

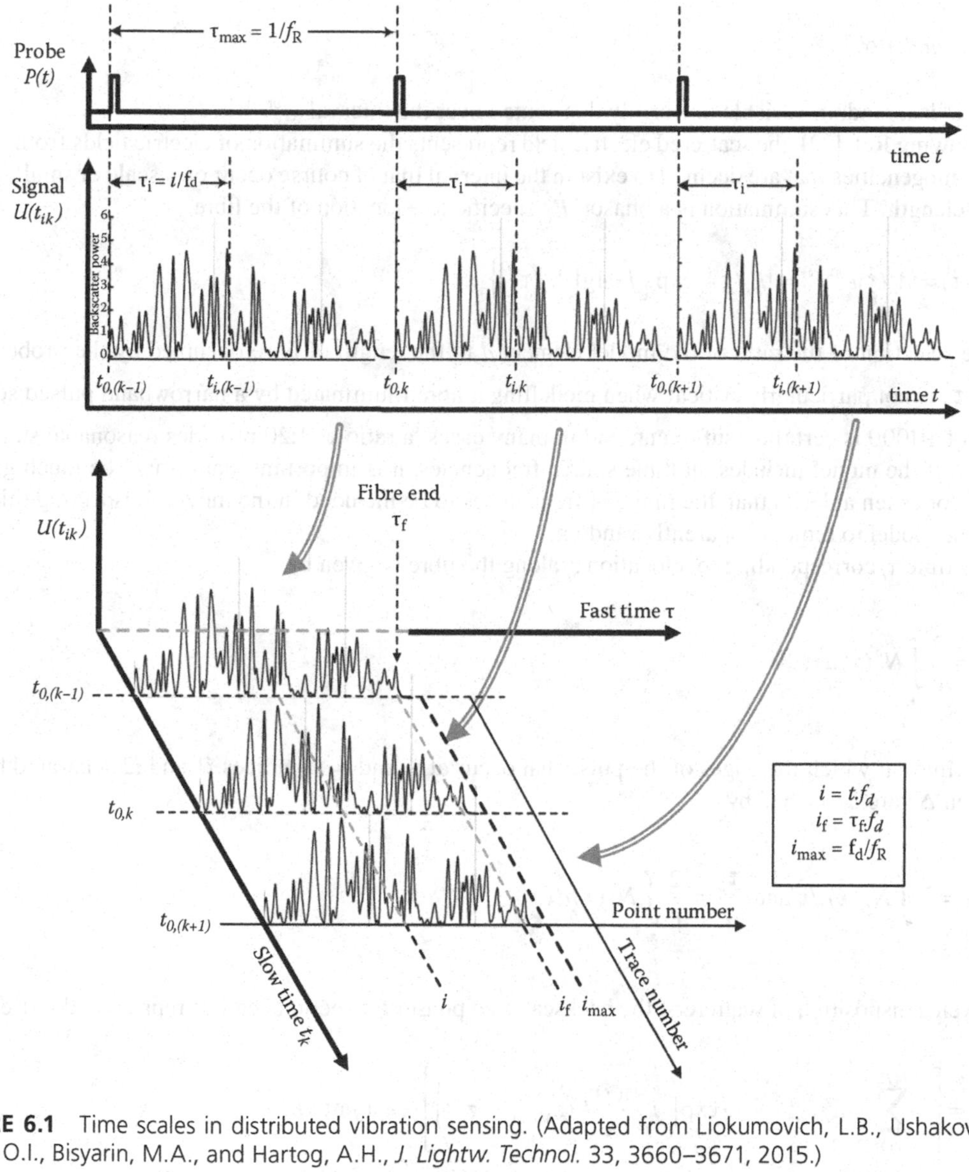

FIGURE 6.1 Time scales in distributed vibration sensing. (Adapted from Liokumovich, L.B., Ushakov, N.A., Kotov, O.I., Bisyarin, M.A., and Hartog, A.H., *J. Lightw. Technol.* 33, 3660–3671, 2015.)

fibre simultaneously, unless they can be distinguished in some way, e.g. by being at a different frequency. The longest fibre that can be measured at pulse repetition rate f_R is therefore $L_{\max} = \frac{1}{f_R}\frac{c}{2 \cdot N_g}$.

6.1.3 Modelling of Coherent OTDR

The signals generated by a coherent OTDR cannot be modelled analytically because they arise from random effects; nonetheless, statistical models can be built showing specific realisations that illustrate the behaviour of signals or the aggregated behaviour of a set of such signals. These can be useful for understanding the statistical properties of the behaviour of coherent backscatter signals.

In the case of a narrowband source, the backscatter signal can be modelled by disposing notional scatterers of amplitude at random locations along the fibre [32]; the mth of these has amplitude a_{sc}^m and is positioned within interval d_{sc}^m. The fibre is thus divided into intervals of length d (much shorter than the spatial resolution) into which a single scatterer is positioned at random. The location of the mth scatterer is given by

$$z_{sc}^m = m.d + \delta\zeta^m, \tag{6.1}$$

where $\delta\zeta^m$ is a random variable uniformly distributed over the interval d_{sc}^m.

Following Ref. [32], the scattered electric field represents the summation of electric fields from all the real inhomogeneities that are deemed to exist in the interval that of course occur on a scale of smaller than one wavelength. This summation is a phasor $\dot{E}_s$ specific to a location of the fibre.

$$\dot{E}_s(\tau) = A_e(\tau)e^{j\phi_s(\tau)} = \left|\dot{E}_s(\tau)\right| \cdot \exp\left\{j \cdot \arg\left[\dot{E}_s(\tau)\right]\right\} \tag{6.2}$$

The selection of the ratio of the model element d to the length of fibre occupied by the probe pulse $\delta z = \frac{v_g}{2}\tau_p$ is not particularly critical when modelling a fibre illuminated by a narrowband pulsed source. A ratio of 1:1000 is certainly sufficient, and in many cases, a ratio of 1:20 provides reasonable statistics. However, if the model includes multiple source frequencies, it is important that $d \cdot v_g/2$ be much greater (by a factor of ten at least) than the range of frequencies to be included in the model. Respecting this rule allows the model to remain apparently random.

The time τ_i corresponding to a location z_i along the fibre is given by

$$\tau_i = \frac{2}{c}\int_0^{z_i} N_g(x)\,dx, \tag{6.3}$$

and the times at which the edges of the pulse that occur at z_1 and z_2 at indices $i1$ and $i2$ separated by the resolution Δz are described by

$$\frac{\tau_p}{2} = \frac{2}{c}\int_{z_{i1}}^{z_i} N_g(x)\,dx \text{ and } \frac{\tau_p}{2} = \frac{2}{c}\int_{z_i}^{z_{i2}} N_g(x)\,dx. \tag{6.4}$$

Given a disposition of scatterers, the backscattered phasor for section i can be represented as Ref. [32]

$$\dot{E}_i = \left[\sum_{m^*=-M/2}^{M/2} a_{M_i+m^*} \cdot w_{m^*} \cdot \exp\left(j\frac{4\pi \cdot n_i}{\lambda}(z_{M_i+m^*} - z_i)\right)\right] \cdot \exp(j\omega\tau_i/K_g), \tag{6.5}$$

which is the summation of the phasors related to each scatterer within the section of fibre, weighted by a window function w that defines the pulse shape and m* spans all elements where the probe pulse in non-zero. The first phase term represents the phase delay for each scatterer, whereas the second, exp $(j\omega\tau_i/K_g)$, is a correction term that accounts for the differences between group velocity and phase velocity.

6.2 AMPLITUDE-BASED COHERENT RAYLEIGH BACKSCATTER DVS

The simplest approach to DVS using coherent Rayleigh backscatter is to use a highly coherent source in an otherwise conventional single-mode OTDR arrangement (Figure 6.2). In this case, the signal that is detected is proportional to the backscatter power returned from each point along the fibre. This approach is sometimes referred to as 'phase-OTDR' or 'Φ-OTDR', which is unfortunate in view of the fact that it is power, not phase, that is measured, and this nomenclature causes confusion with the differential phase OTDR methods that are described later in this chapter (Section 6.3).

The summation in Equation 6.5 is extremely sensitive to disturbances, primarily strain and temperature, that alter the phase relationships between the individual contributions of the scatterers within a resolution cell. The coherent-source OTDR measures intensity only and so ignores the exp $(i\Phi(z))$ term in Equation 6.5; however, it does measure the changes of $A_e(z)$ that result from measurand-related changes in the relative optical distances between scatterers. The amplitude measurement is thus highly localised to where the probe pulse is when the scattering occurs.

An approximate measure of the sensitivity may be gained by considering the relative phase of a scatterer near the beginning of a resolution cell and one near its remote end. If the disturbance alters their relative phase by π, this will lead to a major change in the electric field summation. In Figure 6.3, this is illustrated by splitting the resolution cell into two zones, A and B, deemed each to possess amplitude and phase attributes.* In the case of a strain, the requirement is then to elongate the resolution cell by

$$\delta L = \frac{\lambda}{4n_i\xi} \tag{6.6}$$

in order to effect an additional phase delay of π between Zone A and Zone B. ξ is a correction to the optical path length-change that accounts for the strain-optical effect, i.e. the fact that straining the fibre not only increases its length but also changes its refractive index. $\xi \simeq 0.79$ at wavelengths typically used in DVS [14], and so the strain required for this major change in intensity is about 366 nm at an operating wavelength of 1550 nm. The length change applies over the resolution cell, and so for a pulse duration of 100 ns, i.e. a resolution of 10 m, this corresponds to a strain of 36 nε. In practice, far smaller strains are detectable.

Although the speckle-like behaviour of backscatter intensity in single-mode fibres was known for some time, it was Taylor and Lee [30] who proposed, in a patent application filed in 1991, its use as an intrusion detection system. A little later, Juskaitis et al. [31] demonstrated the concept more or less as described here.

A key issue in amplitude or intensity-based systems is linearity: if the disturbance really was applied at a single point as indicated in Figure 6.3, then its effect would be purely to retard the phase of the light returning from Zone B relative to the phase of the light returned by Zone A, and this would be a linear function of the elongation applied by the disturbance. In practice of course, the disturbance is distributed over a finite length of the sensing fibre.

* In practice, of course, the disturbance is usually distributed along the resolution cell, but Figure 6.3 still illustrates the physical phenomenon.

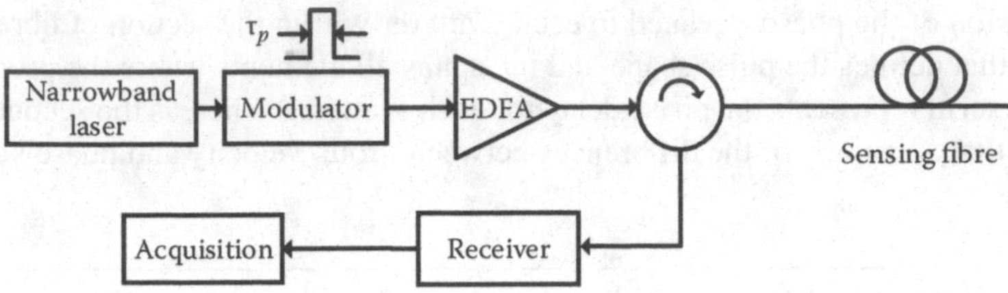

FIGURE 6.2 Schematic diagram of a direct-detection intensity-measuring coherent OTDR.

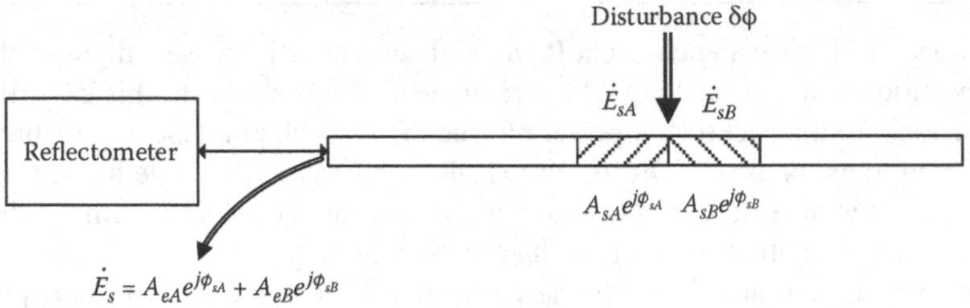

FIGURE 6.3 Illustration of the interference between backscatter signals returning from positions upstream (left) or downstream (right) of a localised disturbance. (Adapted from Juskaitis, R., Mamedov, A.M., Potapov, and V.T., Shatalin. S.V., *Opt. Lett.*, 19, 225, 1994.)

A distributed perturbation affecting an entire resolution cell does not modulate the intensity linearly over any useful range of strain, nor indeed is the transfer function predictable. This is illustrated in Figure 6.4, where the amplitude of the backscatter signal is shown as a composite greyscale image, with distance along the fibre plotted on the vertical axis (m). Along the horizontal axis, short sequences, or slices (corresponding to durations of ~40 ms), have been juxtaposed for 16 different optical carrier frequencies. The variation, for each frequency slice, of the mean intensity along the fibre distance axis is caused by the natural variability of the coherent OTDR signal intensity. In the region from 40 to 80 m from the start of the fibre, a piezo-electric transducer was used to stretch the fibre periodically at a frequency of 105 Hz, leading to about four periods in each frequency slice. The data for all frequencies were acquired simultaneously but processed independently so the strain for each slice is exactly the same in all cases. In this case, the pulse duration was set to ~30 ns and so the pulses occupied about 3 m of fibre at any one time.

The strain applied over a length of fibre corresponding to one pulse duration is about 320 nm, which is equivalent to about 2.7 radian of phase change along that fibre length. Although the strain is applied nominally uniformly along the 40 m that is stretched periodically by the transducer, it can be seen that the response is quite different at different locations along the fibre. This means that the transfer function is completely different in each section of fibre.

This finding also applies at a single location when comparing the response observed for the different frequencies. Figure 6.5 shows a pair of plots taken from the data of Figure 6.4 at locations 50 m (top, solid line) and 70 m (displaced down by four divisions and shown as a dotted line). Several features are immediately apparent. Firstly, the response is different for each frequency and each location. Secondly, the response is usually non-linear, with the signal appearing to be wrapped at one part or other of the waveform. Thirdly, even when the measured signal is roughly sinusoidal, indicating a response that is locally reasonably linear, the peak-to-peak variations in signal level vary widely. Compare, for example, the 1st, 7th and 12th signals in the lower trace; all show a sinusoidal response at the correct frequency, but they vary by a factor of 6 in the magnitude of the measured signal even though the strain applied to the fibre is the same in all cases.

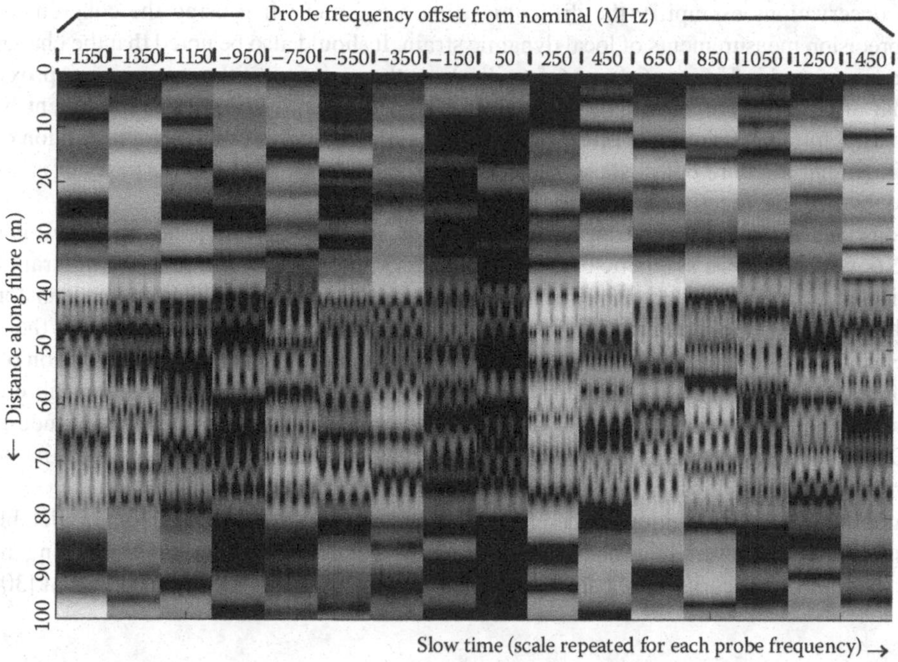

FIGURE 6.4 Greyscale plot of backscatter amplitude with a sinusoidal modulation of the fibre strain over the region delineated by 40 to 80; vertical axis: fast time (distance along the fibre); horizontal axis: slow time. Each of the vertical stripes shows data acquired at a separate optical frequency but at the same time as all other stripes. (Data acquired by Florian Englich and the author at Schlumberger.)

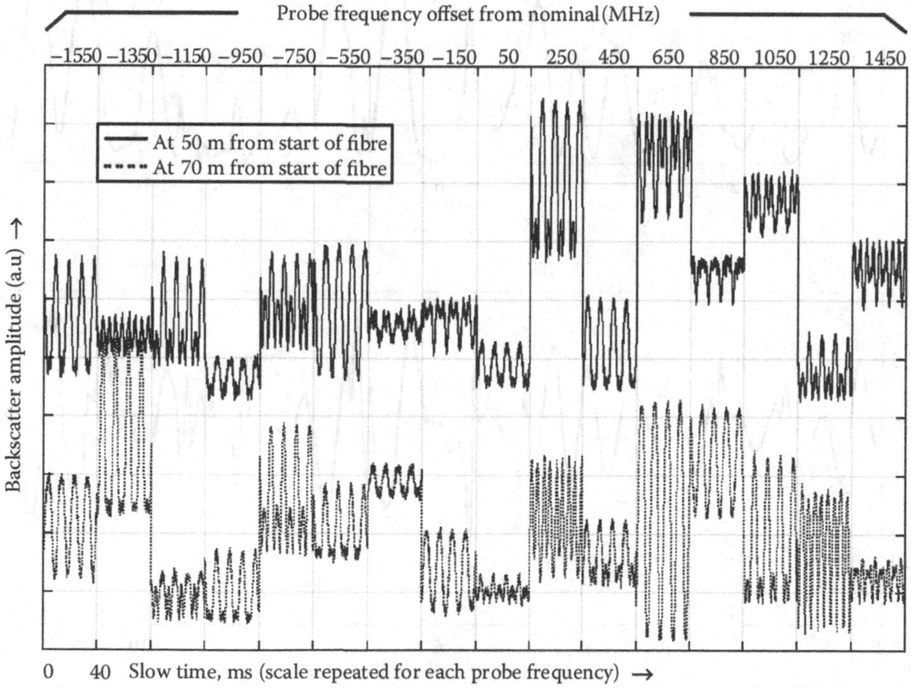

FIGURE 6.5 Slice through two levels (50 m, upper curve; and 70 m, lower, dotted, curve) of Figure 6.4 showing time dependence of backscatter power at various optical frequencies.

These observations exemplify the difficulties that are found when using the coherent backscatter power for precision measurements of local dynamic strain. It should also be noted that the change in probe carrier frequency used in Figures 6.4 and 6.5 to illustrate the variation of the response is a proxy for other changes that affect the fibre, such as temperature or slow strain variations: if the coherent backscatter intensity is used for a precision dynamic strain measurement, not only is the transfer function different at each location along the fibre but it is also subject to drift, and so for practical purposes, it is not possible to calibrate the transfer function in advance of installation.

The non-linearity of the strain transfer function of an amplitude-based coherent OTDR is illustrated in Figure 6.6, which shows a synthetic data set of the backscatter power as a function of strain applied to a 2-m section of fibre (a). The derivative (b) is the local transfer function. Although this transformation is trivial, it illustrates the implications of strain dependence of the backscatter power: clearly, the function is monotonic over ranges of not more than a few hundred nε, but more importantly, each section of fibre will have a different transfer function. Furthermore, the transfer function of each section of fibre also drifts with ambient conditions (temperature, for example). The transfer function of an amplitude-measuring coherent OTDR is therefore utterly unpredictable. However, we shall see in Section 6.5.5 that techniques exist for calibrating or working around these limitations.

The amplitude DVS technique is therefore ill-suited for applications where a predictable transfer function is required. However, it has been used successfully for detecting and even classifying disturbance events, for example, in security applications, and this was the intent of the initial proposal [30]. Most of

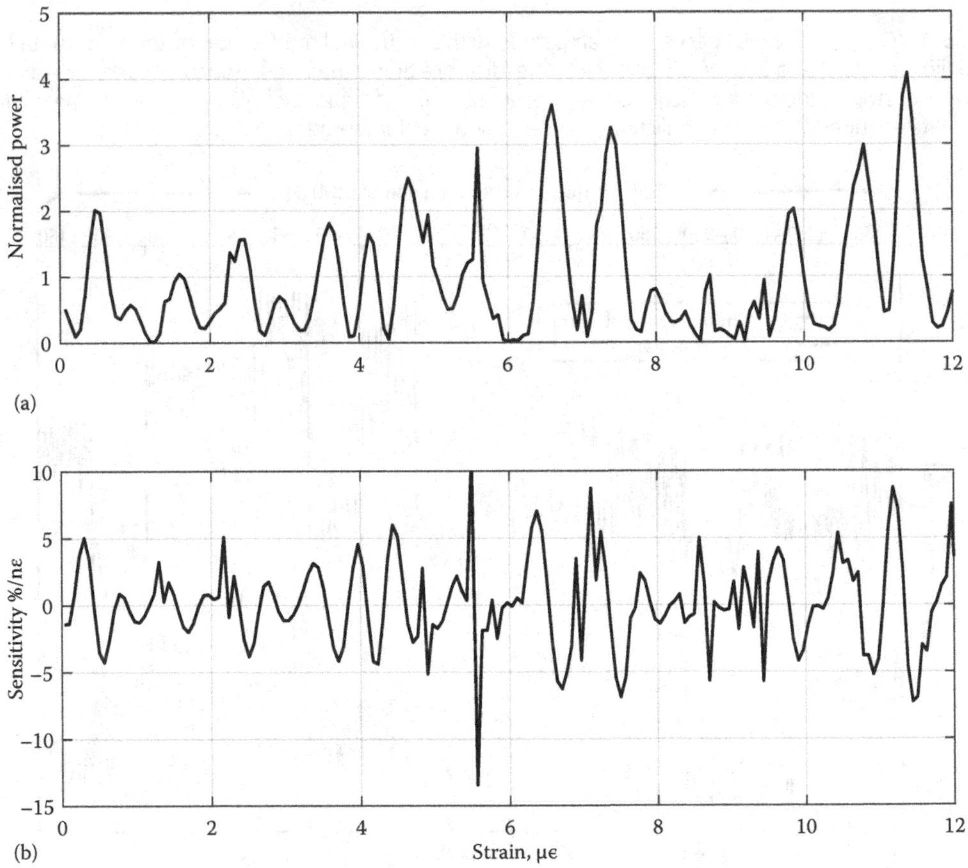

FIGURE 6.6 Example of the strain dependence of the backscatter power (a) and the corresponding sensitivity power with respect to (w.r.t.) strain (b).

the interpretation techniques for amplitude-based DVS use comparisons from pulse to pulse of the measured signal at each location. Although this measurement is totally uncalibrated, it is at least an indication of a disturbance, and it turns out that a surprising level of information can be extracted, including event detection (i.e. confirming within acceptable confidence limits that a disturbance has occurred at a particular location) and even classifying that event, i.e. assigning it to one of several types of disturbance that have been catalogued in the software.

Techniques described in other parts of this book for extending the range have been employed, such as a remotely powered amplification with doped-fibre amplifiers [33], distributed Raman amplification [34] and a combination of Brillouin and Raman amplification [35], in the latter case resulting in a range of 175 km but, of course, a rather limited sampling rate owing to the very long time taken for pulses to travel to the remote end of the fibre and back. Most of the papers on this topic use direct detection, i.e. an arrangement similar to Figure 6.2 with, as appropriate, the addition of components required for providing optical gain in the sensing fibre. However, some publications, e.g. Ref. [36], have used coherent detection, thereby adopting essentially the arrangements discussed in the context of long-range OTDR in Chapter 3, i.e. Figure 3.14.

The main difference between the needs of OTDR and distributed vibration sensing is, of course, that in DVS, the backscatter data from each probe pulse must be acquired and processed location by location as a function of pulse number. In contrast, in OTDR, the backscatter waveforms (after an envelope detection operation in the case of coherent detection OTDR) can be averaged at each location for all shots, and this reduces the volume of data to be stored and processed in OTDR. The data volumes for DVS are typically three to four orders of magnitude higher than in coherent OTDR.

6.3 DIFFERENTIAL PHASE-MEASURING DVS (dΦ-DVS)

If the strain is applied to an infinitesimally small region of the resolution cell, but not elsewhere, the scattered signals from locations upstream and downstream of the strain location are unaffected, other than a delay induced by the strain to signals returning from all downstream locations. This arrangement results in a linear response in the phase of the backscatter signal because the phase delay induced by the strain is indeed linear. This was confirmed experimentally in Ref. [31], which showed a response replicating the input to a piezo-electric transducer with a sinusoidal input. In this case (Figure 6.3), backscatter from the first half of the region occupied by the probe pulse is interfering with that from the second half, and the variation in the relative phase of these backscatter signals provides a signal that follows the strain applied between the two regions. This is the principle applied in phase-sensitive systems, with however, the difference that in more recent systems, the two regions from which the backscatter is interfering are spatially separated, and generally, the contributions in the signals that are acquired from each of the two regions are distinguishable in some way. The specific way in which these contributions are separated is one of the main differences between the different techniques that have been discussed in the literature and that are described later in this section.

Although in the example illustrated in Figure 6.3, a fibre region much smaller than the probe pulse duration is influenced by the measurand, in practice, the interest in distributed systems is primarily where the measurand itself can potentially affect the entire sensing fibre. As we have seen, in this case, the amplitude and – importantly – its derivative w.r.t. the measurand (i.e. the slope of the sensor transfer function) are both random, non-monotonic functions of the measurand.

Just as the amplitude of the backscattered light is determined by the summation of the electric fields re-radiated by the scatterers within the section of the fibre occupied by the probe pulse, so too is its phase. Phase-sensitive methods compare the phase of the backscatter signal between close, but not necessarily adjacent, fibre sections. For that reason, these methods are referred to in this book as differential-phase DVS (dΦ-DVS).

The fibre sections that scatter the probe light on either side of any one location in the fibre may be visualised as reflectors, and their separation, as the path-length imbalance in an interferometer: this arrangement has similarities to a low-finesse Fabry-Pérot interferometer with, however, distributed mirrors and where the mirror separation is the gauge length. The reflector sections may also be thought of as fibre Bragg gratings; these gratings are, however, very weak and formed randomly by the naturally occurring fluctuations of the refractive index that are the basis of Rayleigh scattering.

In some of the implementations, there is no direct, actual interference between the light returning from the reflectors straddling the gauge length; rather, the phase comparison is carried out in the signal processing in these implementations.

6.3.1 Techniques for dΦ-DVS

Systems for dΦ-DVS measurement may be separated (Figure 6.7) into, on the one hand, direct detection techniques, where the phase comparison takes place between two backscatter signals and is conducted in the optical domain and, on the other hand, coherent detection approaches, where the backscatter signal is first mixed with a local oscillator (LO) leading to an electrical signal or signals from which the phase is processed. A further division exists in the direct detection category between systems that create distinguishable backscatter signals resulting from multiple probe pulses that are launched with a slight time separation, the results being mixed on reception (so called 'dual-pulse' systems) and those that process a single backscatter signal to generate copies of the signal with a relative delay (interferometric recovery technique). Amongst coherent detection techniques, the optical local oscillator (OLO) can be at the same frequency as the backscatter signal (homodyne detection) or a different frequency (heterodyne). These approaches will now be reviewed in turn.

In all cases, these approaches determine a change in the length of the fibre between two locations, or scattering regions. This determination is repeated for each resolvable point along the fibre, resulting in a measure of the dynamic strain as a function of time and of location along the fibre. The section of fibre between the centre points of the scattering regions is known as the *gauge length*, L_G (or sometimes the *differentiating interval*).

6.3.2 Phase Unwrapping

The phase-sensitive DVS measurements, being interferometric, are ambiguous in that the response to the measurand is periodic. Thus, from any reading, there is a near-infinite number of values of the measurand that could have resulted in that reading. For small dynamic signals, this is not particularly important. However, when the measurand causes the measured phase to change by more than 2π, the signal returns to a value in the interval $[-\pi, \pi]$ even though the underlying input has continued to move outside that range.

The underlying measurand is proportional to the *principal phase*, whereas what is measured is the *wrapped* phase. The inverse process required to go from wrapped phase to principal phase is known as *phase unwrapping*. With one-dimensional data, such as a time series, the phase unwrapping process is

	Differential phase measurement			
What is measured:	Direct detection		Coherent detection	
Phase of a beat frequency	Dual pulse		Heterodyne	
Quasi-static phase using multiple inputs		Receiving interferometer		Homodyne
Signal detected:	$P_{BS}(z) = E_0\eta(z)\exp(-2\alpha z)$		$\sqrt{P_{BS}(z)\,P_{LO}} = \sqrt{P_{BS}(0)\,P_{LO}}\,\exp(-\alpha z)$	
Differentiation:	Optical domain		Electrical or software domain	

FIGURE 6.7 Summary of the options for dΦ-DVS.

straightforward provided that the signal is sampled sufficiently frequently. Phase unwrapping involves detecting a phase jump and determining whether this is a natural variation of the signal or whether the signal has exceeded the boundaries of the interval $[-\pi, \pi]$, in which case the phase should be corrected by adding or subtracting 2π to the result.

The phase unwrapping process has been formalised elegantly by Itoh [37], who showed that it can be reduced to the following sequence of operations:

1. Forming the successive differences between adjacent samples.
2. Wrapping resulting differences, i.e. replacing the result with its value modulo 2π so that it lies in the interval $[-\pi, \pi]$.
3. Integrating the resulting wrapped sequence of differences.

For one-dimensional data, the Itoh procedure is reliable provided that no two adjacent samples of the principal phase differ by more than π. If this condition is not met, the phase cannot be unwrapped reliably without further information, such as some knowledge of the bandwidth of the underlying principal value.

The unwrapping problem imposes a restriction on the dynamic range of the measurable signal that is inversely proportional to the frequency of the signal. It can be overcome by increasing the sampling rate (probe pulse repetition frequency [PRF]), but ultimately, the PRF is limited by the length of the sensing fibre and the need to ensure that only one probe pulse travels in the fibre at any one time.

The ability to unwrap the data is also limited by noise: the π limit includes any noise that is added to the signal, and so in adverse signal-to-noise ratio conditions, the dynamic range can be severely limited.

The unwrapped phase is a measure of the strain across the gauge length. Some system providers do not unwrap the phase: they simply provide successive differences between probe pulses in the phase differentiated across the gauge length. In this case, they are providing the time derivative of strain, i.e. the strain rate. Although these approaches are nominally equivalent in that one set of data can be derived from another, they do result in different noise properties since the differentiation process tends to enhance the high-frequency noise content. The reasons for some workers in the field choosing one approach and others selecting the alternative approach are not clear; they may simply be a matter of design convenience.

6.3.3 Dual-Pulse DVS

In the dual-pulse DVS (Figure 6.8, upper drawing), a pair of pulses of slightly different frequency is launched with a time separation determined by the required gauge length. This configuration was initially proposed by Dakin and Lamb [29] and may be seen as an extension of the time-division multiplexing of reflective arrays, proposed by Dakin and Wade [38]. The arrangement is similar to a conventional OTDR except that, rather than one pulse of duration τ_p, two such pulses of optical frequency, f_1 and f_2, separated by a time $\Delta T_G = L_G/(2 \cdot c \cdot N_g)$ are launched. As the two probe pulses travel along the sensing fibre, each pulse scatters light at its own carrier frequency. When the backscattered light is collected on its return, it is directed to a photodetector that responds to the square of the electric field. The signal consists of backscatter from each of the probe pulses that is converted to electric current, and they appear at low frequency in the spectrum of the detector output. However, an additional term that results from the light scattered by one pulse mixing with light scattered by the other pulse also appears in the spectrum of the detector output.

Because the pulses are at separate locations when the scattering occurs (regions z_1 and z_2 in the figure), this mixing term is explicitly related to the region between the reflective zones and its phase is a measure of the distance between these reflectors. Of course, because the measurement is of the phase, its response is periodic in strain and so is ambiguous. Nonetheless, using unwrapping techniques, changes in fibre strain can be tracked over signal ranges provided that the signal is sampled sufficiently frequently.

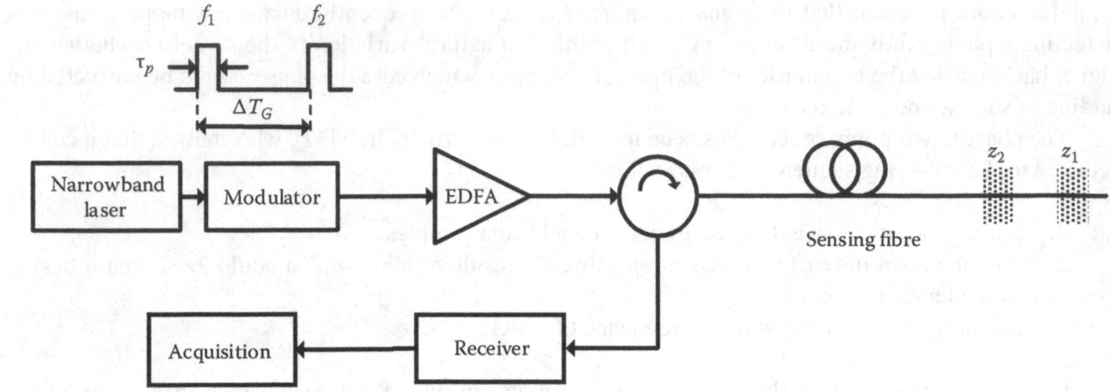

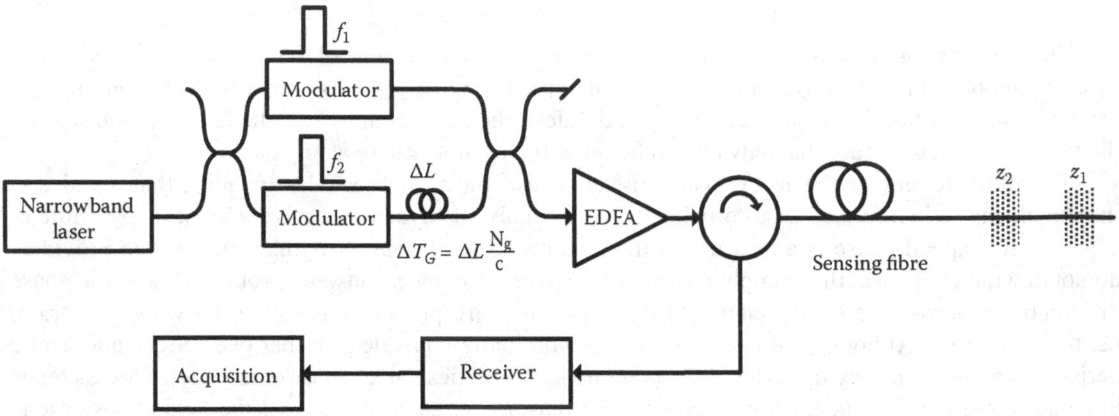

FIGURE 6.8 Two optical arrangements of a dual-pulse phase-measuring coherent OTDRs. (Adapted from Dakin, J.P., Lamb., C. Distributed fibre optic sensor system, GB2222247A, 1990; and Crickmore, R.I., and Hill, D.J., Traffic sensing and monitoring apparatus, US7652245B2, 2010.)

Each of the signals can be modelled according to Equation 6.2, and so the signal current produced at the detector is given by

$$I_{z_1,z_2}(t) = \left|\dot{E}_{s1}\right|^2 + \left|\dot{E}_{s2}\right|^2 + 2\left|\dot{E}_{s1}\right| \cdot \left|\dot{E}_{s2}\right|$$

$$\exp\left[j\left(2\pi\left[\left|f_1 - f_2\right| t + \frac{2n \cdot \xi \cdot \delta l(t)}{\lambda}\right] + \phi_1 - \phi_2\right)\right]. \tag{6.7}$$

So the backscatter signal intensity is a sum of the backscatter at each of the probe frequencies plus a mixing term that has an oscillatory behaviour as a function of the distance between the locations z_1 and z_2 of the two probe pulses. The final term, which can be selected from the backscatter signal owing to its distinct frequency $|f_1 - f_2|$, carries the information of interest for DVS, namely that a change $\delta l(t)$ of $z_1 - z_2$, on a scale comparable to the optical wavelength influences its phase in a similar way to that of a Fabry-Pérot interferometer.

An alternative arrangement, shown in the lower part of Figure 6.8, uses two separate modulators to generate the pulse pair and was originally applied to time-division-multiplexed interferometric arrays [39] but has been applied to the fully distributed vibration sensing technology as well [40]. The output of the source is split into two paths, each of which encounters an acousto-optic modulator (AOM) that creates a pulse at a specific frequency shift in each of the paths. In one of these paths, a delay line then adjusts the timing to provide the required delay between the pulses. This arrangement, although a little more complex than that shown in the upper part of the figure, has the benefit of eliminating the effects of any fluctuation that might occur in the frequency of the laser in the time between the generation of the first and second pulses. This removes one possible source of noise, although variations in the laser frequency in the slow time domain will still result in a noise contribution to the sensor output.

Figure 6.9 shows the typical appearance of a dual-pulse DVS raw signal, in this case, the channel (sample point) separation is 0.34 m as a result of a sampling rate of 300 megasamples per second (MSPS). A modulation is apparent at the intermediate frequency (IF) $\Delta f = f_1 - f_2$ (60 MHz in this example) and this modulation is superimposed on the longer timescale variations of backscatter intensity that relate to the coherent summation of electric fields for each of the probe pulses. After bandpass filtering to select a frequency region around Δf, the signal (Figure 6.10) takes the appearance of a quasi-sinusoidal waveform that is amplitude-modulated by the fading process. It is the phase of this waveform that conveys the vibration information. In Figure 6.10, the difference between two such traces is shown (dot-dash line) when a modulation is applied at around location 230 m. In this example, the pulse durations were 180 ns, and they were separated by 200 ns, thus each occupying 18 m of fibre and separated by 20 m. Other than the contribution of noise, the signals are identical except for the region around 200–260 m, where a disturbance has been applied during the acquisition. The figure illustrates the fact that the modulation induced by a vibration stimulus is truly localised to where the fibre has been disturbed – upstream and, more importantly, downstream of this disturbance, neither the amplitude nor the phase of the IF is affected.

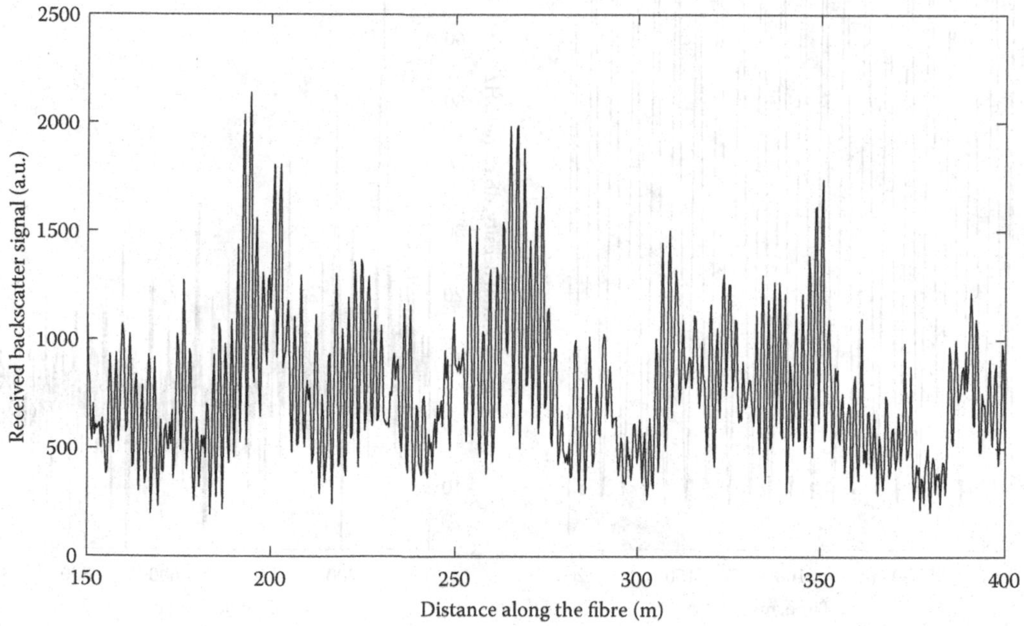

FIGURE 6.9 Typical raw signal acquired with the dual-pulse technique. Data acquired by the author at Schlumberger.

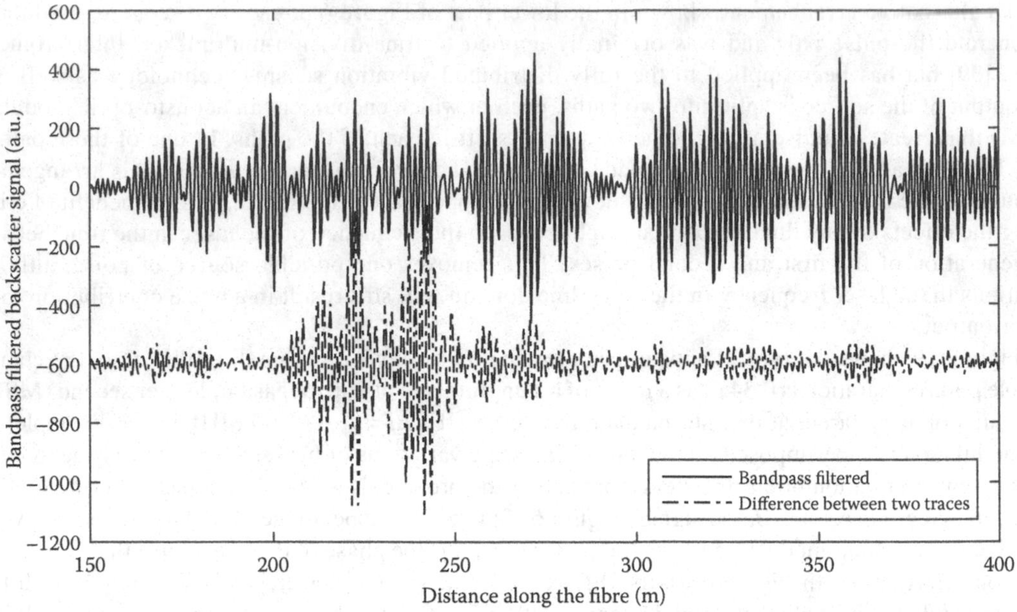

FIGURE 6.10 Typical dual-pulse signal, after bandpass filtering (upper curve), and difference between successive filtered traces (lower curve) from the same data set as is shown in Figure 6.9.

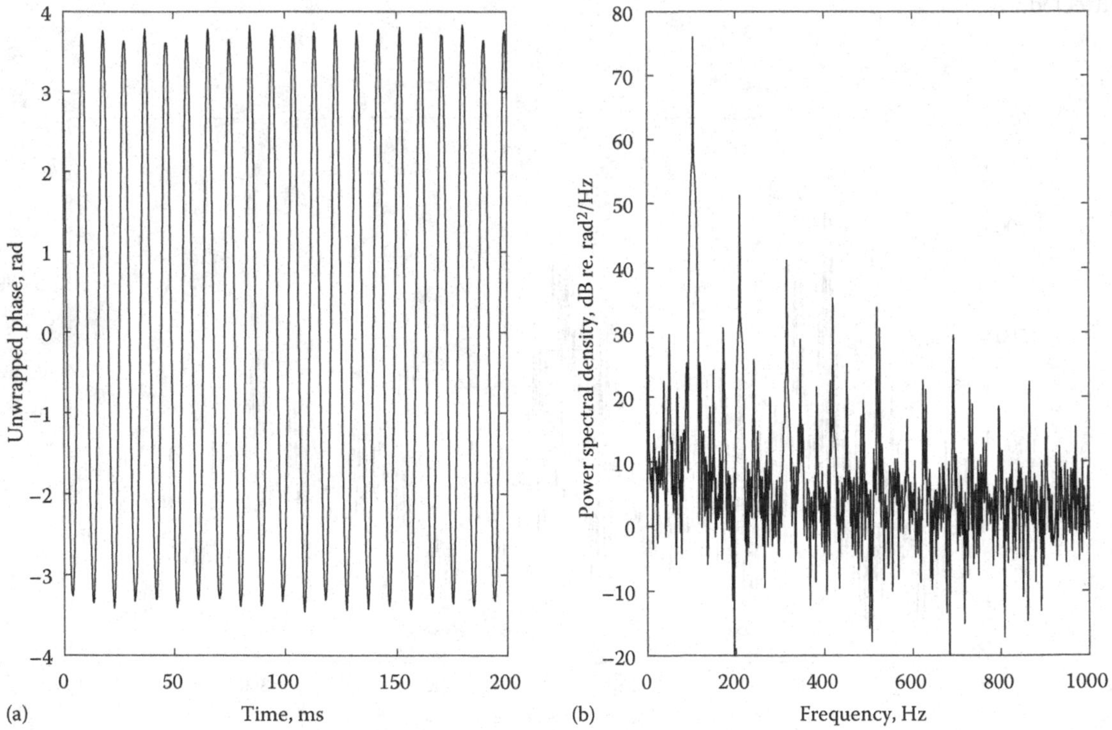

FIGURE 6.11 Phase extracted and unwrapped from the same data set as Figure 6.10 (a); its frequency spectrum is shown in (b).

Figure 6.11 shows the slow-time variation of the phase extracted from the channels around 230 m and unwrapped; this demonstrates a reasonably linear response to the dynamic strain, as confirmed by the power spectral density plot in the right panel of Figure 6.11. In this instance, the sinusoidal modulation that was applied spans about 7 radian; i.e. more than one cycle and phase unwrapping is essential to reproduce the signal faithfully. However, larger signals (by an estimated factor of 3–4) could have been accommodated before the unwrapping algorithm would have failed.

A variant of the dual-pulse technique is described by Alekseev et al. [41]. Here, rather than using a pair of separate pulses, a single pulse is launched into the sensing fibre and collected onto a direct-detection received, similarly to the amplitude-based coherent Rayleigh DVS. The crucial difference compared with the amplitude technique, however, is that a phase modulator is added after the intensity modulator, and it is used to shift the phase of the second half of the probe pulse relative to that of the first half. The arrangement is illustrated in Figure 6.12 together with a diagram of the phase relationships.

In this particular example, the authors chose to shift the phase of the second half in increments of 2/3π, so that a set of three successive probe pulses allows the differential phase between the two sub-pulses to be calculated. If we consider the two halves as separate probe pulses, this technique is similar to the dual-pulse method but differs in the facts that a phase modulation rather than a frequency modulation is applied and that the gauge length is fixed at half the distance occupied by the pulse in the fibre.

The fact that the gauge length is tied to the pulse duration is a drawback of this precise arrangement, but this can be addressed simply by pulsing the intensity modulator twice at times separated $\Delta T_G = L_G/(2 \cdot c \cdot N_g)$. This is what the same team did in Ref. [42], which describes a dual-pulse dΦ-DVS technique where the probe pulses are distinguished by their phase rather than their frequency.

More importantly, the differential phase is measured across the gauge length over a time corresponding to three pulse repetition intervals. The measurement relies on the phase difference (i.e. the dynamic strain that is measured in DVS) not changing substantially over this time period. Whereas this limitation is not serious in some applications, it could be limiting in others.

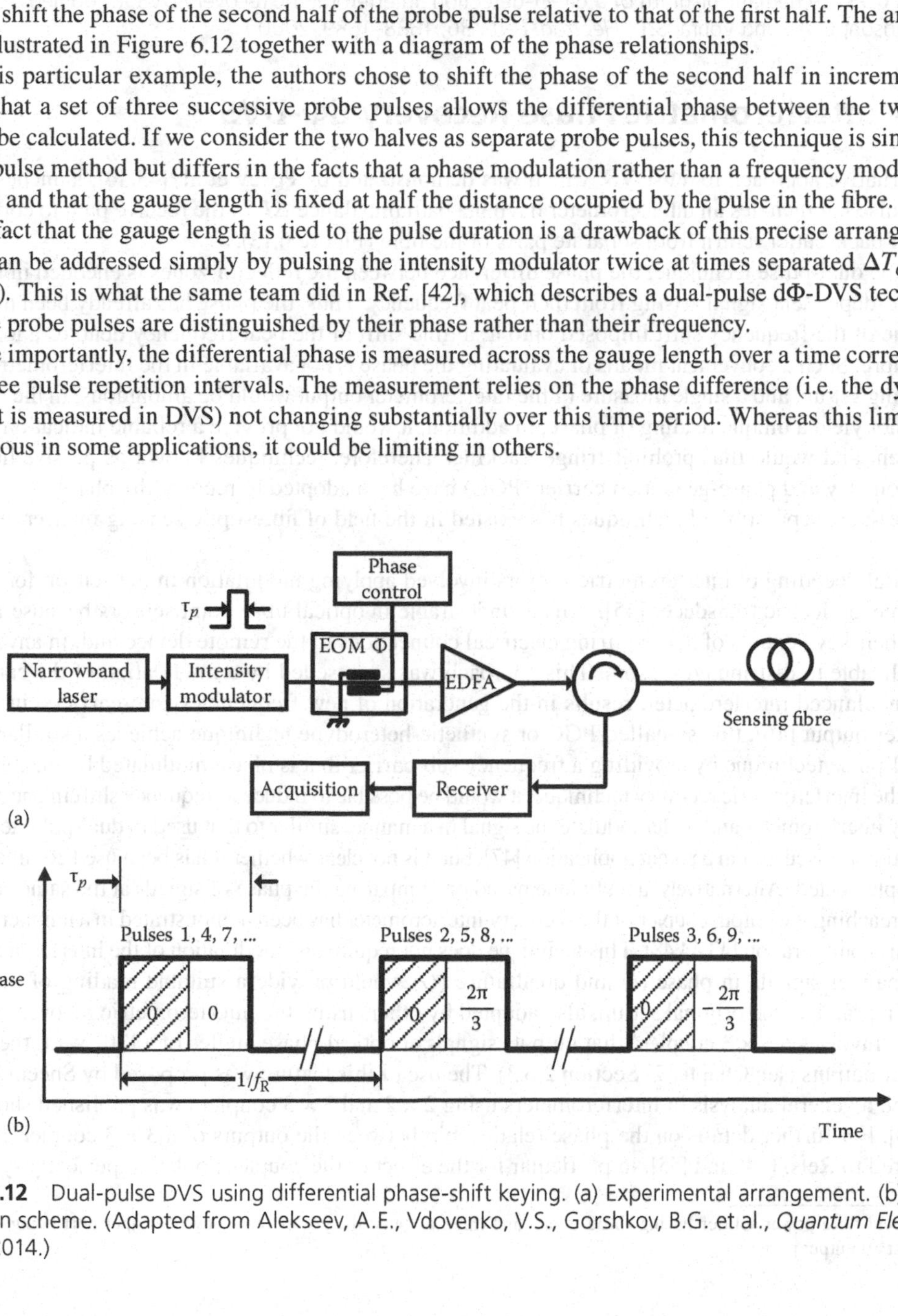

FIGURE 6.12 Dual-pulse DVS using differential phase-shift keying. (a) Experimental arrangement. (b) Phase modulation scheme. (Adapted from Alekseev, A.E., Vdovenko, V.S., Gorshkov, B.G. et al., *Quantum Electron.*, 44, 965, 2014.)

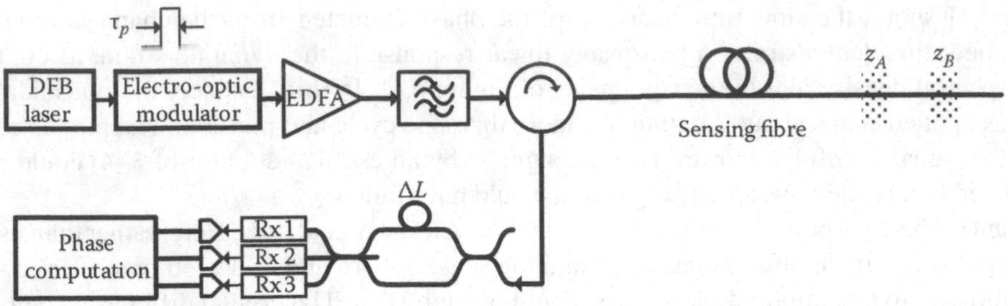

FIGURE 6.13 Schematic diagram of a direct-detection, interferometric recovery DVS. (Modified from Posey, R.J., Johnson, G.A., and Vohra, S.T., *Electron. Lett.*, 36, 1688–1689, 2000.)

6.3.4 Interferometric Phase Recovery dΦ-DVS

An alternative approach to dΦ-DVS, which was demonstrated by Posey et al. [43,44], launches a single probe pulse but includes an interferometer having a path imbalance ΔL in the receive path to compare the phase of backscatter return from separate parts of the fibre (Figure 6.13).

In the dual-pulse technique, the phase difference between the reflector zones is encoded in the phase of a time-dependent signal arising from their beat frequency. Thus, the phase has already been modulated by virtue of the frequency shift imposed onto it; a time shift of the beat frequency denotes a local strain on the fibre. Such a convenient means of evaluating the phase is not available in the interferometric phase-measuring set-up, and a single measure of the interferometer output would be ambiguous, in the sense that it does not yield a unique reading of phase; in addition, it would not provide a reliable indication of fringe movement and would thus prohibit fringe tracking. Therefore, techniques known as passive homodyne interferometry and phase-generated carrier (PGC) have been adopted to recover the phase.

The same separation of techniques has existed in the field of fibre-optic sensing interferometers for decades.

Initial decoding of interferometric sensors involved applying modulation in the sensor, for example, with a piezo-electric transducer [45]. This is undesirable in optical fibre point sensors because it negates one of their key benefits of not requiring electrical connections to the remote device and, in any case, it is not applicable to distributed sensors. This approach was superseded by a modulation of the carrier that, in an unbalanced interferometer, results in the generation of new frequencies (sub-carriers) in the interferometer output [46]; this so-called PGC or synthetic-heterodyne technique achieves a similar result to the dual-pulse technique by providing a frequency sub-carrier that is phase modulated by the measurand.

In the interferometric recovery technique, it would be possible to induce a frequency shift in one arm of the recovery interferometer and so demodulate the signal in a manner similar to that used in dual-pulse techniques. This option is discussed in a patent application [47], but it is not clear whether it has been used commercially or even implemented. Alternatively, a technique based on comparing the phase of signals at the same carrier frequency reaching the output coupler of the recovery interferometer has been demonstrated in a number of papers using this configuration [43,44,48]. This technique does not require any modulation of the interferometer at all.

A pair of signals in phase (I) and quadrature (Q) would provide a suitable reading of the relative phase. In practice, the original set-up, also adopted by others using the interferometric recovery approach [48–50], involves a 3 × 3 coupler* that outputs signals at optical phase angles of ±120° w.r.t. the phase of the other outputs (see Chapter 2, Section 2.3.3). The use of this feature was proposed by Sheem [51,52] in 1980 and a general analysis of interferometers using 2 × 2 and 3 × 3 couplers was published shortly after that [53]. For further details on the phase relationship between the outputs of a 3 × 3 coupler, the reader is referred to Refs. [54] and [55], in particular for the effect of the coupler not being perfectly symmetric.

* An optical fibre power-splitter having three inputs and three outputs, the power from any one input port being split evenly amongst all output ports.

If signals of equal amplitude arrive at two of the input ports of a 3 × 3 coupler, interference occurs within the device and the power emerging from each of the output ports depends on the phase relationship of the two inputs. As the phase-difference between the inputs varies, the power cycles around the three outputs; the relative power of the three output ports is therefore a measure of the relative phase of the inputs.

It turns out that when the splitting ratio is not exactly even between the ports, the phase shift between the ports is also not equally shifted around the unity circle. For the purposes of recovering the phase in an interferometer, however, quite large departures from even splitting can be accommodated; they can be measured and taken into account in the phase extraction algorithm.

Although these signals are not in quadrature, they do provide the information required. In principle, two outputs are sufficient to extract the phase [56]; however, using all three outputs allows the raw signals to be scaled precisely, and this approach is generally more robust.

The relationship between a phase signal (sinusoidal with range ±16 radian, in this case) and the three receiver outputs is illustrated in Figure 6.14: as can be seen, the frequency of the receiver outputs varies in proportion to the rate of change of the phase signal, but the three outputs retain a fixed phase separation that allows the wrapped phase of the input signal to be recovered.

The calculated phase is derived from three separate detector and acquisition channels; the gains and offsets of each of these must be known precisely. In the 3 × 3 case, this can be achieved, for the offset, by measuring the voltage on each channel in the absence of illumination. As to the gains, the signals obtained from a long fibre can be assumed to be equal on average and thus each channel can be normalised by the signal averaged over the entire fibre, or at least a sufficient long fibre section for robust statistics. Alternatively, illuminating the interferometer with broadband light that is expected to provide a signal of equal strength to each of three outputs can be used to calibrate their relative gain [49]. Importantly, once this calibration step is complete, the method is robust to variations in intensity (which are inevitable with coherent OTDR methods) because the sum of the three outputs should be constant. Refs. [57,58] demonstrate one approach to

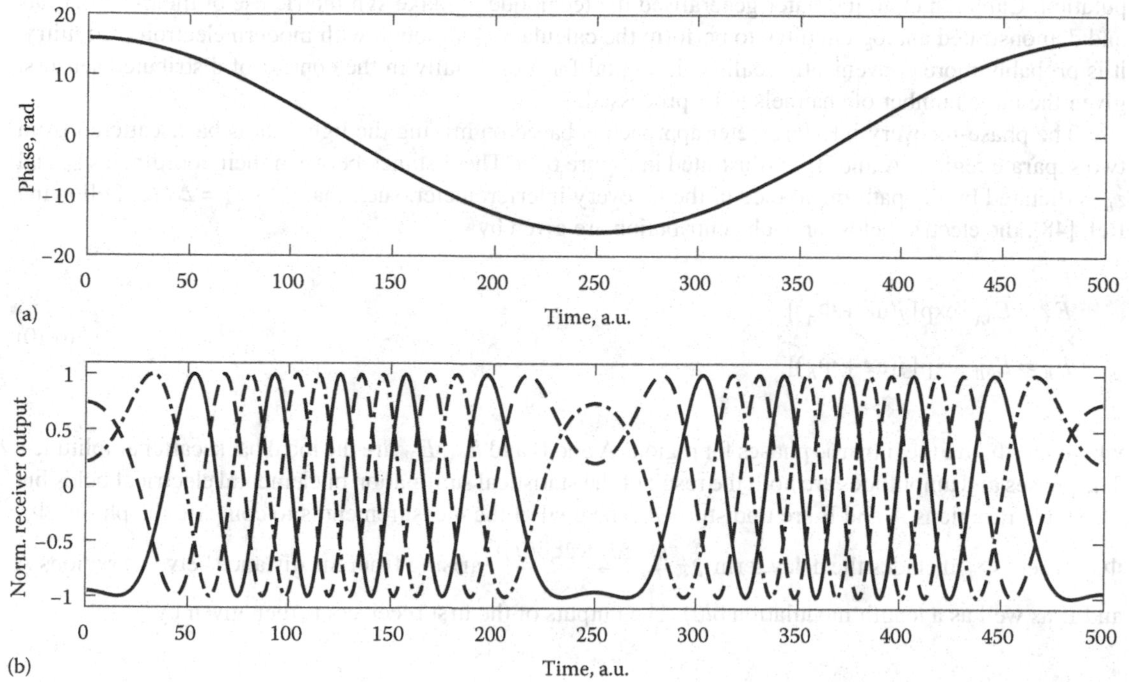

FIGURE 6.14 Slow-time synthesised signals for each of the outputs of the receivers in the setup of Figure 6.13. (a) Presumed phase excursion across the gauge length at the point of interest. (b) Corresponding response of the receivers after DC subtraction and normalisation.

extracting the in-phase (I) and quadrature (Q) components from the three 120° shifted signals, namely given the three signals, where the phase f is the quantity to be determined, and K is a constant

$$S_A = M[1+K\sin(\phi)],\ S_b = M\left[1+K\sin\left(\phi-\frac{2}{3}\pi\right)\right],\ S_C = M\left[1+K\sin\left(\phi+\frac{2}{3}\pi\right)\right]. \tag{6.8}$$

I and Q signals may be derived from S_{A-C} as follows:

$$S_I = -\frac{1}{2}S_A + S_b - \frac{1}{2}S_c,\ S_Q = -\frac{\sqrt{3}}{2}S_A + \frac{\sqrt{3}}{2}S_c. \tag{6.9}$$

S_I and S_Q can then be used to calculate the phase from the four-quadrant arc tangent of their ratio and then unwrapping the result. Although this technique does not make symmetric use of the available data, the process can be repeated by shifting the assignment of the labels to the coupler outputs (i.e. A, B, C becomes B, C, A etc.) and combining the results of all possible combinations to remove possible bias effects from asymmetry in the coupler or differences in the properties of the detection and acquisition channels. Equation 6.9 can be generalised to allow for component imperfections resulting in departure from the ideal 120° phase shift between coupler outputs [56].

An alternative approach is available for demodulating the phase of light emerging from a 3 × 3 coupler, known as 'differentiate and cross-multiply' (DCM) [59], which involves multiplying each component by a time derivative of another, subtracting those intermediate products and integrating the result. This initial approach [59] involved manipulating the outputs to provide a pair of I, Q inputs to the DCM computation. Cameron et al. [60] later generalised the technique to make symmetric use of the three outputs and demonstrated analog circuitry to perform the calculation, although with modern electronic circuitry, it is probably more conveniently realised in digital form especially in the context of distributed sensors, given the large number of channels to be processed.

The phase-recovery interferometer approach is based on mixing the light that is backscattered from two separate regions A and B, as illustrated in Figure 6.13. The distance between their coordinates z_A and z_B is dictated by the path imbalance in the recovery interferometer, such that $z_B - z_A = \Delta L/2$. Following Ref. [48], the electric fields for each contribution are given by

$$\begin{aligned} \dot{E}_A &= E_{0A}\cdot\exp[j(\omega t+\Phi_A)] \\ \dot{E}_B &= E_{0B}\cdot\exp[j(\omega t+\Phi_B)] \end{aligned}, \tag{6.10}$$

where Φ_A, Φ_B are the intrinsic phases for regions A and B and E_{0A}, E_{0B} are the local backscatter amplitudes. The phases and amplitudes are both the result of the statistical summation of elemental electrical fields but are stable if regions A and B are undisturbed. The goal of the measurement is to compare the phases Φ_A, Φ_{Bd}, where Φ_{Bd} includes the delay term $2\pi\frac{n\cdot(\Delta L+2\xi\delta l(t))}{\lambda}$ arising from the distance between regions A and B as well as a length modulation $\delta l(t)$. The outputs of the first receivers is then given by

$$\begin{aligned} I1(t) &= \left(\dot{E}_A+\dot{E}_{Bd}\right)\left(\dot{E}_A+\dot{E}_{Bd}\right)^* \\ I1(t) &= E_{0A}^2+E_{0B}^2+2E_{0A}E_{0B}\exp(j(\Phi_A-\Phi_{Bd})) \end{aligned}. \tag{6.11}$$

Equation 6.11 ignores splitting losses and constant phase terms (such as that arising from the delay line in the interferometer). The outputs of the three receivers are then given by

$$
\begin{aligned}
I1(t) &= E_{0A}^2 + E_{0B}^2 + 2E_{0A}E_{0B}\exp\left(j\left(\Phi_A - \Phi_B - \frac{4\pi n\cdot\xi\cdot\delta l(t)}{\lambda}\right)\right)\\
I2(t) &= E_{0A}^2 + E_{0B}^2 + 2E_{0A}E_{0B}\exp\left(j\left(\Phi_A - \Phi_B - \frac{4\pi n\cdot\xi\cdot\delta l(t)}{\lambda} + \frac{2}{3}\pi\right)\right).\\
I3(t) &= E_{0A}^2 + E_{0B}^2 + 2E_{0A}E_{0B}\exp\left(j\left(\Phi_A - \Phi_B - \frac{4\pi n\cdot\xi\cdot\delta l(t)}{\lambda} - \frac{2}{3}\pi\right)\right)
\end{aligned}
\tag{6.12}
$$

Equation 6.12 can be solved, as discussed previously, to yield $\delta l(t)$, i.e. the dynamic extension between regions A and B. In practice, the outputs of the three receivers are acquired for the entire length of the fibre and the data are converted to a value of $\delta l(t)$ for each part of the fibre.

The original experimental arrangement as demonstrated in Ref. [44] is shown in Figure 6.13; the approach was also adopted by Masoudi et al. [48]. However, it is also possible to use an unbalanced Michelson configuration, using Faraday rotation mirrors [61] that avoid concerns over the relative alignment of the polarisation state of the interfering light [62]. This arrangement is discussed in Ref. [63] for multiplexed sensor arrays and in Refs. [49] and [50] and [64] in the context of DVS. Based on the patent literature, it is quite possible that this approach is the basis of the acquisition systems commercialised by several suppliers.

The PGC demodulation [65] method is well known in interferometric sensors. It involves adding a phase modulation tone to the interferometer; by carefully selecting the phase modulation index, the in-phase and quadrature components may be found at the fundamental and first harmonic of the modulation tone. PGC has recently been demonstrated in relation to the interferometric phase recovery method for DVS [66]. Here, the recovery interferometer is a simple fibre Michelson arrangement using a 2 × 2 coupler. A phase modulator is added to one arm of the interferometer in order to generate specific tones in the detected signal that carry the phase information. Fang et al. [66] clearly demonstrated the viability of the PGC phase recovery approach over a 10-km sensing fibre length with a spatial resolution of order 6 m together with a phase noise of 3 mrad/Hz½. The authors used a modulation tone of 1 kHz in conjunction with a probe repetition rate of 10 kHz. It was not clear how the signal bandwidth and dynamic range trade off with other system parameters and whether the system could handle signals of similar dynamic range to those measured by some of the other approaches.

Yet another variant of the phase-recovery interferometry approach was disclosed in Refs. [67] and [68]. Here, as in Ref. [44], a single probe pulse is launched,* and the backscattered light is split into two paths and recombined after a relative delay that defines the gauge length. However, in this case, the phase difference between the backscattered signals that have followed the direct and delayed paths is estimated by adding a frequency shift with an AOM in one of the paths, which causes the detector output to include a beat-frequency term. This arrangement requires a single detection channel and the phase is derived from that of the beat frequency signal.

6.3.5 Heterodyne DVS (hDVS)

Heterodyne DVS (hDVS) [69,70] differs from the approaches previously discussed in that the phase comparison is not carried out directly in the optical domain.

Figure 6.15 shows a basic optical arrangement for the hDVS technique. The output of a narrowband laser is split into an OLO path and a pulse forming path. In the latter path, the probe pulse is defined from the

* The same patent specification also describes an embodiment where two pulses of differing carrier frequency are launched into the sensing fibre, with again a phase-recovery interferometer with frequency shifters in each arm.

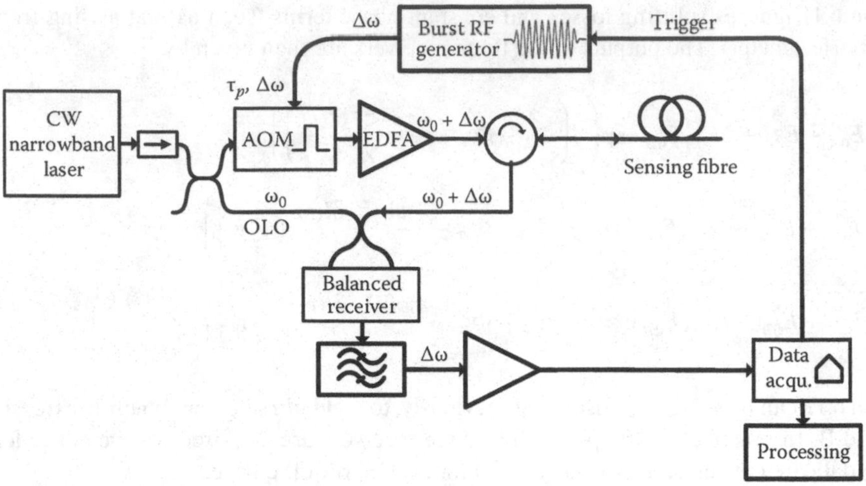

FIGURE 6.15 Schematic diagram of a heterodyne DVS system.

continuous laser output, and, simultaneously, the probe is also frequency shifted. It is convenient to use an AOM for this purpose because it provides both the amplitude modulation and the frequency-shifting functions. AOMs also exhibit excellent extinction ratios. After amplification (and, not shown, possibly optical filtering to eliminate the amplified spontaneous emission of the amplifier), the probe pulses are launched into the sensing fibre. A circulator, as shown, or a directional coupler is used to launch the pulses into the sensing fibre and to direct the backscatter to the balanced receiver via a further coupler that mixes the backscatter with the OLO. A beat signal at the IF $\Delta\omega$ is formed in the mixing process and separated out by the electrical band-pass filter. So far, the system is similar to that used for heterodyne OTDR (as discussed in Chapter 3). However, the crucial difference is that in heterodyne OTDR, the amplitude of the backscatter signal is detected and the phase information is discarded. In contrast, in hDVS, the phase of the backscatter is measured via the down-shifted IF that retains that phase information and yet is at a frequency that can be handled in the electronics.

A typical hDVS raw signal is shown in Figure 6.16. It takes the form of an amplitude-modulated sine wave; its amplitude conveys the same information as would an intensity-based DVS system such as that of

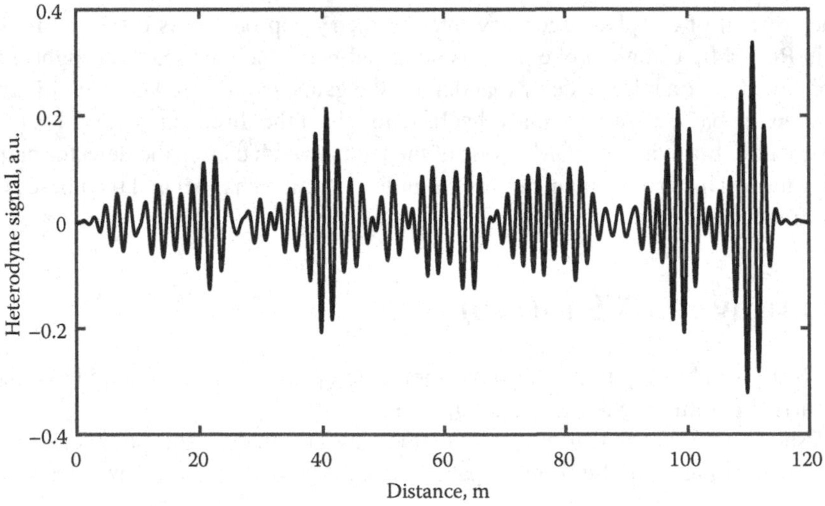

FIGURE 6.16 Sample hDVS data – the beat frequency is 50 MHz in this case.

Figure 6.2. In hDVS, it is the phase that is primarily of interest; the amplitude is used, if at all, to provide an indication of the signal strength and is thus a proxy for signal-to-noise ratio. It is worth noting that, in places, the amplitude is close to vanishing, and this leads to signal fading, a feature that is also apparent in Figure 6.9 for the dual pulse technique.

In the example of Figure 6.16, the frequency of the probe pulse is shifted up by 50 MHz and so the period of heterodyne backscatter signal is equivalent to about 2 m of fibre.

Functionally, the system is required to

a. Select the IF from the receiver output;
b. Detect its phase as a function of time, i.e. distance along the fibre;
c. Differentiate that phase across a selected distance interval (thus forming the gauge length) and
d. unwrap the phase differences between shots of the laser.*

After this processing, a picture of the unwrapped phase as a function of distance along the fibre and of slow time (i.e. on the time scale commensurate with the succession of probe pulses) becomes available.

How these functions are realised is a matter of design decision: the output of the receiver can be captured directly with a sufficiently fast A/D converter. All the processing is then carried out in the digital domain. On the other hand, the filtering, phase detection and even differentiation can be implemented in analog circuitry if the system designer chooses to do so. Both approaches have been tested in the author's laboratory.

The hDVS approach has a number of benefits over the two previously described approaches. Firstly, the signal on the detector is the geometric mean of the LO power and the backscatter power. By choosing a sufficient level of OLO power, the coherent detection process presents to the receiver an optical signal that is strong enough to dominate the receiver noise and so the latter is no longer an issue. In addition, the dynamic range of the signal, being the product of the electric field of the OLO (constant in time) and of the backscatter, falls off more slowly with fibre attenuation than is the case in the techniques that use only the backscatter power, as shown in the comparison in Figure 6.6.

Figure 6.17 (upper box) shows three successive traces in which a disturbance is located part-way along the fibre: as can be seen, the phase of the two signals is highly consistent up to the disturbance – beyond that, there is a constant phase shift for all points downstream of the disturbance. These data, as in Figure 6.16, were acquired with an IF of 50 MHz. The central horizontal box of Figure 6.17 shows again three successive traces acquired at the same time as the data in the upper box, but this time with a down-shift of 150 MHz. As would be expected, the oscillations are at higher frequency, but it can be seen that the phase is also unchanged between the three traces up to the point where the disturbance starts.

The lower section of Figure 6.17 shows enlarged sections of the traces of the upper (left plot, 50 MHz) and central (middle plot, –150 MHz) boxes as well as a further plot (right) acquired with a positive IF (up-shift) of 250 MHz. In all cases, the degree of expansion of the distance axis has been adjusted so that the distance between cycles occupies the same length on the printed page, so the distance covered for each of these plots varies in inverse proportion to the absolute value of the IF. The portion of the trace selected is downstream of the disturbance. It can be seen that phase shift in all cases is the same proportion of a cycle and so it truly is a fixed phase shift that is applied to the signal by the disturbance. Moreover, it can be seen that the sign of the phase shift follows the sign of the IF frequency and so the trace order appears different in the central plot compared with those to the left and right.

The situation may be better visualised using the greyscale image plot of Figure 6.18a, which shows the phase as a function of slow time and distance along the fibre and in which a sinusoidal variation of the

* It is also possible simply to record the change in the phase across a gauge length as a function of the laser shot number. This provides the time derivative of the strain, i.e. the strain rate.

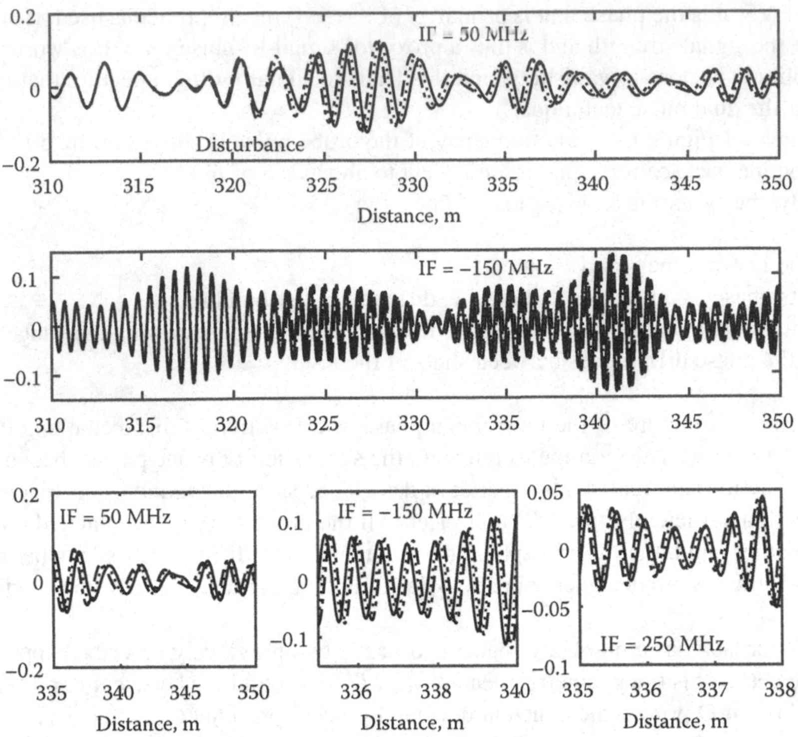

FIGURE 6.17 Triplets of heterodyne backscatter waveforms acquired in quick succession with a phase modulation imposed at a distance of about 323 m. Upper plot is acquired with an IF of 50 MHz (upshifted), whereas the central plot, on the same horizontal scale, has a downshifted IF of −150 MHz. The lower plots are expanded views of a point downstream of the phase modulation: note that the scale each is different and the expansion factor is proportional to the IF. Data acquired by Florian Englich and the author at Schlumberger.

local strain occurs at about 215 m. The phase of the backscatter at all locations beyond the dynamic strain is subject to the same modulation.

In Figure 6.18b, the phase is differentiated, and this restores the localisation of the phase signal to the region surrounding the disturbance. In the case where the strain modulation occurs over a short length of fibre, the distance over which the phase difference varies from the static value is one gauge length. The gauge length is the prime determinant of the spatial resolution in phase-measuring DVS systems.

Figure 6.19 shows an unwrapped differential phase signal obtained in this way: although the range of the phase modulation far exceeds 2π, in this case, the unwrapping process is able faithfully to reconstruct the strain signal because the rate-of-change of the phase is within the unwrapping criteria. The wrapped signal is also shown (dotted curve with markers denoting each sample point). There are at least three data points on each 2π segment of data, and therefore the Itoh criterion, requiring at least two sample points for each 2π phase excursion, is comfortably met in this example. It would be possible to increase the signal range, at the same strain modulation frequency, by approximately 50% before unwrapping errors are expected.

A spectrum of a sinusoidal signal recovered in this way is shown in Figure 6.20, and it illustrates the linearity and the dynamic range that is achievable. However, it should be noted that Figure 6.20 relates to data where the strain modulation operates over a short section of fibre, and, for these data, the phase modulation is contained entirely within one gauge length. In this example, a dynamic range of more than 90 dB, with harmonic distortion at least 60 dB below the signal is demonstrated at the upper end of the seismic frequency band.

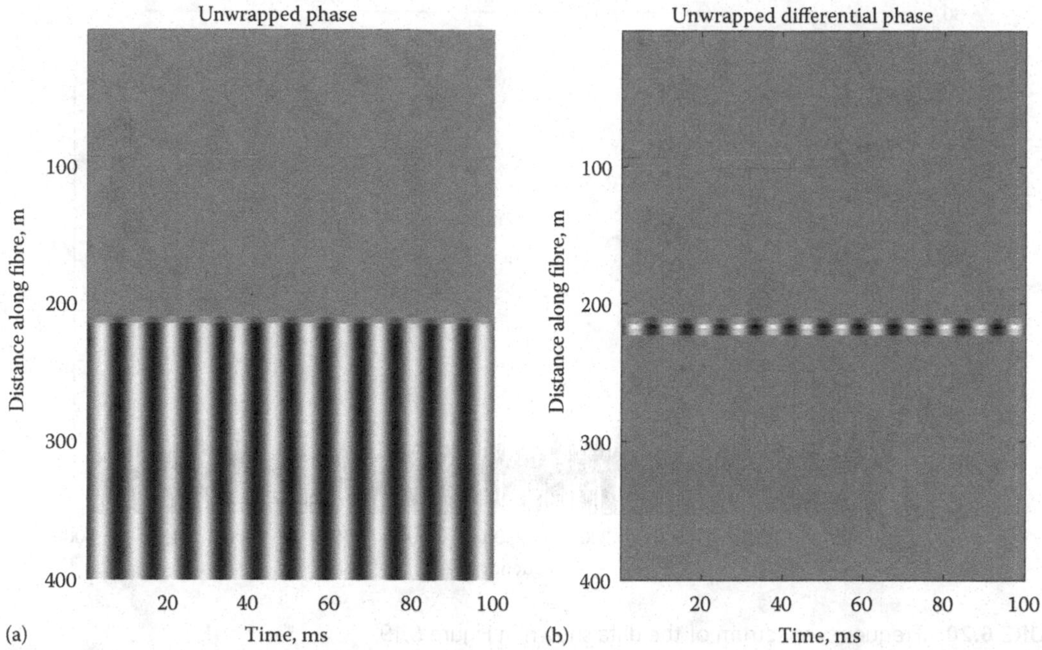

FIGURE 6.18 Phase extracted from a dataset similar to that of Figure 6.16 shown as a function of slow time (*x* axis) and fast time/distance along the fibre (*y* axis), the phase excursion being illustrated on a greyscale (a). (b) Shows the same data after applying a spatial differentiation operation. The distance is truncated at the start of the plot by about 105 m relative to Figure 6.16. Data acquired by Florian Englich and the author at Schlumberger.

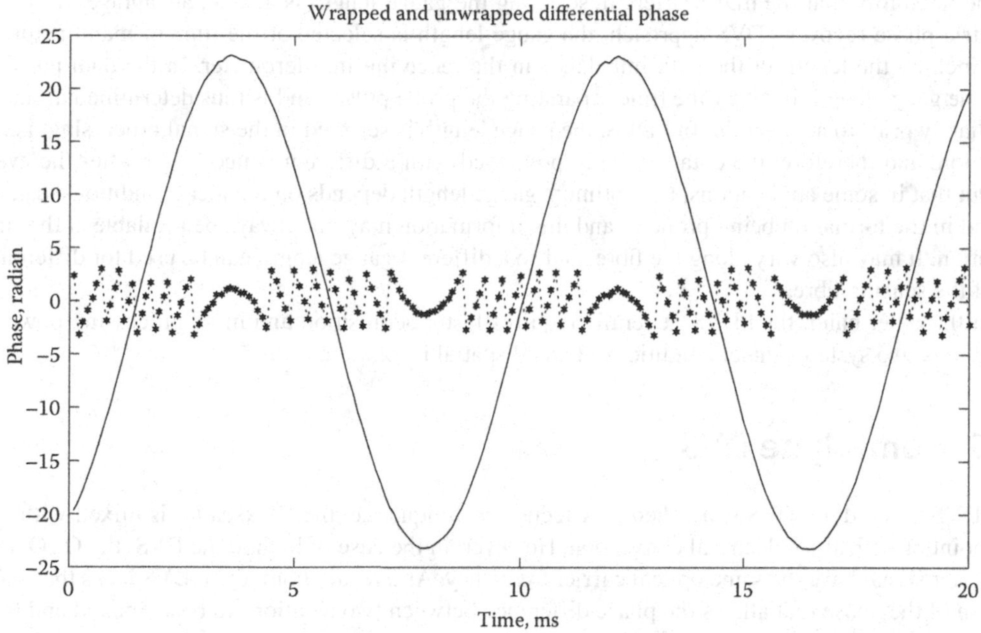

FIGURE 6.19 Differential phase (taken from data at the 218 m location in Figure 6.18) as a function of slow time (dotted curve; the markers represent the actual data points); the solid curve is the unwrapped phase obtained from the differential, wrapped phase.

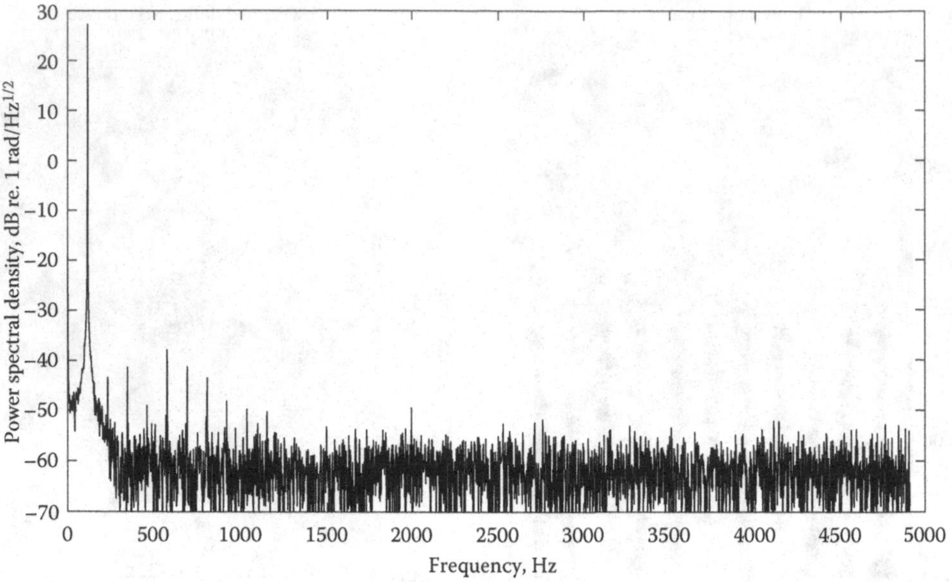

FIGURE 6.20 Frequency spectrum of the data shown in Figure 6.19.

In cases where the strain modulation is on the edge of the gauge, the linearity is not usually as good as illustrated in Figure 6.20. In such cases, the phase of the light scattered in one or other of the regions defining the reflectors is affected by the strain modulation, and this gives rise to a non-linear component to the transfer function.*

The flexibility that the hDVS offers in selecting the gauge length is also an advantage. In the interferometric phase recovery DVS approach, the gauge length is selected at the time of manufacture: it is determined by the length of the path imbalance in the receiving interferometer. In the dual-pulse technique, the gauge length is set by the time separating the probe pulses and is thus determined at the latest immediately prior to acquisition. In hDVS, the gauge length is selected in the signal processing, i.e. post-acquisition, and therefore, the data can be re-processed with a different gauge length after the event. It turns out that in some applications, the optimum gauge length depends on the local conditions (e.g. speed of sound in the formation being probed), and this information may not always be available at the time of acquisition. It may also vary along the fibre, and so a different gauge length can be used for different sections of the sensing fibre.

On the other hand, the hDVS generally requires faster acquisition and more processing power than the previous two systems under conditions of equal spatial resolution.

6.3.6 Homodyne DVS

Like hDVS, homodyne DVS is a coherent detection technique, i.e. the backscatter is mixed with an LO at the point of optical-to-electrical conversion. However, in the case of homodyne DVS, the OLO and the backscatter signal have the same optical carrier frequency. As a result, homodyne DVS lacks the fast-time variation of the phase that allows the phase difference between two locations to be estimated and tracked

* This type of non-linearity affects all types of dΦ-DFOS systems; it is not unique to hDVS.

in slow time. More information is required to estimate the phase, and this comes from phase diversity, i.e. outputs that are naturally at a different phase, preferably in quadrature [69].

In coherent optical communications, the component that provides this functionality is known as an optical 90° hybrid. Examples made from bulk optics were demonstrated in the mid-1980s [71] and used in early coherent phase-shift keying experiments [72,73]. All-fibre techniques, using fibre couplers and fibre polarisers, were also demonstrated at that time [74]. A 3 × 3 coupler has also been used to provide the phase diversity needed for homodyne detection [69,75].

An optical 90° hybrid takes two inputs, namely the LO and the signal, and provides four outputs separated by 90°. Using a pair of balanced receivers with inputs, in the first case, the 0° and 180° ports and the 90° and 270° ports in the second case, provides a pair of electrical outputs in quadrature. Some quarter of a century after the early demonstrations of coherent communications with 90° hybrids, the devices are available commercially and are realised with miniaturised bulk-optics or with integrated optics. A 4 × 4 coupler performs the same function if the splitting ratio is precisely uniform between ports.

The arrangement of a homodyne DVS is shown schematically in Figure 6.21 [69]. As in hDVS, the source laser is split into OLO and probe paths, the latter being modulated to form a pulse, prior to amplification and launching the probe pulses into the sensing fibre. It should be noted that in the homodyne case, the modulator does not shift the frequency of the modulated light. This can be achieved using a different type of modulator or a pair of AOMs with opposite frequency shift, in which case it is convenient to place the second modulator after the amplifier to reduce the ASE accompanying the probe. The backscatter and LO are fed, respectively, to the signal and LO inputs of an optical 90° hybrid. Complementary outputs (0° and 180°, and 90° and 270°) are fed to the inputs of a pair of balanced receivers, the outputs of which are in quadrature and can thus be used to estimate phase simply from a four-quadrant arc tangent calculation.

As in hDVS, the phase calculated for each location along the fibre is differentiated to provide a localised measure of the phase variation at each point and unwrapped. An example of data obtained in this way is shown in Figure 6.22 on the z axis as a function of slow time (x axis) and distance along the fibre (y axis). A piezo-electric transducer was used to apply a sinusoidal strain at a distance of about 60 m from the start of the data shown with a frequency of 105 Hz and a peak-to-peak phase excursion of 10 radian. This signal is clearly distinguishable in the plot. Similar results have been obtained using the 3 × 3 coupler approach [69], also by the author at Schlumberger.

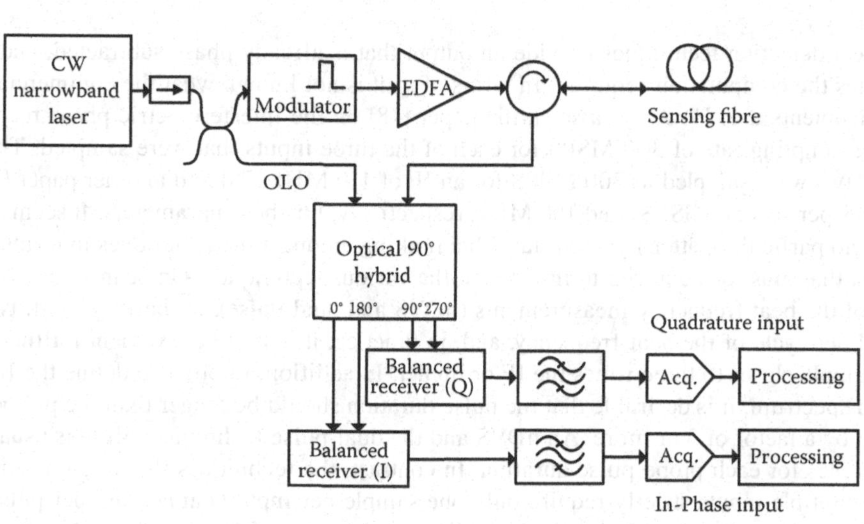

FIGURE 6.21 Schematic diagram of a homodyne DVS system.

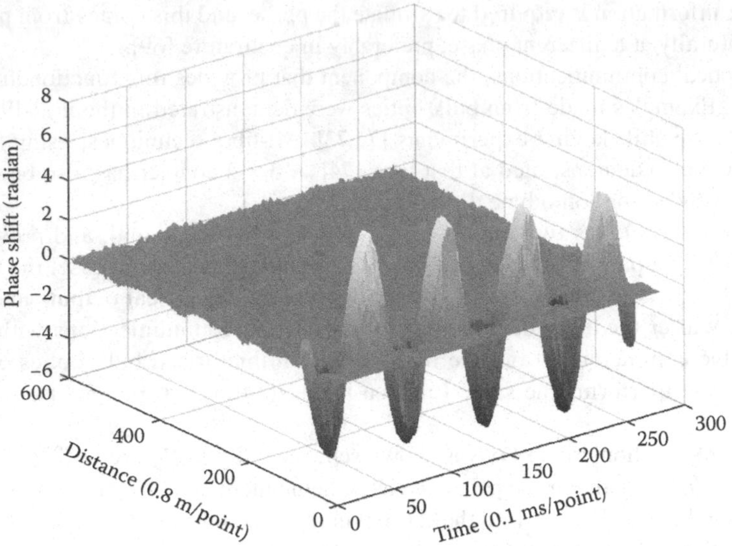

FIGURE 6.22 Homodyne distributed vibration sensing data obtained by the author at Schlumberger with the arrangement of Figure 6.21.

6.4 COMPARISON AND PERFORMANCE OF THE DVS TECHNIQUES

Although all four techniques achieve their goal of measuring dynamic strain and are based on similar principles of estimating a local derivative of the backscatter phase w.r.t. distance along the fibre, major practical differences do arise, and the limitations can be quite different. An exact comparison is difficult because many of the systems have been described as black boxes, with the details of their internal design kept proprietary.

The direct detection techniques provide an output that is already phase subtracted, and this probably simplifies the computation required. In terms of f_d, it is not known what the commercial suppliers of these instruments use. However, a scientific paper [48] on the interferometric phase recovery technique used a sampling rate of 300 MSPS for each of the three inputs that were sampled. The reported results on hDVS were sampled at 300 MSPS for an IF of 110 MHz [76] and another paper [70] reports 3 gigasamples per second (GSPS) and 160 MHz, respectively, for these parameters. It seems, therefore, that there is no particular pattern yet to establish a ranking amongst the techniques in terms of the volumes of data that must be acquired to implement the various techniques. One can state, however, that in the case of the beat-frequency measurements (hDVS and dual pulse), at the very least, two samples are required per cycle of the beat frequency, and, in practice, it is more convenient for the signal processing to sample closer to three times the IF or higher. In addition, in order to define the IF clearly in the acquired spectrum, it is desirable that the pulse duration should be longer than the period of the IF, for example, by a factor of 3 or more. An hDVS and the dual-pulse technique will thus usually acquire some 10 samples for each probe pulse duration. In contrast, the techniques that rely on estimating the phase from multiple inputs strictly require only one sample per input channel and per pulse duration. The data volumes may therefore be somewhat lower (by a factor of 3 to 5) in these types of system for equal probe pulse duration.

In terms of the performance realised, very few figures have been published. From the author's personal experience, the hDVS technique is capable of providing a phase noise of a few (1–5) mrad [76] in the seismic bandwidth (5–150 Hz), depending on the precise conditions, such as the gauge length selected and the PRF as well, of course, of the optical losses between the instrument and the remote end of the fibre. A value of 2 mrad for a probe wavelength of 1550 nm would correspond to a noise on the strain of ~0.23 nm r.m.s. Data have been published for the interferometric phase recovery DVS [77] indicating a noise level of 10 nm, but the bandwidth over which this applied was not specified, nor was the gauge length. In Ref. [48], a value of 160 nm was provided, although this applied to a much wider bandwidth (500–5000 Hz) than is used in seismic applications and so it is not directly comparable. A value of 10 mrad was reported in Ref. [42] using the phase-modulated dual-pulse technique.

The coherent detection approaches circumvent the noise of the receiver, and this allows these techniques to perform without degradation at lower received powers and therefore at higher round-trip optical losses. Thus, whereas the direct detection approaches could be expected to be limited by receiver noise, in the case of coherent detection, hDVS in particular, the fundamental limitations are most likely to reside with the phase noise of the laser. With homodyne detection, component imperfections in 90° hybrids can allow noise sources, such as relative intensive noise in the laser, to impair the system performance.

Unfortunately, no generalised rules covering the sensitivity of all systems can be formulated at present: to understand the system performance, it is necessary to know the details of the system design, the performance of key components as well as the losses in the sensing fibre in order to model the strain noise seen at the system output.

6.5 ISSUES AND SOLUTIONS

6.5.1 Fading

We define *fading** as the phenomenon in coherent Rayleigh backscatter whereby the summation of the electric fields from individual scatterers in the region of the fibre occupied by the probe *occasionally* results in a very small signal that cannot be measured to an acceptable signal-to-noise ratio. This affects all types of DVS measurements. It is illustrated in Figure 3.7, where several instances of weak intensity are shown and similar regions of low signal levels can be seen in Figure 6.9 and Figure 6.16 for the dual-pulse DVS and hDVS approaches, respectively. In the case of amplitude-based DVS, an additional fading mechanism exists, namely that, locally, the intensity of the backscatter is insensitive to disturbance; i.e. the local *derivative* of the intensity vs. strain function is small or zero. This type of fading does not affect phase-based measurements because the linear delay term separating two reflective zones is still modulated and it is this delay term that provides the basis of the measurement.

6.5.1.1 Requirements for statistically independent signals

Fading can be addressed through diversity, i.e. by making multiple measurements under different fading conditions. Because the fading effect is related statistically to the probe frequency, as was illustrated in Figure 3.8, choosing a different frequency provides a measurement with different fading characteristics. According to Mermelstein et al. [78], the frequency change required to achieve a statistically independent measurement is the inverse of the pulse duration. This effect is characterised by the autocorrelation of the

* In the context of optical fibre sensors, fading is often associated with polarisation fading, i.e. the fact that if two light signals are combined with orthogonal polarisation states, no interference occurs. This is a quite different effect to that discussed here.

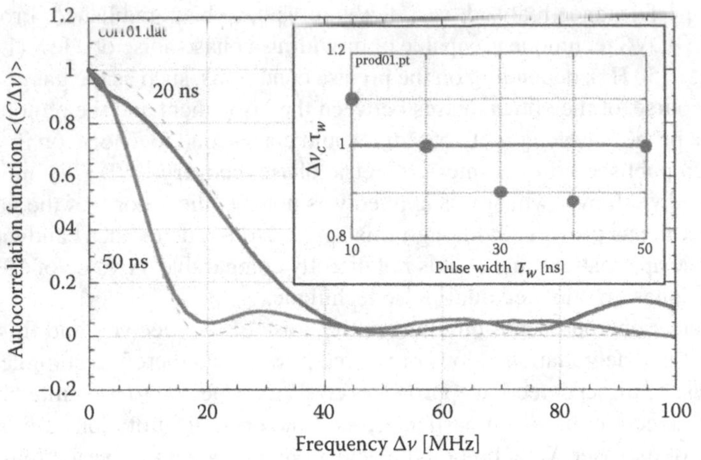

FIGURE 6.23 Autocorrelation of the backscatter power function as a function of carrier frequency. (Reproduced with permission from Mermelstein, M.D., Posey, R.J., Johnson, G.A., and Vohra, S.T. 2001. Rayleigh scattering optical frequency correlation in a single-mode optical fiber. *Opt. Lett.* 26: 2: 58–60 © 2001. Optical Society of America.)

backscattering intensity vs. frequency function. Based on Ref. [78], for a square probe pulse of duration τ_p (note that in Ref. [78], the authors denoted the pulse duration with tw and this appears in the labels of Figure 6.23), the autocorrelation function is given by

$$C(\Delta\nu) = \text{sinc}^2(\pi \cdot \Delta\nu \cdot \tau_p), \tag{6.13}$$

which is zero when $\Delta\nu = 1/\tau_p$ and where the unnormalised definition of the sinc function; i.e. sinc (x) = sin $(x)/x$ has been used.

Although the sinc function has side lobes, in practice, the pulses are seldom square and so the subsidiary regions of correlation in the side lobes are usually substantially lower than would be the case for sharp-edged probe pulses; the relation $\Delta\nu = 1/\tau_p$ can therefore be used as a guide to the minimum frequency separation that is required for statistical independence of the probe signals used for frequency diversity. The autocorrelation function is illustrated in Figure 6.23 for two pulse durations, for square pulses for both experimental data and modelled results, the two being very closely matched. Thus, for a pulse duration of 20 ns, a frequency shift of 50 MHz or more between two probe carrier frequencies provides sufficient separation for two measurements that are statistically independent as regards intensity fading.

When a diversity technique, such as frequency diversity, is employed, several independent measurements are acquired. In the subsequent *aggregation* process, the data from the diverse measurements are fused to provide a global estimate of the phase reading. A number of approaches can be used for this purpose, such as combining the results in proportion to the amplitude of the signal, choosing the median or simply selecting the strongest signal. These are just some of the options available, and more than a dozen such approaches have been identified. It turns out that there is no generally optimal aggregation approach: a technique that maximises the signal-to-noise ratio, for example, may not lead to the best possible linearity.* In addition, some techniques become reliable only with substantial (≫3) numbers of frequencies.

Three independent measurements are sufficient, by and large, for the purpose of eliminating fading. For a larger number of independent measurements, the improvement continues, but at a reduced rate; however, there may be other reasons for increasing the number of independent measurements, such as reducing the noise from sources other than fading or improving the linearity.

* Kotov, O.I., L.B. Liokumovich and N.A. Ushakov, 2014: Personal communication.

6.5.1.2 The effect of diversity on signal quality

The benefit of using multi-frequency interrogation is illustrated in Figure 6.24, where the phase noise recorded along a section of fibre is plotted. The single-frequency results (three panels to the left) show some regions of very high phase noise caused by fading, but they largely appear at different locations along the fibre. The aggregated multi-frequency phase noise is substantially reduced, especially in those places where the single-frequency fading is particularly severe (fourth greyscale panel). The greyscale is the same in all four panels and was adjusted so as just to saturate for the single-frequency data. The r.m.s. noise at each location is shown on a dB scale (20 $\log_{10}$) in the final plot to the right, with the aggregate value shown in black and the single-frequencies in grey (broken line for Frequency 1, dotted line for Frequency 2 and solid line for Frequency 3). Whilst, on average, the improvement due to diversity was less than a factor of 3, the improvement can be more than a factor of 10 in regions of severe fading and, in the most significant locations, a factor of 30.

In order to acquire and process multi-frequency data, it is not sufficient to mix the probe light at the different frequencies and to acquire some aggregate of the resulting backscatter signal: it is necessary to acquire each of the signals generated by each of the probes signals, analyse them independently and then aggregate the results. Of course, this results in a greater computation load and also higher memory requirements, particularly to store intermediate results.

In the discussion on frequency diversity, a lower bound of $\Delta\nu > 1/\tau_p$ has been provided, but no upper bound. It is worth distinguishing frequency diversity based on closely spaced frequencies leading

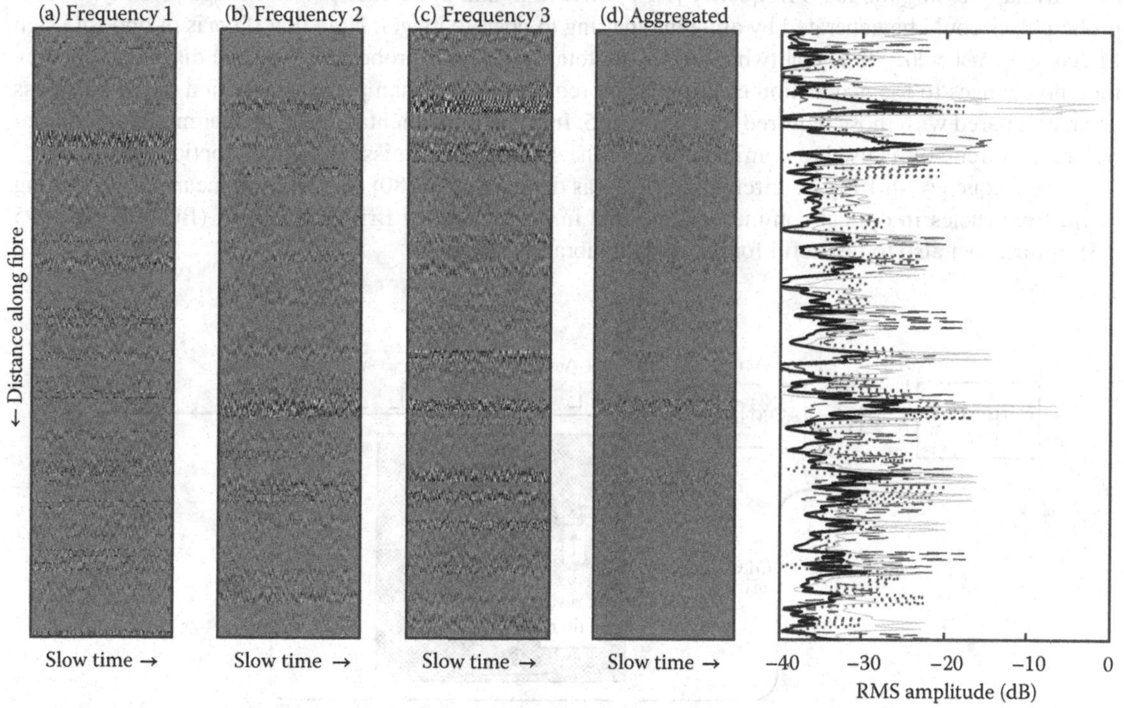

FIGURE 6.24 Measured phase noise for single frequency and aggregated multi-frequency hDVS data. The three leftmost plots show the output signal under quiet conditions at three different IF frequencies. The fourth greyscale plot shows the aggregated output from the three independent measurements. The plot to the right shows an r.m.s. noise level for each location along the fibre for each of the frequencies and the aggregate (the latter in black). Data collected by Alexis Constantinou and processed by Theo Cuny; figure prepared by Tim Dean (all at Schlumberger). (Adapted with permission from Hartog, A.H., Liokumovich, L.B., Ushakov, N.A. et al., The use of multi-frequency acquisition to significantly improve the quality of fibre-optic distributed vibration sensing, 78th EAGE Conference and Exhibition, 2016.)

to signals that can be separated in the electronic domain from those that use separate sources at distinct wavelengths; in the latter case, optical separation, separate receivers and acquisition channels are required. The latter approach might be better described as wavelength diversity. The techniques for adding and splitting optical signals that are well separated in frequency (e.g. by 50 GHz or more) have been well developed in optical communications for dense wavelength-division multiplexing (DWDM), and the components are available commercially.

6.5.1.3 Specific arrangements for frequency diversity

In the case of frequency diversity, the necessary frequency separation can be achieved using a re-circulating optical circuit incorporating a frequency-shifting device. This approach [79] is illustrated for hDVS in Figure 6.25. The schematic is the same as that of Figure 6.15 other than the addition of the functional block labelled 'Ring'. The light from the laser is initially shifted by $\Delta\omega$ in a first AOM that also defines the pulse duration. Half of the power in that pulse is passed through a directional coupler to a pulse amplifier and then launched as a probe into the sensing fibre. The other half of this pulse power is directed, within the ring structure, to a second AOM that adds a frequency shift of $\delta\omega$ MHz to the pulse that is then amplified and filtered in the ring, returning to the directional coupler. At that point, one half of the power of this pulse, now shifted by $\Delta\omega + \delta\omega$ relative to the source laser, provides a second probe pulse, the remainder of its energy re-entering the ring for a further shift to $\Delta\omega + 2\delta\omega$, eventually being launched as a third probe pulse. In a three-frequency system, the AOM in the ring is turned off after the third pulse has been generated (by suppressing the radio frequency [RF] power to it) and no further pulses emerge. Clearly, further probe pulses could be generated by allowing the ring to operate longer. Note that if $\delta\omega$ is of opposite sign to $\Delta\omega$ and $|\Delta\omega| > |(n_{pr} - 1)\,\delta\omega|$ (where n_{pr} is the total number of probe pulses in the diversity scheme), then no changes to the acquisition electronics are required, other than perhaps attention to the bandpass filter, compared with those required for Figure 6.15. In some implementations, a further modulator placed before the circulator is used to eliminate most of the spontaneous emission from the optical amplifiers.

The frequency-shifting, re-circulating ring was demonstrated [80] in 1990 as a means of generating comb frequencies in optical communications and in the context of Brillouin OTDR (BOTDR) in 1995 [81]. It turns out also to be useful for distributed vibration sensing.

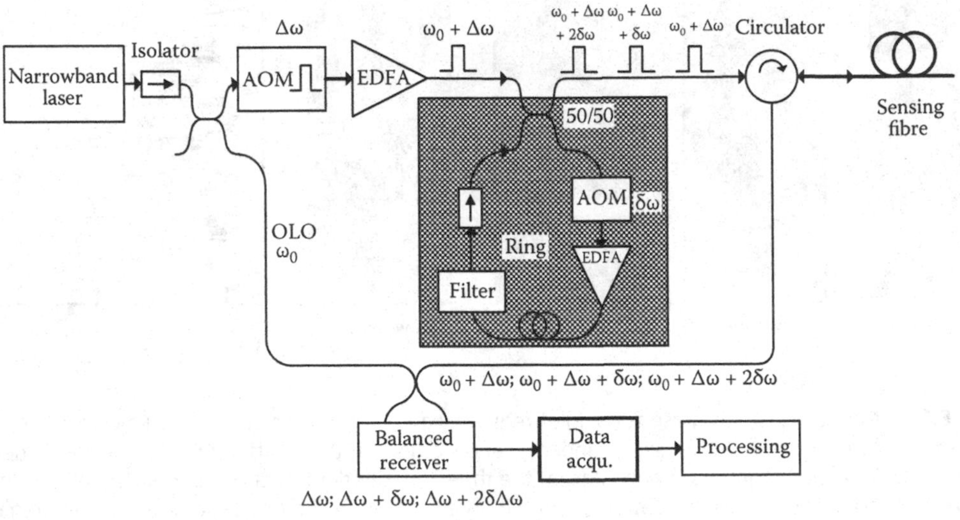

FIGURE 6.25 Schematic arrangement of a frequency-diverse hDVS system. (Adapted with permission from Hartog, A.H., Liokumovich, L.B., Ushakov, N.A. et al., The use of multi-frequency acquisition to significantly improve the quality of fibre-optic distributed vibration sensing, 78th EAGE Conference and Exhibition, 2016.)

Given the limited number of frequencies that are generated in this case, the re-circulating ring is relatively straightforward, compared with the implementation in BOTDR, although attention is required to filtering to avoid amplified spontaneous emission, to ensuring unidirectional operation (by use of an optical isolator) and to providing sufficient delay in the loop to avoid overlapping of multiple pulses in the ring. The operation of a ring with many output pulses becomes more challenging as the number of derived pulses is increased owing to the build-up of spontaneous emission and depletion of the gain of the amplifier that leads to interaction between the powers of neighbouring pulses.

The re-circulating frequency shifting technique has also been proposed in the context of the dual-pulse technique [82]. In this case, the re-circulating ring creates a new pair of probe pulses that must be distinguishable, in the frequency spectrum of the backscatter signal, from other pairs that are generated. This leads to some restrictions in the selection of frequencies because, not only must the frequency differences within each pulse-pair be different, but so too must all frequency differences between any of the pulses launched. This is clearly achievable, but, as mentioned, it is restrictive. In practice, it seems that proponents of the dual-pulse technique have chosen wavelength diversity over frequency in spite of the additional sources and acquisition systems that are required.

6.5.1.4 Alternative approaches to diversity in DVS

Other means of addressing fading through diversity are available. For example, using a pair of orthogonally polarised states in the probe and recovering each polarisation separately will provide *some* diversity. However, in this case, the two measurements are not perfectly independent, given the fact that Rayleigh scattering is somewhat depolarised and that, therefore, some interaction between the backscatter from the two principal polarisation states exists.

A wavelength diverse approach was also illustrated in Ref. [82], which shows two source lasers being multiplexed into the same optical circuitry that is used to generate probe signals and launched simultaneously into the sensing fibre. The resulting backscatter is separated into two receiver channels, the outputs of which are connected to their individual acquisition electronics.

The techniques described so far have involved launching the diverse probes quasi-simultaneously. If it is acceptable to launch independent probes for successive backscatter signals, the range of options is increased. For example, some of the restrictions mentioned in the frequency-diverse dual-pulse technique are lifted. In the case of hDVS, it also allows the use of a phase modulation approach that was demonstrated in Ref. [70]. Here, the probe is separated into two sub-pulses and their relative phase is modulated between successive interrogations. This is illustrated in Figure 6.26, which shows how the second half of the probe pulse is shifted (in this case by π) relative to the first half of the pulse, but only for even-numbered pulses. Because these pulses are coherent, they create a single coherent backscatter signature, including fading. However, when their phase relationship is altered, so too is the electric field summation and, thus, the fading properties are different for odd- and even-numbered probes. In this way, a diverse set of measurements can be constructed, albeit at the cost of a reduction in sampling rate.

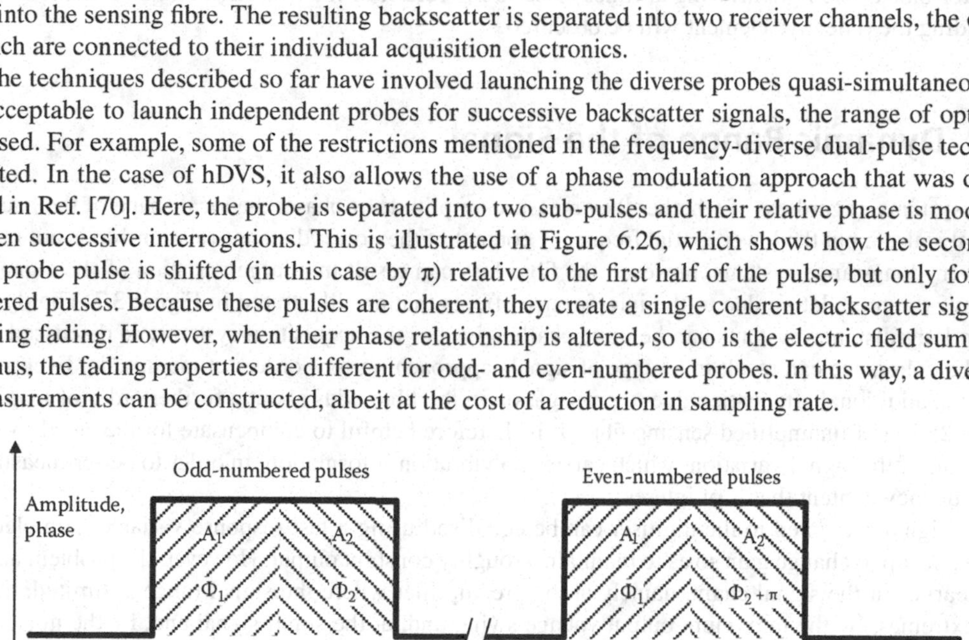

FIGURE 6.26 Timing diagram for a phase-diverse hDVS system. (After Pan, Z., Liang, K., Ye, Q. et al., Phase-sensitive OTDR system based on digital coherent detection, Optical Sensors and Biophotonics III, SPIE, 8311, OS-1, 2011.)

6.5.2 Reflections

Rayleigh OTDR signals are at essentially the same frequency as the probe light, and so they also appear at the same frequency as any reflections that might occur along the fibre; they cannot therefore be eliminated using the optical filtering techniques that are available in Raman and Brillouin distributed sensors. For pulse durations of a few tens of ns, the backscatter is about five orders of magnitude weaker than the reflection that would be caused by a clean break in the fibre normal to the axis. Clearly, this signal will overload a receiver designed to handle the weak backscatter and the problem cannot be alleviated by passive optical solutions. It therefore requires the use of low-reflection connections and also low-reflection terminations at the far end of the sensing fibre. Although it would be desirable to assemble the sensing fibre using only fusion splices that are usually non-reflective, practical systems need to be assembled in the field and maintained; this often forces the use of connectors. In addition, applications where the sensing fibre is installed for a very short time, for example, in borehole surveys, often require the use of an optical rotating joint (also known as an 'optical slip ring' or a 'collector' in various industries) that allows the light to pass with minimum loss whilst the drum holding the un-deployed part of the sensing cable is rotated. The rotating joint is often a source of reflections as well as optical losses, especially after some wear and tear in the field.

The problem of reflections is somewhat less acute in the coherent detection approaches (heterodyne and homodyne) because the unwanted signal, as converted to the electrical domain, is proportional to the electric field rather than the optical power of the reflection.

If strong reflections are inevitable in the system design, an active approach, in which strong reflections are gated out by a fast modulator, may be essential, although this adds complexity to the system and attenuates the signal presented to the receiver even at the most transmissive settings of the gating modulator. It should also be noted that however fast the modulator, some of the sensing fibre will be lost when the modulator blocks the light arriving at times close to the return of the reflection and so a section of fibre surrounding the reflective element will be obscured.

6.5.3 Dynamic Range of the Signal

In DVS systems, regardless of whether they measure amplitude or phase, the local shape of the backscatter signal is used, but the lower spatial frequency content of the power/distance relationship is not of interest. In very-long-range systems, however, the fibre loss can result in a large variation of the signal over and above the modulation due to the interference effects that are illustrated in Figure 3.7. The dynamic range in long-distance systems can be quite challenging for the acquisition systems: as discussed previously, the coherence can result in a range of signals approaching four orders of magnitude. Even in low-loss fibres, additional attenuation due to propagation in the fibre adds about a further order of magnitude for each 25 km of unamplified sensing fibre. It is therefore helpful to compensate for the slowly varying component of the signal variation, which carries no vibration information, in order to better measure the high-frequency content that is of interest.

The signal presented to the digitiser can be equalised using a fast-response variable gain electrical amplifier set up to change gain so as to maintain a roughly constant output. However, the problem actually occurs earlier in the signal chain, namely at the preamplifier where the signal range is limited, at large signal extremes, by the maximum output voltage swing and, at the small signal end, by the noise of the receiver. The solution to this problem [83] is to introduce a variable transmission element that equalises the low-frequency components of the optical signal presented to the preamplifier. Thus, a fast voltage-controlled attenuator can be placed between the circulator and the receiver and driven by a waveform that varies its transmission in approximate inverse proportion to the low-frequency content of the signal including possibly fast increases in attenuation at times corresponding to the location of in-line optical amplifiers.

The dynamic range problem just discussed is less acute in coherent-detection systems owing to the fact that these systems respond in proportion to the amplitude rather than the power of the backscatter signal. In addition, in coherent-detection systems, the variable transmission element can be placed in the LO arm, rather than the signal path, thus avoiding any reduction of the backscatter signal prior to detection.

Rather than equalising the signal by attenuating the strongest elements, a fast-acting variable gain amplifier (e.g. a semiconductor optical amplifier) can be used. However, because the noise figure of these devices is rather high, it is advisable first to preamplify the optical backscatter signal before it is passed to the variable transmission element to avoid degrading its signal-to-noise ratio.

6.5.4 Dynamic Range of the Measurand

The dynamic range of a dΦ-DVS system is limited at the high signal end by the Itoh criterion, as discussed previously. This is a limitation on the rate-of-change of the strain, rather than the change itself. In order to maximise the signal that can be measured, the PRF should be increased to the limit dictated by the requirement that only one probe signal should be present in the fibre at any one time to avoid distance ambiguity, i.e. $f_R < 2N_g c/L$.

To some extent, the PRF limitation can be obviated in combination with diversity: if diverse, e.g. frequency diverse, signals are launched into the sensing fibre for the purposes of overcoming fading; they can also be launched at separate times, for example, evenly dividing the time for any one signal to return from the remote end of the fibre.

This concept was demonstrated using four independent frequencies [84], as illustrated in Figure 6.27. The interrogation arrangement is basically an hDVS arrangement with the addition, however, of an

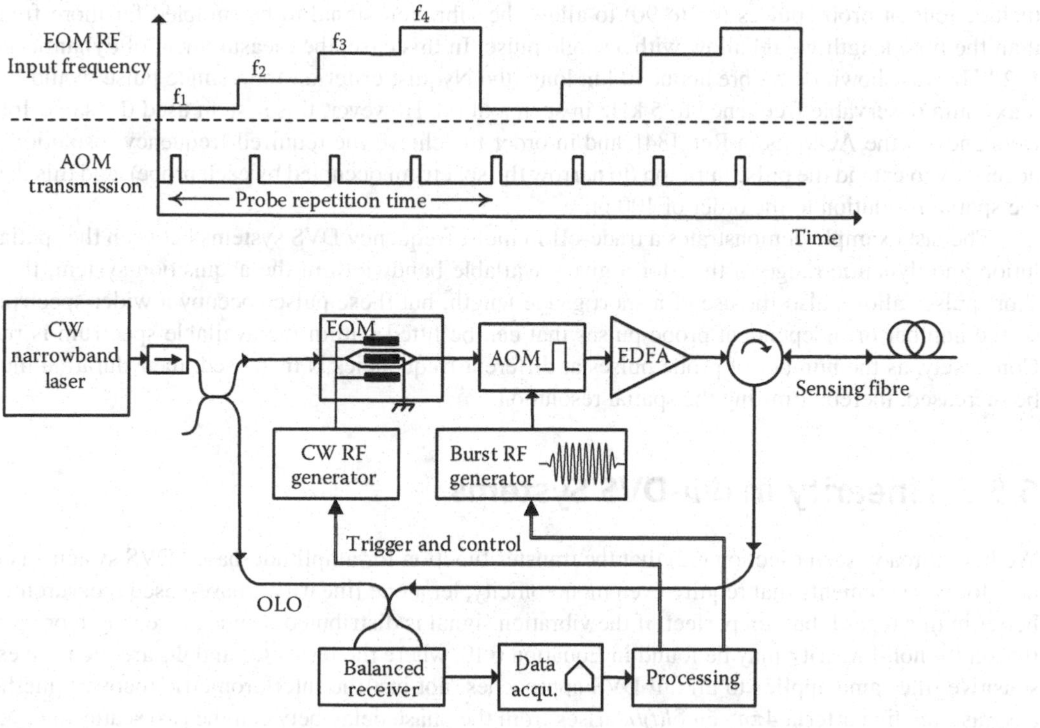

FIGURE 6.27 Multi-frequency measurement for increased dynamic range. (Adapted from Pan, Z., Wang, Z., Ye, Q. et al., High sampling rate multi-pulse phase-sensitive OTDR employing frequency division multiplexing, 23rd Optical Fibre Sensors Conference Santander, Spain, SPIE, 9157, 635, 2014.)

electro-optic modulator (EOM) that serves as a frequency shifter. For a 10-km fibre, as used in this demonstration, the maximum PRF is c. 10 kHz, i.e. a minimum probe interval of 100 μs. However, in this experiment, the frequency was changed at a rate of $4f_R$, i.e. every 25 μs, and a new pulse was launched for each frequency. This does not violate the $f_R < 2N_g c/L$ condition because each signal can be distinguished in the signal processing and handled separately; the limitation on pulse repetition rate applies individually to each frequency in the probe pulse sequence. The result is an increase in the effective sampling rate by a factor of four over and above the limitation defined by the fibre length for a single probe frequency.

In the arrangement, the probe pulse time and duration are determined by the AOM, which opens at a repetition time of 25 μs. However, the optical frequency that is the input to the AOM is varied on a four-pulse cycle so that the overall probe repetition time illustrated in the figure is 100 μs. The authors chose to shift the frequency in steps of 15 MHz from 15 MHz to 60 MHz. An EOM intensity modulator of course creates at least two sidebands (and then only if the modulation depth is carefully controlled). This would normally create an ambiguity between upper and lower sidebands. In this particular arrangement, however, the AOM further shifts the light it transmits (by 160 MHz in the case of Ref. [84]), and so the probe pulses now appear in pairs around 160 MHz separated by multiples of 15 MHz. In this way, the different probes can be distinguished and independent probing of the sensing fibre is possible at a slow-time sampling rate that is four times faster than would be dictated by the fibre length.

A couple of further points should be made in relation to this arrangement. Firstly, the authors operated the EOM in carrier-suppressed mode to ensure that the probe energy is entirely in the useful sidebands (the carrier, being at the same frequency for each probe, would scramble information from the four probes making the information it carries unusable). Secondly, it is necessary to acquire the data continuously, so there is no respite in the data arrival and the acquisition system must be able to cope with the data rate.

The same arrangement, of using multi-frequency probes, was extended by the same team [85] to include tens of probe pulses (up to 90) to allow the vibration signal to be sampled far more frequently than the fibre length would allow with a single pulse. In this case, the measurement of dynamic strain at 122 kHz was shown on a fibre about 10 km long (the Nyquist criterion for a single pulse would limit the maximum observable frequency to 5 kHz in such a fibre). However, this system used the same 160 MHz frequency in the AOM as in Ref. [84], and in order to achieve the required frequency resolution, it was necessary to extend the pulse duration (to narrow the spectrum occupied by each probe), and this degraded the spatial resolution to the order of 100 m.

The last example demonstrates a trade-off in multi-frequency DVS systems between the spatial resolution and dynamic range in that, for a given available bandwidth in the acquisition system, the use of short pulses allows also the use of a short gauge length, but these pulses occupy a wider spectrum, and so the number of independent probe pulses that can be fitted within the available spectrum is reduced. Conversely, as the number of probe pulses at different frequencies is increased, their duration must also be increased, thereby limiting the spatial resolution.

6.5.5 Linearity in dΦ-DVS Systems

We have already seen (Section 6.2) that the transfer function for amplitude-based DVS systems is unsuitable for measurements that require even monotonicity, let alone linearity. Phase-based measurements are better in this regard, but not perfect, if the vibration signal is distributed over a gauge length or more. The reason for non-linearity may be found in Equation 6.10, where the terms Φ_A and Φ_B are themselves strain sensitive (the same applies to all dΦ-DVS approaches, not just the interferometric recovery method). In contrast, the linear term $4\pi n \cdot \xi \cdot \delta l(t)/\lambda$ arises from the phase delay between the two scattering regions A and B, which act as reflectors. The strain dependence of Φ_A and Φ_B is random and, of course, confined to the range $[-\pi, \pi]$.

The intrinsic phases Φ_A and Φ_B cannot usually be observed directly other than that of the first section of fibre, and even then a coherent detection technique is required. Figure 6.28 therefore uses synthetic data to illustrate the strain dependence of the reflector phase for a pulse duration of 20 ns (that therefore occupies a 2-m length of sensing fibre). Although a slowly wandering unwrapped phase underlies the phase that can be observed, in practice, the measurement responds to the wrapped phase, which is shown. The strain range used in the illustration is rather larger than is typically found in geophysical applications, in particular as concerns reflections or even direct source-sensor arrivals. However, this level of strain is not unusual when monitoring tubing vibrations or signals detected in proximity to seismic sources.

The measured phase Φ_m is given by

$$\Phi_m = 4\pi n \cdot \xi \cdot \delta l(t)/\lambda + \Phi_B - \Phi_A. \tag{6.14}$$

Because the reflector intrinsic phases vary randomly with strain, their contributions to Φ_m cause the transfer function to depart from linearity. Assuming that the vibration signal is uniform over a section of fibre that is longer than the gauge length, then the effect on the reflectors is bound, for each reflector, to the interval $[-\pi, \pi]$, whereas the linear term is proportional to gauge length. It follows that the linearity improves as the ratio of gauge length to pulse with increases.

This effect is illustrated in Figure 6.29, which shows how the measured signal departs from linearity as a result of this effect. The DVS output from a single location is simulated for three realisations as a function of strain (these are shown as the thin dotted, broken and dash-dot lines) and contrasted with the linear response term (thick solid line). The gauge length is varied from equal to the pulse width (i.e. the length of fibre occupied by the pulse) to 10 times the pulse width, and the strain range is adjusted for constant elongation over a gauge length, so that the nominal range of the phase is the same in all cases. As can be seen, the linearity improves with increasing gauge length. If one were to use a gauge length shorter than the pulse width, the sensor response can become non-monotonic in some realisations of the random reflector phases.

In Figure 6.29, the results simulate the processing within a DVS, in other words it shows the unwrapped Φ_m.

On the basis of the linearity alone, increasing the gauge is beneficial; however, this also has the effect of degrading the spatial resolution and therefore we are faced again with a trade-off. We shall see in Chapter 9 that there is also an interaction between gauge and frequency response of the measurand in the case of the detection of travelling waves.

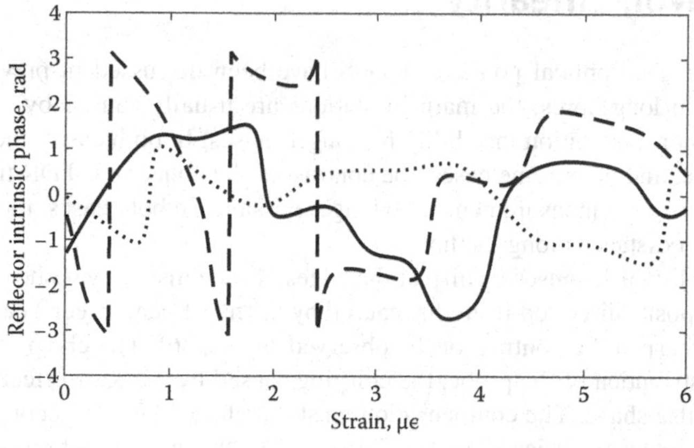

FIGURE 6.28 A few synthetic realisations of the strain dependence of the intrinsic phase, for a 20 ns pulse duration.

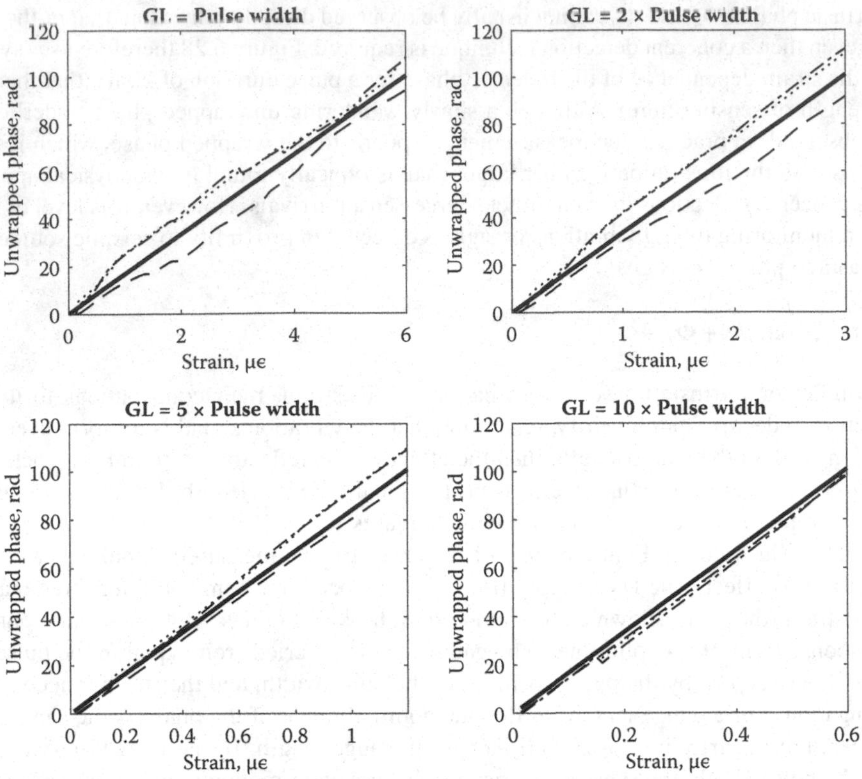

FIGURE 6.29 Strain non-linearity for different ratios of gauge length to pulse width in the case of a uniform vibration signal along the fibre. In each case, the thick solid line is the linear term; the other lines include different realisations of the random reflector phases.

The use of diversity techniques for mitigating fading has the additional benefit of reducing the non-linearity of the response through the averaging of the different non-linear components of the transfer function, in a similar way to the reduction of noise through signal averaging.

6.5.6 Optical Non-Linearity

The limitations on launched optical power in DOFS have been discussed in previous chapters. In the specific case of DVS in long fibres, the main limitations are usually caused by self-phase modulation (SPM) for short pulses or modulation instability for longer pulses. In both cases, the spectral broadening degrades the coherence, and in extreme cases, the contrast of the coherent OTDR trace can be markedly reduced – to a point where the measurement is no longer possible. In both cases, the coherence gradually reduces as a function of distance along the fibre.

In the case of SPM, which causes a chirp at the edges of the pulse, it was shown that a pre-chirping of the probe (in the opposite direction from that caused by the non-linear effects) can be used to compensate the SPM and re-sharpen the contrast of the observed signal [86]. The chirp attempts to reverse, on the time/frequency distribution of the probe, the chirping caused by the Kerr effect acting on changes of intensity within the pulse shape. The compensation must therefore take into account the pulse shape, its intensity and how that intensity varies along the fibre. Strictly, this compensation is effective only at one specific point, but in practice, the correction is useful over a range of several km.

In short sensing fibres, stimulated Brillouin scattering can be the non-linear effect that limits the usable probe power.

6.6 DYNAMIC MEASUREMENTS CONVERTED TO STRAIN

The description of dynamic strain measurements has been focused so far on strain. However, the coherent Rayleigh measurement can be used to estimate any external influence that affects the vector sum of the electric field.

The technique can therefore be applied to a wide range of transducers that convert other measurands to strain. This was demonstrated in Ref. [87], where the sensing fibre was attached to a nickel wire to detect magnetic fields. The transduction principle in this case is magnetostriction, a change in the dimensions of a material when subjected to a magnetic field caused by a re-orientation of grains within the material. Ref. [87] may be viewed as a proof-of-concept, in which a frequency range of [50–5000 Hz] and a resolution of 0.3 G were achieved.

Similar concepts may be applied to electric field measurements via piezo-electric substrates or coatings. Indeed, any process that induces strain on the fibre is a candidate for measurement using coherent Rayleigh scattering.

6.7 STATIC MEASUREMENTS WITH COHERENT RAYLEIGH BACKSCATTER

6.7.1 Static vs. Dynamic Measurements

In the previous sections of this chapter, the measurand is a dynamic signal, i.e. a quantity that varies in time and where the interest is in this variation. The slowest timescale of interest varies between applications but it is usually 1 s or faster. Information that exists on a slower timescale is deemed not to be of interest and is usually filtered out; moreover, the techniques developed for dynamic measurements are ill-suited for acquiring data on a longer timescale, for example, owing to the prevalence of drift in the acquisition. Dynamic measurements are primarily of interest for vibrations, usually mechanical and including, for example, seismic waves.

There are other types of measurements where the long-term stability is crucial; they are described here as static or quasi-static measurements. Here, the user wants to compare data acquired over a long time interval, typically months, and although the measured values will probably change over that time, the key attribute of a quasi-static measurement is the continuity of the data series relative to an initial reference, independently of whether the interrogator is active or not in the intervening period.

As an example, a Raman DTS interrogator can be disconnected from the sensing fibre and reconnected a few months later; the temperature profiles recorded before and after the disconnection can still be compared meaningfully. In contrast, a dΦ-DVS will be unable to track fringes if disconnected; the phase measurements obtained before and after a disconnection will therefore not be related to each other.

6.7.2 Coherent Rayleigh Backscatter in Static Measurements

6.7.2.1 Illustration of the coherent Rayleigh response to static measurements

In the use of Rayleigh backscatter in DVSs, the scattering response to a probe pulse of any section of fibre is seen as a random complex variable, unpredictable in its amplitude and phase owing to the critical relationship between the precise disposition of the scatterers and the probe frequency.

Coherent OTDR, implemented as described in previous sections of this chapter, is unsuited to static measurements. However, we shall show with the aid of synthetic data how the collecting of coherent OTDR traces at multiple source frequencies allows a static measurement to be achieved. It is the additional information obtained by re-measuring the same fibre at different source frequencies and combining the results that transforms the coherent OTDR from a dynamic to a static measurement.

In DVS, the source used is usually essentially single-frequency. In this case, the backscattered signal can be modelled as the summation of the fields scattered by a set of randomly positioned discrete scatterers. The number of scatterers required to model the backscatter response is quite low (a few tens per resolution cell).

However, in the case of a broadband (or frequency-swept) probe, the spacing between the scatterers in the model must be substantially less than the coherence length,* and so the approach of modelling through discrete reflectors becomes rapidly more computationally expensive. Nonetheless, it remains a helpful conceptual model and is used occasionally in the literature when discussing frequency-domain reflectometry.

In order to use the discrete scatterer model of the scattering for a section of fibre with a broadband or multi-frequency source, the frequency resolution in the model must be finer than the reciprocal of the two-way transit time in the section of fibre, i.e. $\ll$100 MHz for 1 m, $\ll$10 MHz for 10 m and so on.

Figure 6.30 (a) shows a synthetic backscatter power measurement returning from a 2-m section of fibre as a function of frequency (offset from a nominal starting value). Here, the frequency is varied in steps of 10 MHz (i.e. roughly one fifth of the inverse transit time in the fibre section) and the frequency range is 800 GHz. A more detailed view of section of the same data is shown in the lower panel over a much narrower frequency range. The frequency separation between adjacent points is still 10 MHz, so there is some correlation between adjacent samples, but nonetheless, the data are very similar in appearance to a single-frequency plot of the measured data as a function of distance along a single-mode fibre with a single-frequency source (Figure 3.7).

Another similarity with time-domain data is that the probability distribution is also close to exponential, as shown in the normalised histogram in Figure 6.30b (cf. Figure 3.10); in this case, the data were decimated to 50 MHz frequency intervals to avoid correlated samples being included in the histogram. This exponential probability density distribution is not surprising because, within the data set of Figure 6.30, there are many independent realisations of the same type of random summation as was discussed in the context of scattering at different locations along the fibre. So for frequencies separated in this case by 50 MHz or more from others in the data set, their electric field summation is statistically independent from summations at different carrier frequencies, just as if the fibre were exchanged for another piece of fibre with a different random arrangement of the scattering centres.

Where the frequency-dependent data for one fibre section (Figure 6.30) *differs* from that of the distance dependence for a long fibre (Figure 3.7), there is an underlying connection between the different points in the frequency data, namely that each frequency is scattered by the same inhomogeneities of the refractive index.

This connection is brought out in Figure 6.31, where two subsets of the spectrum of Figure 6.30c have been expanded to cover a range of 5 GHz each. The lower panels are just subsets of the upper left panel of Figure 6.30. However, in the upper panels of Figure 6.31, the model was recalculated with a decrease in the refractive index in the model of 10^{-5}. The frequency scale in the upper panels has been shifted to higher frequencies by 1.310 GHz, and as can be seen by comparing upper to lower panels, the major features are recognisable between panels, although not totally identical.

What this simple model tells us is that a change to the refractive index is equivalent to a linear shift in the frequency axis. This is further demonstrated in Figure 6.32, where the correlation between the data sets is plotted. The dotted curve is the auto-correlation of the data in the upper left-hand panel of Figure 6.30 and the solid line is the cross-correlation of these data with the frequency distribution

* If this condition on the density of the scatterers is not respected, the model creates a quasi-periodic response that is therefore no longer random as it should be. In this context, the coherence length is that of a source of similar bandwidth to the breadth of the frequency scan.

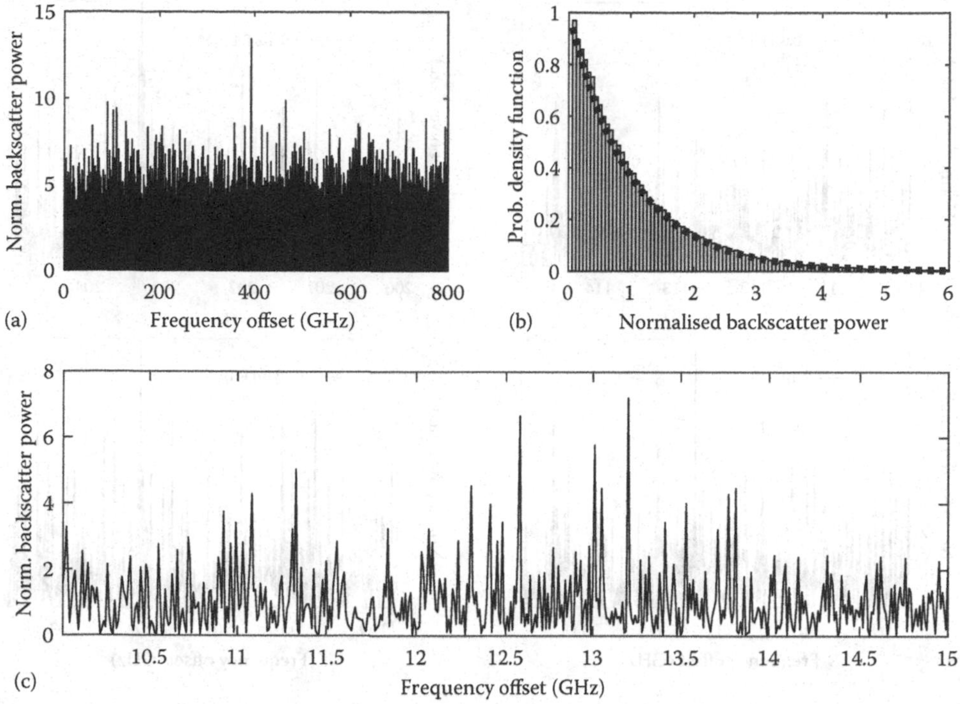

FIGURE 6.30 Frequency dependence of the intensity of the Rayleigh backscatter in a 2 m section of fibre (modelled data). (a) Intensity of the Rayleigh backscatter at one location in a fibre as a function of the frequency of the source (offset from a nominal value); (b) normalised histogram of the intensity value distribution. An enlarged subset of these data (5 GHz span) is shown in (c).

recalculated with a refraction index reduced by 10^{-5}. This curve shows a narrow correlation peak that shifts with a change in the refractive index.

The correlation in the frequency domain illustrated in Figure 6.32 is the basis for static measurements with coherent Rayleigh backscatter. It connects a scaling in the frequency domain to changes in the distance between scatterers in the fibre. Thus, when the refractive index is varied, the optical distance between scatterers is varied in proportion to the refractive index, and this can be detected by cross-correlating a power spectrum with a reference spectrum acquired in a previously known state of the fibre.

For silica-based fibres, a temperature change affects primarily the refractive index (a change to the fibre length from thermal expansion is a second-order effect); the value of 10^{-5} chosen for Figure 6.32 is close to that caused by 1 K change in temperature. In the case of strain, the distance between scatterers is scaled by the strain, but the refractive index is also changed (reduced for elongation), so it is the net effect of elongation and refractive index variation (through the strain-optical effect) that causes a variation of the optical distance between scatterers.

Typical values of the sensitivity of the location of the correlation peak ν_c w.r.t. to temperature and strain are [88] where ν_0 is the centre of the frequency range spanned by the source

$$\frac{\delta \nu_c}{\delta T} = -6.92 \times 10^{-6} \nu_0 \qquad \text{i.e. for } \nu_0 = 193\ THz, \quad \frac{\delta \nu_c}{\delta T} = -1.335 \text{ GHz/K} \tag{6.15}$$

and

$$\frac{\delta \nu_c}{\delta \varepsilon} = -0.78 \nu_0 \qquad \text{i.e. for } \nu_0 = 193\ THz, \quad \frac{\delta \nu_c}{\delta \varepsilon} = -150 \text{ kHz/n}\varepsilon. \tag{6.16}$$

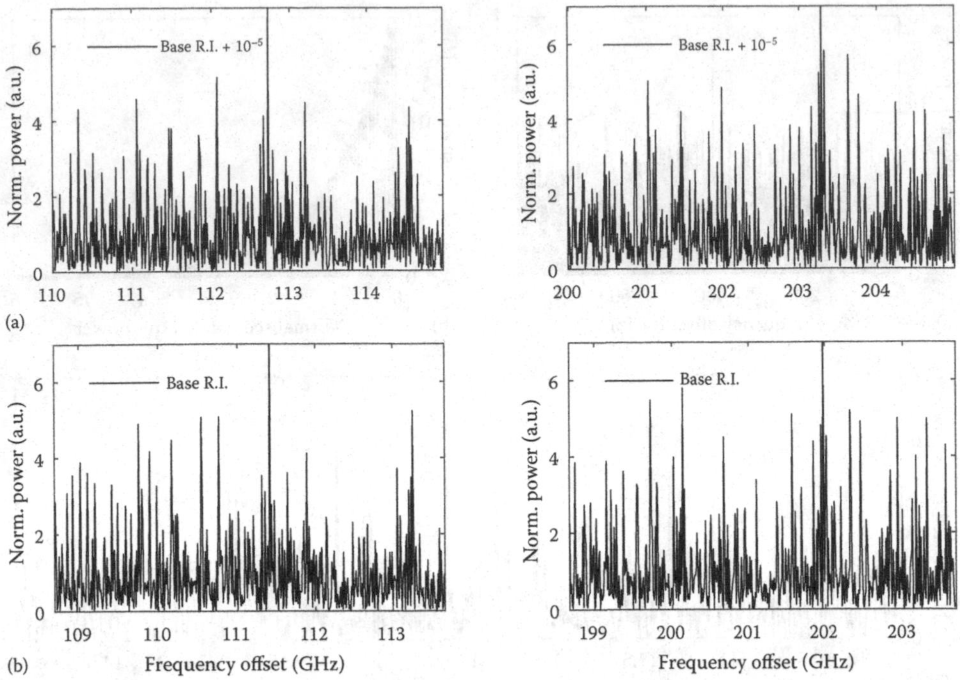

FIGURE 6.31 Further subsets of the data of Figure 6.30, each spanning 5 GHz (b). (a) Were taken from data sets re-calculated with a refractive index reduced by 10^{-5} and the data selection was shifted by 1.310 GHz relative to the lower panels.

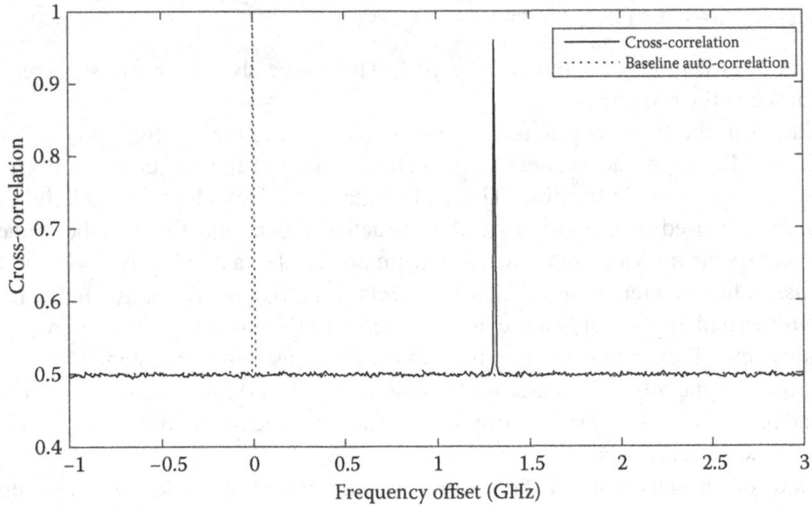

FIGURE 6.32 Correlation of optical frequency-domain data. The auto-correlation of the data of Figure 6.30 is shown (dotted line) along with the cross-correlation between that data and the data recalculated with a refractive index reduced by 10^{-5} (solid line).

This contrasts with typical values for Brillouin sensors of 1 MHz/K and 50 kHz/με, i.e. the frequency shift in the coherent Rayleigh spectrum is about three orders of magnitude greater than that of the Brillouin spectrum, and this reads on to the measurand resolution, provided of course that the frequency resolution is the same for both measurement types.

6.7.2.2 Relationship between the backscatter power spectrum and the refractive index variations

Rather than a collection of randomly placed scatterers, a more detailed representation of the scattered electric field involves a three-dimensional integral of the product of the electric field and a function $\Delta\chi(v_s)$ describing the departure of the permittivity in scattering volume v_s from its mean value [89]. However, in the fibre, we are interested in only the one-dimensional variation of the permittivity, and we therefore take the fluctuations across the surface of the core to be aggregated into a single axial value. The scattering response is then represented by a continuous, complex function of axial position.

Eickhoff and Ulrich [20] showed that the amplitude $\dot{E}_s(\beta)$ of the model element d to the length occupied by the probe pulse $\delta z = \frac{v_g}{2}\tau_p$ of the backscattered wave returning from a section of fibre between coordinates 0 and L, for a wave of propagation constant β and subject to amplitude attenuation $\alpha(x)$, is

$$\dot{E}_s(\beta) = A_0 \int_0^L \sigma(z)\Lambda(z)\exp(2j\beta z)\,dz \text{ where } \Lambda(z) = \exp\left[-\int_0^z \alpha(x)\,dx\right]. \tag{6.17}$$

β is related to the carrier frequency through the phase velocity. The local scatter coefficient $\sigma(z)$ represents the amplitude and phase (relative to the probe) of the re-captured light scattered from elemental fibre section dz.

As can be seen from Equation 6.17, the scattered light field is a Fourier transform of the spatial variations of $\sigma(z)$, so by collecting $E_s(\beta)$ over a wide range of spatial frequencies, the spatial dependence of $\sigma(z)$ can be recovered by measuring the scatter response for the fibre as a function of probe frequency and then transforming it back to the spatial domain through an inverse Fourier transform.

The frequency (or spatial) dependence of the scattering is unique to each piece of fibre, rather like a fingerprint or a DNA profile are unique for humans. It is a direct measure of the axial distribution of the refractive index that is caused by a random process (thermodynamic agitation), the results of which are frozen when the fibre is cooled at the end of the drawing process. When the fibre is stretched or its temperature is varied, the length scale of $\sigma(z)$ is also varied; in the frequency domain, this is a simple scaling operation.

6.7.3 Obtaining Location-Dependent Data from Frequency-Domain Rayleigh Backscatter

In practice, we wish to measure the distribution of the change in the fibre state (temperature or strain) relative to a baseline established, for example, after the fibre was installed, and so the restriction that the change in the state of the fibre must be uniform is not acceptable. Resolving separate locations along the fibre has been achieved principally in two ways. The first is a solution grounded purely in the frequency domain and the second is a mixed time and frequency-domain method.

6.7.3.1 Frequency-domain-only interrogation

The frequency-domain method, initially described in 1999 [90,91], involves sweeping the frequency of a source, typically a tunable laser, and mixing the backscatter return on a photodetector with an OLO derived from the laser, either from a reflector or as a tap from the laser output. This arrangement is essentially a coherent-detection optical frequency domain reflectometer (OFDR). The information on both the spatial distribution and the local state of the fibre is contained in the frequency scan data. The extraction and localisation of the information are achieved through the following steps:

1. Acquire the backscatter signature as a function of source frequency.
2. Convert these data by Fourier transformation from the frequency to the spatial domain; this maps the information from frequency sweep to location along the fibre.
3. Gate the spatial information (by applying a window function) to retain only the information pertaining to a particular fibre section at a time (of course in practice, this gating is applied sequentially to all fibre sections of interest).
4. Apply an inverse Fourier transformation to revert each of the gated data sets to the frequency domain; the transformed data contains information pertinent only to one fibre section for each of the gated segments.
5. Correlate the frequency-domain data from Step 4 with a reference data set acquired with the sensing fibre in a known state.

In principle, this measurement can be conducted by detecting optical *power* as a function of source frequency. In practice, however, most measurements are carried using a heterodyne method. The reference wave is collected either with a reflector in the measurement channel or using a conventional tap to direct a sample of the laser output to the detector where it is mixed with the backscatter. This approach measures the amplitude and phase of the backscatter signal. It also has the usual benefits of coherent detection, such as being able to increase the photocurrent to well above the receiver noise and allowing for narrow-band filtering of the signal.

6.7.3.2 Combined time-/frequency-domain interrogation

In the mixed approach, the position in the fibre is determined using a conventional coherent OTDR approach [88,92,93]. However, the frequency of the source is modulated to provide the change of fibre state information. In one series of articles [88,92], the frequency of the source is varied after acquiring the backscatter trace so as to build up a picture of the backscatter intensity as a function of distance along the fibre and frequency of the source. It is also possible to create a multi-tone probe pulse [93] that contains multiple frequencies that can then be distinguished in the backscatter signal by their offset relative to the laser frequency. Thus, in this mixed approach, the location information is derived from the time-domain data, whereas the measurand value is derived from the carrier-frequency dependence of the backscatter signal.

6.7.4 Limitations in Frequency-Scanned Coherent Rayleigh Backscatter

In both approaches discussed in Section 6.7.3, there is a trade-off between spatial resolution and measurand resolution: reducing the spatial window in the purely frequency-domain approach results in a smoothing of the features of the backscatter spectrum. Likewise, with finite duration probe pulses, the sharpness of the features of the backscatter spectrum are broadened as the resolution cell is made shorter.

In the case of the frequency-domain approach, the spatial resolution is related to the half-range $\Delta\lambda$ of the wavelengths covered in the acquisition or equivalently the frequency spanned Δf by Ref. [90]

$$\delta z = \frac{\lambda^2}{4n \cdot \Delta\lambda} \quad \text{which leads to } \delta z \cdot \Delta f = \frac{v_g}{2}. \tag{6.18}$$

In Ref. [90], it was also shown that the product of δz and the strain resolution $\delta\varepsilon$ in such systems is constant, i.e.

$$\delta z \cdot \delta\varepsilon = \frac{\lambda}{4n}. \tag{6.19}$$

The maximum fibre range is dictated in general by the frequency separation between successive samples of the backscatter signal.

In the case of the mixed time-domain frequency-scanned OTDR using the width δf_c of the autocorrelation function as a criterion for frequency resolution, according to Ref. [78], we can tie the spatial resolution to the frequency resolution through

$$\delta f_c = \frac{1}{\tau_p} = \frac{v_g}{2 \cdot \delta z} \quad \text{i.e.} \quad \delta f_c \cdot \delta z = \frac{v_g}{2}. \tag{6.20}$$

Although these equations (6.18 and 6.20) seem similar, in fact, they are quite different in how they are arrived at; however, they both show how reducing the spatial resolution degrades the resolution to which the shift of the spectrum can be determined and, hence, the measurand resolution.

The limitations on $\delta\varepsilon$ and δz described so far are fundamental; however, they apply on the assumption that the signal-to-noise ratio is sufficient that contributions from receiver noise, for example, can be ignored. With coherent detection, this can be arranged in a wide range of circumstances. The laser linewidth must also be considered, and if it approaches the frequency-domain sample separation, it could further limit the measurand resolution.

In terms of the measurement range, the frequency span must be sufficient to amply cover the variation of measurands expected, based on Equations 6.15 and 6.16, with the addition of a fraction of the range to ensure that the correlation operation can properly identify the shift between the reference measurement and the current one.

It should be remembered that, fundamentally, the frequency-scanned Rayleigh methods determine the distance between scatterers along the fibre. These are altered by temperature and strain, but any other effect that alters the optical path length in the fibre will result in a different value of the measurand. As an example, the ingress of hydrogen and its bonding to the glass can result in a dimensional swelling and changes in the refractive index [94]; although the article cited simply measured the optical path length from one end of the fibre to the other, it is clear that this would also affect frequency-scanned Rayleigh measurements.

6.7.5 Quasi Static Strain Measurements as a Vibration Sensor

If the static strain measurement is carried at a sufficient update rate, i.e. with a sufficiently fast repetition of the laser scan, then of course the technique reverts to one that is sensing a dynamic change, as well as providing a static measurement. However, this concept is quite different from that discussed in previous sections on vibration in that a fast frequency-scanning OFDR yields static data, albeit updated at sub-second rates. In this case, therefore, the issues of phase-tracking do not arise.

This concept was demonstrated in Ref. [95], over a short fibre length (17 m) with a fine spatial resolution (10 cm). The authors were able to measure vibrations at up to 32 Hz. In order to achieve fast scanning, the authors divided the scanned spectra into separate, lower-range scans, each of which carried a snapshop of the strain distribution; this approach allowed the interrogation to be repeated, in effect, at high update rates. The drawback, of course, is that a lower frequency scanning range translated into a coarser spatial resolution.

6.8 LINEARISATION OF THE DYNAMIC STRAIN MEASUREMENT IN INTENSITY COHERENT OTDR

We have already seen (Section 6.2) that the transfer function for amplitude-based DVS systems is unsuitable for measurements that require even monotonicity, let alone linearity. Techniques for alleviating nonlinearity in amplitude-measuring DVS systems have recently become available in specific applications; they are derived from the static measurement concepts described in Section 6.7.

In the first example [96], an intensity-measuring DVS was modified by causing the probe pulses to chirp, i.e. to alter their frequency during the pulse. Specifically, the frequency of the probe pulse was shifted quasi-linearly by about 2 GHz during the 100 ns pulse duration. As a result, the backscatter contains frequency-coded as well as distance-coded information. A comparison of the intensity vs. distance profile for successive probe pulses reveals a distance shift where the temperature and/or strain has changed between the launching of their respective probe pulses.

The change in the state of the fibre causes a local shift of the features of the backscatter intensity vs. distance profile. The magnitude of the shift can be assessed by cross-correlating successive traces. This approach is related to the techniques described in Section 6.7 that used cross-correlation of backscatter acquired at distinct optical frequencies. In Ref. [96], with a chirped probe pulse, probe frequency is mapped locally to distance along the fibre. Therefore, the distance shift obtained from the cross-correlation can be converted to a local frequency translation of the intensity plot. An estimate is thereby obtained of the change in strain or temperature that has occurred locally between the launching of successive probe pulses.

In effect, this method provides a deterministic estimate of the rate-of-change of temperature or strain along the fibre, and this can be integrated over many pulses to estimate the change over a longer timescale. The method is fast and provides a sensitive estimate of the measurand. Of course, it is still a dynamic measurement in the sense that the fibre must be continuously interrogated to track any changes: disconnecting the interrogator from the fibre causes the initial reference to be lost.

An alternative approach [97] is to sum the absolute values of successive differences between backscatter intensity traces. Provided that the changes in temperature or strain are sufficiently small between successive traces, the method provides a reasonably linear estimate of the absolute value of the cumulative change over time. Unlike Ref. [96], it does not provide the sign of the change (in the case of Ref. [96], that information is provided by the sign of the frequency chirp). Nonetheless, the simple method of Ref. [97] provides useful trend analysis that could be used, for example, in fire detection systems, where speed of response is essential.

A third approach [98,99] was demonstrated in the context of vibration sensing for the detection of partial discharge in high-voltage energy transmission cables, particularly at joints. In this case, sections of fibre (200 m), substantially longer than the spatial resolution (50 m) of the coherent OTDR, are packaged as a single sensor and exposed to the vibration. The processing takes advantage of similar exposure of multiple sections in the fibre to dynamic strain from the same source. Although the results are not perfectly linear, a substantial improvement in the transducer response was demonstrated compared with an analysis of a single resolution cell within the transducer.

6.9 SUMMARY

Rayleigh backscatter is rather different from Raman and Brillouin scattering. It relies on subtle variations of the refractive index that are fixed rather than transient, as is the case for inelastic scattering. Thus, interrogating a fibre using coherent Rayleigh backscatter, particularly as a function of carrier frequency, is to probe the inner structure of each section of glass. The glass, although disordered, nonetheless has a random structure that resides permanently within it and it is that structure that the measurements described in this chapter exploit.

In some ways, the refractive index fluctuations that give rise to Rayleigh scattering are similar to a fibre Bragg grating, very weak and random in its phase. The randomness of this pseudo-grating requires it to be probed over a range of frequencies for static measurements, but it provides similar sensitivity to temperature and strain to that of a Bragg grating.

When Rayleigh backscatter is used for distributed vibration sensing, this pseudo-grating provides a pair of reflectors on either side of the disturbance that is measured. The random nature of the grating causes a randomness in the phase of the reflectors, including an unpredictable response to changes in carrier frequency and to strain. In turn this causes an intrinsic non-linearity in the transfer function of DVSs. However, with certain limitations on the design (such as the ratio of pulse duration to gauge length, or the use of multi-frequency interrogation), the linearity of these sensors turns out to be adequate for a number of applications, such as geophysical surveying, in which the signal processing used assumes a linear, stable transfer function.

REFERENCES

1. Bucaro, J. A., H. D. Dardy, and E. F. Carome. 1977. Optical fiber acoustic sensor. *Appl. Opt.* 16: 7: 1761–1762.
2. Culshaw, B., D. E. N. Davies, and S. A. Kingsley. 1977. Acoustic sensitivity of optical-fibre waveguides. *Electron. Lett.* 13: 25: 760–761.
3. Hocker, G. B. 1979. Fiber optic acoustic sensors with composite structure: An analysis. *Appl. Opt.* 18: 21: 3679–3683.
4. Hughes, R., and J. Jarzynski. 1980. Static pressure sensitivity amplification in interferometric fiber-optic hydrophones. *Appl. Opt.* 19: 1: 98–107.
5. Lagakos, N., E. Schnaus, J. Cole, J. Jarzynski, and J. Bucaro. 1982. Optimizing fiber coatings for interferometric acoustic sensors. *IEEE J. Quant. Electron.* 18: 4: 683–689.
6. Nash, P., and J. Keen. 1990. The design and construction of practical optical fibre hydrophones. *Proc. Inst. Acoust.* 12: 201–201.
7. Sudarshanam, V. S., and K. Srinivasan. 1990. Static phase change in a fiber optic coil hydrophone. *Appl. Opt.* 29: 6: 855–863.
8. Wang, C., A. D. Dandridge, A. B. Tveten, and A. Yurek. 1994. Very high responsivity fiber optic hydrophones for commercial applications. 10th International Conference on Optical Fibre Sensors, Glasgow, Scotland, SPIE: 360–363.
9. Yurek, A. M., B. A. Danver, A. B. Tveten, and A. D. Dandridge. 1994. Fiber optic gradient hydrophones. 10th International Conference on Optical Fibre Sensors, SPIE, 2360: 364–367.
10. Knudsen, S., A. B. Tveten, A. Dandridge, and K. Bløtekjær. 1996. Low frequency transduction mechanisms of fiber-optic air-backed mandrel hydrophones. 11th International Conference on Optical Fibre Sensors, Sapporo, Japan, Japan Society of Applied Physics We39.
11. Nash, P. 1996. Review of interferometric optical fibre hydrophone technology. *IEE Proc. Radar Sonar Navigation* 143: 3: 204–209.
12. Kersey, A. D., D. A. Jackson, and M. Corke. 1982. High-sensitivity fibre-optic accelerometer. *Electron. Lett.* 18: 13: 559–561.
13. Pechstedt, R. D., and D. A. Jackson. 1995. Performance analysis of a fiber optic accelerometer based on a compliant cylinder design. *Rev. Sci. Instrum.* 66: 1: 207–214.
14. Cranch, G. A., and P. J. Nash. 2000. High-responsivity fiber-optic flexural disk accelerometers. *J. Lightw. Technol.* 18: 9: 1233–1243.
15. Dakin, J. P., C. A. Wade, and M. Henning. 1984. Novel optical fibre hydrophone array using a single laser source and detector. *Electron. Lett.* 20: 1: 53–54.
16. Sakai, I., R. Youngquist, and G. Parry. 1987. Multiplexing of optical fiber sensors using a frequency-modulated source and gated output. *J. Lightw. Technol.* 5: 7: 932–940.

17. Brooks, J., R. Wentworth, R. Youngquist et al. 1985. Coherence multiplexing of fiber-optic interferometric sensors. *J. Lightw. Technol.* 3: 5: 1062–1072.
18. Hartog, A. H., D. N. Payne, and A. J. Conduit. 1980. Polarisation optical-time-domain reflectometry: Experimental results and application to loss and birefringence measurements in single-mode optical fibres. 6th European Conference on Optical Communication, York, UK, IEE, 190 (post-deadline): 5–8.
19. Healey, P. 1981. OTDR in monomode fibres at 1.3 μm using a semiconductor laser. *Electron. Lett.* 17: 2: 62–64.
20. Eickhoff, W., and R. Ulrich. 1981. Optical frequency domain reflectometry in single-mode fiber. *Appl. Phys. Lett.* 39: 9: 693–695.
21. Healey, P., and D. J. Malyon. 1982. OTDR in single mode fibre at 1.55 μm using heterodyne detection. *Electron. Lett.* 28: 30: 862–863.
22. Healey, P., and D. R. Smith. 1982. OTDR in single-mode fibre at 1.55 μm using a semiconductor laser and PINFET receiver. *Electron. Lett.* 18: 22: 959–961.
23. Healey, P. 1984. Fading rates in coherent OTDR. *Electron. Lett.* 20: 11: 443–444.
24. Healey, P. 1984. Fading in heterodyne OTDR. *Electron. Lett.* 20: 1: 30–32.
25. Healey, P., R. C. Booth, B. E. Daymond-John, and B. K. Nayar. 1984. OTDR in single-mode fibre at 1.55 μm using homodyne detection. *Electron. Lett.* 20: 9: 360–361.
26. Mark, J., E. Bødtker, and B. Tromborg. 1985. Measurement of Rayleigh backscatter-induced linewidth reduction. *Electron. Lett.* 21: 22: 1008–1009.
27. King, J., D. Smith, K. Richards et al. 1987. Development of a coherent OTDR instrument. *J. Lightw. Technol.* 5: 4: 616–624.
28. Mark, J. 1987. Coherent measuring technique and Rayleigh backscatter from single-mode fibers. PhD Technical University of Denmark.
29. Dakin, J. P., and C. Lamb. 1990. Distributed fibre optic sensor system. GB2222247A.
30. Taylor, H. F., and C. E. Lee. 1993. Apparatus and method for fiber optic intrusion sensing. US5194847.
31. Juskaitis, R., A. M. Mamedov, V. T. Potapov, and S. V. Shatalin. 1994. Interferometry with Rayleigh backscattering in a single-mode optical fiber. *Opt. Lett.* 19: 3: 225.
32. Liokumovich, L. B., N. A. Ushakov, O. I. Kotov, M. A. Bisyarin, and A. H. Hartog. 2015. Fundamentals of optical fiber sensing schemes based on coherent optical time domain reflectometry: Signal model under static fiber conditions. *J. Lightw. Technol.* 33: 17: 3660–3671.
33. Strong, A. P., N. Sanderson, G. P. Lees et al. 2008. A comprehensive distributed pipeline condition monitoring system and its field trial. 7th International Pipeline Conference, 26th September–3rd October 2008, Calgary, Alberta, Canada: IPC2008-64549.
34. Martins, H. F., S. Martin-Lopez, P. Corredera et al. 2014. Phase-sensitive optical time domain reflectometer assisted by first-order Raman amplification for distributed vibration sensing over > 100 km. *J. Lightw. Technol.* 32: 8: 1510–1518.
35. Wang, Z. N., J. J. Zeng, J. Li et al. 2014. Ultra-long phase-sensitive OTDR with hybrid distributed amplification. *Opt. Lett.* 39: 20: 5866–5869.
36. Lu, Y., Z. Tao, L. Chen, and X. Bao. 2010. Distributed vibration sensor based on coherent detection of phase-OTDR. *J. Lightw. Technol.* 28: 22: 3243–3249.
37. Itoh, K. 1982. Analysis of the phase unwrapping algorithm. *Appl. Opt.* 21: 14: 2470.
38. Dakin, J. P., and C. A. Wade. 1984. An optical sensing system. GB2126820A.
39. Cranch, G. A., and P. J. Nash. 2001. Large-scale multiplexing of interferometric fiber-optic sensors using TDM and DWDM. *J. Lightw. Technol.* 19: 5: 687–699.
40. Crickmore, R. I., and D. J. Hill. 2010. Traffic Sensing and Monitoring apparatus. US7652245B2.
41. Alekseev, A. E., V. S. Vdovenko, B. G. Gorshkov et al. 2014. Phase-sensitive optical coherence reflectometer with differential phase-shift keying of probe pulses. *Quantum Electron.* 44: 10: 965.
42. Alekseev, A. E., V. S. Vdovenko, B. G. Gorshkov, V. T. Potapov, and D. E. Simikin. 2014. A phase-sensitive optical time-domain reflectometer with dual-pulse phase modulated probe signal. *Laser Phys.* 24: 11: 115106.
43. Posey, R. J., G. A. Johnson, and S. T. Vohra. 2001. Rayleigh scattering based distributed sensing system for structural monitoring. 14th Conference on Optical Fibre Sensors, Venice, Italy, SPIE, 4185: 678–681.
44. Posey, R. J., G. A. Johnson, and S. T. Vohra. 2000. Strain sensing based on coherent Rayleigh scattering in an optical fibre. *Electron. Lett.* 36: 20: 1688–1689.
45. Jackson, D. A., A. Dandridge, and S. K. Sheem. 1980. Measurement of small phase shifts using a single-mode optical-fiber interferometer. *Opt. Lett.* 5: 4: 138–140.
46. Cole, J., B. Danver, and J. Bucaro. 1982. Synthetic-heterodyne interferometric demodulation. *IEEE J. Quant. Electron.* 18: 4: 694–697.
47. Farhadiroushan, M., T. R. Parker, and S. V. Shatalin. 2010. Method and apparatus for optical sensing. WO2010/136810A2.

48. Masoudi, A., M. Belal, and T. P. Newson. 2013. A distributed optical fibre dynamic strain sensor based on phase-OTDR. *Meas. Sci. Technol.* 24: 8: 085204.
49. Farhadiroushan, M., T. R. Parker, and S. V. Shatalin. 2015. Optical sensor and method of use. GB2482641B.
50. Barfoot, D. A. 2015. Interferometric high fidelity optical phase demodulation. WO2015130300A1.
51. Sheem, S. K. 1980. Fiber-optic gyroscope with [3 × 3] directional coupler. *Appl. Phys. Lett.* 37: 10: 869–871.
52. Sheem, S. K. 1981. Optical fiber interferometers with [3 × 3] directional couplers: Analysis. *J. Appl. Phys.* 52: 6: 3865–3872.
53. Priest, R. 1982. Analysis of fiber interferometer utilizing 3 × 3 fiber coupler. *IEEE J. Quant. Electron.* 18: 10: 1601–1603.
54. Gottwald, E., and J. Pietzsch. 1988. Measurement method for the determination of optical phase shifts in 3 × 3 fibre couplers *Electron. Lett.* 24: 5: 265–266.
55. Pietzsch, J. 1989. Scattering matrix analysis of 3 × 3 fiber couplers. *J. Lightw. Technol.* 7: 2: 303–307.
56. Papp, B. 2013. Rock-fibre coupling in distributed vibration sensing. MSc Institut de Physique du Globe de Paris.
57. Johnson, M., and Y. Delaporte. 1993. A passive homodyne fibre vibrometer using a three-phase detector array. *Meas. Sci. Technol.* 4: 8: 854.
58. Mertz, L. 1983. Complex interferometry. *Appl. Opt.* 22: 10: 1530–1534.
59. Koo, K. P., A. B. Tveten, and A. Dandridge. 1982. Passive stabilization scheme for fiber interferometers using (3 × 3) fiber directional couplers. *Appl. Phys. Lett.* 41: 7: 616–618.
60. Cameron, C. B., R. M. Keolian, and S. L. Garrett. 1991. A symmetric analogue demodulator for optical fiber interferometric sensors, Proceedings of the 34th Midwest Symposium on Circuits and Systems, 2: 666–671.
61. Martinelli, M. 1992. Time reversal for the polarization state in optical systems. *J. Mod. Optic.* 39: 3: 451–455.
62. Breguet, J., and N. Gisin. 1995. Interferometer using a 3 × 3 coupler and Faraday mirrors. *Opt. Lett.* 20: 12: 1447–1449.
63. Huang, S. C., and W. W. Lin. 1999. Time-division multiplexing of polarization-insensitive fiber optic Michelson interferometric sensor. US5946429 A.
64. Wang, C., C. Wang, Y. Shang, X. Liu, and G. Peng. 2015. Distributed acoustic mapping based on interferometry of phase optical time-domain reflectometry. *Opt. Commun.* 346: 172–177.
65. Dandridge, A., A. Tveten, and T. Giallorenzi. 1982. Homodyne demodulation scheme for fiber optic sensors using phase generated carrier. *IEEE J. Quant. Electron.* 18: 10: 1647–1653.
66. Fang, G., T. Xu, S. Feng, and F. Li. 2015. Phase-sensitive optical time domain reflectometer based on phase-generated carrier algorithm. *J. Lightw. Technol.* 33: 13: 2811–2816.
67. Crickmore, R. I., and D. J. Hill. 2013. Phase based sensing. US8537345B2.
68. Farhadiroushan, M., T. Parker, and S. Shatalin. 2015. Method and apparatus for optical sensing. GB2518766B.
69. Hartog, A. H., and K. Kader. 2012. Distributed fiber optic sensor system with improved linearity. US9170149B2.
70. Pan, Z., K. Liang, Q. Ye et al. 2011. Phase-sensitive OTDR system based on digital coherent detection. Optical Sensors and Biophotonics III, Shanghai, China, SPIE, 8311: 0S-1.
71. Leeb, W. R. 1983. Realization of 90- and 180 degree hybrids for optical frequencies. *Arch. Elektr. Übertragungstechnik* 37: 203–206.
72. Hodgkinson, T. G., H. R. A., and D. W. Smith 1985. Demodulation of optical DPSK using in-phase and quadrature detection. *Electron. Lett.* 21: 19: 867–868.
73. Hodgkinson, T. G., R. A. Harmon, D. W. Smith, and P. J. Chidgey. 1988. In-phase and quadrature detection using 90° optical hybrid receiver: Experiments and design considerations. *IEE Proc. Pt. J.* 135: 3: 250.
74. Kazovsky, L. G., L. Curtis, W. C. Young, and N. K. Cheung. 1987. All-fiber 90° optical hybrid for coherent communications. *Appl. Opt.* 26: 3: 437–439.
75. Stephens, T. D., and G. Nicholson. 1987. Optical homodyne receiver with a six-port fibre coupler. *Electron. Lett.* 23: 21: 1106–1108.
76. Hartog, A., B. Frignet, D. Mackie, and M. Clark. 2014. Vertical seismic optical profiling on wireline logging cable. *Geophys. Prosp.* 62: 693–701.
77. Parker, T., S. Shatalin, and M. Farhadiroushan. 2014. Distributed Acoustic Sensing – A new tool for seismic applications. First Break 32: February 2014: 61–69.
78. Mermelstein, M. D., R. J. Posey, G. A. Johnson, and S. T. Vohra. 2001. Rayleigh scattering optical frequency correlation in a single-mode optical fiber. *Opt. Lett.* 26: 2: 58–60.
79. Hartog, A. H., and L. B. Liokumovich. 2013. Phase-sensitive coherent OTDR with multi-frequency interrogation. US2013/0113629A1.
80. Coppin, P., and T. G. Hodgkinson. 1990. Novel optical frequency comb synthesis using optical feedback. *Electron. Lett.* 26: 1: 28–30.

81. Horiguchi, T., K. Shimizu, T. Kurashima, M. Tateda, and Y. Koyamada. 1995. Development of a distributed sensing technique using Brillouin scattering. *J. Lightw. Technol.* 13: 7: 1296.
82. Russell, S. J., J. P. W. Hayward, and A. B. Lewis. 2008. Method and apparatus for acoustic sensing using multiple optical pulses. GB2442745B.
83. Hartog, A. H., and K. Kader. 2009. Controlling a dynamic signal range in an optical time domain reflectometry. US7586617B2.
84. Pan, Z., Z. Wang, Q. Ye et al. 2014. High sampling rate multi-pulse phase-sensitive OTDR employing frequency division multiplexing. 23rd Optical Fibre Sensors Conference Santander, Spain, SPIE, 9157: 635.
85. Wang, Z., Z. Pan, Z. Fang et al. 2015. Ultra-broadband phase-sensitive optical time-domain reflectometry with a temporally sequenced multi-frequency source. *Opt. Lett.* 40: 22: 5192–5195.
86. Alekseev, A. E., V. S. Vdovenko, B. G. Gorshkov, V. T. Potapov, and D. E. Simikin. 2016. Contrast enhancement in an optical time-domain reflectometer via self-phase modulation compensation by chirped probe pulses. *Laser Phys.* 26: 3: 035101.
87. Masoudi, A., and T. P. Newson. 2014. Distributed optical fiber dynamic magnetic field sensor based on magnetostriction. *Appl. Opt.* 53: 13: 2833–2838.
88. Koyamada, Y., M. Imahama, K. Kubota, and K. Hogari. 2009. Fiber-optic distributed strain and temperature sensing with very high measurand resolution over long range using coherent OTDR. *J. Lightw. Technol.* 27: 9: 1142–1146.
89. Chu, B. 1974. *Laser light scattering.* Academic Press, London, UK.
90. Froggatt, M. E., and J. Moore. 1998. High-spatial-resolution distributed strain measurement in optical fiber with Rayleigh scatter. *Appl. Opt.* 37: 10: 1735–1740.
91. Froggatt, M., and J. P. Moore. 2003. Apparatus and method for measuring strain in optical fibers using Rayleigh backscatter. US6545760B1.
92. Imahama, M., Y. Koyamada, and K. Hogari. 2008. Restorability of Rayleigh backscatter traces measured by coherent OTDR with precisely controlled light source. *IEICE Trans. Commun.* E91-B: 4: 1243–1246.
93. Hartog, A. H. 2010. Frequency-scanned optical time-domain reflectometry. US7859654B2.
94. Clowes, J. R., J. McInnes, M. N. Zervas, and D. N. Payne. 1998. Effects of high temperature and pressure on silica optical fiber sensors. *IEEE Photon. Technol. Lett.* 10: 3: 403–405.
95. Zhou, D.-P., Z. Qin, W. Li, L. Chen, and X. Bao. 2012. Distributed vibration sensing with time-resolved optical frequency-domain reflectometry. *Opt. Expr.* 20: 12: 13138–13145.
96. Pastor-Graells, J., H. Martins, A. Garcia-Ruiz, S. Martin-Lopez, and M. Gonzalez-Herraez. 2016. Single-shot distributed temperature and strain tracking using direct detection phase-sensitive OTDR with chirped pulses. *Opt. Expr.* 24: 12: 13121–13133.
97. Garcia-Ruiz, A., J. Pastor-Graells, H. Martins, S. Martin-Lopez, and M. Gonzalez-Herraez. 2016. Speckle analysis method for distributed detection of temperature gradients with ΦOTDR. *IEEE Photon. Technol. Lett.* 28: 18: 2000–2003.
98. Rohwetter, P., R. Eisermann, and K. Krebber. 2016. Random quadrature demodulation for direct detection single-pulse Rayleigh C-OTDR. *J. Lightw. Technol.* Early access online. DOI:10.1109/JLT.2016.2557586.
99. Rohwetter, P., R. Eisermann, and K. Krebber. 2015. Distributed acoustic sensing: Towards partial discharge monitoring. 24th International Conference on Optical Fiber Sensors, Curitiba, Brazil, SPIE. 1C.

PART THREE

Applications of Distributed Sensors

Applications of Distributed Temperature Sensors (DTSs)

7

This chapter describes the main applications of distributed temperature sensors (DTSs). In general, they have been addressed using Raman DTS systems (Chapter 4), but in some cases, other technologies such as Brillouin-based distributed sensors (Chapter 5) and coherent optical frequency-domain reflectometry (OFDR) (Chapter 6), that also measure temperature distributions, have been employed in the same application. To avoid repetition, the temperature-related uses of distributed optical fibre sensors (DOFSs) are described here for all distributed sensing technologies.

7.1 ENERGY CABLES: DYNAMIC RATING OF CARRYING CAPACITY

Monitoring buried high-voltage energy cables was probably the first widespread application of DTS. High-voltage cables involve, for each phase, a central conductor, usually made from copper, surrounded by an insulator (typically oil-impregnated paper or cross-linked polyethylene) surrounded by a conductive screen. The whole assembly is usually further sheathed.

Losses in the cable result in heating and, although the cable is designed to operate at somewhat elevated temperature, its life is severely curtailed if operated above the maximum allowable temperature. For example, in some cable designs, the service life is reduced by a factor of two for every 5°C that the cable is operated above 85°C; this is particularly the case with cables insulated by oil-impregnated paper, where the insulation degrades, progressively leading to localised tracking and initially to *partial breakdown*. As in most fields of engineering, the design of such cables involves many compromises such as the conductor area vs. cost.

Buried energy cables are laid in trenches or in ducts within trenches, with a carefully specified backfill that helps dissipate the heat.

Historically, the *ampacity* (i.e. the current carrying capacity) rating of an energy cable was based on the cable design, its backfill and a thermal model based on the seasonal temperature variations and assumptions about soil conditions. The integrity of the transmission network being critical, the cables were generally specified conservatively, particularly when the transmission networks were state owned.

Distributed temperature sensing allows the ampacity of the cable to be estimated more precisely than is the case with the standard models.

Privatisation of the electricity supply industry in many countries has encouraged better asset utilisation. The changes to the industry have also involved trading, with electricity generation companies bidding to deliver power to the network operator for specified time periods, which can be as short as 30-minute slots. The transmission company (network operator) is usually required to compensate the lowest bidder if the supply offered cannot be accepted (for example, because of capacity constraints in their network). This has led to a wish to estimate the ampacity of the cable in real-time and therefore to create a more precise map of the present and short-term future network capacity, based on soil conditions, weather, temperature history of the cable and the recent history of power transmitted through each transmission circuit [1–3]. The loading of a cable varies substantially both during the day and seasonally. Thus, if the ampacity is limited by the temperature of the cable, a circuit that has not been used close to peak rated capacity for some time may be capable of accepting a load in excess of that nominal rating for a short time that depends on its heat capacity and the heating time constant.

DTS is the technology that has enabled real-time thermal rating (or dynamic thermal rating) systems to be deployed. By installing a fibre along an energy cable, the temperatures can be tracked at critical locations that could not be identified prior to commissioning the cable. Sometimes, the ampacity of a circuit is limited by a single hot-spot, which could be caused by poor local backfill or crossing of two cables (that, in effect, heat each other) or interaction with other services, such as district heat pipes. By knowing in real-time the temperature in all parts of the cable, the operator is therefore able to work closer to the actual, rather than the nominal, rating of the cable. In the short-term, this makes a transmission system more resilient, but in the long-term, it also leads to lower capital investment, with DTS technology allowing operators to delay the reinforcement of their network. An example of a temperature profile recorded along a 15-km stretch of live energy cable may be found in Ref. [4] using single-mode fibre as the sensing element. Further temperature profiles and thermal histories of energy transmission cables are shown in Refs. [1–3] and [5].

7.2 TRANSFORMERS

The initial motivation for DTS came from the need to monitor large electrical transformers. Transformers for the transmission network are large (the size of a small bedroom) and expensive and have long manufacturing lead times. As a result, there are generally no directly available spares, and if a unit fails, the circuit that it serves can be out of action for many months. Common causes of transformer failure include lightning strikes and overload, but, like energy cables, the insulation on the windings has limited life at high temperature. Large, high-voltage transformers are monitored with a few point sensors placed in the windings at known critical points. The information is used to control active cooling, a substantial contributor to inefficiency of the apparatus that is therefore turned on only when needed. Reasonably accurate thermal models of transformers have been established by each manufacturer, and so the need for DTS is in fact limited. In addition, the difficulty in laying fibres along the entire set of windings has discouraged their use. Furthermore, the limited spatial resolution available from commercial DTS suppliers has reduced the adoption of the technology in this application. Nonetheless, several manufacturers have used DTS in a limited capacity to validate their thermal models on new transformer designs, and so the technology has been valuable to the industry for qualifying new designs, rather than for monitoring each existing transformer in service.

The application of DTS to transformer design verification was pioneered by ABB (Baden, Switzerland, and Pittsburgh, Pennsylvania), who mastered the means of introducing optical fibres within the transformer windings. They demonstrated the measurement for checking the temporary temperature rise caused by short-term high loads in railway electrical supply systems and also monitored the internal temperature of the transformer on a moving locomotive and showed how its temperature varied as the train moved up and down mountainous terrain [6]. The same reference also shows an example of a temperature profile recorded along the windings of a 22 MW transformer.

In this application (as in many others), the key to a successful DTS measurement is not only the DTS technology but also the expertise in deploying the fibre into the environment in which the measurement must take place. Without a reliable, cost-effective deployment method, DOFS cannot be adopted as a mainstream measurement technique. In the electrical power industry, the sensors cannot be allowed to affect the dielectric performance of the insulation, and there is therefore considerable resistance to adding *anything* to proven designs.

7.3 GENERATORS AND CIRCUIT BREAKERS

Large generators include stator bars that carry high currents and are generally hollow to provide channels for pumping cooling fluid through them. A blockage of a channel would lead very quickly (in seconds) to overheating and failure. This risk, together with the extremely noisy electro-magnetic environment, would suggest useful applications for DTS. In practice, fibre Bragg gratings have been used in this application, perhaps for ease of installation.

The same applies to circuit breakers: when the breaker is activated, an arc usually appears briefly across the terminals; should it persist (perhaps because of a leak of the gas normally present to suppress the arc), damage to the terminals arises. Here again, this application has not seen DTS use.

Although both of these applications benefit from the immunity to electro-magnetic interference brought by optical fibre sensing, neither truly leverages the distributed nature of the measurement and so point sensors are better suited to solving these problems.

7.4 FIRE DETECTION

The use of DTS for fire detection was explored soon after commercial systems became available. However, for cost-effectiveness, the building needs to be large, and installing a fibre cable in each room at a location appropriate to sense the heat of a fire is economically and aesthetically unattractive (the best location for a heat sensor is a few cm below the ceiling, and this is usually unacceptable from an interior design point of view).

It is in transportation infrastructure that DTS technology eventually found applications in heat detection. The fire in King's Cross station on the London Underground on 18 November 1987 [7], in which 31 people died, changed the picture. The cause of the fire was the ignition, probably by a smoker's match, of combustible material (fibre mixed with lubricating grease) that should have been removed by cleaning but had been allowed to accumulate under the escalator over long periods. The very first recommendation (labelled in the 'most important' category) of the report into the disaster was

> All escalator trusses shall be fitted with linear heat detectors and machine rooms with smoke detectors. Priority should be given to escalators with wooden components and consideration given to moving the water fog valves to a protected location outside the machine room. The eventual aim should be for the detection equipment to activate an alarm system, automatic sprinklers or water fog equipment where suitable [7].

London Underground diligently followed the recommendations and installed electrical linear heat detection systems based on a coaxial cable having a heat-sensitive insulation resistance. Unfortunately, in the presence of grease and other contaminants, this technology gave rise to unacceptably high levels of false alarms and in the mid-1990s, the electrical technology was gradually replaced with Raman DTS systems, which, after a trial period, were directly linked to an extinguishing system. The fibres were installed within thin (1/8 in.) stainless steel tubes that provided considerable robustness in a dirty environment that, following the 1987 fire, was subject to periodic rigorous cleaning. A loop measurement arrangement was employed, partly to provide resilience against breaks and changes in fibre loss, but also to provide a so-called 'double-knock' alarm arrangement, i.e. a requirement that both portions of the loop adjacent to each other on the escalator should be in an alarm state for the extinguishing system to be activated, an arrangement that has since been reported for DOFS in another context [8] but is common practice in the fire detection industry.

Fires in road tunnels, for example, in the Mont Blanc tunnel on 24 March 1999 [9], and rail tunnels (e.g. Channel Tunnel fire on 18 November 1996 [10]) unfortunately still occur from time to time. DTS technology is increasingly used as part of the fire detection and fire-fighting solution.

When a fire occurs in a tunnel, smoke quickly obscures any surveillance cameras, and this limits the situational awareness of the fire-fighting crew. A DTS, however, is able to provide information on the location and direction of spread of the heat rising from the fire. This allows the ventilation system to be controlled appropriately, either to starve the fire of oxygen or to give fire-fighting crews a better chance of reaching and dealing with the incident safely.

The thermal profile in tunnels when a fire occurs is the subject of active research. Typically, the sensing fibre is placed near the roof, on the centre line, but stood off the wall. Airflow shifts the apparent location of the fire from vertically above the fire under static air conditions. Radiation and convection both contribute to temperature rise as measured near the roof of the tunnel and so the detection performance depends on whether the fire is open or shielded from the detector; clearly, obstructions to the field of view and airflow affect all types of sensors, but it turns out that linear heat detectors, and DTS sensors in particular, perform particularly well when compared to other technologies (such as smoke detectors, video imaging, infrared [IR] flame detectors) in the case of the fire being hidden from the detector, as would be the case if the fire started inside or under a vehicle [11]. The reason for the difference in response is that many of the other types of detectors rely on a direct optical path between the detector and the fire. Where the fire is hidden, temperature and particularly temperature *with* rate-of-rise are the most sensitive measurands. However, in the case of a fire visible to the detector and when strong air currents are present, video and IR detectors appear to be more effective for detecting the event fast and reliably.

A separate class of applications involves car parks and car parking machines.* Fuel leaks onto hot engines occasionally result in fires, and there is a risk to people as well as to the parked vehicles [12].

European standard EN54 covers fire detection systems in general and linear heat detector systems are covered more particularly by EN54 Part 22, although these are not specific to optical fibre sensors. A DTS-based sensor, quite obviously, uses temperature as an indication of a fire, but in practice, alarm levels can be set based on reaching fixed threshold and/or on a rate-of-rise of temperature. The software that determines whether an alarm should be raised must therefore be able to operate on one or both of these criteria. It is also important to generate the alarm as soon as possible after the criterion has been met and yet to avoid false alarms caused, for example, by noise spikes. The DTS metrology for the fire detection application therefore requires fast but low-noise measurement in order to meet the time-to-alarm required in the industry standards.

7.5 PROCESS PLANT

Vessels containing high-temperature, high-pressure chemical reactions are often operated close to material limits. Failures of the refractory lining, or sometimes of the burner driving the process, have resulted

* Parking systems where the car is driven into a ground-level box that is then moved and stacked by the machine to minimise space usage.

in accidents or near misses and, in all cases, in a loss of production. Although a number of sensors are used to monitor the internal pressure and temperature and the inputs to the chemical process, they do not reliably detect faults that could result in a failure of the containment vessel. Thermal imaging is an effective technique for monitoring the external surface temperature of these vessels, and thermal paint, which changes colour above a critical temperature, is also used. However, these surfaces are often obscured by external insulation or other structures attached to them. A direct measurement of the surface temperature can avoid this problem.

DTS has been applied to monitor refractory-lined reformer vessels [13], blast furnaces [14] and ceramic filters in pressurised, fluidised-bed combustion processes [5,15] to detect hot-spots forming at the surface of the vessel. On detecting such an event, an alarm is raised and used to shut down the process through manual intervention or automatically. The common thread between these applications is the need for high spatial resolution (hot-spots can be far smaller than 1 m), fast response (hot-spots can form fast in the event of the breakdown of the lining) and a complex geometry. In most cases, the fact that a hot-spot is much smaller than the spatial resolution of the DTS cannot be compensated simply by lowering the alarm threshold because the surface temperature is usually non-uniform (see example in Ref. [5]) and liable to change quickly in the event, for example, of a rain shower for vessels exposed to the elements.

With limited spatial resolution, the tendency has been to use innovative fibre patterns to increase the amount of fibre applied per unit area of the vessel, for example, by the close winding of parallel fibre passes, or using serpentine patterns (see illustration in Ref. [5]). Signal processing techniques have been designed to overcome some of the limitations of the hardware, either by deconvolution [16] or by use of dedicated methods to discriminate in favour of small, rapidly forming hot-spots [13]. In the latter case, a thermal history is built up by low-pass filtering and the latest thermal profile is compared against that history to provide a *novelty* profile; this helps overcome the stable non-uniformities due, for example, to deficiencies in the refractory lining. A spatial filter applied to the novelty profile discriminates against sudden temperature changes, affecting large areas of the vessel, such as those due to the weather.

7.6 CRYOGENIC MEASUREMENTS

At the opposite end of the temperature scale, DTS systems have been widely employed to monitor storage tanks and ship-to-shore pipework handling liquefied natural gas (LNG), composed mainly of methane. In this application, the DTS's function is to detect leaks that would result in the build-up of an explosive atmosphere.

The challenge in this case is ensure that the DTS retains its sensitivity at the LNG storage temperature (of order 110 K). It will be recalled that, below −100°C (173 K), the available power in the anti-Stokes Raman band falls very rapidly, although the *relative* sensitivity of this signal to temperature increases. At 110 K, the product of relative sensitivity × anti-Stokes intensity has fallen by a factor of 6 relative to room temperature; at 77 K, this product has fallen by a further factor of 6. As a result, Raman technology can be stretched to monitoring leaks at 77 K (liquid nitrogen) or LNG, but it cannot be used for precise metrology at these temperatures. In addition, attention must be paid, in the design of the DTS, to ensure adequate rejection of the non-temperature-sensitive signals (Rayleigh backscatter and Raman Stokes) that would otherwise swamp the weak anti-Stokes signal with temperature-insensitive signals and thus strongly reduce the DTS sensitivity, and more generally, invalidate the calibration.

Brillouin scattering might be thought to be a more suitable approach, given the very small phonon energy implied by the frequency shift. A simple extrapolation of Equation 5.8 down to cryogenic temperatures would suggest a linear response at least of the anti-Stokes component. Indeed, there is some experimental support in bulk samples [17] for the Brillouin intensity behaving as expected from Equation 5.8. However, the terms that contribute to the Brillouin scattering intensity, such as Young's modulus, behave anomalously in silica at low temperature [18], and so some departure from Equation 5.8 is to be expected. However, there appear to be no publications using Brillouin optical time domain reflectometry (BOTDR) at very low temperature, and the work so far has focused on Brillouin optical time-domain analysis (BOTDA), which does not provide the intensity information.

Cryogenic measurements have been conducted with BOTDA [19–21]. It turns out that the Brillouin frequency is a non-monotonic function of temperature, owing to structural changes taking place in the glass at around 70–110 K (depending on the interrogation wavelength and thus on the hypersonic frequency involved). Fortunately, the structural changes that affect the Brillouin shift also cause the linewidth to vary. The temperature dependence of the linewidth, although also non-monotonic, complements the behaviour of the frequency shift in such a way that, in the region ~1.4–100 K, an unambiguous reading can be obtained.

A third technology, namely Rayleigh OFDR, offers a high sensitivity to temperature, some thousand times that of the Brillouin frequency shift with respect to (w.r.t.) temperature and also no anticipated anomalous behaviour at cryogenic temperature. A resolution of a few mK was demonstrated over the range 77–200 K [22].

The coatings that are applied to the fibre are undoubtedly significant: soft polymer coatings harden below their glass transition temperature (typically below −40°C) and then can induce substantial loss through micro-bending. For very-low-temperature applications, thinner, hard coatings, such as polyimide or organo-ceramic coatings, may be more appropriate. In any case, the thermal expansion of the coating material must be considered as well as its susceptibility to, for example, delamination or cracking at extremely low temperature.

For LNG storage applications, the fibre must be placed where it intercepts any product (liquid or boiled-off gas) that may leak. Sensing fibres are often installed below the tank or pipe and between layers of insulation, which results in a normal temperature that is well below ambient, but higher than that of the LNG. A further benefit of this arrangement is that external damage to the insulation is also detected.

Other applications for distributed measurements at extremely low temperatures (<20 K) include high-energy physics experimental facilities (few in number but very large, complex structures) and, in time, perhaps, energy cables based on high-temperature superconductors, if this technology is ever deployed on a large scale. Another speculative application is the monitoring of hydrogen storage and distribution networks, if this gas displaces hydrocarbons as a transportation fuel (which is far from certain at present).

7.7 PIPELINES: LEAK DETECTION AND FLOW ASSURANCE

Distributed measurements of temperature along gas and liquid pipelines are used to detect leaks, although only a small fraction of the existing pipeline network is fitted with DTS. Leaks, regardless of the fluid transported, are a safety and an environmental hazard.

Other than DTS, one of the main methods for detecting leaks is visual observation from periodic overflights by helicopter; the frequency of these inspections cannot be sufficient to avoid any environmental damage in the event of a leak developing. Mass-balance systems detect leaks by measuring the difference between input and output flow rates. They offer an alternative to visual inspection but the mass-balance technique is capable of detecting only differential flows that are greater than 1% of the nominal flow rate through the pipe; substantial leaks could therefore remain undetected for a significant time. remain undetected for a significant time. Pipeline Instrumented Gauge ('pig') tools are launched into the pipeline and carried along the pipe by the fluid flow. The operator runs pigs at defined time intervals to remove deposits and to detect pipe movement by repeated the in-pipe survey; they also detect corrosion with Eddy current sensors. Acoustic methods have also been attempted.

Generically, it can be readily appreciated that the Joule-Thomson cooling resulting from the reduction of pressure at a leak in a high-pressure gas pipeline will produce a localised cooling that can be detected with an appropriately placed DTS fibre. Likewise, a liquid transport pipeline will generally carry the product at a different temperature from that of the ambient, and again, in principle, a leak can be detected from a change in the thermal profile near the pipe. In spite of these simple concepts, the detection process is not straightforward: long-distance pipelines are large objects: the diameter of a high-pressure pipeline is typically 36–48 in. (0.9–1.2 m), and so a small leak diametrically opposite the sensing fibre would not necessarily be detected.

Each installation must therefore be carefully modelled to ensure that the leak size at which the operator wishes to set an alarm threshold can indeed be detected. The modelling must take into account the nature and temperature of the product, the location of the sensing fibre(s), the ambient temperature of the backfill surrounding the pipe and the permeability of the soil. The results can be surprising. As an example, in the case of a high-pressure gas pipeline, the product is often warmer than the ambient temperature in the soil. Leaking gas is released into the soil and cools as a result of its reduction in pressure (Joule-Thomson effect); one would therefore expect to detect a cold spot. However, under some conditions, the initial effect can be to displace heat from the soil near the pipe (that has been heated by being in proximity to the warm pipe) towards the sensing fibre, and this can result in an *increase*, rather than a *decrease*, in temperature. The contrast in permeability of the backfill relative to the undisturbed soil can also result in containment of the escaped fluid and this increases the probability of detecting an event. However, a detailed model can predict the temperature distribution in time and as a function of position along the sensing fibre; this can be used to optimise the location of the sensing cable and also to refine the algorithms used in the detection process. The model can then be used to estimate the likelihood of failure to detect an event and the false alarm probability.

In the case of pipeline networks collecting fluids in a hydrocarbon production field, the detection problem is even more complex because the pressures and temperatures of the fluid will vary across the network and in time as the production of the field declines. Thus, a leak that might provide a clear signal when the network is brought into service might be challenging to detect 20 years later when the pressures and volumes have fallen.

The deposition of solids is a serious concern for pipeline and flowline operators. Hydrates, waxes and asphaltenes can each form under particular temperature and pressure conditions, and their accumulation can result in a blocked pipeline that is extremely challenging to recover once the process has occurred. Chemicals are used to alleviate the problem but they are expensive, as is heating the pipe. Monitoring the pipeline for signs of deposition is thus attractive to minimise the mitigation measures and to minimise the risk of solids deposition.

Some pipelines carry a product that must be kept in a narrow temperature range through carefully controlled heating. One such example is liquid sulphur, which must be kept above melting temperature (112.8°C) but degrades at elevated temperature, forming increasingly long chains of molecules that are increasingly viscous (gamma sulphur) that clearly could block the pipeline if they are allowed to develop. Sulphur is a by-product of hydrocarbon production in fields that contain large fractions of H_2S, for example, in Western Kazakhstan. In some of these fields, the H_2S is processed to pure sulphur and the product is transported in molten form in pipelines, the temperature of which must be carefully monitored.

Risers are the pipes linking a subsea wellhead to a surface platform or floating production storage and offloading (FPSO) facility. The fluid leaving the seabed undergoes large pressure changes and, in the event of an interruption of the flow, also a large temperature change and dynamic pressure waves (the so-called 'water hammer' effect). This can lead to the deposition of asphaltenes and the formation of hydrates, which are then difficult to clear. Risers with built-in temperature sensing fibres have been deployed in deep-water fields and connected to DTS systems on the FPSO to monitor the temperature of the riser in real time. Heaters and chemical injection can then be used to prevent the deposition of solids that would threaten the free-flow in the riser [23].

7.8 BOREHOLE MEASUREMENT (GEOTHERMAL, OIL AND GAS)

The oil and gas extraction industry is one of the key sectors applying DTS technology; many applications exist for temperature profiles, and some will be described here. At the time of writing, Schlumberger alone has instrumented more than 2000 wells with DTS worldwide. An introduction to the principles of interpreting DTS measurements in hydrocarbon wells is provided by Brown [24].

By October 2014, more than 1000 articles on the use of DTS in the oilfield were available in the OnePetro literature collection, a publication database managed by the Society of Petroleum Engineers but covering the works from several technical societies over many decades. At that time, the total number of entries in OnePetro was of order 200,000, and the fact that 0.5% of them related to the relatively new technology of DTS demonstrates a remarkable interest, given the wide range of topics covered by OnePetro and the long history of the petroleum industry.

7.8.1 Completion of Oil and Gas Wells

A typical well is drilled through a number of formations into the target reservoir. After drilling, a *casing* (heavy gauge tubing joined by threaded terminations at the end of each tubing section) is lowered into the well and cement is pumped through the centre of the casing and back up the annulus between the casing and the formation, providing a barrier to any fluid movement up the well behind the casing. A simple example of a vertical well is shown in Figure 7.1 (left). It is very common for this casing

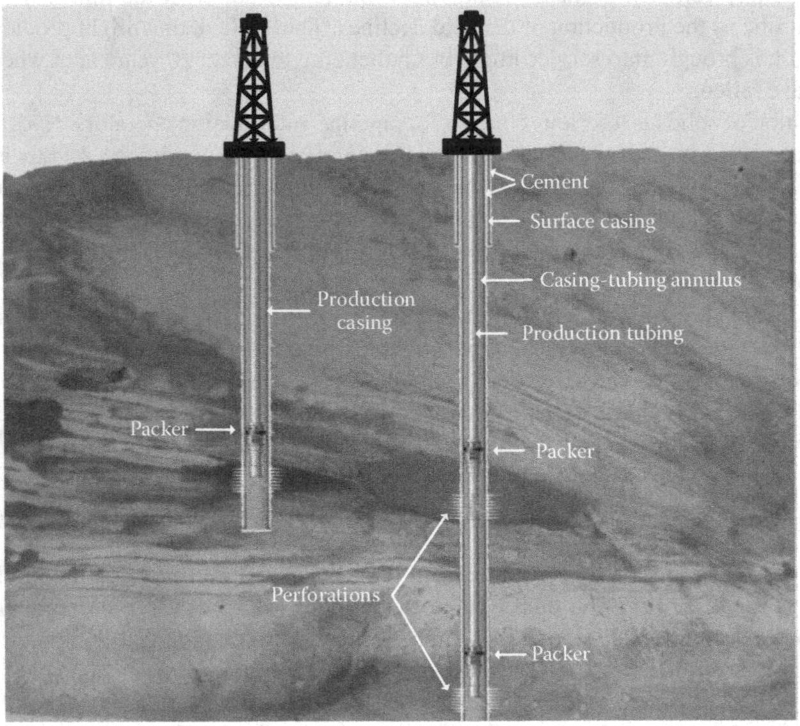

FIGURE 7.1 Illustration of the main features of a cased and perforated vertical well. (Background image courtesy of Christine Maltby.)

operation to be conducted in stages: a section of the well is drilled, cased and cemented; the next section of the well is then drilled through the bore of the first casing string at a smaller diameter and so on as illustrated in the second well (right). One of the reasons for using multiple casings is to allow the drilling fluid density to be adjusted to balance the pressure that must be included in the range between the fluid pressure in the uncased region (i.e. the pressure of the fluids in the rock formation) and the maximum pressure that the rock can sustain (fracture strength).

Production tubing is lowered into the well and the annulus between it and the casing is isolated with a seal, known as a *packer*. A *formation isolation valve* is usually installed just above the production zone to allow maintenance on the well without 'killing' the well with high-density fluids.

The well is designed to cause produced fluids to flow through the production tubing. In the case of a well that is cased through the production zone, the casing is perforated (usually with shaped-charge explosives) to create channels through the casing and the formation; they allow the oil or gas to flow into the well. Alternatively (Figure 7.2), the production interval may be uncased but fitted with a sand screen (a gravel-pack type screen is illustrated in Figure 7.2) that prevents the collapse of the hole and blocks the production of sands and fine particles (*fines*) whilst allowing fluids to flow into the well. Figure 7.2 illustrates the example of a horizontal well that is drilled roughly parallel to the strata to increase the contact area between the borehole and the hydrocarbon-bearing formation. It is also an example of an intelligent completion in which the production interval is divided by packers into multiple zones, the flow into which can be adjusted with inflow-control devices (i.e. valves, usually controlled hydraulically, but sometimes electrically).

Finally, it is worth pointing out that some wells, known as multi-lateral wells, consist of multiple bores, all connected to the main borehole and sharing a common wellhead. This allows, as illustrated in Figure 7.3, independent fluid production from different hydrocarbon-bearing intervals but through the same mother-bore. In other cases, multilateral wells are used to increase the contact area between the wells and the formation; this is particularly important in low-permeability formations.

FIGURE 7.2 Open-hole completion of a horizontal well. (Background image courtesy of Christine Maltby.)

FIGURE 7.3 Illustration of a multi-lateral well. (Image courtesy of Statoil.)

7.8.2 Deployment of Optical Fibres in Wells

Oil and gas wells are a hostile environment. They run at high pressure, commonly to 1,000 bar (wells operating at 2,000 bar do exist), and at high temperature, often at 150°C and beyond, but certain fields, such as in the Gulf of Mexico and certain parts of the UK North Sea, operate above 200°C. When thermal recovery methods are used, for example, in heavy oil production, the temperature can exceed 300°C. The reservoir fluids are often corrosive and toxic (e.g. containing hydrogen sulphide) and frequently also contain hydrogen. More prosaically, it is also a mechanically brutal environment in which large pieces of hardware are used to drill, case and complete the well and these processes are not compatible with the protection of a thin, brittle glass fibre.

An optical fibre cable designed for downhole applications is far more ruggedised than a bare fibre and its design follows that of electrical cables used in the same environment. Firstly, the coating of the fibre is selected for the operating temperature: the acrylate coatings used in the telecommunications industry are not suitable for high temperature; some higher-temperature acrylates are available and silicones are resilient to almost 200°C. For the highest temperatures, polyimide coatings are used and are rated to 300°C. Amorphous carbon coatings are often deposited onto the surface of the fibre prior to applying a polymer coating. The carbon offers some protection against hydrogen ingress, especially at moderate temperatures.

A common cable design (illustrated in Figure 7.4) involves a double-layer of steel tubing, an outer layer of thick steel tubing, usually to a 1/4 in. (6.35 mm) outer diameter together with a thin, inner layer that provides a further barrier to fluid contact. Finally, an outer plastic encapsulation having a square cross-section is often applied to provide further protection against abrasion and to avoid galvanic corrosion from the presence of dissimilar metals. In the case of Figure 7.4, the cable is a hybrid electrical and optical cable allowing both optical fibre sensors and conventional electrical gauges (e.g. pressure and temperature gauges) to be interrogated using a single cable.

As an alternative to installing prepared cables as just described, methods have been developed for installing a sensing fibre in a hydraulic control line that is already installed [25]. Hydraulic control lines are metal conduits typically made out of an alloy suitable for the particular corrosion environment; they are used for actuating valves and other downhole devices by applying pressure or sequences of pressure

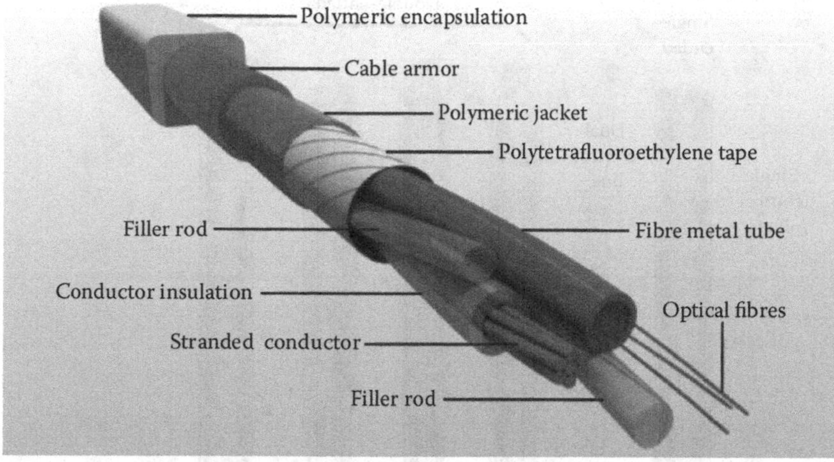

FIGURE 7.4 Hybrid electrical/optical cable for downhole use. (Adapted with permission from Algeroy, J., Lovell, J., Tirado, G., Meyyappan, R., Brown, G., Greenaway, R., Carney, M., Meyer, J., Davies, J., Pinzon, I., *Oilfield Rev.*, 22, 34–41, 2010.)

levels. Their outer diameter is typically 1/4 in., with a wall thickness selected for the pressure they will operate at, e.g. 0.035 in. Optical fibres are installed by pumping a fluid (usually water) through the control line and introducing the fibre in the flow: fluid drag entrains the fibre along the control line. This is quite a gentle process in that the pulling forces are distributed over the entire length of the fibre and the fibre tends to move to the centre of the control line, minimising friction. Lengths of more than 11 km have been pumped successfully through 1/4 in. control lines and more than 15 km through 3/8 in. lines.

The pumping process allows the fibres to be installed after the completion of the well. The control line can be routed through barriers such as packers and well-head outlets and hydraulic connections are more commonly available than optical connections suitable for downhole conditions. Therefore, the entire hydraulic path can be completed and the rig with its crew demobilised; the fibre is installed after the other operations have taken place. This process avoids the need for rig-floor splices, which are time consuming and therefore expensive in rig-time. The pumping process was developed for conditions where the fibre reliability cannot be guaranteed for the required lifetime, and it is common practice, if fibres have degraded, to replace them using pumping methods.

In general, where control lines are used for installing fibres, these are dedicated lines placed in the well for specifically that purpose. The control line is usually looped back to the surface at a '*turn-around sub*' (sub-assembly) so as to form a continuous flow path, which enables a double-ended fibre installation. It also allows the flow to be reversed, so as to remove an existing fibre prior to installing a new, undamaged one. In wells where there is not sufficient space to provide a 180° bend with a sufficiently large radius for a fibre to be pumped around it, the installation is naturally of a single-ended fibre, but a hydraulic loop can still be formed to allow flow reversal. Alternatively, where there is no intention of removing the fibre, the control line can be terminated at the bottom of the well with a one-way valve or *check-valve*. The fluid used for dragging the fibre is then pumped from the surface and flows into the bottom of the well where it is abandoned.

The installation of control lines on production tubing for distributed optical fibre sensing is illustrated in Figure 7.5. To the left, a single control line passes through a packer and is terminated in a check valve. The three central schematics show double-ended control line arrangements; in each case, the turnaround sub is indicated by a short arrow. In centre left, the fibre is looped back above the packer. In the centre, it is attached to a stinger (perforated tube) that hangs below the packer and centre right, and it descends inside the casing. To the right, a two-part completion is illustrated where the lower completion is installed

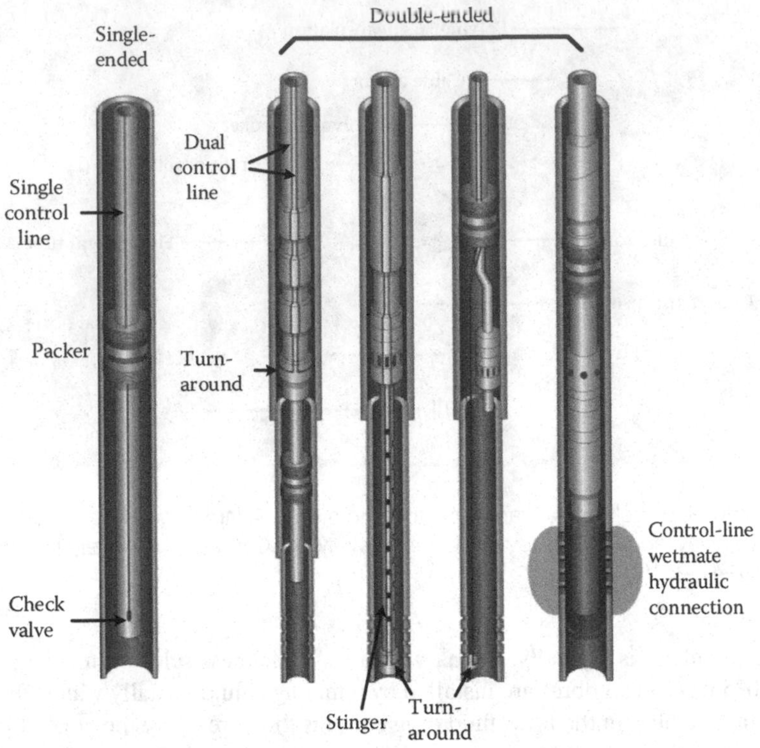

FIGURE 7.5 Examples of control line installations in boreholes for DTS measurements. (Image by S. Siswanto and shown by courtesy of Schlumberger.)

first and the upper completion, including hydraulic connections to the control lines, are mated later at the control-line wet-mate connector (CLWM). This arrangement allows a continuous optical path to be completed by pumping the fibre *after* the completion has been fully installed.

As discussed in Chapter 4, the double-ended method is strongly preferred for the accuracy it provides, particularly in hostile environments, and a looped control-line is ideal for these purposes.

However, there are examples of fibres installed in lines that were intended initially for other purposes such as chemical injection or hydraulic control. Ref. [26] discusses such a case where the fibre was installed in a control line, the primary purpose of which was hydraulic control of downhole valves. Although it was necessary to carefully flush the hydraulic fluid from the control lines using a solvent (isopropyl alcohol in this case), it was found that, after the fibre installation, the lines could be returned to their control function, with the sensing fibre in place and providing data. Of course, the fibre coatings must be compatible with all the fluids involved for pumping, flushing and hydraulic control; likewise, the hydraulic system must also be compatible with these fluids. Although in this example the reason for the fibre sharing a control line with other functions was related to the limited number of penetrators available through various barriers in the well, this is also extremely cost effective, the incremental cost of downhole components being only the fibre itself, together with the service costs for the pumping operations.

The most common location for an optical fibre (protected in a cable or pumped into a control line) is strapped to the production tubing, which is also where cabling for downhole electrical gauges and actuators is usually sited, as are hydraulic control lines. The cables and control lines are held onto the tubing with special protector clamps at each of the joints between tubing sections: these points protrude from the normal tubing diameter, so this is where friction occurs. Further strapping is sometimes applied between clamps to hold the cables tight to the tubing. This is a well-established industry practice. For DTS applications, this means that the temperature measured is close to that of the fluid travelling in the tubing.

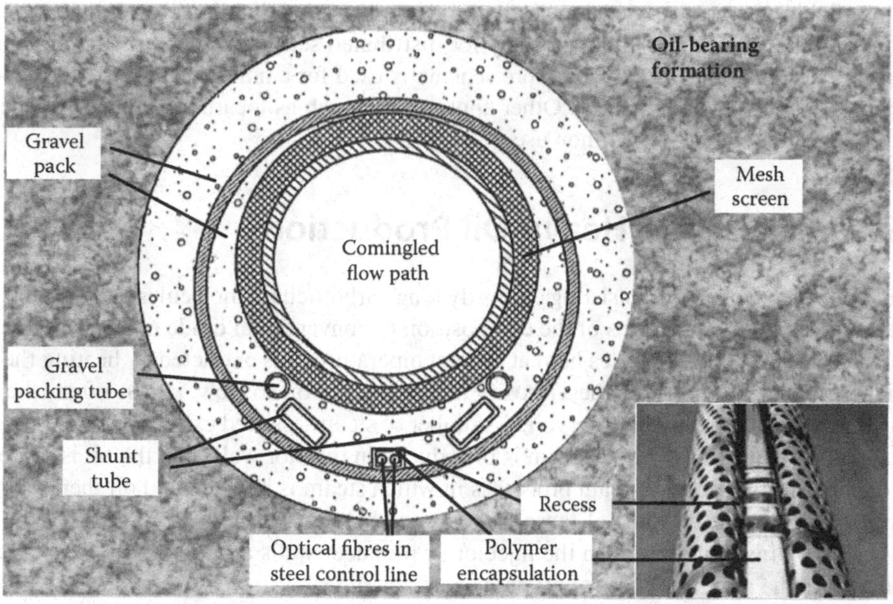

FIGURE 7.6 Dual encapsulated control line installed on the external surface of a sand screen. Main figure: cross-sectional schematic view. (Adapted from Brown, G.A., Method for determining reservoir properties in a flowing well, WO2008/020177 A1, 2007.) The inset shows an oblique view of a complete sandscreen fitted with a dual encapsulated control line.

There are a few examples of fibre installed behind the casing, i.e. in the cemented annulus between the casing and the drilled rock formation: this arrangement provides a measure of the temperature of the formation itself, as opposed to that of the comingled fluid in the production tubing, and has attractions for other sensors, such as distributed vibration sensors. Installing fibres behind casing is more problematic, however: there is a concern about damaging the cables given that the casing is lowered into a hole drilled into rock, in contrast to the tubing that is lowered into the relatively smooth bore of the casing. Although orientated perforation techniques exist, installing cables behind casing still runs the risk of a misplaced perforation charge severing the connection. Finally, there are concerns about the presence of a cable within the cement, in that it could lead to voids in the cement that might result in a leak path for hydrocarbons to reach the surface.

Fibre cables have also been installed on the outside of sand screens, as illustrated in Figure 7.6, where two control lines are encapsulated together in a rectangular cross-section. In this case, an indented grove in the screen allows the cable to be placed in a recess providing protection from abrasion on installation and yet good contact with the formation. Here, the sensor is measuring the temperature of the fluids directly as they enter the well, rather than the temperature of the comingled fluids inside the wellbore. This has important implications for the interpretation of the measurement.

In addition to permanent fibre installations, there are a number of means of introducing fibre on a temporary basis into wells (see Intervention Services, Section 7.8.7). Wireline, an armoured, multi-conductor cable used for electrical measurements for many decades, has now been produced with optical fibres within the cable [27]. Coiled tubing (CT) is a continuous steel tube used to pump fluids and generally operate inside a well. CT strings are now commonly fitted with optical fibres within Schlumberger [28] for communications with a tool at the bottom of the tubing but also for DTS measurements [29].

A type of steel wire, known as slickline, is used for light interventions in wells, including removing and replacing gas-lift valves or actuating sliding-sleeve valves. They are also used for production logging* using battery-powered memory tools that store the data recorded as a function of time for later

* Production logging involves running a suite of measurements whilst a well is flowing to understand which types of fluids are produced from each portion of the well (see Conventional Oil, Section 7.8.4).

interpretation. Schlumberger has developed a special version of a slickline that contains one or more optical fibres [30]; it can be used simultaneously for distributed sensing as well as running conventional slickline tasks. Slickline-conveyed optical fibre sensing is used for a number of tasks such as leak detection and optimising gas-lift systems [31]. Other applications, such as locating the top of the cement [32], have been reported but are not in common use.

7.8.3 Steam-Assisted Heavy Oil Production

Heavy oil is a petroleum product consisting of mostly long carbon-chain molecules and missing the lighter fractions that make up large fractions of the composition of conventional crude oil. As a result, the material is extremely viscous, in some cases hard at room temperature. It is extracted by heating the formation to a sufficient temperature to lower the viscosity and allow the oil to flow.* Most often, the heating is achieved by injecting steam into the reservoir. A number of steam-assisted extraction processes have been developed, such as steam-flood, where steam is introduced in injector wells and the oil is driven laterally towards producing wells or cyclic steam processes in which steam is injected and oil then produced from the same well. DTS is used on most of these processes to help understand which parts of the formation have been heated by installing fibres in the injector or producer wells and often also in specially drilled observation wells. The objective here is optimal use of steam.

Steam-produced oil is cost-sensitive, so as an alternative to permanent monitoring with DTS, often, the fibre is installed in the well and a service company periodically connects a DTS instrument to the sensing fibre to acquire a temperature profile for later analysis. This 'drive-by' approach is suitable for a general monitoring of the effectiveness of the steam injection, but not as an input to the real-time control of the extraction process.

One of the most successful of these steam processes, known as steam-assisted, gravity drainage (SAGD), was developed for the oil sands of Alberta [33]. It involves drilling a pair of horizontal wells one above the other near the bottom of the formation. This is illustrated in Figure 7.7, which also shows fibre for DTS installed in the producing well. The upper well is used to inject steam into the formation along its entire length, typically 1 km. The steam forms a warm chamber in which the viscosity of the oil is lowered and, through gravity, it reaches the lower (producer) well. The injector is typically 5 m above the producer well. The process temperature varies depending on the local composition of the oil, but it is typically in the range 220°C–270°C. SAGD is a very successful technique and has led to an increase of recovery rates from 15% (with earlier processes) to 50% [33,34]; some companies privately claim even higher figures (>70%). The producer well contains a pump, usually electric, and an instrumentation string that was originally an array of thermocouples and a few crude pressure sensors.

SAGD is a rather delicate process in that the formation must be heated sufficiently for the oil to flow, but any non-uniformity in the heating can result in the steam breaking through directly to the producer. If this occurs, a slow and expensive remedial process of heating without flowing the producer is required. In addition, steam is expensive in terms of costs and of energy; some 65%–75% of the operating costs of a SAGD development are related to steam production. The operator tries to maintain a pre-determined difference, known as the *subcool*, between the temperature of the steam leaving the injection tubing and that of the fluid entering the production string.

An example of a series of temperature profiles recorded during the warm-up phase of a SAGD well may be found in Ref. [4], where the progression of the steam may be followed as it extends along the well.

DTS has improved the control of the process by providing a temperature profile along the entire length of the producer well. This means that the temperature of the fluid entering the producer at all points is known at all times. An elevated local temperature can indicate a possible steam breakthrough. Moreover, maintaining the temperature of the entire well close to optimum is important for production efficiency. The operator has some limited control over the steam distribution by varying the steam quality

* With the exception of open-cast mining in cases where the oil-bearing sands are close to the surface.

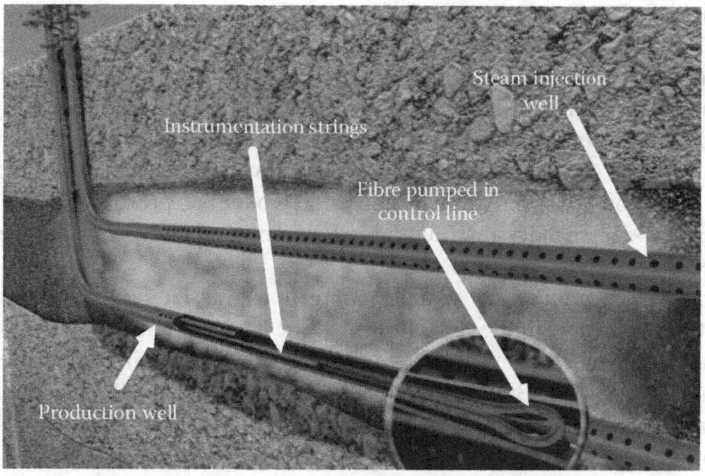

FIGURE 7.7 Illustration of the installation of sensing fibres in the SAGD process. (Adapted with permission from Duey, R. Unfogging the glass, *E&P*, September 2008.)

and usually at least two control valves can be actuated to influence the steam distribution. Active research is underway into using temperature profiles for improved SAGD in automated control loops [34,35].

Double-ended DTS measurements are strongly preferred in SAGD because of the very high temperatures and the related risk of fibre degradation. However, the fibre is usually installed in a 1-in. (25.4 mm) tube (the 'instrumentation string') and the space available is often insufficient for the turnaround in the control line. In the SAGD application, therefore, a loss-compensated single-ended technique (using multiple lasers) is a valuable alternative and in fact was developed for that purpose, at least for multimode DTS systems (see Chapter 4, Section 4.5.1).

7.8.4 Conventional Oil

Conventionally produced oil flows through existing pressure in the reservoir (primary production). Later in the life of the reservoir, production is assisted by a number of techniques to increase recovery, including pressure support, i.e. injecting fluids into the reservoir to maintain the pressure and allow the oil to flow, flooding techniques where fluids are injected to sweep the oil from the proximity of injectors towards the producers and/or injecting fluids that modify the viscosity of the oil and the wettability of the rocks forming the reservoir. A reservoir usually consists of multiple layers with varying permeability, often separated by layers of low or zero permeability. Thus, the reservoir is often naturally partitioned into quite different regions. Producing all the zones in a comingled fashion is seldom optimum and the reservoir engineer needs to understand which fluids (oil, water, gas) are produced from which zone and in what proportion. Producing water is always undesirable because it must be processed and disposed of at surface, e.g. by re-injection after separation. The operator also usually chooses to produce the oil first to allow the gas to continue to drive the production; moreover, in some fields, there are no surface facilities for handling the gas. Production from real formations is complex and causes changes in pressure in each part of the reservoir that in turn change the driving forces on the production process.

In all cases, effective reservoir management requires understanding of the flow profile into the producer. With that knowledge, the reservoir engineer can hope to improve the productivity, by shutting off certain zones that might be producing water, whether from the reservoir or breaking through from injectors, or by stimulating the well (e.g. by hydraulic fracturing or acid treatment) to improve its productivity or by altering the injection regime.

The traditional way of understanding the inflow distribution and nature of the produced fluids is by production logging, which involves lowering a tool that measures fluid velocity (using one or more spinners, i.e. turbine-like devices the rotation rate of which is used to infer fluid velocity), the density (by differential manometry), temperature and sometimes oil/gas ratio by bubble counting [36]. A profile of fluid entry can be derived from the flow rate, density profile and gas/oil ratio by moving the tool down and up the well. However, a production logging operation involves intervening in the well to insert and move the tool, so at best it produces only a snapshot of the production profile, and often, it is necessary to reduce the flow during the log. Understanding the inflow profile under a range of production conditions is lengthy and expensive.

The use of temperature profiles for understanding flow was facilitated by Ramey's approximate model for the heat transfer in a wellbore [37]. It formed the basis for temperature-based analysis of production logs but was largely abandoned, at least in the West, when spinner measurements of sufficient quality became available. In 2000, Brown et al. [38,39] recognised that DTS had changed that situation: by providing a permanently installed, continuous temperature profile of the well, it became possible to obtain *on demand* much of the information provided by a production log and *without intervention* on the well. Also, because of the simplicity of the measurement process, it is possible to repeat the measurement at multiple flow rates or to acquire and interpret additional data when a significant event has occurred.

In practice, the well is instrumented with optical fibre and connected to a DTS system. Logging is generally performed continuously; often, several wells starting from the same platform or pad are interrogated by a single DTS unit, an optical switch being used to scan between fibres. The temperature analysis assumes a quasi-steady-state so the update rate can be relatively slow, favouring high-quality measurements.

The interpretation of these measurements is based on the inversion of a well model: parameters such as the permeability profile (measured during or just after drilling) and geothermal profile are used to calculate a temperature profile that depends on the inflow rate at each of the reservoir intervals: the modelled flow rate profile vs. depth along the well is adjusted so as to match the modelled thermal profile to the actual temperature measurements. The output is a profile of cumulative flow contribution vs. depth and can of course be matched with surface flow meters. In one blind test against spinner measurements, the agreement between the two approaches was better than 2%.

The key to this interpretation is a comparison of the temperature with the local geothermal temperature: fluids entering the well do so at close to* the geothermal temperature and when these fluids flow up the well, they modify the thermal profile above in ways that can be calculated based on Ramey's model [37] or improvements thereof. Software is available to support the workflow of describing the well (or importing an available description), importing the DTS data and providing solutions to match calculated temperature to measured inflow thermal profiles.

This approach works well in single-phase flow conditions with a sufficient geothermal gradient. Although DTS data alone cannot be interpreted unambiguously to provide the multiphase flow in all well types, there are often indications that are nonetheless helpful to the operator. This is particularly the case when the conditions change – a sudden increase in water cut or gas production will be detected by surface flow meters and the DTS often provides clues as to where the fluids entering the well have changed. For example, gas entering the well often breaks out of solution and produces a noisy cooling event above the entry point. Water has a Joule-Thomson coefficient (–0.003°F/psi) that is less than half that of low GOR† oil (–0.0075°F/psi) from oil; in some cases, this can help identify where it is entering. For gas, the coefficient (0.0266°F/psi) is of the opposite sign and ten times the magnitude. If gas is dissolved in the oil, the Joule-Thomson coefficient can vary and even resemble that of water, making the interpretation more difficult.

* There is a change in temperature due to the pressure drop between the formation and the wellbore, caused by the Joule-Thomson effect. For liquids, this effect is small but must be accounted for. For gases, the Joule-Thomson effect is much more significant and results in a strong signal when gases come out of solution.

† Gas-to-oil ratio, the fraction of dissolved gas within the oil.

Another issue arises when the well is perfectly horizontal: the geothermal temperature is then constant and the approach of comparing the measured temperature profile to the geothermal gradient does not yield any useful information; although there is slight Joule-Thomson heating from the pressure drop when the fluid flows from the formation into the well-bore and also from flow along the wellbore, this is usually too small an effect to be used reliably for inferring the inflow profile.

We have to conclude, therefore, that the DTS is a useful tool for continuous production logging in producer wells but, as yet, it is by no means a complete solution. It needs to be complemented with other measurements to provide a full interpretation of the inflow profile. Nonetheless, even the incomplete information that DTS provides is often far better than that already available to the operator; it should also be appreciated that even a coarse (e.g. to within 10%) determination of the inflow profile has considerable value.

Improvements in the temperature resolution and measurement update rate will undoubtedly help interpret more deviated* wells and might ultimately lead to transient analysis where additional parameters, such as fluid heat capacity, can be estimated and used to improve the interpretation. Ideally, quite different measurements, such as pressure distribution, would complete the picture; unfortunately, the precision required is well beyond that which has been demonstrated to date in distributed pressure sensing.

Under special conditions, far more information is available. For example, a sensing fibre cable attached to a sand-screen traversing a geological structure consisting of alternative layers of oil-bearing permeable rock and impermeable layers was found to provide information on the pressure in each producing layer [40]. The reason was that in the impermeable regions, the sensor is measuring the temperature of the comingled fluid in the well, whereas in the regions producing oil, the sensor is measuring primarily the temperature of the fluid entering the wellbore at that point. The temperature is heavily influenced by the Joule-Thomson effect resulting from the pressure drop from the formation and since the pressure in the wellbore can be estimated, the temperature in the permeable regions allows the pressure in these regions to be inferred.

7.8.5 Thermal Tracer Techniques

The interpretation techniques based on referencing the temperature to the geothermal profile are ineffective in the case of oil flowing very slowly. Some fields, especially those yielding rather heavy oil, produce at low flow rates, and the production is often enhanced by drilling lateral wells from a main wellbore. Quite often, these wells are horizontal. In these conditions, a technique based on inducing a sudden temperature event has been devised [41]. The implementation was based on expanding high-pressure nitrogen in a heat exchanger for a relatively brief time. The high-pressure gas is pumped into a down-hole assembly and suddenly released into a larger-diameter return pipe, the first section of which is coiled to form a localised heat exchanger with the borehole fluids. The gas release cools a section of the slowly flowing oil, creating a relatively cool slug of oil that can be tracked using a DTS as it moves towards the well-head. The local flow rate can be estimated from the known pipe diameter and the speed of travel of the temperature feature. This is an example of using a thermal tracer with DTS. It complements the conventional interpretation in that it works particularly well when flow rates are modest, in contrast to the Ramey approach that requires a reasonable flow rate given the temperature resolution of available DTS equipment.

7.8.6 Injector Wells: Warm Back and Hot Slugs

Injector wells are used to insert large volumes of fluids (usually water, but also carbon dioxide, or engineered fluids) into the formation to drive the oil to the producers. The injector well generally flows in

* A deviated well is one that is not vertical. Deviated wells are usually drilled using directional steering techniques and often used to position a long section of the well within a hydrocarbon bearing layer, or to intersect multiple such layers.

single phase and the only aspect of interest to the operator is the distribution of the flow amongst all the layers that are injected: in some cases just one or a few high-permeability layers take most of the fluids. This situation results in early breakthrough of the injected fluid into the producer, which is highly undesirable, and in poor pressure support being provided to other reservoir layers.

A similar approach to that used in producers has been employed in injector wells, namely to compare the well temperature profile with the natural geothermal condition [42]. However, if the well has been flowing for some time with injected fluids, its temperature profile is insensitive to the injectivity profile. In injector wells, it is often more effective to shut-in the well, i.e. to suspend the injection, and allow the temperature to recover towards the natural geothermal gradient using the so-called 'warm-back' technique. The most permeable regions (that have therefore accepted the largest volume of fluids) are more severely cooled by the injection process and so take longer to recover. By analysing the temperature recovery in each location, a profile of the injectivity can be determined. This allows so-called 'thief zones' to be identified and, if appropriate, they can be treated to reduce their permeability [43].

However, even the warm-back technique can be inappropriate if an injector well has been operating continuously for many months: the time required for a shut-in to show an adequate temperature contrast between zones of different injectivity can be prohibitive; given that the primary function of the injector is to provide support to production, the operator is usually reluctant to shut in their injectors for a prolonged period.

An alternative approach, known as the 'hot-slug' technique, has therefore been developed [44]. It involves ceasing the injection for a relatively short time, during which the temperature of the well, above the production interval, warms up relatively quickly because this part of the well has not been cooled by injected fluids. A section of fluid is therefore created that is substantially warmer (by a few to a few tens of degrees) than the fluid in the injection zone. The forming of the hot slug can take a few hours. When injection is resumed, the warm section (the hot slug) travels down the well and is gradually injected into the formation. Its progress can be tracked if the DTS is set to a relatively fast update time (<60 s) and so the local fluid velocity can be inferred. The velocity gradually reduces with increasing depth as more fluid is diverted into the formation at different points. The derivative of velocity w.r.t. depth is a measure of the fraction of the fluid that is injected at that level.

The hot-slug technique is a particular example in borehole measurements where fast but high-temperature resolution measurements are required.

7.8.7 Intervention Services

The foregoing has discussed mainly the permanent deployment of fibre; however, many of the DTS measurements are now performed using temporary deployments with fibre-enabled CT or slickline. This allows DTS measurements to be carried out in wells that have been completed without permanent DTS fibres. The cost of permanently installing an optical cable in an existing well is prohibitive on its own: DTS is sometimes retro-fitted when the completion needs to be removed for some other reason, but never purely to provide the measurement. Having the DTS data available in real-time when intervening on a well provides immediate feedback as to the effectiveness of the operation and in some cases as to where intervention is required.

A few main applications dominate the DTS-enabled intervention services, namely the monitoring of acidising treatment with CT and, on slickline, velocity string gas wells monitoring and the checking of gas-lift valves (both applications are described in the following).

In the case of acid stimulation [43,45], a technique that is used for improving the productivity of oil wells in carbonate reservoirs, a fibre is deployed in the CT. The sequence is then usually as follows:

a. A pre-flush is injected in the space around the CT and the warm-back is used to identify permeable zones;
b. A viscosity-diverting acid (VDA) is pumped through the CT into permeable zones – this is intended to preclude further injection of acid in the zones that are already permeable;

c. DTS measurements are then used to identify or confirm the placement of VDA using the exothermal reaction of the acid with the carbonate rock;
d. Acid is pumped through the CT into zones of poor permeability and DTS warm-back is used to confirm its placement;
e. Final post-treatment flush is pumped and DTS warm-back measurements are used to evaluate the effectiveness of the treatment.

Monitoring the process with DTS in real time allows the service provider to verify the effectiveness of the treatment before leaving the site and to carry out further remedial work if necessary.

7.8.8 Geothermal Energy Extraction

The use of DTS in boreholes was explored by hydrologists and the geothermal community before the oil and gas industry became interested in the subject, with pioneering work from the Geoforschung Zentrum, Potsdam, Germany, initially at near surface [46] and shortly thereafter at elevated temperature [47]. Water at elevated (>300°C) temperature is chemically aggressive, and these applications were amongst the first to demonstrate fibre darkening [48,49].

In temperature measurements in geothermal wells, the local temperature gradient has shown strong correlation with other measurements (gamma ray, sonic) that are characteristic of the detailed layer geology [48]. However, the temperature profile also carries information on the inflow profile and thus fracture locations [50]. This is of particular importance in enhanced geothermal systems (EGSs) that are generally operated at very high temperature (and are therefore very energetic) but also are usually associated with non-sedimentary rocks that exhibit very low permeability unless fractured.

The DTS profile also shows very clearly the presence of steam, which results in essentially constant temperature in the part of the borehole that it occupies [50]. In addition, as a geothermal reservoir is produced, it is important to understand which zones are cooling faster than others, possibly showing higher flow rates or lower contact area with the fractured formation. This information can help manage the production (for example, by adjusting the injection profile).

Geothermal energy extraction is also applied to lower temperature resources from ground water. In this case, the geothermal installation is primarily a heat pump that extracts and upgrades low-grade thermal energy for applications such as district heating or greenhouse heating but not electricity production. An example of the application of DTS for monitoring a ground-coupled heat pumping system is provided in Ref. [51].

7.8.9 CO_2 Sequestration

CO_2 sequestration is the subject of active research as a mitigation for the emission of this greenhouse gas from fossil-fuel burning power stations. DTS and distributed acoustic sensing are two of the techniques used to understand the behaviour of these repositories [52–54], together with a host of conventional techniques, such as wireline logging and repeated surface seismic surveying.

7.8.10 Gas Wells

The principles of logging producing gas wells are similar to those pertaining to conventional oil wells. However, in a gas well, the Joule-Thomson effect is far more important than when oil is produced and so the effect of the pressure in each layer must be taken into account with care [55], especially because in partially produced reservoirs, the depletion of each zone is likely to be different. A short term shut-in of the well is used to detect cross-flow between zones, which allows the pressure distribution in the reservoir to be understood with the support of a thermal simulator.

A particular attraction of DTS in the case of Ref. [55] was the fact that the well was completed with a *velocity string*, a section of tubing designed to concentrate the flow in an annular region to help force water that is co-produced with the gas to move up the well, which otherwise would be 'killed' by the pressure of a large column of water holding the fluids in the formation. The key point here is that the velocity string precludes using conventional spinner logging techniques unless an expensive work-over of the well is conducted to raise the velocity string so that the logging tool can be lowered into the region where the gas from the various entry points is comingling. The use of DTS in these wells simplifies the monitoring task considerably.

7.8.11 Gas Lift Valves

When the reservoir pressure in an oil well becomes insufficient to ensure an adequate flow, it becomes necessary to use artificial lift, such as gas lift. Gas lift valves (GLVs) are devices that are placed in an oil well to inject gas into the fluid column; this reduces the average density, and it therefore also lowers the pressure of the fluid column in the well; this increases the flow rate. GLVs are usually installed in eccentric cavities (side-pockets) and can be retrieved using slickline intervention. GLVs need periodic attention because their optimum setting (the pressure at which they release gas into the column) depends on the flow rate and fluid density. They can also malfunction, for example, by operating intermittently either because of excessive gas pressure, incorrect setting or simply mechanical failure. In addition, if a valve is not operating when it should, the whole operation of the artificial lift system is compromised.

The conventional approach to checking GLVs is production logging on slickline with memory tools [31]. However, with a fibre-enabled slickline attached to a DTS, the data are available immediately. For example, if a valve is found to be operating incorrectly, it can be retrieved right away, brought to surface, adjusted or replaced and then re-positioned. The function of the new valve can then be checked whilst the operators are still on site.

The operation of a valve can be detected based on the slight cooling (c. 1 K) that results from the expansion of gas in the liquid stream.

Permanently installed optical fibre sensor cables have been used in the context of optimising gas lift. In Ref. [56], this was carried out on a sub-sea well where intervention with slickline is more challenging.

7.8.12 Electric Submersible Pumps (ESP)

Electric submersible pumps (ESPs) are commonly used for secondary oil production. DTS monitoring of wells equipped with ESPs was reported as early as 2000 [39], where it was noted that the heat dissipation of the pump creates a distinct change in the temperature profile; under some circumstances, this can help with the flow interpretation because it adds, under steady-state conditions, a known amount of thermal power to the fluid flowing through it and the temperature rise is therefore characteristic of flow rate and heat capacity; it is therefore a proxy for fluid composition.

In addition, ESPs provide the means for varying the production rate in a highly controllable fashion and also often include their own gauges (inlet and outlet pressure and temperature), and the data from these gauges can be used in conjunction with a DTS profile of the entire well to infer the composition as well as the flow rate along the well [57] provided that a surface flowmeter is available to calibrate the total production volume and water cut. In essence, flowing the well at multiple rates whilst logging DTS, pressures and surface production can provide the information required to solve the equations relating temperature and pressure to flow rate and water cut.

7.8.13 Well Integrity Monitoring

The integrity of a well covers many topics including the continuity of the cement behind the casing, structural issues related to ground movement, corrosion and the condition of near-surface equipment.

A prime example of the successful use of DTS in well integrity relates to high-rate gas production wells using vacuum-insulated tubing (VIT). In 1999, the Marlin platform suffered from a leak on well A-2 caused by overheating of VIT, which was located just below the wellhead. This is a deep-water well in the Gulf of Mexico, and apart from environmental risks, a leak can have catastrophic implications for safety, including loss of buoyancy and the possibility of the creation of an explosive atmosphere. The failure was thought to have been caused by thermal effects related to the very large temperature differences that are likely to exist between the production tubing and the surrounding infrastructure [58], in spite of a number of measures that had been taken to limit the effect of the thermal imbalance.

The remedial solution that was adopted after extensive analysis and verification [59] was to include distributed temperature sensing in the re-built completion, which allowed the entire section of the VIT to be monitored in real time [60]. Chemical injection lines were used to pump a sensing fibre into the first 8000 ft. of each of the four wells on the platform. A DTS system was attached to each fibre, and it transmitted temperature profile data to a thermal analysis package that provided real-time safety factors for several critical parts of the completion. This information was fed back to the platform control system. Although this solution may seem simple, it was the culmination of a large remedial program to ensure that operations on the platform could resume safely.

7.8.14 Hydrate Production

Methane hydrates constitute one of the largest reserves of hydrocarbon and the quantity of methane stored in that form is thought to exceed known reserves of natural gas [61]. Methane hydrates are a crystal of water in which molecules of methane are trapped. They are stable under certain conditions of temperature and pressure and are found onshore in polar regions under permafrost and offshore in relatively shallow regions (100–500 m) below the seabed. Estimates of reserves vary, but a consensus estimate at 10 tons of methane is emerging, representing tens of thousands of times the 2012 global consumption of natural gas. Of course, a large fraction of these estimated reserves is inaccessible on technological or economic grounds. Nonetheless, methane hydrates represent a large, essentially untapped, source of energy.

In addition, methane hydrates are a relatively green source of hydrocarbon energy, firstly because burning methane produces far less carbon dioxide (CO_2) than do other hydrocarbons for a given energy output, and secondly, if climate change begins to affect and destabilise the hydrate stores, methane, a far stronger greenhouse gas than CO_2, will be released into the atmosphere in any case. Therefore, exploiting these reserves rather than allowing them to vent naturally might actually reduce greenhouse gas emissions* even in the absence of carbon capture and storage technology.

There are considerable challenges in producing these reserves. Offshore, they exist in relatively unconsolidated sands, often in deep water. No established production method exists, and methane hydrate exploitation is currently a research topic driven by a number of institutions [61]. Test wells have been drilled on a regular basis in recent years to understand the production process.

One production method is to reduce the pressure in the well (and therefore also in the nearby formation). This destabilises the conditions for continuing hydrate existence and therefore the hydrates contained within the pores of the formation then undergo dissociation into methane and water. A number of tests conducted by the Japan Oil, Gas and Metals Corporation have led to a succession of test wells in the Nankai trough off Southern Japan and with other teams in the Canadian Arctic. These tests included

* Here, the timescales are important: methane is a stronger greenhouse gas than carbon dioxide is; however, its lifetime in the atmosphere is shorter and so the relative contributions of a given amount of carbon in one or other of these forms depends on the timescale that is most critical.

a combination of DTS measurements augmented by arrays of point temperature sensors in observation wells near the producing well.

The Nankai trough is in a water depth of c. 950 m. DTS instrumentation was battery powered and operated autonomously for most of a year, probably the first example of a seabed DTS operation in deep water. Specially adapted DTS systems provided by Schlumberger (United Kingdom and Japan) were designed for low power and 'sleep functions' to allow intermittent measurement to be carried out on a regular cycle with equipment shut down between data acquisitions to save power. The results, with many other measurements acquired during drilling and production, provide an understanding of the progression of the dissociation front, and this will feed back into further well tests.

7.8.15 Subsea DTS Deployment

Deploying DTS systems on the seabed is currently of considerable interest in oil and gas production from wells in deep water and in research on production of methane hydrates (discussed earlier).

The earliest DTS systems installed in pressure housings on the seabed were motivated at the Geological Survey of Japan by basic geophysical research into hydrothermal vents at mid-ocean ridges. DTS systems were packaged in pressure housings with battery power and the capacity to operate at pre-determined intervals. Deployments of the system in deep water off the Japanese coast provided both vertical water profiles as well as a line measurement of the seabed showing locations of hot water springs.

In the oil industry, modern subsea well completions usually require the fibre to be installed in at least two sections connected by downhole fibre wet-mate connectors. These are not yet reliable at the time of writing, and so this has limited the application of DTS in these fields, although it is an otherwise ideal technology because it can reduce the number of interventions required on the wells.

7.9 CEMENT AND CONCRETE CURING

The curing of cement is an exothermic chemical reaction that can cause a thermal run-away: the curing process is accelerated at high temperature, and so the warmer the cement, the faster the reaction and, therefore, the more heat is released to stimulate a further acceleration of the curing process. This is a particularly serious problem in thick structures where the heat cannot dissipate naturally and thermal run-away can result in stresses that eventually lead to crack development.

This problem has been recognised for a long time in dam construction; water pipes for cooling the curing material and temperature sensors for monitoring the process are frequently buried in the concrete. DTS allows a high density of sensor points to be installed and the technique has been used in dam construction since the early 1990s, initially in Japan and Turkey and more recently in China [62]. Installing the equivalent number of conventional sensors would be unrealistic on grounds of cost and also because the cables to the individual sensors might result in voids in the structure.

A detailed three-dimensional thermal map of a concrete structure allows a more precise understanding of the curing state, and, in turn, this drives the rate at which the next stage of concrete can be poured safely. In addition, the thermal history, aided by modelling, provides confidence that safety limits on the rate of cure have not been exceeded. Some examples of thermal profiles recorded within a dam after the concrete is poured and as it cures may be found in Ref. [4].

Hydrocarbon wells are generally completed with a steel casing that isolates the formations above the reservoir from the borehole; however, it is also necessary to seal the path behind the casing to prevent leakage behind the casing to the surface or to other sensitive formations, such as potable aquifers.

Conventionally, this sealing is achieved by pumping cement in liquid state through the casing and back up into the annulus between the casing and the drilled formation. At the time of writing, it is uncommon to install sensors in the formation-casing annulus to avoid creating interstices that are not properly filled. However, in those experimental cases where optical fibre cables have been installed behind casing, it has been shown that DTS measurements can easily be used to monitor the state of cure. The heat of the chemical reaction generally results in a temperature rise of 10 K or more. The temperature information can be used firstly to estimate when the curing is sufficiently complete to continue other operations, including safely removing heavy 'kill' fluids that prevent blow-outs when the well is not fully completed. Without monitoring the state of cure in real-time, the well construction 'waits on cement'; i.e. operations are suspended for a prescribed (conservatively judged) time to ensure sufficient curing. Saving even a few hours of rig time is extremely valuable, especially in deep water operations. A more detailed analysis of the temperature, on a somewhat longer timescale that includes clear post-cure cooling, also identifies zones where abnormal amounts of cement have been accepted by the formation, e.g. because of the creation of a cavity during drilling. In time, if installing fibres behind casing for other sensing applications becomes the norm, DTS interpretation is likely to complement, but not replace, ultrasonic cement bond-logging, which is the presently accepted tool for ensuring the integrity of the cement seal.

7.10 HYDROLOGY

7.10.1 Dams and Dykes

Once a dam or dyke is constructed, the ongoing concern is that water seepage within it might lead to increasing flow paths and eventually to failure of the edifice. The materials forming the water barrier consist of a variety of grain sizes, and one slow failure mechanism is the transport of fine particles through the network of coarser material gradually decreasing the resistance to flow. This process allows increasing grain sizes to be removed and leads to the internal erosion of the structure with potentially catastrophic risks to safety and of material damage. The failure can also be caused by external erosion, e.g. from overflowing or slippages of the abutments usually under internal water pressure [63].

DTS technology has been applied to this problem by installing fibres inside dams, dykes and levees, the detection principle being the discovery of a change in temperature due to invading water arriving at a different temperature from that of the monitored structure [63,64].

Using temperature measurements to detect seepage is quite challenging given the small temperature differences involved, the existence of seasonal and diurnal temperature variations as well as faster changes due to exposure to the sun, the perturbative effect of rain and heterogeneities in the edifice caused either by the materials used in the construction or the presence of drainage channels.

The detection problem has been tackled in two ways. In a passive approach, several fibres are installed, some on the reservoir side and some downstream – this allows the temperature contrast to be assessed and the effect of the temperature of the water penetrating the structure to be taken into account. The approach taken in Ref. [63] was firstly to normalise the data (mean subtraction and normalisation to the standard deviation), followed by extracting higher-order statistics that reveal, for example, short-lived changes resulting from precipitation. In a second step, the sources of thermal change are separated using independent component analysis. In a final step, event detection is applied to identify singularities requiring further examination, including site visits. These techniques are still in development but seem promising; they rely on very good temperature metrology (low noise and stability of the calibration over timescales of years) and advanced signal processing adapted to the specific problem.

On the other hand, an active technique involving hybrid sensing/heating cables has been demonstrated [65–68]. The cable contains a heating element able to deliver 10–20 W/m as well as sensing fibres for DTS; the operating mode is usually to pulse modulate the heating so as to establish a contrast between heated and unheated conditions and, in some cases, to observe the transient response.

The heating power that is required clearly depends on the metrology of the DTS available. The authors of Ref. [66] used DTS systems exhibiting a noise level higher by an order of magnitude than was the case in Ref. [63], possibly due in part to the need to make faster measurements for the transient heat pulse detection method. The design of the heating cable is typically the limitation on the range of the system, rather than the DTS measurement. As a result, if the metrology were improved, the range could be extended from the 1–2 km presently achieved in Ref. [66].

The heat pulse measurement is sensitive both to the ground water saturation as well as the heat removal by the fluid flowing past the heated sensor. The first effect provides a measure of the soil's ability to conduct heat away from the heating cable in the absence of fluid movement; the second is a form of hot-wire anemometry (HWA) that measures flow rate indirectly. These effects were separated in Ref. [69] using the contrast between heating and cooling, which is accelerated in the event of fluid flow.

The design of the heating cable is surprisingly significant to the performance of the measurement given the heating times (60–100 minutes) that are typically employed. It is important that the heat can transfer efficiently from the resistive elements to the surrounding medium with minimum thermal impedance.

To summarise, passive measurements using temperature differences between the soil and the retained water (passive tracers) as well as anemometry-based active techniques have been used to attempt to discover and map seepages in dams and dykes. To date, there seem to be no reports of measurements based on active thermal tracer techniques that are well known in other industries such as oil and gas production. This would represent a further dimension in the use of DTS for dam monitoring.

7.10.2 Near-Surface Hydrology, Streams and Estuaries

Ecology research, especially on topics involving water, has benefited from the addition of DTS to the measurement suite. For example, Petrides et al. [70] used DTS to estimate the shade provided by vegetation over a stream in Oregon. An interesting feature of this work is the use of a pair of cables, one black, the other white, which therefore absorb sunlight differentially and can provide a distinction between air temperature and direct solar irradiance. A more detailed thermal analysis of the processes affecting the temperature of a fibre optic cable in water is provided in Ref. [71], and it includes the solar radiation balance as a function of water depth and the dissipative effect of axial and transverse water flow over the cable.

Soil moisture content has been measured using a combination of DTS and heated cable [72], rather similarly to the earlier work on dam leakage [66]. The basic principle of this measurement is that higher moisture improves the thermal conductivity and therefore reduces the temperature rise caused by injecting a defined amount of energy into the heating elements of the cable.

The subtle temperature variations that exist in hydrological applications, particularly in passive sensing (i.e. where no heating is applied) require demanding DTS metrology. Attention needs to be paid to long-term stability and to the effect on the accuracy [73] of environmental changes (temperature particularly) as well as to basic instrument temperature resolution. Careful correction for differential losses is critical in these applications.

DTS metrology is improving, but Ref. [73] shows examples of data that appear to be corrupted by interference between the weak analog signals and the digital processing.

The seepage in stream and polder systems has been studied using DTS [74,75], and the technique was shown to allow the regions of water flow to be identified efficiently; it was noted that even a 1 m resolution was still leaving the finer localisation of seepage to be carried out using traditional, labour-intensive methods.

DTS has found a further related application in understanding inflows into sewage and storm-water systems. In many countries, these are now segregated, but incorrect connections can lead to storm water entering the sewage system or, conversely, foul water being drained into untreated storm water evacuation pipes. Both types of incorrect connection are detrimental to the safe and efficient functioning of the complete water treatment system. Sensing the temperature of a cable placed in the sewer or storm drain or preferably floating in these pipes, together with signal processing techniques designed to detect and classify these events, has allowed a number of incorrect connections to be found and eliminated [76–79].

7.11 ENVIRONMENT AND CLIMATE STUDIES

DTS was used in brackish estuarial waters at Lake Nakaumi in Western Japan [80] to determine the direction of water flow to complement data from acoustic logs and conductivity probes; together, these methods profiled the salinity and the temperature over tidal cycles and helped understand the layering of salt, brackish and fresh waters, which is critical to the ecology of coastal lakes and lagoons.

Growing concerns over changes of the climate and their relationship to greenhouse gases released by human activities has led to a number of applications of DTS in scientific research. For example, DTS systems have been employed [81,82] to monitor the temperature of Antarctic ice shelves. In essence, DTS systems were positioned on the ice shelf and the sensing cable was lowered into the ocean through a borehole specially drilled in the ice. DTS data recorded over a one-year period led to the conclusion that during that time, the interface between the ocean water and the ice was moving up along the cable by ~0.8 m during that time, with rate varying from 8.6 mm/day at the height of the Southern Hemisphere summer to ~1.2 mm/day in the winter, but without any recovery of the base during the Antarctic winter. Of course, this study measured the melting from the base of the shelf only; no comment was reported as to possible growth in thickness from precipitation at the upper surface. It should be noted that, in this case, the DTS system was supplied with batteries recharged with wind and solar energy. This work was thus a useful test of the ability of the technology to work in a hostile remote environment. A four-month gap in the acquisition of data was attributed to a failure of the power supply rather than the DTS equipment.

In an application such as this, the temperature resolution and accuracy are important; a resolution of 0.03 K over ~800 m of sensing cable was reported for averaging times of 5 minutes. The authors used external platinum resistance thermometers to achieve the required calibration accuracy.

A DTS system with a hybrid heating/sensing cable [83] was used to understand the ground thermal history in a region of deep permafrost. The thermal history was reconstructed using the baseline temperature and the thermal conductivity obtained from transient heating. The baseline was obtained by logging the DTS data until a stable value was reached after the completion of the well. The time to reach equilibrium was one month. The well was then perturbed by electrical heating, and the resulting transient temperature changes, measured on cooling, were used to estimate the thermal conductivity at each depth. The thermal conductivity, together with the baseline geothermal profile and geological information gleaned from core samples, was used in a model that was inverted to provide a time dependence of the ground temperature at the surface going back 1000 years.

7.12 BUILDING EFFICIENCY RESEARCH

The drive to improve the thermal efficiency of buildings, including housing, has led to the experimental installation of a Raman DTS system to monitor internal and external walls of two experimental houses constructed specifically for energy efficiency research [84]. A consortium of French research laboratories is conducting long tests using the DTS data to assess the heat and flow in and out of these structures.

7.13 NUCLEAR

The monitoring of temperature in nuclear facilities poses particular problems, in that the fibres can sustain damage from ionising radiation, and, apart from a reduction in the signal-to-noise ratio, it also gives rise to a loss of calibration in single-ended Raman DTS systems because the radiation-induced losses are not the same for both wavelengths [85].

Applications involving the primary cooling loop, especially in experimental reactors using liquid sodium, require operation of the fibre at up to 550°C in a high-radiation environment [86]. In this application, degradation of the fibre is caused by radiation damage.

Another application of DTS in the nuclear industry relates to the safe storage of high-level radioactive waste. It is envisaged to encapsulate this waste in special glasses and store them in tunnels in specially dug repositories. High-level of radioactive waste releases heat and ionising particles and radiation for millions of years and therefore needs to be monitored. The internal radiation can lead to cracking of the glass as a result of internal heating with the potential for leaching of the highly toxic, as well as radioactive, waste [87]. In addition, the entire structure needs to be monitored for self-heating in case the heat concentration leads to damage of the containers and movement of stored waste material.

DTS has been proposed for monitoring certain components of the reactors as well as waste repositories [85,88–90]. Using optical fibres under these conditions has many challenges. Firstly, many of the commonly used coatings degrade in their mechanical strength when exposed to ionisation radiation and this reduces the life of the fibre [91]. However, most attention has been paid to radiation darkening (i.e. an increase in the optical loss) of the fibres when they are exposed to ionising radiation. In pure silica, the dominant effect is that oxygen deficiencies are created and they induce a broad absorption around 630 nm, with effects spreading into the near IR. This is similar to the defects induced in drawing pure-silica core fibres and does not occur in silica fibres containing a high OH content (so-called 'wet fibres') [92]. The addition of common dopants to the core, such as germania and phosphorus pentoxide, further increases the susceptibility to radiation, and they are thus generally avoided in these applications.

Using pure-silica-core fibres with high OH content restricts the wavelength range that can be used and increases the basic attenuation of the fibre even prior to irradiation. The range of a DTS measurement is therefore limited in this application. Because the fibre loss is an exponential function of fibre length, it is usually preferable to deploy several short lengths of sensing fibre in the radiation environment than a single long length. Where an optical switch is used to scan the sensing fibres in turn, the time required for the entire measurement is proportional to (rather than an exponential function of) the number of fibres. In general, therefore, the total time needed for a measurement is far less with multiple, short fibres than that required to average the weaker signal returning from a single, long fibre.

However, it should be recognised that the losses of so-called 'radiation-hardened' fibres deemed suitable for short-distance transmission in high-radiation environments are usually too high for distributed sensing applications. DTS will therefore find applications only under moderate radiation conditions and only if more resilient fibres are developed.

A further application in the nuclear industry concerns the monitoring of insulated steam pipes: very-high-pressure steam pipes are used to convey power from the reactor to the turbines and a leak in the steam could cause serious risk to personnel working on site. Some nuclear plant designs route the high-pressure steam over the control room and a rupture in this area of the pipe could have even more severe consequences if it disables the control systems of the plant. These pipes are heavily insulated and so cannot be visually inspected: the concern is that, by the time leaking steam is apparent in the form of produced water, the pipe could be close to failing completely. DTS was therefore investigated as a monitoring system to detect steam leaks under insulation [93], although it has not yet been used in this application to the author's knowledge.

In the context of measurements in a high-radiation environment, it has recently been shown [94] that the Rayleigh-based coherent OFDR technique (see Chapter 6) is relatively immune in its calibration to the effect of γ-rays. The signal in coherent Rayleigh backscatter relates to the disposition and strength of static scatterers in the glass, and it seems that these are largely unaffected by radiation damage. Of course, it is still necessary for the probe light to reach the point of interest and return and so losses will affect the signal quality, but they should not bias the results, according to Rizzolo et al. [94].

7.14 MINING

The mining industry has used DTS technology for a variety of measurement problems. The use of optical fibre sensors is driven by the potentially explosive atmosphere. If electrical sensors are used in this environment, they must be intrinsically safe, which adds costs and reduces the choice of sensors that can be installed.

Electrical cables to power cutting machines carry considerable power, which leads to obvious risks of overheating in an environment where explosions and fires are a serious hazard. In particular, the energy cables are continuously being reeled in and out and so their ability to dissipate heat is variable; moreover, damage to the cable from constant movement could lead to localised hot-spots [95].

Narrow structures of coal in so-called room-and-pillar mines* lead to stress concentrations, and this, together with airflow through cracks in the pillars, can lead to spontaneous combustion. British Coal installed a DTS on an experimental basis for the purpose of monitoring areas at risk in the Asfordby mine in the early 1990s and it operated successfully for a few years. The pit was shut for economic reasons in 1994 before the trials were finalised.

7.15 CONVEYORS

Conveyor belts are used in mines and in large-scale process plants where the raw material is solid, such as coal-fired power stations, smelters or cement plants. They are a convenient way of moving a steady mass of material from one place (e.g. a dock) to another (e.g. the crushers in a steel plant). The distances covered by these arrangements can amount to several km and they can include thousands of rollers, each one having bearings subject to wear and overheating. In an environment where fine dusk and combustible gas can be present, this is a hazard that can lead to fires, even explosions in some cases. Several sites have been fitted with DTS, and each of the bearings of such conveyors (carrying coal) were fitted with short fibre coils connected optically in series, the entire optical pathway being monitored by a DTS system for overheating. In this way, it was possible to detect indications that a fire might be close to breaking out and undertake preventative maintenance before an incident occurs. Distributed fibre sensors are ideally suited to this long, dirty and potentially hazardous environment.

* In some mines, much of the coal is removed, creating voids that are held up by sections of coal that have been left in place to act as props and hold up the overburden.

7.16 MEASUREMENTS INFERRED FROM TEMPERATURE

7.16.1 Thermal Measurement of Air Flow

The principles of HWA are well understood [96] with electrical sensors. In essence, heat is developed in a sensing wire by passing a current through it and the temperature rise caused by the heating is balanced by dissipation due to flow past the sensor. In HWA, the temperature of the sensing wiring is monitored through its resistance, and from that, the airflow can be deduced. The same concepts can be generalised to optical sensors provided that there is a heat source and a means of measuring temperature. For example. in Ref. [97], the heat is provided by the absorption of light within the sensor and the temperature change is detected interferometrically from the phase difference between light reflecting from a reference surface and light that has travelled through the heated zone. Other examples of optical point anemometers have been published, and in Section 7.10, similar concepts were discussed in the context of hydrology. Although the term is used in a very wide sense, to cover many fluids, of course the etymology of anemometry relates to *air* flow, and this is what is discussed here. Distributed optical fibre hot-wire anemometers using electrical heating and optical heating have been demonstrated [98,99]. The challenge for anemometry is usually to quantify high air velocity because the change in temperature for a given increment of air speed reduces as the air speed is increased. A finer temperature resolution is therefore required at high rates and the technique is increasingly sensitive to bias from other thermal sources.

In the first example cited [98], the distributed sensing technique is coherent Rayleigh OFDR (see Chapter 6), and the fibre was heated by passing an electrical current through a thin (20 μm) coating; a heat rise of ~120°C was observed when dissipating 3.3 W into the coating of a ~0.6 m length of the copper-coated fibre. Given the sensitivity of the Rayleigh measurement (0.1 K for a spatial resolution of 10 mm), a wide range of flow rates should be detectable. The particular experiment was aimed at quite short range and high spatial resolution (as might be required for mapping the airflow in a test rig, such as a wind tunnel), and the Rayleigh measurement was able to detect a 0.2-mm shift in the position of the nozzle delivering the flow to be detected. The same paper used optical heating to circumvent the need for electrical heating.* For optical heating, the sensing fibre was selected deliberately to have high attenuation through incorporation of metal-ion dopants. The temperature rise that was achieved was slightly higher than by electrical heating of a metallic coating. However, a *caveat* is in order here: in the optical case, the heating occurs in the core, whereas in the electrical case, it is the coating that is heated. For anemometry, it is important to heat a region as close as possible to the flow. The difference is subtle in the case of such a thin structure but it might explain the somewhat better (c. 30%) sensitivity observed in the electrical heating in spite of the lower temperature rise generated in this case. It was not clear whether the absorbing fibre in the optical case was coated and, if so, with what material and to what thickness; these practical aspects could have a major influence on the heat transfer from the outer surface of the fibre cladding to the fluid flow.

In Ref. [99], the measurement of airflow was aimed at long distances and more rugged environments and so a Brillouin temperature measurement was used, associated with a more robust stainless steel tube that protected the fibre and acted as a resistive heating element. The results showed a heating relative to ambient of about 70 K; from the curves provided, it can be inferred that flowrates of up to ~2 m/s could be read with reasonable resolution at a heating current of 4.5 A and 3–4 m/s at 6 A. The intention is to cover long distances, and the heating power that was used can be calculated to be about 10 W/m, and so the achievement of the goal of monitoring 10 km of sensing anemometer cable would require some considerable electrical engineering; also, the voltages involved would require substantial insulating coatings that would affect the sensitivity of the sensor, quite apart from the explosion safety issues that they raise.

* This was justified on the grounds of explosion safety, in order to avoid the risk of electrical ignition. It should be noted, however, that optical power is only deemed safe from the point of view of explosion safety at relatively low (<35 mW) average levels, compared with the 1 W used in the work described.

In both the examples discussed, although it is the temperature that was intended to be the prime measurement, in practice, the configurations are likely to lead to strain being applied to the fibre by the heating. It follows that an empirical calibration that allows for the combined result of both effects is likely to be required. From an applications point of view, there is no requirement to interpret the measured value in terms of temperature because of the cross-sensitivity to strain: it is sufficient to calibrate the change in OFDR cross-correlation frequency or Brillouin frequency shift as a function of air velocity. It may, however, be necessary to eliminate changes in ambient temperature by modulating the heating input.

The reports cited are laboratory tests; although they show elegant concepts and demonstrate very good performance, there are substantial difficulties to overcome in practical applications, such as the strain effect* of the airflow, an effect that is not present in electrical hot-wire anemometers, in addition to the question of providing substantial energy over long distances in potentially explosive atmospheres.

7.16.2 Liquid-Level Measurement

The location of the boundary between fluid types is fundamental to the operation of separators.† A liquid-level measurement also assists in understanding the operation of slowly flowing vertical hydrocarbon wells. Passive techniques that rely on the temperature of the different phases can be used in some cases, but it is more reliable to apply heat and then it is the differences in heat capacity of the fluids in the vicinity of the sensing cable that provide the discrimination. In some ways, this concept is similar to HWA, but it works in the absence of fluid flow.

One approach to this problem is in use in the oilfield [100]. In this example, a control line is deployed as a loop in a vertical well and a liquid is left in the control line (whether that was the same liquid that was used to pump the fibre or a replacement fluid that might be more compatible with the fibre coating in the long term). The objective is to identify the interfaces between oil and water and between gas and oil. A double-ended DTS system using a fibre in a control line is used to monitor the steady-state of the system, which settles to the local geothermal profile. In order to discover the location of the interfaces between the fluid phases, the fluid is pumped sufficiently to move the colder (upper) portion of the fluid to the lower part of the control line; alternatively, the fluid is simply exchanged with cold fluid from the surface. This disturbs the equilibrium and the time constant relating to the temperature relaxing back to the steady state is a direct indication of the fluid type. Therefore, by calculating the thermal recovery time as a function of depth from the DTS data, the phase boundary can be located: it corresponds to the depth where the time constant changes.

In this approach, the discrimination of the fluid types is driven directly by the product of heat capacity and thermal conductivity. That product is typically 10 times higher for water than for oil, and there is a further order of magnitude between air or even steam and oil. As a result, the technique of actively disturbing the thermal equilibrium and observing the recovery has proved to be very effective in practical applications. With a reasonable signal-to-noise ratio and with dedicated fitting algorithms, the location of the boundary can be determined to about one tenth of the spatial resolution; this can be improved still further. If better resolution is required, the sensing cable can of course be wound helically to refine the effective depth resolution of the system. The thermal disturbance can also be driven by an external power source [101].

Fluid-level estimation was also demonstrated [102] in a test well for methane hydrate production, where a cable heater used in the lower section of the well was present, as well as a DTS sensing fibre. A thermal transient results when fluids are produced from the bottom, and this can be analysed and interpreted for fluid level.

---

* The strain affects Brillouin or Rayleigh measurements; Raman DTS is immune to this cross-sensitivity.

† Separators are systems that divide an incoming multi-phase flow into single phase streams, for example, gas/oil/water in the petroleum industry.

7.16.3 Performance of Active Fibre Components

The temperature profile of an optical amplifier fibre allows the dissipation of pump energy to be optimised and, more generally, the amplifier design to be optimised.

DTS was used to monitor the temperature distribution within Yb-doped fibre amplifier [103]. Using the BOTDA technique at a wavelength of 1550 nm (well separated from the main absorption bands of the amplifying fibre), the temperature profile of the amplifying fibre (22 m in length) was measured with a spatial resolution of 3 m and a temperature resolution of order 0.1–0.2 K (based on the scatter of the results shown in the paper). The authors were able to show how the temperature rise varied with optical pump power and how it decays along the fibre as a result of the absorption of the pump near the launching end.

A further benefit of the results is that the Brillouin spectrum can be integrated along the fibre to provide a mean linewidth for a given pump level. This linewidth, in turn, dictates the threshold for stimulated Brillouin scattering within the amplifier, which is a serious issue in the design of high-power fibre lasers and amplifiers.

7.17 SUMMARY

The initial applications of DTS in the power and process industries drove the development of the initial technology and commercial offerings. As the technique became established in these applications, and the instrumentation more commonly available commercially, new applications have emerged, some of them seeing considerable uptake, such as in hydrocarbon production and fire protection. In turn, these applications are widening the range of commercial products, opening yet further applications. For the foreseeable future, therefore, DTS technology is likely to continue to develop and be deployed in increasingly numerous applications.

REFERENCES

1. Downes, J., and H. Y. Leung. 2004. Distributed temperature sensing worldwide power circuit monitoring applications, International Conference on Power System Technology, Singapore 2: 1804–1809 vol. 1802.
2. Williams, J. A., J. H. Cooper, T. J. Rodenbaugh, G. L. Smith, and F. Rorabaugh. 1999. Increasing cable rating by distributed fiber optic temperature monitoring and ampacity analysis, IEEE Transmission and Distribution Conference, New Orleans, LA, USA, 1: 128–134 vol. 121.
3. Schmale, M. 2011. Online ampacity determination of a 220-kV cable using an optical fibre based monitoring system, Jicable '11, Versailles, France.
4. Grattan, L., and B. Meggitt. 2000. *Optical fiber sensor technology: Advanced applications-Bragg gratings and distributed sensors*. Springer. Dordrecht, Netherlands.
5. Hartog, A. 1995. Distributed fibre-optic temperature sensors: Technology and applications in the power industry. *IEE Power Eng. J.* 9: 3: 114–120.
6. Ukil, A., H. Braendle, and P. Krippner. 2012. Distributed temperature sensing: Review of technology and applications. *IEEE Sens. J.* 12: 5: 885–892.
7. Fennell, D. Investigation into the King's Cross underground fire. 1988. UK Department for Transport. T022680 0001672 098.
8. Chelliah, P., K. Murgesan, S. Samvel et al. 2010. Looped back fiber mode for reduction of false alarm in leak detection using distributed optical fiber sensor. *Appl. Opt.* 49: 20: 3869–3874.
9. Duffé, P., and M. Marec. Mission administrative d'enquête technique sur l'incendie survenu le 24 mars 1999 au tunnel routier du Mont Blanc [Task force for technical investigation of the 24 March 1999 fire in the Mont-Blanc vehicular tunnel]. 1999. d. T. e. d. L. Ministère de l'Intérieur & Ministère de l'Equipement, France.

10. Channel Tunnel Safety Authority, L. 1997. Inquiry into the fire on heavy goods vehicle shuttle 7539 on 18 November 1996. Stationery Office. ISBN 0115519319.
11. Liu, Z. G., A. H. Kashef, G. D. Lougheed, and G. P. Crampton. 2011. Investigation on the performance of fire detection systems for tunnel applications – Part 2: Full-scale experiments under longitudinal airflow conditions. *Fire Technol.* 47: 1: 191–220.
12. Yamato Protec. Retrieved 4 October 2014, http://www.yamatoprotec.co.jp/english/products/ofds.html.
13. Hartog, A. H., D. P. David, J. J. Hamman, and M. J. Middendorp. 1998. Monitoring wall temperatures of reactor vessels. US5821861.
14. Hironaga, T., O. Iida, and K. Yanagisawa. 1994. Method and apparatus for monitoring temperature of blast furnace and temperature control system using temperature monitoring apparatus. EP0572238B1.
15. Hartog, A. 1995. Fibre-optic temperature sensors monitor Wakamatsu. *Modern Power Syst.* 15: 2: 25–28.
16. Sai, Y. 1997. Temperature distribution measuring apparatus using an optical fiber. US5639162.
17. Vacher, R., and J. Pelous. 1976. Behavior of thermal phonons in amorphous media from 4 to 300 K. *Phys. Rev. B* 14: 2: 823.
18. Spinner, S., and G. W. Cleek. 1960. Temperature dependence of Young's modulus of vitreous germania and silica. *J. Appl. Phys.* 31: 8: 1407–1410.
19. Le Floch, S., and P. Cambon. 2003. Study of Brillouin gain spectrum in standard single-mode optical fiber at low temperatures (1.4–370 K) and high hydrostatic pressures (1–250 bars). *Opt. Commun.* 219: 1: 395–410.
20. Fellay, A. 2003. Extreme temperature sensing using Brillouin scattering in optical fibers. PhD Ecole Polytechnique Fédérale de Lausanne.
21. Fellay, A., L. Thévenaz, J. P. Garcia et al. 2002. Brillouin-based temperature sensing in optical fibres down to 1 K, 15th International Conference on Optical Fiber Sensors, Portland, OR, USA, IEEE, vol. 1: 301–304.
22. Lu, X., M. A. Soto, and L. Thévenaz. 2014. MilliKelvin resolution in cryogenic temperature distributed fibre sensing based on coherent Rayleigh scattering, 23rd International Conference on Optical Fibre Sensors, Santander, Spain, SPIE, 9157: 91573R.
23. Keprate, A. 2014. Appraisal of riser concepts for FPSO in Deepwater. MSc University of Stavanger.
24. Brown, G. A. 2005. *The essentials of fiber-optic distributed temperature analysis.* Schlumberger Educational Services, Sugarland, TX, USA.
25. Clowes, J. R. 2000. Fibre optic pressure sensor for downhole monitoring in the oil industry. PhD University of Southampton.
26. Tolan, M., M. Boyle, and G. Williams. 2001. The use of fiber-optic distributed temperature sensing and remote hydraulically operated interval control valves for the management of water production in the Douglas field. In SPE Annual Technical Conference and Exhibition. New Orleans, LA, USA, Society of Petroleum Engineers. SPE 71676.
27. Varkey, J., R. Mydur, N. Sait et al. 2008. Optical fiber cables for wellbore applications. US7324730 B2.
28. Lovell, J. R., M. G. Gay, S. Adnan, and K. Zemlak. 2009. System and methods using fiber optics in coiled tubing. US7 617 787B2.
29. Jee, V., N. Flamant, H. Thomeer, S. Adnan, and M. Gay. 2012. Use of distributed temperature sensors during wellbore treatments. US8113282B2.
30. Ullah, K. 2010. Fiber optic slickline and tools. US2010/0155059A1.
31. Sanchez, A., G. A. Brown, V. L. S. Carvalho, A. S. Wray, and G. Gutierrez Murillo. 2005. Slickline with fiber-optic distributed temperature monitoring for water-injection and gas lift systems optimization in Mexico, SPE Latin American and Caribbean Peteroleum Engineering Conference, Rio de Janeiro, Brazil, Society of Petroleum Engineers: SPE 94989.
32. Diarra, R., J. Carrasquilla, Y. Gonzalez et al. 2014. Cement evaluation using slickline distributed temperature measurements, AADE Fluids Technical Conference and Exhibition, Houston, Texas, USA: AADE-14-FTCE-41.
33. Butler, R. M. 1998. SAGD comes of AGE! *J. Can. Petrol. Technol.* 37: 7: 9–12.
34. Guo, T. 2014. Automated control for oil sands SAGD operations. University of Calgary, Canada.
35. Stone, T. W., G. Brown, B. Guyaguler, W. J. Bailey, and D. H. S. Law. 2014. Practical control of SAGD wells with dual-tubing string. *J. Can. Petrol. Technol.* January: 32–47.
36. Fordham, E. J., A. Holmes, R. T. Ramos et al. 1999. Multi-phase-fluid discrimination with local fibre-optical probes: I. Liquid/liquid flows. *Meas. Sci. Technol.* 10: 1329–1337.
37. Ramey, H. J. 1962. Wellbore heat transmission. *J. Petroleum Technol.* 14: 04: 427–435.
38. Brown, G. A. 2000. Method and apparatus for determining flow rates. EP1196743B1.
39. Brown, G. A., B. Kennedy, and T. Meling. 2000. Using fibre-optic distributed temperature measurements to provide real-time reservoir surveillance data on Wytch Farm field horizontal extended-reach wells, 2000 SPE Annual Technical Conference and Exhibition, Dallas: TX, USA, SPE 62952.
40. Brown, G. A. 2007. Method for determining reservoir properties in a flowing well. WO2008/020177 A1.

41. Brown, G. A. 2007. Method and apparatus for measuring fluid properties. US7240547B2.
42. Jalali, Y., T. D. Bui, and G. Gao. 2009. System and method for determining a flow profile in a deviated injection well. US7536905B2.
43. Garzon, F., J. R. Amorocho, M. Al-Harbi et al. 2010. Stimulating Khuff gas wells with smart fluid placement. *Saudi Aramco J. Technol.* Summer: 23.
44. Brown, G. A. 2012. Method to measure injector inflow profiles. US8146656B2.
45. Al-zain, A. K., R. Said, S. D. Al-Gamber et al. 2009. First worldwide application of fiber optic enabled coiled tubing with multi-lateral tool for accessing and stimulating a tri-lateral oil producer, Saudi Arabia, Kuwait City, Kuwait SPE 126730.
46. Hurtig, E., J. Schrötter, S. Grosswig et al. 1993. Borehole temperature measurements using distributed fibre optic sensing. *Sci. Drilling* 3: 283–286.
47. Schrötter, J., and H. Hoberg. 1993. Measurements with the distributed fibre optic temperature sensing techniques at temperatures of about 400°C, Symposium on New Developments of Temperature Measurements in Boreholes, Klein Koris, Germany.
48. Wisian, K. W., D. D. Blackwell, S. Bellani et al. 1996. How hot is it? (A comparison of advanced technology temperature logging systems). *Trans. Geotherm. Res. Council*: 427–434.
49. Smithpeter, C., R. Normann, J. Krumhansl, D. Benoit, and S. Thompson. 1999. Evaluation of a distributed fiber-optic temperature sensor for logging wellbore temperature at the Beowave and Dixie Valley geothermal fields 24th Workshop on Geothermal Reservoir Engineering, Stanford University, Stanford, CA.
50. Sakaguchi, K., and N. Matsushima. 2000. Temperature logging by the distributed temperature sensing technique during injection tests, Proc. World Geothermal Congress: Kyushu – Tohoku, Japan 1657–1661.
51. Walker, M. D. 2015. Residential and district-scale ground-coupled heat pump performance with fiber optic distributed temperature sensing. MSc University of Wisconsin-Madison.
52. Daley, T., and B. Freifeld. 2014. Borehole based monitoring of CO_2 storage recent developments in fiber-optic sensing, Rite CCS Workshop, Tokyo, Japan.
53. Freifeld, B. M., T. M. Daley, S. D. Hovorka et al. 2009. Recent advances in well-based monitoring of CO_2 sequestration. *Energy Proc.* 1: 1: 2277–2284.
54. Prevedel, B., S. Martens, B. Norden, J. Henninges, and B. M. Freifeld. 2014. Drilling and abandonment preparation of CO_2 storage wells – Experience from the Ketzin pilot site. *Energy Proc.* 63: 6067–6078.
55. Huebsch, H., M. Moss, T. Trilsbeck et al. 2008. Monitoring inflow distribution in multi-zone, velocity string gas wells using slickline deployed fiber optic distributed temperature measurements, 2008 SPE Annual Technical Conference and Exhibition, Denver, CO, USA: SPE 115816.
56. Costello, C., P. Sordyl, C. T. Hughes et al. 2012. Permanent distributed temperature sensing (DTS) technology applied in mature fields – A Forties field case study, SPE Intelligent Energy International, Utrecht, The Netherlands Society of Petroleum Engineers: SPE 150197.
57. Camilleri, L. A. P., A. H. Akram, and M. E. Quinzo Bravo. 2010. Combining the use of ESPs and distributed temperature sensors to determine layer water cut, PI and reservoir pressure SPE Production and Operations Conference and Exhibition, Tunis, Tunisia: SPE 136057.
58. Vargo, R. F., Jr., M. Payne, R. Faul, J. LeBlanc, and J. E. Griffith. 2003. Practical and successful prevention of annular pressure buildup on the Marlin project. *SPE Drill Completion* 18: 03: 228–234.
59. Ellis, R. C., D. G. Fritchie, Jr., D. H. Gibson, S. W. Gosch, and P. D. Pattillo. 2004. Marlin failure analysis and redesign: Part 2 – Redesign. *SPE Drill Completion* 19: 02: 112–119.
60. Gosch, S., D. Horne, P. Pattillo, J. Sharp, and P. Shah. 2004. Marlin failure analysis and redesign: Part 3 – VIT completion with real-time monitoring. *SPE Drill Completion* 19: 02: 120–128.
61. Birchwood, R., J. Dai, D. Shelander et al. 2010. Developments in gas hydrates. *Oilfield Rev.* 22: 1: 18–33.
62. Su, H., J. Li, J. Hu, and Z. Wen. 2013. Analysis and back-analysis for temperature field of concrete arch dam during construction period based on temperature data measured by DTS. *IEEE Sens. J.* 13: 5: 1403–1412.
63. Khan, A. A. 2009. Séparation de sources thermométriques. PhD Institut National Polytechnique de Grenoble-INPG.
64. Khan, A. A., V. Vrabie, Y.-L. Beck, J. I. Mars, and G. D'Urso. 2014. Monitoring and early detection of internal erosion: Distributed sensing and processing. *Struct. Health Monit.* 1475921714532994.
65. Dornstädter, J. 2013. Leakage detection in dams – State of the art, International Symposium: Dam Engineering in Southeast and Middle Europe, Ljubljana, Slovenia.
66. Aufleger, M., M. Conrad, M. Goltz, S. Perzlmaier, and P. Porras. 2006. Innovative dam monitoring tools based on distributed temperature measurement. *Jordan J. Civ. Eng.* 1: 1: 29–36.
67. Weiss, J. D. 2003. Distributed fiber optic moisture intrusion sensing system. US6581445B1.
68. Dornstädter, J. 2011. Leak detection method for water installations – Using passive temperature sensors to determine temperature changes near leaks. DE19621797B4.

69. Dornstädter, J., A. Fabritius, and P. Heidinger. 2014. Full automatic leakage detection at Ilisu dam by the use of fibre optics, 2nd Baralar Kongresi, Istanbul, Turkey.
70. Petrides, A. C., J. Huff, A. Arik et al. 2011. Shade estimation over streams using distributed temperature sensing. *Water Resour. Res.* 47: 7: W07601.
71. Neilson, B. T., C. E. Hatch, H. Ban, and S. W. Tyler. 2010. Solar radiative heating of fiber-optic cables used to monitor temperatures in water. *Water Resour. Res.* 46: 8.
72. Sayde, C., C. Gregory, M. Gil-Rodriguez et al. 2010. Feasibility of soil moisture monitoring with heated fiber optics. *Water Resour. Res.* 46: W06201: doi:10.1029/2009WR007846.
73. Tyler, S. W., J. S. Selker, M. B. Hausner et al. 2009. Environmental temperature sensing using Raman spectra DTS fiber-optic methods. *Water Resour. Res.* 45: 4.
74. Lowry, C. S., J. F. Walker, R. J. Hunt, and M. P. Anderson. 2007. Identifying spatial variability of groundwater discharge in a wetland stream using a distributed temperature sensor. *Water Resour. Res.* 43: 10.
75. Hoes, O., W. Luxemburg, M. Westhof, N. Van de Giesen, and J. Selker. 2009. Identifying seepage in ditches and canals in Polders in the Netherlands by distributed temperature sensing. *Lowland Technol. Int.* 11: 2: 21–26.
76. Hoes, O., R. Schilperoort, W. Luxemburg, F. Clemens, and N. van de Giesen. 2009. Locating illicit connections in storm water sewers using fiber-optic distributed temperature sensing. *Water Res.* 43: 20: 5187–5197.
77. de Haan, C., J. Langeveld, R. Schilperoort, and M. Klootwijk. 2011. Locating and classifying illicit connections with distributed temperature sensing. In 12th International Conference of Urban Drainage, Porto Alegre, Brazil. 11–16.
78. Schilperoort, R., H. Hoppe, C. de Haan, and J. Langeveld. 2012. Searching for storm water inflows in foul sewers using fibre-optic distributed temperature sensing. In 9th International Conference on Urban Drainage Modelling, Belgrade, Serbia, 4–6 September 2012. IAHR/IWA Joint Committee Urban Drainage, Working Group on Data and Models.
79. Vosse, M., R. Schilperoort, C. de Haan et al. 2013. Processing of DTS monitoring results: Automated detection of illicit connections. *Water Pract. Technol.* 8: 3–4.
80. Nishimura, K., T. Tokuoka, Y. Ueno et al. 2001. Development of aquatic environment measurement systems of estuaries and coastal lagoons. In OCEANS, 2001. MTS/IEEE Conference and Exhibition. Honolulu, HI, USA, IEEE 2: 1342–1347
81. Kobs, S., D. Holland, V. Zagorodnov, A. Stern, and S. Tyler. 2014. Novel monitoring of Antarctic ice shelf basal melting using a fiber-optic distributed temperature sensing mooring. *Geophys. Res. Lett.* 41: 19: 6779–6786.
82. Tyler, S., D. Holland, V. Zagorodnov et al. 2013. Using distributed temperature sensors to monitor an Antarctic ice shelf and sub-ice-shelf cavity. *J. Glaciol.* 59: 215: 583–591.
83. Freifeld, B. M., S. Finsterle, T. C. Onstott, P. Toole, and L. M. Pratt. 2008. Ground surface temperature reconstructions: Using *in situ* estimates for thermal conductivity acquired with a fiber-optic distributed thermal perturbation sensor. *Geophys. Res. Lett.* 35: 14: L14309.
84. Ferdinand, P., M. Giuseffi, N. Roussel et al. 2014. Monitoring the energy efficiency of buildings with Raman DTS and embedded optical fiber cables, 23rd International Conference on Optical Fiber Sensors, Santander, Spain, SPIE, 9157: 9S.
85. Kimura, A., E. Takada, K. Fujita et al. 2001. Application of a Raman distributed temperature sensor to the experimental fast reactor JOYO with correction techniques. *Meas. Sci. Technol.* 12: 966–973.
86. Matsuba, K.-I., C. Ito, H. Kawahara, and T. Aoyama. 2008. Development of fast reactor structural integrity monitoring technology using optical fiber sensors. *J. Power Energy Syst.* 2: 2: 545–556.
87. Weber, W. J., R. C. Ewing, C. A. Angell et al. 1997. Radiation effects in glasses used for immobilization of high-level waste and plutonium disposition. *J. Mater. Res.* 12: 08: 1948–1978.
88. Faustov, A. V., A. Gusarov, M. Wuilpart et al. 2012. Distributed optical fibre temperature measurements in a low dose rate radiation environment based on Rayleigh backscattering, Optical Sensing and Detection II, Brussels, Belgium, SPIE, 8439: 0C-1-8.
89. Fernandez Fernandez, A., P. Rodeghiero, B. Brichard et al. 2005. Radiation-tolerant Raman distributed temperature monitoring system for large nuclear infrastructures. *IEEE Trans. Nucl. Sci.* 52: 6: 2689–2694.
90. Pandian, C., M. Kasinathan, S. Sosamma et al. 2009. Raman distributed sensor system for temperature monitoring and leak detection in sodium circuits of FBR, First International Conference on Advancements in Nuclear Instrumentation Measurement Methods and Their Applications (ANIMMA), Marseille, France: 1–4.
91. Semjonov, S. L., M. M. Bubnov, N. B. Kurinov, and A. G. Schebuniaev. 1997. Stability of mechanical properties of silica based optical fibres under γ-radiation, 4th European Conference on Radiation and Its Effects on Components and Systems. RADECS 97. Cannes France: 472–475.
92. Friebele, E., G. Sigel Jr., and D. Griscom. 1976. Drawing-induced defect centers in a fused silica core fiber. *Appl. Phys. Lett.* 28: 9: 516–518.

93. Craik, N. G. 1996. Detection of leaks in steam lines by distributed fibre-optic temperature sensing (DTS), Specialists' meeting on monitoring and diagnosis systems to improve nuclear power plant reliability and safety, Gloucester. UK International Atomic Energy Agency, Vienna (Austria), 28: 41–53.
94. Rizzolo, S., A. Boukenter, E. Marin et al. 2015. Vulnerability of OFDR-based distributed sensors to high γ-ray doses. *Opt. Expr.* 23: 15: 18997–19009.
95. Kovalchik, P. G., L. W. Scott, F. T. Duda, and T. H. Dubaniewicz. 1997. Investigation of ampacity derating factors for shuttle cars using fiber optics technology, Conference Record of the 32nd IEEE Industry Applications Conference, New Orleans, LA, USA, IEEE, 3: 2074–2077.
96. Bruun, H. H. 1995. *Hot-wire anemometry.* Oxford University Press, Oxford, UK.
97. Giallorenzi, T. 1993. Thermal dilation fiber optical flow sensor. US5208650.
98. Chen, T., Q. Wang, B. Zhang, R. Chen, and K. P. Chen. 2012. Distributed flow sensing using optical hot-wire grid. *Opt. Expr.* 20: 8: 8240–8249.
99. Wylie, M. T. V., A. W. Brown, and B. G. Colpitts. 2012. Distributed hot-wire anemometry based on Brillouin optical time-domain analysis. *Opt. Expr.* 20: 14: 15669–15678.
100. Davies, D. H. 2009. Fluid level indication system and technique. US7472594B1.
101. Hadley, M. R., and D. H. Davies. 2010. Fluid level indication system and technique. US7731421B2.
102. Sakiyama, N., K. Fujii, V. Pimenov et al. 2013. Fluid-level monitoring using a distributed temperature sensing system during a methane hydrate production test, International Petroleum Technology Conference, Beijing, China: IPTC 16769.
103. Jeong, Y., C. Jauregui, D. Richardson, and J. Nilsson. 2009. *In situ* spatially-resolved thermal and Brillouin diagnosis of high-power ytterbium-doped fibre laser by Brillouin optical time domain analysis. *Electron. Lett.* 45: 3: 153–154.

Distributed Strain Sensors

Practical Issues, Solutions and Applications

8

Distributed strain sensors are dominated by Brillouin technology, particularly at long distances; at shorter distances, Rayleigh optical frequency-domain reflectometry (OFDR) has a clear applicability alongside, or independently of, Brillouin scattering. However, both technologies suffer from their intrinsic inability to differentiate the effect on their response of temperature from that of strain. Therefore, this chapter on the applications of distributed strain sensing starts with a discussion of the methods available for temperature/strain discrimination, followed by a description of the main applications of distributed strain that have been demonstrated so far.

8.1 SEPARATION OF TEMPERATURE AND STRAIN EFFECTS

The main weakness of Brillouin distributed sensors, particularly those focused on the Brillouin frequency shift, is their inability to distinguish changes in the Brillouin frequency caused by strain from those caused by temperature. In some applications, contextual information can be used to interpret the results. Nonetheless, in many cases of practical significance, an unambiguous distinction between these measurands is required. In essence, more information is needed, and in general, it must be obtained from additional measurements. In this section, some of the methods that have been used to separate temperature and strain are discussed.

8.1.1 Separation through Cable Design

A conceptually relatively simple approach [1–3] involves a special cable structure that ensures that some fibres, used for temperature determination, are *not strain coupled* and other fibres *are tightly coupled* to the cable structure, the strain on the cable thus being transferred to the latter fibres with a well-calibrated transfer function. The non-strain-coupled fibres are packaged in loose tubes, and validation testing demonstrated that, up to a cable strain of at least 1%, no strain was transferred to the fibres [1]. The Brillouin shift measured on these fibres is therefore reliably representative of their temperature, and so the results can be used to temperature-compensate the strain-coupled fibres. The strain-coupled fibres were attached with a silicone resin to the cable strain member, and the percentage of strain transferred to the fibre was measured in cable qualification testing.

The Brillouin optical time-domain reflectometer (BOTDR) system was modified firstly to interrogate the two fibre types in turn and, secondly, to process the measured frequency shifts into temperature and strain. In this way (and using remote amplification), the temperature and strain distributions were measured independently over >100 km with a resolution of 1 K and 20 με, respectively, with a spatial resolution of 10 m and update times of the order of 20 minutes.

The strain decoupling in a cable cannot always be achieved, and there is also the need to prove that the strain-free state genuinely is strain independent. An alternative is to design a cable with two or more fibres that intrinsically respond differently in their frequency shift to temperature and strain.

8.1.2 Hybrid Measurement

8.1.2.1 Spontaneous Brillouin intensity and frequency measurement

The reader is referred back to Section 5.1.4 on spontaneous Brillouin sensors where the combination of the intensity (usually measured as a Landau-Placzek ratio) and the frequency of the Brillouin shift are used together for temperature/strain discrimination [4–6]. These methods are limited to those Brillouin systems using spontaneous emission.

This option is not available in the approaches based on stimulated Brillouin scattering; in any event, the resulting determination of both temperature and strain is degraded by the intensity measurement.

It should be noted that Brillouin *power* measurement using BOTDR was demonstrated in conjunction with Brillouin optical coherence-domain analysis (BOCDA) [7] for the measurement of the *frequency shift distribution* achieving a 5-cm spatial resolution for the temperature and the strain distributions.

8.1.2.2 Raman/Brillouin hybrid approach

An alternative approach [8,9] is to combine Raman DTS technology with the Brillouin measurement, the latter being temperature compensated by the former. In both Refs. [8] and [9] a heterodyne-detection BOTDR system was complemented by the addition of a detection channel for the anti-Stokes Raman backscatter.

In the case of Ref. [9], a Rayleigh channel was also included for loss compensation. This of course raises the question of the coherent Rayleigh noise (CRN). CRN was addressed by using a Fabry-Pérot laser emitting on 30 lines. In this case, each of the F-P laser modes has its own local oscillator, and so for the BOTDR measurement, the signal is the summation of many separate optical lines. The same measurement was made with a narrow line distributed feedback (DFB) laser and the results were compared.

It was found that the multi-line laser provided better Raman temperature measurements (owing to the reduction of the CRN) but that it compromised the Brillouin frequency measurement. There are two reasons for this degradation, namely the increased shot noise from the multiple local oscillators and also the slight broadening of the Brillouin spectrum caused by the fact that ν_B is different for each of the modes of the F-P laser.* Nonetheless, the F-P laser provided a better resolution of temperature and strain than the DFB laser did. The technique was later enhanced by adding pulse-compression coding [10]; although an improvement over single pulses was demonstrated, insufficient detail in terms of spatial resolution and averaging time was provided and so an objective measure of system performance could not be gleaned from the work.

---

* This follows from the fact that ν_B is a function of incident frequency, which is slightly different for each of the modes of the F-P laser.

8.1.2.3 Coherent Rayleigh/Brillouin hybrid

Certain types of coherent Rayleigh OTDRs can measure the static (i.e. to zero frequency) distribution of temperature and strain (Chapter 6). In fact, the Rayleigh measurement is a far more sensitive (by three orders of magnitude) measurement of these quantities than is the Brillouin frequency shift, and the 2×2 matrix of sensitivity coefficients for the two measurements is well conditioned. The two measurements can therefore be combined to provide independent temperature and strain distributions; this approach was demonstrated in Ref. [11] using a combination of dual-pulse BOTDA (for enhanced spatial resolution) and coherent Rayleigh OTDR. Commercial equipment is available based on this work. Similar work was subsequently reported using a split-wavelength approach with the Brillouin measurement being carried out at around 1310 nm and the Rayleigh measurement performed in the longer wavelength band around 1550 nm [12]. In both reports, the distances covered were rather short, although the commercial equipment is specified to reach up to 25 km [13].

8.1.2.4 Brillouin/fluorescence

Finally, amongst the combinations of Brillouin with other physical measurements for temperature/strain discrimination, the use of fluorescence in a rare-earth-doped fibre [14] should be mentioned. In this case, the temperature measurement is provided by a ratio of the fluorescent power at two wavelengths in an erbium-doped fibre. This measurement, carried out in transmission, provided a temperature calibration corresponding to some average of the temperature in the doped fibre. As pointed out by the authors, the extension of this method to a distributed measurement is problematic for a number of reasons, one of which is the slow fluorescence decay time that would also average the response over the entire length of the sensing fibre. It is therefore not a leading contender for the task of strain/temperature discrimination.

8.1.3 Discrimination Based Solely on Features of the Brillouin Spectrum

The Brillouin spectrum and its sensitivity are determined by the materials making up the fibre as well as the waveguide design and specifically the combination of optical and acoustic guiding structures. The ability to modify the Brillouin response to the measurands leads to a wide range of options in optimising the fibre for sensing.

8.1.3.1 Independent optical cores

Conceptually, the simplest approach is the use of two fibres having sufficiently different (and linearly independent) sensitivity coefficients of the Brillouin frequency shift with respect to (w.r.t.) temperature and strain. This was proposed in Ref. [15], where one fibre having a pure silica core and another having a core doped with germania are spliced together into a loop and laid alongside each other (for example, in the same cable). In this way, each physical point along the cable is measured by both fibres. This technique allows the use of relatively conventional fibres. The drawback was that the strain sensitivities of the two fibres used in the example given differed by only a few percentage; the poor conditioning of the coefficient matrix would result in considerable amplification of the noise. However, far wider ranges of fibre designs have been published recently, and their combinations in a single cable would likely be far more discriminative than the earlier work [15].

Placing fibres of different types in the same cable will ensure that their temperature is equal for all practical purposes, but this is not necessarily the case for their strain, and so a fibre design incorporating multiple cores within a single glass structure might be more suitable. Such a fibre was described in which two cores having radically different Brillouin spectra were included within a single glass cladding structure [16]. Unfortunately, this work reported only the difference in the Brillouin shift between the cores, not their sensitivity, and so little can be deduced as to its effectiveness in measurand discrimination. Nonetheless, the paper demonstrated very low losses (0.23 and 0.29 dB/km for the two cores) and could clearly be combined with some of the designs to be discussed later to provide a solution to the discrimination problem.

8.1.3.2 Multiple Brillouin peaks with distinct properties

Some fibres have more complex Brillouin spectra than the single peak that appears in simple step-index single-mode fibres. Lee et al. [17] showed that a Corning LEAF®* (large effective area) fibre [18] has a multi-peaked Brillouin spectrum and, more importantly for the present purposes, that two of these exhibit a ~9% difference in the temperature sensitivity in their frequency but no difference in their strain sensitivity. In this fibre type, therefore, using the frequency of both of these peaks allows the temperature and strain to be separated. However, the uncertainty on the temperature measurement is increased by an order of magnitude. In spite of the substantial degradation in resolution of the separated measurands, this approach avoids a measurement of the Brillouin intensity, which is unavailable in Brillouin optical time-domain analysis (BOTDA) and often limits the overall metrology even in BOTDR. A related patent [19] claims the concept of multi-peak core compositions more broadly but does not show how to design them for better discrimination.

Alahbabi et al. [20] carried out similar work on long (>22 km), nominally similar fibres and found a rather larger (19%) differential response of the frequency of the two peaks to temperature (and also negligible difference in their strain response). In spite of that more encouraging result than reported earlier, it was still concluded that using the frequency and intensity in a conventional single-mode fibre yielded much lower uncertainty (by a factor of ~7 for temperature and ~4 for strain). This comparison will however be dependent on the precise strengths and weaknesses of each experimental arrangement and in particular on the resolution of the intensity.

Using multi-peak spectra of course requires more sophisticated fitting routines in order to provide separate estimates of the location, strength and linewidth of each of the peaks.

8.1.3.3 Polarisation-maintaining fibres and photonic crystal fibres

It is tempting to consider using, for the separation of temperature and strain, the two orthogonal polarisation states in a polarisation-maintaining fibre [21], which in effect can be thought of as two independent waveguides in the same core. However, measurements on the temperature sensitivity of the Brillouin shift in highly stress-birefringent fibres [22] showed only a trivial difference (a few MHz) in this parameter between the principal axes and no difference at all in their sensitivity. The same work and companion papers [23,24] examined the temperature and strain dependencies of Brillouin gain, linewidth and frequency shift in the hope of using some combination of these parameters. The results suggest that this approach is not a promising candidate for temperature/strain separation Brillouin-based sensors.

Zou et al. [25] explored a similar concept in which the sensing element was a photonic crystal fibre (PCF). In this case, the PCF was a hole-guided structure and it also showed multiple peaks in the gain spectrum. In the work of Zou et al. [25], two peaks (first and third) were identified as candidates for the separation and a relative difference in the temperature sensitivity of their Brillouin frequency shifts of about 30% was reported, but about 15% for the strain sensitivity. The results were obtained on a relatively short (2 m) fibre length albeit with a spatial resolution of 15 cm. Although somewhat more promising than the various reports on LEAF fibres, the limited range covered is a reflection of the immaturity of PCF – it is still a high-cost, high-attenuation fibre that is not available in long lengths, although rapid progress is being made in this field [26].

8.1.3.4 Few-mode fibres

Few-mode fibres [27–29] are another example of independent guidance within the same core. This type of fibre is being actively researched for mode-division multiplexing in optical communications in order to

* Corning and LEAF are registered trademarks of Corning Incorporate, Corning, NY, USA.

extend the information-carrying capacity of individual fibres [30–32] by using each mode, or mode group, as a separate channel. In Ref. [27], such an approach was tested and the coefficient matrix containing the temperature and strain coefficients of the Brillouin for the LP_{01} (first column) and LP_{11} (second column) mode that was $\begin{pmatrix} 1.29\ \text{MHz/K} & 1.25\ \text{MHz/K} \\ 58.5\ \text{kHz}/\mu\varepsilon & 57.6\ \text{kHz}/\mu\varepsilon \end{pmatrix}$. Unfortunately, this matrix is very poorly discriminating. A similar experiment [28] also resulted in a coefficient matrix, $\begin{pmatrix} 1.0169\ \text{MHz/K} & 0.991\ \text{MHz/K} \\ 59.24\ \text{kHz}/\mu\varepsilon & 48.72\ \text{kHz}/\mu\varepsilon \end{pmatrix}$, that is better conditioned than that of Ref. [27] but still amplifies the uncertainty by a very large factor. Finally, on the subject of Brillouin sensing in few-mode fibres, Ref. [29] discloses the use of these fibres in dynamic Bragg gratings where the grating is set up by the Brillouin interaction in one mode and read by a probe beam launched into another mode.

8.1.3.5 Multi-wavelength interrogation

As an alternative, it was proposed to interrogate a fibre at two well-separated wavelengths [33], for example, 1300 nm and 1550 nm. This results in a change of sensitivity to temperature that is roughly inversely proportional to wavelength. Unfortunately, the same applies to the strain sensitivity, and therefore, the sensitivity coefficient matrix is very poorly conditioned and it can be estimated that the discrimination process would amplify the uncertainty by a factor of about 50.

8.1.3.6 Direct measurement of the Brillouin beat spectrum

The authors of Ref. [20] did point out the possibility of measuring the beat frequency between the two peaks in the spontaneous Brillouin backscatter directly, i.e. the Brillouin beat spectrum (BBS). This was accomplished some years later in a BOTDR configuration [34,35] with a LEAF fibre as the sensing element. Here, a pulse of narrowband light is launched into the fibre; the backscatter return is collected by a photodiode (after optical amplification and filtering in the case of Ref. [35]). The spectrum of the photodiode signal is analysed in the frequency domain. In the very-high-frequency/ultra-high-frequency region (100–600 MHz), it consists of the beat between the spontaneous Brillouin scatter (SBS) in the four peaks resulting from the mixing of the Brillouin lines at the detector. Thus, the frequency of each of these peaks represents the frequency difference between pairs of peaks present in the SBS spectrum. In BBS, when interrogated at 1546 nm, peaks appear at 184 and 342 GHz, representing the first Brillouin line at 10.719 GHz beating with the next two at 10.933 GHz and 11.066 GHz [34].

Both of the works [34,35] focused on the relative power of the peaks present in the BBS, rather than the position of the peaks. The reason given was the weak frequency variation with temperature and strain; this conclusion is at variance with the observations of Ref. [20], so there is some lack of agreement in the literature at present. Given the close proximity of the peaks in the Brillouin spectrum and their phase incoherence, the relative intensity of these peaks is expected to be precisely corrected for differential attenuation, and they should not suffer from CRN. In addition, by simply measuring the BBS signal at two pre-defined frequencies, scanning of the spectrum is not required, and this speeds up the measurement. Unfortunately, the changes in power are small and the coefficient matrix is not very well conditioned, as indicated in Equation 8.1, where the superscripts indicate dependence on strain or temperature and subscripts refer to the first or second BBS peak.

$$\begin{pmatrix} C_{P1}^{\varepsilon} & C_{P1}^{T} \\ C_{P2}^{\varepsilon} & C_{P2}^{T} \end{pmatrix} = \begin{pmatrix} -0.75 \cdot 10^{-4}/\mu\varepsilon & 29.5 \cdot 10^{-4}/\text{K} \\ -0.57 \cdot 10^{-4}/\mu\varepsilon & 24.3 \cdot 10^{-4}/\text{K} \end{pmatrix} \tag{8.1}$$

The reported uncertainty of the separated temperature and strain values were 9 K and 380 με, respectively, for a 4.5-km fibre length measured with 20 k averages to a spatial resolution of c. 3 m.

8.1.3.7 Fibres designed for Brillouin measurand discrimination

It should be noted that LEAF fibres are designed for specific dispersion properties, not their Brillouin spectra. In fact, all of the approaches described so far are based on fibres designed for other purposes, primarily for telecommunications, and, in general, the sensing community is dependent on fibres designed for telecommunications where the large production volumes result in attractive costs. To the author's knowledge, at the time of writing, no special fibres with Brillouin spectra optimised for measurand discrimination are available commercially.

However, fibres *could* be designed with multiple lines or features in the Brillouin spectrum that exhibit a different combination of temperature and strain sensitivity and are optimised in some fashion to discriminate temperature from strain.

It should first be noted that in some of the fibres showing multi-peak Brillouin spectra, the stimulated process can dominate, leaving only the strongest mode [36]; in a BOTDR system operating at high probe power, close to the threshold for SBS, the secondary peaks can be eliminated in the competition for pump power. In the case of BOTDA, however, the interaction frequencies are well defined by the pump and probe waves, and this should not be an issue.

There is a continuing research interest in the development of fibres optimised for Brillouin sensing. The route to a successful fibre development must start with a thorough understanding of the dependencies of the frequency shift on the glass properties, which involves not only a modal analysis of the optical waveguide but also of the acoustic waveguide [37–39]. Changes to the dopant affect the refractive index and the acoustic slowness, but they also modify the thermal expansion, and this feeds back (see Section 5.1.4) to optical and acoustic velocity through the strain-optical and strain-acoustic coefficients that are functions of composition [40,41].

Much of the work in the design of fibres with specific Brillouin spectra has been directed at suppressing stimulated Brillouin scattering for high-power laser applications [42–44]. For example, using Al_2O_3 as a core additive results in the acoustic mode experiencing a faster acoustic velocity than the surrounding material and so not being properly guided. GeO_2-doped cores with F-doped cladding can also be designed to prevent guidance of the acoustic modes [45].

If only the frequency shift of the spectral features is considered, then an ideal fibre for discriminating temperature and strain would have well separated peaks (but only by enough to avoid ambiguity in the interpretation); these peaks should also lead to a well-conditioned sensitivity matrix. For example, if two peaks moved in the same direction in response to temperature but in opposite directions for strain, they could be used very effectively in combination to distinguish these two measurands. Preferably, the manipulation of the Brillouin spectrum in the design of the fibre should not broaden the peak because this would degrade the resolution.

A fibre with a multi-peak Brillouin spectrum usually involves multiple acoustic modes, and it was shown that their Brillouin frequency-dependence on doping concentration is increased for very small core diameters [46].

Most of the commonly used additives to silica raise the optical refractive index and reduce the acoustic velocity (i.e. they increase the slowness and thus result in a guided acoustic wave). The major exceptions [47] are Al_2O_3, which raises the refractive index but lowers the acoustic slowness, and B_2O_3 and F, which lower the refractive index but, in common with most other additives, provide an acoustic guiding structure. The designer then has a palette of materials with which to work. Other, less common, dopants have also been used in this context.

None of the designs published based on these principles have yet shown all the characteristics of an ideal measurand-distinguishing Brillouin fibre, but progress has been made. For example, a fibre having a high (25 m%) Al_2O_3 core concentration showed a large, positive-sign strain sensitivity (100 kHz/με, roughly double that of a conventional single-mode fibre) [48]. Unfortunately, in this case, the spectral width of the peak was substantially broadened, which would degrade the ability to measure its centre frequency. Still with Al_2O_3 core doping, the behaviour of two peaks was recently reported [49], where the temperature and strain sensitivity of Peaks 1 and 2 were, respectively, 1.5 MHz/K and 39.2 kHz/με and 1.1 MHz/K and 31.0 kHz/με. In this case, although the response of these peaks to the two measurands is different, the sensitivity coefficient matrix is poorly conditioned.

A fibre doped heavily with P_2O_5 was proposed that is expected to show a Brillouin frequency varying by less than 5 kHz/K over the range −100°C to +100°C [50], and a related study [51] showed a value of 40 kHz/με for the strain coefficient in a similar fibre (about 20% lower than a conventional single-mode fibre and 25% less than a pure-silica-core fibre). A study of the $SiO_2/Al_2O_3/La_2O_3$ system [52] demonstrated that certain concentrations could minimise the temperature sensitivity, and others, the strain sensitivity of at least one of the Brillouin peaks.

An exploration of the properties of baria (BaO)-doped silica showed that this dopant increases the thermo-optic coefficient but reverses the signs of the strain-optical, strain-acoustic and thermal-acoustic coefficients compared with silica. This material system is capable of producing fibres with temperature and strain coefficients that can be either positive or negative. However, there are a few caveats, namely (a) the fibres were produced by melting powder starting materials to form a core within a silica tube, resulting in a relatively impure glass and consequently high loss, (b) the compositions giving rise to negative coefficients have large index-differences and therefore either will be multimode or, if they are designed to be single-mode, will have extremely small cores and (c) the temperature and strain coefficients are quite strongly correlated as functions of BaO concentration and this mitigates against a simple discrimination.

A study of large index-difference fluorine-doped single-mode fibre [53] showed a multi-peak response with increased sensitivity to strain and temperature. The high F-concentration allows several acoustic modes to be guided, given the large (20%) slowness contrast between core and cladding. Unfortunately, these fibres again exhibited a rather poorly conditioned coefficient matrix.

We can conclude that, although attractive, the use of multiple peaks in the Brillouin spectrum has so far been relatively unsuccessful, in that most of the examples would result in significant amplification of the uncertainty. Some progress was however reported [54] with a fibre where the reported sensitivity of the *difference* in Brillouin frequency between two peaks reported to be −1.07 MHz/K, i.e. similar to the frequency shift in conventional single-mode fibres. The interrogation in this case was effected using the BBS technique. Unfortunately, the strain sensitivity was not reported, so even in this case, we cannot be sure that the discrimination problem was solved. The fibre design was not disclosed in this paper, but it appears that this fibre is based on a separate patent [55].

A systematic approach to the design of fibres with separable temperature and strain response was proposed in Ref. [56], where the authors identified a concentric structure for the core, with two zones both having acoustic and optical guidance, the intention being that an internal zone is doped with one additive and causes Brillouin interaction with the lowest order acoustic mode, whereas another zone, further from the axis, is doped with a different material causing Brillouin interaction primarily between the lowest order optical mode and a higher-order acoustic mode. This should provide a dual-peak Brillouin spectrum, and with a suitable choice of materials, it is intended that the strain and temperature sensitivities of each should be non-collinear. According to the calculations provided in this patent application, the most promising combination was a central region occupying the region from the centre of the fibre to a radius 2.3 μm and doped with P_2O_5 to 8.5 w% (yielding an index difference of ~0.0085); a second guiding ring occupying the region between 4 and 8 μm radius would be doped with Al_2O_3 to 0.9 w% (an index difference of ~0.00275). The predicted sensitivity coefficient matrix is $\begin{pmatrix} 1.903\ \text{MHz/K} & 0.73\ \text{MHz/K} \\ -55.6\ \text{kHz}/\mu\varepsilon & -49.18\ \text{kHz}/\mu\varepsilon \end{pmatrix}$, which is far better conditioned than other examples previously given in this subsection. One can predict a degradation of only a factor of 2 for the separated measurands, in contrast to the factor of 10 or more in some of the other work discussed. It must be stressed, however, that the results in Ref. [56] were only modelled and no experimental data were provided.

8.2 INFERRING OTHER MEASURANDS FROM TEMPERATURE OR STRAIN

Distributed Brillouin sensing (in the broad sense of any of the many approaches that have been discussed) have been converted to measurands other than temperature and strain, primarily by use of transduction mechanisms that bring the measurement back sometimes to temperature, but primarily to strain. In this section, a number of measurement techniques that use strain are discussed; a few examples of using temperature as a proxy for other measurands can be found in Chapter 7.

8.2.1 Pressure

From early work on fibre-optic hydrophones, it is known that the coating of an optical fibre converts pressure outside the fibre to strain [57,58], and this provides one route to pressure measurement with distributed strain sensors. This was tested with a number of fibre types [59], using an improved version of the optically separated spontaneous Brillouin arrangement [60]. Although this approach allows the separation of temperature from pressure and/or strain, the sensitivity is in fact quite low. The fibre types used were not disclosed, but the best result showed a Brillouin frequency dependence of about 2.5 MHz/MPa, which is certainly not sufficient for a number of applications, notably in down-hole pressure sensing, where the resolution requirements are of the order of 10 Pa, which would appear to be unachievable with the techniques described. In fact, in these applications, the dynamic range required – of order 10^6 – seems incapable of being addressed with Brillouin technology. On the basis of the 2.5 MHz/MPa mentioned above, the maximum value could be accommodated, but the resolution required at, say, 10 Pa implies determining the frequency shift to a few tens of Hz, i.e. to within 1 part in 10^6 of the linewidth. Based on Section 5.5.3, this implies a signal-to-noise ratio of order 10^5, when most systems only achieve a value of a few hundred.

More recent work in this area describes coating systems (pair of inner and outer coatings) optimised for pressure sensitivity [61], particularly by varying the outer coating applied to a thin, soft inner coating that had a Poisson ratio close to 0.5. Applying a thick outer coating of low Poisson ratio, a substantial enhancement of the Brillouin frequency shift to pressure was predicted. With limited flexibility in the materials, the best result achieved in practice was 1.9 MHz/MPa. However, the analysis showed that a wider control of the materials and even the coating thickness might have led to a further increase in sensitivity by a factor ~5–7.

The work of Guangyu et al. [61] used the BOTDA technique for the Brillouin measurement, which of course raises the issue of separation of temperature and strain. Another team more recently used the differential response of two fibres interrogated in sequence and addressed through fibre switches to provide separation of temperature from pressure [62].

The results of the work lead to a resolution of 0.25 MPa and 0.28 K, although the authors caution against pitfalls they foresee, such as non-linearity of certain polymer materials. They also make the crucial point that the packaging of a distributed pressure sensor is far from trivial: whereas for temperature measurements made in harsh environments, the fibre can be protected reasonably effectively in metallic tubes (even then a number of issues arise with fibre degradation). In the case of distributed pressure measurements, the problem is compounded by the fact that it is necessary to transfer the pressure to the fibre and yet shield it from chemical attack by the fluids that transfer that pressure. This is a central problem in distributed sensing of pressure.

8.2.2 Liquid Flow

The concept of using a distributed measurement for determining build-up of deposits (e.g. waxes, hydrates, asphaltenes) was discussed in Ref. [63] in the context of flow lines (the conduits that connect sub-sea wellhead to other facilities on the seabed). The basic principle of the measurement is to deploy the fibre cable in the flow and to infer the existence of a reduction in the available pipe cross-sectional area from the pressure induced on the cable. The pressure changes that are intended to be detected relate to the Bernoulli effect of pressure changing at a restriction.

It has also been proposed to use fluid drag to convert flow to pressure [64]. As in Ref. [63], the sensing cable is placed in the flow and a change in the Brillouin shift is expected to be observed, this time as a result of the drag caused by the flow past the cable. The inventors of Ref. [64] proposed structuring the coating so as to maximise the drag effect.

Of course, in practice, both the pressure and the strain effect are likely to be present simultaneously. This would complicate the interpretation of the measurement, as would uncertainty as to the radial location of the cable in the pipe, given that the flow is unlikely to be uniform across the pipe diameter, even in single-phase flow conditions. In multi-phase flows, the distribution of the fluids across the pipe is much more complex and depends on, *inter alia*, the flow rate of each phase, the pipe diameter and the angle w.r.t. to vertical.

These approaches have not yet been adopted in practice, to the author's knowledge, possibly owing to objections to the placing in the flow path of unsecured objects that are intrusive even intact and could result in severe restrictions in the event of a cable break.

8.2.3 Chemical Detection

Chemically sensitive coatings that change dimensions in the presence of target species have been applied to fibre Bragg gratings (FBGs) [65], and similar concepts can be applied to the measurement of the resulting strain using Brillouin or Rayleigh OFDR technology, although there is only limited literature on this subject. In Ref. [66], the use of Pd-coated fibres for the detection of hydrogen (demonstrated as point sensors in 1984 [67]) and using Brillouin interrogation was proposed. Conventional fibres also appear [66] to exhibit an intrinsic sensitivity to hydrogen in their Brillouin shift. Experimental results showed a frequency shift of ~20 MHz when a fibre became saturated with hydrogen; the sensing mechanism that caused the shift was not clear, but it seems related to changes in the glass, such as swelling or refractive index changes as a result of the ingress of hydrogen.

The practical issues in exposing a fibre to pressure that were discussed previously are even more acute in chemical sensing because the fibre must be exposed directly to the fluids being sensed. There are, therefore, considerable practical difficulties in the field deployment of this sort of concept, valuable though the monitoring would be.

Another approach, so far demonstrated only as an average measurement on a section of fibre, uses the stimulated Brillouin process to induce acoustic waves, and it uses those acoustic waves to probe the fluid surrounding the fibre [68]. In this case, the stimulated Brillouin process is used in the forward direction (i.e. the stimulated acoustic wave co-propagates with the pump pulse), and this results in much lower-frequency phonons than are observed in Brillouin backscatter (tens to hundreds of MHz for forward scattering, compared to c. 10 GHz for Brillouin backscatter). The lower-frequency phonons have sufficient lifetimes to reach the external boundary of the fibre and return through the core several times.

The measurement consists of detecting (with a separate probing measurement) the phase disturbance caused by these phonons and estimating their lifetime, which can then be related to the acoustic impedance contrast at the external boundary of the fibre cladding. In turn, the acoustic impedance is interpreted to provide an estimate of the external fluid composition, e.g. the salinity of water. The method allows the light to remain in the core (and so avoids effects of light absorption by the fluid that is sensed) and yet it allows a wave generated by that light to probe outside the fibre.

This work emerged only recently, but it provides a deep insight into the physics of stimulated Brillouin scattering and considerable promise as a completely new type of optical fibre sensor.

8.3 POLYMER FIBRE BASED STRAIN MEASUREMENT

Polymer optical fibres (POFs) have been investigated for applications in short-distance communications, where their relatively high attenuation is less of a barrier to use and their flexibility and ease of connection are an advantage. However, their application distributed sensing is relatively recent. The detection of strain using a conventional OTDR with POF was demonstrated in 2004 [69]. These fibres show a localised increase in backscatter under axial strain, and although the exact mechanism was not clarified in Ref. [69], the results are consistent with dimensional changes and tapering that were observed in silica fibres drawn with fibre diameter variations [70,71]. Husdi et al. [69] also showed some discrimination of other types of distortion such as bends and twists. A unique benefit of POFs in the present context is their ability to undergo massive levels of strain (tens of %) without breaking, at least one order of magnitude above the values at which silica fibres would fail.

The technology of POF has progressed and moved from poly-methyl methacrylate (PMMA) to fully fluorine-substituted bonds (replacing C–H bonds), and this has resulted in losses of order 50 dB/km in the spectral region from 850 to 1300 nm [72] with a minimum of 10 dB/km in the 1000–1100 nm region [73]. These perfluorinated graded-index fibres (PFGI) have also demonstrated a bandwidth in excess of 2 GHz over 100 m, with models suggesting the potential for a further order of magnitude in their bandwidth × distance product. These breakthroughs make POF fibres serious contenders for some types of distributed measurement, where the distance range and the ambient temperature are moderate but high localised strain is likely to occur.

In addition to observing the backscatter, the locations of reflective markers caused by imperfections in the sensing fibre were measured with an incoherent (broadband source) OFDR. The results were analysed by cross-correlation with reference traces acquired with the fibre in a known strain state [74]. The measurement itself is similar in concept to the work of Geiger and Dakin [75], but taken much further towards a practical deployment. This work complements backscatter measurements by estimates of the group delay to the reflectors [76,77], and the two approaches were used jointly [78] in geotechnical applications such as the detection of movement in earth works. In particular, the POFs were integrated into geotextiles suitable for burying within the structure to be monitored. In field tests, the opening of a crack over tens of mm was observed, a value that would most likely have resulted in a failure of a silica sensing fibre. Moreover, POFs are more compliant than glass fibres and this allows the material to follow the local strain distribution more closely with weaker attachment methods.

The use of Brillouin scattering in PFGI has appeared in the last few years, probably fuelled by the new perfluorinated materials. Earlier materials, such as PMMA, have lowest-loss transmission windows around 650 nm; in contrast, Brillouin interrogators have usually been designed for 1300 or 1500 nm. Initial BOTDR measurements at 1550 nm [79] showed $C_{\nu_B\varepsilon} = -12.18$ kHz/με and $C_{\nu_B T} = -4.09$ MHz/K, i.e. a temperature sensitivity stronger in silica by a factor of 3.5–4 (but of opposite sign) and a weaker strain coefficient by a factor about 4. These results are supported by the changes in the material properties, notably by the reduction of Young's modulus with increasing temperature. Interestingly, ν_B was found to be about 2.8 GHz (for a 1550 nm probe) with a linewidth of 105 MHz, in contrast with a value of 10–11 GHz and 30 MHz, respectively, in silica-based fibres. It should be remembered that whereas most of the work on distributed sensing using Brillouin techniques in silicate fibres has been conducted on single-mode fibres, the POFs are heavily multimode, the smallest cores that have been reported in this context being 50 μm in diameter in the latest work, with somewhat larger sizes in the initially available fibres. Further work [80] showed that SBS techniques (BOTDA in this case) could also be applied to POFs.

BOTDR work conducted at 1320 nm and 1547 nm on three fibres from different suppliers showed consistent Brillouin spectra in terms of frequency shift but much stronger Stokes signals at 1320 nm, an effect that was attributed to the lower loss at this wavelength (25 vs. 250 dB/km).

Tests over a wider temperature range [81] in PFGI fibres have shown that the material behaves linearly and with little hysteresis from −160°C to about 85°C. From 85°C to 125°C, the response begins to exhibit some hysteresis with increased sensitivity ($C_{\nu_B T}$ = −13.8 MHz/K compared with a value of −3.2 MHz/K for the particular material investigated). It was also found that the linewidth begins to increase sharply above the rated operating temperature (70°C), which presumably is close to the glass transition temperature for this material.

To conclude this section on POFs, the correlation methods developed for silica-based fibres and specifically Brillouin optical coherence domain reflectometry (BOCDR) have been demonstrated on these fibres [82], and a spatial resolution of 7.4 cm was achieved. However, there are particular challenges with Brillouin measurements in POF. In particular, the Brillouin shift is far lower than in silica and the Brillouin gain is lower partially as a result of the multimode character of these fibres and the large mode field diameter. The Brillouin can therefore be swamped by the tail of the Rayleigh scattering that overlaps the Brillouin spectrum. This has been circumvented in part by optimising the relative polarisation of the waves interacting in the Brillouin process [83] so as to minimise the Rayleigh signal relative to the Brillouin one. The need for polarisation optimisation then precludes the use of polarisation scrambling because the signal-to-noise ratio is degraded [84]. Therefore, in order to use polarisation-scrambling, it is necessary to increase the optical powers used to interrogate the sensing fibre.

8.4 APPLICATIONS OF DISTRIBUTED STRAIN MEASUREMENTS

In this section, we discuss some of the applications, emerging or established, of distributed strain sensors. Although most of these have been developed using one form or another of Brillouin scattering; increasingly, the use of frequency-scanned Rayleigh backscatter is finding applications, particularly on fine distance scales and over short measured lengths.

8.4.1 Mechanical Coupling and Limitations on Strain Measurements

Before launching into a description of applications, it is worth considering how strain reaches the sensing fibre. In the case of temperature sensing, exposing the sensing fibre to the measurand is simply a case of allowing the sensor to reach thermal equilibrium with its environment. Layers of materials used to protect the fibre from crushing, bending or chemical attack may extend the time needed to reach equilibrium, and this possibly degrades the useful update rate of the sensor output. However, in the case of strain sensing, there is an additional requirement for mechanical coupling between the object being sensed and the fibre.

Mechanical coupling requires whatever protection is used around the fibre faithfully to transfer the strain to it; it also requires attachment methods between the structure and the sensing cable that transfer the strain between them. The sensing industry has developed methods, such as adhesives, for attaching strain gauges to civil engineering structures. By and large, this knowledge can be applied to distributed sensing, with however a need to re-think the deployment from merely attaching a point sensor to ensuring suitable coupling along the entire sensor length. The attachment methods may not require the sensor to be attached at every point, but the distance between attachments must be less than the targeted spatial resolution of the system.

The nature of optical fibres, and in particular their size, electrical inertness and chemical compatibility with composite technology, makes the option of embedding the sensing fibre into composite structures attractive in applications such as wind turbine blades or airfoils. In this way, the protection of the fibre is provided by the structure that is monitored. However, embedded fibre strain sensors require careful attention to how the fibres are taken out of the structure and protected at the exit points.

Another aspect of strain sensing is that, by definition, in these systems, the sensing fibre is strained. It follows that the maximum strain that the sensor will be subjected to must be considered in relation to the required system life. The mechanical reliability of an optical fibre depends on a number of factors, such as the proof strength to which it was screened, type and quality of the coating and the environment (temperature, humidity) in which it operates. For a fibre protected from humidity and held at room temperature, the maximum strain that can be safely applied and yet cause negligible long term (40 years) failure risk is 1/5 of the proof test strain level [85]. This limits the mean strain that can be applied to the sensing fibre to a few tenths of 1%.

8.4.2 Energy Cables

BOTDR has been proposed for monitoring overhead power lines, partly in connection with heat detection resulting from lightning strikes, but also strain induced by ice loading [86]. The ground wires that run above the phase conductors in high-voltage transmission lines increasingly frequently incorporate optical fibres for communication [87]; there has been limited use of this channel for sensing. Optically enabled ground wires are known as optical ground wire (OPGW). The information that is required is temperature and the related sag caused by thermal expansion; the latter would result in strain on fibres inside the OPGW insofar as they are coupled to the wire.

Subsea energy cables, used to interconnect grids from different countries, e.g. across the North Sea, or to tie offshore wind farms back to the mainland, are now usually equipped with optical fibres that are normally of a single-mode design; the distances involved favour Brillouin technology, although long-range DTS systems are also used to capture the temperature profile independently of strain. In crowded shallow seas, cables are occasionally caught by trawlers or ship anchors, and apart from helping locate the damaged section and thus speed up the restoration of the cable, using real-time distributed strain data allows the precise time and location of the incident and can help assign responsibility for it.

8.4.3 Telecommunications

Given its roots in the laboratories of a major telecommunications utility, NTT in Japan, it is not surprising that distributed Brillouin sensors have found applications in that industry. One of the earliest uses of distributed strain sensing was the monitoring of the installation of multi-fibre cables in ducts [88], where some of the forces imparted to the cable are transferred to the fibres. This work demonstrated a strain on the fibres of order 0.06%, which is still well below the level at which the static strain would impair the life of most fibres; however, it demonstrated that in spite of the use of a slotted liner cable design, intended to isolate the fibres from strain on the cable, some of the forces are still transferred to the fibres.

Another topic of interest to a telecommunications company is the monitoring of branched networks, i.e. configurations where one fibre broadcasts to a number of individual fibres. This was originally carried out with a Rayleigh OTDR systems operating out-of-band at around 1650 nm [89]. The branched network is interrogated from the common end so the backscatter signals from all fibres overlap at the OTDR. However, this approach does not provide the identification of which fibre in the network has failed or suffered increased loss and so BOTDR has been used for this purpose because the Brillouin shifts in the fibres installed in relatively early networks were slightly different [90].

Rather than rely on happenstance, a more recent implementation of this idea uses fibres with deliberately distinct Brillouin shifts [91]. Initially, four fibres were used with Brillouin shifts separated by roughly 200 MHz between neighbouring values. In this case, the BOTDR provides a signal that is characteristic for each of the branches, and so if one branch has deteriorated, it can be identified unambiguously. The interrogation is still performed out of the transmission band.

This work was extended to eight fibres [92]; however, there was found to be little scope for extending this approach further, and besides, it requires differentiated fibres for each branch, an arrangement that poses significant logistical challenges. More recent work [93] has therefore focused on using BOTDA to measure the length difference between each branch, using a variant of the colliding pulse technique demonstrated by Nikles [94] but incorporating coherent detection. This allows the backscatter level to be determined at each location uniquely in relation to the fibre end.

Returning to the origins of BOTDA as an improvement to OTDR, Brillouin measurements in general, benefit from the fact that they occur at a different frequency from the incident light. This results in the behaviour of Brillouin systems being different in the presence of reflections from that of Rayleigh OTDR, and yet they can still operate with narrowband sources without the coherence problems that exist with Rayleigh-based measurements. Although in Brillouin systems reflections still give rise to spurious signals, they can, in general, be distinguished from the backscatter more easily and the receivers are far less susceptible to overload in the presence of reflections. The technique has also been used to assess in-service connector losses in a bi-directional approach (which separates losses from changes in backscatter factor [95]).

8.4.4 Civil Engineering

In civil engineering, monitoring for changes in the dimensions of a structure can provide an early warning of incipient failure. Many existing methods, such as strain gauges and brittle witness plates, are used, but in large-scale structures, the installation of sufficient point sensors is costly and prone to failure; moreover, the parts that are of concern may not be accessible for inspection. Distributed strain sensing therefore has the potential to fill this gap.

8.4.4.1 Tunnels

NTT used BOTDR to monitor the deformation of tunnels used for telecommunication cables [96]; these are normally unattended, and so an automated system able to monitor large structures is ideally suited to ensuring that distortion of the roof is detected early. The sensing cable was stretched between bolts attached to the roof of the tunnel and changes in fibre strain could be interpreted as shifts in the relative positions of the anchor points. The system was able to resolve 0.1 mm elongation or compression in gauge lengths of 2 m.

A detailed study and long-term field test [97,98] of the dimensional changes in concrete-lined tunnels, intended for use in nuclear waste repositories, was conducted by ANDRA (the French radioactive waste management agency) to assess the performance and survival rate of fibre-optic sensors alongside conventional extensometers and thermometers. Installation methods and operation procedures for the fibre sensors were validated.

One of the issues encountered was that when a fibre break does occur, those Brillouin techniques that rely on a loop deployment (e.g. BOTDA) lose all monitoring capability. In the event, sufficient redundancy with additional fibres was in place to provide continuing coverage. Nonetheless, this point must be carefully considered in the design of the monitoring system of a large, critical structure. In particular, the inclusion of additional fibres in one cable will not provide much redundancy in that damage to one cable is likely to result in the failure of all fibres. Many of the failures were the result of operator error in later phases of the construction.

Overall, the trial successfully demonstrated that DOFS can be used for the long-term monitoring of concrete liners with over 95% survival rate after three years' operation. The trial continues.

8.4.4.2 Crack detection

The effective detection of cracks requires a very high spatial resolution to avoid a small dimensional change being lost in a large gauge length; BOCDA, which can achieve mm resolution, was used to demonstrate crack detection in concrete [99]. The smallest crack width that was detected in this study was

0.08 mm, essentially a hairline crack. The change in strain was apparent over c. 10 cm, which is a reflection of the length of fibre over which the strain is transferred from a single point extension. In this context, the buffering materials surrounding the glass part of the sensing fibre (coating, jacket, possibly a cable structure) must be thoroughly understood, as well as the effect of adhesives or other means of attachment. A detailed finite element analysis was reported recently [100], and the model was validated by laboratory testing, and in this case, a dual-pulse BOTDA was able to resolve the opening of a 13-μm gap between two sections to which the fibre had been attached.

Both of these examples were laboratory tests. Field tests on a concrete bridge were reported in Refs. [101] and [102], where fibres were attached to the underside of a steel load-bearing beam, part of a steel-reinforced concrete road bridge. The sensing fibres were monitored using BOTDA, and the response to vehicle loading was recorded. The effect of strain caused by a 40-ton lorry was very clearly identified. In this work, the existence of a crack in the concrete was detected with the BOTDA system. Although its existence had previously been known, changes in the size of the crack caused by thermal cycling allowed the optical system to locate it where it had been expected.

8.4.4.3 Damage in offshore structures

Perfluorinated polymer fibres have been tested as crack detectors for offshore steel structures [103]; when interrogated with a conventional, but high-spatial-resolution, OTDR, these fibres respond with a spike of increased signal when a localised axial strain is applied, as is the case when a crack forms. The plastic nature of the material allows the fibre to be attached continuously on the structure and yet survive a localised dimensional increase that would cause a glass fibre to fail.

8.4.5 Nuclear

The suitability of the distributed Brillouin sensors for ionising radiation environments was reported in Refs. [104–106]. The interest in DOFS includes the monitoring of nuclear waste repository sites, which are hot and are subject to irradiation. Apart from the increase in attenuation caused by ionising radiation, it is also found that the Brillouin frequency shifts change as a result of the radiation-induced damage. This concurs with earlier observations [107] on GeO_2-doped single-mode fibres. The detrimental effects of ionising radiation (γ-rays in particular) can be partly alleviated by using pure-silica core fibres [104–106], in which the shift in Brillouin frequency induced by radiation damage was found to be limited to about 2 MHz, with also reduced radiation-induced attenuation, a factor of two lower than in conventional single-mode fibre. In contrast, higher GeO_2 levels result in a further radiation-induced shift of the Brillouin peak [108].

Shifting the measurement to shorter wavelengths also helps limit the effects of radiation-induced attenuation (RIA), with a reduction by a factor of two in moving from 1550 nm to 1300 nm; the lowest induced loss is around 1000–1200 nm, and it could well be that, if this application warrants the development effort, the measurement might be moved to this wavelength region. That change would however require considerable component development because many of the components that are easily available at 1300 or 1550 nm are either not available or of higher loss at, say 1064 nm, and where available, they are certainly more expensive.

Field tests were conducted using both Raman DTS (for temperature compensation) and BOTDA for monitoring of dimensional changes in a large-scale test site aimed at establishing the viability of long-term underground storage of low and intermediate level waste [108].

8.4.6 Shape Sensing

Shape sensing is desirable in many fields to estimate the 3-D movement of long objects, such as ropes or flexible pipes on a large scale and also in robotics, for example, to gauge the movement of a surgeon's hand when controlling a remote intervention. Shape sensing is commonly achieved by measuring strain at several locations in the cross-section of the body that is sensed so as to be able to determine the bending in at least two axes. This is sufficient in the case of a body that does not twist; for a full picture, torsional strain is also required. In Ref. [109], shape measurement was achieved with coherent OFDR interrogating arrays of point Bragg gratings (so not a fully distributed measurement), which provide very high resolution in distance and strain over moderate distances. In the particular example cited, a three-core fibre was used with the cores arranged at 120° at a radius of 68 μm in an optical fibre having a cladding diameter of 330 μm. The accuracy was tested by inserting the sensing fibre in the grooves of several 3-D templates of known shape. The measured strains were fed into a model in order to invert for the shape. A positional accuracy of ~1.65% (i.e. an error of ~8 mm at a distance of 0.48 m) was achieved. It should be noted that the distance between cores was very small in this case and so a very fine resolution of the strain was required in order to provide reasonably accurate shape estimation.

Similar work was carried out with Brillouin strain sensing [110]; although this technique is less precise, the cores in this case were further apart (c. 2 mm), and generally, the work was aimed at large-scale measurements. It was based on embedding fibres in composite substrates that were then used as shape sensors. It was estimated that a positional error of 0.45 m could be achieved at a distance of 1 km.

8.4.7 Composite Structural Health Monitoring

The deformation of composite structures has been assessed using fibres embedded in the material, for example, in unmanned aerial vehicles. The fibre is reasonably compatible with the manufacturing process. Examples of the deployment of distributed strain sensing in this application may be found using Rayleigh OFDR [111] and high-spatial-resolution BOTDA [112] as well as arrays of FBGs. In this context, the dynamic behaviour of the deformation is critical for detecting damage and so the fast measurement techniques that were described in Chapter 5 are a key enabler for this application of strain measurement.

8.4.8 Casing Deformation

A specific example of shape sensing occurs in the hydrocarbon production industry. In the construction of wells, there is a concern about the mechanical integrity and deformation of the casing or sand-screen that can occur in unstable formations. A major effort was undertaken to instrument a casing [113] with a myriad (5000) of FBGs in a long string wrapped spirally around the tubing at a density of 1 FBG/cm, sufficient to distinguish the major types of deformation (e.g. bending, ovalisation, axial compression, expansion from internal pressure etc.) on the 7-in. and 4.5-in. casings. In this work, a single helical pitch was used, and this can result in ambiguity in the interpretation of some combinations of deformations; however, the authors pointed out that this issue can be addressed using multiple helices of differing pitch.

In principle, similar measurements could be achieved with the high-spatial-resolution variants (BOCDA or BOCDR) of Brillouin interrogation. The first demonstration [114] was carried out using BOTDR in conjunction with FBGs. In practice, the best data were obtained with the FBGs, and so the value Brillouin techniques in this application is still to be proven. The benefit of Brillouin interrogation in this application is that it avoids the need to inscribe thousands of gratings in the sensing fibre. However, it is probably less tolerant of excess losses, particularly the BOCDA/BOCDR approaches. In both cases (Brillouin and FBG), there is a real challenge in installing the sensing fibre so that it is strain coupled to the tubing and yet protected from the massive forces and risk of abrasion that it may experience on installation.

8.4.9 Pipeline Integrity Monitoring

Similar problems exist in the transport of fluids through pipelines. It is a fundamental requirement to ensure the continuing integrity of pipes carrying fluids that could be hazardous or environmentally damaging; however, this is a difficult problem. Tennyson [115] describes methods for detecting damage to pipes by attaching a strain-sensing fibre to it and monitoring changes caused, for example, by wall thinning, which would lead to an increase in strain as a result of internal pressure acting more effectively on the weaker structure; some of the proposals involve wrapping the fibre under tension around the pipe so that dimensional changes will be revealed by variations of the distribution of the Brillouin shift. Sag and buckling were also claimed to be detectable as a reduction of strain on the sensing fibre. An experiment to detect wall-thinning [116,117] by deliberately varying the thickness of the sample and pressurising it internally did indeed show some variations in accordance with known weak points. Similarly, laboratory-scale experiments in which there were sufficient compressive forces to generate buckling showed that this too can be detected with Brillouin distributed sensing [118].

In both cases (thinning and buckling), Brillouin instrumentation with fine (c. 15 cm) spatial resolution was selected in order to resolve the small features. This is a key point: in this type of application, the spatial resolution required is in the cm regime, even for quite large diameter pipes, and the question then arises about the range and sensitivity that can be achieved. Fortunately, in these pipe integrity applications, the measurement time is unlikely to be critical, allowing substantial signal averaging and/or point-by-point measurement, depending on the Brillouin interrogator used.

A further issue in the monitoring of buried pipelines is that generally, they are of large diameter (36 in. being quite common) and trenched. The installation process involves a streamlined operation of digging the trench, inserting some of the backfill material, welding pipe sections into a continuous line and, after weld inspection, lowering the welded pipe stage by stage into the trench and completing the backfilling and soil compaction. There is little room in these operations for an intervention by human operators, on grounds of timescale, personnel safety and possible collateral damage to the pipe.

In addition, most high-pressure gas pipelines are coated with a polymer material to limit corrosion. Many operators will not allow any additional material to be attached to the pipe for any reason; therefore in some cases, attaching a fibre to a pipe is simply not allowed. It will require changes to tried-and-tested operating procedures for the attachment of strain sensors to the pipe to be allowed.

On the other hand, installing a fibre in the trench allows landslides that may cause distortion to the pipe and, thus, potential failures to be predicted.

8.4.10 Compaction and Subsidence in Subterranean Formations

A problem related to integrity of casings is that of the stability of the geological formations surrounding it. The production of hydrocarbons and the injection of fluids in secondary production creates stresses in the rocks that can lead to substantial dimensional changes. The injection of steam can also lead to thermal expansion. Other stability problems arise from dissolution or weakening of a carbonate formation through water injection or simply from the reduction of pore pressure caused by hydrocarbon extraction. At Ekofisk, in the Norwegian North Sea, significant subsidence and formation compaction forced a major remediation programme to be undertaken to raise the platforms by 6 m and eventually replacing the platforms altogether.

On land, there are also concerns for the stability of surface installations. For example, where steam plants are sited above the heated reservoir, the heating could result in heave at the formation, followed by subsidence after extraction. Arrays of FBGs [119] have been deployed to monitor these effects. An alternative based on combined BOTDR (compensated with DTS measurements) [120] has been installed to monitor a steam-injection reservoir situated in the Arabian Peninsula.

8.4.11 Ground Movement and Geomechanics

The prediction and prevention of landslides, the protection of buried pipelines and the detection of erosion in river banks are just a few examples of the need to monitor potentially unstable ground. This monitoring problem is naturally suited to distributed sensors owing to the large distances to be covered, the often isolated locations and the fact that DFOSs do not rely on local power at each sensed location, which would be cumbersome to install and to shield from the environment.

The early prediction of landslides has important benefits by allowing timely evacuation and, possibly, remedial action. Landslides are often caused by heavy rainfall that penetrates and softens unstable soil, causing it to fail under shear stress. In an extreme example, heavy rains in Gansu, China, in 2010 [121] caused landslides that blocked a river with a temporary dam that then broke, flooding three towns. The death toll was between 1000 and 2000 people, and the flood caused 12 million people to be displaced. Therefore, there is a strong incentive to predict the occurrence of such events. Although some preventive measures based on excavation and reinforcements of slopes thought to be at risk is possible, these are often not implemented on the basis of cost–benefit trade-offs. However, at the very least, early detection of movement could allow evacuation whilst the ground movement is still slow, prior to the rapid acceleration that then leads to inescapable injury and loss of life.

Landslides are made more likely by deforestation, mining activity and even the construction of dams [122], where changes in the load to rocks can activate pre-existing weakness in the geology. In the Three Gorges reservoir case, landslides are thought to have been caused by the rising water reducing the load on the toe of the reservoir, thus reducing the frictional forces that had previously held the formations in place.

Many techniques for predicting instability in slopes have been described, ranging from modelling together with recording of potential triggering effects, detailed local surveys to understand the shear stresses present to displacement and tilt measurements using local or remote sensing (e.g. synthetic aperture radar) [123]. BOTDR is one, but only one, of the monitoring techniques that are used to detect early signs of movement. Methods based on temperature measurement, sometimes with DTS, have already been described in Chapter 7 in relation to the detection of water seepage.

When a slope fails, it is often the case (at least in the initial stages of a landslide) that a relatively homogenous section separates from neighbouring sections and moves as a whole. It is therefore of interest to identify the boundaries of the failure as well as the extent of the movement.

Fibre-optic strain measurements used for monitoring ground movement include not only Brillouin distributed sensors (mainly BOTDR) but also arrays of FBGs, as well as interferometric techniques that monitor single points, but where the gauge length can be arranged to cover an extended region [124], if required by the geometry of the problem. The same control of the gauge length has been demonstrated for long-gauge devices incorporating FBGs [125]. Electrical time-domain reflectometers on coaxial cable are also frequently used in this application, primarily to locate regions of high shear distortion, which damages the cable and leads to an impedance contrast that shows up as a spike on the reflectometer display [126].

There are several basic methods for attaching fibre-optic sensors to the soil for the purposes of monitoring movement. Firstly, a series of pickets can be hammered into the ground and the sensor cable attached to it [127]. This is similar to methods used in tunnels. In soil monitoring applications, it leaves the cable exposed to the elements and at risk from weather loading and rodent attack, but it does have the benefit of being specific as to the anchor points that are monitored for displacement.

A second approach consists of burying a sensing in the soil that is monitored. The direct burial of a cable requires planning in the direction of the cable and its layout. It also requires understanding of the friction between soil and cable. When axial forces are exerted on a cable, the resultant strain is distributed over several m of cable, because at the initial portion of the cable, where the force is applied, the frictional forces are quickly overcome and the cable slips, transferring most of the load to the following portions. This mechanism results in a strain and stress gradient of typically a few m depending on the cable surface and the properties of the backfill in which the cable is lying, including aspects such as soil compaction and moisture content. In one test, it was found that fine gravel (5–10 mm) developed substantially more grip on the cable than sand did.

Laboratory tests on the tension applied to a jacketed fibre (0.9 mm Nylon secondary coating) and the related pullout were conducted in laboratory conditions with varying levels of applied pressure (simulating an overburden) [128]. It was shown that the initial tension/pullout displacement initially followed a linear relationship up to the point where the surface friction fails, at which point the fibre slips in the sand and the stress drops dramatically. After the surface friction has failed, the displacement increases for a given pull force as a new equilibrium becomes established; eventually, the axial stress in the fibre tapers smoothly from the full pull force (where the force is applied) to zero at some depth in the sand where the original friction is still intact. When tested in shear, again the forces are redistributed (otherwise the cable would fail by being sheared itself). So in practice, the soil immediately at the shear interface gives way and the stresses on the cable are distributed over a distance that depends on the residual friction. As a result, a BOTDR measurement was able to detect a lateral movement of a few mm of a 2-m simulated trench section [129].

A third method for monitoring soils is to embed the fibre in a 'geotextile', i.e. a mesh that can be buried, usually to provide structural reinforcement but in this case also to serve as support for the fibre sensor [130]. This arrangement, although requiring more extensive effort in the laying of the textile (compared with trenching in a sensing cable), provides a higher level of confidence in the strain coupling of soil to fibre. In geotextiles, the sensor is added to an already planned installation of a soil strengthening element so the incremental cost of including the sensing fibre can be a small fraction of remedial work that is required in any case. These sensor-enabled geotextiles have been used in a number of field installations, such as in gravity dams, dykes, railway embankments and slopes thought to be at risk of sliding. The fibres used include silica as well as polymer fibres. Interrogation methods include Brillouin as well as OTDR, particularly for the polymer fibres. The fibre can be routed to pass through the mesh several times and in different directions, providing a two-dimensional picture of the strain field even with a linear sensor [130]. Krebber et al. [131] provides a review of the use of optical fibres embedded as sensors in geotextiles.

It should be noted that in some landslide monitoring applications, there are roads passing through the areas at risk and it is their integrity that is of prime concern. Field tests have been conducted with cables buried in groves in the road surface [132]. Even in this relatively straightforward application, it is necessary to make corrections to the raw data to take into account the strain transfer from soil (or road) to the fibre.

Successful field tests of BOTDR at extremely long range (100 km) with simulated trench collapse (horizontal landslide) and also simulated subsidence (vertical collapse) are reported in Ref. [1]. In this case, a subsidence of 175 mm resulted in a measured strain of 0.32% – easily detectable, given a resolution of 20 $\mu\varepsilon$, even at 100 km. The simulated landslide showed an even clearer event, at 1.3% strain, and this was validated by a co-located FBG array and a survey before and after the simulated landslide.

8.4.12 River Banks and Levees

River protection systems are intended to prevent unusually high water levels from reaching protected zones on the land side. Modern dykes consist of an internal impermeable layer intended to block the passage of water backed by a supporting mass to provide stability primarily through gravity and, finally, on the land side, a drainage mass of permeable material to allow high water within the structure to exit [133]. Since modern dyke construction techniques often incorporate geotextiles, embedding the sensing fibre within the mesh is an attractive option [133]. Field tests with such an arrangement were conducted in a dyke with a lift cushion placed within the structure to simulate a movement of the drainage mass, the sensor-equipped geotextile being placed near the land-side surface of the drainage mass.

In other developments, a full-scale model of a river levee was built and fitted with sensing fibres interrogated with BOTDR [134]. In this case, the sensing fibres were attached to a specifically designed pierced aluminum plate. The deformation of the plate at 0.02%, following heavy rain, was easily detectable.

The two tests discussed [133,134] and many others that have been reported in the literature have established distributed sensing, primarily with Brillouin techniques, as a valid and valuable method for monitoring waterway flood defences. Their long range, sensitivity and inert character make them ideal candidates for this important task.

8.4.13 Security/Perimeter Fence

Most fibre-optic intrusion detection systems have been based on acoustic signals. However, there are examples of the quasi-static strain imparted to a fibre by the presence of an intruder being used directly to detect an intrusion. The proof of this concept [135] was established using a BOTDA set-up with a sensing fibre attached to a deformable fixture. Fast detection is required for this application, and the measurement was updated at 1.5 s intervals for a strain level assumed to be 0.01% (with the results plotted in arbitrary units, the sensitivity of the set up was not clear). In a real-world application, the sensor might be used to sense ground deformation, for example.

A more specific field test used BOTDA to monitor perimeters using instrumented fences [136]. In this application, a low false alarm rate, together with high probability of detection, is required to maintain the credibility of the system and so the sensing cable was installed at two separate levels on the fence and the coincidence of detection on both fibre cables can be used to cross-validate alarms. The Brillouin approach, being quasi-static, is able to reject transient effects such as wind-induced vibrations, but it does rely on sufficient force being applied so that the sensitivity threshold is crossed.

8.5 SUMMARY

Although one or two decades behind DTS, the development of large-scale applications for distributed strain measurement is gathering pace. By their nature, these applications tend to be site specific, and more detailed engineering is required in their implementation than is generally the case for DTS. However, with increasing build-up of experience in their use, further installations of the technology should become more commonplace. In turn, wide-scale deployment in initial applications, such as ground movement, should make the technology more accessible for further applications.

REFERENCES

1. Strong, A. P., N. Sanderson, G. P. Lees et al. 2008. A comprehensive distributed pipeline condition monitoring system and its field trial, 7th International Pipeline Conference, 26th September–3rd October 2008, Calgary, Alberta, Canada: IPC2008-64549.
2. Inaudi, D., and B. Glisic. 2005. Development of distributed strain and temperature sensing cables, 17th International Conference on Optical Fibre Sensors, Bruges, Belgium, SPIE, 5855: 223.
3. Bao, X., D. J. Webb, and D. A. Jackson. 1994. Combined distributed temperature and strain sensor based on Brillouin loss in an optical fiber. *Opt. Lett.* 19: 2: 141–143.
4. Kee, H. H., G. P. Lees, and T. P. Newson. 2000. All-fiber system for simultaneous interrogation of distributed strain and temperature sensing by spontaneous Brillouin scattering. *Opt. Lett.* 25: 10: 695–697.
5. Maughan, S. M., H. H. Kee, and T. P. Newson. 2001. Simultaneous distributed fibre temperature and strain sensor using microwave coherent detection of spontaneous Brillouin backscatter. *Meas. Sci. Technol.* 12: 17: 834–842.
6. Sakairi, Y., H. Uchiyama, Z. X. Li, and S. Adachi. 2002. A system for measuring temperature and strain separately by BOTDR and OTDR, Advanced Sensor Systems and Applications, Shanghai, China, SPIE, 4920: 274–284.
7. Belal, M., and T. P. Newson. 2011. A 5 cm spatial resolution temperature compensated distributed strain sensor evaluated using a temperature controlled strain rig. *Opt. Lett.* 36: 24: 4728–4730.
8. Alahbabi, M. N., Y.-T. Cho, and T. P. Newson. 2005. Simultaneous temperature and strain measurement with combined spontaneous Raman and Brillouin scattering. *Opt. Lett.* 30: 11: 1276–1278.
9. Bolognini, G., M. A. Soto, and F. D. Pasquale. 2009. Fiber-optic distributed sensor based on hybrid Raman and Brillouin scattering employing multiwavelength Fabry-Pérot lasers. *IEEE Photon. Technol. Lett.* 21: 20: 1523–1525.
10. Bolognini, G., and M. A. Soto. 2010. Optical pulse coding in hybrid distributed sensing based on Raman and Brillouin scattering employing Fabry-Pérot lasers. *Opt. Expr.* 18: 8: 8459–8465.
11. Kishida, K., K. Nishiguchi, C. Li, and A. Guzik. 2009. An important milestone of distributed fiber optical sensing technology: Separate temperature and strain in single SM fiber. Proceeding of the 14th OptoElectronics and Communications Conference, Hong Kong, IEEE: 13–17.

12. Zhou, D.-P., W. Li, L. Chen, and X. Bao. 2013. Distributed temperature and strain discrimination with stimulated Brillouin scattering and Rayleigh backscatter in an optical fiber. *Sensors* 13: 2: 1836–1845.
13. NBX-7000 Specifications. Retrieved 18 May 2015, http://www.neubrex.com/htm/products/pro-nbx7000.htm.
14. Ding, M., Y. Mizuno, and K. Nakamura. 2014. Discriminative strain and temperature measurement using Brillouin scattering and fluorescence in erbium-doped optical fiber. *Opt. Expr.* 22: 20: 24706–24712.
15. Zou, L., and O. M. Sezerman. 2009. Method and system for simultaneous measurement of strain and temperature. US7599047B2.
16. Li, M.-J., S. Li, J. A. Derick et al. 2014. Dual core optical fiber for distributed Brillouin fiber sensors. In Asia Communications and Photonics Conference 2014, Shanghai, China, Optical Society of America AW4I.3.
17. Lee, C. C., P. W. Chiang, and S. Chi. 2001. Utilization of a dispersion-shifted fiber for simultaneous measurement of distributed strain and temperature through Brillouin frequency shift. *IEEE Photon. Technol. Lett.* 13: 10: 1094–1096.
18. Corning LEAF optical fiber. Retrieved 18 May 2015, http://www.corning.com/opticalfiber/products/LEAF_fiber.aspx.
19. Chi, S., C. C. Lee, and P. W. Chiang. 2004. Method of utilizing a fiber for simultaneously measuring distributed strain and temperature. US6698919B2.
20. Alahbabi, M. N., Y.-T. Cho, and T. P. Newson. 2004. Comparison of the methods for discriminating temperature and strain in spontaneous Brillouin-based distributed sensors. *Opt. Lett.* 29: 1: 26–28.
21. MacDougall, T. 2005. Method and apparatus for temperature sensing utilizing Brillouin scattering in polarization maintaining optical fiber. US6910803B2.
22. Yu, Q., X. Bao, and L. Chen. 2004. Temperature dependence of Brillouin frequency, power, and bandwidth in panda, bow-tie, and tiger polarization-maintaining fibers. *Opt. Lett.* 29: 1: 17–19.
23. Yu, Q., X. Bao, and L. Chen. 2004. Strain dependence of Brillouin frequency, intensity, and bandwidth in polarization-maintaining fibers. *Opt. Lett.* 29: 14: 1605–1607.
24. Bao, X., Q. Yu, and L. Chen. 2004. Simultaneous strain and temperature measurements with polarization-maintaining fibers and their error analysis by use of a distributed Brillouin loss system. *Opt. Lett.* 29: 12: 1342–1344.
25. Zou, L., X. Bao, V. S. Afshar, and L. Chen. 2004. Dependence of the Brillouin frequency shift on strain and temperature in a photonic crystal fiber. *Opt. Lett.* 29: 13: 1485–1487.
26. Chen, Y., Z. Liu, S. R. Sandoghchi et al. 2015. Demonstration of an 11 km hollow core photonic bandgap fiber for broadband low-latency data transmission. Optical Fiber Communication Conference, Los Angeles, CA, USA, Optical Society of America. Th5A. 1.
27. Weng, Y., E. Ip, Z. Pan, and T. Wang. 2015. Single-end simultaneous temperature and strain sensing techniques based on Brillouin optical time domain reflectometry in few-mode fibers. *Opt. Expr.* 23: 7: 9024–9039.
28. Li, A., Y. Wang, J. Fang et al. 2015. Few-mode fiber multi-parameter sensor with distributed temperature and strain discrimination. *Opt. Lett.* 40: 7: 1488–1491.
29. Li, M.-J., S. Li, and X. Wang. 2013. Distributed Brillouin sensing systems and methods using few-mode sensing optical fiber. US8493556B2.
30. Berdague, S., and P. Facq. 1982. Mode division multiplexing in optical fibers. *Appl. Opt.* 21: 11: 1950–1955.
31. Richardson, D., J. Fini, and L. Nelson. 2013. Space-division multiplexing in optical fibres. *Nat. Photon.* 7: 5: 354–362.
32. Li, G., N. Bai, N. Zhao, and C. Xia. 2014. Space-division multiplexing: The next frontier in optical communication. *Advances in Optics and Photonics* 6: 4: 413–487.
33. Sanders, P. E., and T. W. Macdougall. 2010. Dual wavelength strain-temperature Brillouin sensing system and method. US7526149B1.
34. Haneef, S. M., P. Prasanth, S. Bhargav, D. Venkitesh, and B. Srinivasan. 2014. Fast distributed temperature sensing using Brillouin beat spectrum of large effective area fiber (LEAF), 23rd International Conference on Optical Fiber Sensors, Santander, Spain, SPIE, 9157: 6Y-4.
35. Lu, Y., Z. Qin, P. Lu et al. 2013. Distributed strain and temperature measurement by Brillouin beat spectrum. *IEEE Photon. Technol. Lett.* 25: 11: 1050–1053.
36. Yeniay, A., J.-M. Delavaux, and J. Toulouse. 2002. Spontaneous and stimulated Brillouin scattering gain spectra in optical fibers. *J. Lightw. Technol.* 20: 8: 1425–1432.
37. Shibata, N., K. Okamoto, and Y. Azuma. 1989. Longitudinal acoustic modes and Brillouin-gain spectra for GeO_2-doped-core single-mode fibers. *J. Opt. Soc. Am. B* 6: 6: 1167–1174.
38. McCurdy, A. H. 2005. Modeling of stimulated Brillouin scattering in optical fibers with arbitrary radial index profile. *J. Lightw. Technol.* 23: 11: 3509–3516.

39. Dragic, P. D., and B. G. Ward. 2010. Accurate modeling of the intrinsic Brillouin linewidth via finite-element analysis. *IEEE Photon. Technol. Lett.* 22: 22: 1698–1700.
40. Mamdem, Y. S., F. Taillade, Y. Jaouën et al. 2012. Optical fiber properties influence on strain coefficient C_ε of Brillouin frequency shift, 22nd International Conference on Optical Fiber Sensors, Beijing, China, SPIE, 8421: 9B-4.
41. Dragic, P., J. Ballato, A. Ballato et al. 2012. Mass density and the Brillouin spectroscopy of aluminosilicate optical fibers. *Opt. Mat. Expr.* 2: 11: 1641–1654.
42. Mountfort, F. H., S. Yoo, A. J. Boyland et al. 2011. Temperature effect on the Brillouin gain spectra of highly doped aluminosilicate fibers, CLEO/Europe and EQEC 2011, Munich, Germany, Optical Society of America: CE-P23.
43. Yoo, S., C. A. Codemard, Y. Jeong, J. K. Sahu, and J. Nilsson. 2010. Analysis and optimization of acoustic speed profiles with large transverse variations for mitigation of stimulated Brillouin scattering in optical fibers. *Appl. Opt.* 49: 8: 1388–1399.
44. Tartara, L., C. Codemard, J.-N. Maran, R. Cherif, and M. Zghal. 2009. Full modal analysis of the Brillouin gain spectrum of an optical fiber. *Opt. Commun.* 282: 12: 2431–2436.
45. Koyamada, Y., S. Sato, S. Nakamura, H. Sotobayashi, and W. Chujo. 2004. Simulating and designing Brillouin gain spectrum in single-mode fibers. *J. Lightw. Technol.* 22: 2: 631–633.
46. Afshar, V. S., V. Kalosha, X. Bao, and L. Chen. 2005. Enhancement of stimulated Brillouin scattering of higher-order acoustic modes in single-mode optical fiber. *Opt. Lett.* 30: 20: 2685–2687.
47. de Oliveira, C. A. S., A. Shang, C. Saravanos, and C. K. Jen. 1993. Stimulated Brillouin scattering in cascaded fibers of different Brillouin frequency shifts. *J. Opt. Soc. Am. B* 10: 6: 969–972.
48. Sikali Mamdem, Y., F. Taillade, Y. Jaouën et al. 2014. Al-doped optical fiber to enhance strain sensitivity of Brillouin based optical fiber sensors, 23rd International Conference on Optical Fibre Sensors, Santander, Spain, SPIE, 9157: 30–34.
49. Mountfort, F. H., M. Belal, and J. K. Sahu. 2012. Frequency dependence of the Brillouin spectrum of an aluminosilicate optical fiber on temperature and strain. *Advanced Photonics Congress, Optical Society of America*, Colorado Springs, CO, USA: JTu5A.56.
50. Law, P.-C., Y.-S. Liu, A. Croteau, and P. D. Dragic. 2011. Acoustic coefficients of P_2O_5-doped silica fiber: Acoustic velocity, acoustic attenuation, and thermo-acoustic coefficient. *Opt. Mat. Expr.* 1: 4: 686–699.
51. Law, P.-C., A. Croteau, and P. D. Dragic. 2012. Acoustic coefficients of P_2O_5-doped silica fiber: The strain-optic and strain-acoustic coefficients. *Opt. Mat. Expr.* 2: 4: 391–404.
52. Dragic, P. D., C. Kucera, J. Ballato et al. 2014. Brillouin scattering properties of lanthano-aluminosilicate optical fiber. *Appl. Opt.* 53: 25: 5660–5671.
53. Zou, W., Z. He, and K. Hotate. 2008. Experimental study of Brillouin scattering in fluorine-doped single-mode optical fibers. *Opt. Expr.* 16: 23: 18804–18812.
54. Dragic, P. D. 2009. Novel dual-Brillouin-frequency optical fiber for distributed temperature sensing, SPIE LASE: Lasers and Applications in Science and Engineering, San Jose, CA, USA, SPIE, 7197: 10.
55. Dragic, P. 2014. Brillouin scattering fiber. US8750655B1.
56. Burov, E., A. Pastouret, and L.-A. de Montmorillon. 2013. Temperature and strain sensing optical fiber and temperature and strain sensor. WO2013108063.
57. Hocker, G. B. 1979. Fiber optic acoustic sensors with composite structure: An analysis. *Appl. Opt.* 18: 21: 3679–3683.
58. Budiansky, B., D. C. Drucker, G. S. Kino, and J. R. Rice. 1979. Pressure sensitivity of a clad optical fiber. *Appl. Opt.* 18: 24: 4085–4088.
59. Parker, T. R., M. Farhadiroushan, E. V. Diatzikis, A. Mendez, and R. L. Kutlik. 2000. Simultaneous optical fibre distributed measurement of pressure and temperature using noise-initiated Brillouin scattering, 14th International Conference on Optical Fibre Sensors, Venice, Italy, SPIE, 4185: 772–775.
60. Parker, T. R., M. Farhadiroushan, V. A. Handerek, and A. J. Rogers. 1997. A fully distributed simultaneous strain and temperature sensor using spontaneous Brillouin backscatter. *IEEE Photon. Technol. Lett.* 9: 7: 979–981.
61. Guangyu, Z., G. Haidong, D. Huijuan, L. Longqiu, and H. Jun. 2013. Pressure sensitization of Brillouin frequency shift in optical fibers with double-layer polymer coatings. *IEEE Sens. J.* 13: 6: 2437–2441.
62. Gu, H., H. Dong, G. Zhang, W. Hu, and Z. Li. 2015. Simultaneous measurement of pressure and temperature using dual-path distributed Brillouin sensor. *Appl. Opt.* 54: 11: 3231–3235.
63. Mendez, A., and E. V. Diatzikis. 2008. Fiber optic sensor and sensing system for hydrocarbon flow. US7397976B2.
64. Watley, D., M. Farhadiroushan, and B. J. Shaw. 2005. Direct measurement of Brillouin frequency in distributed optical sensing systems. WO2005/106396 A2.

65. Chong, L.-K., A. Boersma, and R. Jansen. 2012. Coating based fiber Bragg grating humidity sensor array, 22nd International Conference on Optical Fiber Sensors, Beijing, China, SPIE, 8421: 68.
66. Bertrand, J., S. Delphine-Lesoille, and X. Pheron. 2014. Device for detecting and/or dosing hydrogen and method of detecting and/or dosing hydrogen. US2014/0374578 A1.
67. Butler, M. A. 1984. Optical fiber hydrogen sensor. *Appl. Phys. Lett.* 45: 10: 1007–1009.
68. Antman, Y., A. Clain, Y. London, and A. Zadok. 2016. Optomechanical sensing of liquids outside standard fibers using forward stimulated Brillouin scattering. *Optica* 3: 5: 510–516.
69. Husdi, I. R., K. Nakamura, and S. Ueha. 2004. Sensing characteristics of plastic optical fibres measured by optical time-domain reflectometry. *Meas. Sci. Technol.* 15: 8: 1553.
70. Conduit, A. J., A. H. Hartog, M. R. Hadley et al. 1981. High-resolution measurement of diameter variations in optical fibres by the backscatter method. *Electron. Lett.* 17: 20: 742–744.
71. Conduit, A. J., D. N. Payne, A. H. Hartog, and M. P. Gold. 1981. Optical fibre diameter variations and their effect on backscatter loss measurements. *Electron. Lett.* 17: 8: 307–308.
72. Yoshihara, N. 1998. Low-loss, high-bandwidth fluorinated POF for visible to 1.3-/spl mu/m wavelengths. In Optical Fiber Communication Conference and Exhibit, 1998. Optical Fiber Communication Conference '98, San Jose, CA, USA, Technical Digest. 308.
73. Koike, Y., and M. Asai. 2009. The future of plastic optical fiber. *NPG Asia Mater.* 1: 22–28.
74. Liehr, S., P. Lenke, M. Wendt et al. 2009. Polymer optical fiber sensors for distributed strain measurement and application in structural health monitoring. *IEEE Sens. J.* 9: 11: 1330–1338.
75. Geiger, H., and J. P. Dakin. 1995. Low-cost high-resolution time-domain reflectometry for monitoring the range of reflective points. *J. Lightw. Technol.* 13: 7: 1282–1288.
76. Liehr, D.-I. S. 2015. Fibre optic sensing techniques based on incoherent optical frequency domain reflectometry. PhD Thesis D. Ing. Technische Universität Berlin, Germany.
77. Sascha, L., and K. Katerina. 2010. A novel quasi-distributed fibre optic displacement sensor for dynamic measurement. *Meas. Sci. Technol.* 21: 7: 075205.
78. Liehr, S., P. Lenke, K. Krebber et al. 2008. Distributed strain measurement with polymer optical fibers integrated into multifunctional geotextiles. Photonics Europe, Strasbourg, France, SPIE, 7003: 02.
79. Mizuno, Y., and K. Nakamura. 2010. Potential of Brillouin scattering in polymer optical fiber for strain-insensitive high-accuracy temperature sensing, 21st International Conference on Optical Fiber Sensors, Ottawa, Canada, Optical Society of America, 7753: 775329.
80. Mizuno, Y., M. Kishi, K. Hotate, T. Ishigure, and K. Nakamura. 2011. Observation of stimulated Brillouin scattering in polymer optical fiber with pump-probe technique. *Opt. Lett.* 36: 12: 2378–2380.
81. Kazunari, M., H. Neisei, S. Yuri et al. 2014. Wide-range temperature dependences of Brillouin scattering properties in polymer optical fiber. *Jap. J. Appl. Phys.* 53: 4: 042502.
82. Hayashi, N., Y. Mizuno, and K. Nakamura. 2014. Distributed Brillouin sensing with centimeter-order spatial resolution in polymer optical fibers. *J. Lightw. Technol.* 32: 21: 3397–3401.
83. Mizuno, Y., N. Hayashi, and K. Nakamura. 2013. Polarisation state optimisation in observing Brillouin scattering signal in polymer optical fibres. *Electron. Lett.* 49: 1: 56–57.
84. Neisei, H., M. Kazunari, M. Yosuke, and N. Kentaro. 2015. Polarization scrambling in Brillouin optical correlation-domain reflectometry using polymer fibers. *Appl. Phys. Expr.* 8: 6: 062501.
85. Castilone, R. J. 2001. *Mechanical reliability: Applied stress design guidelines.* Corning, Inc. Corning, NY.
86. Lidong, L., L. Yun, L. Binglin, and G. Jinghong. 2014. Maintenance of the OPGW using a distributed optical fiber sensor. International Conference on Power System Technology (POWERCON), Chengdu, China, 1251–1256.
87. Dey, P., P. Fearns, K. W. Plessner et al. 1982. Overhead electric transmission systems. US 4,359,598.
88. Tateda, M., T. Horiguchi, T. Kurashima, and K. Ishihara. 1990. First measurement of strain distribution along field-installed optical fibers using Brillouin spectroscopy. *J. Lightw. Technol.* 8: 9: 1269–1272.
89. Sankawa, I., S. Furukawa, Y. Koyamada, and H. Izumita. 1990. Fault location technique for in-service branched optical fiber networks. *IEEE Photon. Technol. Lett.* 2: 10: 766–768.
90. Shimizu, K., T. Horiguchi, and Y. Koyamada. 1995. Measurement of distributed strain and temperature in a branched optical fiber network by use of Brillouin optical time-domain reflectometry. *Opt. Lett.* 20: 5: 507–509.
91. Iida, D., N. Honda, H. Izumita, and F. Ito. 2007. Design of identification fibers with individually assigned Brillouin frequency shifts for monitoring passive optical networks. *J. Lightw. Technol.* 25: 5: 1290–1297.
92. Honda, N., D. Iida, H. Izumita, and Y. Azuma. 2009. In-service line monitoring system in PONs using 1650-nm Brillouin OTDR and fibers with individually assigned BFSs. *J. Lightw. Technol.* 27: 20: 4575–4582.
93. Takahashi, H., F. Ito, C. Kito, and K. Toge. 2013. Individual loss distribution measurement in 32-branched PON using pulsed pump-probe Brillouin analysis. *Opt. Expr.* 21: 6: 6739–6748.

94. Nikles, M. 1997. La diffusion Brillouin dans les fibres optiques: Étude et application aux capteurs distribués. PhD Ecole Polytechnique Fédérale de Lausanne.
95. Takahashi, H., K. Toge, and F. Ito. 2014. Connection loss measurement by bi-directional end-reflection-assisted Brillouin analysis. *J. Lightw. Technol.* 32: 21: 4204–4208.
96. Naruse, H., K. Komatsu, K. Fujihashi, and M. Okutsu. 2005. Telecommunications tunnel monitoring system based on distributed optical fiber strain measurement, 17th International Conference on Optical Fibre Sensors, Bruges, Belgium-Post Deadline Paper, SPIE: 168–171.
97. Farhoud, R., S. Delepine-Lesoille, S. Buschaert, and C. Righini-Waz. 2015. Monitoring system design of underground repository for radioactive wastes – *In situ* demonstrator. *IACSIT Int. J. Eng. Technol.* 7: 6: 484–489.
98. Henault, J., J. Salin, G. Moreau et al. 2011. Qualification of a truly distributed fiber optic technique for strain and temperature measurements in concrete structures. In EPJ Web of Conferences. EDP Sciences. 03004.
99. Ong, S., H. Kumagai, H. Iwaki, and K. Hotate. 2003. Crack detection in concrete using a Brillouin optical correlation domain analysis based fiber optic distributed strain sensor. In 16th International Conference on Optical Fiber Sensors, Nara, Japan We3-3 (Oct. 2003).
100. Meng, D., F. Ansari, and X. Feng. 2015. Detection and monitoring of surface micro-cracks by PPP-BOTDA. *Appl. Opt.* 54: 16: 4972–4978.
101. Minardo, A., R. Bernini, L. Amato, and L. Zeni. 2012. Bridge monitoring using Brillouin fiber-optic sensors. *IEEE Sens. J.* 12: 1: 145–150.
102. Minardo, A., G. Persichetti, G. Testa, L. Zeni, and R. Bernini. 2012. Long term structural health monitoring by Brillouin fibre-optic sensing: A real case. *J. Geophys. Eng.* 9: 4: S64.
103. Kuang, K. 2015. Distributed damage detection of offshore steel structures using plastic optical fibre sensors. *Sensor. Actuat. A-Phys.* 229: 59–67.
104. Phéron, X., J. Bertrand, S. Girard et al. 2012. Brillouin scattering based sensor in high gamma dose environment: Design and optimization of optical fiber for long-term distributed measurement, 22nd International Conference on Optical Fibre Sensors, SPIE, 8421: A4.
105. Phéron, X., Y. Ouerdane, S. Girard et al. 2011. *In situ* radiation influence on strain measurement performance of Brillouin sensors, 21st International Conference on Optical Fiber Sensors, Ottawa, Canada, SPIE, 7753: I-4.
106. Phéron, X. 2013. Durabilité des capteurs à fibres optiques sous environnement radiatif. PhD Thesis Université Jean Monnet-Saint-Etienne.
107. Alasia, D., A. Fernandez Fernandez, L. Abrardi, B. Brichard, and L. Thévenaz. 2006. The effects of gamma-radiation on the properties of Brillouin scattering in standard Ge-doped optical fibres. *Meas. Sci. Technol., France* 17: 5: 1091–1094.
108. Delepine-Lesoille, S., X. Phéron, J. Bertrand et al. 2012. Industrial qualification process for optical fibers distributed strain and temperature sensing in nuclear waste repositories. *J. Sensors* 2012: 369375.
109. Moore, J. P., and M. D. Rogge. 2012. Shape sensing using multi-core fiber optic cable and parametric curve solutions. *Opt. Expr.* 20: 3: 2967–2973.
110. Wylie, M. T. V., B. G. Colpitts, and A. W. Brown. 2011. Fiber optic distributed differential displacement sensor. *J. Lightw. Technol.* 29: 18: 2847–2852.
111. Tur, M., I. Sovran, A. Bergman et al. 2015. Structural health monitoring of composite-based UAVs using simultaneous fiber optic interrogation by static Rayleigh-based distributed sensing and dynamic fiber Bragg grating point sensors, 24th International Conference on Optical Fibre Sensors, Curitiba, Brazil, SPIE, 9634: 0P.
112. Stern, Y., Y. London, E. Preter et al. 2016. High-resolution Brillouin analysis in a carbon-fiber-composite unmanned aerial vehicle model wing, Sixth European Workshop on Optical Fibre Sensors (EWOFS'2016), Limerick, Ireland, SPIE: 99162M-99162M-99164.
113. Pearce, J. G., F. H. K. Rambow, W. Shroyer et al. 2009. High resolution, real-time casing strain imaging for reservoir and well integrity monitoring: Demonstration of monitoring capability in a field installation, SPE Annual Technical Conference and Exhibition, New Orleans, LA, USA, SPE, Paper 124932.
114. Zhou, Z., J. He, M. Huang, J. He, and G. Chen. 2010. Casing pipe damage detection with optical fiber sensors: A case study in oil well constructions. *Adv. Civil Eng.* 2010: 9.
115. Tennyson, R. C. 2004. Monitoring of large structures using Brillouin spectrum analysis. US6813403B2.
116. Zou, L., G. A. Ferrier, S. Afshar et al. 2004. Distributed Brillouin scattering sensor for discrimination of wall-thinning defects in steel pipe under internal pressure. *Appl. Opt.* 43: 7: 1583–1588.
117. Gu, G. P., W. Revie, L. Zou, and O. Sezerman. 2009. Pipeline monitoring by Brillouin-scattering-based fibre optic distributed strain sensors: Pipeline wall thickness detection, 20th International Conference on Optical Fibre Sensors, Edinburgh, United Kingdom, SPIE, 7503: 6O.
118. Zou, L., X. Bao, F. Ravet, and L. Chen. 2006. Distributed Brillouin fiber sensor for detecting pipeline buckling in an energy pipe under internal pressure. *Appl. Opt.* 45: 14: 3972–3977.

119. Pearce, J., P. Legrand, T. Dominique et al. 2009. Real-time compaction monitoring with fiber-optic distributed strain sensing (DSS). In SPWLA 50th Annual Logging Symposium, The Woodlands, Texas, USA, Society of Petrophysicists and Well-Log Analysts.
120. Hartog, A. H., and B. Read. 2013. Fiber optic formation dimensional change monitoring. US 2013/0188168 A1.
121. BBC News. 2010. Dozens killed in landslides in China's Gansu province. Retrieved 28 June 2015, http://www.bbc.co.uk/news/world-asia-pacific-10905399.
122. Wang, F., X. Peng, Y. Zhang et al. 2006. Landslides and slope deformation caused by water impoundment in the Three Gorges Reservoir, China, Proceedings of IAEG2006, Nottingham, United Kingdom, Geological Society of London: 137.
123. Thiebes, B. 2011. Landslide analysis and early warning. PhD Diss. Universität Wien.
124. Lloret, S., P. Rastogi, L. Thévenaz, and D. Inaudi. 2003. Measurement of dynamic deformations using a path unbalance Michelson interferometer based optical fiber sensing device. *Opt. Eng.* 42: 3: 662–669.
125. Schulz, W. L., J. P. Conte, and E. Udd. 2001. Long-gage fiber optic Bragg grating strain sensors to monitor civil structures, 8th Annual International Symposium on Smart Structures and Materials, Newport Beach, CA, USA, SPIE: 56–65.
126. Blackburn, J. T., and C. H. Dowding. 2004. Finite-element analysis of time domain reflectometry cable–grout–soil interaction. *J. Geotechnical Geoenvironmental Eng.* 130: 3: 231–239.
127. Higuchi, K., K. Fujisawa, K. Asai, A. Pasuto, and G. Marcato. 2007. Application of new landslide monitoring technique using optical fiber sensor at Takisaka Landslide, Japan, AEG Special Publication. Proceedings of the first North American landslide conference. Vail, Colorado: 1074–1083.
128. Zhang, C.-C., H.-H. Zhu, B. Shi, and J.-K. She. 2014. Interfacial characterization of soil-embedded optical fiber for ground deformation measurement. *Smart Mater. Struct.* 23: 9: 095022.
129. Sugimoto, H. 2001. Landslides monitoring by optical fiber sensor. http://cgsweb.moeacgs.gov.tw/news/1111_dl_data/Lecture%20additional(1)Optical%20fiber.ppt%20Landslides%20monitoring%20by%20optical%20fiber%20sensor.
130. Wang, B.-j., K. Li, B. Shi, and G.-q. Wei. 2009. Test on application of distributed fiber optic sensing technique into soil slope monitoring. *Landslides* 6: 1: 61–68.
131. Krebber, K., S. Liehr, and J. Witt. 2012. Smart technical textiles based on fibre optic sensors. In *Current developments in optical fiber technology*. S. W. Harun and H. Arof, editors. Intech. 978-953-51-1148-1.
132. Iten, M., A. M. Puzrin, and A. Schmid. 2008. Landslide monitoring using a road-embedded optical fiber sensor, 15th International Symposium on: Smart Structures and Materials & Nondestructive Evaluation and Health Monitoring, San Diego, CA, USA, SPIE, 6933: 15.
133. Nöther, N. 2010. Distributed fiber sensors in river embankments: Advancing and implementing the Brillouin optical frequency domain analysis. DEng Dissertation Technische Universität Berlin, Germany.
134. Naruse, H., Y. Uchiyama, T. Kurashima, and S. Unno. 2000. River levee change detection using distributed fiber optic strain sensor. *IEICE Trans. Electron.* E83-C: 3: 462–467.
135. Kwon, I.-B., S.-J. Baik, K. Im, and J.-W. Yu. 2002. Development of fiber optic BOTDA sensor for intrusion detection. Sensor. *Actuat. A-Phys.* 101: 1–2: 77–84.
136. Ferdinand, P., S. Rougeault, N. Roussel et al. 2012. Brillouin sensing for perimetric detection: The SMARTFENCE project, 22nd International Conference on Optical Fiber Sensors, Beijing, China, SPIE, 8421: 9X.

9 Applications of Distributed Vibration Sensors (DVSs)*

A mechanical vibration is a periodic (or quasi-periodic) displacement of the medium in which it takes place. In the case of conventional (electrical) vibration sensors, the response is proportional to particle velocity (for geophones) or acceleration (for accelerometers). These sensors are localised typically to a few cm, sometimes less, they respond independently of each other and they have specific angular and frequency response patterns. Quite separately, hydrophones or microphones measure fluctuations of the ambient pressure; these sensors generally respond to isostatic pressure independently of the direction from which the wave impinges on their surface.

In distributed fibre-optic vibration sensors, the displacement of the medium is converted to a dynamic strain on the sensing fibre and that strain is read remotely by the interrogator. Although a distributed vibration sensor (DVS) responds to vibration, the readings are not directly comparable to those of conventional vibration sensors, such as accelerometers or geophones. The interpretation of DVS data must take these differences into account; they arise partly from the use of strain, rather than particle displacement directly and partly because the fibre sensor integrates the strain over a gauge length. The differences between DVS and conventional point sensors have implications for the angular sensitivity, the directionality and the frequency response of the sensors that depend not only on the optical fibre sensor but also on the properties of the vibrations that it detects as well as the properties of the medium through which the vibration propagates.

The relationship between incoming acoustic or seismic waves and the DVS response has been studied mainly in the context of geophysical measurements, which is why this is the first application to be discussed in this chapter and it is in the section on seismic applications that the properties of the response of a DVS to an acoustic or vibrational wave are discussed. The application background is provided first, and it is followed by a discussion on the specific properties of DVS in these applications and, finally, a few examples are given. In some other applications, the vibration is highly localised and the measurement involves capturing the dynamic strain locally on a structure rather than travelling waves although this distinction is never fully accurate.

In some applications, particularly when a long observation time is required and an indication of the vibrational energy at each location is sufficient, the measurement can be summarised by calculating the variance or standard deviation of the signal at each location over pre-defined timeframes (such as 1 s or 30 s as appropriate to the application). In some cases, the signal is divided into frequency bands that are selected to discriminate between different sources of vibration, a practice known as calculating the

* This chapter is the subject of work that was, at the time of writing, active amongst direct colleagues at the Schlumberger Fibre-Optic Technology Centre (SFTC) and I am grateful to the many colleagues there and wider afield who have provided many insights. I would like to thank especially (in alphabetical order), William Allard, Alexis Constantinou, Theo Cuny, Tim Dean, Paul Dickenson, Tsune Kimura and Bence Papp at SFTC as well as colleagues in other parts of Schlumberger, including Mike Clark (Livingston), Bernard Frignet (Paris), Duncan Mackie (Aberdeen), James Martin (Paris), Alastair Pickburn (Southampton) and Mark Puckett (Paris) and external collaborators, Daniela Donno (Ecole des Mines de Paris) and Artem Khlybov (St. Petersburg, Russia). More recently, Daniele Molteni, Ali Ozbek and Mike Williams (all at the Schlumberger Gould Research Centre, Cambridge, UK) have contributed to the understanding of the application of DVS to geophysics.

frequency band energy (FBE). The technical antecedents for doing this in oilfield applications are to be found in noise logging tools, i.e. single-point acoustic transducers that originally used analogue electronic filters to separate the signal into a few bands for the purposes of identifying small leaks of fluid behind the casing. The FBE calculation reduces by several orders of magnitude the amount of data that are required to be transmitted and displayed in real time.

9.1 ACTIVE SEISMIC APPLICATIONS

Seismic signals are low-frequency (usually <100 Hz) displacement waves travelling through a geological structure. In active seismic acquisition from the surface, vibration stimuli are transmitted into the earth and arrays of sensors are used to capture reflections from rock strata and, in some cases, direct arrivals or refracted waves. From these seismic signals, a picture of the sub-surface can be formed, which is then interpreted in terms of the likely existence of hydrocarbon bearing formations [1,2], or other structures of interest, deep underground. It is the hydrocarbon extraction industry that has led the development of seismic techniques using DVS, and so it has provided most of the examples of its use; nonetheless, geophysical sciences are also applied to improving the understanding of the earth in a wider sense, such as for studying geological faults, or for determining the suitability of potential sites for nuclear waste disposal and carbon sequestration [3]. The initial seismic acquisition for exploration is carried out from the surface, but often, surveys are carried out from within the borehole immediately after exploration and appraisal wells are drilled.

9.1.1 Surface Seismic Acquisition

Seismic acquisition is used in the exploration stage, i.e. to discover prospective locations of hydrocarbon reservoirs. If promising, the same data can be used to plan the drilling of exploration wells. Increasingly, seismic acquisition is also used repeatedly during the production phase of the hydrocarbon extraction to look for changes in the subterranean structures that could indicate where fluids have been produced and where this is not the case indicating, therefore, that reserves could have been by-passed [4]. This is known as time-lapse seismic 4-D seismic or permanent reservoir monitoring (PRM).

In marine seismic exploration, the sensors are towed behind a seismic acquisition vessel in arrays known as streamers that reach some 10 km in length. The sensors have traditionally been hydrophones, but recent developments [5] have added multi-axis accelerometers that allow far better reconstruction of the waves arriving at the receiver arrays and, thus, better processing of the seismic image. Typically, up to 10 streamers are towed in parallel behind a vessel; the streamers are steered with movable control surfaces [6] to maintain a precise depth and lateral position regardless of currents [7]. They also carry equipment such as Global Positioning System (GPS) receivers and heading sensors to inform the positioning control system. These systems are complex and return a huge flow of data (of the order of several Gb/s) that is stored and pre-processed on board the acquisition vessel prior to being processed in dedicated computing centres. The sources for marine seismic exploration are generally compressed-air impulsive sources (known as *air guns*) and are limited in their power by a number of considerations, including damage to cetaceans.

Another category of marine seismic acquisition involves ocean-bottom sensors, either permanently installed in trenches or in the form of retrievable cables or pods.

In land seismic applications, the sensors are generally moving-coil geophones that sense particle velocity, although accelerometers are sometimes used. Seismic land acquisitions involve the mobilisation of large crews and large numbers of sensors, acquisition units, cables and sources. The seismic source on land is sometimes impulsive (dynamite or air guns) for localised surveys, but usually swept-frequency, truck-mounted vibrators are used for large-area surveys.

Naturally, considerable thought has been given to using distributed optical fibre sensor (DOFS) in seismic acquisition. In this application, only the differential phase measurement techniques have so far been proven to be suitable: the intensity measurement of the backscatter signals is not sufficiently linear to meet the requirements of seismic acquisition, where the signal sent to the signal processing must have a known transfer function in relation to incoming seismic waves.

Although optical fibre acoustic sensor arrays are used extensively in naval sonar receivers mounted on or towed behind vessels, this does not yet apply to distributed sensors and, more generally, to seismic receivers used for seismic exploration. The main reason for this absence of a take-up in marine seismic is that the streamers are already highly performing, and they have many functions in addition to towing seismic receivers.

However, arrays of optical point sensors have been developed and installed permanently on the seabed for PRM by companies such as Optoplan [8], PGS [9] and Stingray Geophysical. PRM is still an emerging subject, and it should be pointed out that none of these installations involves fully distributed sensors. Nonetheless, this is perhaps the most likely application for surface seismic surveying using distributed sensing.

9.1.2 Borehole Seismic Surveying

In this subsection, some terms relating hydrocarbon wells are used; the reader is referred to Chapter 7 where they were introduced.

9.1.2.1 Introduction to borehole seismic acquisition

DVS, particularly differential-phase DVS (dΦ-DVS), is rapidly developing into an accepted alternative to conventional, electrical sensors in borehole seismic acquisition. Borehole seismic measurements, also known as *vertical seismic profiling* (VSP), consist of inserting one or more vibration sensors into a borehole and using sources external to the borehole to generate a seismic signal that is detected by the borehole sensors. Only a brief introduction to the subject of borehole seismic survey is provided here; further details may be found in Ref. [10].

There are several motivations for conducting VSP measurements. First and foremost is the calibration of the velocity model. In general, a seismic survey using sources and receivers at the surface will have been conducted before the well is drilled. The surface seismic survey provides a two- or three-dimensional map of the sub-surface using reflections from layers of contrasting acoustic impedance. However, this information is mapped in the time domain, i.e. as a delay from the source energy being coupled to the ground to a reflection being detected. In order to relate the time information to real, physical depth it is necessary to know the velocity of the waves in each of the layers. The seismic picture can then be *migrated* from time to depth. In general, however, the velocity profile can be only estimated from the surface, based on general geological information from the region. A *check-shot* VSP provides the direct arrivals as a function of depth and this allows seismic time to be related to geological depth.

Another driver for borehole seismic surveying is the fact that the earth attenuates high frequencies preferentially and so the seismic bandwidth available at the surface is limited and this restricts the spatial resolution of the seismic image [11]. A VSP, being measured from within the formation, closer to the reflective features, better preserves the high-frequency content of the seismic information and so, when all the features of the seismic response (including reflections, shear conversions etc.) are used to create a seismic image, higher spatial resolution can be achieved, especially in the vicinity of the borehole.

A specific motivation for using optical DVS techniques in borehole seismic applications is the elimination of downhole electronics. Boreholes are hot; they expose the sensor packages to high pressure and often to corrosive fluids. Electronic sensors designed and packaged for these conditions are expensive; moreover, even though the technology of high temperature electronics is improving, there are still conditions in which electrical sensors cannot be operated for the time required for an extensive survey.

Borehole seismic surveying is illustrated in Figure 9.1, which shows a vertical well with an array of sensors and a seismic source that generates waves intended to be detected by the sensor array. In this case, the source is as close as possible to the well-head, a configuration knows as *zero-offset VSP* (ZOVSP). Electrical sensors are usually packaged in *shuttles* separated by a length of wireline cable (typically 50 ft or multiples thereof). In modern borehole seismic tools, each shuttle contains three orthogonally positioned directional sensors, and the sensor package is pushed against the borehole wall using extendible arms, shown in the diagram, in order to achieve good coupling between the borehole wall and the sensors and to decouple the latter from vibrations that could be transmitted along the cable.

Some of the seismic waves arrive at the sensors directly from the source; others arrive after reflecting from a layer below the well. In all cases, a change in the seismic velocity will cause a bending (refraction) of the wave path (in the same way as light rays bend at the interface between regions of differing refractive index), but this effect is minor for ZOVSP given that the waves usually travel close to orthogonally to geological formation.

To the right of Figure 9.1, a time-depth diagram illustrates the signal arrivals at each of the shuttles. A direct arrival from the source to the sensor provides the one-way transit time. This data calibrates the velocity model for the sub-surface conditions close to the well. Waves that reflect from boundaries between formations having different acoustic impedance arrive a little later and can be distinguished in the time/depth data by a reverse slope from the direct arrivals. A time-transformation (delaying the trace at each depth by a time equal to the first, direct, arrival) aligns the waves returning from any given reflector at each of the borehole sensors to a single time for a given depth, and averaging these time-transformed data results in a *corridor stack* that provides the layer structure in a form that can be compared directly with other data (such as logs of sonic velocity recorded as a function of depth) and used to calibrate the depth scale on seismic images.

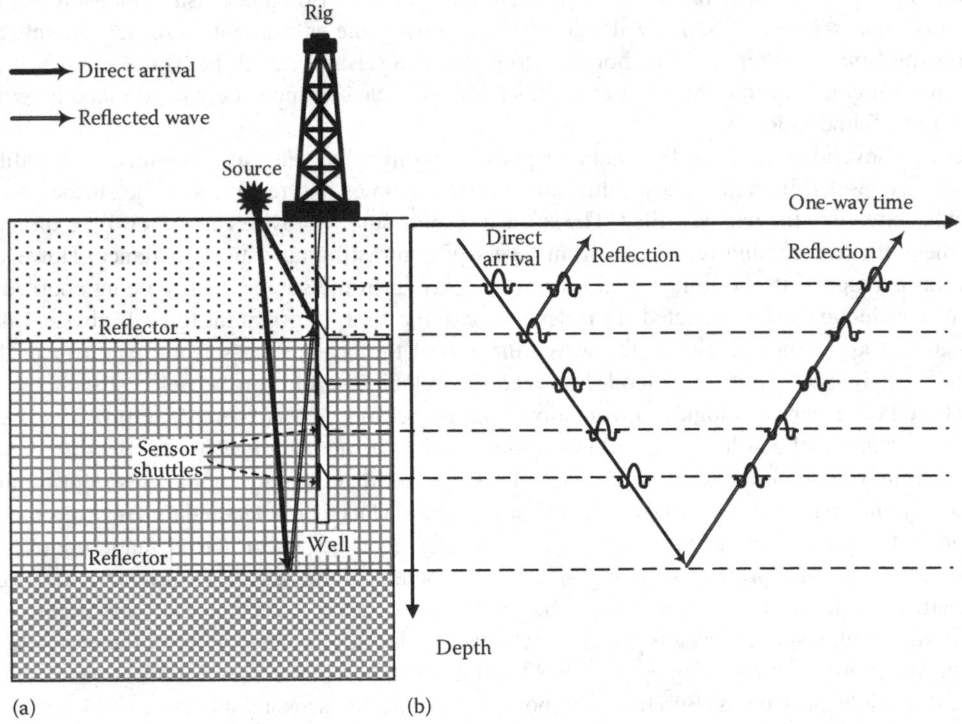

FIGURE 9.1 Illustration of borehole seismic acquisition. (a) Schematic arrangement of the well and geological formations below it. (b) Illustration of the down-going waves and initial reflections.

Implicitly, the waves illustrated are compressional waves (P-waves), but many other types of waves also exist and are routinely detected in VSPs, such as shear waves (S-waves) generated at the source or converted from P-waves at layer interfaces. In addition, tube waves (or Stoneley waves) are guided by the velocity contrast between the fluid in the borehole and the rock formation surrounding the well, in much the same way as modes are guided in an optical fibre. Tube waves are excited by the seismic source through a surface wave that travels to the wellhead and converts at that point to an acoustic mode of the well. It is therefore clear that borehole seismic measurements can provide a wealth of information about the seismic properties of the surrounding formations.

The case of a vertical well illustrated in Figure 9.1 is the simplest, where the source is located near the well-head, i.e. it is a ZOVSP configuration.

Modern wells are often deviated from the vertical or drilled horizontally once the formation of interest is reached. Borehole seismic surveys in these wells are frequently conducted with multiple source points, along the projection of the well trajectory on the surface, a practice termed *walk-above* surveying or vertical-incidence VSP (VIVSP).

The information provided by a ZOVSP is constrained to a region close to the well and has no azimuthal variation. Enhanced surveys, in which the source is moved on a line intersecting the well (*walk-away* VSP), increase the range of angles of arrival of the detected waves and so produce more detailed information in the plane of the well and source line. Azimuthal information can be obtained with a *walk-around* VSP survey and a more complete data set can be obtained using an array of source positions on a two-dimensional grid at the surface (3-D VSP).

If waves that have been reflected multiple times are included in the processing, a better use can be made of the wealth of information available in the seismic data [12] by adding virtual sources arising from reflections that increase the range of angles of arrival. However, the data quality presently available from DVS data may not allow the effective use of these novel imaging techniques until the signal-to-noise ratio of DVS approaches the excellent performance of conventional tools.

VSP surveys are primarily conducted in the exploration phase in order to gain a rapid calibration of the velocity model used for seismic reconstruction. The improved accuracy that this provides on the depth of sub-surface geological structures is then used to adjust the drilling plan and so ensure that the well passes as intended through hydrocarbon-bearing formations. In conventional (electrical) VSPs, tools are lowered on a wireline cable (typically a seven-conductor, steel-armoured cable of 11–12 mm diameter) into the well whilst the survey is shot; modern tools consist of a number of *levels* that allow parallel acquisition at multiple depths. The quality of these measurements is excellent, far exceeding what can be achieved with DVS at the time of writing; however, the number of levels is limited by several factors, such as the time required to deploy the array of tools and the telemetry rates available. As a result, it is often necessary to conduct the survey in several stages, lowering the array, activating the source a few times and then moving the array up the well and repeating the procedure to cover the desired portion of the wellbore. This process is time-consuming, and because rig time is expensive, especially in conditions such as deep-water offshore exploration, sometimes the borehole seismic survey is not conducted at all. In addition, the same constraints often result in compromises on the separation of the shuttles, and this can result in spatially aliasing* some of the slower seismic waves. When the slow waves are aliased, they cannot be accurately separated from other waves, and so the information from fast as well as slow waves is corrupted.

9.1.2.2 Optical borehole seismic acquisition on wireline cable

VSPs using distributed optical fibre sensing is very attractive if it can be shown to provide data of adequate quality because the acquisition can be much faster than with the traditional techniques. To that end, special wireline cables, incorporating optical fibres for sensing as well as optical telemetry, have been constructed

* Not providing sufficiently close spacing of the data points along the well to allow the waveforms of the slower waves (that have shorter acoustic wavelengths than the faster waves) to be properly reconstructed.

and used in conjunction with DVS [13,14]. A comparison of a borehole seismic image obtained with the optical technique on wireline cable with one obtained using a conventional, accelerometer-based tool is shown in Figure 9.2 [14]. Here, the DVS data (acquired using the hDVS technique with a fibre built-into a wireline cable [15] (a) and (b)) is compared with that acquired with the Versatile Seismic Imager (VSI) (c), a Schlumberger tool that includes three-component (x–y–z) accelerometers that are clamped against the borehole wall tool with an extendible arm. This clamping ensures excellent coupling to the seismic waves, assuming the borehole wall itself is well coupled to the formation. In all cases, the source used to acquire these data was a set of airguns in a portable water-filled device [16]. Even the single-shot hDVS image (b) shows the major features of the survey that are visible in the data captured with the conventional tool, namely the first compressional (P) and shear-wave (S) arrivals as well as tube waves guided in the borehole; Figure 9.2a shows the result of averaging nine shots of the source; the averaging process has already removed much of the noise that was visible in the single-shot data. It should be noted that much of the unwanted signals prior to the first arrival in panel (b) are due to tube waves that travel up and down the well and persist for a considerable time (tens of seconds) after the initial shot.

In general, the signal-to-noise ratio produced by the conventional tool is better that that of the optical measurement. The result shown in Figure 9.2 is the first to be recorded with a wireline cable, and the signal-to-noise ratio obtained with DVS is 20–40 dB lower than is obtained with properly clamped

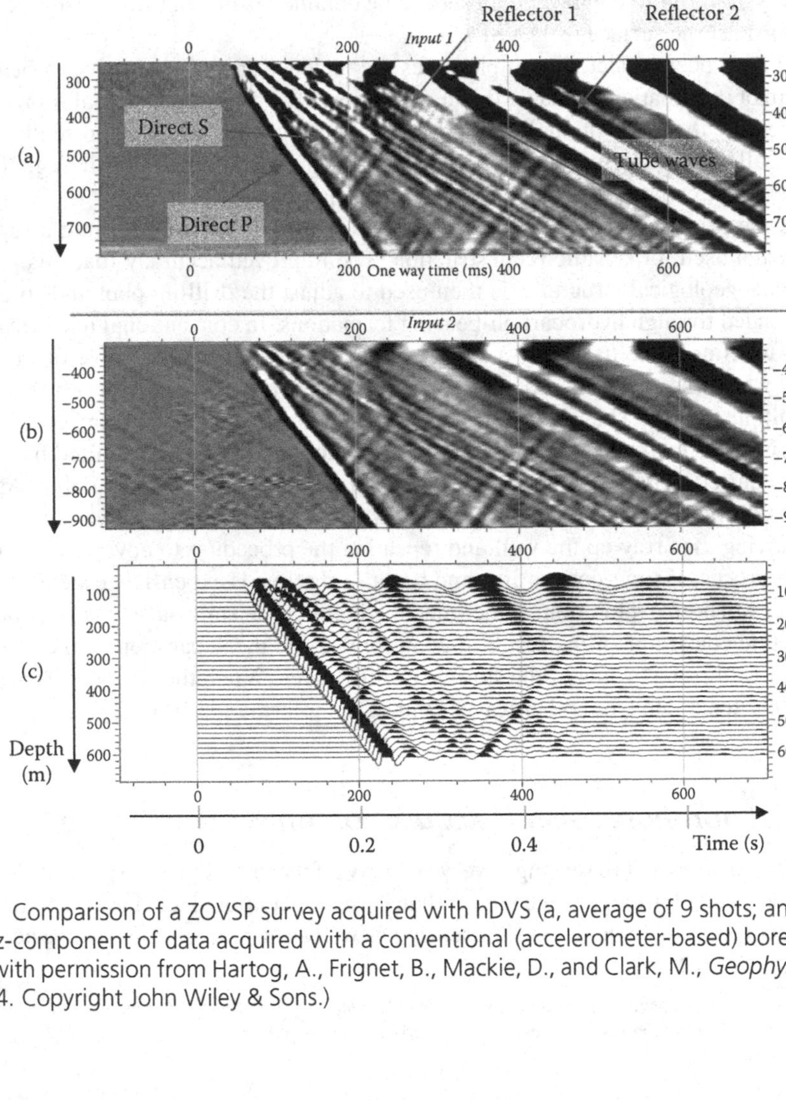

FIGURE 9.2 Comparison of a ZOVSP survey acquired with hDVS (a, average of 9 shots; and b, single source shot) with (c) z-component of data acquired with a conventional (accelerometer-based) borehole seismic tool. (Reproduced with permission from Hartog, A., Frignet, B., Mackie, D., and Clark, M., *Geophys. Prospecting* 62, 693–701, 2014. Copyright John Wiley & Sons.)

electrical sensor array tools. It should also be noted that improvements in dΦ-DVS technology are gradually narrowing the gap between the optical and electrical technologies.

Nonetheless, the optical image shows all the main features of the survey and transit times picked with both approaches match very well. It is also worth noting that the optical image is less granular in the vertical dimensional than the conventional one because the optical data are sampled more densely, in this case the DVS data is down-sampled from intervals of 0.36 m to intervals of 1.34 m, compared with 15 m level spacing for the conventional tool. As a result, some of the slower waves are actually better imaged with the optical data, in spite of its lower signal-to-noise ratio. The denser sampling also allows better performing signal processing techniques to be employed to reject unwanted noise than are available with sparser sampling.

In addition, the optical data are obtained along the entire depth of the well in a single acquisition (or sets of acquisitions, if averaged) without the need to move the wireline tool. This saves a large amount of time, and this can be crucial in considering whether or not to run the measurement. A single-source position borehole seismic survey can take several hours (10–12 hours on typical wells of depth 5000 m), and this is to be compared with just a few minutes with DVS once the cable is lowered (which can be done at a rate of several thousand feet/hour).

Moreover, running a VSP survey using optical techniques on a wireline cable allows the measurement to be combined with other data acquisition taking place in the same borehole, such as sonic, nuclear and resistivity logs, dynamic pressure testing or fluid sampling that are the fundamental measurements acquired in exploration wells. In order to run an optical VSP, the only additional equipment required is a seismic source and the DVS interrogator assuming that a wireline cable containing a sensing fibre is available. If the measurement is run whilst other petro-physical data are collected, little or no additional rig time is required. Above all, it is the rig time required for a conventional VSP that makes the VSP surveys costly and leads to fewer VSP surveys being acquired than would ideally be the case.

9.1.2.3 Borehole seismic acquisition using permanently installed optical cables

Prior to the first DVS borehole seismic measurements on wireline cable, the method had already been demonstrated on fibre permanently installed in wells, using cable cemented behind the casing [17,18] or conveyed on tubing inserted inside the casing [19], which is the conventional way of routing cables, whether electrical or optical, from the surface to downhole sensors.

The use of permanently installed fibres is quite different in its intent from wireline deployment: it is directed at measuring small changes in the seismic image over time, i.e. 4-D seismic, using repeat surveys. Having the fibre permanently installed reduces the cost of such repeat surveys, which would, in most cases, be uneconomic with wireline tools. 4-D seismic, or PRM, is aimed at optimising the extraction from a hydrocarbon formation by understanding which parts of the reservoir are producing and which are by-passed: this knowledge allows the production to be improved, for example, by drilling in-fill wells or altering the enhanced oil recovery (EOR) strategy, e.g. by changing the injection pattern, or the injection fluids.

PRM, being based on estimating differences between successive seismic acquisitions, demands far better (by an order of magnitude) accuracy in the seismic signals compared with a check shot, both in terms of signal-to-noise ratio of the seismic signal and in the timing of the recording of the information relative to the launching of the seismic energy into the ground [20,21]. It is traditionally carried out using permanently installed sensor arrays, electrical or optical [8], and bringing seismic sources back on location, typically at 6-month intervals. Seismic images acquired at separate times are processed to highlight differences that have occurred over time that are interpreted as evidence of effective drainage of the reservoir in the regions where a difference is detected [22] or indeed in the case of injection, e.g. for CO_2 sequestration, of where the injected fluids have migrated to in the reservoir [23].

It is not yet clear whether DVS techniques in the borehole will meet this challenging performance required for PRM, but trials are underway. For example, a large-scale test (involving two long wells and c. 50,000 source shots, with the equivalent of tens of millions of seismic traces acquired) was conducted in the Gulf of Mexico [24] and a successful test was conducted on monitoring steam injection in Oman [25].

In addition to hydrocarbon extraction, DVS techniques have been used in the monitoring of CO_2 storage sites [19,26–28]; here, the interest is not only in understanding the geology but also in tracking changes resulting from CO_2 injection.

DVS technology has also been applied in the monitoring of geothermal wells [29].

9.1.3 Response of Distributed Vibration Sensors To Acoustic or Seismic Waves

Distributed vibration sensors convert the wave, whether a pressure wave or a shear wave, into a strain on the fibre. This has a number of implications for the use of DVS in seismic measurements because the response differs in important ways from that of conventional sensors. Although DVS/DAS systems appear to show seismic images that are similar, for example, when viewed on a greyscale image, to those recorded with conventional sensor arrays, the details of the wavelets recorded by DVS and by electrical signals are different in ways that can affect the interpretation and processing of the data.

It should also be remembered that some designs of DVS systems provide the unwrapped phase at each location, whereas others provide the phase difference between successive probe pulses; the latter is of course the derivative of the former, but the geophysical meaning is subtly different in the two cases.

9.1.3.1 Distance resolution in differential phase DVS systems

Whereas the ability of other types of DOFS to distinguish closely spaced features is characterised by both sampling resolution and spatial resolution (that in turn is defined by the pulse duration and acquisition bandwidth), in DVS, a third parameter is involved, namely the gauge length. Figure 9.3 illustrates the distance scales involved.

The finest scale is usually the (native) sample separation, determined by the sampling rate f_d of the acquisition system. The sample separation is often dictated by the need to estimate the phase. For example, if the data are modulated by a quasi-sinusoidal wave, the phase of which is used to estimate strain (this is the case in hDVS or the dual-pulse technique), then it will be necessary to sample the modulation wave at a rate of at least two samples per cycle (this value is very marginal – a factor of three provides far more resilience to the measurement). However, once the phase is estimated, it is not always necessary to output data at this sample density. Although this criterion is highly dependent on the details of the system design, it is often the case that, after phase estimation, many of the samples are redundant. The output sample separation, illustrated to the right of the figure by the distance between overlapped gauge lengths, is therefore often lower than the native sample separation.

The distance occupied by the probe pulse (a pulse is usually the probe waveform adopted in DVS) is usually a few times longer than the sample separation. The choice of the pulse duration is a compromise between many factors, including the energy launched, the impact of non-linear effects and the relationship between pulse duration and gauge length.

The dependence of the linearity of the response on the ratio of gauge length to pulse length was discussed in Chapter 6 (Section 6.5.5). To summarise, the gauge length should be set to a larger value than the fibre length occupied by the probe pulse in order to ensure a monotonic response; moreover, the longer the gauge length is in relation to pulse width, the more linear the response with respect to strain will be. However, the improved linearity comes at the expense of spatial resolution.

We recall that the response of a dΦ-DVS is the difference between the phase measured at each end of the gauge length. The DVS system will integrate the strain information over the gauge length. A single

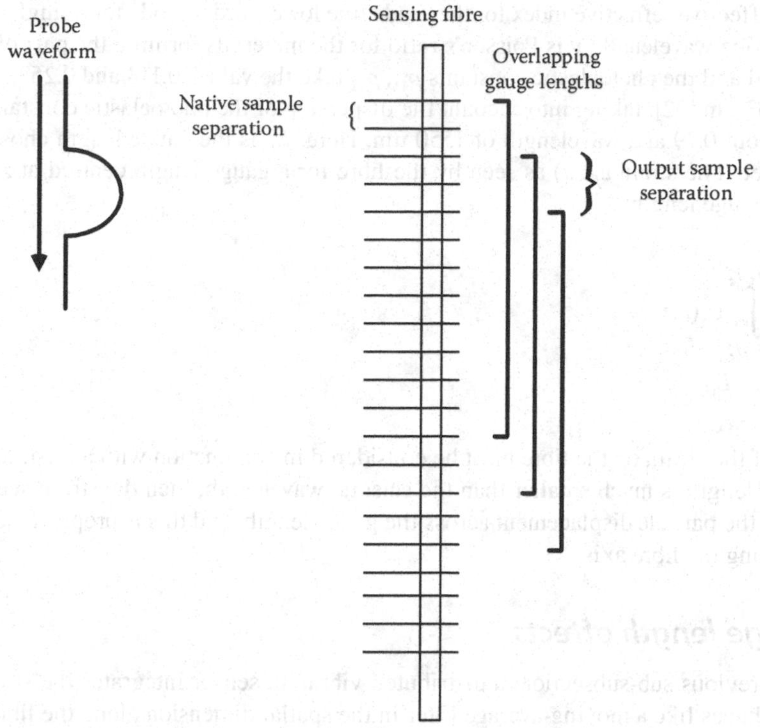

FIGURE 9.3 Distance scales in DVS.

value is returned for each gauge and the response to a localised event anywhere within a gauge length will affect that value. In that sense, the system is unable to distinguish the location of the event to better than a gauge length. However, it is possible in some systems to extract data for overlapping gauge lengths; i.e. there can be some benefit in selecting an output trace separation that is smaller than the gauge length. This provides better clarity of images, as illustrated in Figure 9.2, where the gauge length is similar to the distance between sensors in the conventional survey, but the higher output sampling trace for the optical data shows more detail, particularly for the slower waves.

9.1.3.2 Nature of the measurement

A dΦ-DVS system responds to strain.* More precisely, for each location, it responds to the strain integrated over the gauge length. The change in the phase resulting from a uniform strain ε_z in the longitudinal axis of the fibre is given by Ref. [30]

$$\Delta\phi = \frac{4\pi \cdot n_{eff} \cdot L_G \cdot \varepsilon_z \xi}{\lambda}, \tag{9.1}$$

where [31]

$$\xi = 1 - \frac{n_{eff}^2}{2}[p_{12} - \nu(p_{11} + p_{12})], \tag{9.2}$$

* Some systems natively provide the time derivative of strain, but the underlying physical measurement is still the integral of strain over a gauge length.

where n_{eff} is the effective refractive index for the mode (the lowest order mode for a single-mode fibre), λ is the free-space probe wavelength, ν is Poisson's ratio for the materials forming the part of the fibre where the light is guided and the photoelastic constants p_{11}, p_{12} take the values 0.113 and 0.253, respectively, at a wavelength of 633 nm [32]; taking into account the dispersion of the photoelastic constants, Equation 9.2 ξ evaluates to about 0.79 at a wavelength of 1550 nm. Here, L_G is the gauge length chosen to record the particular data set. The strain $\varepsilon_f(z,t)$ as seen by the fibre for a gauge length centred at z and at time t is averaged over a gauge length:

$$\varepsilon_f(z,t) = \frac{\int_{z-L_G/2}^{z+L_G/2} \varepsilon_z(z,t)}{L_G}. \tag{9.3}$$

The effect of the strain on the fibre must be considered in conjunction with a seismic wave.

If the gauge length is much smaller than the seismic wavelength, then the strain seen by the fibre is the difference in the particle displacement across the gauge length, and this is proportional to the acoustic wave velocity along the fibre axis.

9.1.3.3 Gauge length effects

As seen in the previous sub-subsection, a distributed vibration sensor integrates the strain over a gauge length and so behaves like a moving-average filter in the spatial dimension along the fibre. Locally, at an infinitesimal section of fibre dz at location z, the change in length $\delta\mathcal{L}$ due to a sinusoidal acoustic wave travelling along the fibre with acoustic wavelength λ_A and initial phase ϕ_a will be

$$\delta\mathcal{L}(z) = D_a \sin\left(\frac{2\pi}{\lambda_A} z - \phi_a\right) \cdot dz. \tag{9.4}$$

Here, D_a is the peak displacement due to the acoustic wave. Integrating this over a gauge length yields

$$\Delta\mathcal{L} = -\frac{D_A \cdot \lambda_A}{2\pi} \left\{ \cos\left(\phi_a - \frac{2\pi}{\lambda_A} L_G\right) - \cos\phi_a \right\}. \tag{9.5}$$

It is clear that if L_G is equal to an integral number of wavelengths, the response $\Delta\mathcal{L}$ will be zero. Equivalently, if the spatial response is converted to the spatial frequency (wavenumber) domain, this means that the gauge length introduces notches in the wavenumber response and so it behaves like a wavenumber filter. If an acoustic wave travels along the fibre with *apparent velocity** V_A and frequency f_A, the corresponding wavenumber is $k_A = 1/\lambda_A = f_A/V_A$. Therefore, for a fixed speed of sound, the wavenumber filter applies a frequency filter to the data [33–38].

The filter response for a 5-m gauge is shown in Figure 9.4 (solid curve). The scaling to the frequency domain will depend on V_A, and even for $V_A = 2000$ m/s (at the lower end of the velocities found in subterranean formations, even for sedimentary rocks), this corresponds to a frequency of 400 Hz for the first notch. However, in geophysical applications, significantly longer gauge lengths are desirable for reasons that become apparent in Figure 9.5, i.e. that as the gauge length is increased, the low-frequency response becomes stronger in the sense that the phase shift recorded across the gauge length due to the acoustic wave

* The component of the velocity that is parallel to the fibre axis.

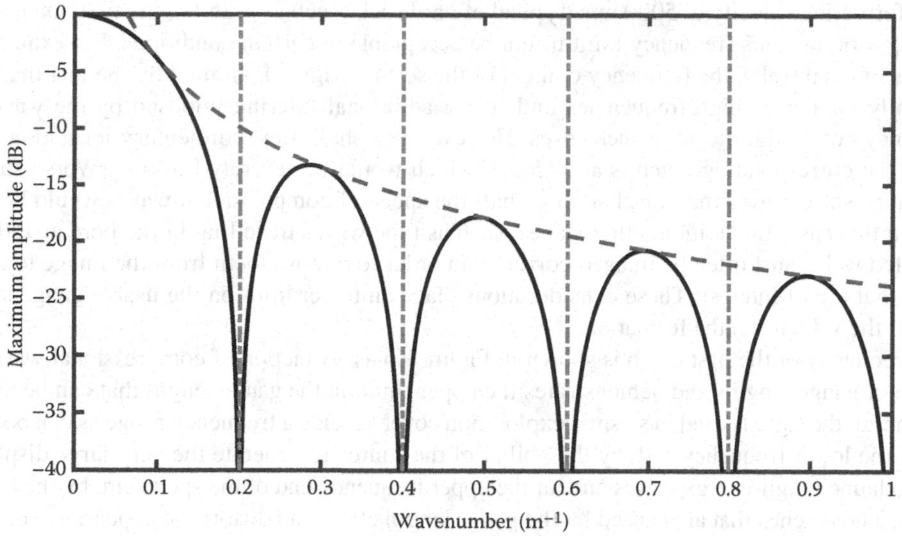

FIGURE 9.4 Wavenumber response for a 5-m gauge length. (Adapted with permission from Dean, T., Papp, B., and Hartog, A., Wavenumber Response of Data Recorded Using Distributed Fibre-optic Systems, 3rd EAGE Workshop on Borehole Geophysics, 2015, and drawn using a script provided by Tim Dean.)

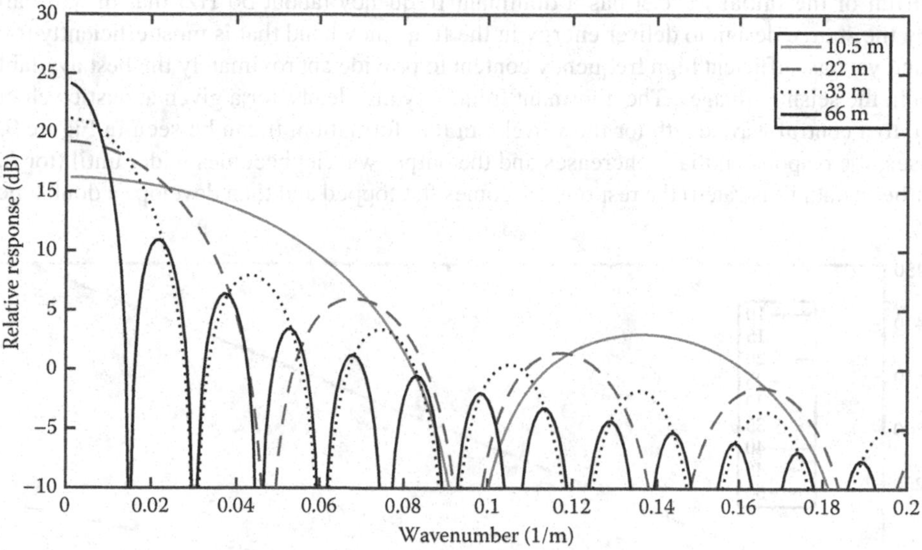

FIGURE 9.5 Response vs. wavenumber for a range of gauge lengths. Drawn using a script provided by Tim Dean.

increases. Of course, if the signals are scaled by gauge length, the strain at low frequency is unchanged by the gauge length effect; the relevance of the phase response (not normalised to L_G) lies in comparing it with the noise level in the acquisition system. In this figure, the wavenumber responses are plotted for a few values of L_G up to 66 m. As can be seen, the amplitude of the signal collected in the gauge length increases with increasing gauge length, but of course, the notch moves to lower frequencies, and even within the first lobe, the higher frequencies are attenuated before the first notch is reached. There is therefore a limit on how long the gauge length can be. The limitation due to the first notch is at $k_A = 0.0166\ \text{m}^{-1}$ for $L_G = 66$ m.

For a fast formation velocity of 5000 m/s (typical of hard rocks such as granite), the first notch appears at 75 Hz if L_G = 66 m. This frequency limit might be acceptable in certain conditions. For example, in the lower parts of deep wells, the frequency content of the seismic signal is limited by the fact that the earth preferentially attenuates high frequencies and so the additional filtering imposed by the wave number response may not be significant in such cases. However, in a shallower sedimentary formation with V_A = 2500 m/s, the corresponding notch is at 37.5 Hz, which would be severely limiting. Worse still, slower waves such as shear waves that travel at about half the speed of compressional waves would be seriously distorted in this case. In addition, other waves, such as tube waves travelling in the bore at 1500 m/s for a water-filled well, must often be imaged correctly in order to remove them from the image to leave only the waves that are of interest. These considerations place an upper limit on the usable gauge length that depends on the velocity in the formation.

The frequency of the first notch is shown in Figure 9.6 as a function of compressional wave velocity for a range of gauge lengths and demonstrates the upper limit on the gauge length that can be used.

In general, the signals used in seismic exploration cover as wide a frequency range as can be achieved, limited, at the lower frequency end, by the ability of the source to generate the very large displacements required to launch high-energy waves and, at the upper frequency end of the spectrum, by the earth filtering effects. The notches that are caused by the gauge length effect can distort the appearance of the signal when shown in the time domain.

A Ricker wavelet is an approximation to the displacement generated by impulsive sources as a function of time that is frequently used in geophysics. Figure 9.7 (left panel) shows one such wavelet transformed through the moving average filter caused by the differentiation across a gauge length, for (an admittedly unrealistic value) V_A = 1000 m/s (left) together with the corresponding signal spectra. The spectrum of the initial wavelet has a dominant frequency (about 50 Hz) that in real hardware is tailored by the source design to deliver energy in the frequency band that is most efficiently transmitted to depth and yet has sufficient high frequency content to provide approximately the best available spatial resolution in the seismic images. The dominant frequency also leads, for a given acoustic velocity in the formation, to a central wavelength for the wavelet in that formation. It can be seen in Figure 9.7 that as L_G increases, the response initially increases and the output wavelet becomes wider until (for L_G = 20 m in the synthetic data illustrated) the response becomes flat-topped and then develops a double peak. This

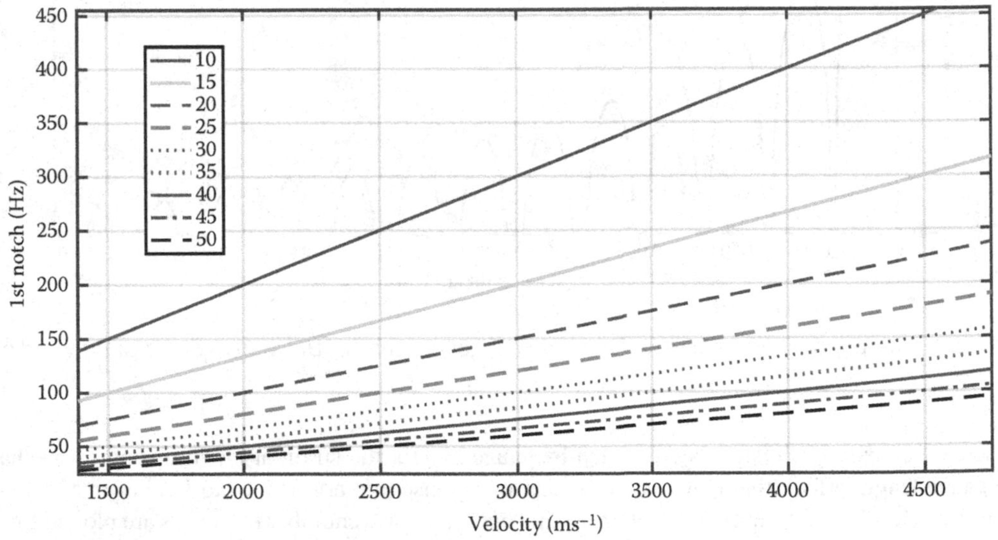

FIGURE 9.6 Frequency of first notch vs. formation velocity for a range of values of the gauge length. (Adapted from Dean, T., Papp, B., and Hartog, A., Wavenumber Response of Data Recorded Using Distributed Fibre-optic Systems, 3rd EAGE Workshop on Borehole Geophysics, 2015; re-plotted using a script provided by Tim Dean.)

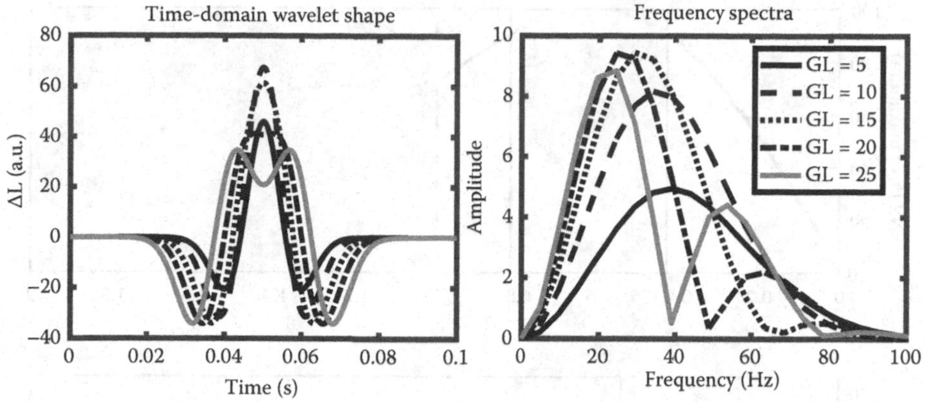

FIGURE 9.7 Time and frequency domain response for a Ricker wavelet with various gauge lengths. (Adapted with permission from Dean, T., Cuny, T., and Hartog, A.H., *Geophys. Prospect.*, in press. Copyright John Wiley & Sons.)

occurs when the dominant spatial wavelength of the wavelet is equal to the gauge length. As the wavelet shape approaches the flat-top position, the resolution of the measurement is degraded, and interpreting the data (for example, picking first arrivals) becomes less accurate. Although the double-peak can be eliminated by filtering in the spatial frequency domain, this still leaves a degraded spatial resolution.

However, as we have also seen, increasing the gauge length increases the low-frequency response of the dΦ-DVS to the wave. Under the assumption that the noise is independent of the gauge length,* then the signal-to-noise ratio will increase with increasing gauge length, so we are faced with a trade-off in choosing the gauge length.

The trade-off is summarised in Figure 9.8. In Figure 9.8a, the signal-to-noise ratio is plotted as a function of the gauge length relative to the dominant wavelength in the wavelet. As can be seen, this is maximised for a ratio in the region of 0.4–0.8 (lightly-shaded box) and peaks at a ratio of 0.6; the dark box indicates the region of multiple peaks and is to be avoided. In contrast (shown in Figure 9.8b), the width of the wavelet, as measured by the DVS, increases with increasing ratio of gauge length to dominant wavelength. This leads to an apparent increase in the wavelength dominant in the wavelet; however, the increase is only substantial at a ratio above about 0.55, and below 0.3, it is trivial.

Figure 9.8 therefore indicates that the gauge length should be selected, if possible, at a factor of 0.3 to 0.6 of the dominant wavelength in the wavelength. Of course, in borehole seismic applications, this ratio is a function of depth in the well because (a) the formation velocity is a function of depth, as the rock properties change in the different strata traversed by the well and (b) because the spectral content of the wave is changing as a result of the attenuation of the highest frequencies by the earth.

It is therefore desirable to vary the gauge length, at least in a few coarse steps, in order to maximise the signal-to-noise ratio of the output at each level. How this is achieved will depend on the DVS system design.

In the case of the interferometric phase recovery techniques (See Chapter 6, Section 6.3.4), the gauge length is fixed by a length of fibre within the instrument, and so this needs to be changed before another data set is recorded, with new source shots. A parallel arrangement of multiple recovery interferometers can be used to allow multiple gauge lengths [39], but each of these interferometers must be connected to a set of receivers and acquisition electronics.

For the dual-pulse techniques (Chapter 6, Section 6.3.3), the gauge length may be set simply by the timing of the acousto-optic modulator (AOM) opening; a different set of data is then acquired for each gauge length setting. In other dual-pulse designs, the gauge length is set, in addition to controlling the

* The validity of this assumption will depend on the design of the dΦ-DVS instrument.

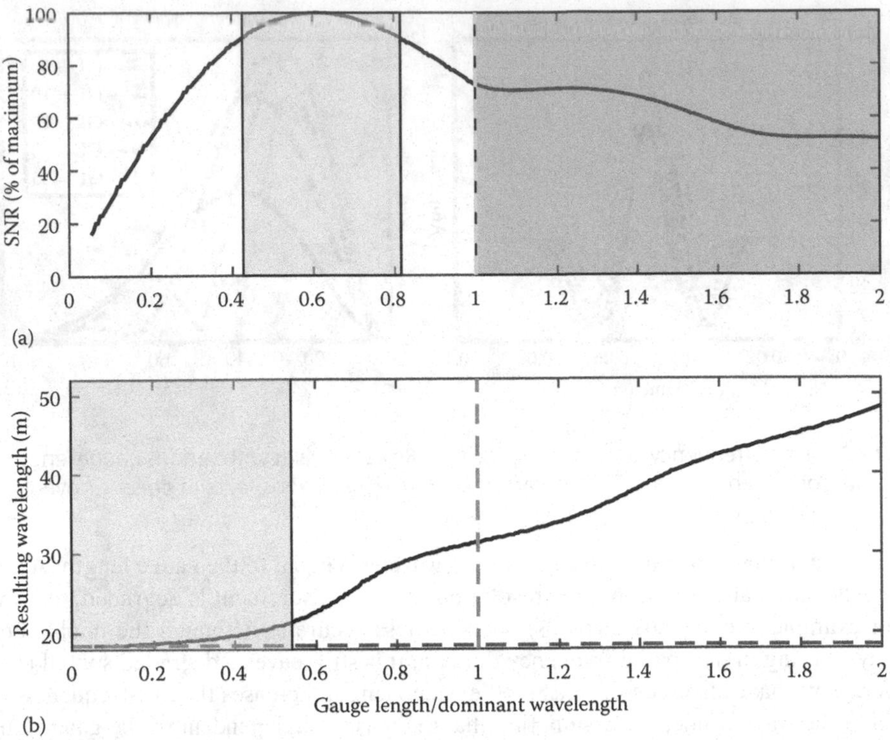

FIGURE 9.8 Optimum gauge length based on spatial resolution and signal-to-noise ratio. (Adapted with permission from Dean, T., Cuny, T., and Hartog, A.H., *Geophys. Prospect.*, in press. Copyright John Wiley & Sons.)

timing of the modulator, by the difference in optical path length in two separate modulator channels.* In these systems, the gauge length can be varied, but the benefits of using separate optical paths are reduced.

Finally, in coherent detection systems (Chapter 6, Sections 6.3.5 and 6.3.6), the gauge length is generally determined in the signal processing, after the data have been acquired. In coherent-detection systems, therefore, it is possible to process the results with a default value of the gauge length, then to estimate the seismic wave velocity for each of the waves of interest† and finally to re-process the data with gauge lengths chosen to broadly optimise the signal-to-noise ratio as a function of depth.

The results of such a re-processing are shown in Figure 9.9a, where a default gauge length of 14 m was used initially. Based on the apparent velocity calculated from the slope of the time/depth curve, the data were then re-processed using just three different values of gauge length, namely 20, 40 and 54 m (Figure 9.9b). This processing has dramatically improved the image quality, reducing the noise and allowing some features such as the reflections visible near the bottom of the well to be seen. It should be noted that the spatial resolution is somewhat degraded in the lower half of the well as would be expected from the foregoing discussion. This gauge optimisation shows that the spatial resolution requirements for borehole seismic acquisition can be relatively modest compared with typical applications of other types of DOFS.

The images in Figure 9.9 are the composite of data acquired at several shots points from a walk-above survey.

* This arrangement is used to cancel some of the effects of laser phase noise (see Figure 6.8, lower diagram).

† Although the main interest is in compressional waves, shear waves also carry valuable information, and their velocity is quite different from that of compressional waves. It might therefore be necessary to select one gauge length for compressional waves and another gauge length for shear waves.

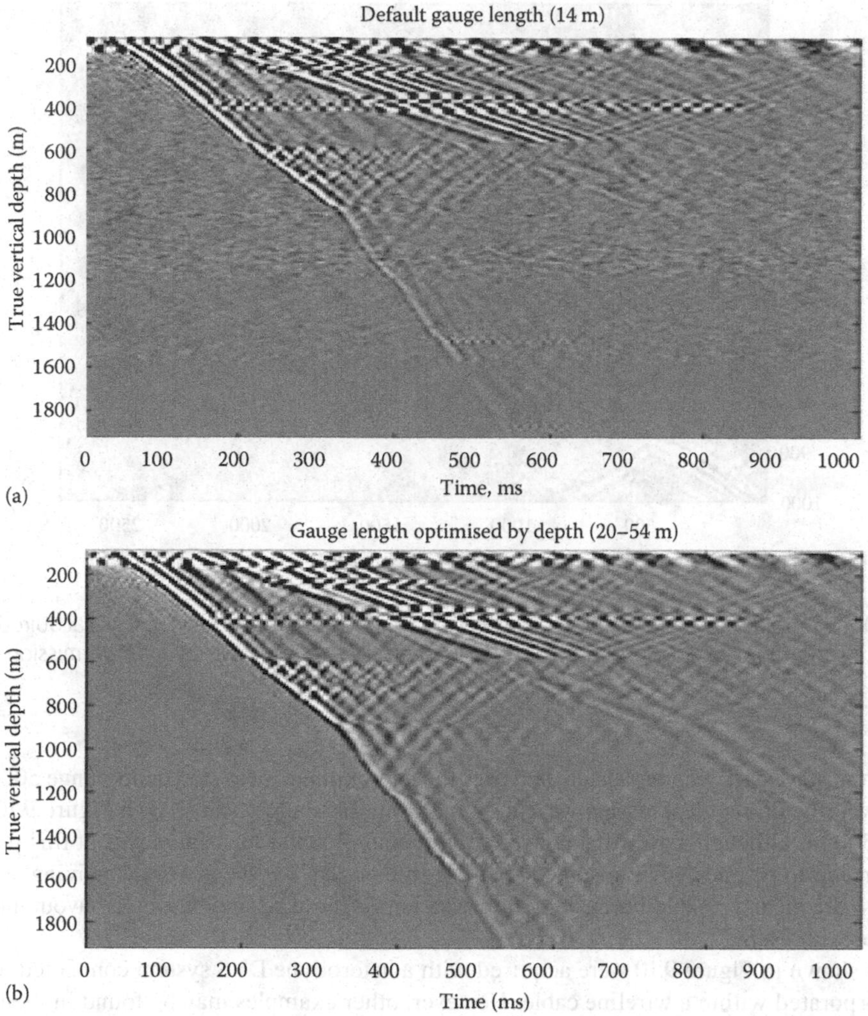

FIGURE 9.9 Borehole seismic survey processed with a fixed 14-m gauge length (a) and optimised by depth (b) Data acquired by Alexis Constantinou, Tsune Kimura and Gordon Milner (Schlumberger) and is shown by kind permission of Lundin Petroleum. (Adapted with permission from Dean, T., Cuny, T., and Hartog, A.H., *Geophys. Prospect.*, in press. Copyright John Wiley & Sons.)

9.1.3.4 Angular and directional response

As mentioned previously, a DVS responds to strain along the fibre. This means that a compressional wave arriving perpendicularly to the fibre axis displaces the fibre sideways, but because there is no relative movement between sections of fibre within a gauge length, the wave does not affect the length of the fibre. DVS systems with straight fibres are therefore insensitive to waves arriving from a direction orthogonal to the fibre axis, an effect sometimes referred to as *broadside insensitivity*. An example of this phenomenon is shown in Figure 9.10, where the phase signal is plotted as a function of arrival time (vertical axis) and measured depth* (horizontal axis). In this case, the well is deviated and the data were acquired by moving the source above the well. For the acquisition of Figure 9.10, the well deviates from the vertical at a depth of about 500 m, and the geometry of the well (including the curvature of its trajectory) and source

* Measured depth is the distance along the well, as opposed to the distance between a location in the well and the surface.

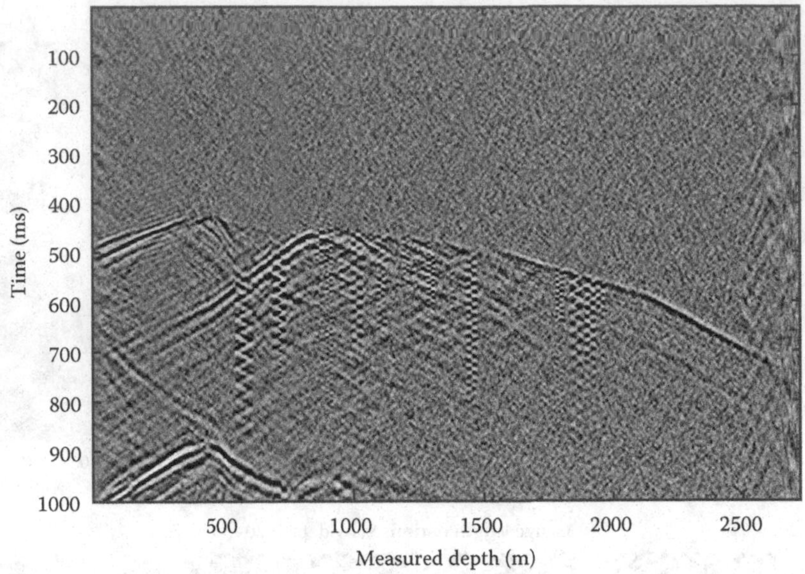

FIGURE 9.10 Example of a walk-above survey showing the broadside insensitivity. Data acquired by Alexis Constantinou, Tsune Kimura and Gordon Milner (Schlumberger) and is shown by kind permission of Lundin Petroleum.

position result in waves arriving broadside data over the approximate measured depth range 500–1500 m. The first arrival (the direct compression wave arriving from the source) vanishes in Figure 9.10 over the range 400–1500 m, although waves that convert from compressional to shear waves at this point allow the time of arrival to be guessed at in some locations. In contrast, the shear waves (arriving at a time of about 900 ms) are clearly visible because the polarisation of these seismic waves is favourable to their being detected with DVS.

The data shown in Figure 9.10 were acquired with a heterodyne DVS system connected to an optical fibre incorporated within a wireline cable. However, other examples may be found in the literature, such as Ref. [18], using the dual-pulse technique with the fibre cable cemented behind the casing or using the interferometric phase recovery technique with the fibre cable attached to production tubing [40].

Broadside insensitivity can be alleviated, to some extent, by shaping the path of the fibre so that it is not perpendicular to the wave arrival, for example, by using helical or serpentine patterns [41–44]. However, for reasons of cable size and design constraints, these techniques have not so far been used in borehole applications.

9.1.3.5 Interaction of wavenumber filtering and angular response

The velocity referred to in Section 9.1.3.3 is the axial component of the seismic wave velocity (the *apparent velocity*), and it is therefore clear that the wavenumber response varies with angle of arrival [35]. The values of Figure 9.4 have been complemented in Figure 9.11 with similar data for 30° incidence – the wavenumber filter is stretched to higher wavenumbers for waves arriving at 30° from the fibre axis.

As the angle of arrival approaches the normal, the notches shift to increasing wavenumber and the peak amplitude is reduced, as illustrated in Figure 9.12 (Figure 9.11 represents the lines extracted from Figure 6.12 at the 0° and 30° angles). The black band near normal incidence indicates the broadside insensitivity.

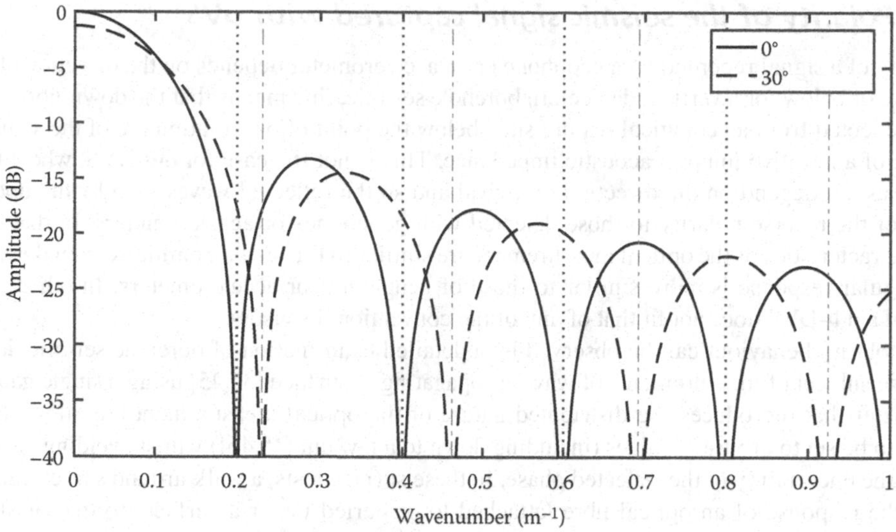

FIGURE 9.11 Wavenumber filter for a 5-m gauge length for wave parallel to and at 30° from the fibre axis. (Adapted with permission from Dean, T., Papp, B., and Hartog, A., Wavenumber Response of Data Recorded Using Distributed Fibre-optic Systems, 3rd EAGE Workshop on Borehole Geophysics, 2015, and drawn using a script provided by Tim Dean.)

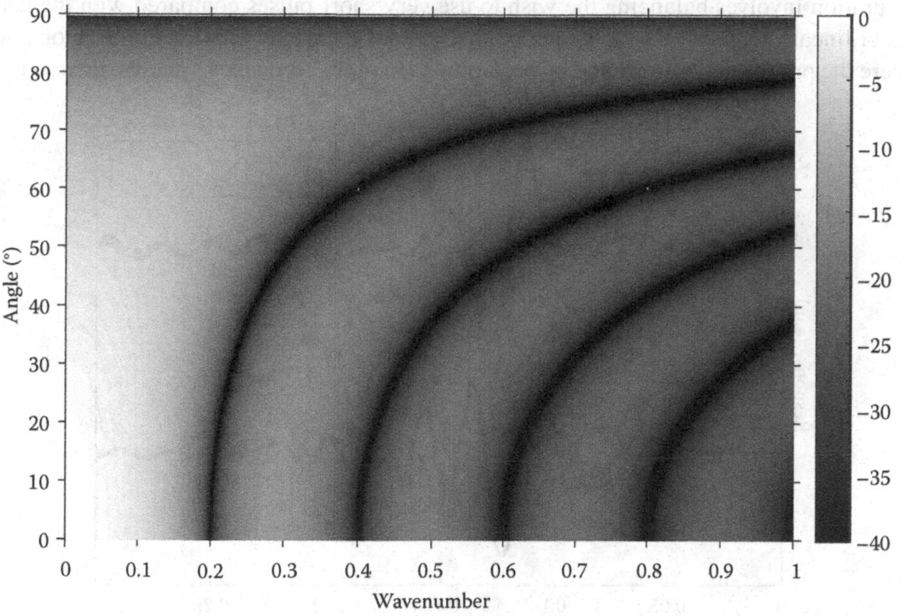

FIGURE 9.12 Maximum strain imposed by a wave as a function of wavenumber and angle of incidence. The greyscale sidebar is graduated in units of dB. (The plot is the output of a script provided by Tim Dean and is adapted with permission from Dean, T., Papp, B., and Hartog, A., Wavenumber Response of Data Recorded Using Distributed Fibre-optic Systems, 3rd EAGE Workshop on Borehole Geophysics, 2015.)

9.1.3.6 Polarity of the seismic signal captured with DVS

The polarity of a signal recorded by a geophone or an accelerometer depends on the direction of the wave (from above or below for a vertical device). In borehole seismic, this means that the down-going wave and the wave reflected from a geological feature sited below the point of observation are of the same polarity in the case of a negative jump in acoustic impedance. This is not the case for dΦ-DVS, where the signal polarity does not depend on the direction of arrival and so the reflected waves seen by the optical technique are of the reverse polarity to those detected with geophones or accelerometers. In that sense, the polarity characteristics of the optical measurement are similar to those of an omnidirectional hydrophone, but the angular response is more similar to those of geophones or accelerometers. In other words, the response of a dΦ-DVS does not fit that of any of the conventional sensors.

This polarity behaviour can be observed by a detailed examination of borehole seismic data [13]. It was also the subject of measurements of waves propagating at surface [35,45] using a single gauge-length interferometer that reproduces the distributed nature of the optical measurement but allows the spatial resolution to be set to arbitrary values (including down to a few cm [45,46]) whilst avoiding issues of fading and of the uncertainty of the reflected phase. In these surface tests, a walk-around survey can compare precisely the response of an optical fibre (attached to or buried under a surface) to that of an array of accelerometers.

A comparison of data collected from opposite ends of a short section of buried optical fibre and from an array of geophones placed just above it is shown in Figure 9.13. The plot demonstrates that the waves detected by the fibre sensor are of the same polarity regardless of the fibre end that they reach first. In contrast, the electrical sensors show a polarity reversal between the 0° and 180° cases.

9.1.3.7 Pulse duration

The finite duration of the pulse results in a further limitation on the spatial resolution [47]. The choice of the pulse duration involves balancing the wish to use very short pulses compared with the gauge length for reasons of linearity against the need to maximise the pulse energy, which requires a long pulse in the regime where the probe energy is limited by peak power. In multi-frequency systems, there is also a desire

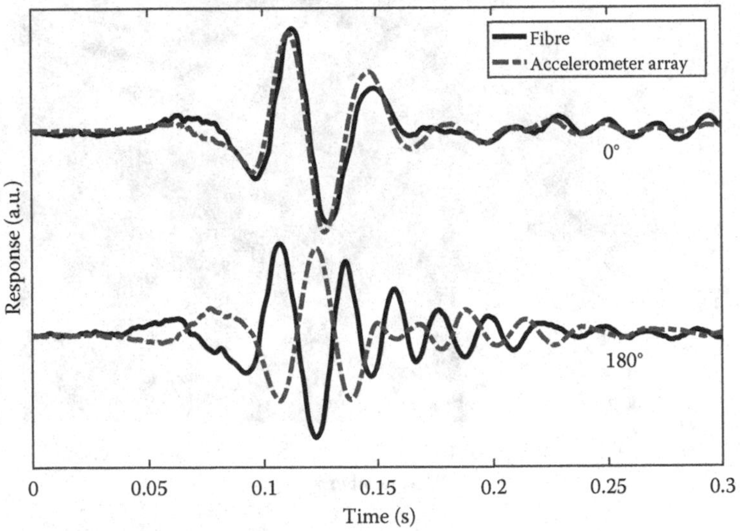

FIGURE 9.13 Response of a straight fibre interferometric sensor to waves arriving from either end (0° and 180°) of the sensor (solid line). The response of an array of reference accelerometers is shown for comparison as a dash-dot line. (Adapted with permission from Dean, T., Hartog, A., Papp, B., and Frignet, B., Fibre optic based vibration sensing: nature of the measurement, 3rd EAGE Workshop on Borehole Geophysics, 2015.)

to limit the spectral width of each probe to avoid overlapping of the probes in the frequency domain; in turn, limiting the spectral usage of each probe pulse places a lower limit on the pulse duration.

The dependence of signal-to-noise ratio on pulse duration is illustrated in Figure 9.14 for a particular laboratory condition on a heterodyne DVS system; the peak power of the probe pulse was largely the same for all pulse durations. The results (square markers) broadly follow a $\sqrt{n}$ curve (solid line) but depart markedly for the 11 and 13 ns data points. For these very short pulse durations, the efficiency of the modulator that defines the probe signal is falling off rapidly with reduced duration and the noise bandwidth is also increasing rapidly.

The wavenumber filtering effect of the pulse width is compared with that of the gauge length in Figure 9.15. The ratio of pulse duration to gauge length (roughly 1:2) is typical of that used in actual field

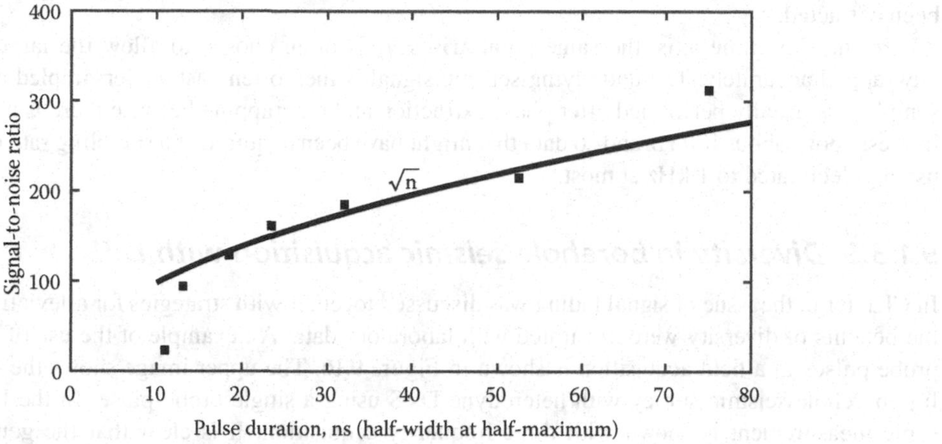

FIGURE 9.14 Measured signal-to-noise ratio as a function of pulse duration (heterodyne DVS). Data acquired by the author and processed by Theo Cuny and Tim Dean at Schlumberger. (Adapted with permission from Dean, T., Hartog, A.H., Cuny, T., and Englich, F.V., The effects of pulse width on fibre-optic distributed vibration sensing data, 78th EAGE Conference & Exhibition, 2016.)

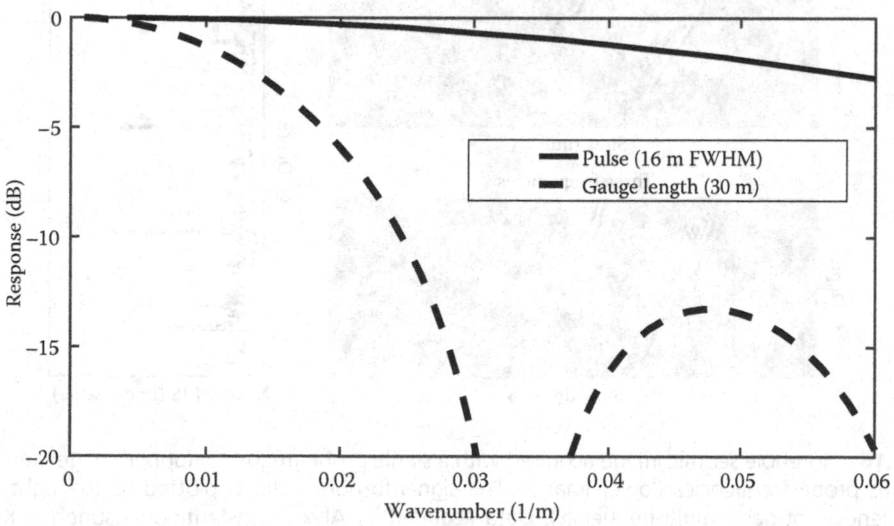

FIGURE 9.15 Wavenumber filter caused by gauge length and that due to pulse duration. (Dean, T., Hartog, A.H., Cuny, T., and Englich, F.V., The effects of pulse width on fibre-optic distributed vibration sensing data, 78th EAGE Conference & Exhibition, 2016.)

use; as can be seen, the pulse width imposes a further wavenumber limitation on the system response, but its effect is minor for the conditions illustrated. The gauge length would need to approach the pulse width for the combined effect to be substantially influenced by the pulse width.

9.1.3.8 Sampling rates

One of the strengths of DVS in borehole seismic acquisition is the very high sampling rate in the distance dimension that it usually offers. Usually, the fast sampling is a by-product of the phase measurement itself and it is far higher than that offered on electrical tools. Although the underlying information does not require the typical sample density available in DVS, it does facilitate subsequent signal processing. A key question for system designers is to what degree the data should be down-sampled after the phase has been extracted.

In the slow time axis, the same point arises: f_R is often chosen to allow the largest signals to be unwrapped accurately. The underlying seismic signal is then often vastly oversampled and some down-sampling is usually performed after phase extraction and unwrapping because there is no information of interest above about 100 Hz and so data that might have been acquired at a sampling rate of a few kHz are usually decimated to 1 kHz at most.

9.1.3.9 Diversity in borehole seismic acquisition with DVS

In Chapter 6, the issue of signal fading was discussed together with strategies for alleviating this problem; the benefits of diversity were illustrated with laboratory data. An example of the use of multi-frequency probe pulses in a field acquisition is shown in Figure 9.16. The upper image shows the result of a wire-line borehole seismic survey with heterodyne DVS using a single probe pulse. In the lower image, the same measurement is shown with three-frequency acquisition. It is clear that the general noise level has improved with the multi-frequency acquisition; in addition, a number of artefacts that are visible in the single frequency as horizontal streaks (sometimes with dashed line appearance) have been almost

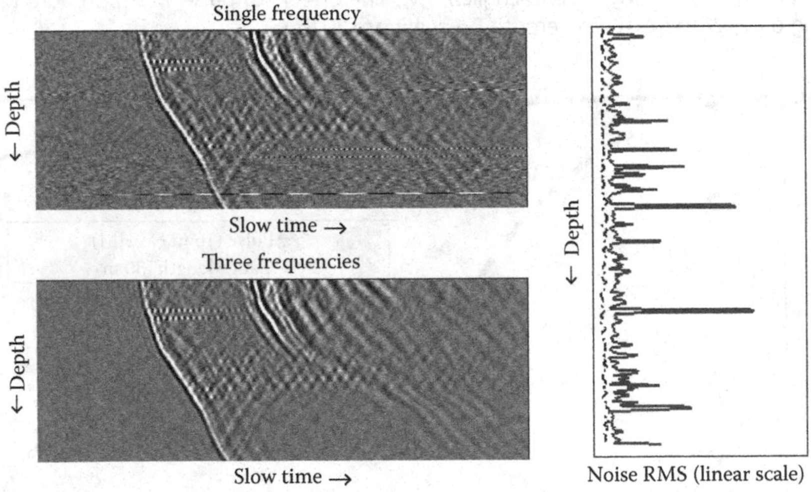

FIGURE 9.16 Borehole seismic image acquired with a single probe frequency (upper image) and with three simultaneous probe frequencies (lower image). The signal-to-noise ratio is plotted to the right (solid line: single-frequency, dot-dash: multi-frequency). Data acquired by Alexis Constantinou, Tsunehisa Kimura and Gordon Milner and processed by Theo Cuny and Tim Dean (all of Schlumberger); data shown by kind permission of Lundin Petroleum. (Adapted with permission from Hartog, A.H., Liokumovich, L.B., Ushakov, N.A. et al., The use of multi-frequency acquisition to significantly improve the quality of fibre-optic distributed vibration sensing, 78th EAGE Conference & Exhibition, 2016.)

completely eliminated in the aggregated multi-frequency data. On the axes used in the figure, a horizontal line indicates noise that is persistent at a single location in the fibre, and this is commonly caused by fading in single-frequency systems. The general improvement of the signal-to-noise level is possibly attributable to averaging of the noise, given the redundancy that is available in the multi-frequency data. To the right of the seismic images, a plot of the noise at each depth of the images compares the single- and three-frequency cases. The noise spikes that were due to fading in the single frequency case (solid line) have been eliminated with the multi-frequency approach, and this improvement is more marked than in the laboratory data comparison that was provided in Chapter 6.

It is clear, therefore, that multi-frequency acquisition provides a major improvement to the signal quality of DVS in borehole seismic surveying. Other forms of diversity (multiple wavelengths, phase diversity) have been demonstrated in the laboratory, but no field test results have been reported to date with these methods. However, Ref. [48] reports a multi-fibre diversity acquisition where a cable containing five sensing fibres was deployed and five separate DVS instruments (dual pulse systems in this case) were used to acquire the same seismic signals simultaneously. The signals from the five instruments were aggregated by summation. An improvement by about 7 dB was demonstrated, as expected. Although the use of a large number of instruments to acquire a single measurement may seem a little profligate, the authors did not have other options for diversity, and in their case, the source was a dynamite charge, which does not lend itself to repeated acquisitions for signal averaging. In any case, in the author's experience, repeating the acquisition with multiple shots improves the general noise level, but it is not very effective for eliminating the artefacts caused by fading owing to the persistence of the fading patterns. The use of multiple fibres or multiple frequencies provides true diversity, unlike repeated measurements on the same fibre that cannot overcome fading without further steps to force the fading pattern to vary (such as changing the frequency of the laser).

9.1.4 Examples of Borehole Seismic DVS Data

Some examples of bore seismic data have already been provided in the discussion of the special features of DVS.

Another example of data acquired in the Ekofisk field in the North Sea is described in a case study that may be found at http://www.slb.com/~/media/Files/seismic/case_studies/acquire-borehole-seismic-data-from-completed-well-norway-cs.pdf (the image cannot be reproduced here for copyright reasons). The physical units have not been shown as this information is deemed to be confidential.

The image in the case study shows 1 of 182 hDVS borehole seismic profiles acquired during a time interval of about 40 minutes, i.e. an acquisition rate of four to five profiles per minute. In this case, the well was offshore and serviced by a platform; the source was a cluster of air guns (total capacity of 3500 in^3) towed behind a dedicated boat. The boat navigated along a pre-planned route and fired the guns on arriving at pre-selected shot points (separated in this case by 25 m). The times and shot positions were recorded using a Global Positioning System (GPS) receiver and this information was used to correlate the hDVS data to the series of shots. In this case, the hDVS data were not only acquired continuously but also time-stamped using GPS and only the data at around the GPS time of each shot are extracted from the voluminous raw data file.

The primary purpose of performing the series of shots was to provide a seismic signal to an array of permanently installed seabed sensors. This arrangement allows the oilfield operator to carry out repeated seismic surveys (in this case at 6-monthly intervals) with the intention of understanding changes to the surface image over time (4D seismic or permanent reservoir monitoring).

The fibre is permanently installed on the production tubing (its original purpose was to connect to a downhole optical pressure sensor) in a cable similar to that shown in Figure 7.4. As a result of this arrangement, the coupling of seismic waves to the sensing fibre is quite good: owing to the inclination of the well, the tubing is pushed against the casing by gravity and so the vibrations pass from rock formation through the cement and casing relatively unhindered to the tubing to which the optical cable is strapped.

The axes are as follows: abscissa – time from the firing of the source; ordinates – measured depth (distance along the well); the section in the riser (between the platform and the seabed) and fibre on the deck of the platform have been truncated from the plot.

In this particular case, the hDVS instrument used was a single-frequency variant and the resulting fading appears as horizontal streaks that can be observed at several locations. This illustrates the need for diversity in DVS acquisition.

The first arrival is very clear for all depths. In this case, where the source is close to the well, the source-well geometry avoids the broadside insensitivity but other plots in the same series, where the source was directly above the well, show the usual broadside insensitivity. Also visible are surface-seabed multiples (reflections of the source from the seabed and back downwards that form a virtual source). In these data, several successive multiples are visible even without further processing. A number of reflections may be seen in the first half of the profile. Ringing (vibration of the casing or production tubing in response to the seismic wave) may be observed at several depths, notably level with the letter 'p' in the ordinate axis label.

The nature of the acquisition is such that only a single shot of the seismic source can be obtained, because the source vessel is moving continuously. In this type of survey, the dΦ-DVS instrument must be capable of recording reliably continuously for the entire recording time without loss of data whilst adding accurate time stamps to the data for a subsequent correlation with the shot-firing times.

Figure 9.17 shows the result of a different survey, also an offshore platform well, referred to here as Well 1.

In this case, the source boat was temporarily stationary and so several source shots were acquired and averaged (stacked in geophysical terminology). In this case, the primary reason for the survey was to acquire an electrical VSP profile with the VSI tool at a few source positions, rather than to acquire a dense source shot line as in the previous example. The VSI survey was recorded in another well (referred here as Well 2, drilled from the same platform as Well 1, with the source positions defined in a line above Well 2). In the case of Figure 9.17, the source–wellhead distance is 1595 m, but because Well 2 deviates at an azimuth of about 100° relative to Well 2, the distance from the source to the well increases to 5 km at

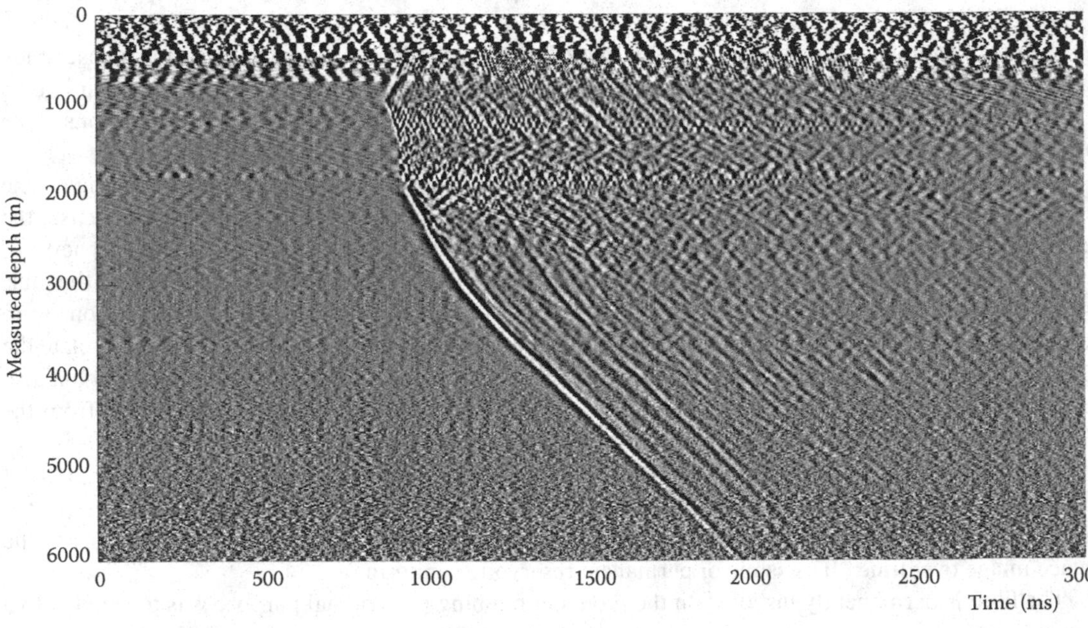

FIGURE 9.17 Example of hDVS survey in an offshore platform well, with source laterally offset from the well. Data acquired by Tsune Kimura and processed by Theo Cuny (both of Schlumberger); it is shown by kind permission of Statoil.

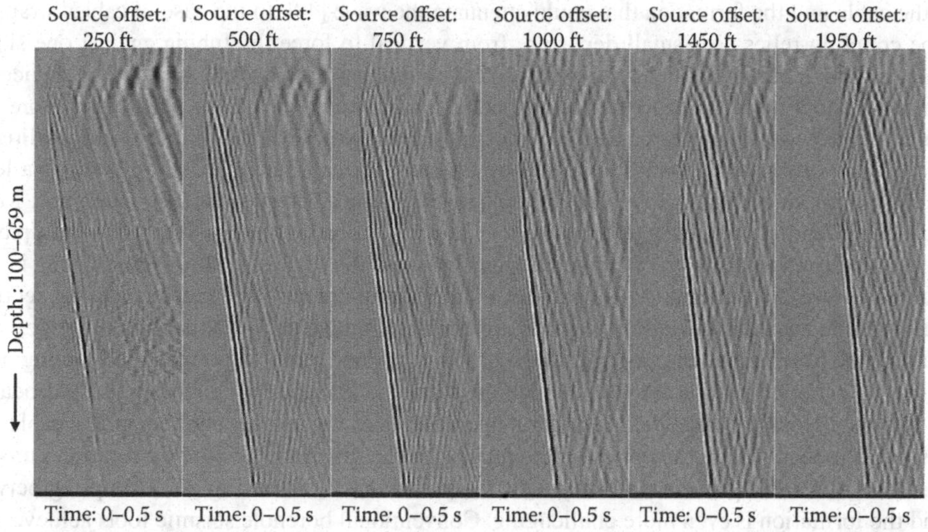

FIGURE 9.18 An example of a multi-offset borehole seismic data set acquired in a vertical well. Data acquired by William Allard, Bernard Frignet, Duncan Mackie, Mike Clark and the author (all of Schlumberger).

the toe of Well 1. In this example, the fibre was also permanently installed on tubing and the well was also deviated (up to 61.9° from vertical).

The data quality is lower than in the example shown in the Ekofisk case study. The main reason for the lesser quality is that the source is weaker (3 × 250 in³) and the averaging of five shots is not sufficient to make up for the lesser source energy; in addition, the source–fibre distance is generally larger than was the case in the Ekofisk example. Nonetheless, the first break is clearly visible all along the well trajectory with the usual exception where the source/well geometry results in the seismic signal arriving broadside onto the sensing fibre (Channels 100–200).

In this case, the hDVS data were recorded continuously and the seismic information was extracted from the resulting file based on GPS time provided by the source boat after the acquisition.

Figure 9.18 shows a composite greyscale seismic image incorporating the dΦ-DVS data recorded in an onshore well. In this case, the well is about 650 m deep, vertical and fitted with just a casing (no production tubing). Each of the panels is a borehole seismic measurement in the same well with the source located at different positions (source to wellhead offset distance is indicated above each panel). The source is a swept-frequency source (Vibroseis) sweeping over the range 8–80 Hz. The data were acquired with a three-frequency hDVS and processed with a gauge length of about 10 m. Several reflections can be observed, as well as a strong shear wave. As the source moves further away from the well, the first arrival is detected a little later. Also, the broadside insensitivity becomes apparent for the 1450 and 1950 ft offsets. It will also be noticed that the earliest arrivals at the offset positions are not at the top of the well: the seismic velocity is usually quite slow near the surface, and the longer path to reach the well at a depth of 1–200 m can actually take less time than the wave travelling close to the surface in the weathered geology.

The data in Figure 9.18 are an example of a walk-away borehole seismic survey (generally, many more shots are acquired than are shown in the figure).

9.1.5 Coupling in Borehole Seismic

Borehole seismic measurements on permanently installed cable are generally reliably coupled to the formation, especially when the cable is cemented behind the casing: in this latter case, the cement ensures a good transmission of the vibration to the sensor and there are no significant heterogeneities in acoustic speed between the cable and the formation that could attenuate the wave [49]. In the case of cables strapped onto tubing, the coupling relies on a small departure from vertical to force the tubing against one side of the well, which is then generally sufficient to ensure good coupling. Alternatively, bow-spring devices can be used to improve coupling of tubing to the casing, but these add cost and make the installation more difficult.

A cable cemented behind the casing provides the most favourable seismic sensing conditions [50]; this approach is, however, not always accepted by operators because of the risk of creating a leak path for fluids to migrate to the surface, bypassing the barriers intended to prevent that occurence. Moreover, installing a cable behind the casing against the formation is probably not possible when multiple casings are used, as is required in many practical well designs because the highest section of the casing is installed and cemented before the next section is drilled. Connecting an optical fibre between casing sections that have been installed sequentially in this way has not been attempted, to the author's knowledge.

Installing the fibre behind the casing into the production zone requires oriented perforating: when the casing is set, an array of holes is created through the steel and into the rock to allow the hydrocarbons to flow. It is important to ensure that the perforation operation does not cut through the optical cable, and this requires special tools to locate the optical cable relative to the orientation of the perforation guns.

In the case of a cable placed in the well on a temporary basis, achieving good coupling between the sensor and the formation is even more challenging. Conventional borehole seismic tools achieve coupling with active clamps that push the sensor pods against the side of the well, a solution that is not very practical with a continuous cable. In deviated wells, the well trajectory, in conjunction with tension on the cable, is often sufficient to ensure the cable touches one side of the well; in other cases, such as in vertical wells, it can be helpful to slack the cable if operating conditions allow this practice. In general, it is sufficient for the cable simply to touch the side of the well: only a small frictional force is sufficient to overcome the extensional forces involved with the very weak seismic signals that are detected. In the case of perfectly vertical wells, providing slack in the cable provides excellent contact [51], although this practice is not always permitted, owing to the risks of the cable becoming stuck.

Figure 9.19 shows a series of borehole seismic acquisitions made in the same vertical well as those shown in Figure 9.18. However, in this case, all measurements were taken from the first source position

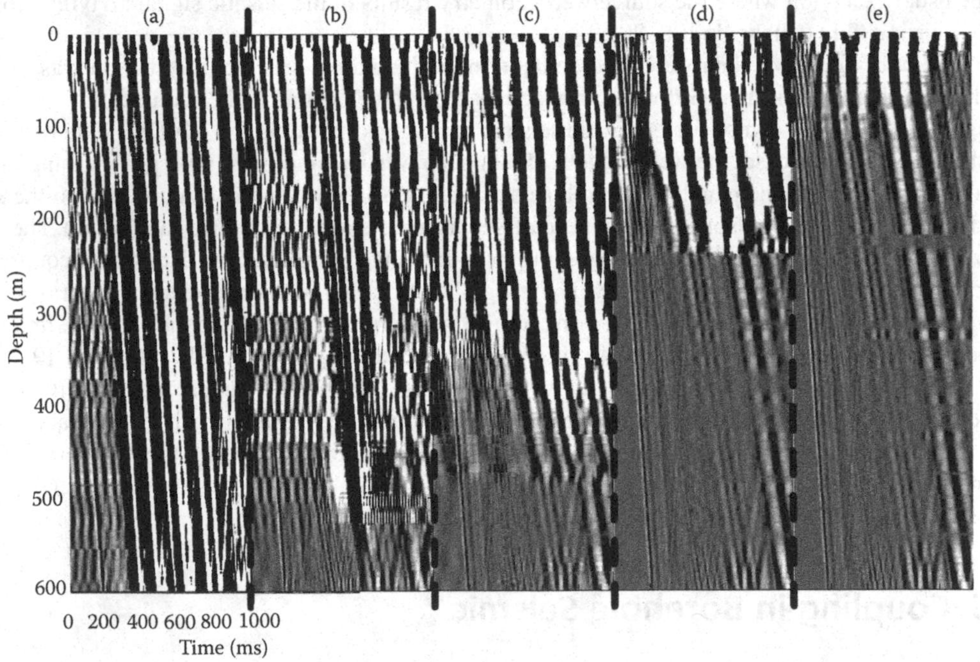

FIGURE 9.19 Borehole seismic profile acquired in a vertical well with various levels of cable slack. Data acquired by William Allard, Bernard Frignet, Duncan Mackie, Mike Clark and the author (all of Schlumberger).

(250 ft offset) but with varying levels of cable slack. The first image, panel (a), was acquired with the cable under the tension applied by a weight that has not quite reached the bottom of the well. Although there are strong travelling vibration signals, these are not geophysical; rather, they represent vibrations of the cable itself induced by movement of the ground shaking the crane from which the cable is suspended. Moving from left to right in Figure 9.19, the tension of the cable is reduced and increasing levels of slack are introduced. As additional cable is fed into the well, the coupling improves and the geophysical signal emerges. The final picture is the left-hand panel of Figure 9.18 (although the greyscale has been adjusted to accommodate the strong cable vibrations in panels (a)–(d) of Figure 9.19); it shows that the cable is coupled to the borehole wall over the entire length of the well.

9.1.6 DVS in Surface Seismic

The use of DVS for surface seismic acquisition is still in the early stages of research; Ref. [52] provides a summary of the applications of DVS in seismic acquisition including land and marine seismic exploration. Field trials of DVS have taken place in recent years, notably in a test at Schoonebeek in the Netherlands that was reported by Hornman et al. [43,53]. One of the limitations of DVS in surface seismic applications is that DVS does not respond natively to compressional waves arriving perpendicularly to the fibre axis (i.e. there is broadside insensitivity, described in Section 9.1.4).

Hornman et al. tackled this problem by developing a cable in which the fibre follows a helical path that allows a more uniform response to waves arriving from all directions; this results in a cable (known as a helically wound cable or HWC) that is largely omnidirectional. The problem of the coupling of incoming seismic waves to a surface cable was addressed by Kuvshinov [54], who showed that the effectiveness of the coupling depends on the mechanical properties of the soil as well as those of the cable materials and its construction, as well as the type of wave (compressional, shear and Rayleigh waves) that is of interest. In some cases, the cable design can cancel the shear or Rayleigh waves whilst improving the response to compressional waves.

The results of Hornman et al. [43,53] are the most detailed of those that are available so far. Although much of the focus of this effort was in demonstrating that the broadside insensitivity can be overcome, Ref. [53] in particular showed seismic reflections from subterranean reflectors. This observation is a milestone in demonstrating the feasibility of DVS for surface seismic applications: waves travelling near the surface are relatively easily observed, but the value of surface seismic acquisition is in gaining an understanding of the sub-surface. A comparison of the HWC and straight fibre results demonstrated, once again, that the broadside insensitivity problem must be addressed for the technology to be applicable because in surface seismic acquisition, the signal broadly arrives at the surface at an angle close to normal incidence.

The HWC is only one approach for reducing the broadside insensitivity. Cables for surface acquisition including curved portions were already discussed in Ref. [41]. Providing azimuthal asymmetry in the cable has been proposed in Ref. [55]. Here, the fibre is wound around a structure in which an internal mass is able to move preferentially in one direction transverse to the fibre axis. The basic idea is that transverse movement in a preferred direction is able to deform the structure on which the fibre is wound and so responds more strongly to arrivals on this azimuth than those arriving on the orthogonal azimuth. The cables described in Refs. [55–57] use inertia of the internal mass and so may be viewed as a distributed version of fibre accelerometers that use inertial masses constrained by a spring force provided by the sensing fibre [31]. In one concept, the fibre is mass loaded and attached to the external structure at internals [57]. A seismic wave arriving from the side moves the cable, and this movement is resisted by the inertial mass, resulting in strain on the fibre. In another approach [56], the fibre is attached to a mass-loaded membrane that has a quite different compliance in orthogonal axes: this arrangement provides differential sensitivity in the two directional directions normal to the fibre axis and could be the basis for a three-component sensor system.

Providing a response with a sensitivity that varies as a function of azimuth can also be achieved by designing the cable with asymmetric compressibility [44]. In addition, by using fibre patterns within a cable that each occupies one plane, a sensing cable providing a three-component response (using at least three fibres) can be achieved [44].

It should also be noted that some of the waves propagating near the surface have elliptical polarisation and that, for example, a HWC wound with a right-hand helix will respond differently from one wound with a left-hand helix, and this feature can be used to distinguish near-surface waves in order to remove them [44]; these waves contribute no seismic information but they can mask the data of interest, and so there is considerable interest in acquiring them to cancel them.

The ultimate objective of applying DVS to buried cables is its application to PRM, where the completely passive and dielectric nature of the sensor is thought to offer advantages of longevity owing to the absence of electrical connections to individual geophones. Permanently buried sensing cables offer a far greater repeatability of the survey (by removing sensor re-positioning errors) than can be achieved using temporarily deployed sensor arrays, in spite of the continuing progress in satellite navigation technology. Repeatability is essential to be able to ascribe to events in the reservoir the small changes in the seismic image that may be obtained between repeat surveys.

The sensing cables can be installed by direct ploughing, which provides a cost-effective approach; however, horizontal drilling techniques have also been employed to install the cables at arbitrary depth – a greater depth reduces the impact of surface waves and also eliminates some of the weathered zone that can degrade the signal quality and make the analysis more challenging.

9.2 SECURITY – PERIMETER, PIPELINE SURVEILLANCE

Some of the earliest work on coherent Rayleigh backscatter suggested perimeter security as an application [58]. Since minute changes in the strain (of order 1 nε) of the fibre can cause substantial changes to the backscatter signature that occur only where the disturbance takes place, the approach has considerable attraction. Sensing cables are buried along boundaries, typically 0.2–0.4 m deep, to detect footsteps or vehicles approaching; the fibres can also be included within fences to alarm on signs of people climbing over, or cutting, them.

In practice, however, it is not sufficient to raise an alarm when the local condition of the fibre changes, indicating a disturbance. The need for intrusion detection is better stated as having a high probability of detecting one of the event types that are pre-defined as threats (a low false-negative probability [FNP]) *whilst* having a very low probability of alarming on conditions that are not threats (low false-positive probability [FPP]). Reliably detecting the existence of an event is primarily an issue of signal-to-noise ratio: given a signal quality, the threshold for detection can be set so as to detect only very strong events, thus probably missing smaller events of interest (i.e. a relatively high FNP but low FPP). Conversely, setting the threshold to lower values will reduce the FNP but increase the FPP. The problem can be alleviated by improving the signal-to-noise ratio.

However, optimising the FPP/FNP balance does not overcome a more fundamental problem, namely that there are a number of events that will genuinely trigger an alarm owing to a sufficient dynamic strain being applied to the fibre, but these events are not seen as threats by the operator. As an example, fences sometime include optical fibre sensors for intrusion detection. The sensors will clearly detect someone climbing the fence, but they will in all likelihood also detect movement caused by wind, animals and other natural phenomena. The detection system must not alarm in such cases: after even a few detections of such events, the credibility of the system is diminished, and ultimately, it will be disregarded. It follows that in all but the simplest applications, the system must be able not only to detect an event characterised by a change in the backscatter signal but also apply higher levels of processing to discriminate between benign events and genuine threats.

Classification of the detected events can help distinguish natural events from threats. It involves extracting features of the event and then applying some form of pattern recognition to assign the event to one of several pre-defined classes, some of which are deemed to be benign and can therefore be ignored, whereas others may represent threats and require further investigation or the immediate raising of an alarm. Feature extraction is usually achieved by transforming the signal into a domain into which all of the events of a particular type exhibit similar features. Fourier and wavelet transforms are examples of transforms that have been used for this purpose [59,60].

The features that have been extracted are compared with the features of a recognition model that has previously been created from events of a known type and the classification process assigns the event to a class within that model. It is therefore a decision-making, non-linear process based on finding the most similar (by some definition) class within the recognition model. A number of techniques have been used for this purpose, such as neural networks, support vector machines and k-nearest neighbour techniques.

Amplitude-based Rayleigh backscatter is the primary approach that has been used for intrusion detection on grounds of simplicity of design and the lower data volumes that it requires for a given coverage and probe repetition frequency.

In this context, it should be remembered that some systems deployed to protect pipelines or national borders interrogated sensing fibres over distances in excess of 100 km. Assuming an interrogation rate of 1 kHz for the case of a 100-km range (the physical limit imposed by round-trip pulse transit times), then for a 2-m sample separation, 50 million data samples are collected in every second. These samples need to be processed in real time at least to detect an event; if no event is detected, in most cases, the data can be discarded. The event detection is necessarily relatively basic owing to the volumes to be treated. However, once an event *is* detected, far more sophisticated signal processing can be applied to the detection and classification on a much smaller data volume selected based on it being acquired close to the event in time and in location along the fibre. For this reason, there is merit in keeping data in temporary storage buffers to allow the algorithms to go back in time to the locations where an event has subsequently been detected: with more processing power applied just to short sections of the sensing fibre, the early stages of the development of the event can sometimes be detected and information gained as to the speed and direction of approach of the threat, for example, and its nature.

Use of the amplitude information in relatively simple systems brings with it a number of issues such as fading and non-predictability of the transfer function. In spite of these limitations, techniques have been evolved that allow reliable classification with a limited number of classes. Once an event has been classified, further levels of processing, such as tracking the event, can be applied.

The problem is, of course, more challenging in fibre sections that are habitually noisy, for example, bordering roads or close to built-up areas: the events that precede a threat might be the approach of a vehicle, followed by people walking near the asset to be protected and finally digging. In populated areas, there are many innocent explanations for signals such as these. In addition, the increased ambient noise from non-threatening events makes detecting and classifying real threats much more difficult, especially with a system with random transfer functions, such as amplitude DVS.

In pipeline applications, the technology is aimed at either deliberate intrusion (for theft of product from the pipe or sabotage) or inadvertent damage caused, for example, by mechanical diggers [61].

The placement of the sensing cable is a critical part of the design: it must be buried sufficiently deep to avoid being uncovered by running surface water or by burrowing animals, and yet the deeper it is sited below the surface, the more high-frequency waves are attenuated and, therefore, the narrower the signal bandwidth. In pipelines, the cable is often buried above the pipe and above the intimate backfill (finer material that is in contact with the pipe). However, where the same cable is used to sense multiple measurands, there is often a conflict between their needs. For example, a cable intended to detect a leak through temperature changes must be relatively close to the pipe and possibly near the bottom of the pipe (given that corrosion is more likely to affect the lower half than the upper half of the pipe); on the other hand, detection of intrusions requires the cable to be quite shallow.

In more general security applications, the installation by direct ploughing often dictates the depth, and in this case, the cable is typically at about 30 cm below ground level.

In the examples that follow and that illustrate the application of DVS in pipeline monitoring, the data are usually shown in false colour, for example, when data are classified, each class (e.g. people walking or digging, or different types of vehicle moving etc.) is shown in a different colour; this option was not available in this book, and so the range of examples has been restricted, and certain features have been identified by annotation for inclusion in this chapter. Figure 9.20 shows pressure transients travelling within a long oil pipeline (the data shown cover a 76-km section of the pipeline). The horizontal lines are examples of fading that occur in this single-frequency amplitude DVS system. At time 20 s, a pressure wave is initiated at a distance of about 70 km from the origin of the plot and the wave travels towards another valve (at distance 84 km, beyond the distance range covered by this particular interrogator) that has shut in the interim and so causes a reflected wave to travel back to the distance origin; as this wave passes through a check-valve at about 62 km, it causes the valve to shut, resulting in a wave that bounces repeatedly between the now shut valves at 62 and 84 km. Meanwhilst, a portion of the wave that caused the check valve at 62 km to shut continues to the distance origin and meets a further valve at 7 km from the origin at time 95 s. The latter valve can be seen to have been shut at time ~70 s, generating its own pressure waves that travel in both directions. Overall, the initial event can be seen to have initiated several pressure waves that travel up and down the pipe (in difference sections) and that can be detected after propagating backward and forward over a distance of some 500 km, even though in this case, the sensing fibre is situated about 15 cm from the pipe itself.

Figure 9.21 shows the signal recorded by a DVS system placed above a buried pipeline at the time of an earthquake. In this case, a single instrument measured a total length of more than 92 km using remotely sited optical amplifiers. The geometry of the event was such that the epicentre of the quake is close to the centre of curvature of the pipeline route, resulting in a near-infinite apparent velocity of the wave for the first 30 km (i.e. the wave reaches all parts of the first section at the same time). Beyond the first 30 km, the wave velocity is about 7000 m/s; however, this is the apparent velocity, which depends on the orientation of the

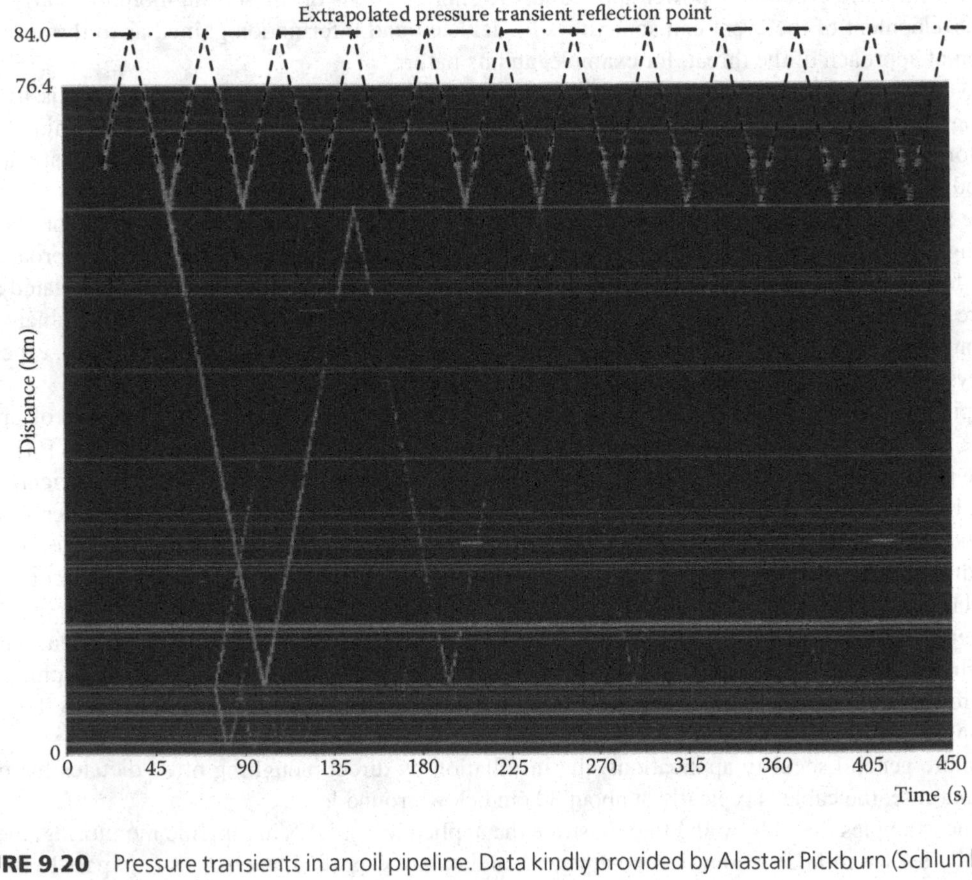

FIGURE 9.20 Pressure transients in an oil pipeline. Data kindly provided by Alastair Pickburn (Schlumberger).

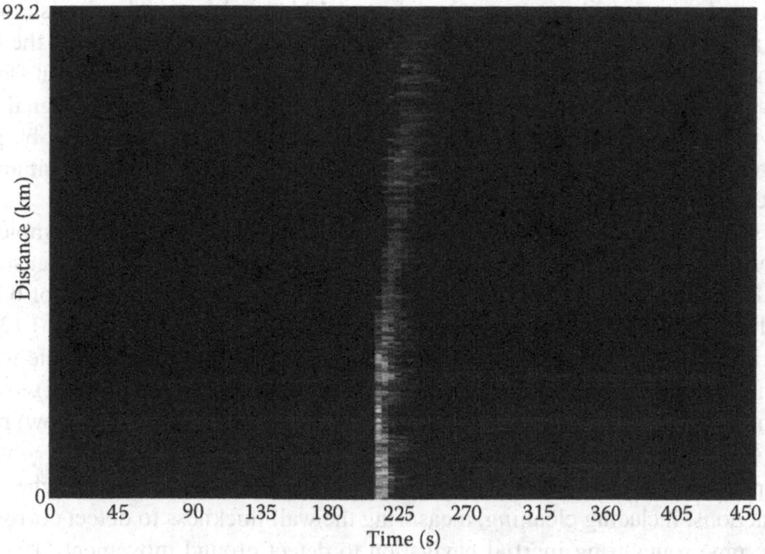

FIGURE 9.21 Earthquake captured by a DVS sensor placed above a pipeline. Data provided by Alastair Pickburn (Schlumberger).

pipe relative to the wave radiation from the epicentre. The fading of the signal is in part due not only to the greater distance from the source of the wave but also due to the classification of the signal that changes over the length of the pipeline (due to the character of the wave changing as it travels, for example, because of earth filtering effects). The classified results, from which the image of Figure 9.21 is derived, encode different types of signal in a colour scheme that the black-and-white format of this book does not capture.

The primary purpose of burying a sensing fibre close to a pipeline is to detect threats, e.g. from human activity. However, the system must also contend with natural events such as animals living in proximity to the pipeline route. Figure 9.22 shows data collected over a long time (35 minutes), but the view is

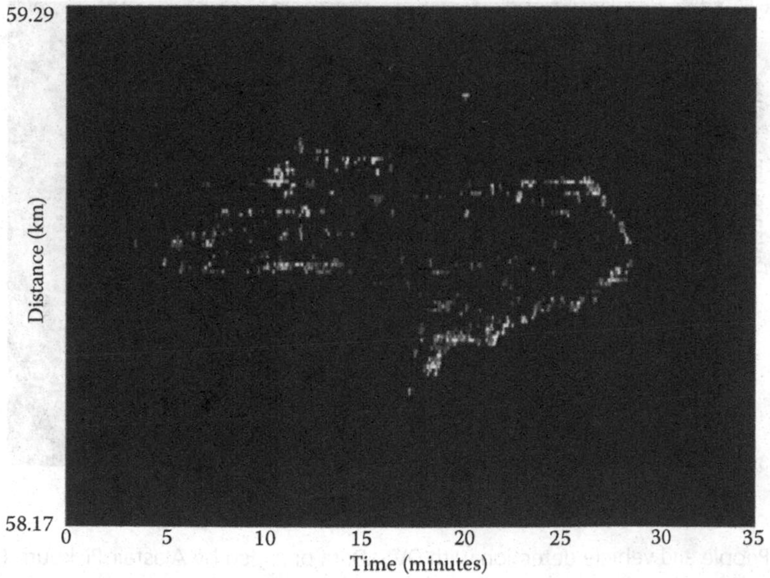

FIGURE 9.22 DVS signals created by cattle grazing near a pipeline. Data provided by Alastair Pickburn (Schlumberger).

expanded from a total span of 92 km to a section of just 1.12 km in length. The dots shown are caused by a herd of cattle grazing near and above the pipeline. The classification (not shown in the figure) identifies these events as mainly human footsteps, but with some element of digging. The threat from human activity can be discounted owing to the meandering, seemingly random, nature of the signal. Although more clarity is visible in the classified image, this example illustrates that the pure point-by-point analysis is best complemented by a higher-level analysis that studies the nature of the movement and the behaviour of the target to separate innocuous events from real threats.

Figure 9.23 shows a 3.2-km section of the same pipeline route used in the example of Figure 9.22; here, two types of events are detected: the near-white events are a wheeled vehicle that starts from a location of about 23.8 km and drives further away from the distance origin at a speed of 6.76 km/h (shown with the broken line annotation) and stops (with engine idling) after travelling about 1.12 km for a whilst before driving a little further. The fainter gray dots (circled) were classified as people walking for a few steps and stopping (a second vehicle is occasionally running near the people walking). Also, the driver of the vehicle that moved can be seen to have got out briefly of their vehicle (white arrow) because the classification changes to that of human footsteps.

Pipeline inspection gauges (known as *pigs*) are a family of tools that are launched in the pipe and serve various functions, including cleaning, measuring the wall thickness to detect corrosion and surveying (mapping the pipe route using inertial navigation to detect ground movement). Pigs have a distinct noise signature, and DVS can be used to track the location of the pig and also to listen to signals emitted by the pig [62,63].

Figure 9.24 shows the track of a pig in a natural gas pipeline that has been classified and all other noises eliminated by disabling the display of the classes to which they were assigned. The total length that was measured by the fibre sensor was about 40 km in this case; only a subset of that data is shown in the figure. Data such as these allow the progress of the pig to be tracked (a study of the slope of the distance/time curve reveals that the rate of progress of the pig is not constant, for example, because of the variations in pressure, friction and slope). The jump in the position of the pig (circled) occurs at a pumping station where excess fibre has been laid so that, locally, the distance marked out by the fibre does not correspond to distance along the pipe.

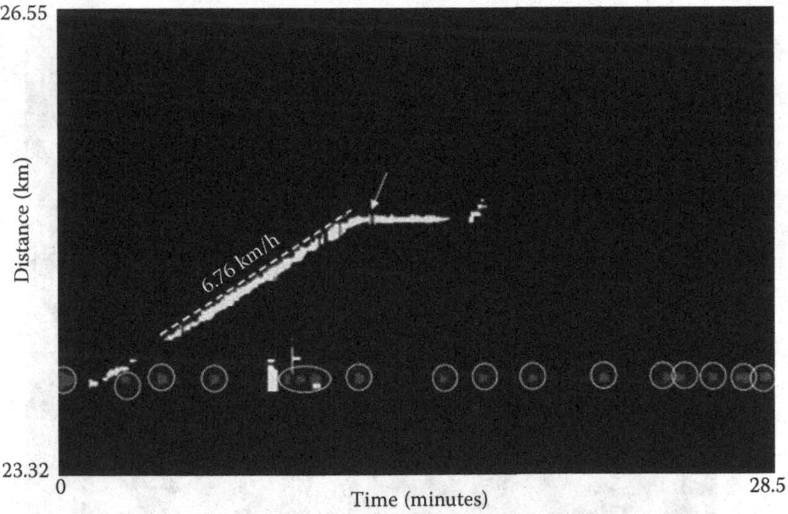

FIGURE 9.23 People and vehicle detection with DVS. Data provided by Alastair Pickburn (Schlumberger).

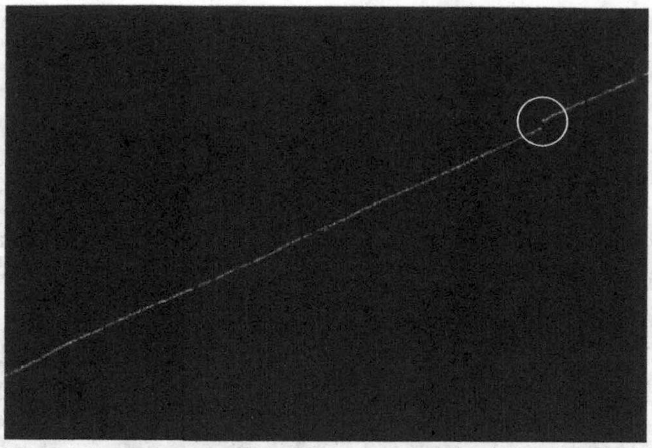

FIGURE 9.24 Classified track of a pipe inspection gauge. Data acquired by Graeme Hilton (Schlumberger).

Apart from re-assuring the operator that the pig is progressing along the pipe, tracking its movements has a number of benefits. Firstly, the DVS signals can be used to estimate precisely the arrival of the pig at its extraction point (the *pig-catcher*). Accurate real-time positioning of the pig from the control room allows the operator to vary the pipeline flow rates to ensure that the pig is received safely. It also informs the operator of the build-up of liquid slugs in gas pipelines that need to be handled by special facilities (slug catchers) to minimise pressure transients.

In addition, pigs are often used to separate fluids transported in pipelines: knowing the time of arrival of the pigs allows fluids of different types to be dispatched into different lines at the pipeline terminal. A particular problem arises when pigs are used to displace liquid from gas in a pipeline: failure to track pigs accurately during these non-routine operations has led to uncontrolled releases of hydrocarbons from downstream facilities with potential environmental and safety consequences. The real-time pig tracking capability offered by the DVS system would enable these process safety incidents to be avoided by alerting the operators when pig movement slows or stops [63].

In addition, in remote and harsh environments, there are health, safety and security benefits from reducing the requirement for pig tracking personnel to spend time on the pipeline right of way.

Another use of DVS in connection with pigging operations is the active marking of locations of interest [62]; for example, if a pig detects a region of serious corrosion, it can be designed to emit a specific acoustic signal that can be detected and located by the DVS – this gives the operator an exact location to start remedial work.

Whilst the pig travels along the pipe, in addition to its normal scraping sound, it can generate 'clunking' noises at each welded joint that launch pressure waves in the pipe, rather like the wheels of railway carriages passing over a joint between rails. An example of this is shown in Figure 9.25 (taken from the same data set as Figure 9.24, but classified differently and expanded). The classification here selects the clunking noise and rejects other sounds (which is why the track of the pig has been added by annotation as a strong white line). A series of chevron-like waves are seen to be generated as the pig hits each weld because the pressure waves travel in both directions in the pipe; the slope of these lines allows the speed of the waves to be estimated (452 m/s in this case, which fits the speed of sound of natural gas). In this case, the welds are spaced by 18 m.

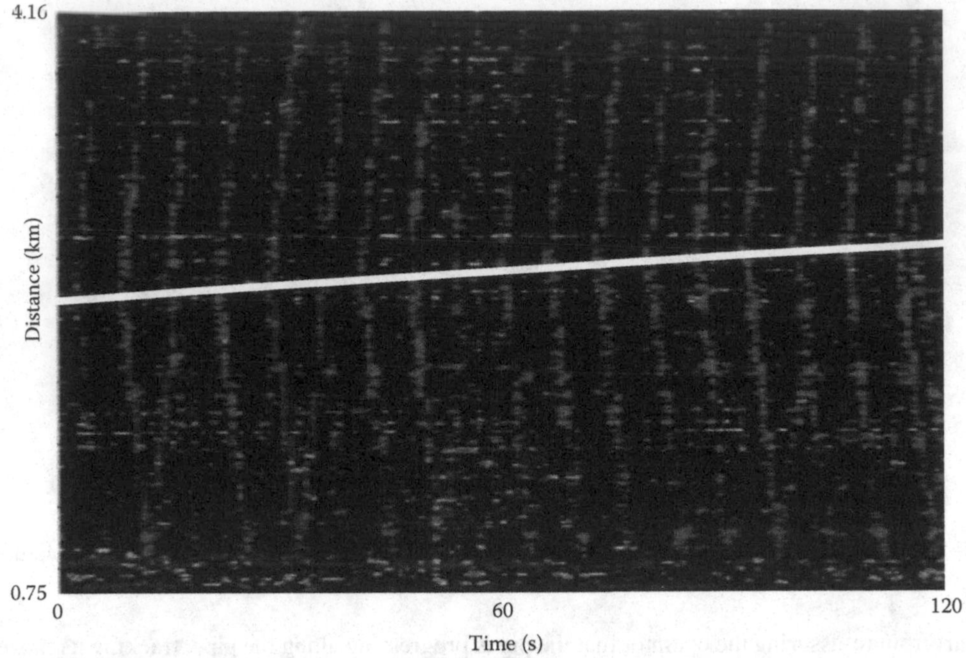

FIGURE 9.25 Noise made by a pipeline inspection gauge as it travels in a welded by pipeline. Data acquired by Graeme Hilton (Schlumberger).

9.3 TRANSPORTATION

DVS has been proposed as a means of monitoring traffic on highways [64], although there seem to be no reports of actual implementation with fully distributed sensors and most of the papers and patents relate to multiplexed sensor arrays. Nonetheless, there are a number of applications in this field, including simply detecting whether traffic is moving, and deducing the speed of vehicles, and in weigh-in-motion applications, i.e. estimating in real time whether each vehicle is within the allowed weight and weight per axle. There are good examples of point optical sensors being used for analysing traffic, for example, on road bridges [65].

The move to distributed sensors would require significant software effort to analyse the many points for which data are provided, but it might allow, for example, the progression of each vehicle to be tracked. Unlike CCTV cameras, fibre-optic sensors must be embedded in the road surface in this application, and this limits the opportunities for deployment.

Trains are relatively easy to detect given their very large mass and consequently the strong seismic waves that they generate and this was noted in an early field trial.* Similar results were published recently [66] in which trains were tracked over a 12-km line using amplitude-based coherent optical time-domain reflectometry (OTDR). In this case, the analysis consisted of estimating the energy in the noise in each location, rather than attempting any more detailed signal processing. By tracking the noise energy, the location, speed and length of the train can be estimated.

One point raised in this paper concerns the pulse repetition rate, i.e. the sampling in slow time. Trains can travel quite fast, c. 110 km/h in this paper [66], corresponding to 30 m/s. On the other hand, long-range DVS systems sample at quite low rates (typically 1 ms between probe pulses), and this leaves only a few

* G. Hilton and S. Mullens, 2008, private communication.

sample points to detect the passing of the train. This effect might limit the usable range in these systems, unless techniques for launching multiple distinguishable probe pulses are used to increase the slow-time sampling rate without compromising range (see Section 6.5.4).

Chapter 10 (Section 10.6.2) includes a description of the use of another type of distributed vibration sensor (based on movement of speckle patterns in multimode fibres) for detecting wheel flats.

9.4 ENERGY CABLES

9.4.1 Subsea Cables

The number of installed subsea electrical transmission cables is increasing rapidly. The growth in installed transmission capacity is driven by changes in the electricity supply, including the building of large offshore wind farms and also of interconnects used to balance electricity demand and supply across national boundaries, in particular under the North Sea.

Subsea cables face a number of hazards, such as trenched sections being uncovered by undersea currents, damage from anchors and fishing gear, in addition to the failures that also exist in onshore cables. Moreover, finding the cable for repairs or to avoid damage when other infrastructure is installed can be challenging because tides and currents sometimes cause the cable to shift significantly away from the position at which it was installed.

Subsea electrical cables are increasingly fitted with optical fibres either for telemetry and control of the offshore wind farm or, in interconnectors, to carry additional telecommunications traffic. The opportunity therefore exists for using spare fibres or spare wavelength-division multiplexing channels for distributed measurements. These fibres are already used in many cases for DTS and static strain measurement, and the distances involved have usually dictated the use of single-mode fibre, a fact that facilitates their re-purposing for DVS.

In this context, distributed vibration measurements can locate a cable by detecting (using a sensing fibre within the cable) sounds emitted by a source boat navigating in the region where the cable is expected to be. For systems that are fitted with permanently installed distributed vibration sensing and real-time event detection and analysis, third-party damage (e.g. dragging anchors, fishing gear or possibly deliberate damage) can also be detected immediately, thereby providing an opportunity to attribute the damage to vessels in the vicinity appropriately.

Distributed vibration sensing has also been shown to detect acoustic emissions caused by applying high-voltage pulses to cables with known faults, a process used to attempt to locate the faulty section prior to undertaking a repair. This technique has long been applied in connection with electrical time-domain reflectometry, but a distributed optical sensing allows a more precise location, thus saving time on the repair.

Figure 9.26 illustrates the measurement of noise energy on a cable connecting an offshore windfarm to a land-based substation. In this case, the fault on the cable had been repaired and all that is visible is the time at which the restored section was lowered back into the sea just after 8 hours from the start of the data set and at distance 11.5 km from the origin. The plot also shows how the noise from the sea state moves closer to the sea wall and recedes with the tides; also, the weather was rather stormy, giving rise to a high level of sea noise at the start and end of the recording, with a quieter time between 2 and 12 hours from the starting time when the storm reduced in strength. Although these observations are not directly related to the task in hand, of locating the cable and the fault, they nonetheless demonstrate the depth of information that can be gleaned. In this case, the picture shows only summaries of the noise level calculated at 30 second intervals; a far deeper level of information is available by studying the time-dependent signals on a millisecond timescale.

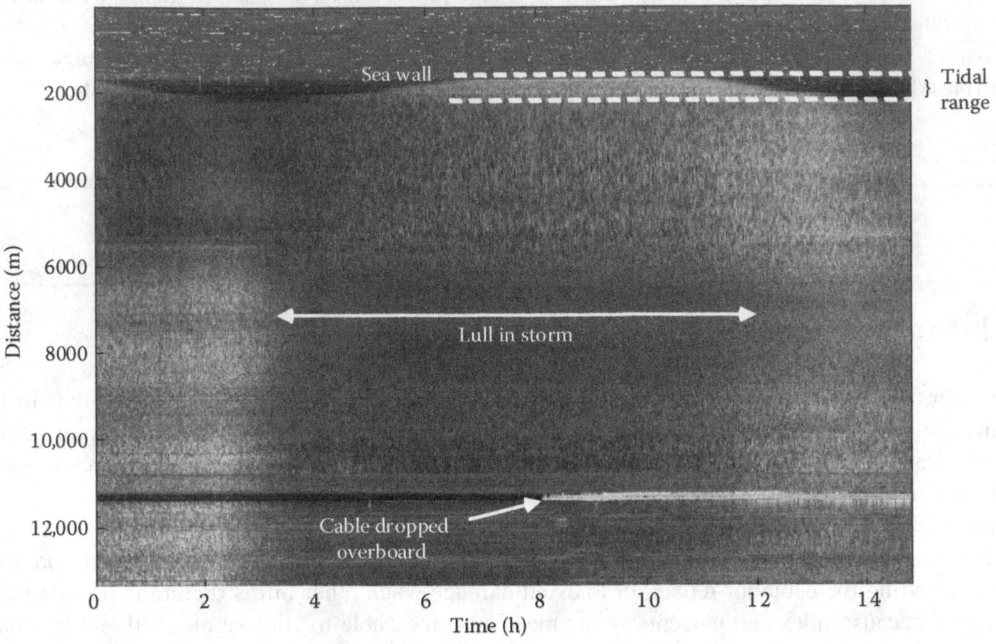

FIGURE 9.26 Plot of noise energy as a function of time and distance along a subsea electrical power transmission cable. Data acquired by Paul Dickenson (Schlumberger) in collaboration with Reece Innovation Ltd.

9.4.2 Detection of Faults and Partial Discharge

Partial discharge is the growth of a conductive track through an insulator, usually in high-voltage systems that eventually causes the failure of the cable or, more often, of a cable joint. Partial discharge is accompanied by acoustic emission at ultrasonic frequencies (sometimes in excess of 1 MHz). The probe repetition rate of distributed vibration sensors cannot usually adequately sample these events; however, even when under-sampled, they still leave a mark on DVS signals. This was recently demonstrated in laboratory models of partial discharge using intensity-detection coherent OTDR [67,68]. A particular feature of this work was a new signal analysis approach using random quadrature demodulation [69], a technique developed for multi-path interferometry. In the DVS context, it allows a more linear strain transfer function to be calculated, under the assumption that the strain is uniform along the section of fibre that is analysed.

9.4.3 Wind and Ice Loading in Overhead Cables

Since the early 1980s, overhead power lines have increasingly included optical fibres integrated in the ground wire (optical ground wires, known as OPGWs). The fibres are usually of the single-mode type; they were primarily installed for telemetry and, with deregulation of the sector, as a telecommunications carrier. However, spare fibres have been used for sensing, mainly for temperature but also for vibration. Reports for vibration sensing have so far are centred on the analysis of state of polarisation in the forward direction [70], which integrates the effect of ambient vibrations over the section length, or polarisation OTDR [71], which provides a true distributed measurement.

One benefit of vibration data in overhead lines is that these provide an understanding of the effects of wind, including resonant effects (sometimes known as the 'galloping wave') that are potentially destructive. Of course, the wind also affects the heat dissipation, and so inferring the wind speed from vibration will allow better usage of the network, although DTS is probably a more effective tool for this purpose.

Vibration sensing might also provide a means of detecting the location of lightning strikes. OPGW cables carry a very large current in the event of a lightning strike, more than 1000 A, and this gives rise to an extremely fast temperature rise (several tens of degrees in less than 1 s), followed by a slower thermal decay. Temperature changes will also vary a DVS signal and a local heating of the order of a lightning strike will engender a phase shift of thousands of radians.

9.5 PRODUCTION AND INTEGRITY APPLICATIONS IN BOREHOLES

The use of DVS in hydrocarbon production (as opposed to exploration) applications is related to performance, i.e. to optimising the flow from production intervals (as is the case for DTS), understanding the hydraulic fracturing process and ensuring that pumps, valves and other in-well devices are operating correctly. Well integrity relates mainly to unwanted leak paths, but also to avoiding problems caused by movement, e.g. compaction and subsidence.

Although PRM is a production activity, it was described with other seismic uses of DVS owing to the similarity of the techniques.

The applications of DVS in production and integrity applications differ from seismic acquisition in that the events are often localised and contain much higher frequencies than is the case for most seismic data. These applications therefore require a DVS system optimised differently from that used in seismic applications, in that there is usually a need for finer spatial resolution, higher frequency recording, but not such an emphasis on signal-to-noise ratio.

9.5.1 Flow Monitoring

The production of fluids from the sub-surface is difficult to monitor, especially on a continuing and distributed basis. The produced fluids can include oil, gas and water; they can also carry solids (sand, usually) with them. The operator wants to know which zone is producing which type of fluid and in what quantity.

Similar needs apply in injection, particularly when complex fluid patterns are pumped, for example, in acidising treatments or hydrofracturing. In these cases, the operators want to understand where the fluids are entering the formation; it is also helpful to track the arrival of different fluid types.

The problem can therefore be summarised as the need for a distributed measurement of flow in one, two, three or even four phases (in the case of solids transport). In general, surface measurements can determine the total flow rate for each phase, and so it is usually sufficient to estimate the fraction of the fluids produced from, or injected into, each zone. The problem is therefore more precisely one of flow allocation. In addition, it is also desirable to understand the flow regime that is established in a wellbore or a pipeline, for example, to predict the arrival of slugs.

DTS provides some of the answers, as described in Chapter 7, but DTS is not usually capable of handling multiphase flow allocation, and it also is unable to estimate flow distribution in most horizontal wells. DVS can complement DTS, and in fact, the combination of both measurements provides new insights that neither technology alone can deliver.

It has been recognised since 2007 [72–74] that DVS techniques can be used in a number of ways to monitor flow in wellbores. Since then, a number of demonstrations of the practical application of DVS to flow monitoring have appeared, and the subject has become very diverse. The range of topics that are involved in flow monitoring with DVS is illustrated in Figure 9.27.

The flow properties at the top of the figure summarise the desired output, which include the flow rate in each phase and also the direction and, in the case of counter-flows, the flow regime and also the path followed by the fluids. The sensor can be sited so as to detect axial flow, in which case the measured value integrates the flow contributions arriving from inlets (in the case of a production well) from the bottom of the well to the point of measurement. In contrast, if the sensor is positioned so as to detect and quantify radial flow, then a direct measure of the local inflow can be obtained. Other aspects of the use of DVS for flow monitoring are discussed in the next few sub-sections.

The DVS technology is used for measuring a wide variety of signals in order to extract flow information. The interpretation methods have used many different physical quantities extracted from the DVS data.

9.5.1.1 Total vibration energy

There is long-standing theoretical and experimental support [75,76] for a linear relationship between flow rate and the root mean square (r.m.s.) value of the dynamic component of the pressure in a pipe for one- and two-phase (liquid-liquid) compositions. In this case, the pressure fluctuations are related to turbulence in the flow and occur at low frequencies (c. 10 Hz).

The most basic approach to flow estimation with DVS is therefore to measure the vibrational energy at each location in the fibre, for example, based on the standard deviation or variance of the dynamic part of the phase signal, evaluated over selected time intervals, such as 1 s or 30 s, depending on the anticipated rate of change of the vibration levels. This is the approach that was used in Figure 9.26, in a different context. Even this relatively simple analysis can provide further insights into the behaviour of the wells; additionally, the data are reduced to a level that can easily be transmitted in real time, for example, over a satellite link or for display on site to the engineers in charge of a process.

However, there is a gap between the early work of Patterson [75,76] and the present processing of DVS data, namely the transduction. In Refs. [75 and 76], a fast-response pressure gauge was connected to the flowline, and its output was processed to provide explicitly the r.m.s. of the dynamic component of the pressure. In DVS, a similar calculation can be made of the phase output, and this is how the results shown in Figure 9.26 were calculated. However, in the case of the DVS data (in this example and others), there is no explicit connection between the DVS output and dynamic pressure. The transfer of dynamic pressure to strain in the fibre will depend on the deployment of the fibre in relation to the pipe, and so the calibration of such a measurement is uncertain in most cases. It is also important to guard against including the effect of mechanical responses in the pipework that could distort the result. Nonetheless, the method provides a useful guide to relative changes in the flow rate.

9.5.1.2 Spectral discrimination in DVS analysis

Spectral analysis of the vibration energy provides a further level of detail. A common practice is to extract the energy in certain frequency bands (typically four to six bands) that, from experience, are thought to provide discrimination of certain features of the flow. For DVS data, the number of bands and the selection of their cut-off frequencies are so far proprietary to each of the companies that provide these measurements, but empirically, it is found that certain types of flow are associated with particular frequency bands.

The study of the acoustic noise spectrum associated with turbulent flow has been documented well before the use of DVS, see Ref. [77], for example. Single-point tools have used a variety of empirical models, including dividing the spectrum in 1/3 octave bands and using fitted relations to associate spectral shape to particular flow regimes. Interpretations of DVS so far have used a coarser spectral division, but the methods that have been long-established in noise log interpretation may well be used in DVS data once the measurement techniques are improved and if a clear benefit in processing the data in finer spectral slices emerges.

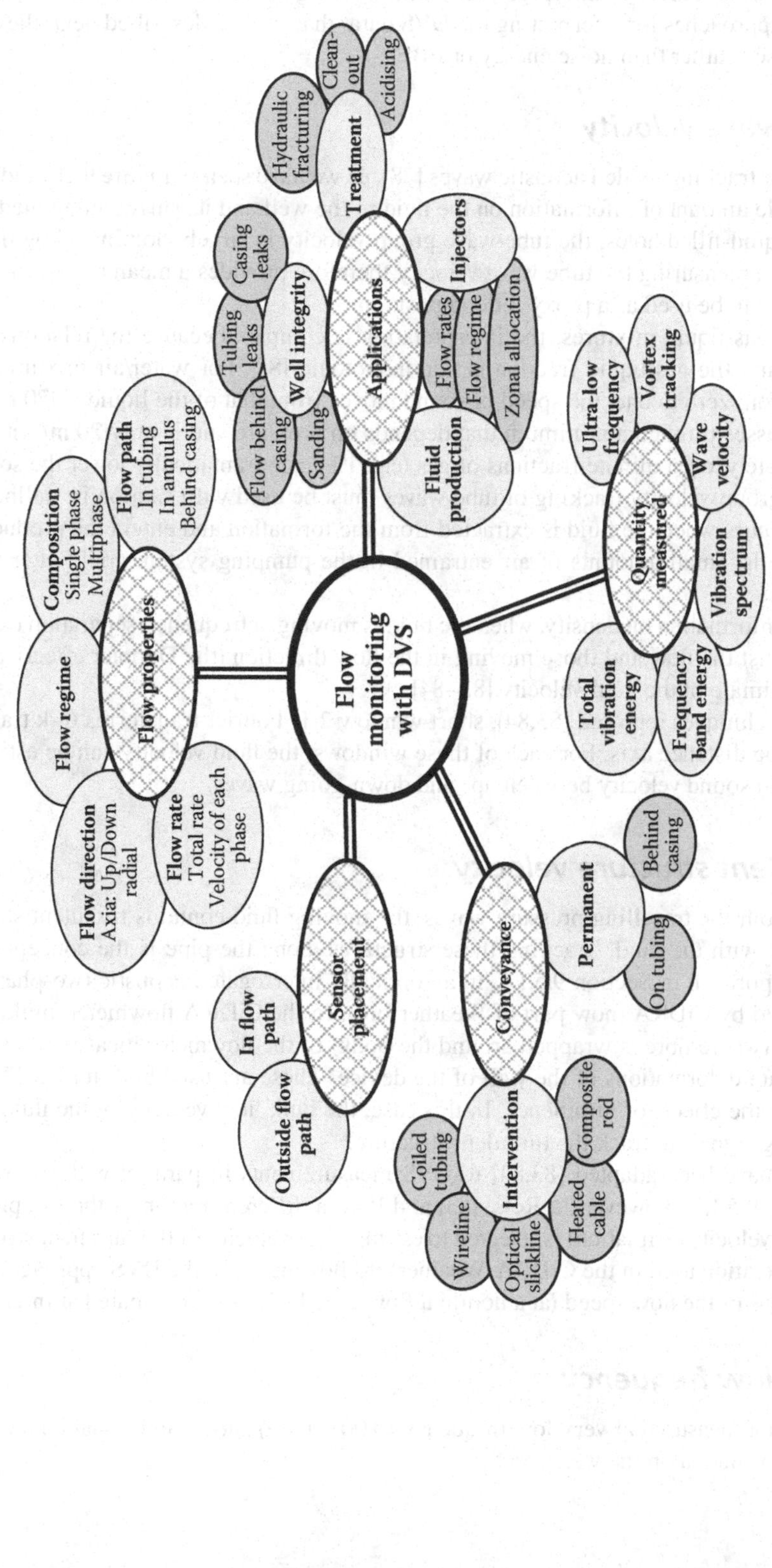

FIGURE 9.27 A pictorial representation of the usage of DVS in flow monitoring.

In terms of improvements to DVS data, it is not simply the signal-to-noise ratio that must be increased: a noise tool has a well-defined transducer, usually sensing acoustic pressure. In the case of DVS, the coupling of the acoustic signal to the sensing cable and its conversion to strain is likely to depend on the well completion, the design of the sensing cable and its placement. The interpretation must allow for these differences and so far, no universal interpretation tools, covering all of these arrangements, are available.

In alternative approaches for interpreting the DVS data, that will be described next, the phase waveform itself that is used rather than noise energy or FBE.

9.5.1.3 Tube wave velocity

DVS can be used for tracking guided acoustic waves [78], as we have seen in Figure 9.20, and these waves reveal a considerable amount of information on the fluid in the well and its surrounding medium [79,80]. In particular, for liquid-filled holes, the tube-wave group velocity is largely dominated by the density of fluid in the borehole; measuring the tube wave velocity therefore provides a means of estimating density. In turn, the density can be used as a proxy for composition.

In the case of gas/liquid mixtures, the interpretation is complex because the relationship between the wave velocity and the gas/liquid fraction is non-monotonic [81]. For water/air mixtures, as the gas fraction is varied from zero to one, the speed of sound moves from that of the liquid (1470 m/s) to that of air (300 m/s) but passes through a minimum that depends on pressure and is only 20 m/s at atmospheric pressure. Unfortunately, even minute fractions of gas (e.g. 1%) can dramatically lower the sound velocity [81], and so using group-velocity tracking of tube waves must be used with caution in wellbores because outgassing often occurs when the fluid is extracted from the formation and enters the production tubing; even in injector wells, small amounts of air entrained in the pumping system could alter the speed of sound significantly.

In addition to information on density, when the fluid is moving, a frequency separation exists between waves moving against the flow and those moving in the flow direction (the Doppler effect), and this provides a separate estimate of the fluid velocity [82–84].

In one of the techniques reported [83,84], short-window 2-D Fourier transforms (f–k transforms) are carried out along the distance axis. For each of these windows, the fluid velocity can be estimated based on the differences in sound velocity between up- and down-going waves.

9.5.1.4 Turbulent structure velocity

Quite separately from the travelling pressure waves, the moving fluid contains turbulent structures (e.g. vortices) that move with the fluid. Tracking these structures along the pipe is the concept that is used, together with the approach in Section 9.5.1.3, in an optically interrogated acoustic two-phase (water/oil) flowmeter developed by CiDRA (now part of Weatherford). In the CiDRA flowmeter, multiplexed interferometric sensors (where fibre is wrapped around the body) of the flowmeter measure pressure fluctuations through minute deformations of the wall of the device. These are used both for tracking travelling acoustic waves and the effects of turbulence. In this case, the bulk flow velocity of the fluid is estimated by cross-correlating signals to track the turbulent structure.

Similar ideas have been adapted [83,84] to DVS measurements in parallel with the wave velocity estimation (Section 9.5.1.2); however, in Refs. [83] and [84], a different region of the f–k plot (from that used for tube wave velocity estimation) is selected to estimate the velocity of the turbulent structure, rather than the cross-correlation used in the CiDRA/Weatherford flowmeter. In the DVS approach, the reported uncertainty was 10% of the flow speed (at a nominal flow speed of 3 m/s), estimated at intervals of 66 m.

9.5.1.5 Ultra-low frequency

DVS data phase data measured at very low frequency (0.001–1 Hz) shows additional characteristics that provide further information on the well.

The origin of the observed structures in the data is not clear. In some cases, it could be related to thermal effects because DVS is sensitive to temperature as well as strain. The time constant of temperature changes in most installations results in thermal effects being filtered in usual frequency ranges (say a few Hz and above); however, at very low frequency, the significance of the thermal can be dominant. It is also possible that these ultra-low-frequency signals, which are clearly visible in some data sets, are indicative of long-term strain changes in the well or reservoir.

The balance of thermal and strain effects differs according to the well construction and the nature of the geological formation. Whatever its origins, ultra-low-frequency structure in the DVS data appears to correlate with events in the well and, pragmatically, it clearly carries information of interest to the operator of the well.

9.5.1.6 Sensor placement

The sensing cable is sometimes placed in the flow path, but often in a separate region, and in this case, it is preferably in mechanical contact with the channel through which the fluid is flowing. The vibration signal of a cable in the flow indicates mainly the interaction between the flow and the cable, and, in general, it sets up its own turbulent structure, which conveys limited information on the fluid. In addition, a very strong signal is typically obtained with a cable vibrating in a turbulent flow, and this often exceeds the dynamic range of the measurement. Even when the measurement is not saturated, careful filtering is required to extract other signals, such as pressure waves travelling in the fluid. Therefore, placing the sensor outside the flow, but in mechanical contact with the pipe in which the flow occurs, is usually the best approach.

9.5.1.7 Conveyance methods for DVS

The conveyance methods used for flow measurements are largely the same as for DTS (see Section 7.8.2). It is common, particularly in production applications, for both measurements to be conducted at the same time using separate fibres in a single cable.

9.5.1.8 Applications of DVS flow monitoring in boreholes

There are many potential uses of DVS for monitoring flow in boreholes; they are summarised in the right-hand part of Figure 9.27 and they apply throughout the life of the well. During the production phase of the field, whether for producer or injector wells, the operator needs to understand the flow distribution, i.e. which zones are producing or accepting fluid and, in the case of producers, which types of fluid are produced at each depth. Ideally, one would obtain a curve of flow rate vs. depth and the composition at each location. Even relatively low accuracy information is usually sufficient to deciding on actions.

The most complex situations arise during treatment, such as hydraulic fracturing, because many different types of fluids are injected in quick succession. The varying viscosity and liquid/gas ratio gives rise to different noise levels; for example, a foam reduces the transmission of vibration signals. In conjunction with the treatment log (sequence of pump rates, pressures and materials injected), good indications of the effectiveness of each fracturing stage have been obtained.*

DVS also has a role in verifying the satisfactory completion of acidising (in conjunction with DTS) and the clean-out stage, where materials used for treating the well are removed prior to production resuming.

DVS also has promising applications in assuring the integrity of the well, for example, by detecting sanding [85] during production. It has also been demonstrated in detecting leaks, particularly low-volume flows after the useful life of the well [86]. In the latter case, the DVS replaces a conventional noise tool with potentially far better efficiency because of the distributed nature of the measurement.

* Paul Dickenson, 2016, personal communication.

9.5.2 Hydrofracturing Monitoring Using Microseismic Events

Hydrofracturing is a technique that has been used since the 1960s to improve the productivity of hydrocarbon wells, but in the last decade, its importance has grown vastly because it is the key to developing unconventional resources held in shale rock. Fracturing involves pumping fluids under sufficiently high pressure to fracture the rock. The operation is usually carried out in separate stages into specific parts of the formation accessed by the well. The fractures can reach hundreds of feet into the formation; they cause new flow paths to be created that are kept open by placing proppant (such as sand particles) carried by the fracturing fluids into the fissures.

Fracturing shale formations has seen a substantial increase in efficiency in recent years by better understanding of all aspects of the process, including improvements in the fluids, proppants and placement techniques. Part of the gain in efficiency has been achieved by monitoring the growth of the fractures with micro-seismic techniques (hydrofracturing monitoring, or HFM). The detection of the fractures may be conducted from the surface, or in-well from neighbouring boreholes used for observation during the fracturing process or the treatment well itself. Installing the sensors for this purpose is expensive, and so a limited number of devices are usually installed, which limits their ability to monitor the process and locate the fracture growth.

Recently, interest has turned to employing DVS for the purposes of HFM.

A field test, combining conventional sensors and DVS, was reported in 2013 by Webster et al. [87] using a dual-pulse dΦ-DVS technique. Those events that were detected by both conventional and DVS techniques showed good agreement. However, the authors reported a difference of 1.75 magnitude units between the better sensitivity of the conventional tools and the optical system. This means that only a few percent of the events captured by the electrical tools were detectable in the optical data.

The sensitivity of DVS will therefore need to improve substantially (about two orders of magnitude depending on the performance of both the optical interrogation system and that of the reference electrical measurement) in order to detect most of the events reported by conventional tools. In addition, the optical technique is hampered by the issue of broadside insensitivity.

In favour of DVS, however, is the fact that the optical techniques can cover the entire length of the well at only a small incremental cost. In addition, the wells that are constructed for extracting shale resources are usually horizontal or highly deviated. As a result, an event detected by the fibre in both the vertical and horizontal sections can be used for triangulation [87], and this mitigates a further disadvantage of the optical technique, namely that at present, it offers a single, axial component of the measurement only. In contrast, the conventional tools are usually three-component devices that provide direction-of-arrival information that is used for locating the detected events.

A further benefit of the optical technique is its wider bandwidth – small micro-seismic events usually contain high-frequency information that is filtered by some electrical tools.

The wide-scale adoption of DVS in HFM will require a significant gain in sensitivity and possibly the development of multi-component cables and acquisition. If these problems can be overcome without interfering with normal fracturing operations, DVS can offer a substantial benefit in measurement productivity, which can be expected to translate into a gain in the efficiency of the treatment process itself.

The value of the data is enhanced by combining it with other measurements, such as a log of the surface equipment (pumping rate, flow rates and nature of the materials injected, pressure readings and many more). It has also been shown [88] that combining DVS, surface readings and DTS data provides a detailed picture of the progress of the fracturing operation.

9.5.3 Artificial Lift Monitoring

The term *artificial lift* covers a number of techniques for enhancing the flow of oil from a well when the natural reservoir pressure is insufficient. It includes rod pumps (the traditional 'nodding donkey' devices

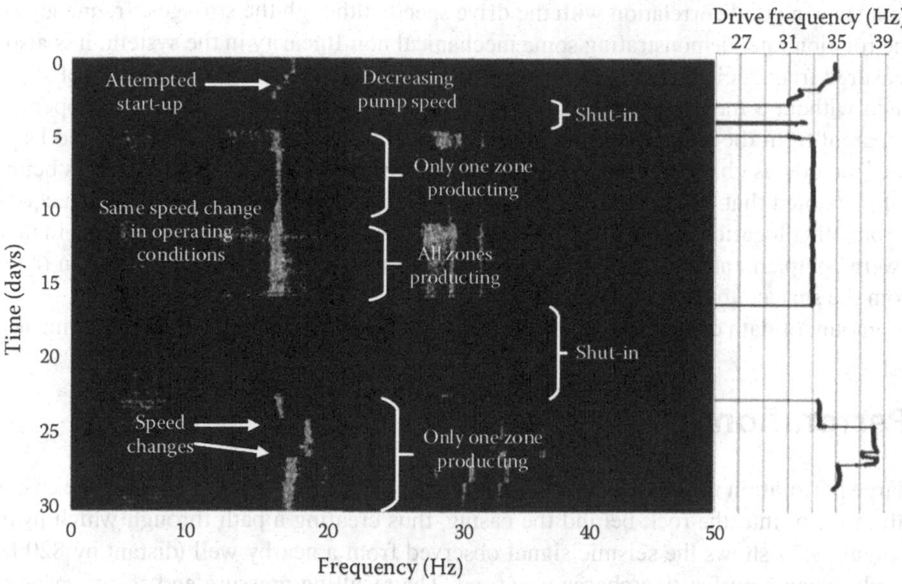

FIGURE 9.28 Long-term DVS recording of the acoustic energy spectrum of an ESP in a SAGD well. Adapted from a figure kindly provided by Paul Dickenson (Schlumberger).

that can be seen in many mature oil fields), gas-lift and electrical submersible pumps (ESPs). DVS has been used to monitor both gas-lift valves (GLVs) and ESPs.

In the case of GLFs, DVS fulfils the same role as was described in Chapter 7 (Section 7.8.11), namely to verify the state of the valve, whether it is leaking and operating when, and only when, it is intended to. DVS has been used in conjunction with DTS to refine and qualify the DTS response: in one particular case,* in a deep-water reservoir, the DTS data seem to indicate some problems with the gas lift system, but the results were a little ambiguous. Given the cost of well intervention in deep water, it was decided to augment the DTS with a DVS (using as sensors the same multimode fibres as had been installed for the DTS measurement). The DVS data, being more sensitive to temperature changes and also able to detect vibrations caused by the entry of gas in the well, were able to determine that, indeed, some valves were malfunctioning. This information allowed the operator to make decisions on intervention in the knowledge of the operational costs of the malfunction in terms of excessive gas use. It also reduced the intervention time because it was known in advance precisely which valves were malfunctioning.

Regarding ESPs, since these are high-power mechanical devices, they emit mechanical vibrations that provide considerable insight into the state of the pump and the well. Many pumps are instrumented, but in the case of very-high-temperature operations, such as in steam-assisted heavy oil extraction, minimal information is available from the pump. A fibre-optic sensing cable conveying fibres for DTS and DVS can inform the operator as to the state of the pump [89]; this can avoid driving the device in conditions where it could be damaged. Figure 9.28 (main panel) shows the data recorded over one month in a steam-assisted, gravity drainage (SAGD) well. The well was heated to about 220°C and it consisted of four zones, each equipped with a flow-control valve. The figure was obtained by calculating the noise spectrum, averaged over the length of the pump. These spectra are then plotted as a function of time (in this case for a continuous period of 30 days).

The thin panel to the right in Figure 9.28 shows the drive frequency of the pump (the control input) recorded by instrumentation at the surface.

* Paul Dickenson, 2015, private communication.

The data show a good correlation with the drive speed, although the strongest frequency component is at half the rotation rate, demonstrating some mechanical non-linearity in the system. It is also observed that the measured frequencies are slightly lower (by ~1 Hz) than the drive, an indication of slip that could not be gained without a measurement local to the motor. A number of changes in the operating conditions are apparent from the data, such as the change in the intensity of the noise at around Day 11, when the control of the flow is changed from allowing just one zone to produce to all the valves being opened.

It should be noted that Figure 9.28 is an analysis of only a small portion of the data acquired in this well: data from other locations carry information on the inflow as a function of depth, and in this case, the DVS data were complemented by data obtained with DTS on a separate fibre conveyed in the same tube running from the surface to the toe of the well.

A vast amount of data can thus be monitored with DFOS in this very hostile environment.

9.5.4 Perforation

Shaped-charge perforation is used in some well designs to form a hole not just through the casing but also reaching almost 2 m into the rock behind the casing, thus creating a path through which hydrocarbons may flow. Figure 9.29 shows the seismic signal observed from a nearby well (distant by 820 ft from the treatment well) when a perforation charge was fired. The resulting pressure and shear waves are clearly visible (identified with white arrows in the figure). A recording from an electrical sensor sited in the observation well below the end of the sensing fibre at a depth of 3700 ft is shown by the grey line in the lower part of the figure. The wavelet recorded by the two sensor types matches very well in time.

The benefit of the DVS in this application is to confirm that the perforation guns have fired; it is also an indication that the microseismic events (e.g. from hydraulic fracturing) can also be detected from nearby observation wells, and this is important for the optimisation of the HFM process. Furthermore, these data are a form of cross-well seismic acquisition that could yield detailed information on the geological structure near the wells. The signal content has far higher frequencies than can be propagated from the surface. Moreover, in the case of shale formations, the geology is often very variable throughout the reservoir and so local high-frequency seismic information can yield information on the formation with

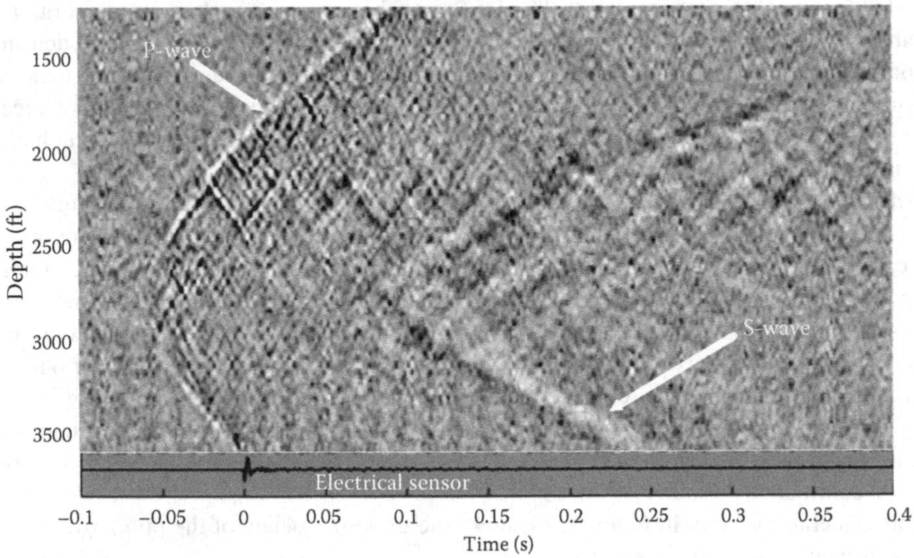

FIGURE 9.29 hDVS measurement of a perforation shot acquired from a neighbouring well. Image shown courtesy of Schlumberger.

very fine spatial detail. Although the same result could be obtained with conventional tools, in practice, the number of sensors that would be required makes this approach uneconomic at present.

9.5.5 Well Abandonment

At the end of its economic life, a hydrocarbon well is sealed (a process known as 'plug and abandon'). There is always a concern as to possible leaks around, or through, the cement used to plug the well: in the event of a leak, well fluids could potentially migrate and reach the surface or potable water aquifers – both eventualities giving rise to serious environmental concerns. It is mandatory in some territories to monitor abandoned wells, and this practice could become more prevalent.

One of the tools available for verifying the integrity of a well is an acoustic tool moved up and down the well that is used to listen for the very faint sounds caused by seeping fluids. DVS has been proposed as a technique for carrying out this work on accessible wells using a provisionally deployed cable; in this case, the DVS provides a faster measurement. Alternately, with a fibre permanently installed at the time of the plugging, DVS allows a frequent, or continuous, monitoring of the well to be carried out.

9.6 MULTIMODE FIBRE IN DISTRIBUTED VIBRATION SENSING

The underlying physics used in dΦ-DVS interrogation involves interference between two light beams, either the backscatter mixing with a local oscillator [90] or the interference between the backscatter signals returning from separate portions of the fibre (in the case of Refs. [91] or [92]). In the case of amplitude-based DVS systems, the interference applies to light returning from different parts of the resolution cell. The interference process compares the phase of the two inputs, and this is effective only when their electric fields are matched spatially. In turn, matching the electric fields requires the two inputs to occupy the same spatial mode. Their polarisations must also be aligned, and the phase distribution across the area where the fields combine must match. The simplest case is for both inputs to have a uniform phase front, in other words to occupy a single transverse mode that is of the same mode field diameter for both waves. This occurs naturally when the interfering optical signals merge within a single-mode fibre, although the question of their state of polarisation still arises.

In the case of a multimode sensing fibre, even if the probe light is launched into the lowest order mode, the backscatter is collected and guided back by many modes [93]. As a result, the backscatter no longer has a uniform transverse electric field distribution. In fact, with the narrowband probes used in DVS, the backscattered electric field distribution on the input fibre face is the interference between the light guided by many modes. Because their relative phase relationship is random, the intensity distribution consists of a speckle pattern, i.e. a set of light and dark spots that move if the fibre is disturbed [94]. The statistics of the intensity distribution (discussed in Chapter 3) change from a number of degrees of freedom $M_f = 1$ for a single-mode fibre to a much higher value of M_f in the case of a multimode fibre, and as a result, most of the contrast is lost from the interference process. Therefore, simply capturing the output of a multimode sensing fibre onto a detector will not yield a DVS measurement.

The solution is to select a single zone within the cross-section of the sensing fibre that has a spatially uniform electric field. This is achieved with a spatial filter that allows a single spatial mode (not necessarily a mode of the fibre, but a mode of free space) and to pass this selected light to the detection system [95]). In traditional optics, the selection is achieved with pin-holes, but it was shown in the field of dynamic light scattering [96,97] that a single-mode fibre acts as a near-perfect aperture for selecting a single speckle in the scattered light in applications such as particle size analysis [98].

It turns out that this spatial filtering concept can also be applied to the rather different case of DVS in multimode fibre. Thus, by interposing a single-mode fibre between a multimode sensing fibre and the detection system, the interference contrast is restored and valid DVS measurements can be obtained.

Under stable conditions, DVS measurements can be obtained on multimode fibres that approach the quality of single-mode data. However, a problem arises if the multimode fibre is disturbed: the relative phase of all modes is altered for all scatter locations beyond the disturbance, and this results in a parasitic signal for all locations downstream of the disruption. This situation arises, for example, in the case of a measurement in a borehole if the portion of the multimode fibre is subject, for example, to vibration from engines or from people moving around the wellhead. The problem of parasitic disturbance of the multimode section is particularly acute for temporary fibre deployments: in permanent installations, the multimode fibre can usually be protected from disturbances.

Measurements can therefore be made with multimode sensing fibres, but the preferred approach is still to use single-mode fibres.

9.7 INTERPRETATION OF DVS DATA

It is apparent from the examples shown previously that DVS generates vast volumes of data, some four orders of magnitude more than does DTS, which itself provides several orders of magnitude more data than is transmitted by a conventional point sensor.

This wealth of data is a marvellous opportunity for improving our understanding of the environment and of processes, for example, in the extractive industries. However, it also presents the users with the serious problem of how to make use of these data in a timely fashion. The data volume far exceeds the capability of human operators to analyse and comprehend the information that it conveys, particularly in applications where the value of the information is time-critical, for example, in controlling a process or determining the action to take when a potential threat has been identified in a security application.

The design of decision-supporting software tools is therefore underway to provide timely and actionable interpretations of the data so that its value can be fully realised. This means understanding the physics linking the DVS data to the source of vibration that it measures, building models that map the physics and creating algorithms that convert the DVS data in near-real-time to a concise formulation of the knowledge that the user or process control software needs to determine the best course of action based on the output of the DFOS and other information. The availability of interpretation software is fundamental to realising the potential of DFOS and DVS in particular.

REFERENCES

1. Kearey, P., M. Brooks, and I. Hill. 2013. *An introduction to geophysical exploration*. Blackwell. Oxford, UK.
2. Sheriff, R. E., and L. P. Geldart. 1995. *Exploration seismology*. Cambridge University Press. Cambridge, UK.
3. Freifeld, B. M., T. M. Daley, S. D. Hovorka et al. 2009. Recent advances in well-based monitoring of CO_2 sequestration. *Energy Proc.* 1: 1: 2277–2284.
4. Lumley, D. E. 2001. Time-lapse seismic reservoir monitoring. *Geophysics* 66: 1: 50–53.
5. Ozdemir, K., M. Vassallo, K. Eggenberger et al. 2013. New marine towed-streamer acquisition technology, Offshore Technology Conference: OTC-24001-MS.
6. Bittleston, S. H. 2003. Control devices for controlling the position of a marine seismic streamer. US6671223B2.
7. Hillesund, O., and S. H. Bittleston. 2005. Control system for positioning of marine seismic streamers. US6932017 B1.

8. Nakstad, H., M. Eriksrud, and S. Valla. 2013. Pressure recordings at Ekofisk – Experience and route forward, 2nd EAGE workshop on permanent reservoir monitoring, Stavanger, Norway, EAGE: Th-01-14.
9. Seth, S., S. Maas, B. Bunn et al. 2013. Full solution deep-water PRM Project in the Jubarte Field in Brazil for petrobras by PGS, 2nd EAGE Workshop on Permanent Reservoir Monitoring, Stavanger, Norway, EAGE: We-01-06.
10. Pereira, A. M., and M. Jones. 2010. *Fundamentals of borehole seismic technology*. Schlumberger. Sugarland, TX, USA.
11. Christie, P., D. Ireson, J. Rutherford et al. 1995. Borehole seismic data sharpen the reservoir image. *Oilfield Rev.* 4: 7: 18–31.
12. Berkhout, A. J., and D. J. Verschuur. 2016. Enriched seismic imaging by using multiple scattering. *The Leading Edge* 35: 2: 128–133.
13. Frignet, B. G., and A. H. Hartog. 2014. Optical vertical seismic profile on wireline cable, SPWLA 55th Annual Logging Symposium, Abu Dhabi, UAE. Society of Petrophysicists and Well Log Analysts.
14. Hartog, A., B. Frignet, D. Mackie, and M. Clark. 2014. Vertical seismic optical profiling on wireline logging cable. *Geophys. Prosp.* 62: 693–701.
15. Varkey, J., R. Mydur, N. Sait et al. 2008. Optical fiber cables for wellbore applications. US7324730 B2.
16. Clark, M., D. Mackie, and B. Frignet. 2013. Portable Airgun Tank – A new approach to Land Borehole Seismic using airguns, 2nd EAGE Borehole Geophysics Workshop, St Julian, Malta.
17. Mestayer, J., B. Cox, P. Wills et al. 2011. Field trials of distributed acoustic sensing for geophysical monitoring, 2011 SEG Annual Meeting, San Antonio, TX, USA: 4253–4257.
18. Mateeva, A., J. Mestayer, B. Cox et al. 2012. Advances in distributed acoustic sensing (DAS) for VSP, SEG: 0739.
19. Barberan, C., C. Allanic, D. Avila et al. 2012. Multi-offset seismic acquisition using optical fiber behind tubing, 74th EAGE Annual conference and exhibition, Copenhagen, Denmark.
20. Lopez, J., P. Wills, J. La Follett et al. 2015. Permanent seismic reservoir monitoring for real-time surveillance of thermal EOR at Peace River, Third EAGE Workshop on Permanent Reservoir Monitoring, Oslo, Norway, EAGE, Th A03.
21. Kasahara, J., Y. Hasada, and T. Tsuruga. 2011. Imaging of Ultra-long-term temporal change of reservoir (s) by accurate seismic source (s) and multi-receivers, EAGE Workshop on Permanent Reservoir Monitoring (PRM)-Using Seismic Data, Trondheim, Norway, EAGE, 10149.
22. Buizard, S., A. Bertrand, K. M. Nielsen et al. 2013. Ekofisk life of field seismic-4D processing, 75th EAGE Conference & Exhibition incorporating SPE EUROPEC 2013, London, United Kingdom, EAGE, Tu 08 02.
23. Chadwick, R. A., R. Arts, O. Eiken et al. 2004. 4D seismic imaging of an injected CO2 plume at the Sleipner Field, central North Sea. Geological Society of London.
24. Mateeva, A., J. Mestayer, Z. Yang et al. 2013. Dual-well 3D VSP in deepwater made possible by DAS, Annual Meeting of Society of Exploration Geophysicists, Houston, TX, USA, SEG, 0667.
25. Kiyashchenko, D., J. Lopez, W. Berlang et al. 2013. Steam-injection monitoring in South Oman from single-pattern to field-scale surveillance. *The Leading Edge* 32: 10: 1246–1256.
26. Daley, T., B. Freifeld, J. Ajo-Franklin et al. 2013. Field testing of fiber-optic distributed acoustic sensing (DAS) for subsurface seismic monitoring. *The Leading Edge* 32: 6: 699–706.
27. Götz, J., S. Lüth, J. Henninges, and T. Reinsch. 2015. Using a fibre optic cable as distributed acoustic sensor for vertical seismic profiling at the Ketzin CO_2 storage site, 77th EAGE Conference and Exhibition 2015, Madrid, Spain, EAGE, Th P2 13.
28. Bettinelli, P., and B. Frignet. 2015. Using optical fiber seismic acquisition for well-to-seismic tie at the Ketzin pilot site (CO_2 storage), 2015 SEG Annual Meeting, New Orleans, LA, Society of Exploration Geophysicists: SEG-5846235.
29. Frignet, B., A. Hartog, A. Constantinou et al. 2014. Hybrid conventional-optical borehole seismic experiment in geothermal well in Alsace, 1st Deep Geothermal Days, Paris, France.
30. Lagakos, N., E. Schnaus, J. Cole, J. Jarzynski, and J. Bucaro. 1982. Optimizing fiber coatings for interferometric acoustic sensors. *IEEE J. Quant. Electron.* 18: 4: 683–689.
31. Cranch, G. A., and P. J. Nash. 2000. High-responsivity fiber-optic flexural disk accelerometers. *J. Lightw. Technol.* 18: 9: 1233–1243.
32. Bertholds, A., and R. Dandliker. 1988. Determination of the individual strain-optic coefficients in single-mode optical fibres. *J. Lightw. Technol.* 6: 1: 17–20.
33. Hartog, A., J. E. Martin, D. Donno, and B. Papp. 2014. Fiber optic distributed vibration sensing with directional sensitivity. WO2014201313.
34. Hartog, A. H., J. E. Martin, D. Donno, and B. Papp. 2014. Fiber optic distributed vibration sensing with wavenumber sensitivity correction. WO2014201316.

35. Dean, T., A. Hartog, B. Papp, and B. Frignet. 2015. Fibre optic based vibration sensing: Nature of the measurement, 3rd EAGE Workshop on Borehole Geophysics, Athens, Greece, EAGE, BG04.
36. Dean, T., T. Cuny, A. H. Hartog, and J.-C. Puech. 2015. Determination of the optimum gauge length for borehole seismic surveys acquired using distributed vibration sensing, 77th EAGE Conference and Exhibition, Madrid, Spain, EAGE, Tu N118 16.
37. Dean, T., T. Cuny, and A. H. Hartog. 2016. The effect of gauge length on axially incident P-waves measured using fibre-optic distributed vibration sensing. *Geophys. Prospect.* In press. DOI: 10.1111/1365-2478.12419.
38. Dean, T., B. Papp, and A. Hartog. 2015. Wavenumber Response of Data Recorded Using Distributed Fibre-optic Systems, 3rd EAGE Workshop on Borehole Geophysics, Athens, Greece, EAGE, BGP07.
39. Farhadiroushan, M., T. R. Parker, and S. V. Shatalin. 2010. Method and apparatus for optical sensing. WO2010/136810A2.
40. Madsen, K. N., M. Thompson, T. Parker, and D. Finfer. 2013. A VSP field trial using distributed acoustic sensing in a producing well in the North Sea. *First Break* 31: 11: 51–55.
41. Kragh, E., E. Muyzert, J. Robertsson, D. E. Miller, and A. H. Hartog. 2012. Seismic acquisition system including a distributed sensor having an optical fibre. US8924158.
42. Den Boer, J. J., A. Mateeva, J. G. Pearce *et al.* 2013. Detecting broadside acoustic signals with a fiber optical distributed acoustic sensing (DAS) assembly. WO2013090544.
43. Hornman, J. C. 2016. Field trial of seismic recording using distributed acoustic sensing with broadside sensitive fibre-optic cables. *Geophys. Prosp.* In press. DOI:10.1111/1365-2478.12358.
44. Martin, J. E., D. Donno, B. Papp, and A. H. Hartog. 2014. Fiber optic distributed vibration sensing with directional sensitivity. WO2014201313.
45. Papp, B., D. Donno, J. E. Martin, and A. H. Hartog. 2016. A study of the geophysical response of distributed fibre optic acoustic sensors through laboratory-scale experiments. *Geophys. Prosp.* In press. DOI: 10.1111/1365-2478.12471.
46. Papp, B. 2013. Rock-fibre coupling in distributed vibration sensing. MSc, Institut de Physique du Globe de Paris.
47. Dean, T., A. H. Hartog, T. Cuny, and F. V. Englich. 2016. The effects of pulse width on fibre-optic distributed vibration sensing data, 78th EAGE Conference & Exhibition, Vienna, Austria, Tu P4 07.
48. Zwartjes, P., and A. Mateeva. 2015. Multi-fibre DAS walk-away VSP at Kapuni, 3rd EAGE Workshop on Borehole Geophysics, Athens, Greece, 3rd EAGE Workshop on Borehole Geophysics, Athens, Greece, EAGE.
49. Mateeva, A., J. Lopez, H. Potters et al. 2014. Distributed acoustic sensing for reservoir monitoring with vertical seismic profiling. *Geophys. Prosp.* 62: 4: 679–692.
50. Mateeva, A., J. Mestayer, B. Cox et al. 2013. Distributed acoustic sensing (DAS) for reservoir monitoring with VSP, 2nd EAGE Borehole Geophysics Workshop, St Julian, Malta, 15685.
51. Hartog, A. H., B. Frignet, D. Mackie, and M. Clark. 2014. Method of borehole seismic surveying using an optical fiber. WO2014190176.
52. Dean, T., T. Brice, A. H. Hartog et al. 2016. Distributed vibration sensing for seismic acquisition. *The Leading Edge* 35: 7: 600–604.
53. Hornman, K., B. Kuvshinov, P. Zwartjes, and A. Franzen. 2013. Field trial of a broadside-sensitive distributed acoustic sensing cable for surface seismic, 75th EAGE Conference & Exhibition incorporating SPE EUROPEC 2013 London, UK, Tu 04 08.
54. Kuvshinov, B. N. 2016. Interaction of helically wound fibre-optic cables with plane seismic waves. *Geophys. Prosp.* 64: 3: 671–688.
55. Crickmore, R. I., and D. Hill. 2014. Fibre optic cable for acoustic/seismic sensing. WO2014/064460A2.
56. Den Boer, J. J., J. M. V. A. Koelman, J. G. Pearce et al. 2012. Fiber optic cable with increased directional sensitivity. WO/2012/177547.
57. Farhadiroushan, M., D. Finfer, D. Strusevich, S. Shatalin, and T. Parker. 2015. Non-isotropic cable. WO2015036735.
58. Taylor, H. F., and C. E. Lee. 1993. Apparatus and method for fiber optic intrusion sensing. US5194847.
59. Qin, Z., L. Chen, and X. Bao. 2012. Continuous wavelet transform for non-stationary vibration detection with phase-OTDR. *Opt. Expr.* 20: 18: 20459–20465.
60. Wu, H., S. Xiao, X. Li et al. 2015. Separation and determination of the disturbing signals in phase-sensitive optical time domain reflectometry (Φ-OTDR). *J. Lightw. Technol.* 33: 15: 3156–3162.
61. Rupture and ignition of a gas pipeline. 30 July 2004. Ghislenghien Belgium 2009 French Ministry for Sustainable Development. DGPR/SRT/BARPI. 27681.
62. Hartog, A. H., K. Forbes, A. P. Strong, and R. Hampson. 2013. Monitoring of the position of a pipe inspection tool in a pipeline. GB2473991B.
63. Espiner, R., and A. Pickburn. 2014. Real-time pig tracking using a fibre-optic pipeline monitoring system, 10th International Pipeline Conference, Calgary, Canada: IPC2014-33398.

64. Crickmore, R. I., and D. J. Hill. 2010. Traffic sensing and monitoring apparatus. US7652245B2.
65. Schulz, W. L., J. P. Conte, and E. Udd. 2001. Long-gage fiber optic Bragg grating strain sensors to monitor civil structures, 8th Annual International Symposium on Smart Structures and Materials, Newport Beach, CA, USA, SPIE: 56–65.
66. Peng, F., N. Duan, Y. J. Rao, and J. Li. 2014. Real-time position and speed monitoring of trains using phase-sensitive OTDR. *IEEE Photon. Technol. Lett.* 26: 20: 2055–2057.
67. Rohwetter, P., R. Eisermann, and K. Krebber. 2015. Distributed acoustic sensing: Towards partial discharge monitoring. 24th International Conference on Optical Fiber Sensors, Curitiba Brazil, SPIE. 1C.
68. Rohwetter, P., R. Eisermann, and K. Krebber. 2016. Random quadrature demodulation for direct detection single-pulse Rayleigh C-OTDR. *J. Lightw. Technol.* Early access online. DOI:10.1109/JLT.2016.2557586.
69. Pouet, B., S. Breugnot, and P. Clémenceau. 2005. Robust laser-ultrasonic interferometer based on random quadrature demodulation, American Institute of Physics Conference Institute of Physics, 820: 233.
70. Wuttke, J., P. M. Krummrich, and J. Rosch. 2003. Polarization oscillations in aerial fiber caused by wind and power-line current. *IEEE Photon. Technol. Lett.* 15: 6: 882–884.
71. Bao, X., J. Snoddy, J. Leeson, and L. Chen. 2009. *Fiber sensor applications in dynamic monitoring of structures, boundary intrusion, submarine and optical ground wire fibers*. Intech Open. Rijeka, Croatia.
72. Hartog, A. H., E. Brown, J. M. Cook et al. 2011. Systems and methods for distributed interferometric acoustic monitoring. US7946341B2.
73. Hartog, A. H., J. E. Brown, J. Cook et al. 2013. Systems and methods for distributed interferometric acoustic monitoring. US8347958B2.
74. Hartog, A. H., J. E. Brown, J. M. Cook et al. 2012. Systems and methods for distributed interferometric acoustic monitoring. US8225867B2.
75. Patterson, M. M. 1976. A flowline monitor for production surveillance, SPE Spring Meeting, Amsterdam, The Netherlands, Society of Petroleum Engineers: SPE 5769.
76. Alford, B. J., M. M. Patterson, and F. A. Womack. 1978. Development and Field Evaluation of the Production Surveillance Monitor. *Journal of Petroleum Technol.* 30: 02: 160–166.
77. Jianrong, W., A. van der Spek, D. T. Georgi, and D. Chace. 1999. Characterizing Sound Generated by Multiphase Flow, Houston, TX, USA, Society of Petroleum Engineers, SPE 56794.
78. Hartog, A. H., D. Miller, K. Kader et al. 2013. Distributed acoustic wave detection. US8408064B2.
79. Paillet, F. E., and J. E. White. 1982. Acoustic modes of propagation in the borehole and their relationship to rock properties. *Geophysics* 47: 8: 1215–1228.
80. Biot, M. A. 1952. Propagation of elastic waves in a cylindrical bore containing a fluid. *J. Appl. Phys.* 23: 9: 997–1003.
81. Kieffer, S. W. 1977. Sound speed in liquid-gas mixtures: Water–air and water–steam. *J. Geophys. Res.* 82: 20: 2895–2904.
82. Johannessen, K., B. K. Drakeley, and M. Farhadiroushan. 2012. Distributed acoustic sensing – A new way of listening to your well/reservoir, Utrecht, The Netherlands, Society of Petroleum Engineers C1-SPE, SPE-149602-MS.
83. Finfer, D. C., V. Mahue, S. Shatalin, T. Parker, and M. Farhadiroushan. 2014. Borehole flow monitoring using a non-intrusive passive distributed acoustic sensing (DAS), Amsterdam, The Netherlands, Society of Petroleum Engineers: SPE-170844-MS.
84. Finfer, D., T. R. Parker, V. Mahue et al. 2015. Non-intrusive multiple zone distributed acoustic sensor flow metering, Houston, TX, USA, Society of Petroleum Engineers, SPE 174916.
85. Mullens, S. J., G. P. Lees, and G. Duvivier. 2010. Fiber-optic distributed vibration sensing provides technique for detecting sand production, Offshore Technology Conference, Houston, TX, USA: OTC-20429.
86. Hull, J., L. Gosselin, and K. Borzel. 2010. Well integrity monitoring & analysis using distributed acoustic fiber optic sensors, IADC/SPE Drilling Conference and Exhibition, New Orleans, LA, USA, SPE: IADC/SPE 128304.
87. Webster, P., J. Wall, C. Perkins, and M. Molenaar. 2013. Micro-seismic detection using distributed acoustic sensing, SEG 2013 Annual Meeting, Houston, TX, USA, Paper 182: 2459.
88. Baihly, J., C. Wilson, A. Menkhaus et al. 2015. Fiber optic revelations from a multistage open hole lateral fracturing treatment, Unconventional Resources Technology Conference (URTEC). URTeC: 2172404.
89. Dickenson, P. 2014. ESP Monitoring using heterodyne distributed vibration sensor (hDVS), European Artificial Lift Forum, Aberdeen, UK.
90. Hartog, A. H., and K. Kader. 2012. Distributed fiber optic sensor system with improved linearity. US9170149B2.
91. Dakin, J. P., and C. Lamb. 1990. Distributed fibre optic sensor system. GB2222247A.
92. Posey, R. J., G. A. Johnson, and S. T. Vohra. 2001. Rayleigh scattering based distributed sensing system for structural monitoring, 14th Conference on Optical Fibre Sensors, Venice, Italy, SPIE, 4185: 678–681.

93. Bisyarin, M. A., O. I. Kotov, A. H. Hartog, L. B. Liokumovich, and N. A. Ushakov. 2016. Rayleigh backscattering from the fundamental mode in multimode optical fibers. *Appl. Opt.* 55: 19: 5041–5051.
94. Rawson, E. G., J. W. Goodman, and R. E. Norton. 1980. Frequency dependence of modal noise in multimode optical fibers. *J. Opt. Soc. Am.* 70: 8: 968–976.
95. Davies, D., A. H. Hartog, and K. Kader. 2010. Distributed vibration sensing system using multimode fibre. US7668411B2.
96. Brown, R. G. W., and A. P. Jackson. 1987. Monomode fibre components for dynamic light scattering. *J. Phys. E Sci. Intrum.* 20: 12: 1503.
97. Brown, R. G. W. 1990. Dynamic light scattering apparatus. US4975237 A.
98. Brown, R. G. W., J. G. Burnett, J. Mansbridge, and C. I. Moir. 1990. Miniature laser light scattering instrumentation for particle size analysis. *Appl. Opt.* 29: 28: 4159–4169.

Other Distributed Sensors 10

Several techniques that do not fit within the bounds of the previous chapters have been demonstrated and, in some cases, commercialised. So far, the commercial interest in these methods has been limited. They are described here mainly for their scientific interest, but it is also possible that technical advances will bring their applicability to the fore and stimulate further research interest and drive the engineering of these concepts into useful products.

We also broaden the scope of distributed optical fibre sensors somewhat from the strict definition given in Chapter 1 to include sensors that provide a reading of an extremum of the measurand. Some of these approaches have relatively simple interrogators that provide no location of the extremum, but they could be adapted to do so using the reflectometric techniques discussed in previous chapters.

10.1 LOSS-BASED SENSORS

It was noted in Chapter 1 that modulating the attenuation of an optical fibre is a possible strategy for a distributed sensor, albeit not that is one suited to a precise measurement at each point of a long sensor. However, there are applications where simply the knowledge of a measurand having passed a threshold at some point along a sensing fibre (and preferably the location of that crossing) is valuable.

In order to be effective in this role, the fibre should present an abrupt, non-linear, response to the measurand to ensure that a moderate increase in its value over extended sections of the fibre is not confused with a localised increase beyond the chosen detection threshold. The interrogation can be particularly simple, for example, a simple power measurement for an alarm without location or an optical distributed optical fibre sensor (OTDR) to also provide location of the closest point above threshold.

Several examples of such sensors have been developed and even commercialised, although the author is not aware of any of these that are still in manufacture.

10.1.1 Micro-Bending-Based Sensors

Distorting the axis of an optical fibre caused losses through bending. In the case of a graded-index (strictly, a parabolic-index) multimode fibre, the propagation constants of the several mode groups are equally spaced in the spatial frequency domain. In other words, a periodic distortion of the axis that couples one mode group to a nearest-neighbour mode group will couple all mode groups to their nearest neighbours, including to mode groups that are marginally guided or indeed are beyond their cut-off. It follows that a deformation at a properly selected pitch-period couples all the nearest-neighbour mode groups to each other and results in a sharp attenuation [1]. At the optimum pitch, a loss of order 1 dB can result from a distortion of the core axis of only 5 μm over a distance of just 50 mm. Admittedly, in the work cited, the index difference for the fibre selected was rather low (0.008), but nonetheless, the principle of a resonant micro-bending loss was established and several devices based on that concept have been published [2,3], and some of the distributed sensor concepts based on this principle are discussed here.

An alternative approach in which the sensing fibre is threaded through an extensible chain provides an adjustable range of strain and a more linear response [4].

An array of short corrugated-jaw sensors interrogated with OTDR was demonstrated in 1982 [5,6]; this was, in effect, a multiplexed array of point displacement sensors, or strain gauges. This is illustrated in Figure 10.1, where the reader will recognise the displacement transducers from Figure 1.3a (based on Ref. [6]) and their multiplexing using a straightforward OTDR arrangement; once the location of the displacement sensors has been mapped along the sensing fibre, their loss can be tracked using built-in software available in many OTDRs. The measurand value is deduced from the size of the point loss at each sensor element. This arrangement requires the sensors to be constructed individually and mounted onto the sensing fibre. A more detailed study of this type of multiplexed sensor array may be found in Ref. [7]. The extensible chain approach was also demonstrated with OTDR interrogation [4].

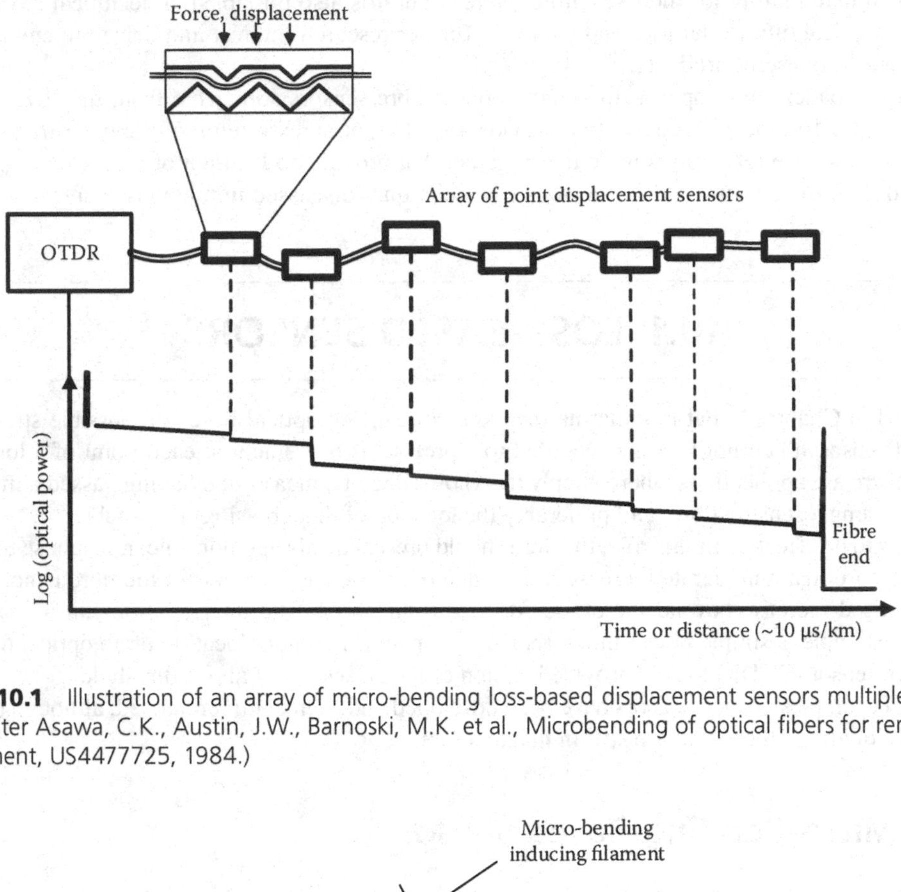

FIGURE 10.1 Illustration of an array of micro-bending loss-based displacement sensors multiplexed using OTDR. (After Asawa, C.K., Austin, J.W., Barnoski, M.K. et al., Microbending of optical fibers for remote force measurement, US4477725, 1984.)

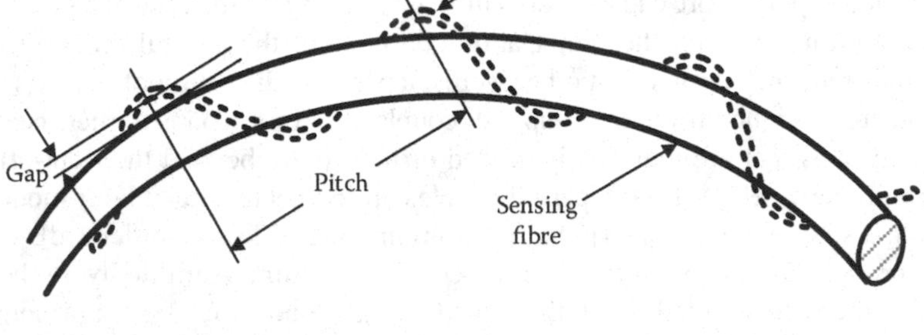

FIGURE 10.2 Continuously sensitive side-pressure sensing fibre. (Re-drawn from Harmer, A.L., Fiber-optical pressure detector, US4618764, 1986.)

The concept was made truly distributed by Alan Harmer [8], who proposed and demonstrated a fibre onto which a plastic helical filament was wound at a pitch suitable for causing the light to be attenuated when side pressure is applied. The functional part of the sensor is shown in Figure 10.2. The gap between the filament and fibre ensures that no excess loss occurs at low side-pressure; as soon as the pressure is sufficient to cause the external layers to deform and force contact between the filament and the fibre to an extent that the latter's axis is deformed, the local attenuation will increase rapidly. Although there appear to be no present commercial uses of this approach, it has been employed for at least three separate applications that will now be described.

10.1.1.1 Safety barriers

In the 1980s, Herga Electric Ltd., a company based in Bury St Edmonds, United Kingdom, manufactured safety barriers to shut down robotic machine tools in the event of anyone stepping onto safety barriers [9]. Their operating principle was that of Ref. [8], i.e. a multimode, graded-index fibre that had been spiral-wound with a thin Nylon thread. These sensing fibres were positioned in mats with simple optical transmitters and receivers and, very simply, any pressure on the mat above a defined threshold caused the received power to drop below a pre-defined threshold, resulting in a shut-down of the machine tool.

Whilst the concept is simple, the devices are required to be extremely reliable (guaranteed to shut down if, but only if, a breach of a safety zone occurs); they must also be cheap and simple to install whilst operating completely without maintenance. A simple optical power level measurement is not sufficiently robust, particularly in view of the high levels of electro-magnetic interference that may be present in machine shops, particularly at the time these sensors were designed, i.e. before regulations on electromagnetic compatibility had been reinforced. In Ref. [9], this was achieved using a pulse-modulation format where the probe pulses are of a specific duration and the output pulses are reduced in duration as the loss in the sensing fibre increases due to activation by lateral pressure. The absence of a detected pulse signifies either a very high loss of the sensor or a break of the sensing fibre or a failure of the emitter; in any of these cases, the system shuts down the equipment that it controls, so ensuring a fail-to-safe mode of operation.

By all accounts [10], the technology was successful but has now been superseded by all-electronic approaches, such as video cameras with real-time image processing.

It should be noted that this sensor was distributed in the meaning of being sensitive at all locations along the fibre but did not report the location of the loss-feature, it simply acted upon the fact that *somewhere*, the safety zone had been breached, which was appropriate to the application.

10.1.1.2 Linear heat detection

Most optical linear heat detectors, as used for fire and overheat alarming, are based on Raman distributed temperature sensor (DTS). However, in the 1990s, the Swedish telecommunications equipment company LM Ericsson produced a fire detection system based on the principle of micro-bending-induced loss [11]. In this case, the fibre is attached using a spiral-wound thread to a tube containing a swellable material. The tube is filled with a wax that melts at a defined temperature and, on melting, expands, causing a periodic distortion of the fibre (at a pitch determined by that of the spiral thread) and, hence, a strong micro-bending loss.

The heat-sensitive cable was interrogated with a dedicated OTDR and so could locate at least the first over-temperature condition as well as alarming on this event. Losses between 0.01 and 0.1 dB/m were reported and a hot-spot typically caused a local drop of the backscatter by 1 dB, easily detectable in a short time scale (response time is critical in fire-detection applications).

Although the system was taken up by a major building alarms company, it fell out of favour partly because the cable tended to 'remember' heat and thus needed to be replaced even after modest excursions above the alarm threshold. This memory property of the sensing medium also precludes testing the alarm system and the ability to test fire alarm systems is a pre-requisite to ensuring their continued effectiveness. In addition, the specialised cable was significantly more expensive than a simple fibre cable as used in Raman systems, and any advantages of simplicity in the interrogation (ability to use the entire backscattered power rather than just the Raman bands and the fact that Raman systems require two acquisition channels) were negated by the higher cable costs. To the author's knowledge, this approach is no longer used commercially.

10.1.1.3 Moisture/hydrocarbon sensing

A similar concept was demonstrated in the form of a coating of swellable material applied to a support structure (a glass-reinforced plastic rod) [12,13]. A thin thread is wound with an appropriate pitch around both the sensing fibre and the support, resulting in micro-bending losses when the swellable material expands. Initial tests [12] were aimed at detecting water and pH using hydrogels as the swelling material. Subsequent work [13] using silicones as the swellable material was aimed at detecting solvents, and the work showed some selectivity between solvent types.

10.1.2 Radiation Detection

The issue of fibre degradation due to ionising radiation was addressed in Chapter 7, where it was shown that the losses induced by exposure to irradiation can seriously limit the performance of distributed optical fibre sensors. In the present context, however, the radiation-induced attenuation (RIA) can be used as an indicator of the radiation dose or dose rate experienced by the fibre.

The RIA depends on a number of factors, even for a single type of radiation. It is a result of defects in the glass network caused by the interaction with high-energy particles or γ-rays [14]. The composition affects the susceptibility of the fibre to radiation damage and also its recovery rate. Some types of damage will gradually reduce with time (annealing), and this recovery is accelerated at high temperature. The annealing rate also depends on the composition [15]; for example, the presence of GeO_2 will slow the recovery, whereas a high OH content drastically accelerates the recovery. P-doping results in damage that, once induced, is permanent (or at least has very high annealing temperature [16,17]) and that increases with the dose and so it could form the basis of a total exposure meter – a witness to the cumulative irradiation undergone by the fibre.

Doping with rare-earth ions can drastically increase the RIA, whereas inclusion of Cerium hardens the fibres against RIA. However, the origins of the RIA in rare-earth-doped fibres appear to be related more to the co-dopants (such as P, Ge or Al) normally associated with the rare-earth elements than these elements themselves [18], and the RIA certainly varies with the dopant type [19].

The recovery of the fibre from RIA can in some cases be accelerated by exposure to light, a process known as photo-bleaching [20].

The operating wavelength also strongly influences the RIA. The damage is predominantly at wavelengths shorter than 1100 nm and increases rapidly as the wavelength is reduced.

All these effects have been well documented in the literature, and the subject is in itself very wide. The use of RIA as a sensor has also been proposed for more than 30 years [21]. Of interest to distributed optical fibre sensors is the use of OTDR to probe fibre samples in a radiation environment specifically to estimate their exposure, which was also shown in Ref. [21]. Apart from this first demonstration, OTDR was further proposed as a more reliable means of performing long term measurements of RIA [22] – it provides better reproducibility than the more standard cut-back method in that it avoids any problems relating to variations in the power launched. It was demonstrated that the OTDR measurements are consistent with the large body of test data acquired under varying conditions. The one *caveat* in this application of OTDR is the high power used in the probe pulses that can induce photo-bleaching, and so the probe power must often be reduced in this application.

The specific use of OTDR to measure radiation levels in particle accelerators was demonstrated at various German particle physics facilities. In Ref. [23], leaks in the particle beam were identified using a fibre sensor interrogated by OTDR, which shows localised increases in the dose that were assessed using the OTDR. In this case, a multi-mode, graded-index fibre doped with both GeO_2 and P_2O_5 was used, ensuring that it is the cumulative dose that is measured. In separate experiments, the clear steps in the OTDR traces due to RIA at specific places along the fibre could be identified [24].

In the work published on estimating of RIA using OTDR, so far, the assumption has been that any detected losses must be related to RIA. Whilst other mechanisms can certainly result in increased loss, a multi-wavelength measurement would in many cases allow the origin of the loss to be confirmed as RIA.

10.1.3 Loss of Numerical Aperture

An optical fibre made from core and cladding materials having markedly different coefficients of refractive index as a function of the measurand will show a variation of its numerical aperture (NA), and hence, it will show a loss or a reduction of the backscatter capture fraction when conditions reduce the NA.

This was the case in the first DTS that used a liquid-core fibre as the sensor [25], where an increase in temperature resulted in a lower refractive index for the core, and so a reduction of the NA. However, in a step-index multimode fibre, a variation of the NA is similar in its consequences to a point loss, and so a localised loss of NA would affect all OTDR readings beyond the point where the NA is restricted. This is undesirable in fully distributed sensor, where independent measurements are required in all locations. Therefore, in Ref. [25], the NA effect was eliminated using a spatial filter that rejected those modes that could potentially be lost in the event of a hot-spot appearing.

However, in a near-parabolic graded-index fibre, the NA modulates the backscatter signal in a way that is local to the modulation, provided that an NA filter is inserted before the receiver. In Ref. [26], it is shown how the backscatter signature measured from both fibre ends allows variations of NA along the fibre to be determined. This seems not yet to have been applied to distributed sensing, probably because there are no known graded-index fibres with a measurand-dependent NA.

The concept of the variation of NA resulting from dissimilar thermal coefficients of the refractive index was explored by Gottlieb and Brandt [27], who investigated a variety of glass pairs to find examples of the NA being reduced at elevated temperature. The specific focus of this work was to design a single-mode fibre, to be used in short lengths, where the NA would be reduced at high temperature, leading to the guided power spreading into the cladding. On the assumption of a cladding glass with higher loss than that of the core, an induced loss of signal would be caused by a hot-spot. So the end goal of this work was in fact a maximum-temperature sensor based on loss rather than a distance-resolved temperature sensor. Nonetheless, the principles explored by Gottlieb and Brandt could be applied to truly distributed sensing if a suitable combination of core and cladding materials were available in low-loss form.

A loss of NA was used in a device designed to detect leaks in cryogenic gas storage tanks [28] in a system that was then manufactured by Pilkington Brothers, United Kingdom.

The sensor concept was based on a fibre constructed of materials with highly dissimilar thermo-optical coefficients, such that the refractive index of core and cladding converge at around –55°C. Although this information was not published, it seems likely that this was achieved using a polymer (silicone)-clad silica fibre. Silicones undergo transitions at around –50°C to –60°C, and their density increases rapidly on the changing of their state, thereby increasing their refractive index. As a result, the fibre loses guidance when its temperature falls below this cross-over point of the refractive indices for core and cladding. The concept was the basis of a sensor that simply monitored the increased loss of the fibre caused by contact with liquefied natural gas (LNG) at any point along its path. By placing the fibre in the secondary confinement container (bund) below the LNG tank, a leak in the liquefied gas will reach the fibre and cause a reduction in transmission and trigger an alarm. Tens of these monitoring systems were installed by British Gas in the early 1980s. Even such a relatively simple sensor requires a thorough design to ensure that (a) it is intrinsically safe, (b) it will trigger alarms reliably and (c) it will not cause false alarms. The product appears to have been withdrawn, probably because its functions have been taken over by Raman DTS systems.

10.1.4 Rare-Earth-Doped Fibre Temperature Sensing

A fibre doped with rare-earth ions exhibits absorption bands, and some combinations of ions and operating wavelengths result in a strong increase of the absorption with increasing temperature, whilst maintaining a very low loss at normal operating conditions. If the loss is monitored with an OTDR technique, the location of the loss increase can be determined.

As with all loss-based techniques, the method is appropriate primarily to applications in which just one, or perhaps a few, hot-spots are expected, and the objective is to determine their location and temperature. There are a number of applications where this information is all that is required. The use of the temperature-dependent absorption in rare-earth-doped fibres was demonstrated in point sensors using Eu-doped and Nd-doped fibre [29].

As illustrated in Figure 10.3, in a rare-earth four-level laser system, such as Nd:glass, Nd^{3+} ions are pumped by absorption of the incident pump wave at wavelength λ_p from a ground state (Level 1, the $^4I_{9/2}$ state for this ion) to an excited state (Level 3, $^4F_{3/2}$) via a higher-level absorption band (Level 4) from which the excited ions decay rapidly to the upper lasing state (Level 3) through a non-radiating process [30]. The system operates as a four-level system when fluorescing from the upper lasing state to a lower lasing state (Level 2, $^4I_{11/2}$) that is distinct from the ground state. The lower lasing state is normally largely unoccupied at room temperature owing to the fast decay time τ_{21} and so photons at the normal lasing wavelength λ_e are not absorbed and losses are therefore normally low at λ_e; this makes this type of system easy to work with (it is not necessary to deplete the ground state for net gain to occur). The contrast of the slow decay rate from Level 3 to Level 2 compared with the rapid decay from Level 2 to Level 1 and from Level 4 to Level 3 causes Level 3 to have a higher population than Level 2 with even small pump power and so exhibit optical gain. For the Nd:glass system, the Level 3 to Level 2 transition results in a broad fluorescence spectrum peaking between 1050 and 1090 nm depending on the precise glass composition.

However, fluorescence also occurs from Level 3 directly to Level 1, in which case the ions are operating as a three-level system with a broad fluorescence peak centred around 938 nm [31]. If pump radiation of slightly shorter wavelength than that corresponding to the Level 1 to Level 3 energy gap is incident of this material, it can be absorbed by the higher-energy tail of the absorption band, a condition

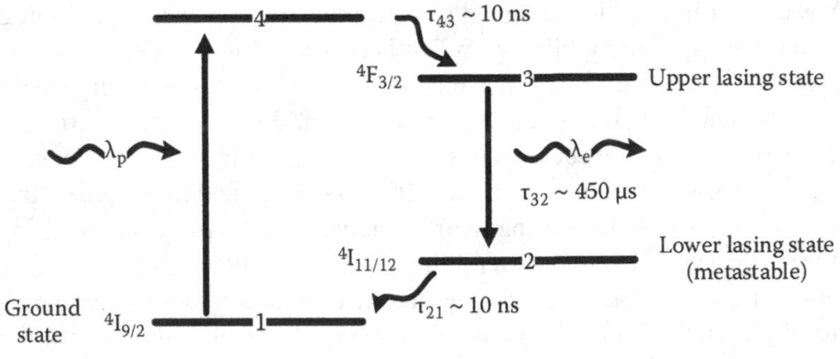

FIGURE 10.3 Energy state diagram for Nd^{3+} in silicate glass. (Adapted from Mears, R.J., Optical fibre lasers and amplifiers, PhD University of Southampton, 1987.)

known as 'quasi-three-level' operation that is common in Er-doped fibres pumped at 1480 nm for optical amplification.

These effects have been investigated as a potential basis for distributed temperature sensing.

In the first example [32], a multimode fibre lightly doped (5 ppm) with Nd ions was probed with an OTDR operating at 904 nm, which is on the long-wavelength edge of an absorption band centred at 880 nm. The location of this edge shifts with temperature as the population of the upper end of the ground state increases with increasing temperature. At this wavelength, therefore, the absorption loss is highly temperature-sensitive. In Ref. [32], it was found that the loss of the fibre (50 dB/km at room temperature) varied by about 0.2%/K. The interrogation was based on an early DTS [33] that had a spatial resolution of 7.5 m when used as a Raman OTDR, and it also provided a Rayleigh scattering output that was used in this experiment. However, the need to differentiate the data along the fibre to estimate the attenuation degraded the spatial resolution to 15 m. Nonetheless, in this first demonstration, a temperature resolution of 2 K was achieved in an acquisition time of 150 s over 140 m of fibre.

The main limitation of the approach just described [32] is that probing on the edge of an absorption band results in a rather large attenuation even when the fibre is at room temperature, if it is to be reasonably sensitive to temperature.

A rather more promising approach is to probe the sensing fibre at λ_e, i.e. at a wavelength corresponding to the transition between Level 2 and Level 3 in Figure 10.3. Given that Level 2 is normally not populated, no absorption occurs and the fibre is relatively transparent. However, Level 2 can be populated thermally from the ground state if the temperature is high enough. The population in Level 2 relative to Level 1 is dictated by the Boltzmann distribution [34], i.e.

$$N_{L2} = N_{L1} \exp\left(-\frac{\Delta\nu_{1,2}}{k_B T}\right). \tag{10.1}$$

Here, N_{L2} and N_{L1} are the population densities of Level 2 and Level 1, respectively, and $\Delta\nu_{1,2}$ is the energy difference between the two levels. k_B is Boltzmann's constant, and T, the temperature. According to Ref. [34], $\Delta\nu_{1,2}$ is 3.97×10^{-20} J, and at room temperature, the ratio N_{L2}/N_{L1} is 54×10^{-6}, growing exponentially with increasing temperature. In the case of the fibre used in Ref. [34], the loss due to the induced absorption at 1064 nm was insignificant at 20°C, increasing to about 15 dB/km at 200°C and rising rapidly thereafter (this fibre had a dopant concentration such as to exhibit a loss of 1100 dB/km at 900 nm at 20°C).

The benefit of using a fibre in a mode where the absorption is low at usual temperatures is, of course, that long distances can be covered and yet hot-spots will be revealed as point losses (or only the nearest hot-spot if its induced loss is sufficient to obscure the remainder of the fibre). This is particularly useful as a hot-spot detector where the non-linearity of the response helps distinguish a small rise occurring over a long length of the sensing fibre from the emergence of a small, very hot anomaly such as a fire or degradation of a cable. The authors of Ref. [34] showed losses increasing to 65 dB/km at 700°C.

The choice of dopant and its concentration can be used to tailor the sensitivity and the threshold at which the temperature causes a noticeable change in attenuation. For example, by selecting a rare earth with a small energy difference between Level 1 and Level 2, the population of Level 2 will start to increase markedly at lower temperature. Thus, Samarium has a value of $\Delta\nu_{1,2}$ about half of that of Neodymium, and this leads to an increase in absorption at around −100°C.

Related work using Holmium as a rare-earth ion [35] demonstrated a room-temperature sensitivity of 2%/K at room temperature in a fibre where the loss had increased to 45 dB/km even at −60°C and reached 130 dB/km at 90°C. In this case, the fibre was only reasonably transmissive at cryogenic temperature with an interrogation wavelength tuned to the region of 900 nm.

In spite of these interesting developments, this work does not seem to have become commercial, partly because of the cost of the very special fibres involved but, more fundamentally, because the range is restricted by the relatively high static losses. The performance of this type of approach has so far not come close to matching that of Raman technology that uses low-cost fibre and is capable of measuring the entire thermal profile, even if the whole fibre is at maximum temperature.

10.2 FLUORESCENCE

The very weak scattering cross-section available with the Raman effect has encouraged the investigation of fluorescence as a source for radiation returned by the sensing process. Similar physics as discussed in the previous section, of thermal population of the Level 2 state, apply to the fluorescence measurement. When probed at the wavelength corresponding to the Level 2 to Level 3 transition (c. 1050–1090 nm for Nd^{3+} ions), a short wavelength fluorescence, also known as anti-Stokes fluorescence [36], can be observed. This effect can be used to judge the temperature at the point where the fluorescence occurs. This possibility was alluded to in Ref. [34]; however, the drawback for rare-earth-doped glasses is the very long fluorescence decay time (many μs), which blurs the spatial resolution and so making this approach of little interest for distributed sensing.

The approach does, however, have important attractions for single-point sensors, and in particular, it was demonstrated [37] that the cross-sensitivity of measured temperature to applied strain is quite low, of the order of 10^{-4} K/με.

Multi-photon fluorescence does not suffer the same limitation on spatial resolution that affects the single-photon fluorescence just discussed, and it will be described amongst the non-linear distributed techniques later in this chapter (Section 10.7.4).

10.3 CHANGES IN CAPTURE FRACTION

The capture fraction in OTDR depends primarily on the NA of the fibre (see Chapter 3, Section 3.1.3).

In the case of a single-mode fibre, the local NA maps directly on to the local capture fraction and, hence, on the strength of the local backscatter signal (rather than its spatial derivative as is the case for loss-based distributed sensors). For multimode fibres, the NA acts as a far-field filter, and specifically for step-index fibres, all locations beyond a restriction of the NA have reduced backscatter signals. Therefore, in a step-index fibre, a dependence of the NA on an external physical quantity provides the maximum or minimum (depending on the sign of the NA sensitivity) of the measurand value along the fibre.

In a graded-index fibre, a gradual variation of the fibre NA (or indeed of the fibre diameter) causes a mode transformation that (provided that a mode filter is installed in front of the detector) allows the value of the NA or fibre diameter to be read all along the sensing fibre [26,38,39].

It is worth mentioning here *pro memoria* the work on strain sensing in polymer fibres, discussed in Section 8.3. Just as NA variations can be observed in backscatter signatures, so too can tapers of the fibre diameter. Conventional, silica-based fibres are brittle and have maximum sustained strain limits below 1%. Polymer fibres, however, can sustain strain levels in excess of 20%, at which point, of course, a significant change in diameter occurs. In graded-index fibres, the strain can be fully resolved to a local disturbance; i.e. there is minimal effect on the measurement results downstream of the strained region. This is probably a unique example of the measurand (strain) being converted to a change in core diameter and then interrogated using an OTDR.

10.4 POLARISATION OPTICAL TIME-DOMAIN REFLECTOMETRY (POTDR)

Polarisation optical time-domain reflectometry (POTDR) is a special case of sensing with OTDR, where the information is conveyed not by the total backscattered power but rather by its state of polarisation. It is, in principle, a very powerful technique in that it can sense many different measurands – however, this is also its weakness, namely poor specificity. The primary practical use of POTDR is the assessment of telecommunication links for polarisation mode dispersion, an effect that can limit the bandwidth of certain types of links. There appears not to have been any commercial application of the concept as originally laid out by Rogers [40]: although at least one company has investigated its use as a pressure sensor; this work seems to have been discontinued.

10.4.1 Basic Principles of POTDR

POTDR is the polarisation analysis of the Rayleigh backscatter signal in a single-mode fibre. It was originally proposed [41] and demonstrated [42] in 1980, and research work on exploiting its capabilities has been sporadic since that time. The basic concept of POTDR is that the Rayleigh backscattering process largely preserves the state of polarisation of the probe pulse. At its most basic, a POTDR measurement involves measuring the backscatter in the parallel and orthogonal states to that of the probe pulse, and a typical arrangement for carrying out this measurement is illustrated in Figure 10.4; it is similar to a conventional OTDR measurement, with the addition of a polariser prior to launching the probe pulse into the sensing fibre and of a polarisation beamsplitter to separate the two backscatter signals returning in orthogonal polarisation states. It is important to ensure that the transition between optical elements does not degrade the polarisation information, so it is common for free-space optics to be used in this measurement. Note also that the beamsplitter used to separate the probe from the backscatter is placed at near-normal incidence (3.5° from normal in the case of Ref. [42]) to avoid any polarisation-dependent losses or splitting ratio in the device.

Figure 10.5 shows a synthetic set of POTDR traces typical of those that are obtained from the arrangement of Figure 10.4 (see Ref. [42] for similar experimental data). The oscillations are due to the linear birefringence of the fibre, indicating a beat-length of order 50 m, which again is typical of modern single-mode fibre. These oscillations in the intensity of the two orthogonal polarisation states are complementary, and

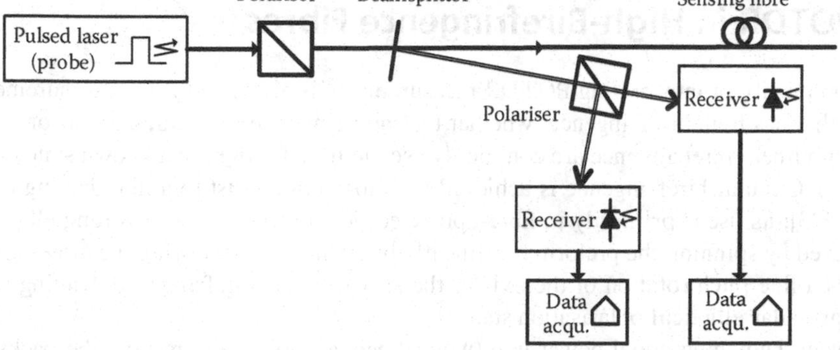

FIGURE 10.4 Basic arrangement for a POTDR measurement.

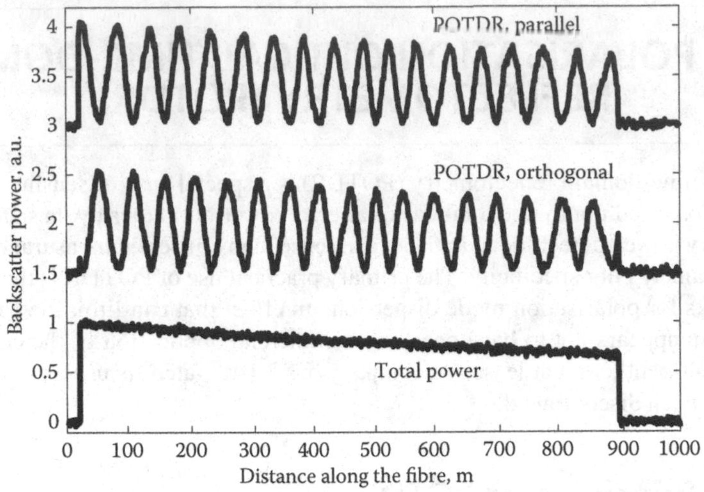

FIGURE 10.5 Typical appearance of POTDR traces.

so the sum of the light measured in the two polarisations sums to the conventional OTDR result to a high accuracy [42]. However, it should be noted that the backscattering process does not fully preserve the incident state of polarisation, and so the minima in Figure 10.5 are non-zero – about 5% of the backscattered light returns in the orthogonal polarisation state from that of the probe at the scattering location.

If the fibre showed only linear birefringence, in a known axis, data such as those of Figure 10.5 would be simple to interpret: the POTDR would measure directly the distribution of the beat length along the fibre that can then be related, for example, to side pressure or to bending that the fibre may be subjected to. Unfortunately, in conventional fibres, matters are not so straightforward. Thus, axis of the birefringence may vary, for example, because in one section the fibre is bent in one axis and in another axis in a different location; twisting of the fibre also has the effect of inducing a change in the birefringence axis. Although twists also cause circular birefringence, this effect is reciprocal, and is therefore cancelled in the two transits undergone by the outgoing pulse and the backscatter return, leaving only a variation in the angle of the linear birefringence. In contrast, linear birefringence adds in the return direction to that experienced by the probe in the forward direction. Finally, Faraday rotation (a rotation of the polarisation in response to a magnetic field parallel to the fibre axis) introduces a non-reciprocal circular birefringence, i.e. a rotation that is cumulative for forward and backward light propagation.

10.4.2 POTDR in High-Birefringence Fibres

The innate complexity of interpreting POTDR measurements is alleviated if the measurement is carried out in a fibre that has high birefringence, whether that birefringence is linear, circular or even elliptical. Fibres with high linear birefringence are commonly used to transfer light in a known state of polarisation (see Chapter 2). Circular birefringence is achieved by imparting a twist to a fibre having otherwise low birefringence [43]; its use is primarily in fibre-optic electrical current sensors. Strong elliptical birefringence is achieved by spinning the preform of a linearly birefringent fibre during the drawing process [44], which induces a fine pitch rotation of the axis of the strong linear birefringence, leading to a fibre that preserves a particular elliptical polarisation state.

In any event, launching equal power into two orthogonal modes and mixing the backscatter power returning in the two orthogonal states leads to a beat frequency that is directly proportional to the birefringence, i.e. the difference in propagation constant between the modes. In the case of a fibre with strong linear birefringence, the usual expression for the backscatter power as a function of distance acquires an additional modulation term as a function of distance z given [45] by $\cos(2\Delta k \cdot z)$, where Δk is the difference in the propagation constants of the two modes. In this case, the impact of other types of birefringence is strongly reduced and the result is a single piece of information, namely the distribution along the fibre of the dominant type of birefringence.

The fibre can be designed to respond in its birefringence to certain measurands. For example, a polarisation-maintaining (PM) fibre based on thermally induced stress birefringence exhibits a strong temperature dependence of birefringence, although some annealing effects can be observed above a few hundred °C [46]. Likewise, side-hole fibres exhibit a dependence of their birefringence on isostatic pressure [47]. Axial strain also modulates the birefringence in stress-birefringent fibres [48].

For fibres with strong linear birefringence, the backscatter signal observed in the orientation parallel (||) and perpendicular (⊥) to the probe light polarisation are given by Ref. [45]

$$P_{BS\|}(z) = \Big[(P_{inx}\cos^2\gamma + P_{iny}\sin^2\gamma) + \sqrt{P_{inx}P_{iny}} \cdot \sin(2\gamma)\left\{\frac{\sin(\Delta k \cdot v_g\tau_p/2)}{\Delta k \cdot v_g\tau_p/2}\right\} \cdot \cos(2\Delta k \cdot z)\Big] \cdot \eta \cdot \tau_p \cdot \exp(-2\alpha z) \tag{10.2}$$

and

$$P_{BS\perp}(z) = \Big[(P_{inx}\cos^2\gamma + P_{iny}\sin^2\gamma) - \sqrt{P_{inx}P_{iny}} \cdot \sin(2\gamma)\left\{\frac{\sin(\Delta k \cdot v_g\tau_p/2)}{\Delta k \cdot v_g\tau_p/2}\right\} \cdot \cos(2\Delta k \cdot z)\Big] \eta \cdot \tau_p \cdot \exp(-2\alpha z). \tag{10.3}$$

Here, γ is the angle between the principal birefringence axes of the fibre and the orientation of the electric field of the probe. If the polarisation of the backscatter were not analysed, the power measured would simply be the sum of Equations 10.2 and 10.3, i.e.

$$P_{BStotal}(z) = [P_{inx}\cos^2\gamma + P_{iny}\sin^2\gamma] \cdot \eta \cdot \tau_p \cdot \exp(-2\alpha z), \tag{10.4}$$

which of course is the case of conventional OTDR. In Equation 10.4, small differences between the capture fractions of the two polarisation modes have been neglected.

Equations 10.2 and 10.3 demonstrate that analysing the polarisation of the backscatter imparts an additional modulation term $\sin(2\gamma)\left\{\frac{\sin(\Delta k \cdot v_g\tau_p/2)}{\Delta k \cdot v_g\tau_p/2}\right\} \cdot \cos(2\Delta k \cdot z)$ for rectangular probe pulses, that has maximum effect for equal power launched into the two polarisation modes, i.e. $\sin(2\gamma) = 1$, e.g. $\gamma = \pi/4$, and that is fully resolved only if $W \ll \Delta k \cdot v_g/2$. The modulation frequency term $\cos(2\Delta k \cdot z)$ carries the birefringence information.

In terms of the observed signal in the time domain, this translates to a beat frequency of

$$f_{beat} = \Delta k \cdot v_g. \tag{10.5}$$

For a beat length of 5 mm (a common value for PM fibres operated at 1550 nm), the resulting beat frequency is of order 40 GHz. Not only is this difficult to detect, but there is also an accompanying requirement on the spatial resolution of the system, namely $\tau_p \ll \Delta k \cdot v_g/2$, which in this case implies a pulse duration $\tau_p \leq 12.5$ ps. This is challenging in terms of the probe energy combined with the extremely wide detection bandwidth that is required.

It should also be appreciated that the fact that the information is conveyed as a frequency limits the product of measurand resolution and spatial resolution: within a certain distance (gauge length) along the fibre over which the beat frequency is estimated, the POTDR signal will contain a certain number of beats – the frequency estimation will be limited by the number of periods within the distance under consideration, so as we increase (degrade) the spatial resolution of the measurement by increasing the gauge length, the ability to estimate the birefringence improves.

Photon counting has been applied to this problem in order to improve the spatial resolution. This allowed the beat length on fibres having beat lengths as low as 60 cm (i.e. having moderate birefringence) to be measured [49].

Two other approaches have been followed to improve the spatial resolution in POTDR. The first approach is to move to the frequency domain, i.e. to polarisation optical frequency-domain reflectometry (POFDR) (see Section 10.4.3). The second approach involves carrying out the measurement at distinct wavelengths and mixing the results.

10.4.3 Polarisation Optical Frequency-Domain Reflectometry

Frequency-domain reflectometry involves operating the source continuously and sweeping its frequency. As discussed in Chapter 3, the frequency range dictates the spatial resolution, and so, for example, a 10 GHz sweep range will result in a spatial resolution of 1 cm. A relatively early paper in POFDR demonstrates clearly that beat lengths of about 3 cm can be resolved with this technique [50].

Provided that the sweep is linear, the processing to convert the results into a distance mapping is straightforward using a Fourier transform. Most commonly, the coherent detection approach has been used, where the present output of the laser is mixed with the backscatter return.

10.4.4 Dual-Wavelength Beat Length Measurements

In order to overcome the demanding requirements on spatial resolution, a differential technique, illustrated in Figure 10.6, has been devised based on using two distinct wavelengths to probe the fibre simultaneously [51]. The detector collects beat signals at high frequency imposed by the beat between the modes; however, this frequency is inversely proportional* to the probe wavelength, and so the beat frequencies further mix on the optical detector to provide an electrical signal at a down-shifted frequency. In this case, the detector provides the optical-electrical conversion, but it also operates as a microwave mixing diode to provide an output corresponding to the difference-frequency between the beat frequencies for the two wavelengths.

The difference-frequency can be controlled by choice of the difference between the wavelengths of the probes.

This approach, however, solves only the issue of detecting very high beat frequencies; the requirement on the pulse duration for each probe pulse that $\tau_p \ll \Delta k \cdot v_g/2$ still holds. Nonetheless, Parvaneh et al. [51] demonstrated that the beat between the POTDR signals at the two probe frequencies exists. Further work by the same team investigated the use of non-linear effects to address this problem, as discussed in Section 10.7.2.

* Neglecting the differential dispersion between core and cladding materials.

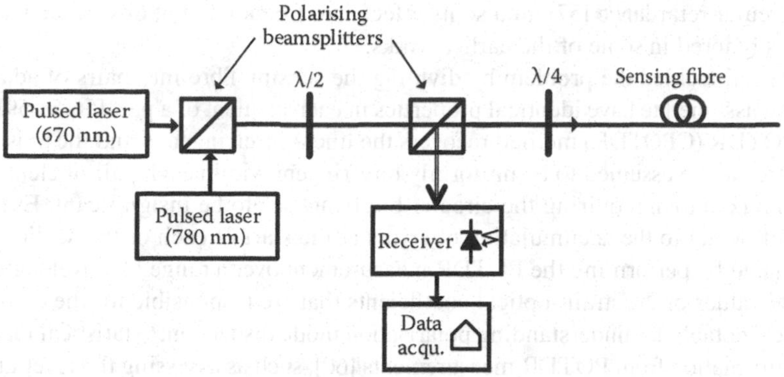

FIGURE 10.6 Schematic diagram of the dual wavelength frequency-derived POTDR experiment. (Adapted from Parvaneh, F., Handerek, V.A., Rogers, A.J., *Electron. Lett.*, 27, 394–396, 1991.)

10.4.5 Full Polarimetric Analysis of POTDR Signals

The approaches discussed so far deal primarily with the simple case of a dominant birefringence, usually linear birefringence. However, the promise (and challenge) of POTDR is to provide a full reconstruction, at any point, of the state of polarisation of the light travelling in the fibre.

The first issue is that a number of input and output state-of-polarisation (SOP) combinations must be recorded. As explained in Ref. [52], the fact that the backscattered light is not fully polarised requires a more complex treatment than Jones calculus and requires more polarisation states to be acquired. The SOP of the backscatter is measured by acquiring two orthogonal states each in a linear state and with a quarter-wave (π/2) phase retardation. Combining these four measurements allows the Stokes parameters to be assessed. In turn, this allows the linear retardation of the fibre and the orientation of the birefringence axis to be estimated. In Ross [52], the angle of the birefringence axis was tracked, showing a full rotation roughly every 10 m, and a beat length of similar order for the particular fibre sample that was tested. Ross achieved his measurement of the Stokes vector using a spinning polariser that was synchronised to the probe pulses. Phase-sensitive amplifiers were then used to capture the in-phase and quadrature components. He also used a phase modulator to flip the detected state by a quarter wave and complete the measurement set.

Ferdinand [53] extended the set-up of Ross [52]. Firstly, his arrangement acquired the Stokes parameters in parallel using separate channels each fitted with a suitable combination of polarising beam-splitters and wave-plates. This avoided the spinning polariser but above all improved the speed of the acquisition, which reduced the degree to which the state of polarisation could change whilst it was analysed. Secondly, Ferdinand used a switched quarter-wave retarder at the input to launch alternately linearly and circularly polarised probes. This is necessary to avoid conditions where the inversion of the Stokes vector is poorly conditioned.

More recently, fully parallel acquisition of the four elements of the Stokes vector has been published [54]; this approach speeds up the measurement by eliminating moving parts and by making the measurements simultaneously and it eliminates errors caused by changes in the measurand during the acquisition. In addition, position-dependent pointing errors in the optics can lead to further errors [55].

The reader is referred to Refs. [52] and [55] for the detailed analysis of POTDR data. However, even the works of Ross and Ferdinand are not sufficient to recover the full birefringence properties of the sensing fibre because circular birefringence (with the exception of Faraday rotation) is cancelled in a two-way transit through any section that is analysed. The problem is that this issue precludes the recovery of even the linear birefringence and its orientation when circular birefringence (e.g. caused by twists) is present [56]. Twist-induced birefringence is a relatively small effect because about 12 turns of twist are required to induce a 2π circular retardance [57], and so its effects are frequently not observed. Therefore, circular birefringence was ignored in some of the earlier works.

Rogers, however, tackled the problem by dividing the sensing fibre into pairs of adjacent small sections that could be assumed to have identical properties in each element of a pair [56,58,59]. This so-called computational POTDR (CPOTDR) method recovers the linear birefringence and the axis orientation provided that the fibre can be assumed to be uniformly birefringent within each pair of elements, a condition that is far more relaxed than requiring the circular birefringence to be insignificant. Even the CPOTDR technique can fail owing to the accumulation of errors at the start of each of the section pairs, although this can be alleviated by performing the POTDR measurement over a range of wavelengths and using the known dispersive values of the strain-optical coefficients that are responsible for the circular retardance.

In some cases, notably for understanding polarisation mode dispersion,* statistical methods have been used to extract information from POTDR measurements [60], such as assessing the level-crossing rate [61], which is adequate for estimating the transmission limitations of an optical link due to birefringence effects.

10.4.6 Alternative Approaches to Estimating the Distribution of the State of Polarisation

The foregoing is based on Rayleigh backscatter, and this is one of the most effective means of measuring the axial distribution of the SOP. However, some of the other measurement techniques discussed elsewhere in this book are also polarisation-sensitive. The Raman gain is one of these (Section 10.7.1). Brillouin gain, measured for example with Brillouin optical time-domain analysis (BOTDA, see Chapter 5), is also sensitive to the relative SOP of the counter-propagating beams; this sensitivity, which is a nuisance in static temperature or strain measurements, can be exploited as a means of inferring the distribution of birefringence along the fibre.

10.4.7 Applications of POTDR

10.4.7.1 Distributed pressure sensing

A demonstration of the use of POTDR for pressure sensing was provided in Ref. [62], where a side-hole fibre [47] was used as the pressure-sensing element. A side-hole fibre, as its name implies, includes two holes, one on either side of the core (the core of these fibres is usually designed for single-mode operation). As a result, isostatic pressure applied to the external surface is transmitted asymmetrically to the core region and so causes stress birefringence that is proportional to the differential pressure between the outside surface and the inner hollow sections.

Some side-hole fibres exhibit substantial intrinsic birefringence (i.e. that which exists in the absence of applied external pressure). If this is the case throughout the pressure range of interest, then an arrangement such as that of Figure 10.4 can be used and the spatial period of the quasi-sinusoidal response provides a direct measure of the differential pressure between side holes and the external surface. In the case of the work reported in Ref. [62], however, the natural birefringence of the side-hole fibre was quite low (beat length of about 13 m), and in fact, it was cancelled by the pressure-induced birefringence partway through the measurand range. A more complex, full polarimetric inversion of the POTDR data was therefore required including the need to evaluate the sign of the birefringence. This report is the only one, to the author's knowledge, to show pressure distribution measurements with POTDR.

Unfortunately, the results are quite limited: based on the scatter of the data shown, the resolution is of order 5% full scale over a range of 50 bar. In contrast, a pressure sensor for downhole use typically

* Pulse broadening caused by birefringence in the fibre and related not only to the birefringence but also the coupling of power between polarisation states.

provides a resolution of order 1 mbar over a dynamic range of approximately 10^6. This application also requires the sensor to drift by less than say 10 mbar per year at temperatures up to 250°C. The packaging issues that this mission profile implies have not been addressed.

The technology might, however, be suitable for coarser measurements at low pressure, such as determining the level of the water table in hydrological studies.

The interpretation of POTDR data involves, explicitly or implicitly, a differentiation step that degrades the resolution of the measurement. Moreover, the calculation of the beat length requires the estimate of the number of beat periods within a spatial resolution, and so there is a direct trade-off between the spatial resolution and measurand resolution, even ignoring how the signal-to-noise ratio varies with spatial resolution.

10.4.7.2 Current/magnetic fields

The Faraday effect introduces a non-reciprocal circular birefringence; i.e. it delays one circularly polarised mode relative to the other; its effect is the same in both directions so that it is doubled in a two-pass arrangement such as POTDR. Equivalently, the Faraday effect causes the orientation of light in a linear polarisation to be rotated by an angle proportional to the magnetic field component parallel to the fibre axis. In a current sensor, the magnetic field H is induced by the current in a conducting coil wrapped around the fibre, giving an angle of rotation ϕ_F for a single pass of Ref. [63]

$$\phi_F = V_F \oint H \cdot dl = V_F \cdot I_{coil} \cdot N_t, \tag{10.6}$$

where V_F, I_{coil} and N_t are, respectively, the Verdet constant of the material making up the core of the fibre, the current flowing in the solenoid through which the fibre is threaded and the number of turns of the solenoid (equally, the fibre can be wrapped around the conductor resulting in the same expression).

The detection of a magnetic field induced by a current driven through a solenoid was reported in Ref. [64] using the arrangement of Figure 10.4 and then in Ref. [65] using a single receiver observing just the orthogonal state of polarisation to that of the probe.

In the case of Ref. [64], the measurement was carried out at a wavelength of 514.5 nm using a mode-locked, cavity-dumped argon ion laser providing 200-ps pulses; the backscatter was captured with photomultiplier tubes that provided a time resolution of 350 ps. For simplicity, the author ensured that the state of polarisation was linear where the fibre entered the test magnetic field by adjusting the SOP at the launching point to compensate the birefringence of the fibre leading up to the coil. The Faraday rotation was measured by forming the ratio $(V_1 - V_2)/(V_1 + V_2)$, where V_1 and V_2 are the detector outputs (a normalisation procedure that is commonly used in optical current sensors). A rotation resolution of 0.25° (corresponding to about 1 mT) was achieved at an update rate of 1 s and a spatial resolution of 0.5 m. Interestingly, the author noted a lower sensitivity than he had calculated. He attributed this to the possible presence of cladding modes. An alternative explanation is that the backscattered light is not fully polarised as we have noted elsewhere in this book.

Following these early reports, little was published on the subject of magnetic field sensing with POTDR until recently.

Polarisation reflectometry, in the form of POFDR, has recently been applied to the measurement of intense magnetic fields within magnetic resonance imaging (MRI) scanners [66]. In this case, a conventional coherent optical frequency-domain reflectometry (OFDR) system was adapted by adding polarisation controllers on the input to the sensing fibre and a polarisation analyser in the return path, thus allowing a sufficient number of polarisation states to be explored. With the combination of a spatial resolution of 3 cm and special fibre layouts allowing the magnetic field to be explored in selected directions, the strength and direction of the field could be determined. The cavity of an MRI scanner is a challenging measurement for all but optical techniques owing to the very strong fields that induce currents and forces in any electrical conductors. Any conductive material in the cavity would, in addition, distort the field to be measured. The polarisation reflectometry approach, with high spatial resolution, has overcome these difficulties. Nonetheless, care still must be applied to the design of the fibre layout and the interpretation of the results.

10.4.7.3 Polarisation mode dispersion

One of the main applications of POTDR/POFDR has been in investigating the birefringence properties of telecommunications fibres owing to the fact that at very high bit rates, the difference in propagation time between orthogonally polarised modes (and coupling of power between polarisation modes) can cause pulse-broadening, an effect known as polarisation mode dispersion. A large body of work on this application polarisation reflectometry was developed from about 1995 onwards [50,54,55,60,61,67–78]. Although this is outside the main topic of this book, the needs of the telecommunications industry have led to a deeper understanding of a technique that is still being refined and enhanced. POTDR/POFDR will find at least niche applications in distributed sensors and possibly some mainstream ones.

10.4.7.4 Vibration sensing

The polarisation of the light travelling in optical cables is influenced by external disturbances and the use of POTDR in distributed sensing has been demonstrated in the laboratory and in field trials. Initially, the polarisation information was seen as a means of supplementing intensity based coherent OTDR.* This was demonstrated in Ref. [79], where it was shown that the information collected in such a coherent OTDR vibration sensor is polarisation dependent, and so using two orthogonal polarisations provides a degree of diversity. Somewhat earlier, the use of POTDR was discussed in the thesis of Seo [80] as an alternative to intensity-based coherent OTDR.

This use of polarisation information for vibration sensing was taken a step further in Ref. [81], where two orthogonal polarisations states of the Rayleigh backscattered light were detected on independent receivers and subtracted in analogue electronics to provide a differential signal. In this work, most of the results were obtained in the forward direction, but preliminary data using Rayleigh backscattering were shown. In particular, it was demonstrated that a vibration signal was localised to 10 m or less, using 100-ns probe pulses. In distributed sensing, demonstrating immunity from cross-talk between locations along the fibre is a key qualifier and this was accomplished in the work cited.

A study of a bidirectional POTDR demonstrated correlation in the inferred disturbance between measurements taken from End 1 and End 2 of a single sensing fibre [82]. This helps clarify the observed disturbance and possibly reject spurious signals.

It is worth considering the relative merits of POTDR and intensity-based coherent OTDR in the distributed vibration application. POTDR provides a signal that is sinusoidal in the strain applied to the fibre, and this is very helpful in interpreting the results and classifying threats, whereas we have seen that the coherent OTDR response is a non-linear function of strain. Both techniques suffer from intensity fading, but it is the fading of the derivative of the intensity with respect to measurand (strain) that the OTDR approach suffers from and that POTDR avoids.

The two techniques measure fundamentally different quantities: coherent OTDR responds to axial strain, whereas POTDR responds to differential strain between the two axes. This makes POTDR insensitive to axial strain† (that is common to both axes) but more sensitive to differential effects such as side-loading. As a result, the POTDR technique measures a different component of the strain field than does coherent OTDR. This can be helpful when multi-axial measurements are required or to disambiguate coherent OTDR measurements. However, the sensitivity of POTDR will, in general, be substantially lower than OTDR, and so this route is unlikely to provide a complete route to three-component vibration

* Discussed in Chapter 6.

† A major exception is found in stress-birefringent PM fibres, where the stress-applying sectors exhibit a different Poisson's ratio from the remainder of the fibre cross-section [48]. Even in this case, the sensitivity to strain of the polarisation is two orders of magnitude lower than the phase change of either mode due to elongation.

sensing. In the case of disturbance sensing, various coupling mechanisms within the cable could contribute to the fibre being sensitive to both axial strain and side loading.

It should be noted that seismic waves generated by a surface disturbance appear in the form of a number of types of wave that travel near the surface, such as Love and Rayleigh waves in addition to compressional waves or shear waves. These near-surface waves have directional components that are likely to result in a polarisation response as well as a phase response in a sensing fibre buried in a shallow trench.

A drawback of the polarisation approach is that the sensitivity will depend on the relative alignment of the mechanical distortion and the light travelling in the fibre: in the worst case, the polarimetric response could be zero. This problem could be alleviated by probing at different input polarisation states (for example at 0° and 45° from some reference orientation).

10.5 BLACKBODY RADIATION

A pyrometer is a temperature sensor that operates by measuring the blackbody radiation that is emitted by all objects; its intensity and spectrum depend on the temperature of the object that is sensed.

Gottlieb and Brandt [83] proposed using an optical fibre as a distributed pyrometer. In contrast to a conventional pyrometer with a target usually selected to be black (and therefore having high emissivity), here, it was proposed to use the fibre itself as a blackbody cavity to capture blackbody emission from within the core and re-captured similarly to the process for OTDR, without, in this case, using a probe pulse.

Optical fibres are designed for very high transparency, and it follows that their emissivity is extremely low. The light emission varies strongly with temperature partly because of the well-known blackbody radiation temperature dependence and also because the emissivity changes with temperature. It follows that the fibre used as a blackbody radiation sensor measures primarily the temperature of the hottest location along the fibre.

The emissivity for a hot-spot is given by $\varepsilon_{bb} = 1 - \exp(-\alpha l)$, where α is the loss per unit length in the hot-spot and is expressed in natural logarithmic units, which approximates to $\varepsilon_{bb} \approx \alpha l$ for short hot-spots. The emissivity can be tailored by adding absorbing dopants, if desired. The analysis in Ref. [83] shows that this type of sensor is best suited to short sensing lengths, up to a few m. Experiments showed that temperatures as low as 125°C could be measured in lengths of some 3 m. This technique was intended to be used within high-temperature power transmission infrastructure (e.g. generator stator bars), where the distances are modest and the electro-magnetic environment precludes the use of electrical sensors. There have been no reports of commercial deployment of this technology probably as a result of other techniques, such as fibre Bragg grating arrays or Raman DTS, becoming available in the years since this ground-breaking work.

10.6 FORWARD-SCATTERING METHODS

Most of the sensors described in other sections of this book rely on reflectometry; i.e. they use the time-of-flight, measured from launching a probe to the time the modulated signal returns to the input end, to relate the measured value to location along the fibre. In contrast, forward-scattering methods rely on coupling probe power at certain points into a separate channel that propagates the light at a different velocity. By comparing the arrival times between the signal in the original channel and that in the coupled channel, a measure is obtained of where the coupling has occurred.

10.6.1 Mode Coupling In Dual-Mode Fibres

The first example of forward-scattering methods exploits the difference in the propagation constants (or equivalently of the light velocity characteristic of the mode) to locate the point at which power is coupled from one mode to another. In this specific example, the two modes in question are orthogonally polarised modes of a high-birefringence fibre that is designed to exhibit a large difference in the effective indices of the two polarisation states [84]. In this approach, power is launched into one mode of the fibre. In the absence of coupling, the power remains in that mode. However, if at some point coupling occurs between the modes, some power is coupled into the second mode, and from that point, it travels to the far end in the second mode at the velocity characteristic of the second mode rather than that of the first mode. The location of the coupling is obtained from the relative timing of the probe light and the coupled light: the difference in the time of the arrival of the coupled light from the remainder of the probe is proportional to the distance between the coupling point and the remote end of the fibre. The time-to-distance calibration is simply related to the difference in propagation speeds for the two modes.

The difficulty with this approach is, of course, that a much finer time resolution is required than in reflectometric approaches, where a distance of 1 m is roughly equivalent to a delay of 10 ns. In the forward-coupling method, assuming a relative difference in velocity of 3×10^{-4} (a typical value for PM fibre), the time difference between the two modes would be 5 ns $\times 3 \times 10^{-4}$, i.e. 1.5 ps/m. Equivalently, if we wanted to resolve a coupling point to within 1 m, the distance resolution for the two optical signals at a particular time would be 0.3 mm. Fortunately, in the forward direction, the power levels are substantial and interferometric techniques can be used. Specifically, low-coherence interferometry was used in Ref. [84] to measure the times of arrival. Figure 10.7 shows schematically the arrangement described in Ref. [84]. A low-coherence source, such as a super-luminescent diode, is launched into one of the polarisation modes of the sensing fibre (the polarisation controller is used to ensure that only one mode is excited). At the far end of the fibre, the light is processed by a compensating interferometer (in the box delineated by a broken line) that delays the light from one polarisation mode relative to the other using the motion of the mirror controlled by a stepper motor. A broadband source has a short coherence length, and a signal (at the frequency of the fine modulation created by the piezo-electric transducer) will be observed only when the path imbalance of the interferometer matches that of two signals that are delayed with respect to each other. Therefore, by scanning the position of the stepper motor, all possible locations of coupling points are identified. In the original paper, coherence length of the source was 50 μm, and for a beat length of 1 mm, this provided a spatial resolution of order 15 cm. Subsequent papers on low-coherence interferometry have shown that the resolution of the coherence peak can be much better than

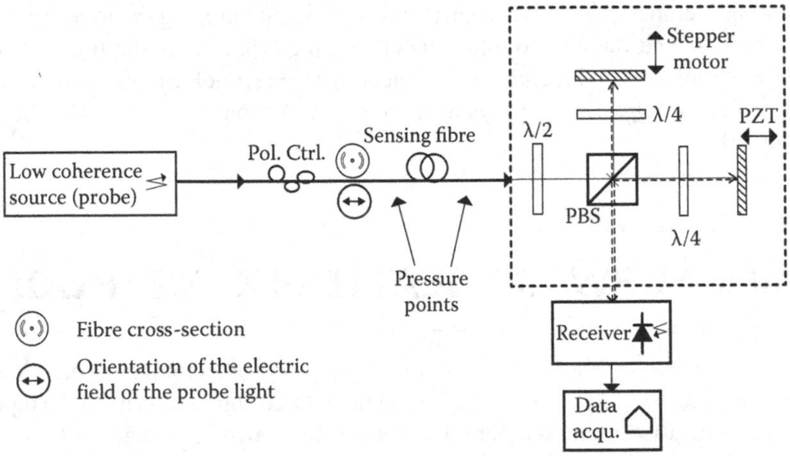

FIGURE 10.7 Optical arrangement for measuring the location of a coupling point in forward-scatter method. (Adapted from Gusmeroli, V., Martinelli, M., Vavassori, P., *Opt. Lett.*, 14, 1330–1332, 1989.)

the coherence length of the source (a resolution of a few tens of nm was demonstrated in Ref. [85]), and so it is quite likely that the pioneering results of Ref. [84] could be improved upon.

The detection of mode coupling in the forward direction has therefore demonstrated the potential for extremely good spatial resolution and a relative simple means of detecting mechanical distortion.

The main limitation of the technique is the difficulty of imparting precisely controlled degrees of cross-coupling in response to particular measurands. In addition, the technique can only cope with relatively few coupling points, each converting small levels of light from one mode to the other. If substantial power is coupled to the second mode, then at additional coupling points along the fibre, the same power could be coupled back from the second mode to the first, leading to a more complex signal to be interpreted.

10.6.2 Modalmetric Sensors

A multimode optical fibre, when illuminated in more than one mode by a coherent source, exhibits at its output a number of spots, whether viewed in the near-field or the far-field. These spots are the result of the interference of the electric field of the several modes that arrive at the far end. In the case of the near-field, the light arriving at each point on the end face is the summation of the electric field at that position from all the modes. Their relative phase depends on the difference in their propagation constants, cumulated over the entire length of the fibre. This is a specific manifestation of the more general phenomenon of laser speckle [86]. If the fibre is distorted in any way (bent, for example), the relative phase between modes is altered, and this can be detected from a change in the speckle pattern at the far end, although the total power carried by the fibre is not changed. By selecting a subset of the speckle pattern with a spatial filter (ideally a single speckle, i.e. a spot having a uniform phase across its surface), its movement can be detected. One approach to the spatial filter is to splice a single-mode fibre to the output of the sensing fibre (the single-mode fibre selects a single region of consistent phase from the complicated near-field present at the output of the sensing fibre) and also to launch a controlled modal power distribution, also by means of a single-mode-to-multimode splice. This is illustrated in Figure 10.8 (adapted from Ref. [87]).

Any disturbance within the multimode sensing fibre alters the near-field distribution (when viewed with visible light, the spots appear to move when the fibre is disturbed even slightly) that is converted to a variation of the power coupled to the receiving single-mode fibre.

This arrangement allows disturbances to be detected, but not their location.

A first step in locating the disturbance is to measure the sensing fibre loop from both ends using a common source but separate detectors, as illustrated in Figure 10.9, based on Ref. [88]. In this arrangement, counter-propagating probes are launched from each end of the sensing loop and detected by a pair

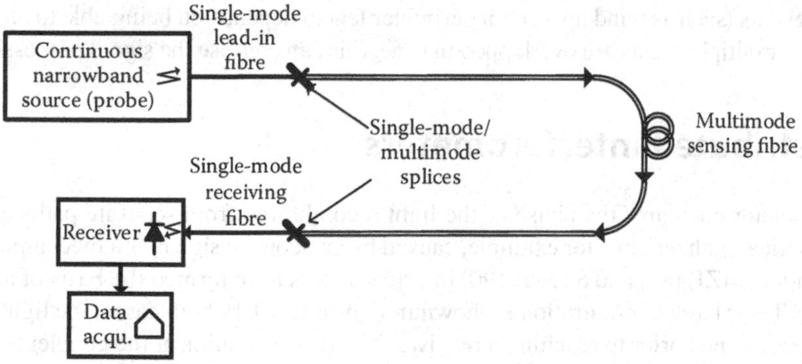

FIGURE 10.8 Modalmetric disturbance sensor. (Adapted from Anderson, D.R., Detecting flat wheels with a fiber-optic sensor. In ASME/IEEE 2006 Joint Rail Conference, American Society of Mechanical Engineers, 25–30, 2006.)

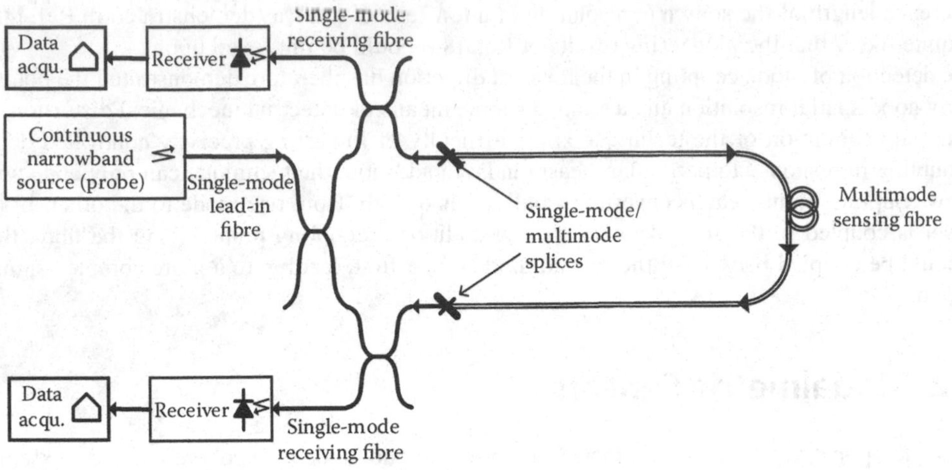

FIGURE 10.9 Modalmetric event locating arrangement. (Adapted from Tapanes, E.E., Goode, J.R., Katsifolis, J., Apparatus and method for monitoring a structure using a counter-propagating signal method for locating events, US6778717 B2, 2004.)

of receivers. Assuming that the event is detected by the signal processors connected to both receivers, the time difference between the detection of the event at the two ends is a direct indication of its location.

The signal is non-linear and non-monotonic, but it can be processed to provide useful information about the disturbance in the same way that intensity-based coherent OTDR can be used for event detection and classification. Its benefit, compared with coherent OTDR, is the relative simplicity of the interrogation system. However, the interpretation of the data is at least as complex as that required for direct-detection coherent OTDR.

This approach is used in a number of distributed sensors that detect movement at any location for intrusion detection, e.g. in perimeter security. Although the techniques used are not always disclosed, it can be inferred from the patent literature that at least three companies, namely Fiber SenSys, Future Fiber Technology and Optellios, use or have used the modalmetric approach in various security applications. They have also been used for preventative maintenance. In one example, a modalmetric sensor laid along a railway line was used to detect the existence of flat wheels that result, for example, from wear during hard braking. A spectral analysis of the sensor output allowed the presence of this problem to be detected [87].

One limitation of the modalmetric approach is that it can be saturated if many effects, whether benign or real threats, occur simultaneously. The ability to classify events and separate serious threats from naturally occurring events (such as wind noise on a perimeter fence) depends on being able to analyse and interpret the signals. If multiple events are overlapped in time, this can confuse the signal processing algorithms.

10.6.3 Distributed Interferometers

A fibre interferometer compares the phase of the light recombining from separate paths and responds to changes in the optical path lengths, for example, caused by an acoustic signal or a mechanical disturbance.

Mach-Zehnder (MZI) [89] and Sagnac [90] interferometers have formed the basis of acoustic sensors for many years.* Their basic configuration is shown in Figure 10.10. In both cases, the light is divided into two paths and recombined prior to reaching a receiver. The recombination at the coupler selects a particular phase relationship between the optical signals. In the case of the MZI, the interferometer responds to a

* MZIs have already been discussed in Chapters 5 and 6 in connection with their wavelength-selective properties; it is their ability to provide phase comparison that is of interest in this section.

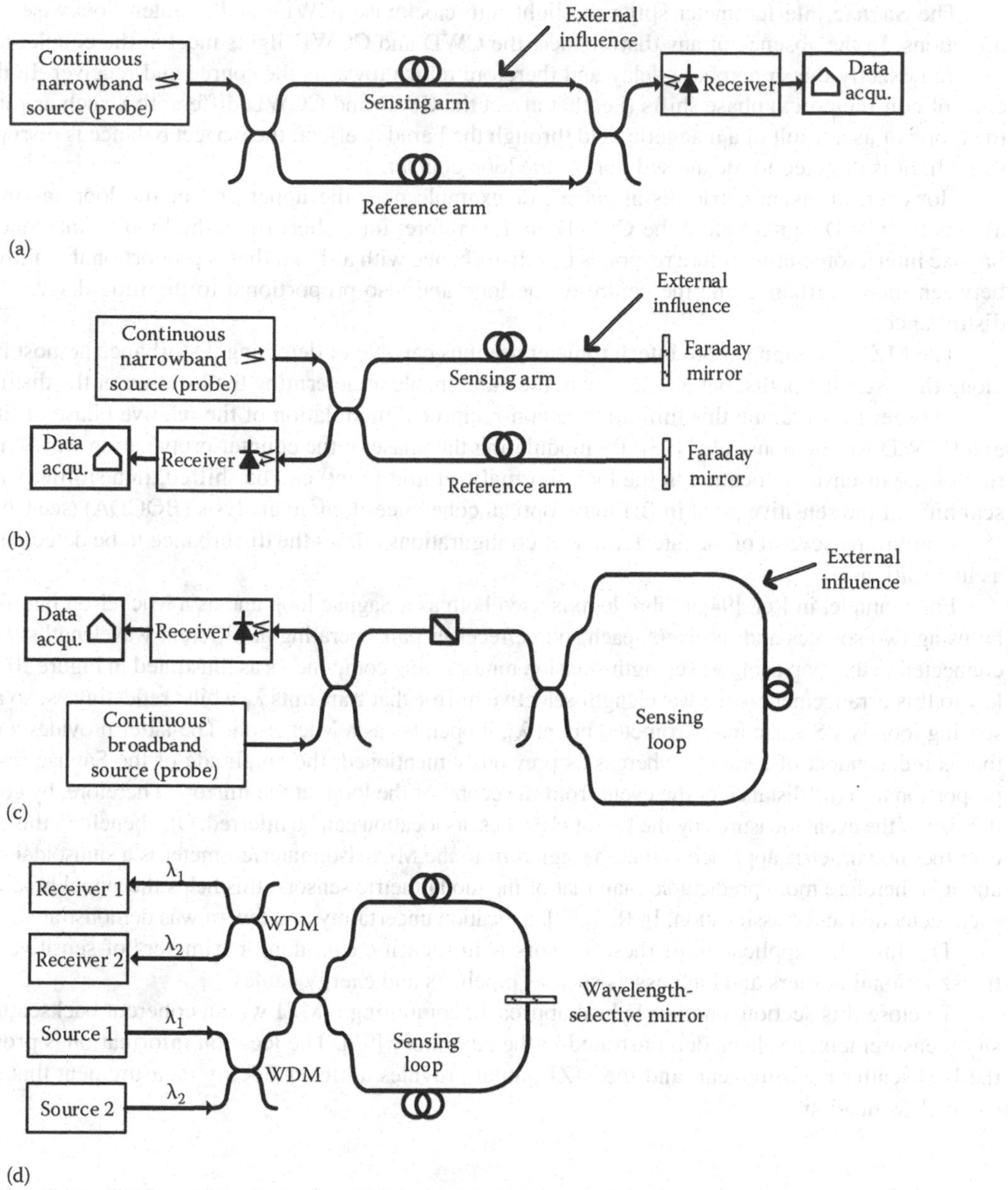

FIGURE 10.10 Interferometer designs used in forward-scattering distributed sensing. (a) Basic Mach-Zehnder fibre interferometer, (b) Basic Michelson fibre interferometer, (c) Basic Sagnac interferometer and (d) Combined Sagnac/Michelson interferometer. (Panel d adapted from Spammer, S.J., Swart, P.L., Chtcherbakov, A.A., *J. Lightw. Technol.*, 15, 972–976, 1997.)

differential change in path length between the two arms. One of these (the reference arm) is often shielded from the measurand, and the other (sensing) arm interacts with the external field. This approach is well established in high-performance point sensors and multiplexed arrays [91]. A Michelson interferometer is similar to the MZI but converted to a reflective configuration, so that the receiver can be sited next to the source. A further benefit of the Michelson configuration is that the use of Faraday mirrors eliminates [92] the problem, that otherwise exists, of needing to ensure that the polarisation states of the two combining optical signals are the same.

The Sagnac interferometer splits the light into clockwise (CWD) and counter-clockwise (CCWD) directions. In the absence of any disturbance, the CWD and CCWD lights meet at the coupler and have suffered exactly the same phase delay and therefore return towards the source and receiver. In the presence of non-reciprocal phase shifts (i.e. that affect the CWD and CCWD differently), such as rotation of the loop* or as a result of a magnetic field through the Faraday effect, the perfect balance is disrupted and some light is directed to the unused port of the loop coupler.†

However, an asymmetric disturbance (for example near the upper end of the loop, as indicated) affects the CWD signal before the CCWD, and therefore, for a short time, the loop is unbalanced. The Sagnac interferometer therefore responds to a disturbance with a signal that is proportional to the distance between the disturbance and the centre of the loop and also proportional to the time-derivative of the disturbance.

The MZI and Sagnac fibre interferometer are thus capable of detecting disturbance at most locations along their sensing paths, but on their own, they are unable to determine the location of the disturbance.

In order to overcome this limitation, a non-reciprocal modulation of the relative phase of the CWD and CCWD was demonstrated [93]. By modulating the phase of the counter-propagating waves, the position of the insensitive location in the loop (normally at mid-point) can be shifted, in a similar way to the scanning of the sensitive point in Brillouin optical coherence-domain analysis (BOCDA) (see Chapter 5).

Combining several of the interferometer configurations allows the disturbance to be detected, as well as its location.

For example, in Ref. [94], a fibre loop is used both as a Sagnac loop and as a Michelson interferometer by using two sources and receivers, each source/receiver pair operating at different wavelengths, λ_1 and λ_2, connected to the loop using wavelength-division multiplexing components as illustrated in Figure 10.10d. The key to this arrangement is the wavelength-selective mirror that transmits λ_1 whilst reflecting λ_2, so at λ_1, the sensing loop is a Sagnac interferometer, but at λ_2, it operates as a Michelson. The latter provides a response that is independent of location, whereas as previously mentioned, the amplitude of the Sagnac response is proportional to the distance of the event from the centre of the loop (at the mirror). Therefore, by comparing the size of the event measured by the two approaches, its location can be inferred. One benefit of this approach over the modalmetric approach is that the signal from the Michelson interferometer is a sinusoidal response, and it is therefore more predictable than that of the modalmetric sensors; this helps the signal processing for event detection and classification. In Ref. [94], a location uncertainty of a few m was demonstrated.

The intended application of these sensors is in security, e.g. along perimeters of sensitive installations, national borders and long assets such as pipelines and energy cables.

To close this section, another hybrid approach, combining a MZI with a coherent backscatter intensity measurement, has been demonstrated in the laboratory [95]. The location information is provided by the backscatter measurement, and the MZI output provides a higher fidelity measurement that supports event classification.

10.7 NON-LINEAR OPTICAL METHODS

Non-linear optical techniques involve either single probes at a power sufficient to exceed the threshold for one of the major non-linear effects or, more commonly, the interaction of probes in a way that responds to a measurand. Stimulated Brillouin scattering, discussed in Chapter 5, is a prime example of non-linear interactions being used in distributed sensing. In this section, we review other non-linear techniques for distributed sensing, excluding stimulated Brillouin scattering, for which the reader is referred to in

* The primary use of the fibre Sagnac interferometer is as a rotation sensor, in inertial navigation.

† In practice, the phase of the light is modulated differentially in the loop in order to bias the sensor – without this biasing modulation, the sensitivity to small signals is zero.

Chapter 5. We also exclude from this chapter the use of non-linear effects for enhancing the range of distributed sensors based on other transduction mechanisms – these have been addressed in the corresponding chapters, notably Chapters 4 and 5.

10.7.1 Stimulated Raman

Stimulated Raman interaction between counter-propagating pump and probe beams was demonstrated by Farries and Rogers [96]. The Raman gain is polarisation sensitive, and so this process is a means of sampling the polarisation state of the probe along the fibre. In that sense, it is similar in its output to POTDR.

The angular dependence of the Raman gain reduces, for interactions between orthogonal polarisation states, to 7% of the peak value achieved with parallel states. Its angular discrimination is therefore slightly worse than that caused by depolarisation of Rayleigh scattering (in the classic POTDR arrangement).

In Ref. [96], the specific experimental arrangement is similar to that for BOTDA (Chapter 5), with a continuous-wave probe at 633 nm propagating in one direction and measured after its forward pass through the fibre, interacting with a pump pulse (in this case up-shifted by one Raman shift to 617 nm) travelling in the opposite direction. The authors demonstrated the angular dependence of Raman gain in this configuration. They also showed some modulation of the time-dependent detected probe when some stress was applied at a particular point along the fibre, although no attempt was made at quantifying the measurand in this case. An analysis of the polarisation optics of this arrangement was provided by Valis et al. [97] in a paper aimed primarily at determining the linear birefringence of the sensing fibre.

The stimulated Raman approach as described in Refs. [96] and [97] does not appear to have any commercial or, indeed, scientific follow-through. This is possibly because it adds little to the POTDR technique (and variants) whilst being substantially more complex. It achieves its objective of providing a stronger signal than is available in linear backscatter-mode POTDR, but the interpretation is more complex given that the interaction depends on the birefringence applying to both pump and probe from their point of launching to their interaction.

10.7.2 The Kerr Effect in Distributed Temperature Sensing

The Kerr effect is a modulation of the refractive index in response to optical intensity (i.e. electric field squared) in contrast to the Pockels effect that, in suitable materials, responds linearly to the electric field. The Kerr effect is orientation dependent; i.e. it modulates the birefringence of the medium that it affects. It could therefore be an attractive measurement of the magnitude of the electric field, but measurements of its temperature dependence [98] suggested that it could also be a means of providing a DTS.

The principle of the use of the Kerr effect was described in Refs. [99] and [100], and it is shown conceptually in Figure 10.11. Like BOTDA and the stimulated Raman process described in Section 10.7.1, the arrangement involves counter-propagating continuous-wave (CW) probe light and pulse pump. However, in this case, there is no need for a frequency relationship between pump and probe, and in fact, it would be advisable to avoid there being a frequency difference equal to the Raman or Brillouin shift between the counter-propagating waves. The path from the probe laser to the detector is a MZI (between Couplers 1 and 2) that compares the phase of the light that has travelled through the sensing fibre to that emerging from the reference fibre. The pump pulse is launched in the opposite direction from the probe into the sensing fibre through Coupler 3. Any change in the optical path length in the sensing fibre will be detected as a phase change that is converted by the interferometer to a signal variation at the balanced receiver. Through the Kerr effect, the pump pulse induces a phase change on the probe signal that is local to the region of interaction between pump and probe. Insofar as the Kerr effect responds to a measurand, the phase variation is an indication of the spatial distribution of that measurand.

The motivation for this work was to provide a faster measurement of temperature; in the Kerr effect measurement, the probe light is detected directly rather than a far weaker backscatter signal, and it was

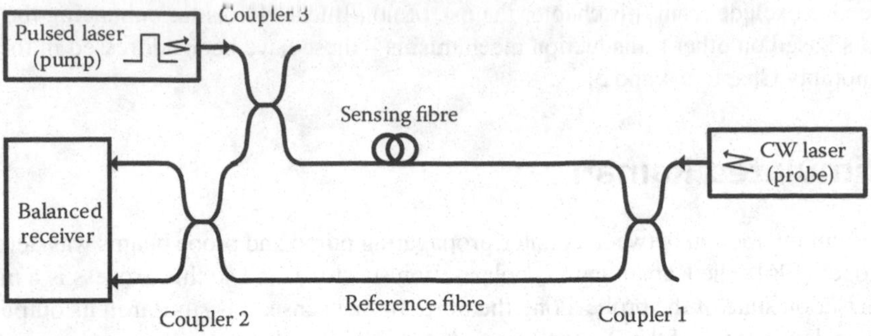

FIGURE 10.11 Conceptual arrangement for Kerr-based DOFS. (Based on Dakin, J., Pratt, D., Edge, C., Goodwin, M., Bennion, I., A distributed fibre temperature sensor using the optical Kerr effect, Fibre Optics Sensors II, The Hague, The Netherlands, SPIE, 798, 149–157, 1987; and the description in Dakin, J.P., Improvements relating to optical fibre sensing arrangements, GB2207236B, 1991.)

expected that this would provide a much higher signal-to-noise ratio, possibly leading to a complete temperature profile with a single pump pulse.

In the actual experimental verification of this technique, the interferometer function was realised by sensing the differential phase change between two modes of a PM fibre. This has the benefit of minimising unwanted differential sensitivity between the reference and sensing channels (i.e. the two modes of the PM fibre in this case) to environmental changes. It also overcomes the problem of ensuring that the polarisation state of the pump is consistent throughout the fibre. In this arrangement, the pump is launched on one axis of the sensing fibre at one end and the probe is launched evenly into both modes from the opposite end.

The expected phase change $\Delta\phi$ is [99]

$$\Delta\phi = \frac{2\pi}{\lambda} K \frac{P_p}{A_{eff}} \frac{\tau_p c}{N_g}, \tag{10.7}$$

where λ is the probe wavelength, K is the Kerr coefficient, P_p/A_{eff} is the pump intensity and τ_p is its duration. K was measured at low frequencies [98] – from 1 kHz to 1 MHz – to be ~5.3×10^{-16} m V^{-2} c is the speed of light in free space and N_g is the group velocity. This would lead to a phase change of several radians at a few hundred W of pump power. However, it is not possible to propagate this sort of power level over any reasonable fibre length owing to, for example, stimulated Raman scattering.

The experimental results in Ref. [99] were obtained with a pump pulse power of 1 W and a pulse duration of 25 ns; a Kerr-induced phase change of 10 mrad was observed, slightly below the expected value. This can easily be detected in an interferometric arrangement. However, the problem with the results was that no significant temperature sensitivity was observed, in contrast to the results of Ref. [98]. The contradiction was explained by considering that the Kerr effect has a fast component that is temperature *insensitive* and a slow component that is temperature *sensitive*, and the slow component was measured in Ref. [98]. The electric field that applies in the case of the measurements of Ref. [99] is that caused by the pump, i.e. in the hundreds of THz and so unable to excite the slow component. As a result, the Kerr effect can be seen to be *ineffective* in the prime purpose for which it was tested, namely a fast DTS measurement.

10.7.3 The Kerr Effect for POTDR in High-Birefringence Fibres

The use of the Kerr effect to map the birefringence of an optical fibre was demonstrated by Parvaneh et al. [101]. As in the previous section, counter propagating pump pulses at wavelength λ_1 and a continuous-wave

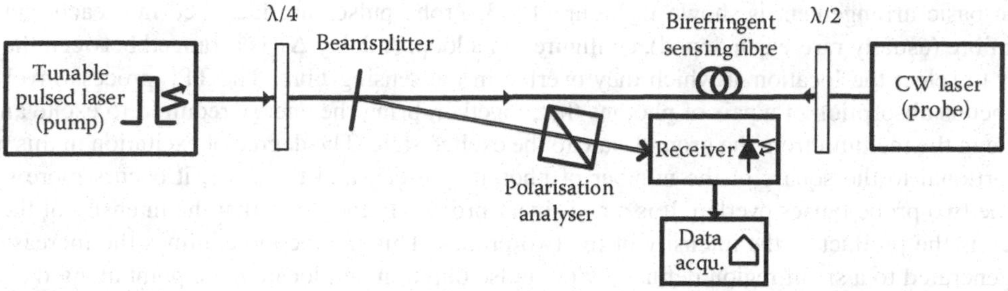

FIGURE 10.12 Experimental arrangement for measuring the beat length in a high-birefringence fibre using the Kerr effect. (Adapted from Parvaneh, F., Handerek, V.A., Rogers, A.J., *Opt. Lett.*, 17, 1346–1348, 1992.)

probe at wavelength λ_2 are launched into opposite ends of a sensing fibre and interact. However, in the case of the birefringence measurement, the pump pulse (having a duration selected so as to occupy less than half a beat length in the fibre) is launched at 45° from the principal axes of a sensing fibre. The polarisation of the pump pulse passes periodically through linear and circular states; when the pump is in the linear state, it is again polarised at 45° from the principal axes, but alternates in direction between −45° and +45°.

In the Kerr-effect arrangement, the probe is launched in a state of polarisation corresponding to a principal axis of the fibre. The arrangement is illustrated in Figure 10.12. Where the probe light encounters the pump in a linear state of polarisation, the Kerr effect modulates the refractive index in a direction parallel to the electric field of the probe, i.e. at 45° from the principal axes. This has the effect of 'rocking' the principal axes very slightly. This couples a small fraction of the probe into the orthogonal state with a spatial period of $\Delta\beta$, which is related to the fibre birefringence and the difference between the probe and pump wavelengths. Rather like the arrangement of Section 10.4.4, this has the effect of down-shifting the frequency produced by the measurement. The frequency produced at the receiver is given by Ref. [102]

$$F_{Kerr} = \frac{C}{n_1} B_F \cdot \Delta\lambda \frac{\Delta\lambda}{\lambda_1 \cdot \lambda_2} \quad \text{where } \Delta\lambda = |\lambda_1 - \lambda_2|. \tag{10.8}$$

Here, B_F is the intrinsic birefringence of the fibre. So at equal probe and pump wavelengths, no beat is observed and the beat frequency increases linearly with increasing wavelength difference (neglecting dispersion effects).

Experimental results in Ref. [102] demonstrated the basic principles, although the distance covered (24 m) and the spatial resolution (0.4 m) were modest (given the need to keep the pulse duration shorter than half a beat length).

However, in subsequent work [103], the technique was refined to provide a temperature resolution of 1.2°C with a single laser shot with a spatial resolution of order 0.5 m over a distance of 100 m.

Although based on very elegant physics, the practical application of measuring the fibre birefringence using the Kerr effect has been limited so far owing to the large optical powers required for a reasonable output signal; these powers bring with them a number of other non-linear effects, in addition to the Kerr interaction, that limit the permissible length of the sensing fibre. The technique does not seem to have attracted any commercial developments.

10.7.4 Two-Photon Fluorescence

Single-photon fluorescence (see Section 10.2) is limited in its spatial resolution owing to the long decay time of the luminescence. However, an alternative approach [104,105] based on two-photon fluorescence (TPF) avoids this problem.

The basic arrangement is shown in Figure 10.13. Probe pulses are launched into each end of the sensing fibre (usually rare-earth doped), configured in a loop. A delay ΔT is arranged between the pulses in order to select the location at which they overlap in the sensing fibre. The TPF process involves the simultaneous absorption of a pair of photons that, together, bring the energy required to excite an ion or molecule in the medium from the ground state to the excited state. The degree of excitation in this process is proportional to the square of the number of photons present, and therefore, it occurs more strongly where the two probe pulses overlap. It is a non-linear process in the sense that the intensity of the signal observed is the product of the intensity of the two probes. This interaction confines the increase in the signal generated to a small region defined by the pulse duration and located at a point along the sensing fibre that can be controlled by adjusting the delay ΔT (which could be negative in order to explore the upper half of the loop illustrated in Figure 10.13). This approach therefore circumvents the blurring issues that are present in single-photon fluorescence.

A background fluorescence level will exist owing to the absorption of a portion of each of the two pulses. If the probe pulses are at the same wavelength, when the pulses meet, the fluorescence is doubled, giving rise to a temporary increase in signal that is uniquely related to the region where the pulses overlap. However, by choosing dissimilar wavelengths for Probe 1 and Probe 2, it can be arranged for neither single-photon fluorescence nor TPF to occur at the wavelength as the emission from TPF. The background fluorescence generated individually by each probe pulse can therefore be reduced, and it can also be distinguished if different transitions are excited compared with that caused by the combination of Probe 1 and Probe 2. The ability to discriminate against the background fluorescence of the individual probes will of course depend on the absorption spectrum of the dopant ions.

Moreover, the change in signal can be analysed in a number of ways, including not only its intensity but also its decay time. It is known that the fluorescence decay time is a function primarily of the local temperature.

The technique was studied in detail at Strathclyde University [106–111], including experiments on short samples of Pr-doped fibre, which was selected for its large two-photon excitation cross-section and relatively short decay time (25 μs at room temperature), coupled with a relatively strong sensitivity of the decay time to temperature (64 ns/K). The fibre was excited at 935 nm and the emission collected at around 600 nm. It was demonstrated that a measurable change in the decay time exists.

This work is still far from commercial, and a number of practical issues remain to be solved. It is also not clear that the technique, in spite of the innovative physics on which it is based, is able to perform

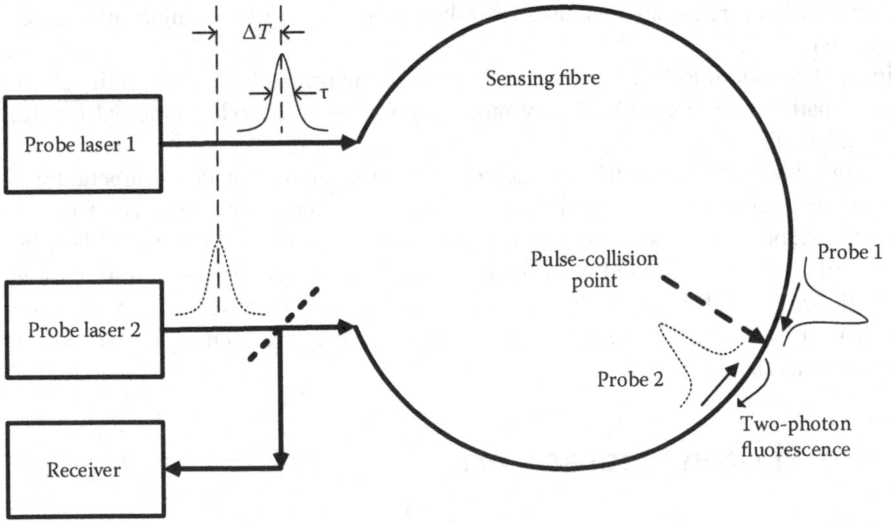

FIGURE 10.13 Arrangement of a two-photo fluorescence DTS. (Adapted from Han, T.P.J., Ruddock, I.S., Fibre optic sensing system, WO2006027613A2, 2006.)

better than better established approaches such as Raman or Brillouin backscatter, nor what issues will arise in producing the rare-earth doped fibre that are required. It was pointed out, however, that two-photon-fluorescence techniques might be applicable to other measurands, such as magnetic fields using the Zeeman effect [108].

10.8 CLOSING COMMENTS

We have described a wide variety of physical mechanisms, probing approaches and interrogation systems that have extended the scope of DFOS beyond the well-developed Raman, Brillouin and coherent Rayleigh backscatter techniques. Some of these have demonstrated elegant physics but have seen no commercial applications. Others, such as polarisation reflectometry, in spite of tens of years of development, are only just reaching the threshold of performance and applications demonstration that might allow their commercial development. Finally, some of the techniques have seen commercial development but been superseded by other technologies; the competing technologies have sometimes been other types of DFOS, but in other instances, the displacing technology is provided by electronic sensors.

It is worth bearing these less developed approaches in mind because their relative merit can be radically changed by advances in component technology novel implementations.

REFERENCES

1. Fields, J. N. 1980. Attenuation of a parabolic-index fiber with periodic bends. *Appl. Phys. Lett.* 36: 10: 799–801.
2. Horsthuis, W., and J. Fluitman. 1983. The development of fibre optic microbend sensors. *Sens. Actuators* 3: 99–110.
3. De Fornel, F., J. Arnaud, and P. Facq. 1983. Microbending effects on monomode light propagation in multimode fibers. *J. Opt. Soc. Am.* 73: 5: 661–668.
4. Marvin, D. C., and N. A. Ive. 1984. Wide-range fiber-optic strain sensor. *Appl. Opt.* 23: 23: 4212–4217.
5. Asawa, C. K., S. K. Yao, R. C. Stearns, N. L. Mota, and J. W. Downs. 1982. High-sensitivity fibre-optic strain sensors for measuring structural distortion. *Electron. Lett.* 18: 362–364.
6. Asawa, C. K., J. W. Austin, M. K. Barnoski et al. 1984. Microbending of optical fibers for remote force measurement. US4477725.
7. Mickelson, A., O. Klevhus, and M. Eriksrud. 1984. Backscatter readout from serial microbending sensors. *J. Lightw. Technol.* 2: 5: 700–709.
8. Harmer, A. L. 1986. Fiber-optical pressure detector. US4618764.
9. Botting, C. 1995. Signal sensing and measuring in fibre optic systems. EP0419267 B1.
10. Herga Electric Limited. 2014. Personal communication.
11. Bergqvist, E. A. 1993. Optical fibre cable for detecting a change in temperature. US5231681.
12. Michie, W. C., B. Culshaw, M. Konstantaki et al. 1995. Distributed pH and water detection using fiber-optic sensors and hydrogels. *J. Lightw. Technol.* 13: 7: 1415–1420.
13. MacLean, A., C. Moran, W. Johnstone et al. 2003. Detection of solvents using distributed fibre optic sensor. *Electron. Lett.* 39: 17: 1237–1238.
14. Griscom, D. L., M. E. Gingerich, and E. J. Friebele. 1994. Model for the dose, dose-rate and temperature dependence of radiation-induced loss in optical fibers. *IEEE Trans. Nucl. Sci.* 41: 3: 523–527.
15. Friebele, E., D. Griscom, and G. Sigel Jr. 1974. Defect centers in a germanium-doped silica-core optical fiber. *J. Appl. Phys.* 45: 8: 3424–3428.
16. Griscom, D. L., E. Friebele, K. Long, and J. Fleming. 1983. Fundamental defect centers in glass: Electron spin resonance and optical absorption studies of irradiated phosphorus-doped silica glass and optical fibers. *J. Appl. Phys.* 54: 7: 3743–3762.
17. Paul, M. C., D. Bohra, A. Dhar et al. 2009. Radiation response behavior of high phosphorous doped step-index multimode optical fibers under low dose gamma irradiation. *J. Non-Cryst. Solids* 355: 28: 1496–1507.

18. Lezius, M., K. Predehl, W. Stöwer et al. 2012. Radiation induced absorption in rare earth doped optical fibers. *IEEE Trans. Nucl. Sci.* 59: 2: 425–433.
19. Fox, B. P. 2013. Investigation of ionizing-radiation-induced photodarkening in rare-earth-doped optical fiber amplifier materials. PhD The University of Arizona.
20. Vassilopoulos, C., A. Kourtis, and C. Mantakas. 1993. Conception of an ionising radiation detection scheme based on controlled light induced annealing of silica fibres. *IEE Proc. J.* 140: 4: 267–272.
21. Gaebler, W., and D. Bräunig. 1983. Application of optical fibre waveguides in radiation dosimetry, 1st International conference on optical fibre sensors, London, UK. IEE, 185–189.
22. West, R., H. Buker, E. Friebele, H. Henschel, and P. Lyons. 1994. The use of optical time domain reflectometers to measure radiation-induced losses in optical fibers. *J. Lightw. Technol.* 12: 4: 614–620.
23. Schmidt, G., E. Kasl, K. Wille et al. 2002. Optical fibre beam loss monitors for storage rings at DELTA. In Proc. EPAC. 1969–1971.
24. Henschel, H., O. Köhn, M. Körfer, T. Stegmann, and F. Wulf. Preliminary trials with optical fiber dosimeters at TTF. 2000. TESLA-Report No. 2000-25.
25. Hartog, A. H. 1983. A distributed temperature sensor based on liquid-core optical fibers. *J. Lightw. Technol.* 1: 3: 498–509.
26. Eriksrud, M., and A. Mickelson. 1982. Application of the backscattering technique to the determination of parameter fluctuations in multimode optical fibers. *IEEE J. Quant. Electron.* 18: 10: 1478–1483.
27. Gottlieb, M., and G. B. Brandt. 1981. Temperature sensing in optical fibers using cladding and jacket loss effects. *Appl. Opt.* 20: 22: 3867–3873.
28. Pinchbeck, D. 1986. The optical fibre cryogenic leak detection system. *Meas. Control* 19: 7: 46–47.
29. Snitzer, E., W. Morey, and W. Glenn. 1983. Fiber optic rare earth temperature sensors. 1st International Conference on Optical Fibre Sensors, London. IEE, 79–82.
30. Mears, R. J. 1987. Optical fibre lasers and amplifiers. PhD University of Southampton.
31. Reekie, L., I. M. Jauncey, S. B. Poole, and D. N. Payne. 1988. Continuous-wave tunable and Q-switched operation at 938 nm of a diode-laser-pumped Nd3+-doped fiber laser, Conference on Lasers and Electro-optics, Anaheim, CA, USA, Optical Society of America, THM46.
32. Farries, M. C., M. E. Fermann, R. I. Laming et al. 1986. Distributed temperature sensor using Nd3+-doped optical fibre. *Electron. Lett.* 22: 8: 418–419.
33. Hartog, A. H., A. P. Leach, and M. P. Gold. 1985. Distributed temperature sensing in solid-core fibres. *Electron. Lett.* 21: 23: 1061–1062.
34. Farries, M. C., and M. E. Fermann. 1987. Temperature sensing by thermally-induced absorption in a neodymium doped optical fibre. Fibre Optic Sensors II, SPIE, The Hague, The Netherlands, 0798, 115–120.
35. Farries, M. C., M. E. Fermann, S. B. Poole, and J. E. Townsend. 1987. Distributed temperature sensor using holmium-doped optical fibre, Optical Fiber Conference/International Conference on Integrated Optics and Optical Fiber Communication '87.
36. Garlick, G. F. J., and R. E. Nwokedi. 1978. Anti-stokes effects for Nd^{3+} ions in solids. *Phys. Lett. A* 67: 3: 205–206.
37. Wade, S. A., S. F. Collins, G. W. Baxter, and G. Monnom. 2001. Effect of strain on temperature measurements using the fluorescence intensity ratio technique (with Nd3+- and Yb3+-doped silica fibers). *Rev. Sci. Instrum.* 72: 8: 3180–3185.
38. Mickelson, A. R., and M. Eriksrud. 1982. Theory of the backscattering process in multimode optical fibers. *Appl. Opt.* 21: 11: 1898–1909.
39. Kotov, O. I., L. B. Liokumovich, A. H. Hartog, A. V. Medvedev, and I. O. Kotov. 2008. Transformation of modal power distribution in multimode fiber with abrupt inhomogeneities. *J. Opt. Soc. Am.* B 25: 12: 1998–2009.
40. Rogers, A. J. 1981. Polarisation-optical time domain reflectometry: A technique for the measurement of field distributions. *Appl. Opt.* 20: 6: 1060–1073.
41. Rogers, A. J. 1980. Polarisation optical time-domain reflectometry. *Electron. Lett.* 16: 13: 489–490.
42. Hartog, A. H., D. N. Payne, and A. J. Conduit. 1980. Polarisation optical-time-domain reflectometry: Experimental results and application to loss and birefringence measurements in single-mode optical fibres, 6th European Conference on Optical Communication, York, UK, IEE, 190 (post-deadline): 5–8.
43. Barlow, A. J., J. J. Ramskov-Hansen, and D. N. Payne. 1981. Birefringence and polarisation mode-dispersion in spun singlemode fibers. *Appl. Opt.* 20: 17: 2962–2968.
44. Laming, R., and D. N. Payne. 1989. Electric current sensors employing spun highly birefringent optical fibers. *J. Lightw. Technol.* 7: 12: 2084–2094.
45. Nakazawa, M. 1983. Theory of backward Rayleigh scattering in polarisation-maintaining single-mode fibers and its application to polarisation optical time domain reflectometry. *IEEE J. Quant. Electron.* 19: 5: 854–861.

46. Ourmazd, A., M. P. Varnham, R. D. Birch, and D. N. Payne. 1983. Thermal properties of highly birefringent optical fibers and preforms. *Appl. Opt.* 22: 15: 2374–2379.
47. Xie, H. M., P. Dabkiewicz, R. Ulrich, and K. Okamoto. 1986. Side-hole fiber for fiber-optic pressure sensing. *Opt. Lett.* 11: 5: 333–335.
48. Varnham, M. P., A. J. Barlow, D. N. Payne, and K. Okamoto. 1983. Polarimetric strain gauge using high birefringence fibre. *Electron. Lett.* 19: 17: 699–700.
49. Wegmuller, M., F. Scholder, and N. Gisin. 2004. Photon-counting OTDR for local birefringence and fault analysis in the metro environment. *J. Lightw. Technol.* 22: 2: 390–400.
50. Wegmuller, M., M. Legre, and N. Gisin. 2002. Distributed beatlength measurement in single-mode fibers with optical frequency-domain reflectometry. *J. Lightw. Technol.* 20: 5: 828–835.
51. Parvaneh, F., V. A. Handerek, and A. J. Rogers. 1991. Frequency-derived distributed optical fibre sensing: Signal-frequency downshifting. *Electron. Lett.* 27: 5: 394–396.
52. Ross, J. N. 1982. Birefringence measurement in optical fibers by polarisation optical time-domain reflectometry. *Appl. Opt.* 21: 19: 3489–3495.
53. Ferdinand, P. 1990. Etat du domaine des capteurs à fibre optiques et réseaux associés: Capteurs répartis par analyse de la polarisation de la lumière rétrodiffusée. PhD Université de Nice. France.
54. Ellison, J. G., and A. S. Siddiqui. 1998. A Fully polarimetric optical time-domain reflectometer. *IEEE Photon. Technol. Lett.* 10: 246–248.
55. Schuh, R. E. 1997. Characterisation of birefringence in optical fibres. PhD University of Essex.
56. Rogers, A. 2002. Distributed fibre measurement using backscatter polarimetry, 15th International Conference on Optical Fiber Sensors, IEEE, 1: 367–370.
57. Barlow, A., and D. Payne. 1983. The stress-optic effect in optical fibers. *IEEE J. Quant. Electron.* 19: 5: 834–839.
58. Rogers, A. J. 2002. Optical fibre backscatter polarimetry. WO2002/095349.
59. Rogers, A. J., S. V. Shatalin, and S. E. Kannellopoulos. 2005. Optical-fibre backscatter polarimetry for the distributed measurement of full strain fields. In Fiber Optic Sensor Technology and Applications IV. SPIE, Boston, MA, USA, 6004: 600404–600414.
60. Corsi, F., A. Galtarossa, and L. Palmieri. 1998. Polarisation mode dispersion characterization of single-mode optical fiber using backscattering technique. *J. Lightw. Technol.* 16: 10: 1832–1843.
61. Galtarossa, A., L. Palmieri, M. Schiano, and T. Tambosso. 2000. Measurements of beat length and perturbation length in long single-mode fibers. *Opt. Lett.* 25: 6: 384–386.
62. Rogers, A. J., S. V. Shatalin, and S. E. Kanellopoulos. 2005. Distributed measurement of fluid pressure via optical-fibre backscatter polarimetry. In 18th International Conference on Optical Fibre Sensors. SPIE, Bruges, Belgium, 5855: 230–233.
63. Papp, A., and H. Harms. 1980. Magnetooptical current transformer. 1: Principles. *Appl. Opt.* 19: 22: 3729–3734.
64. Ross, J. N. 1981. Measurement of magnetic field by polarisation optical time-domain reflectometry. *Electron. Lett.* 17: 17: 596–597.
65. Kim, B. Y., P. Dann, and C. Sang. 1982. Use of polarisation-optical time domain reflectometry for observation of the Faraday effect in single-mode fibers. *IEEE J. Quant. Electron.* 18: 4: 455–456.
66. Palmieri, L., and A. Galtarossa. 2012. Distributed fiber optic sensor for mapping of intense magnetic fields based on polarisation sensitive reflectometry, 3rd Asia Pacific Optical Sensors Conference, SPIE, Sydney, Australia, 8351: 31.
67. Schuh, R., J. Ellison, A. Siddiqui, and D. Bebbington. 1996. Polarisation OTDR measurements and theoretical analysis on fibres with twist and their implications for estimation of PMD. *Electron. Lett.* 32: 4: 387–388.
68. Schuh, R., and A. Siddiqui. 1996. Measurement of SOP evolution along a linear birefringent fibre with twist using polarisation OTDR. NIST/IEEE/OSA Symposium on Optical Fibre Measurements, Boulder, CO, USA, NIST Special Publication: SP 159–162.
69. Thévenaz, L., M. Facchini, A. Fellay, M. Nikles, and P. Robert. 1997. Evaluation of local birefringence along fibres using Brillouin analysis, Optical Fibre Measurement Conference OFMC '97, Teddington, UK, National Physical Laboratory: 82–85.
70. Zhang, J., V. A. Handerek, I. Cokgor, V. Pantelic, and A. J. Rogers. 1997. Distributed sensing of polarisation mode coupling in high birefringence optical fibers using intense arbitrarily polarised coherent light. *J. Lightw. Technol.* 15: 5: 794–801.
71. Huttner, B., J. Reecht, N. Gisin, R. Passy, and J. P. Von der Weid. 1998. Local birefringence measurements in single-mode fibers with coherent optical frequency-domain reflectometry. *IEEE Photon. Technol. Lett.* 10: 10: 1458–1460.
72. Corsi, F., A. Galtarossa, and L. Palmieri. 1999. Beat length characterization based on backscattering analysis in randomly perturbed single-mode fibers. *J. Lightw. Technol.* 17: 7: 1172–1178.

73. Corsi, F., A. Galtarossa, and L. Palmieri. 1999. Analytical treatment of polarisation-mode dispersion in single-mode fibers by means of the backscattered signal. *J. Opt. Soc. Am.* A 16: 3: 574–583.
74. Huttner, B., B. Gisin, and N. Gisin. 1999. Distributed PMD measurement with a polarisation-OTDR in optical fibers. *J. Lightw. Technol.* 17: 18: 1843–1848.
75. Galtarossa, A., L. Palmieri, A. Pizzinat, M. Schiano, and T. Tambosso. 2000. Measurement of local beat length and differential group delay in installed single-mode fibers. *J. Lightw. Technol.* 18: 10: 1389–1394.
76. Wuilpart, M., P. Megret, M. Blondel, A. J. Rogers, and Y. Defosse. 2001. Measurement of the spatial distribution of birefringence in optical fibers. *IEEE Photon. Technol. Lett.* 13: 8: 836–838.
77. Galtarossa, A., and L. Palmieri. 2002. Measure of twist-induced circular birefringence in long single-mode fibers: Theory and experiments. *J. Lightw. Technol.* 20: 7: 1149.
78. Wuilpart, M., G. Ravet, P. Megret, and M. Blondel. 2002. Polarisation mode dispersion mapping in optical fibers with a polarisation-OTDR. *IEEE Photon. Technol. Lett.* 14: 12: 1716–1718.
79. Juarez, J. C., and H. F. Taylor. 2005. Polarisation discrimination in a phase-sensitive optical time-domain reflectometer intrusion-sensor system. *Opt. Lett.* 30: 24: 3284–3286.
80. Seo, W. 1994. Fiber optic intrusion sensor investigation. PhD, Texas A&M.
81. Zhang, Z., and X. Bao. 2008. Continuous and damped vibration detection based on fiber diversity detection sensor by Rayleigh backscattering. *J. Lightw. Technol.* 26: 7: 832–838.
82. Franciscangelis, C., C. Floridia, and F. Fruett. 2015. Optical fibre intrusion detection method based on Lefevre-loop and bidirectional polarisation optical time-domain reflectometer-C technique. *J. Eng.* 1: 1.
83. Gottlieb, M., and G. B. Brandt. 1981. Fiber-optic temperature sensor based on internally generated thermal radiation. *Appl. Opt.* 20: 19: 3408–3414.
84. Gusmeroli, V., M. Martinelli, and P. Vavassori. 1989. Quasi-distributed single-length polarimetric sensor. *Opt. Lett.* 14: 23: 1330–1332.
85. Danielson, B. L., and C. Y. Boisrobert. 1991. Absolute optical ranging using low coherence interferometry. *Appl. Opt.* 30: 21: 2975–2979.
86. Goodman, J. W. 1976. Some fundamental properties of speckle*. *J. Opt. Soc. Am.* 66: 11: 1145–1150.
87. Anderson, D. R. 2006. Detecting flat wheels with a fiber-optic sensor. In ASME/IEEE 2006 Joint Rail Conference. American Society of Mechanical Engineers. 25–30.
88. Tapanes, E. E., J. R. Goode, and J. Katsifolis. 2004. Apparatus and method for monitoring a structure using a counter-propagating signal method for locating events. US6778717 B2.
89. Cole, J. H., R. L. Johnson, and P. G. Bhuta. 1977. Fiber-optic detection of sound. *J. Acoust. Soc. Am.* 62: 5: 1136–1138.
90. Cahill, R. F., and E. Udd. 1983. Optical acoustic sensor. US4375680 A.
91. Giallorenzi, T., J. Bucaro, A. Dandridge et al. 1982. Optical fiber sensor technology. *IEEE J. Quant. Electron.* 18: 4: 626–665.
92. Martinelli, M. 1992. Time reversal for the polarisation state in optical systems. *J. Mod. Optic.* 39: 3: 451–455.
93. Kurmer, J. P., S. A. Kingsley, J. S. Laudo, and S. J. Krak. 1992. Distributed fiber optic acoustic sensor for leak detection, Distributed and Multiplexed Fiber Optic Sensors, SPIE, 1586: 117–128.
94. Spammer, S. J., P. L. Swart, and A. A. Chtcherbakov. 1997. Merged Sagnac-Michelson interferometer for distributed disturbance detection. *J. Lightw. Technol.* 15: 6: 972–976.
95. Qian, H., Z. Tao, X. Xianghui et al. 2013. All Fiber distributed vibration sensing using modulated time-difference pulses. *IEEE Photon. Technol. Lett.* 25: 20: 1955–1957.
96. Farries, M. C., and A. J. Rogers. 1984. Distributed sensing using stimulated Raman interaction in a monomode optical fibre. 2nd International Conference on Optical Fibre Sensors. SPIE/VDE Verlag, Stuttgart, Germany, 0514: 121–132.
97. Valis, T., R. D. Turner, and R. M. Measures. 1989. Distributed fiber optic sensing based on counterpropagating waves. *Appl. Opt.* 28: 11: 1984–1990.
98. Farries, M. C., and A. J. Rogers. 1983. Temperature dependence of the Kerr effect in a silica optical fibre. *Electron. Lett.* 19: 21: 890–891.
99. Dakin, J., D. Pratt, C. Edge, M. Goodwin, and I. Bennion. 1987. A distributed fibre temperature sensor using the optical Kerr effect, Fibre Optics Sensors II, The Hague, The Netherlands, SPIE, 798: 149–157.
100. Dakin, J. P. 1991. Improvements relating to optical fibre sensing arrangements. GB2207236B.
101. Parvaneh, F., L. Valente, V. Handerek, and A. Rogers. 1992. Forward-scatter frequency-derived distributed optical fibre sensing using the optical Kerr effect. *Electron. Lett.* 28: 12: 1080–1082.
102. Parvaneh, F., V. A. Handerek, and A. J. Rogers. 1992. Frequency-derived remote measurement of birefringence in polarisation-maintaining fiber by using the optical Kerr effect. *Opt. Lett.* 17: 19: 1346–1348.
103. Parvaneh, F., M. Farhadiroushan, V. A. Handerek, and A. J. Rogers. 1997. Single-shot distributed optical-fiber temperature sensing by the frequency-derived technique *Opt. Lett.* 22: 5: 343–345.

104. Han, T. P. J., and I. S. Ruddock. 2006. Fibre optic sensing system. WO2006027613A2.
105. Ruddock, I. S., and T. P. J. Han. 2006. Potential of two-photon-excited fluorescence for distributed fiber sensing. *Opt. Lett.* 31: 7: 891–893.
106. Dalzell, C. J., T. P. J. Han, I. S. Ruddock, and D. B. Hollis. 2012. Two-photon excited fluorescence in rare-earth doped optical fiber for applications in distributed sensing of temperature. *IEEE Sens. J.* 12: 1: 51–54.
107. Dalzell, C., T. Han, and I. Ruddock. 2011. The two-photon absorption spectrum of ruby and its role in distributed optical fibre sensing. *Appl. Phys. B* 103: 1: 113–116.
108. Dalzell, C. J. 2011. Two-photon excited fluorescence and applications in distributed optical fibre temperature sensing. PhD University of Strathclyde.
109. Ruddock, I., C. Dalzell, and T. Han. 2010. Nonlinear spectroscopy of doped glass and crystal for applications in distributed fibre sensing. The Central Laser Facility Annual Report, Science & Technology Facilities Council.
110. Dalzell, C., T. Han, and I. Ruddock. 2008. Distributed optical fibre sensing of temperature using time-correlated two-photon excited fluorescence: Theoretical overview. *Appl. Phys. B* 93: 2–3: 687–692.
111. Dalzell, C. J., T. P. Han, and I. S. Ruddock. 2009. Distributed optical fibre sensing of temperature using time-correlated two-photon excited fluorescence. In 20th International Conference on Optical Fibre Sensors. SPIE, Edinburgh, United Kingdom, 7503: 75036K.

Conclusions

11

Distributed optical fibre sensors (DOFSs) are at the intersection of many disciplines, including optical fibre technology and photonics, the instrumentation techniques that are applied to provide distance resolution, the physics that is used to encode the measurand onto the return signal and the knowledge used in each of the applications to implement a DOFS installation and extract value from the data.

Figure 11.1 summarises these aspects. The figure is not a complete classification scheme for distributed sensors, but it provides a picture of the relationships between the functional elements of a DOFS system and a hierarchy in the process of defining the requirements and the system that will meet them.

The top left-hand quadrant illustrates considerations relating to the fibre sensor itself and its protection and deployment, i.e. the input end of the sensing system. The lower part of Figure 11.1 relates to the interrogation technology, namely how the distance dimension is captured (to the right of the centre) and how the information concerning the local value of the measurand is converted to a modulation of the optical signal. Most of the latter approaches developed to date relate to modulation of a form of scattered light, which dominated the technology descriptions in this book (Chapters 4–6). However, some of the other techniques, described in Chapter 10, are included in the lower left-hand corner. Also shown are the major measurands, tied to the interrogation techniques that are able to provide them. The bottom right-hand corner of the figure shows the measurands that have been demonstrably derived indirectly (inferred) from a DOFS output.

The top right-hand corner illustrates the issues around storing and using the data from a DOFS. The volume of data generated by DOFS is very large by comparison with conventional sensors. In the case of distributed vibration sensing (DVS), the raw data can amount to several GB/s, and even after pre-processing to extract the usable information, the data rate can still exceed 10 MB/s per fibre. This data volume poses a number of challenges.

Firstly, it requires adequate internal bandwidth to handle the raw data flows (and sometimes this implies transfers outside the acquisition system in real time to large capacity storage). Although these data rates are common in optical telecommunications, they are unusual in the sensing world, and so usually, new infrastructure is required for data transfer and storage. Secondly, acquiring data in itself fulfils no useful role: it is only when it is interpreted and acted upon that the investment in the acquisition can generate a return. This implies the presence of a data repository (sometimes known as a data historian) facilitating the storage, retrieval and comparison of data sets.

Interpretation tools are required to convert the data into actionable information. This usually means constructing a model of the system that is monitored and populating that model, not just with the DOFS data, but also metadata on the asset that is surveilled and complementary data from other sources.

As an example, a distributed temperature sensor (DTS) installed on an energy transmission cable provides regularly updated temperature profiles. In order to convert that data into useful knowledge (e.g. a dynamic rating for the cable section), the thermal history is also required as well as information on the weather forecast, the cable route and backfill quality (that may, in part, have been inferred from earlier DTS data), the cable design and its ageing history. All of these data form the input to a model of the system consisting of the cable and its environment. The model may also include financial information to trade off the present value of ampacity compared with the future value of ampacity at a time when the transmission capacity demand might be higher, given that a high present load will heat the cable and leave the transmission system with reduced capacity until it cools; these comparisons depend, of course, on the anticipated loading state of the network, the expected availability of power especially from intermittent sources and

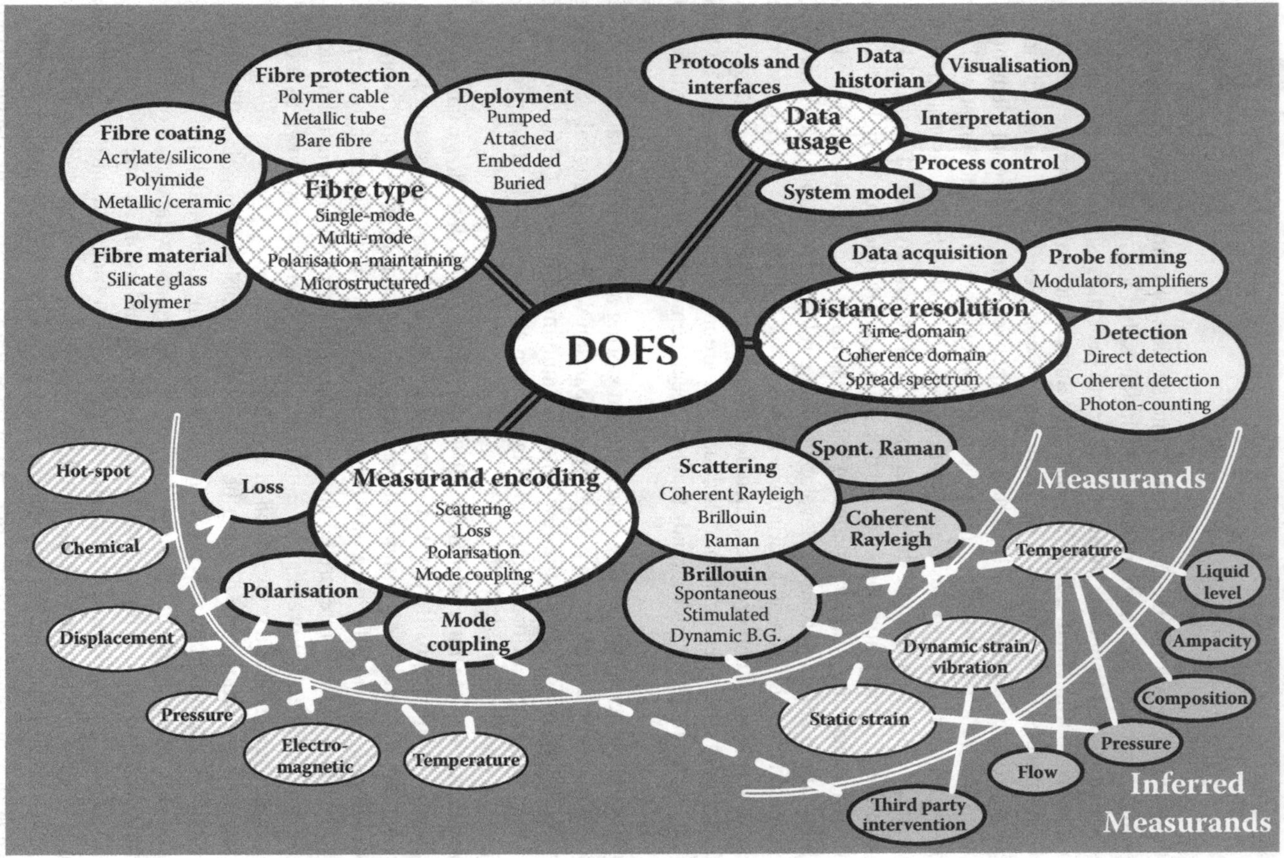

FIGURE 11.1 The interrelationships between the main elements of a *DOFS* system, the measurands it addresses and the underlying physical principles.

the cost/benefit of a faster usage of the cable lifespan if its temperature is increased. A system model is a practical means of transforming the data into actionable information.

Where human beings interact with the data, the problem of visualising it arises: the data emerging from even relatively low-update rate DOFS (such as DTS) are such that simple 2-D graphs are not adequate. The number of sample points provided by most DOFS is well beyond the pixel count of even the highest-resolution displays or the visual acuity of most operators. Therefore, innovative approaches have been devised to present the DOFS results accessibly. These depend of course on the application, but they include 3-D images emphasised with colour (or moving 3-D objects), videos showing the time evolution of the measurand together with other data thought to be relevant (such as inputs to the process) or mimic displays that automatically switch the focus to areas that the model has identified as being important. These visualisation techniques are all intended to minimise the workload of the person making decisions, sometimes in real time, based on the output generated by the DOFS and other sensors.

Finally, data standards are gradually being evolved to provide the outputs in forms that can be read by interpretation software regardless of the specifics of the interrogation. The rapid evolution of DOFS is challenging the industry standardisation bodies to keep up with the innovation in the interrogation technology and applications development.

11.1 SELECTION PROCESS

Assuming that the measurand is known, the selection of the sensor involves choosing a technology that is suited to the performance required, in particular in relation to the four key parameters of spatial resolution, range, measurement resolution and update time. Most often, the precise requirements cannot all be provided in commercially available systems and the skill of the specifier is in deciding how best to trade off the performance of one parameter against another. In the author's experience, the measurand distribution can be very different (e.g. far less uniform) from that which was anticipated at the time of specifying the project. As a result, the expectations of the system might not be fulfilled; it is difficult to guard against these issues, but some due diligence in understanding the application and exploring certain scenarios may help (e.g. 'What if the backfill of this cable is not to the standard required?' 'Are there any known services crossing my cable route?' 'How do I know that the lining of my process vessel is built correctly?' 'Is the strain really decoupled from the sensing fibre so that I can consider temperature effects only?'). In addition, undertaking modelling of the expected measurand distribution in cases of interest will help reduce the risk and allow realistic targets to be set for system performance.

The specifier must also understand the limitations of each approach to interrogation, such as the relative benefits and drawbacks of time-domain vs. frequency-domain or pulse-compression coding. The particular capabilities of each means of encoding the data, i.e. usually which type of scattering is used, are also relevant: whilst there is some overlap between the approaches, Raman backscatter, for example, has some limitations in terms of range, but it is unique in being sensitive to temperature only and to reject other measurands.

The environment in which the sensing fibre will be installed and operated will dictate the type of fibre to be used, its protection and the installation methods that are appropriate. When DOFS technology fails to fulfil the client expectations, it is more often than not due to failure of the fibre or the installation process, rather than the interrogation technology. Not all fibres that are used in DOFS are necessarily available in forms that will survive each environment, whether the limiting factor is temperature, static strain or chemical attack on the coating or hydrogen ingress. It follows that one technology that is otherwise selected on grounds of measurement performance could be rejected because the sensing fibres that it requires are not available in a form that guarantees long-term operation within specified limits.

Of course, in general, it is not the measurement *per se* that is of interest, it is the deductions that will be made from it and the decisions that will be taken based on the DOFS data and other inputs. This

TABLE 11.1 Overview of the main types of DOFSs

	MEASURANDS	*SEPARABLE*	*STRENGTHS*	*LIMITATIONS*	*BEST SPATIAL RESOLUTION*	*MAX NUMBER OF POINTS*	*MAX RANGE ([a]WITH OPTICAL AMPLIFICATION)*
Spontaneous Raman							
Analog direct detection	Temperature	Not needed	Temperature only	Weak signal. Cannot optically pre-amplify	0.4 m	10,000	50 km
Single-photon counting	Temperature	Not needed	Temperature only	Weak signal. Biased results at high signals	1 cm	100	10 m
Brillouin							
BOTDR	Temperature and strain	Yes, but loss of precision	Single ended. Can use optical amplification	Spatial resolution limited for strain	5 cm	20,000–50,000	100 km[a]
BOTDA	Temperature and strain	Difficult	Can use optical amplification	Double-ended. Vulnerable to fibre breaks	5 cm	1,000,000	>300 km[a]
BOCDA	Temperature and strain	Difficult	Very fine spatial resolution	Double-ended. Locations addressed serially	<1 cm	c. 1000. Serial	
BOCDR	Temperature and strain	Difficult	Single ended. Very fine spatial resolution	Weak signal. Locations addressed serially	~1 cm	Serial	
Coherent Rayleigh							
Dynamic	Temperature and strain	Not demonstrated	Single ended. Can use optical amplification			10,000–50,000	>150 km[a]
Static	Temperature and strain	Possible with PM fibres	Single ended. Can use optical amplification	Fundamental trade-off with measurand resolution	2 mm	~1,000	15 m
POTDR	Temperature, strain, pressure magnetic field	Under some conditions	Magnetic fields	Complex interrogation and interpretation Measurand discrimination	6 cm		~100 m
Modalmetric	Dynamic strain	N/A	Simple optical implementation	Poor discrimination of multiple events	Not applicable	Not applicable	5 km. Remote by up to 20 km
Loss	Temperature, chemical, side pressure, strain	Requires special fibres	Simple optical implementation. Good for detecting a single anomaly	Measurand-induced loss limits power available to more distant locations	~3 m	1,000 potential hot spots	~3 km

means that the planning also involves the systems to which the DOFS will interface, whether this be a data repository attached to a modelling package to provide long-term answers, alerts sent to operators or process engineers based on the data or even a real-time control system that will select the inputs to provide to the process.

The applications engineering (planning, installation, downstream use of the data) is therefore fundamental to making DOFS a success; it is usually more important than the selection of the DOFS technology itself.

Table 11.1 provides an overview of the main physical effects that are used in distributed sensors, listed according to the physical principles used to gain sensitivity. The main measurands that are addressed directly with each technique are listed together with a qualitative judgement on the relative strengths and limitations in each case. Some indication of the best published results in terms of spatial resolution, number of sensed points and maximum range that have been reported is also provided. However, these performance indications are merely a snapshot at the time of writing and progress in this field is rapid, meaning that the reader should look to current publications and commercial offerings when making quantitative comparisons. The context is also important: as an example, POTDR measurements have been achieved over very long distances, but mainly to extract statistical properties of transmission fibres, rather than localising the value of a measurand. As a further note of caution, the interdependence between performance criteria requires that the potential user should verify that the important parameters in their requirements can be achieved together, rather than in isolation. The three right-hand columns therefore do not provide a full picture of the system performance.

The table is not specific as to the interrogation methods (e.g. single-pulse time-domain reflectometry, frequency-domain reflectometry or pulse-compression coding) that are available. Ultimately, it is the performance that matters to the user, not the means by which it has been achieved. Nonetheless, the user should be conscious of the subtle differences between the techniques – such as the allowable loss in the sensing fibre that can depend on the specifics of system design.

11.2 CHOICE OF DISTRIBUTED SENSORS VS. MULTIPLEXED POINT SENSORS

DOFS and multiplexed point sensors are both highly capable sensing approaches. The question then arises as to which of these options to choose in any given application. Given the vast number of sensor types and applications, there can be no single answer to that question. One technology in particular, that of fibre Bragg gratings (FBGs), provides arrays of tens of sensors per fibre if multiplexed in the wavelength domain and hundreds per fibre if multiplexed in the time domain (with several thousand points per fibre with combined time and wavelength multiplexing). FBG technology is close to fully distributed sensors in its attributes.

DOFSs map the measurand over the entire sensing fibre,* whereas point sensors address a number of individual locations. Monitoring problems where the *entire* measurand distribution must be acquired at every point naturally lend themselves to the distributed approach. On the other hand, where a few small selected points are of interest, then the use of multiplexed point sensors may be more appropriate.

The value of point sensors in this comparison is not so much that they do not waste the unused sensing capability between points of interest. Rather, point sensors can be highly specific in the location of the measurand they sense – down to a few mm in some cases – this provides a spatial resolution that is

* Here, we should flag the exception of Brillouin coherence domain approaches (BOCDA and BOCDR) where a single, addressable point is sampled at a time.

difficult to achieve in DOFS in combination with long range. So it is the ability to provide, at reasonable cost, very local measurements at selected positions spaced over a wide range of distances that can be highly attractive in multiplexed point sensors.

In contrast, where a relatively uniform spatial resolution is required that includes many (≫100) points, then DOFS tends to be more attractive.

Even then, the distinction can be blurred: several successful DTS installations have involved packaging small coils of fibre to form point sensors (the length of coiled fibre being chosen exceed the spatial resolution by typically a factor of 2 to 3). One example of such an installation is in monitoring the bearings of coal conveyors, where many hundreds of highly localised measurements were conducted using a conventional Raman OTDR. Here, the benefits of a long-distance coverage were combined with achieving highly localised measurements by compressing the spatial resolution into small devices (c. 50 mm × 50 mm × 8 mm) that were rugged and could be placed at will along the sensing fibre. In addition, in an application such as this, the regular spacing between point probes allowed short sections of ruggedised sensing cable incorporating a number of local probes to be assembled and tested in a factory environment and the sections merely installed and concatenated on site.

Conversely, point sensors – capable of measuring on a mm scale – are sometimes used to determine the cumulative value of the measurand over an extended distance. Well-documented examples of this practice occur in strain measurement with Bragg gratings [1] and interferometric point sensors [2]. In both cases, the intrinsically short spatial resolution is extended by calculating the mean strain over a gauge length defined between anchoring points.

The boundaries of the applications of multiplexed sensors and distributed sensors are further shifted by recent advances in very-high-spatial-resolution, but long-range, Brillouin sensors [3–5]. Optical frequency-domain reflectometry (OFDR) has for some time demonstrated mm resolution as has Brillouin optical coherence-domain analysis (BOCDA)/Brillouin optical coherence domain reflectometry (BOCDR). However, it is the recent developments that combine 1–2 cm resolution with ranges of order 10 km that are competing very effectively with FBG in some applications. More commonly, however, the multiplexed and distributed approaches are being combined [6,7] in the same application to gain the benefits of each technology.

The technologies themselves are becoming blended with approaches such as faint long gratings (FLOGs) [8] that consist of very weak, but continuously inscribed, gratings in a sensing fibre. FLOGs have the effect of enhancing the sensor return without increasing the loss in proportion: whereas all of the scattering processes discussed in this volume are inefficient in terms of the power that is extracted vs. that returned to the interrogator because they scatter more or less uniformly into 4π steradian, FBGs reflect most of the light that they extract from the probe back to the interrogator. The FLOG technology therefore combines the inscription process of an FBG with a fully distributed interrogation and an efficient use of the probe energy.

11.3 TRENDS IN DOFS

In spite of a history of more than 30 years, the pace of development of distributed sensors shows no sign of abating. Even in the relatively mature area of Raman DTS, progress is still being made in basic interrogation, accuracy improvements and of course in novel applications.

Although the concept of Brillouin-based sensing can be traced back more than 25 years, the last five years has seen massive improvements in the design of the probe waveforms (particularly for very fine spatial resolution) and increased range with a variety of distributed amplification approaches and in fast measurements, for example, using the slope-assisted techniques and sweep-free approaches.

Rayleigh-based techniques for static measurements (OFDR) and DVS, although conceived around 20–25 years ago and having been used in limited numbers since that time, are now very rapidly growing in numbers deployed, in applications development and in the scientific progress underpinning these technologies.

The rate of publications and patent filings is very high, with tens of articles appearing each month across the major, peer-reviewed journals in the field of optics as well as those of similar status in applications fields, such as geophysics, electrical power systems and civil engineering.

Looking back on the last 30 years, progress in this field has been helped by the new scientific concepts, but above all by improvements in the supporting technologies: in the mid-1980s, when the present author was involved in the design of the first commercial DTS system, the fastest available A/D converters sampled their inputs at 20 megasamples per second to a resolution of 8 bits. A/D converters with sampling rates two orders of magnitude higher and having an order of magnitude better resolution of the signal (12 bit, rather than 8 bit) are now available from several suppliers. The technologies for handling fast digital signals, analogue signal conditioning and embedded processing have all progressed enormously. These technological developments mean that interrogation approaches that would have been dismissed as impractical in the mid-1980s now seem quite reasonable to implement; they have 'freed the spirit' of DOFS developers by removing many practical constraints on their ideas.

Of course, the same applies in the field of photonics, with narrowband, high-power lasers now easily available, enabling far simpler Brillouin sensing, for example, than the pioneers of this approach had to employ to prove their concepts. Likewise, the fibre amplifier has provided huge flexibility in the design of interrogators by boosting the probe waveforms and sometimes pre-amplifying the return signals. The availability of low-cost passive components such as splitters, circulators, in-line filters and wavelength-division devices has made the assembly of new, experimental types of interrogator simpler and faster. This has increased the pace of innovation in DOFS, and this acceleration was facilitated in large part by developments in the much larger market of optical communications.

Although DOFS can, and usually do, use relatively conventional fibre designs, specific coatings must be applied to these fibres to provide some defence against harsh environments, and ultimately, the glass itself must be optimised for the most resistant fibre types. Although the high-temperature, hydrogen-rich environment is a relatively small fraction of all DOFS deployments, it attracts a price premium that supports improvements to the fibres; conversely, the availability of specialty optical fibres in particular for sensing in harsh environments has facilitated the growth of DOFS.

Another recent development is the recognition that the output of a distributed sensor is fundamentally a function of two variables, namely the measurement time and distance along the fibre. This has led to the application of techniques derived from image processing [9–11] to filter the signals; these approaches have led to substantial improvements in measurement quality in DVS, Brillouin optical time-domain analysis (BOTDA) and DTS, respectively, in the specific references cited.

The history of DOFS is therefore one of a virtuous circle of new concepts in sensing that are facilitated by rapidly improving enabling technologies that then feed back into new applications, requiring yet further developments in interrogation concepts and instrument design.

In spite of some parts of the technology maturing, there is no reason to expect the field as a whole to slow down in its progress. New ideas are being proposed at increasing frequency, and they will drive applications that presently are excluded either for lack of the interrogation techniques or because the existing solutions are too complex or costly. The field of DOFS has shown over the years that complicated proof-of-concept demonstrations can be simplified and cost reduced through refinement of the ideas and the development of enabling technologies.

A decade or two of further rapid progress can therefore be expected, but we can only guess at the precise form of the technologies to be developed.

It is hoped that this book will guide newcomers to the field and will be a useful reference to those of us who have already enjoyed the stimulation and challenge of the development of this field of science and technology.

REFERENCES

1. Schulz, W. L., J. P. Conte, and E. Udd. 2001. Long-gage fiber optic Bragg grating strain sensors to monitor civil structures, 8th Annual International Symposium on Smart Structures and Materials, SPIE, 4330: 56–65.
2. Lloret, S., P. Rastogi, L. Thévenaz, and D. Inaudi. 2003. Measurement of dynamic deformations using a path unbalance Michelson interferometer based optical fiber sensing device. *Opt. Eng.* 42: 3: 662–669.
3. Denisov, A., M. A. Soto, and L. Thévenaz. 2014. 1'000'000 resolved points along a Brillouin distributed fibre sensor, 23rd International Conference on Optical Fibre Sensors, Santander, Spain, SPIE, 9157: D2.
4. London, Y., Y. Antman, N. Levanon, and A. Zadok. 2015. Brillouin analysis with 8.8 km range and 2 cm resolution, 24th International Conference on Optical Fibre Sensors, Curitiba, Brazil, SPIE, 9634: 0G.
5. Song, K. Y., Y. H. Kim, and K. Lee. 2015. Brillouin optical correlation domain analysis with more than 1 million effective sensing points, 24th International Conference on Optical Fibre Sensors, Curitiba, Brazil, SPIE, 9634: 0I.
6. Chiuchiolo, A., L. Palmieri, M. Consales et al. 2015. Cryogenic temperature monitoring in superconducting power transmission line at CERN with hybrid multi-point and distributed fiber optic sensors, 24th International Conference on Optical Fibre Sensors, Curitiba, Brazil, SPIE, 9634: 1U.
7. Tur, M., I. Sovran, A. Bergman et al. 2015. Structural health monitoring of composite-based UAVs using simultaneous fiber optic interrogation by static Rayleigh-based distributed sensing and dynamic fiber Bragg grating point sensors, 24th International Conference on Optical Fibre Sensors, Curitiba, Brazil, SPIE, 9634: 0P.
8. Thévenaz, L., S. Chin, J. Sancho, and S. Sales. 2014. Novel technique for distributed fibre sensing based on faint long gratings (FLOGs), 23rd International Conference on Optical Fibre Sensors, Santander, Spain, SPIE, 9157: 6W.
9. Zhu, T., X. Xiao, Q. He, and D. Diao. 2013. Enhancement of SNR and spatial resolution in ϕ-OTDR system by using two-dimensional edge detection method. *J. Lightw. Technol.* 31: 17: 2851–2856.
10. Soto, M. A., J. A. Ramirez, and L. Thevenaz. 2016. Intensifying the response of distributed optical fibre sensors using 2D and 3D image restoration. *Nat. Commun.* 7 in press. doi: 10.1038/ncomms10870.
11. Soto, M. A., J. A. Ramírez, and L. Thévenaz. 2016. Reaching millikelvin resolution in Raman distributed temperature sensing using image processing. Sixth European Workshop on Optical Fibre Sensors (EWOFS'2016), Limerick, Ireland, SPIE, 9916: 2A.

Index

Page numbers followed by f, t and n indicated figures, tables and footnotes, respectively.

A

ABB, 23
Absorption, 37
Accelerometers, 344
Accuracy, of sensor, 22
Acoustic noise spectrum, 374
Acoustic phonons, 162
Acoustic/seismic waves, response of DVS to
  angular/directional response, 353–354
  distance resolution in differential phase DVS systems, 346–347
  diversity in borehole seismic acquisition with DVS, 358–359
  gauge length effects, 348–352
  nature of measurement, 347–348
  polarity of seismic signal captured with DVS, 356
  pulse duration, 356–358
  sampling rates, 358
  wavenumber filtering/angular response, interaction of, 354–355
Acousto-optic modulators (AOMs), 48–49, 89, 99, 162, 175, 179, 180, 195, 243, 250, 260, 264, 351
Acquisition; *see also* Data acquisition
  in BOTDA
    parallel (multi-tone) approaches, 208–210, 209f
    slope-assisted fast acquisition, 210
  of OTDR, 86
  parallel multi-frequency, 192–193
Active phase/frequency feedback, semiconductor DFB lasers with, 44
Active seismic applications; *see also* Distributed vibration sensors (DVSs)
  borehole seismic surveying
    about borehole seismic acquisition, 341–343
    borehole seismic acquisition using permanently installed optical cables, 345–346
    optical borehole seismic acquisition on wireline cable, 343–345
  coupling in borehole seismic measurements, 361–363
  examples of bore seismic DVS data, 359–361
  response to acoustic/seismic waves
    angular and directional response, 353–354
    distance resolution in differential phase DVS systems, 346–347
    diversity in borehole seismic acquisition with DVS, 358–359
    gauge length effects, 348–352
    nature of measurement, 347–348
    polarity of seismic signal captured with DVS, 356
    pulse duration, 356–358
    sampling rates, 358
    wavenumber filtering/angular response, interaction of, 354–355
  surface seismic acquisition, 340–341, 363–364
Air flow, thermal measurement of, 308
Air guns, 340
Alumina (Al_2O_3), 41, 320, 321
Ampacity, 281
Amplified OTDR systems, 101
Amplified spontaneous emission (ASE), 45
Amplifiers
  optical, 45–46, 101
    doped fibre amplifiers, 45–46
    semiconductor, 46
    symbol for, 43f, 46
  remotely pumped rare-earth-doped fibre amplifiers, 190–192
Amplitude-based coherent Rayleigh backscatter DVSs, 235–239, 236f–238f
Amplitude-based Rayleigh backscatter, 365
Anti-Stokes Raman scattering, 109, 111
Anti-Stokes Raman with Rayleigh loss correction, 147–148
Anti-Stokes scattering, 12
Anti-Stokes signal, 285
AOMs, *see* Acousto-optic modulators (AOMs)
APDs, *see* Avalanche photodiode detectors (APDs)
Artificial lift monitoring, 378–380
Asahi Glass, 23
Avalanche photodiode detectors (APDs), 51–52, 73–74, 120
  SNR as function of, 81–83, 85, 81f–84f
Azimuthal asymmetry, 363

B

Backfill, 7
Background noise suppression, in BOCDA, 216
Backscatter curves, 58f
Backscatter factor, 58, 61, 62, 62f, 65
Backscatter signals; *see also specific entries*
  Raman OTDR, 114–115
Bandwidth, optical fibres, 40–41
Bessel functions, 36
Birefringence, 36, 219
  elliptical, 396
Blackbody radiation, 154, 403
BOCDA, *see* Brillouin optical coherence domain analysis (BOCDA)
BOCDR, *see* Brillouin optical coherence domain reflectometry (BOCDR)

BOFDA, *see* Brillouin optical frequency-domain analysis (BOFDA)
Borehole, production/integrity applications in; *see also* Distributed vibration sensors (DVSs)
  artificial lift monitoring, 378–380
  flow monitoring
    about, 373–374
    conveyance methods for DVS, 377
    DVS flow monitoring in boreholes, 377
    sensor placement, 377
    spectral discrimination in DVS analysis, 374–376
    total vibration energy, 374
    tube wave velocity, 376
    turbulent structure velocity, 376
    ultra-low frequency, 376–377
  hydrofracturing monitoring using microseismic events, 378
  perforation, 380–381
  well abandonment, 381
Borehole measurement; *see also* Distributed temperature sensors (DTS)
  conventional oil, 295–297
  CO_2 sequestration, 299
  electric submersible pumps (ESP), 300
  gas lift valves (GLV), 300
  gas wells, 299–300
  geothermal energy extraction, 299
  hydrate production, 301–302
  injector wells, 297–298
  intervention services, 298–299
  oil and gas wells, completion of, 288–289
  optical fibres in wells, deployment of, 290–294
  steam-assisted heavy oil production, 294–295
  subsea DTS deployment, 302
  thermal tracer techniques, 297
  well integrity monitoring, 301
Borehole seismic acquisition; *see also* Active seismic applications
  about, 341–343
  using permanently installed optical cables, 345–346
Borehole seismic surveying, *see* Active seismic applications
Bore seismic DVS data, examples of, 359–361
BOTDA, *see* Brillouin optical time-domain analysis (BOTDA)
BOTDR, *see* Brillouin optical time-domain reflectometer (BOTDR)
Bow-spring devices, 362
Bragg reflectors, 44
Brillouin and fluorescence measurement, combined, 317
Brillouin backscatter, 154–155, 161
Brillouin-based distributed sensing, 14–15
  vs. Raman technology, 15
Brillouin-based distributed strain and temperature sensors (DSTSs), *see* Distributed strain and temperature sensors (DSTSs), Brillouin-based
Brillouin beat spectrum (BBS), direct measurement of, 319
Brillouin frequency shift, sensitivity of, 167–168
Brillouin measurand discrimination, fibres designed for, 320–321
Brillouin optical coherence domain analysis (BOCDA), 162, 213–217, 316
  arrangement of, 213–214, 213f
  background noise suppression, 216
  low coherence, 217
  range extension, 215–216
  simplified approach, 215
  using phase coding, 216
Brillouin optical coherence domain reflectometry (BOCDR), 162, 213, 217–218, 217f, 325
Brillouin optical frequency-domain analysis (BOFDA), 162, 208
Brillouin optical time-domain analysis (BOTDA), 14, 162, 197–198, 310, 318, 400
  basic arrangement, 197, 198f
  BOFDA, 208
  composite pulse methods in, 203–206
    differential pump interrogation, 205–206
    simultaneous gain/loss interrogation, 206
    single shaped pump interrogation, 203–205, 204f
  optical arrangements, 199–202, 200f–203f
  overview, 197–199
  performance and limitations
    distance enhancements through remote amplification, 210–211
    parallel (multi-tone) acquisition approaches, 208–210, 209f
    power limitations due to SPM, 211
    slope-assisted fast acquisition, 210
  phase measurement in, 211–213, 212f
  pulse compression coding
    as extension of composite pulse techniques, 207
    for range extension, 206–207
  sensitivity, 203
  slow light correction, 207–208
  spatial resolution, 203
  trade-offs in, 203
Brillouin optical time-domain reflectometer (BOTDR), 14–15, 161, 174n, 177, 286, 316, 324
  dual-pulse method in, 195–196, 195f–196f
  heterodyne measurement, 177–179, 178f
  limitations and benefits, 184–186
  microwave heterodyne-detection arrangement, 180–181, 180f
  pulse compression coding in, 194
  remote amplification for range extension, 187–192
    distributed Raman amplification, 188–190
    hybrid Raman/EDFA amplification, 192
    remotely pumped rare-earth-doped fibre amplifiers, 190–192
  simultaneous multi-frequency measurement in, 194, 194f
  spectral efficiency, 197
Brillouin scattering, 12, 87n, 285
  overview, 161
  spontaneous, 14
  stimulated, 14, 87
  vs. Raman scattering, 161
Brillouin shift measurement, 315, 316, 409
Brillouin spectrum, 12–13, 12f; *see also* Distributed strain sensors
  direct measurement of, 319
  few-mode fibres, 318–319
  fibres designed for Brillouin measurand discrimination, 320–321
  independent optical cores, 317–318
  multiple Brillouin peaks with distinct properties, 318
  multi-wavelength interrogation, 319
  polarisation-maintaining fibres/photonic crystal fibre, 318

Broad contact semiconductor lasers, 42–43, 43f
Broadside insensitivity, 353, 354
Brugg Cable, 23
Building efficiency research, 305; *see also* Distributed temperature sensors (DTS)

C

Calibration, Raman OTDR DTS systems, 149–150
Capture fraction
  changes, 394
  defined, 57
  for graded-index multimode fibres, 64
  OTDR, 64–65
Carbon dioxide sequestration, 299; *see also* Borehole measurement
Casing, 289
Casing deformation, 329
CD (chromatic dispersion), 40, 41
Cement and concrete curing, 302–302
Check-valve, 291
Chemical detection, 323
Chemical injection, 287, 301
Chemical vapour deposition, 41, 42
Chromatic dispersion (CD), 40, 41
Circuit breakers, 283
Circulators, 47
Civil engineering; *see also* Distributed strain sensors
  crack detection, 327–328
  offshore structures, damage in, 328
  tunnels, 327
Climate studies, 305
Coating, spontaneous Brillouin scattering and, 168–169
C-OFDR, *see* Coherent OFDR (C-OFDR)
Coherence effects, in OTDR, 66–72, 67f–71f; *see also* Coherent OTDRs (C-OTDRs)
  backscatter power and probability density function, 69–72, 70f–71f
  backscatter power *vs.* distance measured, 67–68, 67f–68f
  backscatter signals, 67–68, 69f
Coherent-detection systems, 352
Coherent FMCW, 128–129
Coherent OFDR (C-OFDR), 92–93
Coherent OTDRs (C-OTDRs), 67, 88–91, 88f, 95, 177
  dynamic strain measurement linearisation, 274
  frequency-diverse, 99–101, 100f
  heterodyne detection, 89–91
  homodyne detection, 89, 91
  modelling of, 234–235
Coherent Rayleigh backscatter, 232
  amplitude-based DVSs, 235–239, 236f–238f
  frequency-scanned, 272–273
  limitations in frequency-scanned, 272–273
  in static measurements, 267–271
Coherent Rayleigh/Brillouin hybrid, 317
Coherent Rayleigh noise (CRN), 175, 316
Coiled tubing (CT), 293
Composite pulse methods, in BOTDA, 203–206
  differential pump interrogation, 205–206
  pulse compression coding as extension of, 207
  simultaneous gain/loss interrogation, 206
  single shaped pump interrogation, 203–205, 204f
Composite structural health monitoring, 329
Compressional waves (P-waves), 343, 344
Computational POTDR (CPOTDR) method, 400
Connectors, 50
Continuous-wave (CW) probe light, 409
Control line installations in boreholes, 291, 292f
Conventional oil, 295–297; *see also* Borehole measurement
Conveyance methods for DVS, 377
Conveyors, 307; *see also* Distributed temperature sensors (DTS)
Cossor Electronics, 23
Costs
  multiplexed sensor systems and, 6–7, 6f, 8f
  of optical fibre sensors, 6
C-OTDRs, *see* Coherent OTDRs (C-OTDRs)
Counter-clockwise (CCWD) signal, 408
Coupling, in borehole seismic measurements, 361–363
Coupling effects, spontaneous Brillouin scattering and, 168–169
Crack detection, 327–328; *see also* Civil engineering
Critical angle, 31
Cross-talk, between sensing positions, 22
Cryogenic measurements, 285–286
Curing of cement/concrete, 302–303; *see also* Distributed temperature sensors (DTS)
Cut-off frequency, 33
CWD signal, 408

D

Dakin, John, 13
Dam, 303–304
DAS, *see* Distributed acoustic sensing/sensor (DAS)
Data acquisition, in OFDR, 129–130, 129f
dB, in optical fibre, 74n
DBGs, *see* Dynamic Bragg gratings (DBGs)
Dead time, 134
  correction for, 136
Decision-supporting software tools, 382
Degeneracy, 36
Dense point sensor arrays, 25–26
Depolarisation, 221
Detectors, 50–52; *see also* Avalanche photodiode detectors (APDs)
  imperfections, corrections for, 136
  in OTDR, 73–74
  photomultiplier tubes (PMT), 133
  photon number resolving, 137
  single-photon avalanche detector (SPAD), 133
  superconducting nanowire single-photon detectors (SNSPD), 133
Deviated wells, 297
dΦ-DVS, *see* Differential phase-measuring DVS (dΦ-DVS)
Differential mode attenuation (DMA), 64n
Differential phase-measuring DVS (dΦ-DVS), 239–256; *see also* Dual-pulse DVSs; Heterodyne DVS (hDVS); Homodyne DVS
  interferometric phase recovery, 246–249, 247f
  linearity in, 264–266, 265f–266f
  phase unwrapping, 240–241
  techniques for, 240, 240f
Differential pump interrogation, BOTDA and, 205–206
Dispersion, 40
  intermodal, 40
  intramodal, 40

Distance-step resolution, 21
Distributed acoustic sensing/sensor (DAS), 24, 232
Distributed feedback (DFB) lasers, 175, 316
  fibre, 44–45
  semiconductor, 44
    with active phase/frequency feedback, 44
Distributed interferometers, 406–408; *see also* Forward-scattering methods
Distributed optical fibre hot-wire anemometers, 307
Distributed optical fibre sensors (DOFSs)
  advantages, 3–4, 7
  background, 3–4
  commercialisation history, 22–25
    Brillouin-based distributed strain and temperature sensors, 24
    distributed vibration sensing, 24–25
    Raman DTS, 22–24
  concept, 3–9
  connectors, 50
  defined, 7
  described, 7–9
  dynamic sensing, 231
  elements of, 419, 420f
  historical development, 9–16
    Brillouin-based distributed sensing, 14–15
    distributed temperature sensors, 11
    elastic and inelastic scattering, 11–13, 12f
    optical TDR (OTDR), 9–11, 9f
    Raman OTDR, 13
    with Rayleigh backscatter, 15–16
  lasers, 42–45
  Lidar systems, 8
  optical amplifiers, 45–46
  PDs in, 51
  performance criteria, 16–22
    estimation of parameters, 21–22
    interplay between, 20–21
    measurand resolution, 16, 17
    measurement time, 19–20
    range, 16, 17
    sampling resolution, 19
    spatial resolution, 17–18, 17f–19f
  static (or quasi-static) measurement, 231
  TDR, 8–9
  trends in, 424–425
  types of, 422t
Distributed pressure sensing, 400–401
Distributed Raman amplification, 188–190
Distributed sensing, 25–26
  fibre reflectometry application to, 102–103
  with Rayleigh backscatter, 15–16
Distributed sensing systems
  circulators, 47
  components used in, 42–52
  connectors and splices, 50
  detectors, 50–52
  Faraday rotation mirror (FRM), 47
  fibre Bragg gratings (FBGs), 47–48
  fused-taper couplers, 46
  isolators, 47
  lasers, 42–45
  micro-bulk optics, 47
  modulators, 48–49
  optical amplifiers, 45–46
  polarisation controllers, 50
  switches, 49
Distributed sensors, 11, 25–26; *see also* Distributed optical fibre sensors (DOFSs); *specific entries*
  vs. multiplexed point sensors, 423–424
Distributed strain and temperature sensors (DSTSs), Brillouin-based
  based on coherence-domain reflectometry-based
    BOCDA, 213–217; *see also* Brillouin optical coherence domain analysis (BOCDA)
    BOCDR, 217–218; *see also* Brillouin optical coherence domain reflectometry (BOCDR)
  commercial, 24
  dynamic Bragg gratings, 218–220, 220f
  overview, 161–162
  polarisation effects, 220–221
  spontaneous Brillouin scattering; *see also* Spontaneous Brillouin scattering (SpBS)
    coating and coupling effects, 168–169
    effect, 162
    frequency shift, 162–163
    loss compensation for intensity measurements, 166–167
    sensitivity coefficients, 164–165
    sensitivity of Brillouin frequency shift, 167–168
    sensitivity of intensity to strain, 166
    sensitivity of intensity to temperature, 165–166
    spectral width, 163–164
  spontaneous Brillouin scattering, recent advances in
    dual-pulse method in BOTDR, 195–196, 195f, 196f
    fast measurement techniques, 192–193
    pulse compression coding, 194
    remote amplification for range extension, 187–192, 188f
    simultaneous multi-frequency measurement, 194, 194f
    spectral efficiency, 197
  spontaneous Brillouin scattering, systems based on, 171–187
    electrical separation, 177–183, 178f, 180f, 182f–183f
    frequency, gain and linewidth estimations, 183–184, 184f
    limitations and benefits, 184–186
    optical separation, 171–177, 172f, 173f, 175f–177f
  stimulated Brillouin scattering, 169–171, 170f; *see also* Stimulated Brillouin scattering (SBS)
  stimulated Brillouin scattering, system based on
    BOFDA, 208; *see also* Brillouin optical frequency-domain analysis (BOFDA)
    BOTDA, 197–208; *see also* Brillouin optical time-domain analysis (BOTDA)
    performance and limitations, 208–211
    phase measurement, 211–213
    pulse compression coding, 206–207
    sensitivity and spatial resolution, 203
    slow light correction, 207–208
Distributed strain sensing, SpBS-based, 187–197
  dual-pulse method in BOTDR, 195–196, 195f, 196f
  fast measurement techniques, 192–193
  pulse compression coding, 194
  remote amplification for range extension, 187–192
  simultaneous multi-frequency measurement, 194, 194f
  spectral efficiency, 197

Distributed strain sensors, 324–325
application
casing deformation, 329
civil engineering, 327–328
composite structural health monitoring, 329
energy cables, 326
ground movement/geomechanics, 330–332
mechanical coupling/limitations on strain measurements, 325–326
nuclear, 328
pipeline integrity monitoring, 329–330
river banks and levees, 332
security/perimeter fence, 332–333
shape sensing, 328–329
subterranean formations, compaction/subsidence in, 330
telecommunications, 326–327
measurands using strain
chemical detection, 323
liquid flow, 322–323
pressure, 322
polymer optical fibres (POF)-based strain measurement, 324–325
separation of temperature/strain effects
Brillouin spectrum and, 317–321
hybrid measurement, 316–317
through cable design, 315–316
Distributed temperature sensors (DTS), 11, 13, 35, 58
active fibre components, performance of, 310
borehole measurement
conventional oil, 295–297
CO_2 sequestration, 299
electric submersible pumps (ESP), 300
gas lift valves (GLV), 300
gas wells, 299–300
geothermal energy extraction, 299
hydrate production, 301–302
injector wells, 297–298
intervention services, 298–299
oil and gas wells, completion of, 288–289
optical fibres in wells, deployment of, 290–294
steam-assisted heavy oil production, 294–295
subsea DTS deployment, 302
thermal tracer techniques, 297
well integrity monitoring, 301
building efficiency research, 305
cement and concrete curing, 302–303
conveyors, 307
cryogenic measurements, 285–286
energy cables, 281–282
environment and climate studies, 305
fire detection, 283–284
generators/circuit breakers, 283
hydrology
dam/dyke, 303–304
near-surface, streams and estuaries, 304–305
liquid-level measurement, 309
market turnover, 23
mining industry, 307
in nuclear facilities, 306–307
pipelines, leak detection/flow assurance, 286–287
process plant, 284–285
Raman DTS, commercialisation history of, 22–24
Raman OTDR DTS systems
alternative approaches, 128–137
backscatter signals, 114–115
design, 115–121, 116f
fundamental limitations, 122–125, 124f
high-performance systems operating at around 1064 nm, 121
long-range systems, 121
performance of, 121–128
practical issues and solutions, 137–155
principles, 109–115
Raman scattering, temperature sensitivity of, 111–112, 113f
reference coil, 116–117
short-range systems, 120
signal-to-noise ratio, 125
spatial resolution, 121–122
spontaneous Raman scattering, 109–111, 111f
temperature resolution, 122
vs. high-spatial resolution systems, 125–128, 126t–127t
walk-off effect, 123–125, 124f
thermal measurement of air flow, 308
transformers, 282–283
Distributed vibration sensing (DVS), 24
applications, 24–25
in security, 25
Distributed vibration sensors (DVSs), 232–235
active seismic applications
borehole seismic surveying, 341–346
coupling in borehole seismic measurements, 361–363
examples of bore seismic DVS data, 359–361
response to acoustic/seismic waves, 346–359
surface seismic acquisition, 340–341
surface seismic acquisition, DVS in, 363–364
amplitude-based coherent Rayleigh backscatter, 235–239, 236f–238f
comparison and performance, 256–257
differential phase-measuring DVS (dΦ-DVS), 239–256
interferometric phase recovery, 246–249, 247f
phase unwrapping, 240–241
techniques for, 240, 240f
dual-pulse, 241–245, 242f–245f
energy cables
fault detection/partial discharge, 372
subsea cables, 371
wind and ice loading in overhead cables, 372–373
heterodyne (hDVS), 249–254, 250f, 252f–254f
homodyne, 254–255, 255f–256f
interpretation of DVS data, 382
issues and solutions
dynamic range of measurand, 263–264
dynamic range of signal, 262–263
fading, 257–261
linearity in dΦ-DVS systems, 264–266, 265f–266f
optical non-linearity, 266
reflections, 262
multimode fibre in distributed vibration sensing, 381–382
overview, 339–340
perimeter security/pipeline surveillance, 364–370
production/integrity applications in boreholes
artificial lift monitoring, 378–380
flow monitoring, 373–377

hydrofracturing monitoring using microseismic events, 378
perforation, 380–381
well abandonment, 381
timescales, 232–234, 233f
transportation, 370–371
Diversity, in borehole seismic acquisition with DVS, 358–359
DOFSs, *see* Distributed optical fibre sensors (DOFSs)
Dopants, 39
Doped fibre amplifiers, 45–46
Doping with rare-earth, 390
Double-ended measurements
OTDR, 60–64
configuration, 61f
folded-path configuration, 63–64, 63f
separation of traces, 61–62, 62f
Raman OTDR DTS systems, 143–144
DTS, *see* Distributed temperature sensors (DTS)
Dual-mode fibres, mode coupling in, 404–405; *see also* Forward-scattering methods
Dual-pulse DVSs, 241–245, 242f–245f
Dual-pulse method, in BOTDR, 195–196, 195f–196f
Dual-wavelength beat length measurements, 398
DVS, *see* Distributed vibration sensing (DVS)
DVSs, *see* Distributed vibration sensors (DVSs)
Dyke, 303–304, 332
Dynamic Bragg gratings (DBGs), 162, 218–220, 220f
Dynamic measurements
Rayleigh backscatter
strain measurements, 267
vs. static, 267
of strain, linearisation in intensity C-OTDR, 274
Dynamic range
of measurand, 263–264, 263f
of signal, 262–263
Dynamic sensing, 231

E

Elastic scattering, 11–13
Electrical cables, 307
Electrical local oscillator (ELO), 179, 181
Electrical seismic sensors, 342
Electrical separation, spontaneous Brillouin scattering and, 177–183, 178f, 180f, 182f–183f
Electrical submersible pumps (ESP), 300, 379; *see also* Borehole measurement
Electrical time-domain reflectometers, 331
Electronic absorptions, 37
Electronic downhole sensors, 341
Electro-optic modulators (EOMs), 48, 49, 180
Elliptical birefringence, 396
ELO, *see* Electrical local oscillator (ELO)
Energy cables, 281–282; *see also* Distributed temperature sensors (DTS); Distributed vibration sensors (DVSs)
fault detection/partial discharge, 372
subsea cables, 371
wind/ice loading in overhead cables, 372–373
Enhanced geothermal systems (EGS), 299
Enhanced oil recovery (EOR) strategy, 345
Environment and climate studies, 305
EOMs, *see* Electro-optic modulators (EOMs)
Er-doped fibre amplifier (EDFA), 46, 175, 179, 180, 188, 190
Estuaries, streams and, 304–305
European standard EN54 (fire detection systems), 284
Evanescent wave, 33
Excess noise, 76
External cavity feedback semiconductor lasers, 44
Extrinsic sensors, 5, 5f
vs. intrinsic sensors, 5

F

Fabry-Pérot interferometer (FPI), 172, 172f
Fabry-Pérot laser, 316
Fabry-Pérot single-mode laser diodes, 43–44
Fading, 66
in Rayleigh backscatter, 257–261
alternative approaches, 261, 261f
diversity effect on signal quality, 259–260, 259f
specific arrangements for frequency diversity, 260–261, 260f
statistically independent signals requirements, 257–258, 258f
False-negative probability (FNP), 364
False-positive probability (FPP), 364
Faraday effect, 401
Faraday rotation, 396
Faraday rotation mirror (FRM), 47
Fast measurement techniques
frequency discriminators, 192
parallel multi-frequency acquisition, 192–193
signal tracking, 193
Fault detection, 372
Few-mode fibres, 318–319; *see also* Brillouin spectrum
Fibre Bragg gratings (FBGs), 44, 47–48, 323, 329, 423
Fibre cables, 293
Fibre components, active, 310
Fibre degradation, in loss compensation methods, 138–142, 139f–142f
Fibre DFB lasers, 44–45
Fibre-optic intrusion detection systems, 332–333, 364, 365
Fibre-optic strain measurements, 331
Fibre reflectometry, application to distributed sensing, 102–103
Fibres
for Brillouin measurand discrimination, 320–321
for Raman OTDR DTS systems, 117–119, 119f
Fictive temperature (T_f), 37, 166
Field-effect transistor (FET), 77–78, 120
Field tests, 328, 332, 359, 378
Filters, Raman OTDR DTS systems and, 152–153
Fire detection, 283–284; *see also* Distributed temperature sensors (DTS)
Fires in road tunnels, 284
Flame hydrolysis, 41–42
Floating production storage and offloading (FPSO) facility, 287
Flow assurance, 286–287
Flowmeter by CiDRA, 376
Flow monitoring, 373–377, 374–377; *see also* Boreholes, production/integrity applications in
Fluid-level estimation, 309
Fluorescence, 317, 392, 394
FMCW, *see* Frequency-modulated, continuous-wave (FMCW)

Folded-path OTDR approach, 63–64, 63f
Formation isolation valve, 289
Forward Raman scattering, 153–154, 153f
Forward-scattering methods
  about, 403
  distributed interferometers, 406–408
  modalmetric sensors, 405–406
  mode coupling in dual-mode fibres, 404–405
Fourier transformed data, 130
Fourier transforms, 376
FPI, *see* Fabry-Pérot interferometer (FPI)
Fracturing, 378
Free spectral range (FSR), 172
Frequency
  estimation, spontaneous Brillouin scattering and, 183–184, 184f
  ultra-low, 376–377
Frequency band energy (FBE), 340
Frequency discriminators, 192
Frequency-diverse C-OTDR, 99–101, 100f
Frequency-domain-only interrogation, Rayleigh backscatter, 271–272
Frequency-modulated, continuous-wave (FMCW), 128
  coherent, 128–129
Frequency-scanned coherent Rayleigh backscatter, 272–273
Frequency shift
  Brillouin, sensitivity of, 167–168
  spontaneous Brillouin scattering, 162–163
Fujikura, 23
Full polarimetric analysis of POTDR signals, 399–400
Furukawa, 23
Fused-taper couplers, 46
Fusion splicers, 50

G

Gain, spontaneous Brillouin scattering and, 183–184, 184f
Gas-lift valves (GLV), 300, 379; *see also* Borehole measurement
Gas wells, 299–300
Gauge length effects, 348–352
Generators, 283
GeO_2-doped graded-index multimode fibres, 72
Geomechanics, 330–332
Geotextile, 332
Geothermal energy extraction, 299; *see also* Borehole measurement
Germania (GeO_2), 41, 111, 306
Global Positioning System (GPS), 340, 359
Graded-index fibre, 394
Graded-index multimode fibres, 34–36, 34f
  capture fraction for, 64
  GeO_2-doped, 72
Ground movement, 330–332
Group velocity, defined, 33n

H

hDVS, *see* Heterodyne DVS (hDVS)
Heaters, 287
Heat pulse measurement, 304
Heat-sensitive cable, 389
Heavy oil production, steam-assisted, 294–295
Helically wound cable (HWC), 363, 364
Hermetic coatings, 139
Heterodyne detection
  BOTDR, 177–179, 178f
  C-OTDRs, 89–91
Heterodyne DVS (hDVS), 249–254, 250f, 252f–254f, 255, 256, 257, 344, 346, 360
  frequency-diverse hDVS system, 260, 260f
  phase-diverse hDVS system, 261, 261f
High-birefringence fibres
  Kerr effect for POTDR in, 410–411
  POTDR in, 396–398
High-electron mobility transistors (HEMT), 78
High-performance systems, operating at around 1064 nm, 121
High-spatial resolution systems *vs.* Raman OTDR DTS systems, performance, 125–128, 126t–127t
High-temperature operation, of Raman sensors, 154
Hitachi Cable, 23
Holmium, 393
Homodyne detection, C-OTDRs, 89, 91
Homodyne DVS, 254–255, 255f–256f
Hot-slug technique, 298
Hot-wire anemometry (HWA), 304, 308
Hybrid measurement; *see also* Distributed strain sensors
  Brillouin/fluorescence, 317
  coherent Rayleigh/Brillouin hybrid, 317
  Raman/Brillouin hybrid approach, 316
  spontaneous Brillouin intensity/frequency measurement, 316
Hybrid Raman/EDFA amplification, 192
Hydrate production, 301–302
Hydrocarbon sensing, 390
Hydrocarbon wells, 302
Hydrofracturing monitoring using microseismic events, 378
Hydrogels, 390
Hydrology; *see also* Distributed temperature sensors (DTS)
  dam/dyke, 303–304
  near-surface, streams and estuaries, 304–305

I

Ice loading in overhead cables, 372–373
IF (intermediate frequency), 89
Ignition, 283, 308n
Incoherent OFDR, 94, 94f
Incoherent OTDR, 95
Independent optical cores, 317–318; *see also* Brillouin spectrum
Index-modifying additives, 39
Inelastic scattering, 11–13
InGaAs devices, 76, 85, 121
Injector wells, 297–298; *see also* Borehole measurement
Interferometers, distributed, 406–408
Interferometric phase recovery, dΦ-DVS, 246–249, 247f
Intermediate frequency (IF), 89
Intermodal dispersion, 40
Intervention services, 298–299
Intramodal dispersion, 40
Intrinsic sensors, 5, 5f
  vs. extrinsic sensors, 5
Ionising radiation, 328
Isolators, 47

J

'J-fibre' configuration, Raman OTDR DTS systems, 144–145
Johnson noise, 77, 78, 79
Johnson noise current density, 77
Joule-Thomson cooling, 287
Joule-Thomson effect, 296, 297

K

Kerr effect, 87
in distributed temperature sensing, 409–410
for POTDR in high-birefringence fibres, 410–411
Kramers-Kronig relations, 207, 219

L

Landau-Placzek ratio (LPR), 165, 166–167, 316
measurement of, 175f
Land seismic applications, sensors in, 340
Landslides, 331
Lasers, 42–45
broad contact semiconductor lasers, 42–43, 43f
external cavity feedback semiconductor lasers, 44
Fabry-Pérot single-mode laser diodes, 43–44
fibre DFB lasers, 44–45
Raman OTDR DTS systems, 117–119, 119f
semiconductor DFB lasers with active phase/frequency feedback, 44
semiconductor distributed feedback lasers, 44
symbols for optical sources, 43f, 45
tunable lasers, 45
Laser speckle, 405
LEAF fibres, 320
Leak detection, 286–287
Levenberg-Marquart non-linear least squares algorithm, 183
Lidar (light radar) systems, 8
Light, attributes, 4
Light-emitting diode (LED), 86
Linear birefringence, 396
Linear heat detection, 389–390
Linearity, in dΦ-DVS, 264–266, 265f–266f
Linewidth estimations, spontaneous Brillouin scattering and, 183–184, 184f
Liquefied natural gas (LNG), 286, 391
Liquid flow, 322–323
Liquid-level measurement, 309; *see also* Distributed temperature sensors (DTS)
Logging, 293n, 296
Longitudinal acoustic (LA) waves, 162
Long-range systems, 121
Lorentzian spectral distribution, 183
Loss-based sensors
loss of numerical aperture, 391–392
micro-bending-based sensors
about, 387–389
linear heat detection, 389–390
moisture/hydrocarbon sensing, 390
safety barriers, 389
radiation detection, 390–391
rare-earth-doped fibre temperature sensing, 392–394
Loss compensation methods
intensity measurements of Brillouin scattering, 166–167
Raman OTDR DTS systems and
anti-Stokes Raman with Rayleigh loss correction, 147–148
doubled-ended measurements, 143–144
dual sources separated by two Stokes shifts, 148
fibre degradation, 138–142, 139f–142f
'J-fibre' configuration, 144–145
multi-wavelength measurements, 145
Raman ratio with separate sources, 145–147, 146f
Loss mechanisms, optical fibres
absorption, 37
other mechanisms, 39–40
scattering, 37–39, 38f
Low coherence BOCDA, 217
Low-coherence interferometry, 404
Low-loss optical fibres, 41

M

Mach-Zehnder interferometer (MZI), 49, 173–175, 173f, 406–407
interferometer, rejection of, 174n
refinement, 176–177, 177f
Magnetic field detection, 401
Magnetic resonance imaging (MRI) scanners, 401
Marine seismic exploration, 340
Mass-balance systems, 286
Master oscillator, power amplifier (MOPA), 121
Maxwell's equations, 33
MC-PC, *see* Multi-channel photon counting (MC-PC)
Measurand distribution, 18
Measurand resolution, 16, 17, 22
Measurands
defined, 3n
dynamic range of, 263–264, 263f
Measurement time, 19–20
Mechanical coupling, 325–326
Mechanical splices, 50
Meridional ray, 31
Methane hydrates, 301
Michelson interferometer, 407, 408
Micro-bending-based sensors; *see also* Loss-based sensors
about, 387–389
linear heat detection, 389–390
moisture/hydrocarbon sensing, 390
safety barriers, 389
Micro-bulk optics, 47
Microchip Nd:YAG Q-switched lasers, 121
Micro-electro-mechanical (MEMS) devices, 49
Microwave heterodyne-detection BOTDR arrangement, 180–181, 180f
Mie scattering, 39
Mining industry, 307
Modalmetric sensors, 405–406
Modal power distribution, 35
Mode coupling in dual-mode fibres, 404–405; *see also* Forward-scattering methods
Mode groups, 35
Modified chemical vapour deposition process (MCVD), 42
Modulation instability (MI), 185
Modulators, 48–49
acousto-optic (AOMs), 48–49
electro-optic (EOMs), 48, 49

Moisture sensing, 390
m-sequences, 95
Multi-channel photon counting (MC-PC), 132, 136
Multi-frequency measurement, simultaneous, in BOTDR, 194, 194f
Multi-frequency measurements in DVS, 264, 356, 358
Multimode, short-wavelength OTDR, 78–85, 79t, 80f–84f
Multimode fibres
in distributed vibration sensing, 381–382
graded-index, 34–36, 34f
step-index, 33–34, 34f
Multi-photon fluorescence, 394
Multiple Brillouin peaks with distinct properties, 318
Multiplexed point sensors *vs.* distributed sensors, 423–424
Multiplexed sensor systems, 6–7, 6f, 8f; *see also* Distributed optical fibre sensors (DOFSs)
Multi-wavelength interrogation, 319; *see also* Brillouin spectrum
Multi-wavelength measurements, Raman OTDR DTS systems, 145
MZI, *see* Mach-Zehnder interferometer (MZI)

N

Near-surface hydrology, 304–305
Noise
background noise suppression, in BOCDA, 216
in optical receivers, 75–85
multimode, short-wavelength OTDR, 78–85, 79t, 80f–84f
single-mode, long-wavelength OTDR, 85
Non-linear optical methods
about, 408–409
Kerr effect for POTDR in high-birefringence fibres, 410–411
Kerr effect in distributed temperature sensing, 409–410
stimulated Raman interaction, 409
two-photon fluorescence (TPF), 411–413
Non-strain-coupled fibres, 315
Normalised frequency, 33
Nuclear facilities, temperature monitoring in, 306–307
Nuclear waste repository, monitoring of, 328
Numerical aperture (NA), loss of, 391–392

O

OFDR, *see* Optical frequency-domain reflectometry (OFDR)
Offshore structures, damage in, 328; *see also* Civil engineering
Oil and gas wells, completion of, 288–289
OLO, *see* Optical local oscillator (OLO)
Optical amplification, 187
Optical amplifiers, 45–46, 101
doped fibre amplifiers, 45–46; *see also* Er-doped fibre amplifier (EDFA)
semiconductor, 46
symbol for, 43f, 46
Optical borehole seismic acquisition on wireline cable, 343–345
Optical fibre acoustic sensor arrays, 341
Optical fibres; *see also* Distributed optical fibre sensors (DOFSs)
bandwidth, 40–41
construction, 41–42
graded-index multimode fibres, 34–36, 34f
loss mechanisms
absorption, 37
other mechanisms, 39–40
scattering, 37–39, 38f
propagation in, 31–41, 32f
single-mode optical fibres, 34f, 36–37
step-index multimode fibres, 33–34, 34f
structure, 31
types of, 33–37
in wells, deployment of, 290–294
Optical fibre sensors, 3–6; *see also* Distributed optical fibre sensors (DOFSs)
advantages, 3–4
background, 3–4
cost of, 6
intrinsic *vs.* extrinsic, 5, 5f
multiplexing strategies, 6–7, 6f, 8f
patch panel, 3
system components, 3, 4f
transit cable, 3
Optical frequency domain analysis, 25
Optical frequency-domain reflectometry (OFDR), 92, 128–131, 286, 308, 401
coherent (C-OFDR), 92–93
data acquisition in, 129–130, 129f
incoherent, 94, 94f
signal recovery, 130
vs. OTDR methods, 130–131
Optical ground wire (OPGW), 326, 372, 373
Optical heating, 308
Optical local oscillator (OLO), 88–89, 177, 179, 180, 181, 182, 240, 249–250, 254, 255, 271
Optical non-linearity, 145, 151, 189, 266
Optical phonons, 162
Optical receivers, noise in, 75–85
multimode, short-wavelength OTDR, 78–85, 79t, 80f–84f
single-mode, long-wavelength OTDR, 85
Optical separation, spontaneous Brillouin scattering and, 171–177, 172f–173f, 175f–177f
Optical time-domain reflectometry (OTDR), 9–11, 9f, 370, 372, 388
amplified systems, 101
Brillouin OTDR (BOTDR), 14–15
capture fraction, 64–65
coherence effects, 66–72, 67f–71f
concept, 55
cross-sensitivity of location to environmental effects, 57
double-ended measurement, 60–64, 61f–63f
historical development, 55
multimode, 74
multimode, short-wavelength, 78–85, 79t, 80f–84f
photodiodes, 73–74
process overview, 55–56, 56f
pulsed laser, 74
Raman OTDR, *see* Raman OTDR
range, enhancement of
detection improvement, 88–91
photon counting, 91–92
spread-spectrum techniques, 92–101
Rayleigh OTDR, 15
re-capture process, 55, 56f
signal levels, 72–73
single-ended measurement, 56–60, 58f

single-mode, long-wavelength, 85
spatial resolution of, 59–60
system design
acquisition and further processing, 86
common design, 74–75, 75f
detectors, 73–74
limitations to power launched, 87–88
noise in optical receivers, 75–85
overload considerations, 85–86
sources, 73–74
trends in, 102
vs. OFDR methods, 130–131
Opto-electronic system, 3
OTDR, *see* Optical time-domain reflectometry (OTDR)
Outside vapour phase oxidation (OVPO), 41
Overhead cables, wind/ice loading in, 372–373
Overhead power lines, 372
Overload considerations, in OTDR, 85–86
OVPO (outside vapour phase oxidation), 41

P

Packer, 289
Parallel (multi-tone) acquisition approaches, BOTDA and, 208–210, 209f
Parallel multi-frequency acquisition, 192–193
Parameters, estimation (DOFSs), 21–22
Partial breakdown, 281
Partial discharge, 372
Passive techniques, 309
Patch panel, optical fibre sensors, 3
PCC, *see* Pulse compression coding (PCC)
Perfluorinated graded-index fibres (PFGI), 324
Perfluorinated polymer fibres, 328
Perforation, 380–381
Perimeter fence, 332–333
Perimeter security, 364–370
Permanent reservoir monitoring (PRM), 340, 341
Phase coding, BOCDA using, 216
Phase measurement, in BOTDA, 211–213, 212f
Phase-OTDR, 235
Phase-sensitive amplifiers, 399
Phase unwrapping, dΦ-DVS, 240–241
Phosphorus pentoxide (P_2O_5), 41, 306
Photodiodes (PDs), 50
in OTDR, 73–74
Photomultiplier tubes (PMT), 133
Photon counting, 91, 132, 398
limitations, 133–134
multi-channel (MC-PC), 132, 136
OTDRs, 91–92
for Raman DTS, 133–134
recent improvements in, 136–137
time-correlated single-photon counting (TC-SPC), 132, 134–136, 134f
Photonic crystal fibre, 318; *see also* Brillouin spectrum
Photonic crystal fibres (PCFs), 37, 318
Photon number resolving detectors, 137
Pig-catcher, 369
Pipeline
inspection gauges, 368
integrity monitoring, 329–330
leak detection/flow assurance, 286–287
surveillance, 364–370
Pipeline Instrumented Gauge, 286
Planck-Einstein relation, 51
Plane wave, 32
Plasma deposition, 41, 42
PMT, *see* Photomultiplier tubes (PMT)
Point Bragg gratings, 329
Point sensor arrays, 25–26
Polarisation controllers, 50
Polarisation diversity, 90, 221
Polarisation effects, 220–221
Polarisation-maintaining (PM) fibres, 318; *see also* Brillouin spectrum
Polarisation-preserving fibres, 34f, 36
Polarisation scrambling, 90, 221
Polarity of seismic signal captured with DVS, 356
Polarisation mode dispersion, 402
Polarisation optical frequency-domain reflectometry (POFDR), 398
Polarisation optical time-domain reflectometry (POTDR), 11, 25
application
current/magnetic fields, 401–402
distributed pressure sensing, 400–401
polarisation mode dispersion, 402
vibration sensing, 402–403
basic principles of, 395–396
distribution of state of polarisation, estimating, 400
dual-wavelength beat length measurements, 398
full polarimetric analysis of POTDR signals, 399–400
in high-birefringence fibres, 396–398
Kerr effect for, in high-birefringence fibres, 410–411
polarisation optical frequency-domain reflectometry (POFDR), 398
Polarisation reflectometry approach, 401
Polymer fibres, 394
Polymer optical fibres (POF)-based strain measurement, 324–325; *see also* Distributed strain sensors
Poly-methyl methacrylate (PMMA), 324
POTDR, *see* Polarisation optical time-domain reflectometry (POTDR)
Pound-Drever-Hall technique, 44
Power, launched, in OTDRs, 87–88
Pressure, 322
Primary cooling loop, 306
Privatisation of electricity supply, 282
Probe power
Raman OTDR DTS systems and, 150–152
linear power limitation, 151–152
wavelength selection role in extending linearity, 152
spontaneous Brillouin scattering, 185–186
Process plant, 284–285
Production logging, 293n
Profile dispersion effect, 117
Pulse compression coding (PCC), 94–99, 96f–97f, 131–132
in BOTDA
as extension of composite pulse techniques, 207
for range extension, 206–207
in BOTDR, 194
Pulsed laser, in OTDR, 74
Pulsed Raman OTDR system, 122

Pulse duration, 356–358
Pulse repetition frequency (PRF), 187
Pump pulse, 409
Pure-silica-core fibres, 306

Q

Q-switched fibre lasers, 121
Quantum efficiency (*Q*), 50–51
Quasi-static (static) measurement, 231
 Rayleigh backscatter
  coherent, 267–271
  as vibration sensor, 273
  vs. dynamic measurements, 267
 vs. vibration, 231

R

Radiation detection, 390–391
Radiation-induced attenuation (RIA), 328, 390
Raman amplification, distributed, 188–190
Raman backscatter signals, 114–115
Raman/Brillouin hybrid approach, 316
Raman DTS system, 25, 39
 commercialisation history, 22–24
Raman gain, 189
Raman interaction, stimulated, 409
Raman OTDR, 13, 23
Raman OTDR DTS systems
 alternative approaches, 128–137
 backscatter signals, 114–115
 design, 115–121, 116f
  receiver, 120
  reference coil, 119–120
  wavelength, laser and fibre selection, 117–119, 119f
 fundamental limitations, 122–125, 124f
 high-performance systems operating at around 1064 nm, 121
 long-range systems, 121
 performance of, 121–128
  comparisons, 125–128, 126t–127t
 practical issues and solutions, 137–155
  Brillouin backscatter, 154–155
  calibration, 149–150
  forward Raman scattering, 153–154, 153f
  high-temperature operation of Raman sensors, 154
  insufficient filter rejection, 152–153
  loss compensation, 137–149
  probe power limitations, 150–152
 principles, 109–115
 Raman scattering
  spontaneous, 109–111, 111f
  temperature sensitivity of, 111–112, 113f
 reference coil, 116–117
 short-range systems, 120
 signal-to-noise ratio, 125
 spatial resolution, 121–122
 temperature resolution, 122
 vs. high-spatial resolution systems, 125–128, 126t–127t
 vs. OFDR methods, 130–131
 vs. with analog direct detection, 137
 walk-off effect, 123–125, 124f
Raman scattering, 12, 165
 anti-Stokes, 109, 111
 elastic, 39
 forward, 153–154, 153f
 frequency shift in, 14
 spontaneous, 109–111, 111f
 Stokes, 109, 111
 temperature sensitivity of, 111–112, 113f
 vs. Brillouin scattering, 161
Raman sensors, high-temperature operation of, 154
Raman signal, 128
Raman spectra, 12, 12f
Raman technology, 285
 vs. Brillouin-based distributed sensing, 15
Raman thermometry, 13
Ramey's approximate model, 296
Range
 extension, in BOCDA, 215–216
 extension in BOTDA, pulse compression coding for, 206–207
 of OTDR, enhancement of
  detection improvement, 88–91
  photon counting, 91–92
  spread-spectrum techniques, 92–101
 remote amplification for extension, BOTDR systems, *see* Remote amplification, for range extension, BOTDR systems
 of sensor, 16, 17
Rare-earth-doped fibre amplifiers, remotely pumped, 190–192
Rare-earth-doped fibre temperature sensing, 392–394
Rayleigh backscatter, 10, 128, 307
 coherent, 232
  amplitude-based DVSs, 235–239, 236f–238f
  limitations in frequency-scanned, 272–273
  in static measurements, 267–271
 combined time-/frequency-domain interrogation, 272
 distributed sensing with, 15–16
 distributed vibration sensors, *see* Distributed vibration sensors (DVSs)
 dynamic range
  of measurand, 263–264, 263f
  of signal, 262–263
 dynamic strain measurements, 267
 fading in, 257–261
  alternative approaches, 261, 261f
  diversity effect on signal quality, 259–260, 259f
  specific arrangements for frequency diversity, 260–261, 260f
  statistically independent signals requirements, 257–258, 258f
 frequency-domain-only interrogation, 271–272
 overview, 231
 power spectrum and refractive index variations, 271
 quasi static strain measurements as vibration sensor, 273
 reflections, 262
 static *vs.* dynamic measurements, 267
Rayleigh optical frequency-domain reflectometry, 286, 308
Rayleigh OTDR, 15
Rayleigh scattering, 12, 37, 55, 64, 393
Rayleigh scattering-loss coefficient, 57
Rayleigh scattering spectrum, 91
Re-capture process, in OTDR, 55, 56f

Receivers
design, 120
optical, noise in, 75–85
multimode, short-wavelength OTDR, 78–85, 79t, 80f–84f
single-mode, long-wavelength OTDR, 85
Reference coil, 116–117
design, 119–120
Reflections, Rayleigh OTDR signals, 262
Refractive index, 12, 31, 57
Refractory-lined reformer vessels, 285
Refractory lining, failure of, 284
Relative index-difference, 32
Remote amplification
distance enhancements through, BOTDA, 210–211
for range extension, BOTDR systems, 187–192
distributed Raman amplification, 188–190
hybrid Raman/EDFA amplification, 192
remotely pumped rare-earth-doped fibre amplifiers, 190–192
Remotely pumped rare-earth-doped fibre amplifiers, 190–192
Repeatability, 22
Resolution cell, 15, 66
Resonance Raman scattering, 110n
Ricker wavelet, 350
Risers, 287
River banks and levees, 332
Rogers, Alan, 11
Room-and-pillar mines, 307

S

SA-BOTDA (slope-assisted BOTDA), 210
Safety barriers, 389
Sagnac interferometer, 407, 408, 408n
Samarium, 393
Sampling resolution, 19
vs. spatial resolution, 19
Scattering, 37–39, 38f; *see also specific entries*
elastic, 11–13
inelastic, 11–13
Schlumberger Fibre-Optic Technology Centre (SFTC), 231n
SEAFOM®, 21
Security, 332–333
Seismic signals, 340
Self-phase modulation (SPM), 87, 186
power limitations due to, 211
Semiconductor lasers
broad contact, 42–43, 43f
distributed feedback (DFB), 44
with active phase/frequency feedback, 44
external cavity feedback, 44
Semiconductor optical amplifiers (SOAs), 46
Sensitivity
of BOTDA, 203
of Brillouin frequency shift, 167–168
of spontaneous Brillouin scattering
intensity to strain, 166
intensity to temperature, 165–166
Sensitivity coefficients, 164–165
Sensors; *see also specific entries*
defined, 3
industrial use, 3, 4f
placement, 377
selection process, 421
Separation of temperature and strain effects; *see also* Distributed strain sensors
Brillouin spectrum
direct measurement of, 319
few-mode fibres, 318–319
fibres designed for Brillouin measurand discrimination, 320–321
independent optical cores, 317–318
multiple Brillouin peaks with distinct properties, 318
multi-wavelength interrogation, 319
polarisation-maintaining fibres/photonic crystal fibres, 318
hybrid measurement
Brillouin/fluorescence, 317
coherent Rayleigh/Brillouin hybrid, 317
Raman/Brillouin hybrid approach, 316
spontaneous Brillouin intensity/frequency measurement, 316
through cable design, 315–316
SF-BOTDA (sweep-free BOTDA), 209–210
Shape sensing, 328–329; *see also* Distributed strain sensors
Shear waves (S-waves), 343, 344
Short-range systems, 120
Short-wavelength edge (SWE), 138
Signal levels, in OTDR, 72–73
Signal recovery, in OFDR, 130
Signal-to-noise ratio (SNR), 20, 59, 344, 351, 357
defined, 75
as function of APD gain, 81–83, 81f–84f
Raman OTDR DTS systems, 125
Signal tracking, 193
Silica (SiO_2), 41, 138, 154, 163, 168, 306, 324
Simultaneous gain/loss interrogation, BOTDA and, 206
Simultaneous multi-frequency measurement, in BOTDR, 194, 194f
Single-ended measurement, OTDR, 56–60, 58f
Single-mode, long-wavelength OTDR, 85
Single-mode fibres, 72
Single-mode optical fibres, 34f, 36–37
Single-photon avalanche detector (SPAD), 133
Single-photon fluorescence, 411
Single-point sensor, 16
Single shaped pump interrogation, BOTDA, 203–205, 204f
Slickline, 293
Slope-assisted BOTDA (SA-BOTDA), 210
Slope-assisted fast acquisition, BOTDA, 210
Slow light
correction, in BOTDA, 207–208
defined, 207
Snell-Descartes law of refraction, 31
SNR, *see* Signal-to-noise ratio (SNR)
SNSPD, *see* Superconducting nanowire single-photon detectors (SNSPD)
Soil moisture, 304
Soil monitoring, 330–332
SPAD, *see* Single-photon avalanche detector (SPAD)
Spatial resolution, 17–18, 17f–19f
of BOTDA, 203
of OTDR, 59–60
Raman OTDR DTS systems, 121–122

SpBS sensor, 186, 187f
very high, 132–134
vs. sampling resolution, 19
Spatial temperature resolution, 21
SpBS, *see* Spontaneous Brillouin scattering (SpBS)
Spectral discrimination in DVS analysis, 374–376
Spectral efficiency, BOTDR, 197
Spectral width, spontaneous Brillouin backscatter, 163–164
Splices, 50
SPM, *see* Self-phase modulation (SPM)
Spontaneous Brillouin intensity/frequency measurement, 316
Spontaneous Brillouin scattering (SpBS), 14, 161, 319; *see also* Stimulated Brillouin scattering (SBS)
coating and coupling effects, 168–169
distributed strain sensing based on, 187–197
dual-pulse method in BOTDR, 195–196, 195f, 196f
fast measurement techniques, 192–193
pulse compression coding, 194
remote amplification for range extension, 187–192
simultaneous multi-frequency measurement, 194, 194f
spectral efficiency, 197
effect, 162
frequency shift, 162–163
limitations and benefits, 184–186
power launched, 185–186
spatial resolution, 186, 187f
loss compensation for intensity measurements, 166–167
sensitivity
of Brillouin frequency shift, 167–168
of intensity to strain, 166
of intensity to temperature, 165–166
sensitivity coefficients, 164–165
spectral width, 163–164
stress-optical effect, 162
systems based on
electrical separation, 177–183, 178f, 180f, 182f–183f
frequency, gain and linewidth estimations, 183–184, 184f
optical separation, 171–177, 172f–173f, 175f–177f
vs. spontaneous Raman scattering, 161
Spontaneous Raman scattering, 109–111, 111f
vs. spontaneous Brillouin scattering, 161
Spread-spectrum techniques, 92–101
coherent OFDR (C-OFDR), 92–93
frequency-diverse C-OTDR, 99–101, 100f
incoherent OFDR, 94, 94f
OFDR, 92
overview, 92
pulse compression coding, 94–99, 96f–97f
SRS, *see* Stimulated Raman scattering (SRS)
State-of-polarisation (SOP), 11, 399
Static (or quasi-static) measurement, 231
Rayleigh backscatter
coherent, 267–271
as vibration sensor, 273
vs. dynamic measurements, 267
vs. vibration, 231
Steam-assisted, gravity drainage (SAGD), 294, 295, 379
Steam-assisted heavy oil production, 294–295
Steam-produced oil, 294
Step-index multimode fibres, 33–34, 34f
Stimulated Brillouin scattering (SBS), 14, 87, 101, 161–162, 169–171, 170f; *see also* Spontaneous Brillouin scattering (SpBS)
system based on
BOFDA, 208
BOTDA, 197–199, 198f
composite pulse methods in BOTDA, 203–206
optical arrangements for BOTDA, 199–202, 200f–202f
performance and limitations, 208–211
phase measurement in BOTDA, 211–213, 212f
pulse compression coding, 206–207
sensitivity and spatial resolution, 203
slow light correction, 207–208
trade-offs in BOTDA, 203
Stimulated Raman interaction, 409
Stimulated Raman scattering (SRS), 87, 110
Raman OTDR systems and, 123, 150
Stokes lines/bands, 12
Stokes Raman scattering, 109, 111
Strain measurements, limitations on, 325–326
Strain sensing, 326
Strain sensitivity, to intensity of Brillouin scattering, 166
Streams and estuaries, 304–305
Stress-birefringent fibres, 396
Stress-optical effect, 162
Subcool, 294
Subsea cables, 371
Subsea DTS deployment, 302
Subsea energy cables, 326
Subterranean formations, compaction/subsidence in, 330
Sumitomo Electric, 23
Superconducting nanowire single-photon detectors (SNSPD), 133
Surface seismic acquisition, 340–341, 363–364
Sweep-free BOTDA (SF-BOTDA), 209–210
Switches, 49
Symbols
for optical amplifiers, 43f, 46
for optical sources, 43f, 45

T

TAC, *see* Time-to-amplitude converter (TAC)
TC-SPC, *see* Time-correlated single-photon counting (TC-SPC)
TDC, *see* Time-to-digital converter (TDC)
TDR, *see* Time-domain reflectometry (TDR)
Telecommunications, 326–327
Temperature resolution, Raman OTDR DTS systems, 122
Temperature sensitivity
to intensity of Brillouin scattering, 165–166
of Raman scattering, 111–112, 113f
Thermal tracer techniques, 297; *see also* Borehole measurement
Time-correlated single-photon counting (TC-SPC), 132, 134–136, 134f
Time-domain reflectometry (TDR), 8–9
optical TDR (OTDR), 9–11, 9f
Time-/frequency-domain interrogation, combined, Rayleigh backscatter, 272
Timescales, in DVSs, 232–234, 233f
Time-to-amplitude converter (TAC), 134–135
Time-to-digital converter (TDC), 134, 135

Toshiba, 23
Total internal reflection, 31
Total vibration energy, 374
Transformers, 282–283; *see also* Distributed temperature sensors (DTS)
Transimpedance preamplifier, 86, 120
Transit cable, 3
Transportation, 370–371; *see also* Distributed vibration sensors (DVSs)
Transverse acoustic (TA) waves, 162
Tube wave velocity, 376
Tunable lasers, 45
Tunnels, 327; *see also* Civil engineering
Turbulent structure velocity, 376
Two-photon fluorescence (TPF), 411–413

U

Ultraviolet (UV) absorption bands, 37
US Naval Research Laboratory, 16

V

Vacuum-insulated tubing (VIT), 301
Vapour-phase axial deposition (VAD) process, 41
Velocity string, 300
Versatile Seismic Imager (VSI), 344
Vertical-incidence vertical seismic profiling (VIVSP), 343
Vertical seismic profiling, 24
Vertical seismic profiling (VSP), 341, 343
Very-high-resolution systems, 132–134
Vibration; *see also* Distributed vibration sensors (DVSs)
  vs. quasi-static measurement, 231
Vibration sensing, 402–403
Viscosity-diverting acid (VDA), 298

W

Walk around VSP, 343
Walk away VSP, 343
Walk-off effect, 123–125, 124f
Warm-back technique, 298
Waveguide dispersion, 40–41
Wavelength
  Raman OTDR DTS systems, 117–119, 119f
  selection, role in extending linearity, 152
Wavelength-division multiplexing (WDM) coupler, 188
Wavenumber filtering, 354–355, 357
Well abandonment, 381
Well integrity monitoring, 301
Wind in overhead cables, 372–373
Wireline, 293

Y

York Sensors Ltd., 23

Z

Zero-offset vertical seismic profiling (ZOVSP), 342